Designed Experiments for Science and Engineering

Designed Experiments for Science and Engineering is a versatile and overarching toolkit that explores various methods of designing experiments for over 20 disciplines in science and engineering.

Designed experiments provide a structured approach to hypothesis testing, data analysis, and decision-making. They allow researchers and engineers to efficiently explore multiple factors, interactions, and their impact on outcomes, ultimately leading to better-designed processes, products, and systems across a wide range of scientific and engineering disciplines. Each discipline covered in this book includes the key characteristics of the steps in choosing and executing the experimental designs (one factor, fractional factorial, mixture experimentation, factor central composite, 3-factor+central composite, etc.) and reviews the various statistical tools used as well as the steps in how to utilize each (standard deviation analysis, analysis of variance [ANOVA], relative standard deviation, bias analysis, etc.).

This book is essential reading for students and professionals who are involved in research and development within various fields in science and engineering, such as mechanical engineering, environmental science, manufacturing, and aerospace engineering.

Designed Experiments for Science and Engineering

Michael D. Holloway

CRC Press
Taylor & Francis Group
Boca Raton London New York

CRC Press is an imprint of the
Taylor & Francis Group, an **informa** business

Designed cover image: "Ancient Wagon Wheel found in the Ljubljana Slovenia Marshes" by Michael D. Holloway

First edition published 2025
by CRC Press
2385 NW Executive Center Drive, Suite 320, Boca Raton FL 33431

and by CRC Press
4 Park Square, Milton Park, Abingdon, Oxon, OX14 4RN

CRC Press is an imprint of Taylor & Francis Group, LLC

© 2025 Michael D. Holloway

Reasonable efforts have been made to publish reliable data and information, but the author and publisher cannot assume responsibility for the validity of all materials or the consequences of their use. The authors and publishers have attempted to trace the copyright holders of all material reproduced in this publication and apologize to copyright holders if permission to publish in this form has not been obtained. If any copyright material has not been acknowledged please write and let us know so we may rectify in any future reprint.

Except as permitted under U.S. Copyright Law, no part of this book may be reprinted, reproduced, transmitted, or utilized in any form by any electronic, mechanical, or other means, now known or hereafter invented, including photocopying, microfilming, and recording, or in any information storage or retrieval system, without written permission from the publishers.

For permission to photocopy or use material electronically from this work, access www.copyright.com or contact the Copyright Clearance Center, Inc. (CCC), 222 Rosewood Drive, Danvers, MA 01923, 978-750-8400. For works that are not available on CCC please contact mpkbookspermissions@tandf.co.uk

Trademark notice: Product or corporate names may be trademarks or registered trademarks and are used only for identification and explanation without intent to infringe.

ISBN: 978-1-032-85441-0 (hbk)
ISBN: 978-1-032-86658-1 (pbk)
ISBN: 978-1-003-52853-1 (ebk)

DOI: 10.1201/9781003528531

Typeset in Times
by codeMantra

This work is dedicated to my twin brother Christopher L. Holloway (Topper), for your continuous support, love, and friendship. Any place you are is truly the place to be!

Michael D. Holloway

Contents

Preface

Early human survival required hunting, gathering, and observing, as well as a team effort to communicate. When ancient humans tried different combinations for favorable outcomes, they were using designed experiments. This could have been a recipe for bread or mortar, a process to domesticate an animal, a selection process for stone or timber. While trial and error was the method, documentation remained essential. Prior to written language, the methods were passed down through actions and then verbally. When symbols, numbers, and letters were developed, accuracy and precision increased.

As societies grew and became more complex, there was a need to communicate information over long distances and across time. Writing provided a means to record thoughts, ideas, and events, enabling communication between individuals who were not physically present or who lived in different time periods. With the advent of agriculture and the rise of civilizations, there arose a need to keep records of transactions, inventories, laws, and other important information. Writing allowed for the preservation of such records, facilitating administrative tasks and the organization of societies.

Writing enabled cultures to preserve their history, myths, legends, and religious beliefs for future generations. It provided a means for passing down knowledge and traditions, contributing to the continuity and cohesion of societies. It facilitated the development and dissemination of knowledge by allowing individuals to document their ideas, theories, and discoveries. It played a crucial role in the advancement of fields such as science and engineering.

PERFECTING EXPERIMENTATION

The origins of using math in designed experiments can be traced back to ancient civilizations, though its development evolved in distinct phases across different historical periods. Ancient civilizations like the Babylonians and Egyptians used basic mathematics for measurements and record-keeping in agricultural practices, but not in formal experiments. The Greeks and Romans employed descriptive statistics (averages and ratios) in astronomy and demographics, but not yet in controlled experiments. Mathematicians like Pierre de Fermat, Blaise Pascal, and Isaac Newton laid the groundwork for probability theory and statistical analysis.

With the rise of scientific experimentation, scientists like Galileo Galilei and William Harvey started incorporating basic statistical methods (e.g., averages and percentages) into their studies. While math entered scientific investigations, formal statistical methods for designed experiments were still in their infancy. Another one of the pioneers in this field was Sir Francis Galton, a British polymath who lived in the 19th century. Galton made significant contributions to the development of statistical methods and their application in various fields, including engineering and science.

Galton's work in statistics laid the foundation for modern statistical techniques and their use in scientific inquiry. He developed methods for analyzing and interpreting data, such as regression analysis and correlation. Galton's ideas and methodologies greatly influenced later generations of scientists and engineers, shaping the way data are collected, analyzed, and interpreted in these fields.

Mathematicians like Karl Pearson and Sir Ronald Fisher developed formal statistical methods like hypothesis testing, analysis of variance (ANOVA), and regression analysis, making them vital tools for designed experiments. Fisher pioneered the design of experiments (DOE) with techniques like randomized block designs and factorial designs, revolutionizing how experiments are conducted and analyzed.

Statistical methods became essential in fields like agriculture, biology, and engineering, driving advancements in designed experiments across various disciplines. New statistical methods and

DOE approaches continue to emerge, catering to specific research needs and fields like Taguchi methods and Six Sigma methodologies.

While early civilizations used basic math in record-keeping and observations, the formal integration of mathematics into designed experiments really began in the 19th century with the statistical revolution.

It was with the work of pioneers like Galton, Pearson, and Fisher and the development of statistical inference and experimental design theory that math truly became a cornerstone of designed experiments across various scientific and engineering disciplines.

Let us consider experiments to be reactions and reactions to be qualified into the following types. When experimenting, look to identify what reaction is occurring:

$1+2=3$ *Additive*: a known and predictable outcome

$1+2+n>3$ *Potentiometric*: an unknown variable n results in a favorable response

$1^n+2^n>3$ *Synergistic*: an unknown power n results in a favorable response

$1+2+\text{-}n<3$ *Antagonistic*: an unknown variable $\text{-}n$ results in an unfavorable response

$1+n+\text{-}X!=0$ *Nihilistic*: regardless of any variable, the result is a disaster

Understanding these reactions helps us develop and perfect. Designing experiments brings us closer to understanding these reactions. Using designed experiments has proven to be exceptionally beneficial in scientific discovery, product development, and refinement. A working knowledge of statistics will be helpful to gain the greatest insight and usefulness from this work. As with any endeavor, scientific or engineering, the outcome is as important as the journey. There is much to learn along the way. Keep good notes.

Michael D. Holloway
April 13, 2024

Author

Michael D. Holloway is the President of 5th Order Industry (https://5thorderindustry.com/leadership/), which provides competency development and training programs for many industries and offers standard, customized, and certification preparation courses. He has over 38 years of experience in the industry, starting with research and development for Olin Chemical and WR Grace, product management for Rohm & Haas, technical marketing and application engineering for GE Plastics, and business development and management for NCH, ALS, and SGS. He is a subject matter expert in failure analysis, reliability engineering, and designed experiments for science and engineering. He holds 16 professional certifications, a patent, a Master of Science degree in polymer engineering from the University of Massachusetts, and a Bachelor of Science degree in chemistry and a Bachelor of Arts degree in philosophy from Salve Regina University, Newport, Rhode Island. He has authored 8 books and contributed to several others, and he has been cited in more than 1000 manuscripts and several hundred master's theses and doctoral dissertations.

1 Introduction

There are many types of designed experimental approaches and statistical tools you can use to research and develop. Designed experiments provide a structured approach to hypothesis testing, data analysis, and decision-making. They allow researchers and engineers to efficiently explore multiple factors, interactions, and their impact on outcomes, ultimately leading to better-designed processes, products, and systems across a wide range of scientific and engineering disciplines. The following are key characteristics of experimentation:

Controlled Conditions: Experiments involve manipulating and controlling certain variables or conditions to observe their effects on a particular phenomenon. This control is essential for drawing meaningful conclusions.

Hypothesis Testing: Experiments are often conducted to test hypotheses or predictions. A hypothesis is a statement or educated guess about the relationship between variables, and experiments aim to confirm or refute these hypotheses.

Randomization: In many experiments, randomization is used to assign subjects or samples to different experimental groups or conditions. This helps minimize bias and ensure that the results are representative of the larger population.

Data Collection: Experiments involve collecting data through systematic observations, measurements, or other means. These data are then analyzed to draw conclusions.

Replication: To ensure the reliability of findings, experiments should be replicable. This means that other researchers should be able to repeat the experiment and obtain similar results.

Independent and Dependent Variables: Experiments typically involve at least one independent variable (the variable that is manipulated) and one dependent variable (the variable that is measured or observed as a result of the manipulation). The relationship between these variables is a central focus of experimentation.

Experimental and Control Groups: In many experiments, there is an experimental group that receives the treatment or manipulation and a control group that does not. This allows researchers to compare the effects of the treatment to a baseline condition.

Ethical Considerations: Ethical guidelines and principles should be followed in conducting experiments, especially when they involve human subjects or animals. Ensuring the well-being and rights of participants is a fundamental ethical concern.

The choice of experimental design depends on several factors:

- **Number of Factors**: More complex designs handle a larger number of factors.
- **Desired Information**: Comprehensive analysis vs. initial screening or optimization.
- **Resource Constraints**: Experiment number and complexity limitations.
- **Experimental Context**: Specific needs of the analysis being conducted.

By understanding these different types of designed experiments and considering the specific needs of your analysis, you can make an informed decision and leverage the power of experimentation to extract valuable insights and optimize your analysis processes.

DOI: 10.1201/9781003528531-1

1.1 THE FIRST EXPERIMENT

The most significant example of experimentation is the discovery of creating fire. However, pinpointing the exact group or individual who first discovered fire and the specific date is very challenging. While there is evidence for fire use which comes from archaeological finds like burned animal bones or changes in tool types, these don't provide a specific date or identify the exact humans involved. Estimates for the earliest controlled fire use range from 1 million to 1.5 million years ago which is a vast timeframe that makes it difficult to pinpoint a single group. Simply stated, we don't know how or when and we can only speculate as to why; however, we do know the outcome of this pivotal discovery.

Homo erectus were considered the earliest humans. The earliest evidence for fire use is associated with *Homo erectus* or early human ancestors living around 1 million to 1.5 million years ago. The use of fire was a gradual process. Fire discovery and management likely weren't a single "eureka" moment. It might have involved a series of observations and experiments over a long period. From the first onset of the use and management of fire to today's complex technologies, it would seem as though humans took their time in the development process. Perhaps a drastic change wasn't required? Or perhaps humans developed a greater skill set that made for advancement happen at an exponential pace? That skill was experimentation caulked with observation and a logical path of trial and error. Sometimes the goal was not defined. Perhaps it was an individual just trying out various combinations? Dry leaves vs. green leaves? Which burn easier, which smell better? Do rocks or bones burn? What happens if I stick my hand in the flame?

While we can't identify the specific individuals who discovered and experimented, the ability to control fire is considered a major turning point in human evolution. It had a profound impact on early humans, allowing them to illuminate and to stay warm during colder nights as well as cook food, making it easier to digest and extract nutrients. The addition of animal protein into the early human diet is thought to be a catalyst for increased neurodevelopment. It is also known that cooking meat will destroy many pathogens; therefore, human survival and advancement most likely was due to a steak dinner. It is also suggested that fire helped scare away predators. There is evidence that fire was used to create tools by hardening wood and eventually clay leading to early pottery and eventually metallurgy. The discovery of fire, even if accidental at first, stands as a testament to the ingenuity and adaptability of our early ancestors.

When engaged in the process of science, there are a few basic aspects to consider: discovery, learning, and limitations of resources. A designed experiment early humans may have used to figure out how to create and manage fire would consider the following:

THE DISCOVERY ASPECT

Imagine early humans with limited tools and knowledge. It is plausible that someone tasked with sharpening flint tools for the hunt may have vigorously struck a piece of flint against a rough, reddish stone (iron pyrite) to create sharp edges. This action may have generated sparks. If a spark landed on the dry tinder, it's plausible that the spark would engage with the tinder igniting a small flame. It is also possible that this scenario would bring about adding more tinder to the fire. This is considered accidental management. The early humans may have observed how different materials burn – leaves providing a quick but fleeting flame, while small twigs offer a more sustained burn.

Through this accidental series of events, early humans may have stumbled upon the concept of fire creation. They may have learned that sparks from flint and pyrite can ignite dry tinder, and by adding fuel, they can control the fire's size and duration. This newfound knowledge provides warmth, protection from predators, a way to cook food, and a source of light during dark nights.

This is a simplified scenario, but it highlights the potential role of curiosity, experimentation, and observation in early human fire discovery. The exact methods used by early humans might have varied depending on the environment and available materials. This scenario showcases how a series of seemingly unrelated actions could have led to the accidental discovery of fire. Early humans, driven by curiosity and a need for survival, might have unwittingly stumbled upon this crucial advancement through trial and error.

THE LEARNING PROCESS

Early humans likely learned about fire by trial and error. A possible scenario outlining the observations and experiments is early humans might have gone through to understand and control fire through a series of stages, observation, unintended experiments, deliberate experiments, and utilization.

Observation:
- **Natural Fire Events:** Early humans likely witnessed wildfires caused by lightning strikes or volcanic eruptions. They would have observed the destructive power of fire but also its ability to provide warmth and light.
- **Animal Behavior:** They might have noticed animals being cautious around burning areas or using fire-affected zones for warmth. This could have sparked curiosity about fire's properties.

Unintentional Experiments:
- **Tool Sharpening:** As mentioned previously, sharpening tools with flint and stone could have accidentally produced sparks. If these sparks landed on dry tinder (like leaves or grass) near their campsites, it might have occasionally ignited a small fire.
- **Cooking with Hot Rocks:** Heating rocks in fires created by natural events might have been used for cooking food. Observing how hot rocks transferred heat to food could have led them to experiment with placing food near naturally occurring fires.

Deliberate Experimentation:
- **Recreating Sparks:** After witnessing that sparks ignite tinder accidentally, early humans might have started deliberately striking flint and stone together near dry materials to see if they could recreate fire at will.
- **Fueling the Flames:** Once a fire was started, they would have observed how adding more tinder or small twigs kept the flames going for a longer duration. This would have led to them understanding the concept of adding fuel to maintain the fire.
- **Controlling Fire Spread:** Early humans likely learned to control the spread of fire by creating fire pits or clearing flammable materials around the desired burning area. Witnessing how fire consumed everything in its path would have led them to develop methods for containment.
- **Extinguishing Fire:** They might have unintentionally extinguished small fires by throwing water or sand at them. This would have given them a basic understanding of how to put out unwanted fires.

Utilization:
- **Cooking Food:** Fire would have revolutionized food preparation. Early humans could now roast or boil meat and vegetables, making food easier to digest and extracting more nutrients.
- **Tool Production:** Fire hardening techniques could be developed for wooden tools, making them stronger and more durable.
- **Warmth and Light:** Fire provided a vital source of warmth during cold nights, especially in colder climates. The light from the fire also allowed them to extend their activities beyond daylight hours.
- **Protection:** Fire could be used to scare away predators or create smoke barriers for added security.

This is pure conjecture and without actual physical evidence, we are making assumptions, yet it stands to reason that these are the scenarios that unfolded. The observations and experiments likely occurred over a vast period, with knowledge slowly accumulating across generations. Early humans, through their ingenuity and adaptability, gradually shifted from simply witnessing fire to understanding its potential and developing methods to control and utilize it.

THE LIMITATIONS OF RESOURCES

The following is an example of a designed experiment where early humans could have stumbled upon fire creation and developed basic fire management skills using readily available materials:

Materials:
    Flint: A hard, knappable stone commonly found in nature.
    Iron Pyrite ("Fool's Gold"): A common mineral that produces sparks when struck against harder materials like flint.
    Dry Tinder: A pile of dry leaves, moss, or shredded bark collected near the campsite.
    Small Twigs and Branches: Gathered from nearby trees or bushes.

The Experiment (Accidental Discovery):
    Tool Sharpening: A member of the group uses a piece of flint to sharpen a wooden spear or another tool and vigorously strikes the flint against a rough-looking stone (unbeknownst to him, iron pyrite).
    Sparks and Observation: The forceful strikes generate a shower of sparks. One spark lands on a pile of dry tinder (accumulated from previous nights spent at this location) and ignites a small flame.
    Fueling the Flame: Intrigued, someone (maybe a child – they tend to do things like this!) cautiously pokes the small flame with a stick. This action dislodges some tinder, causing a few sparks to fly. These sparks land on nearby dry leaves, causing the fire to grow slightly larger.

Gradual Development of Fire Management Skills:
    Maintaining the Fire: As night deepens and the temperature drops, the fire starts to dwindle. Instinctively, someone throws more dry leaves and small twigs on the flames, keeping them from going out entirely. This accidental addition of fuel demonstrates the concept of maintaining the fire. (This action seems instinctual to humans as witnessed by countless times that someone at a campsite keeps feeding the fire. There is also a direct correlation between the amount of fermented beverage one consumes and the amount of materials and attention one pays to the management of the fire.)
    Fuel Selection: Through repeated attempts to keep the fire going, they observe how different materials burn. Leaves provide a quick but short-lived flame, while small twigs offer a more sustained burn. This leads to a preference for using twigs as fuel.
    Fire Pit Creation: Over time, they might notice how the fire damages the ground beneath it. To prevent this and potentially contain the fire better, they start digging a shallow pit in the ground before creating the fire. This is a rudimentary form of fire pit creation.
    Extinguishing the Fire: When it's time to move camp or the fire becomes too large, they might try throwing water or sand on the flames to extinguish it. This would give them a basic understanding of how to put out unwanted fires. There is also a distinct possibility that urinating on the fire would prove to extinguish; it was witnessed by those at campsites who seem to want to instinctually urinate on the campfire which also correlates to the number of fermented beverages they imbibe.

Outcomes:
- Through this series of accidental observations and trial-and-error actions, early humans stumble upon fire creation and begin to develop basic fire management skills.
- They learn that sparks from flint and pyrite can ignite dry tinder.

- Adding fuel (twigs) keeps the fire going for longer periods.
- Creating a fire pit helps contain the fire and protect the ground.
- Throwing water, sand, or urine extinguishes unwanted flames.

This is a simplified experiment, and the exact sequence of events might have varied depending on the specific environment and available materials. Repetition and observation over generations likely played a crucial role in solidifying their understanding of fire. This scenario showcases how readily available materials and a series of seemingly unrelated actions could have led to the discovery of fire and the development of basic fire management skills by early humans. Their curiosity, adaptability, and trial-and-error approach would have paved the way for this critical advancement.

1.2 EARLY DESIGNED EXPERIMENTS IN ENGINEERING

I once visited eastern Europe for holiday and found myself in a museum where on display was an ancient wagon wheel. This was the oldest wooden wheel ever discovered and was found in the Ljubljana Slovenia Marshes in 2002. Radiocarbon dating places its age at around 5,100–5,350 years old. It's made from ash and oak and belonged to a prehistoric two-wheeled cart. The oldest wooden wheels differed quite a bit from the spoked wheels we see on carriages today. They are an engineering wonder, and it sparked the question "How did they know how to do this?" This wheel was not a single piece of wood or even made of pegs. It is only speculated that a thick wooden disc was cut from a tree trunk, with a hole bored through the center for the axle, and it was the first wheel, yet we have no physical proof, only conjecture. It is believed that this was a common design for the earliest wheels. However, there is actually no physical evidence to suggest that was the design of the first wooden wheel. This wheel was different. This was an engineering marvel!

The Ljubljana marsh wheel was made from several planks of wood joined together. They used the technique of mortise and tenon joints where a protruding peg fits into a carved slot to hold the pieces securely without nails or metal.

There are two main reasons why mortise and tenon joints were a good choice for constructing ancient wooden wheels, especially when compared to other joining methods available at the time:

Strength and Stability: The mortise and tenon joint creates a very strong and stable connection between two pieces of wood. This is crucial for a wheel, which needs to withstand significant stress as it rolls and carries loads. The tenon (projecting peg) interlocks with the mortise (carved slot) to prevent the pieces from pulling apart under pressure. This is especially important when dealing with the radial forces acting on a wheel as it turns.

Simple Construction without Metal: During the period when the earliest wheels were being built, metalworking wasn't widespread or well developed. The mortise and tenon joint rely solely on woodworking techniques, making it ideal for the tools and materials available at the time. Lashes and bindings might have been used to further secure the joint, but the mortise and tenon itself provided a strong foundation.

Simply carving notches or holes wouldn't be strong enough, and the pieces could easily break apart under stress. While pegs (like dowels) could be used, they might just pop out under pressure. The mortise provides a locked-in fit for the tenon.

The mortise and tenon joint offered a strong, reliable method for joining wooden components in early wheel construction, relying on well-established woodworking techniques without the need for advanced metalworking. Other considerations include

Material Selection: Since metalworking wasn't widespread in the very beginning, early wheelwrights favored hardwoods like oak, ash, or elm. These woods are strong, durable, and can withstand the stress of carrying loads.

Shaping Techniques: Stone tools like axes and adzes were likely used to shape the wood into the desired form. Fire might have also been used for hardening or bending the wood in some cases.

Spoked Wheels Weren't Invented Until Much Later: The innovation of spoked wheels, which are lighter and stronger than solid discs, came about 2,000 years after the first wheels.

Regional Variations Likely Existed: While the basic principles remained similar, the specific construction techniques and materials might have differed depending on the location and available resources.

If one was to produce a designed experiment for such a wheel, it may look like this:

DESIGNED EXPERIMENT: IMPACT OF MORTISE-AND-TENON-JOINT DESIGN ON ANCIENT WOODEN WHEEL STRENGTH

Objective: Investigate how variations in mortise-and-tenon-joint design affect the breaking strength of a wooden wheel constructed with this technique.

Materials:

- Seasoned hardwood boards (e.g., oak) with consistent thickness.
- Woodworking tools (saw, chisel, mallet).
- Glue (optional, depending on chosen design).
- Testing machine capable of applying a concentrated load to the wheel rim.

Design Matrix:

The experiment will explore two key variables in the mortise-and-tenon-joint design:

- **Tenon Size (diameter):** This will be varied across three levels (e.g., thin, medium, thick).
- **Mortise Depth (as a percentage of wood board thickness):** This will be tested at three depths (e.g., shallow, medium, deep).

An example of a design matrix illustrating the experiment setup is shown in Table 1.1.

Each group (e.g., Group 1A) will represent a unique combination of tenon size and mortise depth. A minimum of 3–5 replicates should be made for each group to account for natural variations in wood strength.

Procedure:

1. Prepare wooden boards with the specified thickness.
2. Following the design matrix, create mortise and tenon joints on each board with the designated dimensions. Glue may be used for additional reinforcement in some groups, depending on the research question.

TABLE 1.1

Wooden Wheel Design Matrix

Tenon Diameter	Shallow Mortise (Depth %)	Medium Mortise (Depth %)	Deep Mortise (Depth %)
Thin	Group 1A	Group 2A	Group 3A
Medium	Group 1B	Group 2B	Group 3B
Thick	Group 1C	Group 2C	Group 3C

3. Assemble the wooden boards to form complete wheels, ensuring consistent construction across all groups.
4. Mount the wheel on the testing machine with the axle secured.
5. Apply a gradually increasing load to the rim of the wheel until it breaks.
6. Record the breaking load (maximum weight sustained) for each wheel.

Data Analysis:

- Calculate the average breaking load for each group (e.g., Group 1A, 1B, 1C).
- Analyze the data statistically to identify any significant differences in breaking strength between the various joint designs.

Expected Outcome:
The experiment should reveal how the size of the tenon and the depth of the mortise affect the overall strength of the wheel. We might expect:

- **Stronger Joints with Thicker Tenons:** A larger tenon provides a better interlocking area within the mortise, potentially increasing breaking strength.
- **Optimal Mortise Depth:** There might be a "sweet spot" for mortise depth. A shallow mortise might not provide enough engagement, while a mortise that goes too deep could weaken the surrounding wood.

This experiment provides a basic framework. You can modify it further by

- Including additional factors like the length of the tenon.
- Testing the wheels under dynamic loading conditions (simulating real-world use).
- Analyzing the failure modes of the broken wheels for a more detailed understanding.

By implementing this designed experiment, a Wainwright could gain valuable insights into how mortise-and-tenon-joint design influences the strength of these ancient wooden wheels.

1.3   EARLY DESIGNED EXPERIMENTS IN MATERIAL SCIENCE

I visited a castle in Kilkenny Ireland. I was taken aback when told that the ceilings and walls used mortar that was reinforced with thin tree branches. Looking at the ceiling the first thought was "Composite". I wondered how someone would have come up with such an idea, and furthermore, what sort of work was done to perfect it? The ceilings in Kilkenny castle varied depending on the specific area and its construction period. The castle has undergone many changes over its 800-year history. In the medieval sections in the oldest parts of the castle, particularly the base of the west tower, you might find evidence of a wattle-and-daub ceiling. This technique uses interwoven willow branches as a framework and then covers it with a clay mixture for a finished surface. Wattle and daub was a widespread building technique used in ancient times for constructing walls, fences, and even entire houses.

A composite building method combines a wooden framework ("wattle") with a packed-in filling ("daub"). The wattle is typically made of interwoven strips of flexible wood like hazel or willow. The daub is a sticky mixture usually made from a combination of wet soil, clay, sand, and even animal dung (for added binding). Plant materials, such as straw (for insulation and structure), as well as thin willow branches for strength and flexibility.

The wooden strips were woven together to create a lattice framework. This framework would be secured to a foundation or existing posts. The daub mixture would then be applied by hand or even trampled by people or animals to pack it in firmly between the woven wood. The daub would dry to form a hard and relatively weatherproof surface. In some cases, a lime plaster might be applied on top of the dried daub for a smoother finish.

Wood and earth were easily obtainable in most locations. The woven wood frame provided structural support, while the daub filled in the gaps and offered some insulation. The flexibility of the materials helped the structure withstand movement from wind or earthquakes. Wattle and daub was a low-cost building method, and repairs could be done relatively easily by replacing damaged sections.

Archaeological evidence suggests wattle-and-daub construction dates back at least 6,000 years. It has been found in various cultures worldwide, including ancient Egyptians; Romans; people of the linear pottery; and Rössen cultures in central Europe, North American Mississippian culture, and South American cultures. While less common in modern construction, wattle and daub is still used in some parts of the world and can be a sustainable and low-impact building method. The following is an example of a designed experiment for the effectiveness of the method.

Designed Experiment: Effectiveness of Different Wattle-and-Daub Wall Constructions

Objective: Investigate how variations in wattle spacing and daub composition affect the strength and water resistance of wattle-and-daub walls.

Materials:

- Coppiced wood poles (e.g., willow, hazel) for wattle.
- Clay soil.
- Sand.
- Straw (or other appropriate fiber).
- Animal dung (optional, for binding).
- Water.
- Panels or frames for constructing test walls (depending on chosen size).
- Testing equipment:
  - Spray nozzle for water resistance test.
  - Load-bearing capacity tester (or weights for a simpler setup).

Design Matrix:
This experiment will explore two key variables in wattle-and-daub construction:

- **Wattle Spacing:** This will be tested at three levels (e.g., close spacing, medium spacing, wide spacing).
- **Daub Composition:** We will investigate two daub mixes (e.g., clay rich, clay+straw).

An example of a design matrix illustrating the experiment setup is shown in Table 1.2.

Each group (e.g., Group 1A) will represent a unique combination of wattle spacing and daub composition. A minimum of 3 replicates should be made for each group to account for natural variations in materials and construction.

TABLE 1.2
Wattle-and-Daub Design Matrix

Wattle Spacing	Clay-Rich Daub	Clay+Straw Daub
Close (e.g., 2 cm gaps)	Group 1A	Group 1B
Medium (e.g., 5 cm gaps)	Group 2A	Group 2B
Wide (e.g., 8 cm gaps)	Group 3A	Group 3B

Procedure:

1. Prepare the wattle by cutting the coppiced wood poles to a consistent length and removing bark if necessary.
2. Construct test wall frames of a designated size using appropriate materials.
3. Following the design matrix, weave the wattle onto the frames at the designated spacing for each group.
4. Prepare the daub mixtures according to the matrix (clay rich or clay with added straw).
5. Apply the chosen daub mix to the wattle frame for each group, ensuring a consistent thickness across the surface. Allow the daub to dry completely.
6. Water Resistance Test: Subject each dried wall to a controlled water spray for a set duration (e.g., 30 minutes) and measure the amount of water penetration through the wall (moisture meters or visual inspection of the interior surface).
7. Strength Test: After water resistance testing, use a load-bearing capacity tester (or apply a measured weight for a simpler setup) to determine the maximum load each wall can support before failure.

Data Analysis:

- Calculate the average water penetration and maximum load for each group (e.g., Group 1A, 1B, etc.).
- Analyze the data statistically to identify any significant differences in water resistance and strength between the various wattle-and-daub wall constructions.

Expected Outcome:
The experiment should reveal how wattle spacing and daub composition influence the overall effectiveness of the wall. We might expect:

- **Closer Wattle Spacing**: Provides better support for the daub, potentially leading to increased strength and water resistance.
- **Clay-Rich Daub**: May offer higher initial water resistance, but might be prone to cracking as it dries.
- **Daub with Straw**: Straw fibers can improve the flexibility and crack resistance of the daub, potentially leading to better long-term performance.

This is a basic framework, and you can modify it further by

- Adding a surface-treatment group (e.g., lime plaster) to see if it affects water resistance.
- Testing the walls for thermal insulation properties.
- Using destructive testing methods to analyze the internal structure of the daub after testing.

By implementing this designed experiment, a mason in medieval times could gain valuable insights into the effectiveness of different wattle-and-daub wall construction techniques.

1.4 EARLY DESIGNED EXPERIMENTS IN CHEMISTRY

Prehistoric humans might not have had the formal scientific understanding of chemistry we have today, but there's evidence of them engaging in activities that involved unknowingly manipulating and experimenting with chemical reactions. The very act of creating fire by manipulating materials like flint and tinder involves a chemical reaction. Additionally, early humans used pigments like ochre for creating cave paintings. Ochre itself is a naturally occurring iron oxide compound, and processing it might have involved heating or grinding, which could unintentionally alter its chemical properties. There are several theories as to why prehistoric humans painted on cave walls. The most reasonable explanation is that the paintings could have served as a way to document the animals they encountered, record hunting strategies, or communicate knowledge across generations. These early artists may have tried various combinations of clays and water percentages in order to get the most favorable effects.

Another example of early humans using chemistry is fermentation to create alcohol. The process of fermentation likely arose from early attempts at food preservation. Storing fruits, grains, or even animal products could have led to the discovery of alcoholic beverages or fermented foods like yogurt, or it may have been through witnessing drunkenness when eating fruit that was fermented. It is well documented that **moose** are particularly fond of fallen apples that have begun to ferment. Sweden even has a season were police deal with reports of drunken moose. **Elephants will eat** marula fruit; when fermented, it is a suspected culprit behind wobbly elephant behavior seen from time to time. Some monkey species have been observed deliberately raiding bars and stealing sugary drinks, likely for the fermented content. Even birds can get in on the action, with parrots and others sometimes indulging in fermented fruit and becoming a bit disoriented.

It would not be inconsistent with human behavior for them to witness the behavior of other animals and look to replicate. Gathering fruits and keeping them in a container, such as an animal's gut in an earthen pot or even the ground, would contain the batch for further analysis. Over time, it would be determined that fruit with a high sugar content led to the best hootch! Fermentation involves the action of microorganisms breaking down sugars and starches, producing alcohol or lactic acid as byproducts. This process would come in handy when making cheese and yogurt which provided to be a valuable, portable protein source. This would become essential in various parts of the world which would be seasonally deficient in games that could be scavenged or killed. Choice of the right materials and perfecting the process would result in being fed or starving, or in the case of fermented fruit, the choice of dealing with your unhygienic little brother, a mad wolf pack, a feuding clan or a good buzz…

An example of a designed experiment with a matrix for someone in prehistoric times trying fruit fermentation for the first time, keeping in mind their limited knowledge and resources is shown in Table 1.3.

Variables:
- Fruit: They might test readily available fruits like berries or grapes.
- Storage Container: Different natural containers like animal skins, gourds, or hollowed logs could be used.
- Sealing: Leaving the container open versus covering it with leaves or animal hide.

Design Matrix

After the first pass is completed, a second pass may look to switch the container and sealing process according to fruit type. It is essential that good notes be kept even if through spoken langue or kinetic depictions. The odds of remembering the right combination diminish with the greater success of the brew.

TABLE 1.3

Fermentation and Container Design Matrix

Fruit	Storage Container	Sealing
Raspberries	Animal Skin	Open
Raspberries	Animal Skin	Covered (leaves)
Grapes	Gourd	Open
Grapes	Gourd	Covered (animal hide)

Limited Knowledge and Resources:
- In prehistoric times, an understanding of yeast wouldn't be present. Fermentation would occur naturally from wild yeasts on the fruit skins.
- Sugar wouldn't be readily available as a separate ingredient. The experiment relies on the natural sugars present in the fruits.
- Temperature control would be limited. They might choose a location with consistent warmth like a sunny spot near a cave entrance.

Measurements and Observations:
- Visual: Observing bubbling activity, changes in fruit texture, and color.
- Taste: Periodically tasting the fermenting mixture to assess changes in sweetness and potential intoxicating effects (be very cautious with tasting!).

Analysis:

By comparing the different combinations, they could observe:
- Does the container type or sealing affect bubbling activity?
- Does covering the container impact taste or spoilage?
- Which fruit ferments more readily based on visual and taste changes?
- Which fruit and process produces great hootch with the best buzz?

The Outcome

Through trial and error, they might discover that covered containers promote fermentation (trapping CO_2) and some fruits ferment better than others. This knowledge could be passed down and lead to further experimentation with fermentation techniques.

Early humans used various plant materials for medicine, tools, and other purposes. Soaking or boiling plants could have yielded extracts with specific properties due to the presence of active chemical compounds within them. The right combinations and processes would produce the best product. This was only possible with experimentation as well as time to carry out the work. It is theorized that the advancements made were due to additional time made available because early humans did not have to spend a large amount of time foraging, scavenging, or hunting. Eventually, tool making and pottery ensued producing greater efficiency in daily tasks. Eventually, heating stones to create tools or shaping and firing clay to make pottery involved altering the physical and chemical properties of the materials occurred. Trial and error along with a good memory would lead to advancements. These were the first designed experiments.

These examples represent a rudimentary understanding of chemistry, but they involved manipulating materials and observing the resulting changes. Over time, these observations likely contributed to the development of more advanced techniques and a deeper understanding of the world around them. It's important to remember that evidence from this period is often indirect. However, by studying tools, artifacts, and even cave paintings, archaeologists can glean insights into the ingenuity and experimentation of our prehistoric ancestors.

1.5 FIELDS OF STUDY FOR DESIGNED EXPERIMENTS

The following chapters examine several different experimental methods through

- Definition of the designed experimental method.
- Steps on how to perform the methods.
- Potential pitfalls and remedies for the methods.
- Examples of the methods, including a sample matrix.
- Writing the report (statistics, outline, and critique example).

Covered in the following chapters are some examples of engineering, science, and specific industry segments where designed experiments play a crucial role.

Engineering:
- **Chemical Engineering**: Design experiments optimize various chemical processes, like refining fuels, synthesizing materials, and developing new catalysts.
- **Mechanical Engineering**: Experiments test and improve the efficiency and performance of machines, vehicles, and structures.
- **Civil Engineering**: Design experiments optimize building materials, assess structural integrity, and analyze the impact of construction projects on the environment.
- **Electrical Engineering**: Experiments play a role in developing new electronics, optimizing communication systems, and testing power grids.
- **Aerospace Engineering**: Designed experiments are used to test aircraft performance, develop new fuels and engines, and ensure safety in space travel.

Science:
- **Chemistry**: Experiments study the properties and interactions of matter, develop new materials, and investigate chemical reactions.
- **Physics**: Design experiments to explore fundamental physical laws, test new theories, and develop innovative technologies like lasers and semiconductors.
- **Biology**: Designed experiments to explore genetics, cell behavior, drug efficacy, and disease mechanisms.
- **Psychology**: Researchers use experiments to understand human behavior, assess the effectiveness of therapies, and study cognitive processes.
- **Environmental Science**: Design experiments help understand the impact of human activities on the environment, develop sustainable solutions, and monitor pollution levels.

Designed experiments are also a fundamental tool used in subcategories in the various fields of science and engineering to systematically study and optimize processes, products, and systems.

Specific Industry Segments:
- **Agriculture**: In agriculture, designed experiments help optimize crop yields, study the effects of fertilizers and pesticides, and improve farming practices.
- **Biotechnology**: Designed experiments are used in biotechnology to optimize bioprocesses, such as fermentation and cell culture, and develop biopharmaceuticals and biofuels.
- **Energy and Renewable Resources**: In energy research, designed experiments help optimize energy production processes, study energy efficiency, and develop renewable energy technologies.
- **Food Science**: In food science and technology, designed experiments are used to develop new food products, optimize food processing methods, and ensure food safety and quality.
- **Healthcare and Medical Research**: Designed experiments are used to study the effects of medical treatments, optimize healthcare processes, and conduct clinical trials.
- **Information Technology**: In software development and IT, designed experiments can be used to optimize software performance, test system reliability, and improve user experience.
- **Lubrication**: Designed experiments are used during the formulation and manufacturing stages to understand the influences of chemistry and processes as they affect the performance.
- **Manufacturing**: Designed experiments are essential in manufacturing to improve product quality, reduce defects, optimize production processes, and minimize variability in production lines.

- **Materials Science**: Material scientists employ designed experiments to study material properties, optimize material compositions, and develop new materials for various applications.
- **Pharmaceuticals**: In the pharmaceutical industry, designed experiments are used to optimize drug formulations, investigate drug interactions, and ensure product quality and consistency.
- **Social Sciences**: Designed experiments are increasingly used in social sciences, economics, and psychology to conduct controlled experiments and study human behavior and decision-making.
- **Transportation and Aerospace**: The transportation and aerospace industries use designed experiments to optimize vehicle and aircraft design, study aerodynamics, and improve safety.

These are just a few examples, and the list truly spans countless subfields within science and engineering. Any study that seeks to control variables, understand cause-and-effect relationships, and draw conclusions based on data benefits from implementing designed experiments.

Ultimately, the specific types of designed experiments used vary depending on the research question, available resources, and the nature of the field. Some common designs include factorial experiments, block designs, response surface methods, and Taguchi methods, each offering unique advantages for specific situations.

1.6   STEPS IN CHOOSING AND DOING THE BASIC EXPERIMENTAL DESIGNS

The following are basic steps for determining the best-suited experimental design for a desired outcome. The following section covers creating a decision process for selecting a designed experiment and involves considering various factors and criteria to determine the most suitable experimental design for a particular research or problem. Here's a simplified decision tree to guide you in selecting a designed experiment.

Designed Experiment Selection Process:

Step 1: Determine your research objective:
- Is it screening and identifying key factors?
- Are you looking to estimate the main effects and interactions?
- Do you want to build a response surface model and optimize the process?

Step 2: Consider the number of factors involved:
- One Factor: One factor at a time (OFAT): Simple but inefficient for exploring interactions.
- Two Factors:
  - Full Factorial: Efficient for exploring all combinations and interactions but can be costly for many factors.
  - 2-Factor Central Composite Design (CCD): More advanced, allows for model building and optimization with curvature exploration.
- Three or More Factors:
  - 3-Factor CCD: More advanced, allows for model building and optimization with curvature exploration.
  - Box–Behnken Design (BBD): Efficient for exploring quadratic terms without requiring as many runs as a full factorial.
  - Plackett–Burman Design (PBD): Useful for screening many factors with limited resources, but only provides information about main effects.
  - Fractional Factorial Design: Efficient for screening and identifying key factors, requiring fewer runs than a full factorial.

     – Derringer Design: Useful for optimizing multiple responses simultaneously when interactions are important.

Step 3: Analyze your budget and resource constraints:
- Limited resources: Consider PBD, fractional factorial, or even OFAT if interactions are not a major concern. Choose smaller designs with fewer runs.
- Ample resources: Full factorial, CCD, or BBDs can be beneficial for detailed analysis and model building.

Step 4: Assess the expected relationship between factors and response:
- Linear relationship: Factorial, BBD, or fractional factorial designs might suffice.
- Non-linear relationship: CCD (Derringer design) might be better to capture curvature and complex interactions.
- Unknown relationship: Start with a fractional factorial design (PBD) to identify key factors; then, follow up with a more specific design based on the findings.

Step 5: Consider experimental error and accuracy needs:
- High accuracy: Include replicates in your design, especially at the center point. Choose designs with inherent replication or error estimation capability.
- Initial exploration and rough estimates: Fewer replicates might be acceptable, depending on the tolerance for error.

After the type of designed experiment is chosen, here are the elements for structuring your designed experiments:

Designed Experiment Structure Process:

Step 1: Define Your Objective: Start by clearly defining your research objective and the specific questions you want to answer through the experiment.

Step 2: Identify Your Factors: Determine the factors (independent variables) that you want to investigate. Consider both the qualitative and quantitative factors involved.

Step 3: Choose the Response Variable: Identify the response variable (dependent variable) that you want to measure or observe to assess the impact of the factors.

Step 4: Consider Other Experimental Design Types as a Backup: Experimental design types are categorized based on different considerations. Choose accordingly but have a backup plan.

Step 5: Isolating a Single Factor: If you want to investigate the effect of a single factor while keeping others constant, use a one-factor-at-a-time (OFAT) design.

Step 6: Balancing Factors: If you have multiple factors and want to balance the effects of each factor, consider using factorial design.

Step 7: Finding Optimal Conditions: If your goal is to optimize a process or system, consider using response surface methodology (RSM).

Step 8: Screening Many Factors: If you have many factors and want to identify the most significant ones efficiently, use screening designs (e.g., PBD or Taguchi).

Step 9: Handling Constraints: If you have constraints on your experimental setup or factors, consider using robust design or constrained optimization techniques.

Step 10: Explore Interaction Effects: If you suspect interactions between factors, use higher-order factorial designs (e.g., 2^3 or 2^4) or RSM.

Step 11: Resource Constraints: If you have limited resources or time, consider using fractional factorial designs to explore factor interactions with fewer runs.

Step 12: Experimental Replicates: Determine whether you need experimental replicates (repeating the experiment multiple times) to account for variability.

Step 13: Data Analysis Approach: Decide on the statistical techniques and software tools you'll use to analyze the experimental data.

Step 14: Document Your Experiment: Ensure you have a well-documented experimental plan, including the chosen design, factor levels, response variable measurements, and data analysis methods.

Step 15: Conduct the Experiment: Execute the designed experiment according to your plan, ensuring proper data collection and control of variables.

Step 16: Analyze the Results: Analyze the experimental results using appropriate statistical methods to draw conclusions and make recommendations.

Step 17: Validate and Implement: Validate the findings from your designed experiment and implement any recommended changes or optimizations in your process or system.

These methods are used in science and engineering for several important reasons. Designed experiments allow researchers to control and manipulate variables systematically, ensuring that only the factors of interest are influencing the outcomes. By controlling for extraneous variables, researchers can isolate the effects of specific factors and draw more reliable conclusions about cause-and-effect relationships. They are often more efficient than observational studies or ad hoc experimentation because they allow researchers to achieve statistically significant results with fewer resources (such as time, materials, or subjects). By carefully planning the experimental design, researchers can optimize the use of resources and minimize waste. Well-designed experiments are replicable, meaning that they can be repeated by other researchers to verify the results. Additionally, designed experiments are often designed with the goal of generalizability, allowing researchers to make valid inferences about a broader population or phenomenon beyond the specific conditions of the experiment.

The work should be structured in a way that allows for rigorous statistical analysis, enabling researchers to make valid inferences and draw meaningful conclusions from the data. Statistical techniques, such as analysis of variance (ANOVA), regression analysis, and hypothesis testing, are commonly used to analyze experimental data and assess the significance of observed effects. In engineering and industrial settings, designed experiments are often used for process optimization and quality improvement. By systematically varying process parameters and measuring the resulting outcomes, engineers can identify the optimal settings for maximizing efficiency, minimizing waste, and meeting quality specifications. The method can help mitigate risks associated with decision-making by providing empirical evidence and quantifying uncertainty.

By conducting experiments to test hypotheses and evaluate potential outcomes, researchers and engineers can make more informed decisions and reduce the likelihood of costly errors or failures. They also play a pivotal role in both scientific research and engineering endeavors, offering a structured and systematic approach to investigating complex phenomena and optimizing processes. One primary advantage of designed experiments is their ability to control and manipulate variables with precision, allowing researchers and engineers to isolate the effects of specific factors and elucidate cause-and-effect relationships. By controlling for extraneous variables, designed experiments enhance the reliability and validity of findings, enabling researchers to draw more accurate conclusions and make informed decisions based on empirical evidence.

Designed experiments are instrumental in maximizing efficiency and minimizing resource utilization. Through careful planning of experimental designs, researchers can achieve statistically significant results with minimal time, materials, or subjects, thus optimizing resource allocation and reducing waste. This efficiency is particularly crucial in engineering applications, where experiments may involve costly materials, complex processes, or time-sensitive operations. By streamlining experimentation and maximizing the information gleaned from each trial, designed experiments contribute to the advancement of knowledge and innovation in science and engineering.

This work will facilitate rigorous statistical analysis and inference, enabling researchers and engineers to derive meaningful insights and draw valid conclusions from experimental data. Statistical techniques, such as ANOVA, regression analysis, and hypothesis testing, are commonly employed to analyze experimental results, assess the significance of observed effects, and make informed decisions based on quantitative evidence. By applying rigorous statistical methods, designed experiments enhance the credibility and robustness of research findings, fostering confidence in the validity of scientific discoveries and engineering solutions. The designed experiments serve as a cornerstone of scientific inquiry and engineering practice, offering a systematic and rigorous approach to hypothesis testing, process optimization, and evidence-based decision-making.

1.7 SPECIFIC DESIGNS

1.7.1 ONE FACTOR

1.7.1.1 OFAT: Simple but Inefficient for Exploring Interactions

OFAT experimentation is a simple and straightforward approach to experimental design where you vary one factor (variable) at a time while keeping all other factors constant. While it has limitations compared to more advanced experimental design methods like factorial or response surface design, OFAT can still provide useful insights in some situations. Here are the steps to develop a OFAT experiment:

1. **Identify the Problem or Objective**: Clearly define the problem you want to solve or the objective you want to achieve through experimentation. This step is crucial to ensure that your OFAT experiment addresses the right questions.
2. **Select the Factor to Investigate**: Determine which single factor (independent variable) you want to investigate. This is the variable you will vary during the experiment to observe its effect on the response variable.
3. **Define the Range of Levels**: Determine the range or levels at which you will vary the chosen factor. You need to decide the minimum and maximum values or levels that will be tested. These levels should be relevant to your problem or objective.
4. **Choose a Response Variable**: Identify the response variable (dependent variable) that you will measure to evaluate the effect of the factor under investigation. The response variable should be relevant to your objective and should provide meaningful data.
5. **Hold Other Factors Constant**: To maintain simplicity and isolate the effect of the chosen factor, keep all other relevant factors or variables constant. Ensure that they do not vary during the experiment.
6. **Plan the Experimental Runs**: Create a plan for conducting the experiment, specifying the order in which you will conduct the experimental runs. For OFAT, you typically conduct a series of experiments, each with a different level of the chosen factor while keeping all other factors constant.
7. **Perform the Experiments**: Execute the experimental runs according to your plan. Measure and record the values of the response variable for each run.
8. **Analyze the Data**: Examine the collected data to understand the relationship between the chosen factor and the response variable. This analysis typically involves plotting the data and looking for trends or patterns.
9. **Draw Conclusions**: Based on the data analysis, draw conclusions about the impact of the chosen factor on the response variable. Assess whether the factor had a significant effect and, if so, in what direction.
10. **Report Findings**: Document the results of your OFAT experiment, including the methodology, data collected, and the conclusions drawn. Be clear about the limitations of the OFAT approach, as it may not account for interactions between factors.
11. **Consider Further Experiments**: Depending on the results and the nature of your problem, you may need to conduct additional experiments or use more advanced experimental design methods to explore interactions between factors and optimize your process or product.

It's important to note that while OFAT experimentation is easy to conduct, it has limitations in terms of its ability to uncover complex relationships and interactions between factors. More advanced experimental designs, such as factorial designs or response surface methodologies, are often preferred when possible, as they provide a more comprehensive understanding of the system being studied.

In OFAT design experimentation, you vary one factor at a time while keeping all other factors constant. Here's an example of a matrix for conducting an OFAT experiment to optimize the cooking time for a pasta dish:

Objective: To determine the optimal cooking time for a pasta dish in terms of taste and texture.

Factor: Cooking Time (minutes)

- Levels: 8, 10, 12, 14, 16 minutes

Response Variables:

1. Taste rating (on a scale of 1 to 10, with higher values indicating better taste).
2. Texture rating (on a scale of 1 to 10, with higher values indicating better texture)

You'll conduct five experiments, each with a different cooking time while keeping all other factors constant (Table 1.4).

In this matrix, each row represents an experiment with a specific cooking time (the factor under investigation). You will cook the pasta for the specified time and then have taste testers rate the taste and texture of the dish. Record the taste and texture ratings in the respective columns.

After conducting these experiments, you can analyze the data to determine which cooking time resulted in the best taste and texture, based on the taste and texture ratings provided by the testers. Keep in mind that OFAT experimentation only explores the effect of one factor (cooking time) while assuming that all other factors remain constant. It may not capture potential interactions between factors, so more advanced experimental designs may be needed for a comprehensive optimization process.

1.7.2 Two Factors

1.7.2.1 Full Factorial: Efficient for Exploring All Combinations and Interactions but Can Be Costly for Many Factors

Developing a full factorial design experiment involves systematically planning and conducting experiments to investigate the effects of multiple factors on a response variable. In a full factorial design, all possible combinations of factor levels are tested. Here are the steps to develop a full factorial designed experiment:

TABLE 1.4
OFAT Example

Experiment	Cooking Time (Minutes)	Taste Rating	Texture Rating
1	8	—	—
2	10	—	—
3	12	—	—
4	14	—	—
5	16	—	—

1. **Define the Objective**: Clearly define the research objective or problem that you want to address through the experiment. Determine the response variable (dependent variable) that you want to optimize or understand better.
2. **Identify the Factors**: Identify all the factors (independent variables) that may influence the response variable. These factors could be process variables, ingredients, or any other relevant parameters.
3. **Determine the Factor Levels**: Define the low and high levels for each factor. These levels should encompass the expected operating range of each factor.
4. **Create a Factorial Matrix**: Generate a full factorial matrix that represents all possible combinations of factor levels. Each row in the matrix corresponds to a unique combination of factor levels, and each column represents a factor.
5. **Conduct Randomization (Optional)**: Depending on your experimental design, you may choose to randomize the order in which experiments are conducted to reduce the impact of uncontrolled factors.
6. **Execute the Experiments**: Conduct the experiments as specified in the factorial matrix. Ensure that you control any sources of variability other than the factors under investigation.
7. **Measure and Record Responses**: Measure and record the response variable for each experimental run. Ensure consistency and accuracy in data collection.
8. **Analyze the Data**: Perform statistical analysis, such as ANOVA, to analyze the data and identify significant main effects and interactions between factors.
9. **Interpret Results**: Interpret the results to understand the relationships between factors and the response variable. Identify which factors have the most significant impact and how they interact with each other.
10. **Optimize and Draw Conclusions**: If the objective is optimization, use the results to determine the optimal factor settings that maximize or minimize the response variable, depending on your objective. If the objective is understanding, draw conclusions based on the observed effects.
11. **Verify and Validate**: Conduct additional experiments or validations, if necessary, to confirm the robustness and reliability of the results obtained from the full factorial design.
12. **Report and Document**: Document the experimental design, data collected, analysis methods, results, and conclusions in a clear and comprehensive report. Ensure that the report includes all relevant details for transparency and reproducibility.

Full factorial designs are powerful tools for understanding the impact of multiple factors on a response variable and optimizing processes or products. They provide a comprehensive view of the factor effects and interactions, making them suitable for a wide range of applications in fields, such as engineering, manufacturing, chemistry, and more.

A full factorial design experiment is characterized by testing all possible combinations of factor levels for multiple factors. Here's an example of a matrix for a 2×2 full factorial design, which means there are two factors, each with two levels:

Objective: Determine the effect of two factors, A (temperature) and B (pressure) on the yield of a chemical reaction.

Factors:

1. Factor A: Temperature (T).
   - Level 1: Low temperature (T1) 150°C.
   - Level 2: High temperature (T2) 400°C.
2. Factor B: Pressure (P).
   - Level 1: Low pressure (P1) 2 atm.
   - Level 2: High pressure (P2) 100 atm.

TABLE 1.5
Full Factorial Example

Run	Factor A (Temperature)	Factor B (Pressure)	Yield
1	150°C	2 atm	—
2	400°C	2 atm	—
3	150°C	100 atm	—
4	400°C	100 atm	—

Factorial Matrix: In a 2×2 full factorial design, there are four experimental runs, corresponding to all possible combinations of factor levels (Table 1.5).

In this matrix, each row represents a unique experimental run. The levels of factor A (temperature) are labeled as T1 (low temperature) and T2 (high temperature), while the levels of factor B (pressure) are labeled as P1 (low pressure) and P2 high pressure).

Conduct the Experiments: Execute the experiments as specified in the factorial matrix, varying both temperature and pressure levels. Ensure that all other experimental conditions are consistent.

Measure and Record Responses: Measure and record the yield of the chemical reaction for each experimental run.

Analyze the Data: Perform statistical analysis, such as ANOVA, to analyze the data and determine the main effects of temperature and pressure, as well as any potential interaction effects.

Interpret Results: Based on the analysis, interpret the results to understand how temperature and pressure levels affect the yield of the chemical reaction. Identify which factor(s) have a significant impact and whether there are any interactions between the factors.

Optimize and Draw Conclusions: Depending on the objective, you can use the results to optimize the conditions for maximum yield or draw conclusions about the relationship between factors and the response variable.

In this example, a 2×2 full factorial design allows for a systematic exploration of the effects of temperature and pressure on the chemical reaction's yield, covering all possible combinations of the two factors. In practice, full factorial designs can be expanded to include more factors and levels, providing a comprehensive understanding of complex systems and processes.

1.7.2.2 2 Factor CCD: More Advanced, Useful for Exploring Curvature in the Response Surface

Developing a CCD for an experiment involves a series of steps to systematically plan and conduct the experiment. CCD is a commonly used experimental design methodology that allows for exploring both linear and quadratic effects of factors on a response variable, as well as the assessment of factor interactions. Here are the steps for developing a CCD designed experiment:

1. **Define the Objective**: Clearly define the research objective or problem that you want to address through the experiment. Determine the response variable (dependent variable) that you want to optimize or understand better.
2. **Identify the Factors**: Identify the factors (independent variables) that may influence the response variable. These factors could be process variables, ingredients, or any other relevant parameters.

3. **Determine the Factor Levels**: Define the high and low levels for each factor. These levels should encompass the expected operating range of each factor. In a CCD, there are typically three levels: low, high, and a central point.
4. **Decide on the Number of Center Points**: Determine the number of center points you want to include in your CCD. Center points are experimental runs conducted at the central values of all factors. Including center points helps assess the curvature of the response surface.
5. **Choose the Alpha (α) Value**: Select the α value, which represents the distance of the axial points from the center of the design space. Common values for α are 1 or $\sqrt{2}$, depending on the desired balance between axial points and center points.
6. **Create a Factorial Design**: Generate a full factorial design matrix for your factors, including the center points. This matrix represents all possible combinations of factor levels for a 2^k factorial design, where "k" is the number of factors. The center points are placed at the center of the design matrix.
7. **Calculate Axial Points**: Calculate the axial points (corner points) by applying the α value to the high and low levels of each factor. These axial points are placed at a distance α from the center of the design.
8. **Plan the Experimental Runs**: Combine the center points, axial points, and factorial design points to create a list of experimental runs. Each row in this matrix represents a unique combination of factor levels.
9. **Conduct the Experiments**: Execute the planned experimental runs in a randomized order. Ensure that you control for any sources of variability other than the factors under investigation.
10. **Measure and Record Responses**: Measure and record the response variable for each experimental run. Ensure consistency and accuracy in data collection.
11. **Analyze the Data**: Perform statistical analysis, such as regression analysis, to fit a response surface model to the data. This model helps quantify the relationship between the factors and the response variable, including any linear, quadratic, or interaction effects.
12. **Optimize and Interpret Results**: Use the response surface model to identify optimal factor settings that maximize or minimize the response variable, depending on your objective. Interpret the results to gain insights into the factors' effects and interactions.
13. **Verify and Validate**: Conduct additional experiments or validations, if necessary, to confirm the robustness and reliability of the optimized settings.
14. **Report and Document**: Document the experimental design, data collected, analysis methods, results, and conclusions in a clear and comprehensive report.

CCD designed experiments are powerful tools for optimizing processes, products, or systems while considering the effects of multiple factors and interactions. They are widely used in various fields, including engineering, chemistry, and manufacturing to efficiently explore the design space and achieve desired outcomes.

CCD is a widely used experimental design methodology for optimizing processes or products by exploring factor effects and interactions. Here's an example of a matrix for a CCD designed experiment to optimize the baking time and temperature of a cake:

Objective: Optimize the baking time and temperature to achieve a cake with the best taste and texture.

Factors:

1. Baking Time (minutes).
   - Low Level: 20 minutes.
   - High Level: 40 minutes.

2. Baking Temperature (°C).
   - Low Level: 160°C.
   - High Level: 180°C.

Center Points:

- Center point for baking time = 30 minutes.
- Center point for baking temperature = 170°C.

Alpha (α) Value: Typically set to $\sqrt{2}$ (approximately 1.414) for balanced axial points.

Factorial Design: Example of a full factorial design matrix for the two factors, including the center points (Table 1.6).

Axial (Corner) Points: Calculate the axial points using the α value and the high and low levels of each factor.

For Baking Time:

- Low Axial Point: $20 - \alpha * (30 - 20) = 20 - 1.414 * 10 \approx 5.86$ minutes.
- High Axial Point: $40 + \alpha * (40 - 30) = 40 + 1.414 * 10 \approx 51.4$ minutes.

For Baking Temperature:

- Low Axial Point: $160 - \alpha * (170 - 160) = 160 - 1.414 * 10 \approx 145.86$°C.
- High Axial Point: $180 + \alpha * (180 - 170) = 180 + 1.414 * 10 \approx 194.14$°C.

Experimental Runs (Including Axial Points): Combine the center points, axial points, and factorial design points to create the list of experimental runs (Table 1.7).

TABLE 1.6

Example of a 2-Factor CCD

Run	Baking Time (Minutes)	Baking Temperature (°C)	Taste Rating	Texture Rating
1	20	160	—	—
2	40	160	—	—
3	20	180	—	—
4	40	180	—	—
5	30	170	—	—

TABLE 1.7

Example of a 2-Factor CCD Axial Points

Run	Baking Time (Minutes)	Baking Temperature (°C)	Taste Rating	Texture Rating
1	5.86	145.86	—	—
2	51.4	145.86	—	—
3	5.86	194.14	—	—
4	51.4	194.14	—	—
5	30	170	—	—

Conduct the Experiments: Execute the planned experimental runs, ensuring that all measurements are accurate and consistent.

Measure and Record Responses: Measure and record the taste and texture ratings for each cake in each experimental run.

Analyze the Data: Use statistical analysis, such as regression analysis or response surface methodology, to fit a response surface model to the data. This model helps identify the optimal combination of baking time and temperature for the best taste and texture.

Interpret Results: Interpret the results to determine the optimal settings for baking time and temperature that yield the desired taste and texture for the cake.

By following this CCD designed experiment, you can efficiently optimize the baking process for the cake while considering the effects of baking time and temperature and their potential interactions.

1.7.3 Three or More Factors

1.7.3.1 3-Factor + CCD: More Advanced, Useful for Exploring Curvature in the Response Surface with 3 or More Factors

A 3-Factor+CCD is a type of response surface design used to explore the effects of multiple factors (3+) on a response variable while considering both linear and quadratic relationships. It includes three sets of experimental runs: factorial points, axial points, and center points. Here's an example of a matrix for a CCD experiment with three factors:

Objective: Optimize the yield of a chemical reaction by varying three factors: temperature (T), pressure (P), and concentration (C).

Factors:

Factor A: Temperature (T)
- Low Level: 150°C.
- High Level: 250°C.

Factor B: Pressure (P)
- Low Level: 20 atm.
- High Level: 40 atm.

Factor C: Concentration (C)
- Low Level: 0.5 M.
- High Level: 1.5 M.

Center Points:

- The center points represent the midpoint of each factor's range.

Axial Points:

- Axial points are experiments conducted at distances from the center equal to $\pm\alpha$, where α is a scaling factor.

Alpha (α) Value: Typically set to $\sqrt{2}$ (approximately 1.414) for balanced axial points.

Factorial Matrix: Here's an example of the CCD matrix, which includes the center points, factorial points, and axial points (Table 1.8).

TABLE 1.8

Example of a 3-Factor+ CCD

Run	Factor A (T)	Factor B (P)	Factor C (C)	Yield
1	−1	−1	−1	—
2	+1	−1	−1	—
3	−1	+1	−1	—
4	+1	+1	−1	—
5	−1	−1	+1	—
6	+1	−1	+1	—
7	−1	+1	+1	—
8	+1	+1	+1	—
9	0	0	0	—
10	$-\alpha$	0	0	—
11	$+\alpha$	0	0	—
12	0	$-\alpha$	0	—
13	0	$+\alpha$	0	—
14	0	0	$-\alpha$	—
15	0	0	$+\alpha$	—

In this matrix,

- Each row represents a unique combination of factor levels for temperature (T), pressure (P), and concentration (C).
- The levels of each factor are represented as −1 for the low level and +1 for the high level.
- The center points (runs 9–15) are at the midpoints of each factor's range.
- The factorial points (runs 1–8) represent combinations of factors at specified distances from the center.
- The axial points (runs 10–15) are experiments conducted at distances α ($\alpha \approx 1.414$) from the center along each factor's axis.

Conduct the Experiments: Execute the experiments as specified in the CCD matrix, varying temperature, pressure, and concentration levels. Measure and record the yield of the chemical reaction for each experimental run.

Analyze the Data: Perform statistical analysis, such as response surface modeling, to analyze the data and identify the optimal factor settings that maximize or minimize the yield while considering interactions between factors.

1.7.3.2 BBD: Efficient for Exploring Quadratic Terms without Requiring as Many Runs as a Full Factorial

Developing a BBD for an experiment involves a series of steps to systematically plan and conduct the experiment. BBD is a response surface design that allows for the exploration of factor effects and interactions with a reduced number of experimental runs compared to a full factorial design. Here are the steps to develop a BBD experiment:

1. **Define the Objective:** Clearly define the research objective or problem that you want to address through the experiment. Determine the response variable (dependent variable) that you want to optimize or understand better.

2. **Identify the Factors**: Identify the factors (independent variables) that may influence the response variable. These factors could be processing variables, ingredients, or any other relevant parameters.
3. **Determine the Factor Levels**: Define the low and high levels for each factor. These levels should encompass the expected operating range of each factor.
4. **Choose the Number of Factors:** Determine the number of factors you want to include in your BBD. BBD is typically used for experiments involving three or more factors.
5. **Generate the Design Matrix**: Use statistical software or tables to generate a BBD matrix. This matrix represents a subset of all possible combinations of factor levels, including center points.
6. **Plan the Experimental Runs**: Combine the center points and the non-center points (factorial points) from the design matrix to create a list of experimental runs. Each row in this matrix corresponds to a unique combination of factor levels.
7. **Conduct Randomization (Optional):** Depending on your experimental design, you may choose to randomize the order in which experiments are conducted to reduce the impact of uncontrolled factors.
8. **Execute the Experiments**: Conduct the experiments as specified in the BBD matrix. Ensure that you control for any sources of variability other than the factors under investigation.
9. **Measure and Record Responses**: Measure and record the response variable for each experimental run. Ensure consistency and accuracy in data collection.
10. **Analyze the Data**: Perform statistical analysis, such as regression analysis or response surface methodology, to analyze the data and fit a response surface model to the observed responses.
11. **Interpret Results**: Interpret the results to understand the relationships between factors and the response variable. Identify significant main effects and potential interactions between factors.
12. **Optimize and Draw Conclusions**: If the objective is optimization, use the response surface model to determine the optimal factor settings that maximize or minimize the response variable, depending on your objective. If the objective is understanding, draw conclusions based on the observed effects.
13. **Verify and Validate**: Conduct additional experiments or validations, if necessary, to confirm the robustness and reliability of the results obtained from the BBD.
14. **Report and Document**: Document the experimental design, data collected, analysis methods, results, and conclusions in a clear and comprehensive report. Ensure that the report includes all relevant details for transparency and reproducibility.

BBDs are valuable tools for optimizing processes, products, or systems by efficiently exploring the design space and assessing the effects of multiple factors and their interactions. They strike a balance between full factorial designs and simpler experimental designs, allowing for practical experimentation with reduced resource requirements.

A BBD is typically used when you have three continuous factors (variables) and you want to optimize a response variable while exploring the effects and interactions of these factors within a specified range. Here's an example of a BBD matrix for a hypothetical experiment:

Objective: Optimize the yield of a chemical reaction by varying three factors: temperature (T), pressure (P), and concentration (C).

Factors:

1. Factor A: Temperature (T)
   - Low Level: 150°C.
   - High Level: 200°C.

2. Factor B: Pressure (P)
   - Low Level: 20 atm.
   - High Level: 40 atm.
3. Factor C: Concentration (C)
   - Low Level: 0.5 M.
   - High Level: 1.5 M.

Center Points:

- The center points represent the midpoint of each factor's range.

Factorial Points:

- The BBD includes a set of factorial points, which are experimental runs for combinations of factors at specified distances from the center.

Axial Points:

- Axial points represent experiments conducted at distances from the center equal to $\pm\alpha$, where α is a scaling factor.

Alpha (α) Value: Typically set to $\sqrt{2}$ (approximately 1.414) for balanced axial points.

Factorial Matrix:

- Generate the BBD matrix, which includes the center points, factorial points, and axial points (Table 1.9).

In this matrix,

- Each row represents a unique combination of factor levels for temperature (T), pressure (P), and concentration (C).
- The center points (run 9 and run 10) are at the midpoints of each factor's range.
- The factorial points (runs 1–8) represent combinations of factors at specified distances from the center.

TABLE 1.9

Example of a BBD Matrix

Run	Factor A (T)	Factor B (P)	Factor C (C)	Yield
1	200°C	20 atm	0.5 M	—
2	200°C	20 atm	1.5 M	—
3	200°C	40 atm	0.5 M	—
4	200°C	40 atm	1.5 M	—
5	150°C	20 atm	0.5 M	—
6	150°C	20 atm	1.5 M	—
7	150°C	40 atm	0.5 M	—
8	150°C	40 atm	1.5 M	—
9	175°C	30 atm	1.0 M	—
10	175°C	30 atm	1.0 M	—

- The axial points (runs 1–4 and runs 5–8) are experiments conducted at distances α ($\alpha \approx 1.414$) from the center along each factor's axis.

Conduct the Experiments: Execute the experiments as specified in the BBD matrix, varying temperature, pressure, and concentration levels. Measure and record the yield of the chemical reaction for each experimental run.

Analyze the Data: Perform statistical analysis, such as response surface modeling, to analyze the data and identify the optimal factor settings for maximizing the yield of the chemical reaction while considering interactions between factors.

BBDs allow you to efficiently explore the design space and optimize complex processes or systems by conducting a relatively small number of experiments while still capturing the important effects and interactions of the factors involved.

1.7.3.3   PBD: Useful for Screening Many Factors with Limited Resources, but Only Provides Information about Main Effects

Developing a PBD for an experiment involves a series of steps to systematically plan and conduct the experiment. A PBD is a type of screening design used to identify the most significant factors affecting a response variable. It is particularly useful when you have several factors to consider and want to quickly identify the key factors for further investigation. Here are the steps to develop a PBD experiment:

1. **Define the Objective**: Clearly define the research objective or problem that you want to address through the experiment. Determine the response variable (dependent variable) that you want to optimize or understand better.
2. **Identify the Factors**: Identify all the factors (independent variables) that may influence the response variable. These factors could be process variables, ingredients, or any other relevant parameters.
3. **Determine the Factor Levels**: For each factor, define the high and low levels. These levels should encompass the expected operating range of each factor.
4. **Choose the Number of Factors**: Determine the number of factors you want to include in your PBD. A PBD is typically used for experiments involving a larger number of factors, often more than ten.
5. **Generate the Design Matrix**: Use statistical software or tables to generate a PBD matrix. This matrix represents a subset of all possible combinations of factor levels.
6. **Plan the Experimental Runs:** The PBD matrix includes a series of experimental runs, each representing a unique combination of factor levels. Ensure that you follow the order specified in the matrix.
7. **Conduct Randomization (Optional):** Depending on your experimental design, you may choose to randomize the order in which experiments are conducted to reduce the impact of uncontrolled factors.
8. **Execute the Experiments:** Conduct the experiments as specified in the PBD matrix. Ensure that you control any sources of variability other than the factors under investigation.
9. **Measure and Record Responses:** Measure and record the response variable for each experimental run. Ensure consistency and accuracy in data collection.
10. **Analyze the Data:** Perform statistical analysis to determine the most significant factors affecting the response variable. This analysis often involves identifying factors with p-values below a specified significance level.
11. **Interpret Results:** Interpret the results to understand which factors have the most significant impact on the response variable. Focus on the key factors that are identified as significant.

12. **Select Factors for Further Investigation:** Based on the results, select the most significant factors for further investigation or optimization. These factors are candidates for more in-depth experimentation or process improvement.
13. **Report and Document:** Document the experimental design, data collected, analysis methods, results, and conclusions in a clear and comprehensive report. Ensure that the report includes all relevant details for transparency and reproducibility.

PBDs are valuable tools for efficiently screening a large number of factors to identify the most important ones affecting a response variable. They are particularly useful in the early stages of experimentation when you want to quickly narrow down the list of factors for more detailed investigations.

A PBD is typically used to screen a large number of factors to identify the most significant ones that affect a response variable. Here's an example of a matrix for a PBD experiment with eight factors:

Objective: Determine the most significant factors affecting the yield of a chemical reaction.

Factors: You have eight factors that may influence the reaction yield.

Factor Levels: For a PBD, you typically use two levels for each factor: a high level (+1) and a low level (–1).

Design Matrix: The matrix below represents a PBD with eight factors (Table 1.10).
In this matrix,

- Each row represents a unique combination of factor levels for the eight factors.
- The factors are denoted as factor 1, factor 2, and so on.
- For each factor, +1 represents the high level, and –1 represents the low level.
- The last column (yield) is where you would record the response variable (e.g., the yield of the chemical reaction) for each experimental run.

The PBD matrix systematically varies the factors at their high and low levels to assess their impact on the response variable. After conducting these eight experimental runs and measuring the response (yield) for each run, you can perform statistical analysis to identify which factors have the most significant effects on the yield. Factors with significant effects may warrant further investigation or optimization in subsequent experiments.
Consider another example with specific factors.

TABLE 1.10

Example of an 8-Factor PBD Matrix

Run	Factor 1	Factor 2	Factor 3	Factor 4	Factor 5	Factor 6	Factor 7	Factor 8	Yield
1	+1	–1	–1	–1	+1	–1	+1	+1	—
2	–1	+1	–1	–1	+1	+1	–1	+1	—
3	–1	–1	+1	–1	–1	+1	+1	+1	—
4	–1	–1	–1	+1	–1	–1	–1	+1	—
5	+1	+1	+1	+1	–1	–1	–1	–1	—
6	–1	–1	+1	+1	+1	+1	–1	–1	—
7	+1	–1	+1	+1	–1	+1	+1	–1	—
8	+1	+1	–1	+1	+1	–1	+1	–1	—

Objective: Identify the most significant factors influencing the yield of a chemical reaction.

Factors: For this example, we have six factors, each with two levels (high and low):

1. Factor A: Temperature (T) $+1 = 150°C$, $-1 = 10°C$.
2. Factor B: Pressure (P) $+1 = 30$ atm, $-1 = 1$ atm.
3. Factor C: Catalyst concentration (C) $+1 = 0.35/L$, $-1 = 0.075/L$.
4. Factor D: Catalyst type (CT) $+1 = Ni$, $-1 = Sn$.
5. Factor E: Stirring speed (SS) $+1 = 1000$ rpm, $-1 = 100$ rpm.
6. Factor F: Gravity (G) $+1 = 9.8$ m/s^2, $-1 = 0.001$ m/s^2.

Design Matrix: A PBD matrix typically consists of experimental runs, each representing a unique combination of factor settings. In PBDs, each factor is examined at two levels (+1 for high and −1 for low), and all possible combinations of low and high levels are considered to evaluate the main effects.

For a 12-run PBD matrix, here's a partial example of the design (Table 1.11).
In this matrix,

- Each row represents an experimental run with a specific combination of high (+1) and low (−1) levels for each factor.
- The response variable (yield) is recorded for each experimental run to determine the effect of the factor settings on the yield of the chemical reaction.

After conducting these experiments and recording the yield values, statistical analysis, such as calculating the main effects of each factor, is performed to identify the most significant factors influencing the response variable. The goal is to screen and prioritize the factors for further, more detailed experimentation or optimization.

1.7.3.4 Fractional Factorial Design: Efficient for Screening and Identifying Key Factors, Requiring Fewer Runs Than a Full Factorial

Developing a fractional factorial design for an experiment involves a series of steps to systematically plan and conduct the experiment. Fractional factorial designs are used when you have multiple

TABLE 1.11

Example of a 6-Factor PBD Matrix

Run	A (T)	B (P)	C (C)	D (CT)	E (SS)	F (G)	Yield
1	+1	−1	+1	−1	+1	−1	—
2	−1	+1	+1	−1	−1	+1	—
3	+1	+1	−1	+1	−1	−1	—
4	−1	−1	−1	+1	+1	+1	—
5	+1	−1	−1	+1	+1	+1	—
6	−1	+1	−1	+1	−1	−1	—
7	+1	+1	+1	−1	−1	−1	—
8	−1	−1	+1	−1	+1	+1	—
9	+1	+1	−1	−1	+1	−1	—
10	−1	−1	−1	−1	−1	+1	—
11	+1	−1	+1	+1	−1	+1	—
12	−1	+1	+1	+1	+1	−1	—

factors, but you want to reduce the number of experimental runs required while still capturing the main effects and some interactions between factors. Here are the steps to develop a fractional factorial design experiment:

1. **Define the Objective**: Clearly define the research objective or problem that you want to address through the experiment. Determine the response variable (dependent variable) that you want to optimize or understand better.
2. **Identify the Factors:** Identify all the factors (independent variables) that may influence the response variable. These factors could be process variables, ingredients, or any other relevant parameters.
3. **Determine the Factor Levels**: Define the low and high levels for each factor. These levels should encompass the expected operating range of each factor.
4. **Choose the Fraction:** Determine the fraction of the full factorial design that you want to use. Common fractions include half-fraction (1/2), quarter-fraction (1/4), and eighth-fraction (1/8). The fraction represents the reduction in the number of experimental runs compared to a full factorial design.
5. **Generate the Fractional Factorial Matrix:** Use statistical software or tables to generate the fractional factorial design matrix. This matrix represents a subset of all possible combinations of factor levels.
6. **Plan the Experimental Runs:** The fractional factorial design matrix includes a series of experimental runs, each representing a unique combination of factor settings. Ensure that you follow the order specified in the matrix.
7. **Conduct Randomization (Optional):** Depending on your experimental design, you may choose to randomize the order in which experiments are conducted to reduce the impact of uncontrolled factors.
8. **Execute the Experiments**: Conduct the experiments as specified in the fractional factorial design matrix, varying the factor levels accordingly. Ensure that you control for any sources of variability other than the factors under investigation.
9. **Measure and Record Responses**: Measure and record the response variable for each experimental run. Ensure consistency and accuracy in data collection.
10. **Analyze the Data**: Perform statistical analysis to analyze the data and identify the main effects of factors. Fractional factorial designs are designed to capture the main effects but may not fully account for interactions between factors.
11. **Interpret Results**: Interpret the results to understand which factors have the most significant impact on the response variable based on their main effects. If interactions are of interest, further experimentation may be needed.
12. **Select Factors for Further Investigation**: Based on the results, select the most significant factors for further investigation or optimization. If interactions are important, consider conducting additional experiments to study them in more detail.
13. **Report and Document**: Document the experimental design, data collected, analysis methods, results, and conclusions in a clear and comprehensive report. Ensure that the report includes all relevant details for transparency and reproducibility.

Fractional factorial designs are useful for efficiently exploring factor effects while reducing the number of experimental runs. They are commonly used in situations where resources or time constraints make it impractical to conduct a full factorial experiment.

A fractional factorial design matrix represents a subset of all possible combinations of factor levels, allowing you to conduct experiments while reducing the number of runs compared to a full factorial design. Here's an example of a 1/2 fractional factorial design matrix for a hypothetical experiment with four factors:

TABLE 1.12

Example of a Fractional Factorial Design Matrix ½ Fraction

Run	Factor A	Factor B	Factor C	Factor D	Yield
1	−1	−1	−1	−1	—
2	+1	−1	−1	−1	—
3	−1	+1	−1	−1	—
4	+1	+1	−1	−1	—
5	−1	−1	+1	−1	—
6	+1	−1	+1	−1	—
7	−1	+1	+1	−1	—
8	+1	+1	+1	−1	—
9	−1	−1	−1	+1	—
10	+1	−1	−1	+1	—
11	−1	+1	−1	+1	—
12	+1	+1	−1	+1	—

Objective: Investigate the effects of four factors (A, B, C, and D) on the yield of a chemical reaction.

Factors: For this example, we have four factors, each with two levels (low and high):

1. Factor A.
2. Factor B.
3. Factor C.
4. Factor D.

Fractional Factorial Matrix (1/2 Fraction): A 1/2 fractional factorial design reduces the number of runs to half of a full factorial design. The matrix for a 1/2 fractional factorial design is shown in Table 1.12.

In this matrix,

- Each row represents an experimental run with a specific combination of factor settings.
- The levels of each factor are represented as -1 for the low level and +1 for the high level.
- The response variable (yield) is recorded for each experimental run to determine the effects of the factor settings on the yield of the chemical reaction.

With a 1/2 fractional factorial design, you can study the main effects of the four factors while conducting only half of the experiments compared to a full factorial design. This approach is useful when you want to efficiently identify the most influential factors in a complex system while conserving resources and time.

1.7.3.5 Derringer Design: Useful for Optimizing Multiple Responses Simultaneously When Interactions Are Important

Derringer's desirability function is a methodology used to optimize processes or products by considering multiple responses or quality characteristics simultaneously. Developing a Derringer design involves several steps:

1. **Define the Objective:** Clearly define the objective of your experiment. Determine what you want to optimize or improve and identify the responses or quality characteristics you need to consider.
2. **Identify the Factors**: Identify the factors (independent variables) that may influence the responses or quality characteristics. These factors could be process variables, ingredient quantities, or any other relevant parameters.
3. **Determine the Factor Levels**: Define high and low levels for each factor. These levels should encompass the expected operating range of each factor.
4. **Specify the Target Values**: Determine the target values or desired ranges for each response or quality characteristic. You should have a clear understanding of what constitutes an ideal outcome for each response.
5. **Create a Response Surface Model**: Develop a mathematical model that relates the factor levels to the responses or quality characteristics. This model can be linear, quadratic, or more complex, depending on the nature of the relationship.
6. **Define the Desirability Function**: Define a desirability function for each response or quality characteristic. The desirability function assigns a desirability score to different levels or ranges of each response, indicating how desirable they are. The function should reflect your preferences and objectives, aiming to maximize desirability.
7. **Calculate the Overall Desirability**: Combine the desirability scores for all responses to calculate an overall desirability score. This score provides a measure of how well the combination of factor settings meets your objectives across multiple responses.
8. **Optimize the Factors**: Use optimization techniques to find the factor settings that maximize the overall desirability score. Optimization algorithms can help you identify the best combination of factor levels that yield the most desirable outcomes across all responses.
9. **Conduct Experiments**: Execute experimental runs based on the factor settings identified in the optimization step. Ensure that you control for any sources of variability other than the factors under investigation.
10. **Measure Responses**: Measure and record the responses or quality characteristics for each experimental run. Ensure consistency and accuracy in data collection.
11. **Analyze the Data:** Analyze the experimental data and compare the observed responses to the target values or desired ranges. Calculate the overall desirability score to assess how well the optimized factor settings meet your objectives.
12. **Interpret Results**: Interpret the results to determine whether the factor settings identified through optimization effectively optimize the multiple responses or quality characteristics. Assess whether the objectives are met across all responses.
13. **Verify and Validate**: Conduct additional experiments or validations, if necessary, to confirm the robustness and reliability of the optimized factor settings and their impact on the responses.
14. **Report and Document**: Document the experimental design, data collected, analysis methods, results, and conclusions in a clear and comprehensive report. Ensure that the report includes all relevant details for transparency and reproducibility.

Derringer's desirability function approach is particularly useful when you need to optimize multiple responses simultaneously, considering both desirable and undesirable characteristics, to achieve an overall optimal outcome. It allows for a holistic approach to optimization, taking into account the trade-offs between different quality characteristics.

Derringer's desirability function design experimentation involves optimizing multiple responses or quality characteristics simultaneously. While there is no specific matrix for Derringer's desirability function, you can illustrate the concept with an example using hypothetical responses and their associated desirability functions.

Example 1

Objective: Optimize the formulation of a food product with multiple quality characteristics: taste, texture, and shelf life.

Factors:

- Factor A: Amount of spice.
- Factor B: Sugar content.
- Factor C: Storage temperature.

Responses and Desirability Functions: For each response (taste, texture, shelf life), you define a desirability function that quantifies how desirable different levels or ranges of that response are. The desirability function typically ranges from 0 (undesirable) to 1 (ideal). Here's a simplified example:

1. Taste:
   - Desirability function:
     - Taste is rated on a scale from 1 to 10.
     - Desirable taste (rating 8–10) is assigned a desirability of 1.
     - Acceptable taste (rating 6–7) is assigned a desirability of 0.7.
     - Undesirable taste (rating 1–5) is assigned a desirability of 0.
2. Texture:
   - Desirability function:
     - Smooth texture is desirable (rated 1).
     - Slightly grainy texture is acceptable (rated 0.7).
     - Coarse texture is undesirable (rated 0).
3. Shelf Life:
   - Desirability function:
     - Longer shelf life is desirable (rated 1).
     - Moderate shelf life is acceptable (rated 0.7).
     - Short shelf life is undesirable (rated 0).

Overall Desirability: To combine the desirability functions for all three responses, you can calculate an overall desirability score. This score reflects how well a particular combination of factor settings optimizes taste, texture, and shelf life simultaneously.

For each experimental run, you calculate the overall desirability as the geometric mean of the desirability values for individual responses. The geometric mean ensures that you consider all responses equally and penalizes combinations that perform poorly in any one response.

Experimental Runs: Conduct a series of experimental runs using different factor settings for the amount of spice, sugar content, and storage temperature. Measure the responses (taste, texture, and shelf life) for each run.

A simplified example with two experimental runs for illustration is shown in Table 1.13.

TABLE 1.13

Example of a Derringer Design Matrix

Run	Factor A (Spice)	Factor B (Sugar)	Factor C (Temp)	Taste	Texture	Shelf Life	Overall Desirability
1	Low	High	Low	9	0.8	0.9	Calculated
2	High	Low	High	7	1	0.7	Calculated

Analysis:

- Calculate the desirability for each response based on the desirability functions.
- Calculate the overall desirability for each run using the geometric mean.
- Identify the factor settings that result in the highest overall desirability score. These settings represent the optimal formulation for the food product that balances taste, texture, and shelf life according to the desirability functions.

In practice, Derringer's desirability function allows you to optimize processes or products considering multiple responses and their associated desirabilities, providing a holistic approach to quality improvement or product formulation. The specific desirability functions and factor settings depend on your objectives and the nature of the responses.

Example 2

Derringer's desirability function approach focuses on optimizing processes or products by considering multiple responses or quality characteristics simultaneously. Unlike traditional experimental designs with a fixed matrix, the design matrix in this approach is not predefined but rather a result of optimization based on the desirability function. Here's a simplified example to illustrate how Derringer's desirability function works:

Objective: Optimize the properties of a pharmaceutical tablet with three quality characteristics: hardness, dissolution rate, and friability.

Factors: For this example, let's consider two factors:

1. Factor A: Compression force (low and high levels).
2. Factor B: Binder type (binder X and binder Y).

Responses: Each response or quality characteristic should be optimized. Target values or ranges are defined for each response:

1. Hardness: Target range (60–70 N).
2. Dissolution rate: Target range (90%–95%).
3. Friability: Target range (0.1%–0.2%).

Desirability Function: Create a desirability function for each response that quantifies how close the observed values are to the target values. You may use a sigmoid function, linear function, or other suitable functions for each response. The desirability scores range from 0 (undesirable) to 1 (perfectly desirable).

Overall Desirability: Combine the desirability scores for all three responses to calculate an overall desirability score for each combination of factor settings. The overall desirability score reflects the quality of the tablet with respect to all three responses.

Here is a simplified example of how the optimization process might work for a Derringer's desirability function:

1. Define the desirability functions for each response (hardness, dissolution rate, friability).
2. Create a grid of factor settings for compression force (factor A) and binder type (factor B), considering the specified levels.
3. For each combination of factor settings, calculate the desirability score for each response.
4. Calculate the overall desirability score for each combination of factor settings by combining the individual desirability scores using a weighted geometric mean or other aggregation method.

5. Use optimization techniques (e.g., numerical optimization algorithms) to find the factor settings that maximize the overall desirability score. These settings represent the best combination of compression force and binder type that optimizes all three tablet properties simultaneously.
6. Conduct experimental runs based on the optimized factor settings to produce tablets that meet the desired quality characteristics.

Remember that this example is simplified for illustration purposes. In practice, the desirability functions, response targets, and factor ranges would be determined based on scientific knowledge and experimental data. The optimization process would involve more complex mathematical calculations, and the design space may include a larger number of factors and levels.

Example 3

This example looks at optimizing multiple responses or quality characteristics simultaneously. It involves defining desirability functions for each response and combining them into an overall desirability score. While there isn't a specific matrix associated with Derringer's desirability function, I can provide a simplified example to illustrate how it works.

Objective: Optimize a baking process to simultaneously improve three quality characteristics: taste, texture, and color of a baked product.

Responses (Quality Characteristics):

1. Taste (Response 1): Higher values are better.
2. Texture (Response 2): Higher values are better.
3. Color (Response 3): Higher values are better.

Factor: Oven temperature (factor A)

Factor Levels:

- Low Level (A1): 350°F.
- High Level (A2): 400°F.

Desirability Functions: For each response, you need to define a desirability function that quantifies how desirable a particular outcome is. Desirability functions can take various forms, such as linear, quadratic, or custom shapes, depending on your preferences. Here, we'll use a simple linear desirability function:

- Taste (Response 1):
  - Desirability is 0 if the taste is below a certain threshold (e.g., 5 on a scale of 1 to 10).
  - Desirability linearly increases from 0 to 1 as taste improves from the threshold to a maximum desired value (e.g., 9 on a scale of 1 to 10).
  - Desirability is 1 if taste equals or exceeds the maximum desired value.
- Texture (Response 2):
  - Similar desirability function as taste.
- Color (Response 3):
  - Similar desirability function as taste.

Matrix for Experiment: You would conduct a series of experimental runs with different combinations of oven temperatures (low and high levels) and measure the taste, texture, and color for each run. A simplified representation is shown in Table 1.14.

TABLE 1.14

Example of a Derringer Optimization Design Matrix

Run	Factor A (Oven Temp)	Taste (Response 1)	Texture (Response 2)	Color (Response 3)	Desirability
1	A1	7	8	7	--
2	A2	8	7	8	--
3	A1	6	9	8	--
4	A2	9	6	9	--
...	...	...	...	...	...

Desirability Calculation:

Using the defined desirability functions for taste, texture, and color, you would calculate the desirability score for each response for each run. Then, you'd combine these scores to calculate an overall desirability score.

For instance, if taste, texture, and color have desirability scores of 0.8, 0.7, and 0.9, respectively, for a particular run, the overall desirability for that run could be calculated as a weighted average:

$$\text{Overall desirability} = (0.8 * \text{weight for taste} + 0.7 * \text{weight for texture} + 0.9 * \text{weight for color})/(\text{total weights})$$

The weights reflect the relative importance of each response. A higher overall desirability score indicates a better compromise between the multiple quality characteristics.

The optimization process would involve finding the factor settings (oven temperature in this case) that maximize the overall desirability score while simultaneously improving taste, texture, and color.

2 Science Designed Experiments

2.1 BIOLOGY

Designed experiments are essential in biology for conducting controlled studies, optimizing laboratory protocols, and understanding biological processes. In biology designed experiments, various types of variables are considered to investigate biological processes, phenomena, and organisms. These variables can be categorized into several common types, including:

IVs:

Treatment Conditions: Variables related to the experimental treatments or interventions applied to biological systems, such as drug dosages, exposure to environmental factors, or genetic modifications.

Environmental Factors: Variables related to environmental conditions, such as temperature, light, humidity, and nutrient availability, which can impact biological organisms and processes.

Genetic Factors: Variables related to genetic traits, mutations, or gene expression levels, particularly relevant in genetics and molecular biology experiments.

Time: Variables related to the timing and duration of events or processes in biological systems.

DVs:

Biological Responses: Variables related to the biological responses or outcomes under investigation, including growth rates, enzyme activity, gene expression levels, cell viability, and behavioral responses.

Phenotypic Traits: Variables related to observable traits or characteristics of organisms, such as size, color, shape, and morphological features.

Biochemical Data: Variables related to biochemical measurements, including concentrations of metabolites, proteins, and biomarkers.

Molecular Data: Variables related to molecular data, such as deoxyribonucleic acid (DNA) sequences, ribonucleic acid (RNA) expression profiles, or protein concentrations.

Categorical Variables:

Species or Organisms: Categories of biological species or organisms used in experiments, particularly relevant in ecology and biodiversity studies.

Experimental Groups: Categories of experimental groups or conditions, such as control groups, treatment groups, and different genetic strains.

Life Stages: Categories representing different life stages of organisms, such as larval, juvenile, or adult stages.

Control Variables (Covariates): These are variables that are held constant or controlled during experiments to eliminate their influence on the DV. For example, maintaining a constant temperature or pH in a cell culture experiment.

Random Variables: In some cases, random variables may be introduced to account for variability or uncertainty in biological measurements, such as individual differences in response.

Interaction Variables: Interaction variables are used to investigate the combined effects of two or more IVs on the DV. For example, the interaction between genetic factors and environmental conditions on phenotype expression.

DOI: 10.1201/9781003528531-2

Noise Variables (Error Variables): These are uncontrolled or unmeasured variables that can introduce variability or errors into the experimental results. Techniques like statistical analysis and experimental controls are used to minimize the impact of noise variables in biology experiments.

Biology designed experiments are essential for understanding biological processes, disease mechanisms, ecological interactions, and genetic functions. They play a critical role in advancing knowledge in fields, such as genetics, ecology, physiology, and molecular biology. Careful control and manipulation of variables are crucial for obtaining reliable and meaningful data in biology experiments.

The following are examples of designed experiments for biological investigation. The common designs for experiments include:

Full Factorial Design: Examines all possible combinations of factor levels.

Fractional Factorial Design: Studies a subset of factor-level combinations to reduce the number of **experimental runs.**

Taguchi Method: Focuses on robust parameter design by considering factors and their interactions.

Central Composite Design (CCD): Investigates both linear and quadratic effects of factors.

And customized design based on specific research needs.

2.1.1 BIOLOGY FACTORIAL EXPERIMENTS

Factorial design experiments are a powerful tool in biology, enabling researchers to efficiently explore complex biological systems, optimize experimental conditions, and gain insight into the underlying mechanisms governing biological processes. By systematically varying multiple factors and their interactions, factorial experiments provide a robust and flexible framework for hypothesis testing and knowledge discovery in biology.

Factorial design experiments are widely used in biology for several reasons:

Efficient Exploration of Multiple Factors: Biology is often characterized by complex interactions between multiple factors, such as genetic variation, environmental conditions, and physiological processes. Factorial experiments allow researchers to systematically explore the effects of multiple factors and their interactions in a single experiment. By varying multiple factors simultaneously, factorial designs enable researchers to identify synergistic or antagonistic effects that may not be apparent when studying individual factors in isolation.

Optimization of Experimental Conditions: Factorial experiments are valuable for optimizing experimental conditions and identifying optimal parameter settings. By systematically varying factors, such as temperature, pH, nutrient concentrations, or treatment dosages, researchers can identify the conditions that maximize desired outcomes, such as cell growth, enzyme activity, or gene expression. Factorial designs allow researchers to efficiently explore a wide range of conditions and identify the most influential factors and their optimal levels.

Statistical Efficiency and Power: Factorial experiments are statistically efficient, meaning they allow researchers to estimate main effects and interactions with high precision using a relatively small number of experimental runs. By fully exploiting the factorial structure of the design, researchers can maximize the information obtained from each experimental observation, leading to more precise estimates of treatment effects and higher statistical power for detecting significant effects. This is particularly important in biology, where experiments often involve variability and noise inherent in biological systems.

Insight into Complex Biological Systems: Biological systems are inherently complex, with numerous interacting components and feedback loops. Factorial experiments provide a systematic framework for studying these complexities and elucidating the underlying mechanisms governing biological processes. By manipulating multiple factors simultaneously, researchers can uncover synergistic or antagonistic interactions between factors and gain insight into the underlying biology of the system.

Flexibility and Adaptability: Factorial designs offer flexibility and adaptability, allowing researchers to modify and expand experimental designs as new questions arise or new factors need to be considered. Researchers can easily add or remove factors, change factor levels, or introduce additional treatments within the factorial framework, enabling iterative experimentation and hypothesis testing.

Biologists may conduct factorial experiments to study the growth of plants under different environmental conditions. They can vary factors, such as light intensity, temperature, and soil nutrients to assess their effects on plant growth.

Factorial design experiments are used in biology in various research areas and applications to systematically investigate the effects of multiple factors or variables on biological processes, organisms, or systems. Factorial design experiments are used across various fields of biology for studying complex interactions between multiple factors and elucidating underlying biological mechanisms. Here are some specific areas where factorial experiments are commonly employed in biology:

Genetics and Genomics: Factorial experiments are used to study gene–environment interactions, gene–gene interactions, and the effects of genetic mutations or variations on phenotypic traits. For example, researchers may use factorial designs to investigate how different gene mutations interact with environmental factors to influence disease susceptibility, drug response, or phenotypic variation.

Ecology and Environmental Science: In ecology, factorial experiments are used to study the effects of multiple environmental factors (such as temperature, precipitation, nutrient availability, and species interactions) on ecosystem dynamics, species abundance, and community composition. Factorial designs enable researchers to assess how combinations of factors interact to shape ecological processes and biodiversity patterns.

Physiology and Pharmacology: Factorial experiments are employed in physiology and pharmacology to investigate the effects of multiple treatments or interventions on physiological responses, drug efficacy, and toxicity. Researchers may use factorial designs to study how different drugs, dosages, and treatment regimens interact with genetic or physiological factors to influence disease progression or treatment outcomes.

Cellular and Molecular Biology: In cellular and molecular biology, factorial experiments are used to study signaling pathways, gene regulation, and protein interactions. Researchers may use factorial designs to investigate how multiple factors, such as signaling molecules, transcription factors, and post-translational modifications, interact to regulate cellular processes and molecular pathways.

Agriculture and Plant Science: In agriculture and plant science, factorial experiments are used to study the effects of agronomic practices, environmental conditions, and genetic factors on crop productivity, disease resistance, and stress tolerance. Factorial designs enable researchers to optimize agricultural practices, breed resilient crop varieties, and develop sustainable farming systems.

Behavioral Ecology and Animal Behavior: In behavioral ecology and animal behavior, factorial experiments are employed to study the effects of multiple factors, such as social interactions, environmental cues, and genetic predispositions, animal behavior, mating strategies, and social organization. Factorial designs enable researchers to manipulate and control variables to test hypotheses about the adaptive significance of behavior.

Factorial design experiments are a versatile and powerful tool in biology, allowing researchers to systematically investigate complex biological systems, identify interactions between multiple factors, and advance our understanding of the underlying mechanisms driving biological processes.

Factorial design experiments allow researchers in biology to systematically vary and analyze multiple factors to understand complex biological phenomena, interactions, and responses. They help uncover patterns, relationships, and dependencies that would be challenging to identify through single-variable experiments.

STEPS FOR FACTORIAL DESIGN EXPERIMENTS

Developing a factorial design experiment in biology involves several systematic steps to investigate the effects of multiple factors on a biological response or outcome. Here are the typical steps for developing a factorial design experiment in biology:

1. **Define the Research Objective:** Clearly articulate the research question or objective of your study. What specific biological phenomenon or response are you trying to understand or optimize?
2. **Identify the Factors of Interest:** Identify the IVs or factors that you want to study. These factors could be biological variables, treatments, conditions, or interventions that may influence the biological response.
3. **Determine Factor Levels:** Specify the different levels or values for each factor. These levels represent the range of conditions or treatments you want to investigate. Ensure that the levels are biologically relevant and meaningful.
4. **Determine the Number of Experimental Runs:** Calculate the total number of experimental runs required for your selected factorial design type. This depends on the number of factors and their levels.
5. **Randomize Experimental Runs (Optional):** To minimize the impact of uncontrolled variables or systematic biases, consider randomizing the order of experimental runs.
6. **Conduct the Experiments:** Perform the experiments according to the designed factorial matrix, varying the factor levels for each run systematically.
7. **Measure and Record Data:** Collect accurate and consistent data for the biological response or outcome of interest. Ensure that data collection protocols are standardized.
8. **Data Analysis:** Analyze the data using appropriate statistical methods, such as analysis of variance (ANOVA) or regression analysis, to assess the main effects of factors and their interactions. Determine which factors significantly influence the biological response.
9. **Interpret the Results:** Interpret the statistical results to understand how each factor and their interactions affect the biological response. Identify the most influential factors and the direction of their effects.
10. **Optimize Conditions (if applicable):** If your objective is to optimize biological conditions or outcomes, use the results to determine the optimal combination of factor levels that maximize or minimize the response.
11. **Validation and Replication (if necessary):** Consider validating the findings through replication or additional experiments to ensure the robustness and reliability of the results.
12. **Report and Document:** Prepare a comprehensive report that includes details of the experimental design, data collected, statistical analysis, results, conclusions, and any implications for biological understanding or practical applications.

Factorial design experiments in biology allow researchers to explore complex relationships between multiple factors and biological responses systematically. Properly designed and executed experiments can provide valuable insights into biological processes, interactions, and optimization strategies.

PITFALLS AND REMEDIES

Factorial design experimentation is a powerful approach in biology, but it can also present certain pitfalls. Here are some common pitfalls and their remedies when using factorial design experiments in biology:

Insufficient Understanding of Biological Context
Pitfall: Conducting factorial experiments without a thorough understanding of the biological system or processes involved can lead to incorrect interpretations and conclusions.

Remedy: Before designing experiments, invest time in gaining a deep understanding of the biological context. Consult relevant literature, collaborate with domain experts, and consider preliminary studies to gather essential background knowledge.

Poorly Defined Research Objectives
Pitfall: Ambiguous or vaguely defined research objectives can result in experiments that lack focus, leading to inconclusive or irrelevant outcomes.

Remedy: Clearly define your research objectives, specifying the biological response(s) of interest and the specific factors and their levels that you intend to investigate. Be explicit about the hypotheses you want to test.

Inappropriate Factor Selection
Pitfall: Selecting factors without considering their biological relevance or significance may lead to experiments that produce uninformative results.

Remedy: Choose factors based on biological knowledge and prior research. Prioritize factors that are likely to have a meaningful impact on the biological response of interest. Consult with experts if needed.

Neglecting Factor Interactions
Pitfall: Ignoring or overlooking potential interactions between factors can result in an incomplete understanding of the biological system.

Remedy: Consider the possibility of interactions between factors. Use factorial designs specifically to investigate interactions. Analyze and interpret main effects and interaction effects to capture the full picture of factor influence.

Inadequate Sample Size
Pitfall: Small sample sizes can lead to low statistical power, making it challenging to detect significant effects or interactions.

Remedy: Conduct a power analysis to determine an appropriate sample size based on the expected effect sizes and significance levels. Ensure that you have enough replicates for each combination of factor levels.

Nonrandom Experimental Design
Pitfall: Failing to randomize the order of experimental runs can introduce bias and confound the results.

Remedy: Randomize the order of experimental runs to minimize the impact of uncontrolled variables or systematic biases. This ensures that the effects observed are more likely to be attributed to the manipulated factors.

Data Collection Errors
Pitfall: Inconsistent or inaccurate data collection procedures can lead to unreliable results.

Remedy: Implement rigorous and standardized data collection protocols. Train experimenters to minimize measurement errors and ensure consistency in data collection procedures.

Overcomplicating the Experimental Design

Pitfall: Overly complex factorial designs with many factors and levels may result in impractical experiments, increased resource demands, and difficulties in data analysis.

Remedy: Keep the experimental design as simple as possible while addressing the research objectives. Consider fractional factorial designs if you have a large number of factors to reduce the number of runs.

Ignoring Biological Variability

Pitfall: Failing to account for inherent biological variability can lead to misleading results and overgeneralization.

Remedy: Recognize and account for biological variability by including replicates and using appropriate statistical methods, such as ANOVA, to distinguish between true effects and random variation.

Lack of Post-experimental Validation

Pitfall: Neglecting to validate the findings of the factorial experiment through follow-up experiments or independent studies can leave the results unverified.

Remedy: Consider conducting follow-up experiments or studies to confirm the observed effects and conclusions, especially if the findings have practical implications or require further validation.

Addressing these pitfalls with careful planning, adequate preparation, and rigorous execution of factorial design experiments in biology can enhance the reliability and meaningfulness of the results, ultimately advancing scientific understanding and applications in the field.

EXAMPLE

Factorial experiments in biology are used to study the effects of multiple factors on a biological response variable. Here's an example of a matrix for a 2^2 full factorial experiment in biology:

Objective: To investigate the effects of two factors (nutrient type and light exposure) on the growth of a specific plant species.

Factors:

1. **Nutrient Type (Factor A):**
   - Nutrient A.
   - Nutrient B.
2. **Light Exposure (Factor B):**
   - Low light.
   - High light.

Response Variable: Plant growth (measured in centimeters).

Full Factorial Experiment Matrix (2^2):

In this 2^2 full factorial experiment, all possible combinations of the two factors (nutrient type and light exposure) are studied. The matrix is shown in Table 2.1.

In this matrix, each row represents a specific experimental run with a unique combination of nutrient type and light exposure levels. Plant growth measurements are taken for each combination.

TABLE 2.1
Biology Full Factorial Example

Run	Nutrient Type (A)	Light Exposure (B)	Plant Growth (cm)
1	Nutrient A	Low light	...
2	Nutrient B	Low light	...
3	Nutrient A	High light	...
4	Nutrient B	High light	...

The full factorial design allows biologists to systematically investigate how changes in nutrient type and light exposure affect the growth of the plant species. By analyzing the results, they can draw conclusions about the main effects and interactions of these factors on plant growth, providing insights into the plant's growth requirements.

This is a simplified example, and in actual biology experiments, additional factors and experimental runs may be considered to address more complex research questions.

Reporting

Factorial design experimentation in biology often involves complex statistical analyses to understand the effects of multiple factors on biological responses. Here are some of the statistical tools commonly used in factorial design experiments in biology:

ANOVA: ANOVA is a fundamental tool for assessing the significance of main effects and interactions between factors. It helps determine whether observed differences in responses are statistically significant.

Regression Analysis: Regression models can be used to model the relationship between factors and responses. Linear, multiple, and nonlinear regression models are applied to fit data and make predictions.

Response Surface Methodology (RSM): RSM is used to model and optimize responses when multiple factors interact in a nonlinear manner. It involves fitting mathematical models to experimental data to identify optimal factor settings.

Design of Experiments (DOE) Software: Specialized software packages like Minitab, JMP, or R with DOE libraries are often used for experimental design, data analysis, and optimization of factorial experiments.

Factorial ANOVA: Factorial ANOVA is used to analyze experiments with two or more factors. It assesses the main effects of each factor and their interactions, helping researchers understand how multiple factors influence the response.

Analysis of Covariance (ANCOVA): ANCOVA extends ANOVA by including covariates (additional variables that might affect the response). It helps account for covariate effects while analyzing the main factors.

Contrast Analysis: Contrast analysis is used to test specific hypotheses about the differences between factor levels. It is particularly useful for post-hoc analysis of factor effects.

Multivariate Analysis: Multivariate techniques like principal component analysis (PCA) and canonical correlation Analysis (CCA) are applied when multiple responses are involved. They help explore relationships between multiple variables and identify patterns.

Logistic Regression: Logistic regression is used when the response variable is binary or categorical, as opposed to continuous. It is commonly used in biology to model binary outcomes, such as presence/absence or disease/no disease.

Mixed-Effects Models: When dealing with hierarchical data or repeated measures, mixed-effects models (or hierarchical linear models) account for both fixed effects (factors) and random effects (e.g., subject or experimental site).

Nonparametric Tests: Nonparametric tests like the Kruskal–Wallis test, Wilcoxon signed rank test, or Friedman test are used when the data distribution is not normal or when assumptions for parametric tests are violated.

Bayesian Analysis: Bayesian approaches, including Bayesian ANOVA and Bayesian regression, allow for a probabilistic framework to estimate parameters and make inferences, incorporating prior knowledge into the analysis.

Survival Analysis: Survival analysis techniques, such as Kaplan–Meier survival curves and Cox proportional hazards models, are employed in studies involving time-to-event data, such as disease progression or survival times.

Machine Learning and Data Mining: Advanced statistical and machine learning methods, including decision trees, random forests, and neural networks, are increasingly applied to analyze complex biological data and identify patterns and predictors.

The choice of statistical tools depends on the specific research question, the nature of the data, and the complexity of the biological system under investigation. Researchers often use a combination of these tools to gain a comprehensive understanding of the factors influencing biological responses and to draw meaningful conclusions from factorial design experiments in biology.

A well-structured and comprehensive report for factorial design experimentation in biology should include the following key elements:

Title: A clear and concise title that reflects the focus and purpose of the study.

Abstract: A brief summary of the experiment's objectives, methods, major findings, and conclusions. It should provide a concise overview of the entire report.

Table of Contents: An organized list of sections and subsections within the report, along with their page numbers.

List of Figures and Tables: A separate list that enumerates all figures and tables included in the report, along with their respective page numbers.

Introduction: Provide context for the study by explaining the background, rationale, and significance of the research. Clearly state the research objectives, hypotheses, and the factors to be investigated.

Literature Review: Review relevant literature and previous studies to establish the scientific basis for your research. Discuss any existing knowledge related to the factors and biological responses under investigation.

Materials and Methods:
- Describe the experimental design, including the factorial design matrix, factor levels, and the number of replicates.
- Detail the procedures, protocols, and equipment used for data collection.
- Explain any statistical methods or software tools employed for data analysis.
- Provide information on how factors were manipulated and controlled.

Results:
- Present the data collected during the experiments in a clear and organized manner.
- Use tables, figures, graphs, and charts to illustrate key findings.
- Include statistical analyses, such as ANOVA, regression, or other relevant tests.
- Discuss any significant main effects and interactions among factors.

Discussion:
- Interpret the results, emphasizing the biological implications of the findings.
- Compare your results with previous research and discuss any discrepancies or agreements.

- Address the limitations of the study and potential sources of error or bias.
- Consider the broader implications of the research and its relevance to the field of biology.

Conclusion: Summarize the main findings and their significance in the context of the research objectives. Restate the key conclusions and their implications for the biological system under study.

Recommendations: Suggest any practical recommendations or further research directions based on your findings. Highlight areas where additional studies or investigations may be needed.

Acknowledgments: Acknowledge individuals or organizations that provided support, funding, or resources for the research.

References: List all references, citations, and sources used in the report following a consistent citation style (e.g., American psychological association (APA), modern language association (MLA), or Chicago). Include both primary research articles and relevant literature.

Appendices: Include supplementary information that supports the main text but is not essential for understanding the report (e.g., raw data, experimental protocols, additional figures or tables).

Figures and Tables: Ensure that all figures and tables are correctly labeled, with informative captions that explain their content. Number figures and tables sequentially.

Glossary (Optional): If specialized terminology or abbreviations are used, provide a glossary to define key terms for readers' reference.

Author Information: Include the names and affiliations of the authors, as well as contact information.

Date and Version Information: Specify the date of the report and indicate if it is a final version or a revision.

Ethical Considerations (if applicable): If the research involves human subjects, animals, or any ethical considerations, provide a section detailing ethical approvals and compliance with relevant guidelines.

Ensure that the report is well organized, logically structured, and written in a clear, concise, and scientific manner. Properly cited references and consistent formatting are essential for a professional and credible presentation of your factorial design experimentation in biology.

2.1.2 RSM

RSM designed experiments are a valuable tool in biology for optimizing experimental conditions, exploring complex response surfaces, identifying optimal conditions and factor interactions, achieving statistical efficiency and precision, and adapting experimental designs to address the complexity of biological systems. RSM enables researchers to systematically investigate biological processes, optimize experimental outcomes, and advance our understanding of the underlying biology of living organisms. RSM is applied in microbiology to optimize the culture conditions for bacterial fermentation. Scientists vary factors like pH, temperature, and nutrient concentrations to maximize the yield of a specific microbial product. RSM designed experiments are widely used in biology for several reasons:

Optimization of Experimental Conditions: RSM allows researchers to optimize experimental conditions and identify optimal parameter settings for biological processes, treatments, or interventions. By systematically varying multiple factors and observing the response, RSM enables researchers to identify the combination of factors that maximize desired outcomes, such as cell growth, enzyme activity, or gene expression. This optimization process is crucial in biology for maximizing research efficiency, resource utilization, and the effectiveness of experimental interventions.

Exploration of Complex Response Surfaces: Biological systems often exhibit complex, nonlinear responses to changes in experimental factors. RSM is well-suited for exploring these complex response surfaces and characterizing the relationships between multiple factors and the response of interest. By fitting mathematical models to experimental data, RSM allows researchers to visualize and understand the shape, curvature, and interactions of response surfaces, providing valuable insights into the underlying biology of the system.

Identification of Optimal Conditions and Factor Interactions: RSM enables researchers to identify optimal conditions and factor interactions that maximize or minimize the response of interest. By systematically varying factors and observing the response within a defined experimental region, RSM allows researchers to identify critical factors, assess their individual and combined effects, and optimize experimental protocols or process parameters accordingly. This capability is particularly valuable in biology for identifying key factors influencing biological processes, designing effective interventions, and optimizing experimental outcomes.

Statistical Efficiency and Precision: RSM is statistically efficient, meaning it allows researchers to estimate response surfaces and factor effects with high precision using a relatively small number of experimental runs. By strategically selecting experimental points within the design space, RSM maximizes the information obtained from each experimental observation, leading to more accurate parameter estimates and more reliable predictions of response behavior. This statistical efficiency is crucial in biology, where experiments may involve variability and noise inherent in biological systems.

Robustness and Adaptability: RSM is robust and adaptable, allowing researchers to modify and expand experimental designs as new questions arise or additional factors need to be considered. Researchers can easily add or remove factors, change factor levels, or introduce additional treatments within the RSM framework, enabling iterative experimentation, hypothesis testing, and model refinement. This flexibility is essential in biology for accommodating the complexity and variability of biological systems and adapting experimental designs to address evolving research questions and hypotheses.

RSM designed experiments are widely used in biology for various applications where researchers need to optimize multiple factors or conditions to achieve desired biological responses. Here are some areas in biology where RSM experiments find applications:

Enzyme Kinetics: RSM can be employed to optimize reaction conditions for enzyme-catalyzed reactions, such as determining the optimal pH, temperature, and substrate concentrations to maximize enzyme activity.

Cell Culture and Fermentation: RSM is used to optimize culture conditions for cell lines or microorganisms, such as optimizing nutrient concentrations, pH, and oxygen levels to maximize cell growth, protein production, or biofuel yield.

Pharmaceutical Formulation: In pharmaceutical research, RSM helps optimize drug formulations, such as identifying the best combination of excipients, pH, and temperature to enhance drug stability and release profiles.

Agricultural and Crop Science: RSM is applied to optimize agricultural practices, such as determining the ideal combination of fertilizers, irrigation, and soil amendments to maximize crop yield and quality.

Drug Delivery Systems: In drug delivery, RSM helps optimize drug release kinetics from various delivery systems (e.g., nanoparticles, liposomes) by adjusting factors like particle size, drug loading, and release mechanisms.

Bioprocess Engineering: RSM plays a crucial role in bioprocess optimization, such as designing and controlling bioreactors for the production of biopharmaceuticals, biofuels, or biochemicals.

Protein Purification: RSM is used to optimize purification processes for proteins and biomolecules, such as chromatography conditions to maximize protein yield and purity.

Environmental Science: In ecological studies, RSM helps optimize conditions for plant growth, pollution remediation, or aquatic ecosystems, considering factors like nutrient levels, light exposure, and water quality.

Toxicology and Pharmacology: RSM can be used to optimize dosages, drug combinations, and treatment protocols in preclinical and clinical studies to achieve desired therapeutic outcomes.

Food Science: RSM is applied to optimize food processing and formulation, including factors like ingredient proportions, cooking temperatures, and storage conditions to improve taste, texture, and shelf life.

Biotechnology: RSM is used for optimizing biotechnological processes, such as bioremediation, biosensors, and bioconversion processes for bio-based products.

Vaccine Development: In vaccine research, RSM helps optimize vaccine formulations and production processes, including antigen concentrations, adjuvants, and storage conditions.

Environmental Monitoring: RSM can be applied in environmental monitoring to optimize the DOE and sensor networks for efficient data collection and analysis.

Neuroscience: RSM can assist in optimizing the parameters of brain stimulation protocols for therapeutic purposes or for studying neural activity.

Molecular Biology: RSM can be used to optimize conditions for molecular biology techniques like polymerase chain reaction (PCR), quantitative reverse transcription polymerase chain reaction (RT-qPCR), and DNA sequencing to improve assay sensitivity and specificity.

In these and many other areas of biology, RSM allows researchers to systematically explore and optimize multiple factors and their interactions to achieve desired biological responses or outcomes efficiently. It helps reduce experimentation time and resources while enhancing the quality of research and product development.

STEPS FOR DEVELOPING A RSM DESIGNED EXPERIMENT

Developing an RSM designed experiment in biology involves several steps to optimize multiple factors or conditions and understand their impact on a biological response. Here are the key steps to develop an RSM experiment in biology:

1. **Define the Research Objective:** Clearly state the research goal or objective. What specific biological response or outcome are you trying to optimize or understand?
2. **Identify the Factors:** Determine the factors (IVs) that you suspect may influence the biological response. These factors can be physical parameters, chemical concentrations, or other relevant variables.
3. **Determine Factor Levels:** Define the range of values or levels for each factor that will be studied in the experiment. Consider both lower and upper bounds for each factor.
4. **Select the Response Variable:** Identify the biological response variable (DV) that will be measured or observed as the outcome of interest. This could be a growth rate, enzyme activity, protein yield, cell viability, etc.
5. **Choose the Experimental Design:** Select the appropriate experimental design, such as a CCD or BBD, based on the number of factors and levels, as well as your knowledge of potential interactions.
6. **Generate the Experimental Matrix:** Create a matrix of experimental runs that includes all combinations of factor levels according to the chosen experimental design. This matrix will guide the execution of experiments.

7. **Conduct the Experiments:** Execute the experiments systematically, following the combinations of factor levels specified in the experimental matrix. Ensure that conditions are controlled and replicates are collected.
8. **Collect Data:** Measure and record the responses for each experimental run. Ensure data accuracy and consistency.
9. **Perform Data Analysis:** Use statistical analysis tools and software (e.g., regression analysis, ANOVA, RSM software) to analyze the collected data. Fit response surface models to the data.
10. **Model Validation:** Validate the response surface models to ensure that they accurately represent the observed biological responses. Use techniques like cross-validation or model diagnostics.
11. **Interpret Results:** Interpret the statistical results to understand the effects of each factor and their interactions on the biological response. Identify significant factors and their optimal levels.
12. **Optimization:** If the goal is optimization, determine the factor settings (combinations of levels) that maximize or minimize the biological response. Use the response surface models to guide this process.
13. **Validation Experiments:** Conduct validation experiments to confirm the predicted optimal conditions and verify that the desired biological response is achieved.
14. **Report and Documentation:** Prepare a comprehensive report that includes details of the experimental design, data collected, statistical analysis, results, and conclusions. Document all steps and findings.
15. **Discussion and Conclusion:** Discuss the implications of the results in the context of the research objectives and the biological system under study. Draw conclusions and suggest areas for further research.
16. **Recommendations:** If applicable, provide recommendations for practical applications or strategies based on the optimized conditions or biological insights gained.
17. **References:** Cite relevant literature and sources used in your research and experimental design.
18. **Appendices:** Include any supplementary information, such as raw data, graphs, tables, and detailed protocols, in the appendices.

Throughout the entire process, rigorously document all experimental procedures, data, and analytical methods to ensure transparency and reproducibility. Collaboration with statisticians or experts in experimental design may also be beneficial to design and analyze RSM experiments effectively in biology.

PITFALLS AND REMEDIES

RSM design experiments are a powerful tool in biology, but they can also be susceptible to certain pitfalls. Here are common pitfalls and remedies when using RSM in biology:

Insufficient Biological Understanding
Pitfall: Conducting RSM experiments without a solid understanding of the biological system, its complexities, and underlying mechanisms may lead to misinterpretation of results.

Remedy: Invest time in gaining a deep understanding of the biological system under study. Collaborate with domain experts, conduct preliminary studies, and consult relevant literature to build a strong biological foundation for your RSM design.

Overfitting Response Surface Models
Pitfall: Overfitting occurs when response surface models are overly complex and fitted too closely to the experimental data, resulting in poor predictive performance.

Remedy: Use model validation techniques, such as cross-validation or split-sample validation to assess the model's predictive accuracy. Choose models that balance complexity and fit while maintaining good predictive power.

Neglecting Factor Interactions

Pitfall: Ignoring or underestimating the importance of factor interactions can lead to incomplete or inaccurate models.

Remedy: Explicitly account for factor interactions in your RSM design. Include interaction terms in the response surface models and thoroughly analyze their significance. Plot response surfaces to visualize interactions.

Inadequate Model Validation

Pitfall: Failing to validate response surface models or using inappropriate validation techniques can lead to unreliable predictions.

Remedy: Employ proper validation techniques, such as cross-validation, holdout samples, or external validation, to assess the model's accuracy and generalizability. Validate models with new experimental runs if possible.

Failure to Consider Biological Constraints

Pitfall: Optimized conditions suggested by RSM may not always be biologically feasible or practical.

Remedy: Incorporate biological constraints and practical considerations into the optimization process. Constraints may relate to toxicity limits, resource availability, or regulatory requirements.

Inadequate Sample Size

Pitfall: Small sample sizes can result in low statistical power and may make it challenging to detect significant effects or accurately estimate model parameters.

Remedy: Conduct a power analysis to determine an appropriate sample size based on expected effect sizes and desired confidence levels. Ensure that you have a sufficient number of replicates for each experimental run.

Poor Experimental Design

Pitfall: A poorly designed experiment can lead to biased or confounded results.

Remedy: Carefully plan the experimental design by choosing appropriate factor levels, ensuring randomization, and using a balanced design. Consult with statisticians or experimental design experts if needed.

Lack of Replicates

Pitfall: Insufficient replicates can reduce the reliability and reproducibility of experimental results.

Remedy: Include an adequate number of replicates for each combination of factor levels to reduce random variation and enhance the statistical validity of your results.

Misinterpreting Statistical Significance

Pitfall: Overemphasizing statistical significance without considering practical significance or biological relevance.

Remedy: Always interpret statistical significance in the context of the biological system. Focus on effect sizes and practical implications when making conclusions.

Not Accounting for Uncertainty

Pitfall: Failing to account for uncertainty in factor settings and predictions can lead to unrealistic expectations.

Remedy: Use confidence intervals or prediction intervals to quantify and communicate the uncertainty associated with model predictions and factor settings.

Addressing these pitfalls with careful planning, robust experimental design, thorough statistical analysis, and a strong understanding of the biological context can help ensure the success and reliability of RSM experiments in biology. Collaboration with interdisciplinary teams can also be valuable in navigating these challenges effectively.

EXAMPLE

In RSM design experiments in biology, a matrix is used to plan and organize the experimental runs, specifying the combinations of factor levels to be tested. Here's an example of a matrix for an RSM experiment in a biological context, where we want to optimize the production of a specific protein in a bacterial fermentation process:

Objective: To maximize the yield of a target protein produced by a bacterial strain in a fermentation process by optimizing three key factors: temperature (°C), pH, and agitation speed (rpm).

Factor Levels:

- Temperature (Factor A): Low (25°C) and high (37°C).
- pH (Factor B): Low (6.0) and high (7.5).
- Agitation Speed (Factor C): Low (200 rpm) and high (300 rpm).

Central Points for RSM:

- Center point runs are included to estimate the curvature of the response surface. These values are often set at the midpoint of the factor levels.
  - Temperature (Factor A): Midpoint (31°C).
  - pH (Factor B): Midpoint (6.75).
  - Agitation Speed (Factor C): Midpoint (250 rpm).

Experimental Matrix (a CCD example with 2 levels and 5 center points) (Table 2.2):

For this example, the experimental matrix consists of 13 runs, with a combination of low and high levels for each of the three factors, as well as center point runs. The center points are typically included to estimate the curvature of the response surface, helping to fit a quadratic model.

During the experiment, each combination of factor levels is tested, and the yield of the target protein is measured for each run. The data collected from these runs are then used for statistical analysis, modeling, and optimization to find the optimal conditions for maximizing the protein yield while considering the interactions between temperature, pH, and agitation speed.

TABLE 2.2

RSM Example for Biology

Run	Temperature (A)	pH (B)	Agitation Speed (C)
1	−1	−1	−1
2	+1	−1	−1
3	−1	+1	−1
4	+1	+1	−1
5	−1	−1	+1
6	+1	−1	+1
7	−1	+1	+1
8	+1	+1	+1
9	0	0	0
10	0	0	0
11	0	0	0
12	0	0	0
13	0	0	0

REPORTING

RSM design experiments in biology involve the use of various statistical tools and techniques to model, analyze, and optimize the relationships between multiple factors and a biological response. Here are some of the common statistical tools used in RSM for biology:

Regression Analysis: Linear and nonlinear regression models are used to fit response surface models to experimental data. These models describe how the biological response varies with changes in factor levels.

ANOVA: ANOVA is used to assess the significance of the main effects of factors and their interactions. It helps determine which factors have a significant impact on the response.

ANOVA for Quadratic Models: In RSM, quadratic models are often used to capture curvature in the response surface. ANOVA for quadratic models evaluates the significance of quadratic terms and interactions.

DOE Software: Specialized software tools like JMP, Minitab, or Design-Expert are used to design experiments, analyze data, and build response surface models. These tools automate many aspects of RSM.

Response Surface Plots: These graphical representations show the three-dimensional response surface, making it easier to visualize how changes in factor levels affect the response.

Contour Plots: Contour plots are used to visualize response surfaces in two dimensions, highlighting regions where the response is optimized.

Desirability Function: Desirability functions are used to simultaneously optimize multiple responses. They assign a desirability value to different levels of factors and help find the best compromise.

Model Diagnostics: Various diagnostics, such as residual plots, normal probability plots, and leverage plots, are used to assess the adequacy of the response surface models and identify outliers or influential data points.

Model Validation: Cross-validation and holdout validation are used to validate the response surface models, ensuring they accurately represent the underlying biological relationships.

Optimization Algorithms: Numerical optimization algorithms like gradient-based methods or genetic algorithms are used to find the optimal factor settings for maximizing or minimizing the response.

Confidence Intervals and Prediction Intervals: These intervals provide a measure of uncertainty around the predicted responses and optimal factor settings, helping to assess the reliability of predictions.

Sequential Experimentation: Sequential experimental designs, such as the RSM guided by sequential experimentation (RSMGSE), are used to iteratively refine models and optimize responses based on initial experimental results.

PCA: PCA can be applied to identify underlying patterns and relationships in multivariate data, especially when multiple responses are involved.

Canonical Analysis: Canonical analysis explores the relationships between factor settings and multiple responses, revealing how changes in factors impact multiple biological outcomes.

Monte Carlo Simulation: Monte Carlo simulations can assess the robustness of optimized conditions and predictions by considering the effects of variability and uncertainty in factors.

Bayesian Analysis: Bayesian approaches allow the incorporation of prior knowledge and uncertainty into response surface modeling and optimization.

The choice of statistical tools and techniques depends on the specific objectives of the RSM experiment and the complexity of the biological system under investigation. Proper statistical analysis is critical for drawing meaningful conclusions, optimizing biological processes, and making informed decisions in biological research and development.

A well-structured and comprehensive report for RSM design experiments in biology should include the following key elements:

Title: A clear and concise title that reflects the focus and purpose of the study.

Abstract: A brief summary of the RSM experiment's objectives, methods, major findings, and conclusions. It should provide a concise overview of the entire report.

Table of Contents: An organized list of sections and subsections within the report, along with their page numbers.

List of Figures and Tables: A separate list that enumerates all figures and tables included in the report, along with their respective page numbers.

Introduction: Provide context for the study by explaining the background, rationale, and significance of the RSM experiment. Clearly state the research objectives, hypotheses, and the factors to be investigated.

Literature Review: Review relevant literature and previous studies to establish the scientific basis for your research. Discuss any existing knowledge related to the factors and biological responses under investigation.

Materials and Methods:
- Describe the experimental design, including the RSM design matrix, factor levels, and the number of replicates.
- Detail the procedures, protocols, and equipment used for data collection.
- Explain any statistical methods or software tools employed for data analysis.
- Provide information on how factors were manipulated and controlled.

Results:
- Present the data collected during the RSM experiments in a clear and organized manner.
- Use tables, figures, graphs, and charts to illustrate key findings.
- Include statistical analyses, such as ANOVA, regression, or other relevant tests.
- Discuss any significant main effects and interactions among factors.

Response Surface Analysis: Present response surface plots or contour plots to visualize the relationship between factors and responses. Explain the significance of these plots and how they help understand the response surfaces.

Discussion:
- Interpret the results, emphasizing the biological implications of the findings.
- Compare your results with previous research and discuss any discrepancies or agreements.
- Address the limitations of the study and potential sources of error or bias.
- Consider the broader implications of the research and its relevance to the field of biology.

Conclusion: Summarize the main findings and their significance in the context of the research objectives. Restate the key conclusions and their implications for the biological system under study.

Optimization Recommendations: If applicable, provide recommendations for optimizing biological processes or conditions based on the RSM findings. Specify the optimal factor settings and their practical implications.

References: List all references, citations, and sources used in the report, following a consistent citation style (e.g., APA, MLA, or Chicago). Include both primary research articles and relevant literature.

Appendices: Include supplementary information that supports the main text but is not essential for understanding the report (e.g., raw data, experimental protocols, additional figures, or tables).

Figures and Tables: Ensure that all figures and tables are correctly labeled, with informative captions that explain their content. Number figures and tables sequentially.

Glossary (Optional): If specialized terminology or abbreviations are used, provide a glossary to define key terms for readers' reference.

Author Information: Include the names and affiliations of the authors, as well as contact information.

Date and Version Information: Specify the date of the report and indicate if it is a final version or a revision.

Ethical Considerations (if applicable): If the research involves human subjects, animals, or any ethical considerations, provide a section detailing ethical approvals and compliance with relevant guidelines.

Ensure that the report is well organized, logically structured, and written in a clear, concise, and scientific manner. Properly cited references and consistent formatting are essential for a professional and credible presentation of your RSM design experiments in biology.

2.1.3 FRACTIONAL FACTORIAL EXPERIMENTS

Fractional factorial design experiments are a valuable tool in biology for efficiently screening factors, reducing experimental burden, identifying key factors and interactions, achieving statistical efficiency and precision, and adapting experimental designs to address the complexity of biological systems. Fractional factorial designs enable researchers to systematically investigate biological processes, identify influential factors, and advance our understanding of the underlying biology of living organisms. Fractional factorial design experiments are utilized in biology for several compelling reasons:

Efficient Screening of Factors: Biology often involves studying systems with numerous variables, some of which may interact in complex ways. Fractional factorial designs enable researchers to efficiently screen a large number of factors while running fewer experiments compared to full factorial designs. This allows for the identification of the most influential factors or interactions, which can then be further investigated in more detail, saving time and resources.

Reduction of Experimental Burden: Conducting experiments in biology can be labor-intensive and resource demanding, especially when dealing with living organisms or complex biological processes. Fractional factorial designs allow researchers to reduce the number of experimental runs required while still capturing essential information about factor effects and interactions. This reduction in experimental burden is particularly valuable in biology, where experimental conditions may be challenging to control or manipulate.

Identification of Key Factors and Interactions: Fractional factorial designs help identify key factors and interactions that significantly influence biological outcomes. By systematically varying a subset of factors at multiple levels, researchers can assess the main effects of individual factors as well as interactions between factors. This information is crucial for understanding the underlying mechanisms driving biological processes and for prioritizing further investigation of the most influential factors.

Statistical Efficiency and Precision: Fractional factorial designs are statistically efficient, allowing researchers to estimate main effects and certain interactions with high precision using a relatively small number of experimental runs. This statistical efficiency maximizes the information obtained from each experimental observation, leading to more accurate estimates of factor effects and interactions. This is particularly important in biology, where experiments often involve variability and noise inherent in biological systems.

Flexibility and Adaptability: Fractional factorial designs offer flexibility and adaptability, allowing researchers to modify and expand experimental designs as new questions arise or additional factors need to be considered. Researchers can easily add or remove factors, change factor levels, or introduce additional treatments within the fractional factorial framework, enabling iterative experimentation, hypothesis testing, and model refinement. This flexibility is essential in biology for accommodating the complexity and variability of biological systems and adapting experimental designs to address evolving research questions and hypotheses.

Fractional factorial design experiments are used in biology across various research and application areas to efficiently explore and understand the effects of multiple factors or variables on biological systems. Some common areas where fractional factorial design experiments find applications in biology include:

Genetics and Genomics: Fractional factorial designs are used to study the genetic basis of complex traits, such as identifying the factors influencing gene expression, epistatic interactions, or gene–gene interactions.

Molecular Biology: These designs help optimize conditions for molecular biology techniques, including PCR, DNA sequencing, and DNA microarray experiments, by assessing the effects of factors like reagent concentrations and reaction conditions.

Protein Expression and Purification: Researchers use fractional factorial designs to optimize protein expression systems, growth conditions, and purification protocols to maximize protein yield and purity.

Cell Culture: Fractional factorial designs are applied to optimize cell culture conditions for growth, viability, and productivity of cell lines or microorganisms used in biotechnology, pharmaceuticals, and research.

Pharmacology and Drug Development: Fractional factorial experiments help optimize drug formulations, drug screening assays, and pharmacological studies by assessing the effects of factors like drug concentrations, solvents, and assay conditions.

Enzyme Kinetics: Researchers use fractional factorial designs to investigate enzyme-catalyzed reactions, determining optimal conditions for factors, such as pH, temperature, substrate concentrations, and enzyme concentrations.

Ecology and Environmental Science: Fractional factorial experiments are used to study the effects of multiple environmental factors on ecosystems, species interactions, and biodiversity.

Agriculture and Crop Science: Fractional factorial designs help optimize agricultural practices by assessing the impact of various factors, such as soil nutrients, irrigation, and pest management on crop yield and quality.

Toxicology and Risk Assessment: These designs are applied to assess the effects of multiple factors on toxicity and risk profiles in toxicology studies, including the evaluation of chemical exposure.

Neuroscience: Fractional factorial experiments are used to optimize experimental conditions for neuroscience research, including neuroimaging studies and electrophysiological recordings.

Microbiology and Microbial Ecology: Researchers use these designs to investigate factors affecting microbial growth, diversity, and metabolic activities in various environments.

Environmental Monitoring: Fractional factorial experiments are applied to design efficient sampling and monitoring strategies in environmental studies, assessing the impact of multiple variables on data collection.

Bioinformatics and Systems Biology: Fractional factorial experiments can be used to optimize algorithms and parameters in bioinformatics analyses or to study the interactions of multiple biological molecules.

Bioprocess Engineering: These designs help optimize bioprocesses for the production of biopharmaceuticals, biofuels, or biochemicals by assessing the effects of various factors on product yield and quality.

Virology and Vaccine Development: Researchers use fractional factorial experiments to optimize vaccine formulations, viral growth conditions, and vaccine production processes.

Fractional factorial design experiments offer a cost-effective and time-efficient way to explore the effects of multiple factors on biological systems, making them valuable in a wide range of biological research and applied fields. They allow researchers to efficiently identify critical factors and interactions, ultimately leading to better understanding and optimization of biological processes.

STEPS

Developing fractional factorial design experiments in biology involves careful planning and execution to efficiently explore multiple factors and their interactions with minimal experimental effort. Here are the steps to develop such experiments in biology:

1. **Identify the Research Question or Hypothesis**: Clearly define your research question or hypothesis. Determine the specific biological factors or variables you want to investigate and understand how they might interact.
2. **Select the Factors of Interest**: Identify the factors (IVs) that are relevant to your research question. These could include biological parameters, treatments, or conditions that you want to study.
3. **Determine Factor Levels**: Decide on the different levels or values at which each factor will be tested. Ensure that these levels are biologically relevant and cover the range of interest.
4. **Choose the Fractional Factorial Design**: Decide on the fraction of the full factorial design you want to use. Common fractions are 1/2, 1/4, or 1/8, depending on the number of factors and their interactions you want to investigate.
5. **Generate the Experimental Design Matrix**: Use statistical software or tools specifically designed for experimental design to generate the matrix that outlines the experimental runs. This matrix will indicate which combinations of factor levels to test in your experiment.
6. **Randomize Experimental Runs**: To minimize the influence of uncontrolled variables, randomize the order of experimental runs. This helps ensure that the results are not biased by external factors.
7. **Conduct the Experiments**: Execute the experiments according to the design matrix, carefully following the protocols and procedures you've established for each run.
8. **Record Data**: Collect and record data meticulously for each experimental run, ensuring accuracy and consistency in data collection.
9. **Perform Statistical Analysis**: Analyze the data using appropriate statistical techniques, such as ANOVA or regression analysis. Evaluate the main effects of factors and any interactions that may be significant.
10. **Draw Conclusions**: Based on the statistical analysis, draw conclusions about the effects of the factors on the biological response of interest. Determine which factors are significant and understand any interactions between them.
11. **Optimize and Fine-Tune**: If your goal is optimization, you can use the results to fine-tune conditions or treatments to achieve desired outcomes more efficiently.
12. **Report Findings**: Document and report your findings in a clear and comprehensive manner, including statistical results, conclusions, and implications for your research question or hypothesis.

13. **Repeat or Validate**: Depending on the complexity of your research, you may need to repeat the experiments or conduct additional validation studies to ensure the robustness of your findings.

Fractional factorial design experiments in biology can be a powerful tool for efficiently exploring multiple factors while conserving resources. However, careful planning, rigorous execution, and thorough data analysis are essential to ensure the validity and reliability of the results obtained.

PITFALLS AND REMEDIES

Fractional factorial design experiments in biology offer significant advantages, but they can also present pitfalls that researchers need to be aware of. Here are some common pitfalls and their corresponding remedies:

Inadequate Factor Selection
Pitfall: Choosing the wrong factors or omitting important ones can lead to incomplete or inconclusive results.

Remedy: Conduct a thorough preliminary investigation to identify and select the most relevant factors. Consult with experts in the field to ensure you're not overlooking critical variables.

Insufficient Factor Levels
Pitfall: Using a narrow range of levels for your factors may limit your ability to capture the full spectrum of biological responses.

Remedy: Ensure that the selected levels for each factor encompass the biologically relevant range. Pilot experiments or literature reviews can help you determine appropriate levels.

Over-reliance on Fractional Factorial Designs
Pitfall: Relying solely on fractional factorial designs may overlook potential interactions among factors that are not included in the design.

Remedy: Consider complementing fractional factorial experiments with additional follow-up experiments to explore interactions or factors that were excluded in the initial design.

Neglecting Randomization
Pitfall: Failing to randomize the order of experiments can introduce bias or confounding factors into your results.

Remedy: Randomize the order in which experiments are conducted to minimize the impact of uncontrolled variables. Use a randomized complete block design (RCBD) if applicable.

Inadequate Replication
Pitfall: Conducting experiments without sufficient replication can lead to unreliable or uninterpretable results.

Remedy: Include appropriate levels of replication to account for experimental variability and ensure the robustness of your conclusions. Calculate sample size requirements based on statistical power analysis.

Ignoring Assumptions
Pitfall: Failing to check or meet the assumptions of the statistical model (e.g., normality, homoscedasticity) can affect the validity of your results.

Remedy: Perform statistical diagnostics to verify that the assumptions are met. If not, consider data transformation or use alternative analysis methods suited to your data distribution.

Neglecting Data Validation
  Pitfall: Not checking for data errors or outliers can compromise the quality of your results.
  Remedy: Conduct data validation and verification to identify and address errors, outliers, or anomalies in your data. Outliers can significantly impact the analysis and should be investigated or removed if appropriate.
Misinterpretation of Results
  Pitfall: Misinterpreting the main effects and interactions can lead to incorrect conclusions.
  Remedy: Thoroughly understand the statistical analysis techniques used in your fractional factorial design and seek assistance from statisticians or data analysis experts if necessary. Interpret results in the context of your biological knowledge.
Failure to Communicate Findings
  Pitfall: Not effectively communicating your results and their implications can hinder the dissemination of valuable insights.
  Remedy: Prepare clear and concise reports or presentations to communicate your findings to colleagues, collaborators, or the broader scientific community. Highlight the practical implications and areas for further research.

By being aware of these potential pitfalls and implementing the suggested remedies, researchers can conduct more robust and meaningful fractional factorial design experiments in biology, leading to improved understanding and advancements in the field.

Example

Fractional factorial experiments in biology involve studying the effects of factors while conducting only a fraction of the full set of experimental runs. Here's an example of a matrix for a 2^3-1 fractional factorial experiment in biology:

Objective: To investigate the effects of three factors (temperature, nutrient concentration, and pH) on the growth of a specific microorganism.

Factors:

1. **Temperature (Factor A):**
   - Low temperature (25°C).
   - High temperature (35°C).
2. **Nutrient Concentration (Factor B):**
   - Low concentration (0.5 g/L).
   - High concentration (2 g/L).
3. **pH (Factor C):**
   - Acidic (pH 5).
   - Neutral (pH 7).
   - Alkaline (pH 9).

Response Variable: Microorganism growth (measured in colony-forming units per milliliter, CFU/mL).

Fractional Factorial Experiment Matrix (2^3-1):
   In this fractional factorial experiment, a fraction of the full factorial design (2^3) is conducted, specifically the one-half fraction, which reduces the number of experimental runs. The matrix might look like this (Table 2.3):

TABLE 2.3

Biology Fractional Factorial Design Experiment Matrix Example

Run	Temperature (A) (°C)	Nutrient Concentration (B) (g/L)	pH (C)	Microorganism Growth (CFU/mL)
1	25	0.5	5	...
2	35	0.5	5	...
3	25	2.0	5	...
4	35	2.0	5	...
5	25	0.5	7	...
6	35	0.5	7	...
7	25	2.0	7	...
8	35	2.0	7	...

In this matrix, only eight experimental runs are conducted, which is a fraction of the 2^3 (8-run) full factorial design. The "1" in 2^{3-1} signifies that one-half of the full design is carried out. Each row represents a specific experimental run with a unique combination of temperature, nutrient concentration, and pH levels. Microorganism growth is measured for each combination.

Fractional factorial experiments help reduce the number of experiments required while still allowing researchers to understand the main effects and some interactions between factors. This is particularly useful when conducting large-scale experiments is impractical or resource intensive.

REPORT

Fractional factorial design experiments in biology utilize various statistical tools and techniques to analyze data, assess factor effects, and interpret results. Here are some common statistical tools used in fractional factorial design experiments for biology:

ANOVA: ANOVA is a fundamental statistical tool used to partition the variation in experimental data into different sources, including main effects and interactions between factors. It helps assess the significance of each factor and their interactions on the observed biological response.

Main Effects Plots: Main effects plots visualize the impact of each factor on the biological response while holding all other factors constant. These plots are useful for identifying the relative importance of individual factors.

Interaction Plots: Interaction plots illustrate the interactions between two or more factors. They help researchers understand how the effects of one factor depend on the levels of another factor, which can provide insights into complex biological systems.

Residual Analysis: Residual analysis involves examining the differences between observed data and the values predicted by the statistical model. Residual plots and tests are used to check the assumptions of the model and identify outliers or unusual observations.

Fractional Factorial Design Matrices: The design matrix is a crucial tool for setting up and organizing fractional factorial experiments. It specifies which factor combinations are included in the design and which are omitted, helping researchers plan and conduct experiments efficiently.

RSM: RSM is a statistical technique often used in conjunction with fractional factorial designs. It helps researchers optimize experimental conditions by modeling the relationship between multiple factors and the response, allowing for the identification of optimal factor settings.

Normal Probability Plots and Box–Cox Transformation: These tools are used to assess the normality of data distributions. If data are not normally distributed, Box–Cox transformations may be applied to make the data conform to normality assumptions.

Factorial ANOVA: Factorial ANOVA extends the basic ANOVA to analyze the effects of multiple factors simultaneously. It allows for the assessment of main effects and interactions between factors in a factorial design.

Post-Hoc Tests: In cases where significant factor interactions are observed, post-hoc tests (e.g., Tukey's honestly significant different (HSD), Bonferroni) can help identify specific differences between factor levels.

Statistical Software: Statistical software packages like R, statistical analysis system (SAS), statistical package for social sciences (SPSS) or specialized tools for experimental design and analysis are commonly used to perform statistical analyses, generate plots, and interpret results.

Power Analysis: Power analysis helps researchers determine the required sample size to achieve adequate statistical power, ensuring that the experiment can detect meaningful effects.

These statistical tools are essential for designing, conducting, and analyzing fractional factorial experiments in biology. They help researchers gain insights into the relationships between factors and biological responses, optimize experimental conditions, and draw meaningful conclusions from their research.

A comprehensive report for fractional factorial design experiments used in biology should provide a clear and organized account of the experimental design, procedures, results, and conclusions. Here are the essential elements to include in such a report:

Title Page:
- Title of the report.
- Names of the authors and their affiliations.
- Date of submission or completion.

Abstract: A concise summary of the experiment's objectives, methods, key findings, and conclusions.

Table of Contents: List of sections, subsections, and page numbers for easy navigation.

List of Figures and Tables: Enumeration of all figures and tables along with their respective titles and page numbers.

Introduction:
- Clear statement of the research question or hypothesis.
- Background information to provide context for the experiment.
- Objectives and rationale for conducting the fractional factorial design experiment in the context of biology.

Experimental Design:
- Explanation of the factors and levels investigated in the experiment.
- Description of the fractional factorial design matrix, including which factor combinations were included and which were omitted.
- Details on the randomization and replication methods employed.

Materials and Methods:
- Detailed description of the experimental procedures, including sample preparation, data collection, and data analysis.
- Mention of any specialized equipment, software, or reagents used.
- Justification for the chosen methods.

Results:
- Presentation of the data in a clear and organized manner, often using tables, figures, and charts.

- Main effects and interaction effects, if applicable, should be reported.
- Statistical analyses, including ANOVA tables or relevant statistical tests, should be included.
- Any significant findings, trends, or patterns should be discussed.

Discussion:
- Interpretation of the results, including the biological implications of the observed effects.
- Consideration of potential sources of error, limitations of the study, and any unexpected findings.
- Comparison of the results with existing literature or relevant studies.
- Discussion of the practical relevance and applications of the findings.

Conclusion: A concise summary of the key findings and their significance. Restate the research objectives and whether they were achieved. Implications for future research or practical applications in biology.

Acknowledgments: Acknowledgment of individuals, organizations, or funding sources that contributed to the experiment or provided support.

References: Citation of all relevant sources, including research articles, books, and scientific literature, using a consistent citation style (e.g., APA, MLA).

Appendices: Supplementary information, such as raw data, calculations, or additional figures and tables. Any additional details or documentation necessary for reproducing the experiment.

Ethical Considerations (if applicable): Discussion of any ethical considerations, including compliance with research ethics, animal welfare, or human subjects' protection.

Glossary (if necessary): Definitions of technical terms and abbreviations used in the report.

Tables and Figures: Clear and labeled tables and figures that support the text and help illustrate key points or findings.

A well-structured report ensures that the details of the fractional factorial design experiment are conveyed effectively, allowing readers to understand the research process, outcomes, and their biological relevance.

2.1.4 CCD

CCD designed experiments are a valuable tool in biology for exploring response surfaces, optimizing experimental conditions, assessing factor interactions, achieving statistical robustness, and adapting experimental designs to address the complexity of biological systems. CCD enables researchers to systematically investigate biological processes, optimize experimental outcomes, and advance our understanding of the underlying biology of living organisms. CCD designed experiments are used in biology for several important reasons:

Exploration of Response Surfaces: Biology often involves studying complex systems where the relationship between factors and responses is nonlinear. CCD allows researchers to explore the response surface, which represents the relationship between experimental factors and the response of interest. By systematically varying factors at multiple levels and observing the corresponding responses, researchers can gain insights into the shape, curvature, and interactions of the response surface, helping to elucidate the underlying biology of the system.

Optimization of Experimental Conditions: CCD facilitates the optimization of experimental conditions by identifying the optimal combination of factor levels that maximize or minimize the response of interest. By fitting mathematical models to experimental data collected from CCD experiments, researchers can predict the optimal factor settings that achieve desired outcomes, such as maximizing cell growth, enzyme activity, or gene expression. This optimization process is crucial in biology for maximizing research efficiency, resource utilization, and the effectiveness of experimental interventions.

Assessment of Factor Interactions: CCD allows researchers to assess the interactions between factors, including linear, quadratic, and interaction effects. By systematically varying factors at different levels and observing changes in the response, researchers can identify significant interactions between factors that influence biological processes. Understanding these interactions is essential for elucidating the complex relationships between factors and for designing effective interventions or treatments in biology.

Robust Statistical Analysis: CCD experiments are designed to provide robust statistical analysis, allowing researchers to estimate factor effects and interactions with high precision. By strategically selecting experimental runs within the design space, CCD maximizes the information obtained from each experimental observation, leading to more accurate parameter estimates and more reliable predictions of response behavior. This statistical robustness is crucial in biology, where experiments often involve variability and noise inherent in biological systems.

Flexibility and Adaptability: CCD offers flexibility and adaptability, allowing researchers to modify and expand experimental designs as new questions arise or additional factors need to be considered. Researchers can easily add or remove factors, change factor levels, or introduce additional treatments within the CCD framework, enabling iterative experimentation, hypothesis testing, and model refinement. This flexibility is essential in biology for accommodating the complexity and variability of biological systems and adapting experimental designs to address evolving research questions and hypotheses.

CCD experiments are used in biology for a variety of research applications. CCD is a type of RSM that enables researchers to study the relationships between multiple factors and optimize experimental conditions. Here are some common areas where CCD experiments are applied in biology:

Drug Formulation and Optimization: CCD is often used in pharmaceutical and biotechnology research to optimize drug formulations. Researchers can investigate factors, such as drug concentration, pH, temperature, and excipient composition, to maximize drug stability, release, or bioavailability.

Enzyme Kinetics and Bioprocess Optimization: In enzymology and bioprocess engineering, CCD is employed to optimize reaction conditions, enzyme concentrations, and reaction kinetics. This helps enhance enzyme efficiency, yield, or selectivity in various biotechnological applications, such as biofuel production or enzyme-catalyzed reactions.

Environmental Toxicology and Ecotoxicology: CCD can be applied to study the effects of various environmental factors (e.g., temperature, pH, pollutant concentrations) on the growth, survival, or behavior of organisms in ecological and toxicological experiments. It helps assess the impact of environmental changes on biological systems.

Microbial Fermentation and Bioreactor Optimization: Bioprocessing and fermentation industries utilize CCD to optimize bioreactor conditions, including nutrient concentrations, temperature, agitation, and aeration. This optimization aims to maximize microbial growth, product yield, or bioproduct quality.

Nutritional Studies: In nutritional and dietary research, CCD can be used to investigate the influence of dietary components, such as nutrients or additives, on biological responses. Researchers can optimize diets or nutrient compositions to achieve specific health or growth outcomes in animals or microorganisms.

Plant Growth and Crop Yield Optimization: Agricultural and horticultural researchers apply CCD to optimize factors like soil composition, irrigation, and light exposure to maximize crop yield, quality, or resistance to environmental stressors.

Biological Response Surface Modeling: CCD is valuable for modeling and predicting biological responses, such as cell growth, enzyme activity, or gene expression, as a function of multiple experimental factors. This modeling aids in understanding complex biological systems and designing experiments for specific outcomes.

Experimental Design in Toxicology Studies: CCD can be used to design toxicity studies to investigate the effects of different chemical concentrations and exposure durations on living organisms. It helps determine the dose–response relationships and toxicological thresholds.

Vaccine Development and Optimization: In vaccine research, CCD can assist in optimizing antigen concentrations, adjuvant formulations, and vaccine delivery methods to enhance immune responses and vaccine efficacy.

Biological Response Surface Mapping: CCD experiments are used to map biological response surfaces, helping researchers identify optimal operating conditions, critical factors, and interactions among variables in complex biological systems.

CCD experiments are versatile tools in biology, offering a systematic approach to understanding and optimizing various biological processes, environmental interactions, and experimental conditions across a wide range of applications.

STEPS

Developing CCD experiments in biology involves a systematic approach to studying the effects of multiple factors on a biological response. Here are the steps to develop CCD experiments for biology:

1. **Define the Research Objectives**: Begin by clearly defining the research objectives and the specific biological response you aim to study or optimize. This could include aspects, such as enzyme activity, cell growth, protein expression, or any other relevant biological outcome.

2. **Identify Factors and Levels**: Determine the factors (IVs) that may influence the biological response. Factors can include environmental conditions, chemical concentrations, nutrient compositions, or other relevant variables. For each factor, specify the range or levels to be investigated.

3. **Choose a Response Surface Model**: Select the appropriate response surface model to represent the relationship between the factors and the biological response. Common models include linear, quadratic, or cubic models, depending on the expected nature of the response.

4. **Determine the Design Resolution**: Decide on the resolution of the CCD. A resolution of III is commonly used as it allows for the estimation of main effects, two-way interactions, and quadratic terms. Higher resolutions are more complex but can capture additional interactions.

5. **Generate Factorial Points**: Generate the factorial points (2^k) based on the number of factors and levels. These points represent the combinations of factors at extreme levels and form the basis of the CCD. Include center points for assessing curvature and replication for assessing experimental error.

6. **Choose Axial Points**: Add axial points ($\pm\alpha$) to the factorial points to explore the design space further. These axial points are placed at a specified distance (α) from the center of the design and are used to estimate the linear effects and potential curvature.

7. **Randomize the Experimental Run Order**: Randomize the order in which experiments are conducted to minimize the impact of uncontrolled variables or systematic errors. Ensure that experiments are conducted in a randomized and balanced manner.

8. **Conduct Experiments**: Perform the experiments according to the predetermined factor combinations, including the factorial points, center points, and axial points. Ensure that experimental conditions are carefully controlled and standardized.
9. **Collect and Record Data**: Record the data obtained from each experiment, including the values of the biological response and the corresponding factor settings. Maintain detailed records to ensure data accuracy.
10. **Analyze Data and Fit the Response Surface Model**: Analyze the collected data using regression analysis or appropriate statistical software. Fit the chosen response surface model to the data to estimate coefficients for main effects, interactions, and quadratic terms.
11. **Assess Model Adequacy**: Evaluate the adequacy of the fitted model using statistical criteria, such as the coefficient of determination (R^2), lack-of-fit tests, and residual plots. A good model fit should adequately describe the relationship between factors and the biological response.
12. **Interpret and Optimize Results**: Interpret the coefficients of the response surface model to understand the effects of factors on the biological response. Use the model to optimize factor settings that maximize or minimize the desired response while considering biological and practical constraints.
13. **Report and Document Findings**: Prepare a comprehensive report that includes the experimental design, procedures, results, statistical analyses, and conclusions. Document all findings and insights obtained from the CCD experiments in a clear and organized manner.

By following these steps, researchers can systematically develop and execute CCD experiments in biology to gain a deeper understanding of biological systems, optimize experimental conditions, and achieve specific biological outcomes.

PITFALLS AND REMEDIES

CCD experiments in biology are powerful tools for studying complex relationships between factors and biological responses. However, they can be prone to several pitfalls. Here are some common pitfalls and their corresponding remedies:

Inadequate Factor Selection
Pitfall: Choosing the wrong factors or overlooking important ones can lead to incomplete or misleading results.

Remedy: Conduct thorough preliminary research and consultation with domain experts to identify and select the most relevant factors for your biological study. Carefully consider the biological context.

Insufficient Factor Range
Pitfall: Narrow or inappropriate factor ranges may limit the ability to capture the full biological variability.

Remedy: Ensure that the selected factor ranges encompass the biologically relevant values. Pilot experiments, literature reviews, or prior knowledge can help establish appropriate factor ranges.

Violation of Model Assumptions
Pitfall: Fitting a response surface model to data that does not meet underlying assumptions (e.g., linearity, normality) can result in inaccurate predictions.

Remedy: Conduct diagnostic checks to verify model assumptions and consider data transformations or alternative models if necessary. Be cautious of extrapolating beyond the observed data.

Overfitting

> **Pitfall:** Overcomplicated models with too many terms can lead to overfitting and poor generalization.
>
> **Remedy:** Strike a balance between model complexity and goodness of fit. Use techniques like cross-validation or stepwise regression to identify the most important terms while avoiding excessive model complexity.

Lack of Replication

> **Pitfall:** Insufficient replication may lead to unreliable estimates of factor effects and interactions.
>
> **Remedy:** Include adequate replication for each factor combination to account for variability and assess experimental error. Calculate sample size requirements based on statistical power analysis.

Neglecting Model Validation

> **Pitfall:** Fitting a response surface model without proper validation can lead to misleading results.
>
> **Remedy:** Validate the model by conducting additional experiments or using holdout data not used in model fitting. Assess the model's predictive performance to ensure its reliability.

Ignoring Biological Variability

> **Pitfall:** Failing to account for biological variability, such as genetic differences or natural variation, can lead to overly optimistic conclusions.
>
> **Remedy:** Recognize and incorporate biological variability into the experimental design and interpretation. Consider using biological replicates or nested experimental designs.

Misinterpreting Results

> **Pitfall:** Misinterpreting the biological significance of model coefficients or response surface contours can lead to incorrect conclusions.
>
> **Remedy:** Carefully interpret the model coefficients and response surface plots in the biological context. Seek input from experts to validate the biological relevance of your findings.

Neglecting Practical Constraints

> **Pitfall:** Optimized factor settings may not be practically achievable or biologically meaningful.
>
> **Remedy:** Consider practical constraints and biological constraints when interpreting and applying optimization results. Ensure that optimized conditions are feasible in a real-world context.

Failure to Communicate Findings

> **Pitfall:** Failing to effectively communicate and share findings with colleagues or stakeholders can limit the impact of your research.
>
> **Remedy:** Prepare clear and concise reports or presentations that convey the experimental design, results, and implications in an accessible manner. Engage with relevant audiences to discuss the practical applications of your findings.

By being aware of these potential pitfalls and implementing the suggested remedies, researchers can conduct more robust and meaningful CCD experiments in biology, leading to better understanding and optimization of biological systems.

EXAMPLE

A CCD is used to study the response of a system to various factors, especially when you want to explore the optimal conditions within a specified range. A CCD matrix for experiments in biology typically represents the combinations of factor levels used in the experimental design. The matrix

TABLE 2.4

Biology 3-Factor CCD Experiment Matrix Example

Run	Factor 1 (A)	Factor 2 (B)	Factor 3 (C)
1	−1	−1	−1
2	+1	−1	−1
3	−1	+1	−1
4	+1	+1	−1
5	−1	−1	+1
6	+1	−1	+1
7	−1	+1	+1
8	+1	+1	+1
9	0	0	0
10	0	0	0

allows researchers to systematically study the effects of multiple factors on a biological response. Here's a simplified example of a CCD matrix for a biology experiment with three factors (A, B, and C), each at two levels (−1 and +1) (Table 2.4):

Assuming three factors (A, B, and C) with two levels (−1 and +1) for each factor:

Run A (factor 1) B (factor 2) C (factor 3)
  In this example,

- Factors A, B, and C are varied at two levels: −1 and +1, representing low and high settings, respectively.
- There are a total of 10 experimental runs, including eight factorial points (combinations of −1 and +1 for each factor), and two center points (at level 0 for all factors).
- The center points are included for estimating the curvature of the response surface.
- Researchers would perform experiments at each of these factor combinations and measure the biological response of interest.

Please note that in a real-world experiment, the number of factors, factor levels, and design resolution may vary depending on the complexity of the system and research objectives. Researchers may also use software or statistical tools to generate the CCD matrix with more factors and levels to suit their specific research needs.

Here's another example of a matrix for a CCD designed experiment in biology:

Objective: To optimize the growth of a specific strain of bacteria by studying the effects of two factors (temperature and nutrient concentration) using a CCD.

Factors:

1. **Temperature (Factor A):**
   - Coded Low (−1): 25°C.
   - Coded Medium (0): 30°C.
   - Coded High (+1): 35°C.
2. **Nutrient Concentration (Factor B):**
   - Coded Low (−1): 0.5 g/L.
   - Coded High (+1): 2 g/L.

TABLE 2.5

Biology 2-Factor CCD Experiment

Run	Coded Temperature (A)	Coded Nutrient Concentration (B)	Bacterial Growth (CFU/mL)
1	−1	−1	...
2	+1	−1	...
3	−1	+1	...
4	+1	+1	...
5	0	0	...
6	−1.68	0	...
7	+1.68	0	...
8	0	−1.68	...
9	0	+1.68	...

Response Variable: Bacterial growth (measured in CFU/mL).

CCD Experiment Matrix:

In a CCD, the matrix is designed to explore the response surface by conducting experiments at the central point, as well as at the corners and axial points of the experimental space. The matrix is shown in Table 2.5.

MATRIX EXAMPLE

In this matrix, each row represents a specific experimental run with a unique combination of coded temperature and nutrient concentration levels. The central point (run 5) represents the midpoint of the range for both factors, while runs 6 to 9 explore the corners and axial points. Bacterial growth is measured for each combination.

The 2-factor CCD allows biologists to fit response surface models to the data and find the optimal conditions for bacterial growth by studying the effects of temperature and nutrient concentration while considering interactions. This design is useful for efficiently exploring the experimental space and identifying optimal conditions for biological processes.

REPORTING

CCD experiments in biology involve various statistical tools and techniques for data analysis, model fitting, and optimization. Here are some common statistical tools used in CCD design experiments for biology:

Regression Analysis: Regression analysis is used to fit mathematical models to the experimental data. In CCD experiments, researchers typically use polynomial regression models to describe the relationship between factors and the biological response. These models can include linear, quadratic, and cubic terms to capture the effects of factors and their interactions.

ANOVA: ANOVA is employed to assess the significance of factors and their interactions on the biological response. It helps partition the variability in the data into contributions from main effects and interactions. ANOVA tables are often used to summarize the analysis.

Response Surface Analysis: Response surface analysis involves the graphical representation of the response surface, which depicts how the biological response changes as factors vary within their specified ranges. Researchers use response surface plots to visualize the relationships and identify optimal factor settings.

Lack-of-Fit Tests: Lack-of-fit tests assess whether the fitted response surface model adequately describes the experimental data. If the lack-of-fit test is significant, it suggests that the model may not fit the data well, and further investigation is needed.

Model Selection Techniques: Researchers may use model selection techniques, such as stepwise regression or model simplification, to identify the most appropriate and parsimonious model that adequately explains the relationship between factors and the response.

Optimization Algorithms: Optimization techniques, such as gradient-based or response surface optimization algorithms, are applied to identify the optimal factor settings that maximize or minimize the biological response. These techniques help researchers find the conditions that yield the desired outcome.

Confidence Intervals and Prediction Intervals: Confidence intervals are used to quantify the uncertainty associated with model coefficients, while prediction intervals estimate the range within which future observations are likely to fall. Both intervals provide valuable information about the reliability of predictions.

Diagnostic Plots and Residual Analysis: Diagnostic plots, such as residual plots and normal probability plots, are used to assess the adequacy of the regression model. Residual analysis helps identify outliers, heteroscedasticity, or other deviations from model assumptions.

ANOVA for Model Validation: Researchers may perform ANOVA on the regression model to validate its adequacy. This involves comparing the residual sum of squares to the pure error, which can help determine whether the model adequately represents the data.

Statistical Software: Statistical software packages, such as R, SAS, JMP, or specialized optimization software are often used to perform the statistical analyses, generate response surface plots, and optimize factor settings.

These statistical tools and techniques are essential for analyzing CCD experiments in biology. They help researchers understand the relationships between factors and biological responses, identify optimal conditions, and draw meaningful conclusions from their experiments.

A well-structured report for CCD experiments used in biology is crucial for effectively communicating the experimental design, procedures, results, and conclusions to the scientific community. Here are the key elements that should be included in such a report:

Title Page:
- Title of the report.
- Names of the authors and their affiliations.
- Date of submission or completion.

Abstract: A concise summary of the research objectives, experimental design, key findings, and implications.

Table of Contents: A list of sections, subsections, and page numbers for easy navigation.

List of Figures and Tables: Enumeration of all figures and tables included in the report, along with their respective titles and page numbers.

List of Abbreviations and Symbols (if necessary): Definitions and explanations of abbreviations, symbols, or technical terms used in the report.

Introduction:
- Clear statement of the research objectives and the specific biological response under investigation.
- Background information to provide context for the CCD experiment.
- Rationale for using CCD and the significance of the study in the field of biology.

Materials and Methods:
- Detailed description of the experimental setup, including factors studied, their levels, and any specialized equipment or procedures.
- Information on how the biological response was measured or observed.

- Explanation of the statistical methods used for analysis, including the response surface model and regression techniques.

Experimental Design: Presentation of the CCD matrix, including factor combinations and corresponding factor levels. Explanation of the rationale behind the chosen factor ranges and levels.

Results:
- Presentation of the experimental data, including the observed biological responses for each experimental run.
- Statistical analysis of the data, including ANOVA tables, regression coefficients, and significance levels.
- Response surface plots or contour plots to visualize the relationships between factors and the response.

Discussion:
- Interpretation of the results, including the biological significance of factor effects and interactions.
- Comparison of the findings with prior research or relevant literature.
- Discussion of any limitations or sources of variability in the experiment.
- Implications of the results for the biological system under study.

Conclusion:
- Summary of the key findings and their implications.
- Reiteration of the research objectives and whether they were achieved.
- Recommendations for further research or practical applications in biology.

Acknowledgments: Acknowledgment of individuals, organizations, or funding sources that contributed to the experiment or provided support.

References: Citation of all relevant sources, including research articles, books, and scientific literature, using a consistent citation style (e.g., APA, MLA).

Appendices (if necessary): Supplementary information, such as raw data, statistical calculations, or additional figures and tables. Any relevant details or documentation that support the experiment's methodology or findings.

Graphs, Figures, and Tables: Clear and labeled graphical representations, figures, and tables that support the text and help illustrate key points or findings.

A well-organized and informative report ensures that the details of the CCD experiment are effectively conveyed to the scientific community, allowing for the proper evaluation, understanding, and utilization of the research outcomes in the field.

2.1.5 Taguchi Methods in Biology

Taguchi method designed experiments offer a systematic and efficient approach to studying complex biological systems, optimizing experimental conditions, and enhancing the reliability and reproducibility of experimental results. By leveraging orthogonal arrays (OAs), robust parameter design principles, and statistical analysis techniques, the Taguchi method enables researchers to efficiently explore factor effects, optimize multiple responses, and advance our understanding of the underlying biology of living organisms. The Taguchi method designed experiments are utilized in biology for several compelling reasons:

Robustness to Variability: Biology is inherently variable due to factors, such as genetic diversity, environmental fluctuations, and stochastic processes. The Taguchi method is designed to be robust to variability, allowing researchers to identify optimal conditions or treatments that are less sensitive to variations in factors, such as genetic background, environmental conditions, or experimental procedures. This robustness is particularly valuable in biology, where reproducibility and reliability of experimental results are essential.

Efficiency in Screening Factors: Biology often involves studying systems with numerous variables, some of which may interact in complex ways. The Taguchi method enables efficient screening of factors by using OAs to systematically vary factors at multiple levels with a minimal number of experimental runs. This allows researchers to identify the most influential factors or interactions that significantly impact biological outcomes, saving time and resources compared to traditional factorial designs.

Optimization of Multiple Responses: In biology, researchers often need to optimize multiple responses simultaneously, such as maximizing cell growth while minimizing resource consumption or toxicity. The Taguchi method allows for the optimization of multiple responses using a single experimental design. By considering the trade-offs between different responses and identifying the factor settings that simultaneously optimize multiple criteria, researchers can design experiments that achieve balanced and desirable outcomes.

Reduction of Experimental Variation: The Taguchi method emphasizes the reduction of experimental variation by considering the effects of noise factors that may influence experimental outcomes. By incorporating noise factors into the experimental design and analyzing their effects on response variability, researchers can identify and control sources of variation that may otherwise confound experimental results. This reduction of variation enhances the reliability and reproducibility of experimental findings in biology.

Statistical Efficiency and Precision: The Taguchi method is statistically efficient, allowing researchers to estimate main effects and interactions with high precision using a relatively small number of experimental runs. By strategically selecting experimental settings and incorporating statistical analysis techniques, such as signal-to-noise ratios (SNRs), researchers can maximize the information obtained from each experimental observation, leading to more accurate estimates of factor effects and interactions.

The Taguchi method, developed by Genichi Taguchi, is a robust design methodology commonly used in engineering and manufacturing to improve product quality and performance while minimizing variation and cost. While the Taguchi method originated in engineering disciplines, it has also found applications in biology and life sciences. Here are some areas where the Taguchi method designed experiments are used in biology:

Bioprocess Optimization: In biotechnology and bioengineering, the Taguchi method is used to optimize bioprocess parameters, such as fermentation conditions, cell culture media composition, and enzyme reaction conditions. By systematically varying factors, such as pH, temperature, nutrient concentrations, and agitation speed, researchers can improve the yield, productivity, and quality of bioprocesses.

Environmental and Ecological Studies: The Taguchi method can be applied in environmental and ecological studies to optimize experimental conditions and control factors affecting biological responses. For example, researchers may use Taguchi experimental designs to investigate the effects of pollutants, temperature, pH, and other environmental variables on the growth, survival, and behavior of organisms in aquatic or terrestrial ecosystems.

Pharmaceutical Formulation Development: In pharmaceutical research, the Taguchi method is employed to optimize drug formulations, dosage forms, and drug delivery systems. Researchers use Taguchi experimental designs to systematically vary formulation factors, such as excipient composition, drug-to-excipient ratio, and processing parameters, to enhance drug stability, bioavailability, and therapeutic efficacy.

Gene Expression Studies: In molecular biology and genetics, the Taguchi method can be used to optimize experimental conditions for gene expression studies, such as PCR assays or gene transfection protocols. By optimizing factors, such as primer concentrations, annealing temperatures, and reaction times, researchers can improve the efficiency and reproducibility of gene expression experiments.

Plant Breeding and Crop Improvement: In agriculture and plant sciences, the Taguchi method is applied to optimize agronomic practices, breeding techniques, and crop management strategies. Researchers use Taguchi experimental designs to systematically vary factors, such as planting density, irrigation frequency, fertilizer application rates, and genotype combinations to maximize crop yield, quality, and resistance to biotic and abiotic stresses.

Toxicology and Drug Screening: In toxicology and drug screening studies, the Taguchi method can be used to optimize experimental conditions for assessing the toxicity, efficacy, and safety of chemical compounds or drug candidates. Researchers use Taguchi experimental designs to systematically vary factors, such as drug concentrations, exposure durations, and cell culture conditions, to identify optimal assay conditions and minimize variability in experimental outcomes.

Microbiology and Microbial Ecology: The Taguchi method is utilized in microbiology and microbial ecology to optimize experimental conditions for studying microbial growth, metabolism, and community dynamics. Researchers use Taguchi experimental designs to systematically vary factors, such as nutrient availability, temperature, pH, and substrate concentrations to elucidate the factors influencing microbial responses and interactions in various environments.

The Taguchi method designed experiments find applications in a wide range of biological disciplines, offering a systematic and efficient approach to optimizing experimental conditions, improving process performance, and enhancing understanding of biological systems.

STEPS

Developing Taguchi method designed experiments in biology involves a systematic approach to optimizing experimental conditions and factors affecting biological responses. Here are the steps to develop Taguchi method designed experiments in biology:

1. **Define the Objective**: Clearly define the research objective or problem statement. Determine the specific biological response or outcome that you want to optimize or improve.
2. **Identify Factors and Levels**: Identify the factors (IVs) that may influence the biological response. These factors could include environmental conditions, chemical concentrations, genetic parameters, or experimental parameters. Determine the range or levels for each factor to be studied.
3. **Select OAs**: Choose an appropriate OA based on the number of factors and levels. OAs are predefined experimental designs that ensure efficient and balanced experimentation. Select the OA that best suits the number of factors and levels in your experiment.
4. **Assign Factors to Columns**: Assign each factor to a column in the OA. Ensure that the factors are evenly distributed across the columns to minimize the confounding effects of interactions.
5. **Determine SNRs**: Select appropriate SNRs to evaluate the performance of the experimental factors. SNRs quantify the variability in the response data relative to the experimental factors and noise factors. Choose SNRs that are relevant to the specific biological response being studied.
6. **Conduct Experimental Runs**: Perform the experimental runs according to the combinations of factor levels specified in the OA. Ensure that the experimental conditions are accurately controlled and standardized across all runs.
7. **Collect Data and Calculate SNRs**: Measure the biological response or outcome for each experimental run. Calculate the SNR for each response using the specified SNR calculation method. The SNR represents the quality or performance of the experimental factors relative to noise factors.

8. **Analyze Data and Identify Optimal Conditions**: Analyze the experimental data to identify the optimal factor levels that maximize or minimize the SNR for the biological response. Use statistical methods, such as ANOVA, regression analysis, or graphical methods, to identify significant factors and their interactions.

9. **Perform Confirmation Runs**: Conduct confirmation runs to validate the optimized factor levels and ensure reproducibility of the results. Verify that the observed improvements in the biological response are consistent with the predicted optimal conditions.

10. **Interpret Results and Draw Conclusions**: Interpret the results of the Taguchi method designed experiments in the context of the research objective. Draw conclusions regarding the effect of experimental factors on the biological response and the optimized conditions for achieving the desired outcome.

11. **Document and Report Findings**: Prepare a comprehensive report documenting the experimental design, procedures, results, and conclusions of the Taguchi method designed experiments. Clearly communicate the insights gained from the study and their implications for further research or practical applications in biology.

By following these steps, researchers can systematically develop and execute Taguchi method designed experiments in biology to optimize experimental conditions, improve process performance, and enhance understanding of biological systems.

PITFALLS AND REMEDIES

Implementing Taguchi method designed experiments in biology can encounter various pitfalls, which, if not addressed, may lead to inaccurate conclusions or suboptimal outcomes. Here are some common pitfalls along with their corresponding remedies:

Inadequate Factor Selection:
> **Pitfall:** Selecting incorrect or insufficient factors may result in incomplete understanding or optimization of the biological system.
> **Remedy:** Conduct thorough preliminary research and consultation with domain experts to identify all relevant factors that could influence the biological response. Ensure that factors selected represent the key variables affecting the system.

Improper Factor Range Determination:
> **Pitfall:** Narrow or inappropriate ranges for experimental factors may limit the ability to capture the full range of variation in the biological response.
> **Remedy:** Conduct pilot studies or literature reviews to establish suitable ranges for each factor that encompass biologically relevant values. Ensure that the ranges cover both low and high levels of each factor.

Ignoring Factor Interactions:
> **Pitfall:** Neglecting interactions between factors may lead to an incomplete understanding of the biological system or suboptimal experimental designs.
> **Remedy:** Consider potential interactions between factors when designing the experiment. Use OAs to ensure balanced a representation of interactions. Analyze interaction effects along with main effects to identify significant factors.

Insufficient Replication:
> **Pitfall:** Inadequate replication of experimental runs may result in unreliable estimates of factor effects and variability in the biological response.
> **Remedy:** Include sufficient replication for each experimental condition to account for variability and assess experimental error. Use statistical methods to determine the appropriate sample size based on desired power and precision.

Violation of Assumptions:
> **Pitfall:** Violating assumptions of the Taguchi method, such as the assumption of linear effects or homogeneity of variance, can lead to inaccurate results.
>
> **Remedy:** Validate model assumptions through diagnostic checks, such as residual analysis or normality tests. Transform data if necessary to meet assumptions or consider alternative analysis methods.

Overlooking Noise Factors:
> **Pitfall:** Failing to account for noise factors or sources of variability outside the experimental control may obscure the true effects of experimental factors.
>
> **Remedy:** Identify and control or minimize sources of noise through proper experimental design, randomization, and blocking. Use SNRs to assess the robustness of the experimental factors to noise.

Inadequate Model Validation:
> **Pitfall:** Failing to adequately validate the predictive accuracy of the model may result in unreliable optimization or prediction of factor settings.
>
> **Remedy:** Perform validation experiments using holdout data or additional experimental runs. Compare predicted responses to observed responses to assess the model's predictive performance.

Misinterpretation of Results:
> **Pitfall:** Misinterpreting the significance or practical implications of experimental results may lead to erroneous conclusions.
>
> **Remedy:** Ensure clear and accurate interpretation of results by considering biological context and practical relevance. Seek input from domain experts to validate interpretations and conclusions.

By being aware of these potential pitfalls and implementing appropriate remedies, researchers can conduct Taguchi method designed experiments in biology more effectively, leading to robust optimization of experimental conditions and improved understanding of biological systems.

EXAMPLE

Taguchi methods, also known as the Taguchi robust design, are used to optimize processes or systems by identifying factors that affect performance and determining the optimal factor settings. Here's an example of a matrix for a Taguchi method designed experiment in biology:

Objective: To optimize the conditions for the growth of a specific plant species by studying the effects of two factors (light intensity and watering frequency) using a Taguchi L9 OA.

Factors:

1. **Light Intensity (Factor A):**
   - Low light (level 1).
   - Medium light (level 2).
   - High light (level 3).
2. **Watering Frequency (Factor B):**
   - Low frequency (level 1).
   - Medium frequency (level 2).
   - High frequency (level 3).

Response Variable: Plant growth (measured in height in centimeters).

TABLE 2.6
Taguchi Method Designed Experiment Matrix Example for Biology

Run	Light Intensity (A)	Watering Frequency (B)	Plant Growth (cm)
1	1	1	...
2	1	2	...
3	1	3	...
4	2	1	...
5	2	2	...
6	2	3	...
7	3	1	...
8	3	2	...
9	3	3	...

Taguchi L9 OAs:

Taguchi methods often use OAs to efficiently explore factor settings. In this case, a Taguchi L9 OA is chosen to perform nine experimental runs, which is suitable for studying two factors at three levels each. The matrix is shown in Table 2.6.

In this matrix, each row represents a specific experimental run with a unique combination of light intensity and watering frequency levels. Plant growth is measured for each combination.

Taguchi methods focus on finding the optimal factor settings that minimize variation and produce consistent results. The use of OAs and a systematic approach helps researchers efficiently identify the best conditions for plant growth while considering interactions between factors.

REPORTING

In Taguchi method designed experiments for biology, various statistical tools are employed to analyze experimental data, optimize factor settings, and interpret results effectively. Here are some common statistical tools used in Taguchi method designed experiments for biology:

ANOVA: ANOVA is used to partition the variation in the biological response into components attributed to different factors and their interactions. It helps assess the significance of factors and identify the most influential ones affecting the response.

SNRs: SNRs are calculated to quantify the signal (desired response) relative to the noise (undesired variation) in the experimental data. Different types of SNRs, such as smaller-the-better (STB), larger-the-better (LTB), and nominal-the-best (NTB), are used depending on the nature of the response.

OAs: OAs are predefined experimental designs used to systematically vary factors and levels while ensuring efficient experimentation. They facilitate the exploration of factor effects and interactions with a minimal number of experimental runs.

Main Effects Plots: Main effects plots visually display the estimated effects of individual factors on the biological response. They provide insights into the relative importance of different factors in influencing the response.

Interaction Plots: Interaction plots illustrate the interactions between pairs of factors on the biological response. They help identify synergistic or antagonistic effects between factors, which may not be evident from main effects alone.

RSM: RSM involves fitting mathematical models to experimental data to describe the relationship between factors and the biological response. It allows for the optimization of factor settings to achieve desired response levels.

Regression Analysis: Regression analysis is used to develop predictive models, relating factors to the biological response. Linear, quadratic, or higher-order regression models may be fitted to the data depending on the complexity of the relationship.

Diagnostic Plots: Diagnostic plots, such as residual plots, normal probability plots, and scatterplots of predicted versus observed responses, are used to assess the adequacy of the fitted models and identify any deviations from model assumptions.

ANOVA for Model Validation: ANOVA is performed on the fitted regression models to assess their goodness of fit and significance. It helps determine whether the models adequately represent the variability in the data and whether they can be used for prediction and optimization.

Optimization Algorithms: Optimization algorithms, such as gradient-based methods or heuristic search algorithms, are used to identify optimal factor settings that maximize or minimize the biological response. They help find the best combination of factors for achieving desired performance levels.

These statistical tools play a crucial role in analyzing Taguchi method designed experiments for biology, facilitating the optimization of experimental conditions, and enhancing understanding of biological systems.

A comprehensive report for Taguchi method designed experiments used in biology should effectively communicate the experimental design, procedures, results, and conclusions to the scientific community. Here are the key elements that should be included in such a report:

Title Page:
- Title of the report.
- Names of the authors and their affiliations.
- Date of submission or completion.

Abstract: A concise summary of the research objectives, experimental design, key findings, and implications.

Table of Contents: A list of sections, subsections, and page numbers for easy navigation.

List of Figures and Tables: Enumeration of all figures and tables included in the report, along with their respective titles and page numbers.

List of Abbreviations and Symbols (if necessary): Definitions and explanations of abbreviations, symbols, or technical terms used in the report.

Introduction:
- Clear statement of the research objectives and the specific biological response under investigation.
- Background information to provide context for the Taguchi method designed experiments.
- Rationale for using the Taguchi method and the significance of the study in the field of biology.

Materials and Methods:
- Detailed description of the experimental setup, including factors studied, their levels, and any specialized equipment or procedures.
- Information on how the biological response was measured or observed.
- Explanation of the statistical methods used for analysis, including the Taguchi method, OAs, and SNRs.

Experimental Design:
- Presentation of the Taguchi experimental design, including the selected OA and factor combinations.
- Explanation of how factors were assigned to columns and the rationale behind factor selection and levels.

Results: Presentation of the experimental data, including the observed biological responses for each experimental run. Statistical analysis of the data, including ANOVA tables, SNRs, and graphical representations of main effects and interactions.

Discussion:
- Interpretation of the results, including the biological significance of factor effects and interactions.
- Comparison of the findings with prior research or relevant literature.
- Discussion of any limitations or sources of variability in the experiment.
- Implications of the results for the biological system under study.

Conclusion:
- Summary of the key findings and their implications.
- Reiteration of the research objectives and whether they were achieved.
- Recommendations for further research or practical applications in biology.

Acknowledgments: Acknowledgment of individuals, organizations, or funding sources that contributed to the experiment or provided support.

References: Citation of all relevant sources, including research articles, books, and scientific literature, using a consistent citation style (e.g., APA, MLA).

Appendices (if necessary): Supplementary information, such as raw data, statistical calculations, or additional figures and tables. Any relevant details or documentation that support the experiment's methodology or findings.

Graphs, Figures, and Tables: Clear and labeled graphical representations, figures, and tables that support the text and help illustrate key points or findings.

A well-organized and informative report ensures that the details of the Taguchi method designed experiments are effectively conveyed to the scientific community, allowing for the proper evaluation, understanding, and utilization of the research outcomes in the field of biology.

2.1.6 MIXTURE EXPERIMENTS FOR BIOLOGY

Mixture designed experiments offer a systematic and efficient approach for studying complex mixtures in biology, optimizing experimental conditions, and understanding the interactions between components. By systematically varying the proportions of components in a mixture and observing changes in biological responses, researchers can optimize the composition of mixtures for specific outcomes, enhance our understanding of complex biological systems, and advance biomedical research and applications. Mixture designed experiments are valuable in biology for several reasons:

Studying Complex Mixtures: Many biological systems involve complex mixtures of components, such as cell culture media, drug formulations, or environmental samples. Mixture designed experiments allow researchers to systematically vary the proportions of different components in a mixture and study their effects on biological responses. This enables researchers to optimize the composition of mixtures for specific outcomes, such as maximizing cell growth, enhancing drug efficacy, or minimizing toxicity.

Understanding Synergistic or Antagonistic Effects: In biology, the interactions between components in a mixture can have synergistic or antagonistic effects on biological responses. Mixture designed experiments enable researchers to investigate these interactions by systematically varying the proportions of components and observing changes in response. By identifying synergistic combinations or mitigating antagonistic effects, researchers can optimize the composition of mixtures to achieve desired biological outcomes.

Optimization of Experimental Conditions: Mixture designed experiments facilitate the optimization of experimental conditions by identifying the optimal composition of mixtures that maximize or minimize biological responses. By systematically varying the proportions of components and fitting mathematical models to experimental data, researchers can predict the optimal mixture composition that achieves desired outcomes, such as maximizing enzyme activity, enhancing protein expression, or improving cell viability.

Reduction of Experimental Runs: Mixture designed experiments offer the advantage of reducing the number of experimental runs required compared to full factorial designs. By focusing on the proportions of components in a mixture rather than their absolute values, researchers can create efficient experimental designs that require fewer experimental runs while still capturing essential information about mixture effects and interactions. This reduction in experimental runs saves time, resources, and labor, making mixture designed experiments particularly valuable in biology, where experiments can be labor-intensive and costly.

Statistical Efficiency and Precision: Mixture designed experiments are statistically efficient, allowing researchers to estimate mixture effects and interactions with high precision using a relatively small number of experimental runs. By strategically selecting experimental settings and incorporating statistical analysis techniques, researchers can maximize the information obtained from each experimental observation, leading to more accurate estimates of mixture effects and interactions.

Mixture designed experiments are widely used in biology for studying complex systems where multiple components or factors interact to produce a biological response. Here are several areas within biology where mixture designed experiments find application:

Pharmaceutical Formulation Development: In pharmaceutical research and development, mixture designed experiments are used to optimize the composition of drug formulations, including tablets, capsules, creams, and injectables. Researchers investigate the effects of different excipients, such as binders, disintegrants, and lubricants, on drug stability, solubility, bioavailability, and release kinetics.

Nutrition and Food Science: Mixture designed experiments are employed in nutrition and food science to optimize the composition of food products and dietary supplements. Researchers study the effects of various ingredients, such as vitamins, minerals, proteins, fats, and carbohydrates, on nutritional content, flavor, texture, and shelf life of food products.

Cell Culture Media Optimization: In cell biology and biotechnology, mixture designed experiments are used to optimize the composition of cell culture media for growing cells in vitro. Researchers investigate the effects of different nutrients, growth factors, salts, and pH levels on cell growth, viability, productivity, and phenotype.

Ecology and Environmental Science: Mixture designed experiments find applications in ecology and environmental science for studying species interactions, community dynamics, and ecosystem functions. Researchers investigate the effects of various environmental factors, such as temperature, humidity, pH, and nutrient availability, on species diversity, population dynamics, and ecological processes.

Plant Breeding and Crop Improvement: In agriculture and plant sciences, mixture designed experiments are used to optimize the composition of crop fertilizers, soil amendments, and growth media. Researchers study the effects of different nutrient combinations, soil properties, and environmental conditions on crop yield, quality, and resistance to biotic and abiotic stresses.

Microbiology and Microbial Ecology: Mixture designed experiments are employed in microbiology and microbial ecology to study microbial growth, metabolism, and interactions. Researchers investigate the effects of various nutrients, carbon sources, energy substrates, and environmental factors on microbial community structure, function, and activity.

Chemical and Biochemical Engineering: In chemical and biochemical engineering, mixture designed experiments are used to optimize the composition of reaction mixtures, catalysts, and process conditions. Researchers investigate the effects of different reactants, solvents, temperatures, and pressures on reaction kinetics, yield, selectivity, and product purity.

Medicinal Plant Extract Optimization: Mixture designed experiments are applied in herbal medicine and phytochemistry to optimize the composition of medicinal plant extracts. Researchers investigate the effects of different extraction solvents, extraction techniques, and extraction parameters on the bioactive compound content, potency, and pharmacological activities of plant extracts.

Mixture designed experiments are used in various branches of biology to optimize the composition of complex systems; understand the interactions between components; and improve the quality, performance, and efficiency of biological processes and products.

STEPS

Developing mixture designed experiments in biology involves a systematic approach to optimizing the composition of complex systems where multiple components interact to produce a biological response. Here are the steps to develop mixture designed experiments in biology:

1. **Define the Objective**: Clearly define the research objective or problem statement. Determine the specific biological response or outcome that you want to optimize or improve by varying the composition of the mixture.
2. **Identify Components and Constraints**: Identify the components (ingredients or factors) that will make up the mixture. Consider the biological relevance of each component and any constraints or limitations on their composition or availability.
3. **Select the Mixture Design**: Choose an appropriate mixture design methodology based on the number of components and constraints. Common mixture design approaches include simplex-centroid designs, simplex-lattice designs, and extreme vertices designs.
4. **Determine the Design Space**: Define the feasible region or design space within which the mixture components can vary. Consider any constraints or limitations on the composition of the mixture, such as upper and lower bounds, proportionality constraints, or physical/chemical constraints.
5. **Assign Factor Levels**: Determine the levels or proportions of each component to be studied in the experiment. Ensure that the chosen factor levels are within the feasible region defined by the design space.
6. **Generate Experimental Design**: Generate experimental design using software or statistical tools that implement the chosen mixture design methodology. The design should specify the combinations of component levels to be tested in the experiment.
7. **Conduct Experimental Runs**: Perform the experimental runs according to the combinations of component levels specified in the experimental design. Ensure that the experimental conditions are accurately controlled and standardized across all runs.
8. **Measure Biological Response**: Measure the biological response or outcome of interest for each experimental run. This could include quantitative measurements, qualitative observations, or subjective assessments depending on the nature of the response.
9. **Analyze Data**: Analyze the experimental data to determine the effects of mixture components on the biological response. Use statistical methods, such as regression analysis, ANOVA, or RSM, to model the relationship between mixture composition and response.
10. **Optimize Mixture Composition**: Use optimization techniques to identify the optimal composition of the mixture that maximizes or minimizes the biological response.

This may involve response surface optimization, numerical optimization algorithms, or graphical methods.

11. **Validate and Confirm**: Validate the optimized mixture composition through additional experiments or validation studies. Confirm that the observed improvements in the biological response are consistent with the predicted optimal conditions.

12. **Interpret Results and Draw Conclusions**: Interpret the results of the mixture design experiments in the context of the research objectives. Draw conclusions regarding the effects of mixture components on the biological response and the optimized composition for achieving the desired outcome.

13. **Document and Report Findings**: Prepare a comprehensive report documenting the experimental design, procedures, results, and conclusions of the mixture design experiments. Clearly communicate the insights gained from the study and their implications for further research or practical applications in biology.

By following these steps, researchers can systematically develop and execute mixture design experiments in biology to optimize the composition of complex systems and improve understanding of the relationships between mixture components and biological responses.

PITFALLS AND REMEDIES

Implementing mixture design experiments in biology may encounter various pitfalls, which, if not addressed, can lead to inaccurate conclusions or suboptimal outcomes. Here are some common pitfalls along with their corresponding remedies:

Improper Selection of Components:
> **Pitfall:** Choosing inappropriate or irrelevant components for the mixture may lead to ineffective experimentation or misleading results.
>
> **Remedy:** Conduct thorough preliminary research and consultation with domain experts to identify relevant components that significantly influence the biological response. Ensure that the selected components are biologically meaningful and have a plausible impact on the outcome of interest.

Failure to Consider Component Interactions:
> **Pitfall:** Neglecting interactions between mixture components may overlook important synergistic or antagonistic effects that influence the biological response.
>
> **Remedy:** Investigate potential interactions between components during experimental design and analysis. Use statistical techniques, such as two-way ANOVA, interaction plots, or RSM, to assess and account for component interactions.

Inadequate Design Space Exploration:
> **Pitfall:** Failing to explore a sufficient range of component levels within the design space may limit the ability to identify optimal mixture compositions.
>
> **Remedy:** Define an appropriate design space that encompasses a wide range of plausible component levels. Ensure that the experimental design includes factor levels that span the entire design space to adequately explore the effects of each component on the biological response.

Overfitting of Models:
> **Pitfall:** Overfitting occurs when a statistical model fits the noise in the data rather than the underlying biological relationships, leading to poor predictive performance.
>
> **Remedy:** Use parsimonious modeling techniques and avoid including unnecessary complexity in the model. Validate the model using independent data or cross-validation methods to assess its predictive accuracy and generalizability.

Ignoring Nonlinear Relationships:

Pitfall: Assuming linear relationships between mixture components and the biological response may oversimplify the true underlying relationships, leading to inaccurate predictions.

Remedy: Consider nonlinear modeling techniques, such as quadratic or higher-order models, to capture nonlinear effects and curvature in the relationship between mixture components and the biological response. Use diagnostic plots and residual analysis to assess the adequacy of the chosen model.

Insufficient Replication:

Pitfall: Conducting experiments with inadequate replication may result in unreliable estimates of component effects and variability in the biological response.

Remedy: Include sufficient replication for each experimental condition to account for variability and assess experimental error. Use statistical methods to determine the appropriate sample size based on desired power and precision.

Biological Complexity Oversimplification:

Pitfall: Oversimplifying the biological system may lead to unrealistic or misleading conclusions about the relationship between mixture components and the biological response.

Remedy: Consider the complexity of the biological system and incorporate relevant biological knowledge and expertise into the experimental design and analysis. Take into account potential confounding factors, biological mechanisms, and ecological interactions that may influence the response.

By addressing these pitfalls and implementing appropriate remedies, researchers can conduct robust and informative mixture designed experiments in biology, leading to a better understanding of complex biological systems and the optimization of mixture compositions for desired outcomes.

EXAMPLE

Mixture designed experiments in biology are used when a response variable depends on the proportions of multiple components that make up a mixture. Here's an example of a matrix for a mixture designed experiment in biology:

Objective: To optimize the growth medium for a specific microorganism by studying the effects of three components (A, B, and C) on microbial growth.

Components:

1. **Component A (Proportion in Mixture)**: Proportion ranging from 0% to 100%.
2. **Component B (Proportion in Mixture)**: Proportion ranging from 0% to 100%.
3. **Component C (Proportion in Mixture)**: Proportion ranging from 0% to 100%.

Response Variable: Microbial growth (measured in CFU/mL).

Mixture Designed Experiment Matrix:

In a mixture designed experiment, the matrix is designed to vary the proportions of the components within a specified range. The matrix might look like this (Table 2.7):

In this matrix, each row represents a specific experimental run with a unique combination of component A, component B, and component C proportions. Microbial growth is measured for each combination.

TABLE 2.7

Mixture Designed Experiment Matrix Example for Biology

Run	Component A (%)	Component B (%)	Component C (%)	Microbial Growth (CFU/mL)
1	10	30	60	...
2	50	20	30	...
3	25	45	30	...
4	70	10	20	...
5	30	60	10	...
6	40	40	20	...
7	20	20	60	...
8	60	30	10	...

Mixture designed experiments help biologists optimize the composition of growth media or solutions by systematically varying the proportions of components to achieve the desired biological response. This approach is particularly valuable when the response depends on the interactions between the components in the mixture.

REPORTING

In mixture designed experiments for biology, various statistical tools are employed to analyze experimental data, optimize mixture compositions, and interpret results effectively. Here are some common statistical tools used in mixture designed experiments for biology:

1. **ANOVA**: ANOVA is used to partition the variation in the biological response into components attributed to different mixture components and their interactions. It helps assess the significance of mixture components and identify the most influential ones affecting the response.
2. **RSM**: RSM involves fitting mathematical models to experimental data to describe the relationship between mixture components and the biological response. It allows for the optimization of mixture compositions to achieve desired response levels. RSM typically includes techniques, such as polynomial regression, surface plots, and contour plots.
3. **Regression Analysis**: Regression analysis is used to develop predictive models relating mixture components to the biological response. Linear, quadratic, or higher-order regression models may be fitted to the data depending on the complexity of the relationship between mixture components and the response.
4. **Mixture Model Analysis**: Mixture model analysis is specifically designed for analyzing data from mixture experiments. It accounts for the constrained nature of mixture components (i.e., the sum of component proportions equals one) and provides estimates of component effects on the response.
5. **ANOVA for Mixture Models**: ANOVA techniques adapted for mixture experiments are used to assess the significance of mixture components and their interactions. These techniques may include traditional ANOVA, as well as analyses tailored to the constraints of mixture experiments, such as Scheffé's method or sum of squares decomposition.
6. **Optimization Algorithms**: Optimization algorithms are used to identify optimal mixture compositions that maximize or minimize the biological response. These algorithms search for the combination of mixture components that yields the best response using techniques, such as gradient-based methods, evolutionary algorithms, or response surface optimization.

7. **Desirability Functions**: Desirability functions provide a comprehensive approach to optimizing multiple responses simultaneously. They combine individual response variables into a single desirability index, allowing researchers to identify mixture compositions that meet multiple criteria or constraints.

8. **Diagnostic Plots**: Diagnostic plots, such as residual plots, normal probability plots, and scatterplots of predicted versus observed responses, are used to assess the adequacy of fitted models and identify any deviations from model assumptions.

9. **Model Validation Techniques**: Model validation techniques, such as cross-validation, bootstrapping, or holdout validation, are used to assess the predictive performance of fitted models and ensure their generalizability to new data.

10. **PCA**: PCA can be used to explore patterns and relationships among mixture components and responses in high-dimensional datasets. It helps identify underlying trends, correlations, and sources of variability in the data.

By leveraging these statistical tools, researchers can effectively analyze experimental data from mixture designed experiments, optimize mixture compositions, and gain insights into the relationships between mixture components and biological responses in diverse biological systems.

A comprehensive report for mixture designed experiments used in biology should effectively communicate the experimental design, procedures, results, and conclusions to the scientific community. Here are the key elements that should be included in such a report:

Title Page:
- Title of the report.
- Names of the authors and their affiliations.
- Date of submission or completion.

Abstract: A concise summary of the research objectives, experimental design, key findings, and implications.

Table of Contents: A list of sections, subsections, and page numbers for easy navigation.

List of Figures and Tables: Enumeration of all figures and tables included in the report, along with their respective titles and page numbers.

List of Abbreviations and Symbols (if necessary): Definitions and explanations of abbreviations, symbols, or technical terms used in the report.

Introduction:
- Clear statement of the research objectives and the specific biological response under investigation.
- Background information to provide context for the mixture design experiments.
- Rationale for using mixture design and the significance of the study in the field of biology.

Materials and Methods:
- Detailed description of the experimental setup, including mixture components, their levels, and any specialized equipment or procedures.
- Information on how the biological response was measured or observed.
- Explanation of the statistical methods used for analysis, including mixture design methodology, regression techniques, and model validation.

Experimental Design: Presentation of the mixture design used in the experiment, including the selected design methodology, component levels, and constraints. Description of how the experimental runs were generated and conducted.

Results: Presentation of the experimental data, including the observed biological responses for each experimental run. Statistical analysis of the data, including regression analysis, ANOVA tables, and graphical representations of response surfaces.

Discussion:
- Interpretation of the results, including the biological significance of mixture components and their effects on the biological response.
- Comparison of the findings with prior research or relevant literature.
- Discussion of any limitations or sources of variability in the experiment.
- Implications of the results for the biological system under study.

Conclusion: Summary of the key findings and their implications. Reiteration of the research objectives and whether they were achieved. Recommendations for further research or practical applications in biology.

Acknowledgments: Acknowledgment of individuals, organizations, or funding sources that contributed to the experiment or provided support.

References: Citation of all relevant sources, including research articles, books, and scientific literature, using a consistent citation style (e.g., APA, MLA).

Appendices (if necessary): Supplementary information, such as raw data, statistical calculations, or additional figures and tables. Any relevant details or documentation that support the experiment's methodology or findings.

Graphs, Figures, and Tables: Clear and labeled graphical representations, figures, and tables that support the text and help illustrate key points or findings.

A well-organized and informative report ensures that the details of the mixture designed experiments are effectively conveyed to the scientific community, allowing for the proper evaluation, understanding, and utilization of the research outcomes in the field of biology.

2.1.7 SEQUENTIAL EXPERIMENTATION FOR BIOLOGY

Sequential design experiments provide a systematic and adaptive approach for studying complex biological systems, refining hypotheses, optimizing experimental conditions, and efficiently allocating resources. By conducting experiments sequentially and iteratively refining experimental designs based on previous results, researchers can advance our understanding of biological processes, address research questions, and make meaningful contributions to biomedical science and applications. Sequential design experiments are utilized in biology for several compelling reasons:

Progressive Refinement of Hypotheses: Biology often involves complex systems with numerous variables and interactions. Sequential design experiments allow researchers to progressively refine hypotheses and experimental designs based on the results of previous experiments. By conducting experiments sequentially, researchers can iteratively test hypotheses, modify experimental conditions, and build upon previous findings, leading to a deeper understanding of biological processes and phenomena.

Adaptation to Dynamic Systems: Biological systems are dynamic and can exhibit variability over time or in response to external stimuli. Sequential design experiments enable researchers to adapt experimental designs to the dynamic nature of biological systems by incorporating new information and adjusting experimental conditions in real time. This flexibility allows researchers to respond to unexpected findings, refine experimental protocols, and address emerging research questions or hypotheses.

Efficient Resource Utilization: Conducting experiments in biology can be resource intensive, requiring time, labor, and materials. Sequential design experiments enable researchers to optimize resource utilization by prioritizing experiments based on previous results and allocating resources to the most promising avenues of investigation. By focusing resources on experiments with the highest potential for yielding meaningful results, researchers can maximize the efficiency and impact of their research efforts.

Iterative Optimization of Experimental Conditions: Sequential design experiments facilitate the iterative optimization of experimental conditions by allowing researchers to systematically explore parameter space and identify optimal conditions over multiple iterations. By adjusting experimental parameters based on previous results and incorporating statistical analysis techniques, researchers can refine experimental protocols, optimize conditions for specific outcomes, and achieve more robust and reproducible experimental results in biology.

Enhanced Statistical Power and Precision: Sequential design experiments offer enhanced statistical power and precision by allowing researchers to adaptively allocate samples and resources based on observed variability and effect sizes. By focusing resources on experiments with the highest potential for detecting significant effects, researchers can maximize the statistical power of their experiments and obtain more precise estimates of biological parameters and relationships. This adaptive allocation of resources enhances the reliability and reproducibility of experimental findings in biology.

Sequential experimentation design (SED) experiments are used in biology in various contexts where iterative or adaptive experimental approaches are required to efficiently explore complex biological systems or optimize experimental procedures. Here are some areas where SED experiments find application in biology:

Drug Discovery and Development: In pharmaceutical research, SED is used to optimize drug candidates, dosage formulations, and treatment protocols. Researchers iteratively explore the effects of different drug compounds, concentrations, and delivery methods on biological targets, pharmacokinetics, and therapeutic outcomes.

Clinical Trials: In clinical research, SED is used to optimize study protocols, patient recruitment strategies, and treatment regimens. Researchers iteratively adjust study parameters, sample sizes, and randomization procedures based on accumulating data and interim analyses to enhance the efficiency and validity of clinical trials.

Genomics and Functional Genomics: In genomics research, SED is used to explore gene functions, regulatory networks, and genetic interactions. Researchers iteratively design and perform experiments, such as gene knockout studies, RNA interference screens, or clustered regularly interspaced short palindromic repeats (CRISPR)-based perturbations, to elucidate the roles of genes and genetic pathways in biological processes and diseases.

Proteomics and Metabolomics: In proteomics and metabolomics research, SED is used to identify and quantify proteins, metabolites, and biomarkers associated with physiological states, diseases, or environmental exposures. Researchers iteratively refine experimental protocols, sample preparation methods, and analytical techniques to improve sensitivity, accuracy, and throughput of omics analyses.

Systems Biology and Computational Biology: In systems biology and computational biology, SED is used to model, simulate, and predict complex biological systems and their emergent properties. Researchers iteratively refine computational models, parameterize simulations, and validate predictions using experimental data to gain insights into biological networks, signaling pathways, and cellular dynamics.

Ecology and Environmental Science: In ecology and environmental science, SED is used to study ecosystem dynamics, species interactions, and environmental responses to perturbations. Researchers iteratively design field experiments, monitoring programs, and modeling studies to assess the effects of climate change, pollution, habitat fragmentation, and species invasions on biodiversity, ecosystem services, and ecological resilience.

Bioprocess Engineering and Biomanufacturing: In bioprocess engineering and biomanufacturing, SED is used to optimize fermentation processes, bioreactor configurations, and downstream purification methods. Researchers iteratively optimize process parameters, media formulations, and scale-up strategies to maximize product yields, quality, and efficiency in biopharmaceutical and biofuel production.

Neuroscience and Brain Imaging: In neuroscience and brain imaging research, SED is used to investigate brain structure, function, and connectivity in health and disease. Researchers iteratively design neuroimaging studies, cognitive tasks, and clinical interventions to probe neural circuits, identify biomarkers, and develop therapeutic interventions for neurological and psychiatric disorders.

The SED experiments are used in various areas of biology to iteratively explore complex biological systems, optimize experimental procedures, and generate actionable insights for advancing scientific knowledge and addressing real-world challenges.

STEPS

Developing SED experiments in biology involves a systematic approach to iteratively explore complex biological systems, optimize experimental procedures, and generate actionable insights. Here are the steps to develop SED experiments in biology:

1. **Define Research Objectives**: Clearly define the research objectives or questions that the SED experiments aim to address. Identify the specific biological system, process, or phenomenon of interest, and articulate the desired outcomes or hypotheses to be tested.
2. **Review Existing Knowledge**: Conduct a comprehensive literature review to identify relevant prior research, experimental methods, and theoretical frameworks related to the research objectives. Gain insights into the current state-of-the-art methodologies and potential gaps or limitations in existing knowledge.
3. **Identify Experimental Variables**: Identify the key variables, factors, or parameters that may influence the biological system under investigation. Consider both IVs (e.g., experimental treatments, environmental conditions) and DVs (e.g., biological responses, outcomes).
4. **Design Initial Experiment**: Develop an initial experimental design that systematically varies the identified variables to explore their effects on the biological system. Choose appropriate experimental techniques, protocols, and measurements to collect relevant data and observations.
5. **Perform Initial Experiment**: Conduct the initial experiment according to the planned design, procedures, and protocols. Ensure that experimental conditions are accurately controlled, standardized, and replicated to generate reliable and reproducible data.
6. **Analyze Initial Results**: Analyze the data and results obtained from the initial experiment using appropriate statistical methods, visualization techniques, and data interpretation approaches. Identify trends, patterns, and relationships between experimental variables and biological responses.
7. **Iterative Experimentation**: Based on the analysis of initial results, iteratively refine the experimental design, modify experimental parameters, and plan additional experiments to further explore or validate hypotheses, optimize procedures, or address emerging questions.
8. **Adaptive Experimental Design**: Incorporate adaptive experimental design principles to dynamically adjust experimental parameters, sample sizes, or study protocols based on accumulating data and interim analyses. Optimize resource allocation, experimental efficiency, and scientific yield by adapting experimental plans in real time.
9. **Evaluate and Refine Models**: Continuously evaluate and refine mathematical models, computational simulations, or conceptual frameworks that describe the relationships between experimental variables and biological responses. Incorporate new data, insights, or constraints to improve model accuracy and predictive performance.

10. **Validate Findings**: Validate the findings, conclusions, and implications derived from SED experiments through independent replication, cross-validation, or external validation studies. Ensure that the observed effects are robust, reproducible, and generalizable across different experimental conditions or contexts.

11. **Document and Communicate Results**: Document the experimental design, procedures, data, results, and conclusions of SED experiments in a comprehensive and transparent manner. Communicate findings through scientific publications, presentations, or reports to share knowledge and contribute to the advancement of biological research.

By following these steps, researchers can systematically develop and execute SED experiments in biology to explore complex biological systems, optimize experimental procedures, and generate actionable insights for scientific discovery and innovation.

PITFALLS AND REMEDIES

Implementing SED experiments in biology may encounter various pitfalls, which, if not addressed, can lead to inefficient experimentation, biased results, or misinterpretation of findings. Here are some common pitfalls along with their corresponding remedies:

Biased Sampling:
   Pitfall: Biased sampling occurs when the selection of experimental units or study subjects is not representative of the population under study, leading to skewed results.
   Remedy: Randomize the selection of experimental units or study subjects to minimize bias and ensure the representativeness of the sample. Use stratification or blocking techniques to account for known sources of variation and ensure a balanced representation across experimental conditions.

Overfitting Models:
   Pitfall: Overfitting occurs when a statistical model fits the noise in the data rather than the underlying biological relationships, leading to poor generalization and predictive performance.
   Remedy: Use parsimonious modeling techniques and avoid including unnecessary complexity in the model. Regularize the model parameters, cross-validate the model using independent data, or use model selection criteria to prevent overfitting and improve model robustness.

Ignoring Sequential Dependencies:
   Pitfall: Ignoring sequential dependencies between experimental observations or treatments may violate assumptions of independence and lead to biased estimates or inflated type I error rates.
   Remedy: Account for sequential dependencies using appropriate statistical methods, such as time-series analysis, repeated measures models, or hierarchical modeling approaches. Incorporate autocorrelation structures or random effects to model temporal or spatial dependencies in the data.

Inadequate Control of Type I Error Rate:
   Pitfall: Failing to control the type I error rate (false-positive rate) in sequential experimentation may result in spurious findings or erroneous conclusions.
   Remedy: Apply multiple comparison adjustments, such as Bonferroni correction, false discovery rate (FDR) control, or sequential testing procedures (e.g., alpha spending functions), to account for the accumulation of type I errors over multiple comparisons or interim analyses.

Limited Generalizability:
   Pitfall: Limited generalizability occurs when findings from SED experiments are not applicable beyond the specific experimental conditions or study population.

Remedy: Ensure that experimental designs and protocols are sufficiently diverse and representative to capture the variability and heterogeneity of the target population or biological system. Validate findings through external replication or cross-validation studies in independent samples or contexts to assess their generalizability.

Uncontrolled Confounding Factors:

Pitfall: Uncontrolled confounding factors, such as environmental variables, genetic background, or experimental artifacts, may introduce bias or spurious associations in sequential experimentation.

Remedy: Implement appropriate experimental controls, randomization procedures, or matching techniques to minimize the influence of confounding factors and isolate the effects of experimental variables of interest. Use factorial designs, covariate adjustment, or sensitivity analyses to assess and address potential confounding effects.

Inefficient Resource Allocation:

Pitfall: Inefficient resource allocation occurs when experimental resources (e.g., time, budget, sample size) are not optimally allocated to maximize scientific yield or statistical power.

Remedy: Use adaptive experimental design principles to dynamically adjust sample sizes, treatment allocations, or study protocols based on accumulating data and interim analyses. Apply sequential stopping rules or futility criteria to terminate ineffective or unpromising experiments early and reallocate resources to more promising research avenues.

Publication Bias:

Pitfall: Publication bias arises when only statistically significant or positive findings are reported, while nonsignificant or negative results are omitted, leading to an overestimation of effect sizes or an inaccurate portrayal of the scientific evidence base.

Remedy: Promote transparency and reproducibility in research by preregistering study protocols, outcomes, and analysis plans. Publish study protocols, raw data, and negative results in open-access repositories to mitigate publication bias and ensure the integrity and reliability of scientific evidence.

By addressing these pitfalls and implementing appropriate remedies, researchers can conduct robust and informative SED experiments in biology, leading to reliable scientific findings and actionable insights for advancing biological knowledge and addressing real-world challenges.

EXAMPLE

Sequential experimentation in biology involves a series of experiments where the results of each experiment inform the design of subsequent experiments. Here's a simplified example of a matrix for a sequential designed experiment in biology:

Objective: To identify the optimal conditions for the germination of a specific plant species by studying the effects of various factors over multiple stages.

Factors (Varied Sequentially):

1. **Light Exposure:**
   - Stage 1: Low light (L).
   - Stage 2: Medium light (M).
   - Stage 3: High light (H).

2. **Temperature:**
   - Stage 1: Low temperature (T).
   - Stage 2: Medium temperature (M).
   - Stage 3: High temperature (H).

3. **Watering Frequency:**
   - Stage 1: Low frequency (F).
   - Stage 2: Medium frequency (M).
   - Stage 3: High frequency (H).

Response Variable: Germination rate (measured as a percentage).

Sequential Experimentation Matrix:
In sequential experimentation, the matrix is designed to vary the factors sequentially, with each stage building upon the results of the previous stage. The matrix is shown in Tables 2.8–2.10.

Stage 1:
Stage 2:

Based on the results of stage 1, you adjust the factors for stage 2:

Stage 3:

TABLE 2.8
Sequential Design Experiments Stage 1 for Biology

Run	Light Exposure	Temperature	Watering Frequency	Germination Rate (%)
1	L	T	F	...
2	M	T	F	...
3	H	T	F	...

TABLE 2.9
Sequential Design Experiments Stage 2 for Biology

Run	Light Exposure	Temperature	Watering Frequency	Germination Rate (%)
4	L	M	F	...
5	M	M	F	...
6	H	M	F	...

TABLE 2.10
Sequential Design Experiments Stage 3 for Biology

Run	Light Exposure	Temperature	Watering Frequency	Germination Rate (%)
7	L	M	H	...
8	M	M	H	...
9	H	M	H	...

TABLE 2.11

Sequential Design Experiments Iterative Adjustments for Biology

Experiment	Treatment A	Treatment B	Treatment C	Outcome 1	Outcome 2	Outcome 3
1	High	Low	Low			
2	High	High	Low			
3	Low	High	High			
...	...	...	...	...	...	...
N	Low	Low	High			

Based on the results of stage 2, you adjust the factors for stage 3:

In this example, each stage represents a set of experimental runs with specific combinations of factors. The results of each stage guide the adjustments made in subsequent stages to systematically identify the optimal conditions for plant germination. Sequential experimentation allows biologists to efficiently refine their experimental conditions based on the evolving understanding of the system being studied.

Another example of a SED experiment in biology can take various forms depending on the specific research objectives, experimental variables, and study design. Sequential experimentation designs often involve iterative adjustments to experimental parameters based on accumulating data and interim analyses. A simplified example of a matrix for a SED in biology is shown in Table 2.11.

In this simplified matrix,

- Each row represents a single experimental run or iteration.
- Columns represent experimental variables, such as different treatments or conditions applied in the experiment.
- "Outcome" columns indicate the measured responses or outcomes for each experimental run.
- The specific treatments or conditions applied in each experiment may vary based on the sequential design strategy, such as adaptive allocation, response-adaptive randomization, or dynamic treatment assignment.

During the course of the experiment, researchers may update the matrix based on interim analyses, adjust treatment allocations or experimental parameters, and collect additional data to refine the experimental design and optimize outcomes. Sequential experimentation designs allow for flexible and adaptive exploration of complex biological systems, enabling researchers to efficiently learn from ongoing experiments and iteratively refine their research approach.

Reporting

Sequential experimentation design experiments in biology often involve complex data analysis techniques to adaptively adjust experimental parameters, optimize study protocols, and interpret evolving results. Here are some statistical tools commonly used in SED experiments for biology:

1. **Bayesian Methods**: Bayesian statistical methods are commonly used in SED to update prior knowledge with new data and iteratively refine probability distributions for model parameters or hypotheses. Bayesian techniques allow for flexible modeling of complex biological systems, incorporation of prior information, and adaptive decision-making based on accumulating evidence.

2. **Adaptive Design Methods**: Adaptive design methods enable researchers to dynamically adjust study protocols, treatment allocations, or sample sizes based on interim analyses of accumulating data. Adaptive designs include response-adaptive randomization, group sequential designs, and adaptive dose-finding algorithms. These methods optimize resource allocation, statistical power, and efficiency in sequential experimentation while maintaining statistical validity and controlling error rates.

3. **Sequential Hypothesis Testing**: Sequential hypothesis testing methods allow for ongoing monitoring of study outcomes and iterative hypothesis testing based on interim analyses. Sequential testing procedures, such as sequential probability ratio tests (SPRT) or sequential likelihood ratio tests (SLRT), enable researchers to make efficient decisions regarding treatment efficacy, dose–response relationships, or biomarker validation in real time.

4. **Sequential ANOVA**: Sequential ANOVA techniques are used to analyze data from SED experiments, accounting for dependencies between experimental observations or treatments. Sequential ANOVA allows for the detection of treatment effects, temporal trends, or interaction effects while adapting to evolving experimental conditions and accumulating evidence over time.

5. **Optimization Algorithms**: Optimization algorithms, such as sequential Monte Carlo methods, genetic algorithms, or simulated annealing, are used to iteratively optimize experimental parameters, study designs, or treatment regimens based on objective functions or performance criteria. Optimization algorithms enable researchers to explore large experimental spaces, identify optimal solutions, and adaptively refine experimental strategies in real time.

6. **Time-Series Analysis**: Time-series analysis techniques are employed to analyze longitudinal data collected from SED experiments. Time-series models, autoregressive integrated moving average (ARIMA) models, or state–space models allow for the characterization of temporal patterns, trends, and dependencies in biological responses over time.

7. **Survival Analysis**: Survival analysis methods, such as Kaplan–Meier estimation, Cox proportional hazards regression, or accelerated failure time models, are used to analyze time-to-event data in SED experiments, such as patient survival times, disease progression, or treatment outcomes. Survival analysis enables researchers to assess treatment effects, prognostic factors, and risk factors in dynamic study settings.

8. **Machine Learning Techniques**: Machine learning techniques, including classification, regression, clustering, and dimensionality reduction algorithms, are applied to analyze complex, high-dimensional data generated from SED experiments. Machine learning methods facilitate pattern recognition, predictive modeling, and knowledge discovery in biological systems, allowing researchers to uncover hidden relationships, identify biomarkers, or predict treatment responses.

By leveraging these statistical tools, researchers can effectively analyze data, adaptively adjust experimental parameters, and interpret results in SED experiments for biology. These approaches enable efficient exploration of complex biological systems, optimization of experimental procedures, and generation of actionable insights for scientific discovery and innovation.

Sequential experimentation design experiments in biology often involve complex data analysis techniques to adaptively adjust experimental parameters, optimize study protocols, and interpret evolving results. Here are some statistical tools commonly used in SED experiments for biology:

Bayesian Methods: Bayesian statistical methods are commonly used in SED to update prior knowledge with new data and iteratively refine probability distributions for model parameters or hypotheses. Bayesian techniques allow for flexible modeling of complex biological systems, incorporation of prior information, and adaptive decision-making based on accumulating evidence.

Adaptive Design Methods: Adaptive design methods enable researchers to dynamically adjust study protocols, treatment allocations, or sample sizes based on interim analyses of accumulating data. Adaptive designs include response-adaptive randomization, group sequential designs, and adaptive dose-finding algorithms. These methods optimize resource allocation, statistical power, and efficiency in sequential experimentation while maintaining statistical validity and controlling error rates.

Sequential Hypothesis Testing: Sequential hypothesis testing methods allow for ongoing monitoring of study outcomes and iterative hypothesis testing based on interim analyses. Sequential testing procedures, such as SPRT or SLRT, enable researchers to make efficient decisions regarding treatment efficacy, dose–response relationships, or biomarker validation in real time.

Sequential ANOVA: Sequential ANOVA techniques are used to analyze data from SED experiments, accounting for dependencies between experimental observations or treatments. Sequential ANOVA allows for the detection of treatment effects, temporal trends, or interaction effects while adapting to evolving experimental conditions and accumulating evidence over time.

Optimization Algorithms: Optimization algorithms, such as sequential Monte Carlo methods, genetic algorithms, or simulated annealing, are used to iteratively optimize experimental parameters, study designs, or treatment regimens based on objective functions or performance criteria. Optimization algorithms enable researchers to explore large experimental spaces, identify optimal solutions, and adaptively refine experimental strategies in real time.

Time-Series Analysis: Time-series analysis techniques are employed to analyze longitudinal data collected from SED experiments. Time-series models, ARIMA models, or state–space models allow for the characterization of temporal patterns, trends, and dependencies in biological responses over time.

Survival Analysis: Survival analysis methods, such as Kaplan–Meier estimation, Cox proportional hazards regression, or accelerated failure time models, are used to analyze time-to-event data in SED experiments, such as patient survival times, disease progression, or treatment outcomes. Survival analysis enables researchers to assess treatment effects, prognostic factors, and risk factors in dynamic study settings.

Machine Learning Techniques: Machine learning techniques, including classification, regression, clustering, and dimensionality reduction algorithms, are applied to analyze complex, high-dimensional data generated from SED experiments. Machine learning methods facilitate pattern recognition, predictive modeling, and knowledge discovery in biological systems, allowing researchers to uncover hidden relationships, identify biomarkers, or predict treatment responses.

By leveraging these statistical tools, researchers can effectively analyze data, adaptively adjust experimental parameters, and interpret results in SED experiments for biology. These approaches enable efficient exploration of complex biological systems, optimization of experimental procedures, and generation of actionable insights for scientific discovery and innovation.

Sequential experimentation design (SED) experiments in biology typically involve iterative testing and refinement of hypotheses or experimental conditions. The elements of a report for such experiments may include:

Title: Clearly states the topic or purpose of the experiment.

Abstract: Provides a concise summary of the experiment, including the objectives, methods, key findings, and implications.

Introduction: Describes the background information, rationale, and objectives of the experiment. It may include a literature review to contextualize the study within existing research.

Hypotheses: States the specific hypotheses or research questions being investigated.

Materials and Methods: Details the experimental design, including the organisms or systems used, treatments applied, measurements taken, and statistical methods employed. For SED experiments, it should explain the iterative nature of the design and how decisions were made based on previous results.

Results: Presents the data collected during the experiment, typically in tables, figures, or graphs. It may include descriptive statistics, such as means and standard deviations, as well as any significant findings or trends observed.

Discussion: Interprets the results in the context of the hypotheses and previous research. It should address the significance of the findings, potential limitations of the study, and suggestions for future research directions. For SED experiments, it should also discuss how the sequential nature of the design influenced the results and conclusions.

Conclusion: Summarizes the main findings of the experiment and their implications for the broader field of biology.

References: Lists the sources cited in the report, following a specific citation style (e.g., APA, MLA).

Appendices: Includes any additional supplementary information, such as raw data, detailed experimental protocols, or supporting analyses.

These elements provide a structured framework for reporting the methods, results, and conclusions of SED experiments in biology, ensuring clarity and transparency for readers and facilitating replication and further research in the field.

2.1.8 ROBUST PARAMETER DESIGN (RPD)

RPD experiments provide a systematic and robust approach to studying complex biological systems, optimizing experimental conditions, identifying critical factors and interactions, and enhancing the reproducibility and reliability of experimental findings in biology. By systematically evaluating parameter effects and interactions, RPD enables researchers to design experiments that are less susceptible to variability, more robust to changes in experimental conditions, and more informative for advancing our understanding of biological processes. RPD experiments are utilized in biology for several compelling reasons:

Reduction of Variability: Biological experiments often face challenges related to variability arising from factors, such as genetic differences, environmental fluctuations, and stochastic processes. RPD experiments enable researchers to identify experimental conditions that are less sensitive to variability, leading to more robust and reliable results. By systematically evaluating the effects of key parameters and their interactions on experimental outcomes, RPD allows researchers to design experiments that are less susceptible to variability, enhancing the reproducibility and reliability of experimental findings in biology.

Optimization of Experimental Conditions: RPD experiments facilitate the optimization of experimental conditions by identifying parameter settings that maximize or minimize biological responses under varying conditions. By systematically varying experimental parameters and observing their effects on response variability, researchers can identify optimal parameter settings that achieve desired outcomes, such as maximizing cell growth, enhancing protein expression, or minimizing experimental noise. This optimization process is crucial in biology for maximizing research efficiency, resource utilization, and the effectiveness of experimental interventions.

Identification of Critical Factors and Interactions: RPD experiments help identify critical factors and interactions that significantly influence biological responses. By systematically varying parameters and quantifying their effects on response variability, researchers can identify the most influential factors and interactions that drive biological processes. Understanding these critical factors and interactions is essential for elucidating the underlying mechanisms governing biological systems and for designing effective interventions or treatments in biology.

Enhanced Statistical Power and Precision: RPD experiments offer enhanced statistical power and precision by allowing researchers to systematically evaluate parameter effects and interactions using robust statistical analysis techniques. By strategically selecting experimental settings and incorporating statistical analysis methods, such as ANOVA or regression analysis, researchers can obtain more accurate estimates of parameter effects and interactions, leading to more reliable and interpretable experimental results in biology.

Adaptation to Dynamic Systems: Biological systems are dynamic and can exhibit variability over time or in response to external stimuli. RPD experiments enable researchers to adapt experimental designs to the dynamic nature of biological systems by incorporating new information and adjusting experimental conditions in real time. This flexibility allows researchers to respond to changes in experimental conditions, refine experimental protocols, and address emerging research questions or hypotheses in biology.

RPD experiments in biology are used to optimize experimental conditions or parameters to achieve robust and reliable results. While RPD is more commonly associated with engineering and manufacturing processes, it can also be applied in biological research settings where variability and uncertainty need to be minimized for accurate and reproducible outcomes. Here are some areas in biology where RPD experiments may be utilized:

Biomedical Research: RPD experiments can be employed in biomedical research to optimize experimental protocols, such as cell culture conditions, assay parameters, or drug delivery systems. By identifying and controlling critical parameters, researchers can enhance the reliability and reproducibility of their results.

Genetics and Genomics: In genetic studies, RPD experiments may be used to optimize PCR conditions, sequencing protocols, or gene editing techniques. By minimizing variability in experimental procedures, researchers can improve the accuracy of genetic analyses and reduce the risk of false-positive or false-negative results.

Ecology and Environmental Science: RPD experiments can be applied in ecology and environmental science to optimize field sampling protocols, monitoring techniques, or data analysis methods. By controlling environmental variables and experimental conditions, researchers can improve the reliability of their observations and enhance the comparability of data collected across different studies or locations.

Microbiology: In microbiology, RPD experiments may be used to optimize growth conditions for microbial cultures, fermentation processes, or bioreactor systems. By controlling factors, such as temperature, pH, nutrient concentrations, and agitation rates, researchers can maximize the yield, purity, and consistency of microbial products or biomolecules.

Pharmacology and Drug Development: RPD experiments are crucial in pharmacology and drug development to optimize formulation parameters, drug delivery systems, and manufacturing processes. By minimizing variability in drug properties and performance, researchers can enhance the efficacy, safety, and reproducibility of pharmaceutical products.

Agriculture and Crop Science: RPD experiments can be employed in agriculture and crop science to optimize agricultural practices, crop breeding programs, or crop protection strategies. By controlling factors, such as soil conditions, irrigation regimes, and pest management strategies, researchers can improve crop yields, quality, and resilience to environmental stressors.

Overall, RPD experiments play a valuable role in biology by helping researchers optimize experimental conditions, reduce variability, and enhance the reliability and reproducibility of their results across diverse biological disciplines.

STEPS

Developing an RPD experiment in biology involves several key steps to optimize experimental conditions and minimize variability. Here's a generalized outline of the process:

1. **Define the Problem**: Clearly identify the research question or problem that the RPD experiment aims to address. This could involve optimizing a biological process, improving the reliability of experimental results, or minimizing variability in a specific outcome.
2. **Identify Critical Parameters**: Determine the key factors or parameters that influence the outcome of interest. These may include environmental conditions, experimental variables, sample characteristics, or assay parameters.
3. **Design Experimental Plan**: Develop a structured experimental plan that systematically evaluates the effects of different parameter settings on the outcome of interest. Consider factors, such as sample size, experimental design (e.g., factorial design, RSM), and randomization to ensure robustness and statistical validity.
4. **Select Response Variables**: Define the response variables or metrics that will be used to assess the outcome of the experiment. These should be relevant to the research question and measurable with sufficient precision and accuracy.
5. **Conduct Preliminary Studies**: Conduct preliminary studies or pilot experiments to assess the range of variability in the selected parameters and identify potential sources of variation. This will help refine the experimental design and determine appropriate levels for each parameter.
6. **Optimize Parameter Settings**: Use statistical methods, such as DOE, ANOVA, or regression analysis, to analyze the experimental data and identify optimal parameter settings that maximize the desired outcome while minimizing variability.
7. **Perform Sensitivity Analysis**: Conduct sensitivity analysis to assess the robustness of the optimized parameter settings to variations in environmental conditions or experimental factors. Identify potential sources of variability that may affect the reliability of the results.
8. **Validate Results**: Validate the optimized parameter settings through additional experiments or validation studies. Verify that the observed improvements in the outcome are reproducible under different conditions and robust to variations in experimental parameters.
9. **Implement Quality Control Measures**: Implement quality control measures to ensure consistency and reproducibility in future experiments. This may involve standardizing protocols, monitoring key parameters, and implementing corrective actions to address deviations from optimal conditions.
10. **Document and Report Findings**: Document the experimental procedures, results, and conclusions in a comprehensive report or research paper. Clearly communicate the rationale, methods, and outcomes of the RPD experiment to facilitate replication and dissemination of the findings.

By following these steps, researchers can systematically develop and implement RPD experiments in biology to optimize experimental conditions, minimize variability, and improve the reliability and reproducibility of their results.

Pitfalls and Remedies

In RPD experiments in biology, several pitfalls may arise that can affect the reliability and validity of the results. Here are some common pitfalls and potential remedies to mitigate them:

Insufficient Understanding of System Dynamics:
>**Pitfall:** Inadequate knowledge of the underlying biological system or processes may lead to incorrect assumptions about critical parameters or their interactions.
>
>**Remedy:** Conduct thorough preliminary studies or literature reviews to gain a comprehensive understanding of the biological system. Consult with domain experts to identify relevant factors and their potential interactions.

Inadequate Experimental Design:
>**Pitfall:** Poorly designed experiments with inadequate sample size, inappropriate factor levels, or lack of randomization can lead to biased or unreliable results.
>
>**Remedy:** Utilize robust experimental designs, such as factorial designs, RSMs, or optimal designs tailored to the specific research question. Ensure proper randomization and replication to account for variability and minimize bias.

Failure to Identify Critical Parameters:
>**Pitfall:** Overlooking important factors or parameters that significantly influence the outcome can result in suboptimal experimental conditions.
>
>**Remedy:** Conduct comprehensive sensitivity analyses or screening experiments to identify critical parameters. Use statistical methods, such as ANOVA or regression analysis, to assess the relative importance of different factors.

Inadequate Control of Experimental Conditions:
>**Pitfall:** Inconsistent or poorly controlled experimental conditions can introduce variability and confound the results.
>
>**Remedy:** Implement rigorous quality control measures to standardize experimental protocols and minimize sources of variation. Monitor key environmental factors, assay parameters, and sample characteristics throughout the experiment.

Ignoring Nonlinear Effects or Interactions:
>**Pitfall:** Focusing solely on main effects while ignoring nonlinear effects or interactions between factors can lead to incomplete or misleading conclusions.
>
>**Remedy:** Use appropriate statistical techniques, such as regression analysis or nonlinear modeling, to capture and assess the effects of interactions between factors. Conduct thorough data analysis to identify nonlinear relationships and optimize parameter settings accordingly.

Overfitting Models to the Data:
>**Pitfall:** Overfitting statistical models to the experimental data can result in overly complex models that do not generalize well to new data.
>
>**Remedy:** Use parsimonious models with a sufficient number of parameters to adequately describe the data without unnecessary complexity. Validate the model using independent datasets or cross-validation techniques to ensure robustness.

Lack of Reproducibility and Generalizability:
>**Pitfall:** Inability to reproduce experimental results under different conditions or generalize findings to other biological systems can limit the applicability of the research.
>
>**Remedy:** Validate the experimental findings through independent replication studies or validation experiments conducted under diverse conditions. Document experimental procedures and conditions in detail to facilitate replication by other researchers.

By being aware of these potential pitfalls and implementing appropriate remedies, researchers can enhance the robustness and reliability of RPD experiments in biology, leading to more accurate and reproducible results.

RPD experiments in biology aim to optimize factors to make a system less sensitive to variations and external factors. Here's a simplified example of a matrix for an RPD experiment in biology:

Objective: To optimize the conditions for a fermentation process to produce a specific compound in a microorganism while making the process robust against variations in pH and temperature.

Factors:

1. **Fermentation Time (Factor A):**
   - Low time (12 hours).
   - Medium time (18 hours).
   - High time (24 hours).
2. **pH (Factor B):**
   - Low pH (5.5).
   - Medium pH (6.5).
   - High pH (7.5).
3. **Temperature (Factor C):**
   - Low temperature (30°C).
   - Medium temperature (35°C).
   - High temperature (40°C).

Response Variable: Compound yield (measured in grams per liter, g/L).

RPD Experiment Matrix:

In an RPD experiment, the matrix is designed to optimize the factors while considering robustness to variations. The matrix is shown in Table 2.12.

In this matrix, each row represents a specific experimental run with a unique combination of fermentation time, pH, and temperature levels. Compound yield is measured for each combination.

The RPD approach involves systematically optimizing the factors while considering robustness. Statistical tools are used to assess the sensitivity of the system to variations in pH and temperature and to identify factor settings that minimize this sensitivity, making the fermentation process more robust and reliable.

Another example of a RPD design experiments in biology could be a factorial design matrix. Factorial designs allow researchers to systematically vary multiple factors or parameters

TABLE 2.12

RPD Matrix Example for Biology

Run	Fermentation Time (A)	pH (B)	Temperature (C)	Compound Yield (g/L)
1	12	5.5	30	...
2	18	6.5	35	...
3	24	7.5	40	...
4	18	7.5	30	...
5	12	6.5	40	...
6	24	5.5	35	...
7	18	6.5	30	...
8	12	7.5	35	...
9	24	5.5	40	...

TABLE 2.13
RPD Matrix Example for Biology

Run	Temperature (T)	pH Level (pH)	Nutrient Concentration (N)
1	+	+	+
2	+	+	−
3	+	−	+
4	+	−	−
5	−	+	+
6	−	+	−
7	−	−	+
8	−	−	−

simultaneously to assess their individual and combined effects on the outcome of interest. Here's a hypothetical example of a factorial design matrix for an RPD experiment in biology:

Let's consider an experiment aimed at optimizing the growth conditions for a particular species of bacteria. The factors of interest are temperature (T), pH level (pH), and nutrient concentration (N). Each factor is tested at two levels, high (+) and low (−), resulting in a 2^3 full factorial design.

Factorial Design Matrix: See Table 2.13

In this design matrix,

- Each row represents a unique combination of factor levels, known as a "run" or "experimental condition."
- The columns represent the factors being investigated (temperature, pH level, and nutrient concentration), with their respective levels (+ for high, − for low).
- Each factor is varied independently of the others, resulting in $2^3 = 8$ total experimental conditions.

Researchers would conduct experiments for each combination of factor levels, measuring the growth rate or biomass yield of the bacteria under each condition. Statistical analysis would then be used to identify optimal parameter settings that maximize growth while minimizing variability across different environmental conditions.

This factorial design matrix allows researchers to efficiently explore the effects of multiple factors on the biological response of interest and identify robust parameter settings that optimize experimental outcomes in a systematic and rigorous manner.

REPORT

In RPD experiments for biology, several statistical tools and techniques are commonly used to analyze experimental data, optimize parameter settings, and assess the robustness of experimental outcomes. Here are some statistical tools frequently employed in RPD experiments:

ANOVA: ANOVA is a widely used statistical technique for comparing the means of multiple groups to determine whether there are significant differences between them. In RPD experiments, ANOVA is used to assess the effects of different parameters or factors on the outcome of interest and identify significant sources of variability.

Factorial Designs: Factorial designs allow researchers to systematically vary multiple factors or parameters simultaneously to assess their individual and combined effects on the outcome variable. Full factorial designs, fractional factorial designs, and mixed factorial designs are commonly used in RPD experiments to efficiently explore parameter space and identify optimal settings.

RSM: RSM is a statistical technique used to model and optimize response variables as a function of multiple IVs (factors). RSM involves fitting mathematical models, such as quadratic or cubic regression models, to experimental data to identify optimal parameter settings that maximize or minimize the response variable while minimizing variability.

DOE: DOE is a systematic approach to planning, conducting, and analyzing experiments to optimize process parameters and improve product quality. DOE techniques, such as Taguchi methods, CCDs, and BBDs, are commonly used in RPD experiments to efficiently explore parameter space and identify optimal settings with minimal experimental effort.

Sensitivity Analysis: Sensitivity analysis involves assessing the impact of variations in input parameters on the output or response variable. Sensitivity analysis helps identify critical parameters that have a significant influence on the outcome and evaluate the robustness of optimal parameter settings to variations in input conditions.

Optimization Algorithms: Optimization algorithms, such as gradient descent, genetic algorithms, simulated annealing, or particle swarm optimization, are used to search for optimal parameter settings that maximize or minimize an objective function. These algorithms are particularly useful for complex optimization problems with multiple constraints and nonlinear relationships.

Statistical Modeling: Statistical modeling techniques, such as regression analysis, generalized linear models (GLMs), or machine learning algorithms, are used to develop predictive models of biological systems based on experimental data. These models can help identify key factors influencing the outcome variable and optimize parameter settings for robust experimental outcomes.

By leveraging these statistical tools and techniques, researchers can effectively analyze experimental data, optimize parameter settings, and enhance the reliability and reproducibility of RPD experiments in biology.

When reporting the results of RPD experiments used in biology, it's essential to provide a clear and comprehensive account of the experimental procedures, findings, and implications. Here are the key elements typically included in a report for RPD experiments in biology:

Title: Clearly states the topic or purpose of the experiment.

Abstract: Provides a concise summary of the experiment, including the objectives, methods, key findings, and implications.

Introduction: Describes the background information, rationale, and objectives of the experiment. It should also include a brief overview of the RPD approach and its relevance to the study.

Materials and Methods: Details the experimental design, including the factors or parameters studied, levels of each factor, experimental procedures, data collection methods, and statistical analysis techniques. It should also explain how the RPD methodology was applied to optimize parameter settings and minimize variability.

Results: Presents the data collected during the experiment, typically in tables, figures, or graphs. It should include descriptive statistics, such as means, standard deviations, and confidence intervals, as well as the results of statistical analyses, such as ANOVA, regression analysis, or optimization algorithms. Highlight any significant findings or trends observed.

Discussion: Interprets the results in the context of the research question and objectives. Discusses the implications of the findings, including the optimized parameter settings and their robustness to variations in experimental conditions. Addresses any limitations of the study and suggests directions for future research.

Conclusion: Summarizes the main findings of the experiment and their implications for the broader field of biology. Reinforces the importance of the RPD approach in optimizing experimental outcomes and enhancing the reliability of biological research.

References: Lists the sources cited in the report, following a specific citation style (e.g., APA, MLA).

Appendices: Includes any additional supplementary information, such as raw data, detailed experimental protocols, or supporting analyses.

Acknowledgments: Acknowledges any individuals or organizations that contributed to the research, including funding sources, collaborators, and technical support.

By including these elements in the report, researchers can effectively communicate the methods, results, and implications of RPD experiments in biology, facilitating replication, and further research in the field.

2.1.9 RANDOMIZED EXPERIMENTS FOR BIOLOGY

(RD) experiments provide a robust and reliable approach for studying biological systems, controlling for confounding variables, making valid statistical inferences, minimizing bias, enhancing precision and power, and accommodating various experimental designs and research questions in biology. By randomly allocating experimental units to treatment groups, randomized experiments ensure the validity and reliability of experimental findings, advancing our understanding of biological processes and informing evidence-based decision-making in biology. RD experiments are widely used in biology for several important reasons:

Control of Confounding Variables: In biology, experiments often involve complex systems with multiple factors that can influence experimental outcomes. RD experiments help control for confounding variables by randomly assigning experimental units (e.g., subjects, samples, or treatments) to different treatment groups. This randomization process ensures that any observed differences between treatment groups are due to the treatments themselves rather than systematic differences in the characteristics of experimental units, leading to more valid and reliable conclusions.

Statistical Inference and Generalizability: RD experiments enable researchers to make valid statistical inferences and generalize their findings to larger populations or biological systems. By randomly allocating experimental units to treatment groups, randomized experiments ensure that the sample is representative of the population of interest, allowing researchers to draw unbiased conclusions about treatment effects and their generalizability to broader contexts. This is essential in biology for extrapolating experimental findings to real-world applications or populations.

Minimization of Bias: RD experiments help minimize bias in experimental results by ensuring that treatment assignments are not influenced by researcher preferences, preconceptions, or systematic biases. By randomly assigning experimental units to treatment groups, randomized experiments distribute potential sources of bias evenly across treatment groups, reducing the risk of bias influencing the interpretation of experimental results. This objectivity enhances the credibility and reliability of experimental findings in biology.

Enhanced Precision and Power: RD experiments offer enhanced precision and statistical power by reducing the variability attributable to extraneous factors or sources of bias. By randomly allocating experimental units to treatment groups, randomized experiments distribute variability evenly across treatment groups, allowing researchers to detect treatment effects with greater precision and accuracy. This enhanced precision and power enable researchers to draw more reliable conclusions from experimental data and make more informed decisions in biology.

Flexibility and Versatility: RD experiments are flexible and versatile, allowing researchers to accommodate various experimental designs and research questions in biology. Whether investigating the effects of treatments, interventions, or environmental factors, randomized experiments can be adapted to different study designs, including parallel-group designs, crossover designs, factorial designs, and repeated-measures designs. This flexibility makes randomized experiments a valuable tool for studying diverse biological phenomena and addressing a wide range of research questions in biology.

RD experiments are widely used in biology across various subfields for conducting controlled experiments, particularly when studying the effects of treatments or interventions on biological systems. Here are some areas in biology where RD experiments are commonly applied:

Clinical Trials: Randomized controlled trials (RCTs) are essential in clinical research to evaluate the safety and efficacy of new drugs, treatments, or medical interventions. Patients are randomly assigned to different treatment groups (e.g., experimental treatment, placebo, standard care) to minimize bias and ensure the validity of the results.

Agricultural Research: RD experiments are used in agricultural science to evaluate the effects of different treatments, such as fertilizers, pesticides, or crop varieties, on crop yields, quality, and resilience to environmental stressors. Randomized field trials help researchers assess the effectiveness of agricultural practices and technologies under real-world conditions.

Ecology and Environmental Science: RD experiments are employed in ecology and environmental science to study the effects of environmental factors; habitat manipulations; or conservation interventions on ecosystems, species diversity, and ecosystem services. Randomized field experiments help researchers understand ecological processes and inform management and conservation strategies.

Genetics and Genomics: RD experiments are used in genetics and genomics research to study gene expression, genetic variation, and gene function. Randomized experimental designs, such as randomized block designs or complete RDs, are employed in gene knockout studies, gene expression profiling, and genetic mapping experiments to control for confounding variables and ensure the validity of genetic analyses.

Behavioral Studies: RD experiments are utilized in behavioral studies to investigate the effects of interventions, treatments, or environmental manipulations on animal behavior, social interactions, and cognitive processes. RCTs are commonly used in ethology, psychology, and neuroscience research to assess the efficacy of behavioral interventions or drug treatments in animal models.

Microbiology and Immunology: RD experiments are employed in microbiology and immunology research to study the effects of antimicrobial agents, vaccines, or immune-modulating compounds on microbial growth, pathogen virulence, and host immune responses. RCTs are used in clinical microbiology and immunology to evaluate the efficacy of new antimicrobial drugs or vaccine candidates in human or animal subjects.

Overall, RD experiments are essential in biology for conducting rigorous and unbiased experiments to test hypotheses, evaluate treatments, and advance scientific knowledge across diverse biological disciplines.

STEPS

To develop RD experiments in biology, researchers typically follow a systematic approach to design and implement controlled experiments while minimizing bias and ensuring the validity of the results. Here are the steps to develop RD experiments in biology:

1. **Define the Research Question**: Clearly articulate the specific research question or hypothesis that the experiment aims to address. This could involve investigating the effects of a treatment, intervention, or manipulation on a biological system or process.
2. **Identify Experimental Variables**: Determine the key variables or factors that will be manipulated or measured in the experiment. This may include IVs (e.g., treatments, conditions) and DVs (e.g., outcomes, responses).
3. **Design Experimental Treatments**: Determine the different levels or categories of each experimental variable that will be tested in the experiment. Randomize the assignment of experimental treatments to subjects or experimental units to minimize bias and ensure balance across treatment groups.
4. **Select Experimental Units**: Identify the units or subjects on which the experimental treatments will be applied or tested. This could involve selecting biological organisms, samples, tissues, or individuals that are representative of the population under study.
5. **Randomize Experimental Design**: Randomly assign experimental treatments to experimental units using a randomization procedure, such as simple randomization, stratified randomization, or block randomization. Randomization helps ensure that treatment groups are comparable and that any observed differences are not due to systematic biases.
6. **Control Extraneous Variables**: Identify potential sources of variability or confounding factors that could affect the outcome of the experiment. Implement control measures, such as standardization of experimental procedures, environmental controls, or blocking, to minimize the impact of extraneous variables on the results.
7. **Implement Experimental Protocol**: Conduct the experiment according to the predetermined experimental protocol, ensuring that all experimental treatments are applied or tested consistently and accurately. Record relevant data and observations during the experiment.
8. **Analyze Experimental Data**: Use appropriate statistical techniques to analyze the experimental data and assess the effects of experimental treatments on the outcome variable. This may involve descriptive statistics, inferential statistics (e.g., t-tests, ANOVA), or multivariate analyses depending on the experimental design and research question.
9. **Interpret Results**: Interpret the results of the statistical analysis in the context of the research question and objectives. Determine whether the observed differences between treatment groups are statistically significant and biologically meaningful.
10. **Draw Conclusions and Communicate Findings**: Draw conclusions based on the results of the experiment and discuss their implications for the broader field of biology. Communicate the findings through scientific publications, presentations, or reports, ensuring clarity and transparency in reporting the methods, results, and conclusions of the experiment.

By following these steps, researchers can develop RD experiments in biology that are rigorously designed, properly controlled, and statistically valid, allowing for robust conclusions and scientific advances in the field.

PITFALLS AND REMEDIES

In RD experiments used in biology, several pitfalls may arise that can affect the validity and reliability of the results. Here are some common pitfalls and potential remedies to mitigate them:

Selection Bias:
Pitfall: Nonrandom selection of experimental units or subjects can introduce bias and compromise the validity of the experiment.

Remedy: Randomly assign experimental treatments to subjects or experimental units using appropriate randomization procedures to ensure that treatment groups are comparable and representative of the population under study.

Lack of Randomization:

Pitfall: Inadequate randomization of experimental treatments can lead to imbalances between treatment groups and confound the interpretation of the results.

Remedy: Implement robust randomization procedures, such as simple randomization, stratified randomization, or block randomization, to ensure that each experimental unit has an equal chance of receiving each treatment.

Confounding Variables:

Pitfall: Failure to control for confounding variables or extraneous factors can obscure the true effects of experimental treatments on the outcome variable.

Remedy: Identify potential confounding variables and implement control measures, such as blocking, matching, or statistical adjustment, to minimize their influence on the results. Design experiments with appropriate factorial designs or control groups to isolate the effects of interest.

Measurement Error:

Pitfall: Inaccurate or imprecise measurement of the outcome variable can introduce variability and bias into the results.

Remedy: Use reliable and validated measurement techniques to ensure accurate and precise data collection. Implement quality control measures, such as calibration, standardization, or repeated measurements, to minimize measurement error.

Sample Size and Power:

Pitfall: Inadequate sample size or statistical power can limit the ability to detect meaningful differences between treatment groups.

Remedy: Conduct power analyses to determine the appropriate sample size needed to detect the expected effect size with sufficient statistical power. Ensure that the sample size is large enough to detect meaningful differences while balancing practical constraints and ethical considerations.

Noncompliance or Dropout:

Pitfall: Noncompliance with experimental protocols or dropout of subjects can compromise the integrity of the experiment and bias the results.

Remedy: Implement strategies to minimize noncompliance and dropout, such as clear communication of experimental procedures, incentives for participation, and follow-up procedures to track and address missing data.

Multiple Comparisons:

Pitfall: Conducting multiple statistical tests without appropriate adjustments can increase the likelihood of false-positive results.

Remedy: Correct for multiple comparisons using methods, such as Bonferroni correction, FDR control, or permutation testing, to maintain the overall type I error rate at the desired level.

By being aware of these potential pitfalls and implementing appropriate remedies, researchers can enhance the validity, reliability, and interpretability of RD experiments in biology, leading to more robust and trustworthy scientific conclusions.

EXAMPLE

Randomized experiments in biology involve random assignment of treatments or conditions to experimental units to eliminate bias and draw valid statistical conclusions. Here's an example of a matrix for a randomized designed experiment in biology:

Objective: To evaluate the effectiveness of three different fertilizers (fertilizer A, fertilizer B, and fertilizer C) on the growth of a specific crop plant.

Treatments:

1. **Fertilizer A.**
2. **Fertilizer B.**
3. **Fertilizer C.**

Response Variable: Crop plant growth (measured in height in centimeters).

Randomized Experiment Matrix:
In a randomized experiment, the treatments are randomly assigned to experimental units (e.g., pots or plots) to ensure that the results are not influenced by any systematic biases. The matrix is shown in (Table 2.14).

In this matrix, each row represents an experimental unit (e.g., a pot or plot) to which one of the three fertilizers is randomly assigned. Crop plant growth is measured for each experimental unit.

The random assignment of treatments ensures that any differences in crop plant growth are more likely to be due to the effects of the fertilizers rather than other extraneous variables or biases. Statistical analysis can then be used to compare the effectiveness of the fertilizers in promoting plant growth. Randomized experiments are fundamental in drawing causal relationships in biological research.

Another example of a matrix for RD experiments in biology is a RCBD matrix. RCBD is a type of experimental design that incorporates both randomization and blocking to control for variability and increase the efficiency of the experiment. Here's a hypothetical example of an RCBD matrix for a biology experiment:

Let's consider a study investigating the effects of different fertilizers on plant growth, with three types of fertilizers (A, B, and C) being tested. The experiment is conducted in four different locations (blocks), and each location represents a different soil type or environmental condition.

Randomized Complete Block Design Matrix (Table 2.15)**:**
In this RCBD matrix,

- Each row represents an experimental unit (e.g., a plot of land, a pot) within a specific block (location).
- The columns represent the blocks and the treatment (fertilizer) applied to each experimental unit.

TABLE 2.14
Randomized Experiment Matrix Example for Biology

Experimental Unit	Treatment (Fertilizer)	Crop Plant Growth (cm)
1	Randomly assigned	...
2	Randomly assigned	...
3	Randomly assigned	...
4	Randomly assigned	...
5	Randomly assigned	...
6	Randomly assigned	...
7	Randomly assigned	...
8	Randomly assigned	...
9	Randomly assigned	...
10	Randomly assigned	...

TABLE 2.15
Randomized Complete Block Experiment
Matrix Example for Biology

Block	Treatment (Fertilizer)
1	A
2	B
3	C
4	B
1	C
2	A
3	B
4	C
1	B
2	C
3	A
4	A

- The treatments (fertilizers) are randomly assigned to experimental units within each block, ensuring that each treatment is tested in every block.
- Randomization within blocks helps control for variation due to environmental factors or other sources of variability specific to each location.

Researchers would collect data on plant growth (e.g., height, biomass) for each experimental unit and analyze the results using appropriate statistical techniques, such as (ANOVA), to assess the effects of the fertilizers while accounting for the variability between blocks.

This RCBD matrix allows researchers to efficiently compare the effects of different fertilizers while controlling for potential sources of variability due to environmental conditions, leading to more robust and interpretable results in biology experiments.

REPORTING

In RD experiments in biology, a variety of statistical tools and techniques are utilized to analyze experimental data, assess treatment effects, and draw conclusions. These tools help researchers make valid inferences and understand the significance of their findings. Here are some common statistical tools used in RD experiments for biology:

ANOVA: ANOVA is a widely used statistical technique for comparing means among multiple groups. In RD experiments, ANOVA is employed to analyze the effects of different treatments or interventions on the outcome variable while accounting for variability within and between treatment groups.

Regression Analysis: Regression analysis is used to model the relationship between one or more IVs (e.g., treatments) and a DV (e.g., outcome). Linear regression, logistic regression, and GLMs are common regression techniques used in biology to assess the effects of treatments and identify significant predictors of the outcome.

Post-Hoc Tests: Post-hoc tests are conducted following ANOVA to determine specific differences between treatment groups when a significant overall effect is detected. Tukey's HSD, Bonferroni correction, Dunnett's test, and Scheffé's method are examples of post-hoc tests used in RD experiments to identify pairwise differences between treatment means.

Multiple Comparison Correction: When conducting multiple statistical tests or comparing multiple treatment groups, it's important to adjust for multiple comparisons to control the familywise error rate or FDR. Bonferroni correction, FDR control, and the Benjamini–Hochberg procedure are common methods used to correct for multiple comparisons in RD experiments.

Nonparametric Tests: Nonparametric tests, such as the Wilcoxon rank sum test, Kruskal–Wallis test, or Mann–Whitney U test, are used when the assumptions of parametric tests (e.g., normality, homogeneity of variance) are violated or when analyzing ordinal or non-normally distributed data in RD experiments.

Survival Analysis: Survival analysis techniques, such as Kaplan–Meier curves and Cox proportional hazards models, are used in biology to analyze time-to-event data, such as survival times or disease recurrence rates, when studying the effects of treatments on patient outcomes or disease progression.

Factorial ANOVA and Interaction Analysis: Factorial ANOVA is used to analyze the effects of multiple factors (IVs) and their interactions on the outcome variable in RD experiments. Interaction analysis helps identify whether the effects of one factor depend on the levels of another factor.

Multivariate Analysis: Multivariate analysis techniques, such as PCA, canonical correlation analysis (CCA), or multivariate analysis of variance (MANOVA), are used to analyze relationships among multiple variables simultaneously and identify patterns or clusters in complex biological datasets.

By utilizing these statistical tools and techniques, researchers can effectively analyze data from RD experiments in biology, interpret the results, and draw valid conclusions about the effects of treatments on biological systems.

When reporting the results of RD experiments used in biology, it's important to provide a clear and comprehensive account of the experimental procedures, findings, and implications. Here are the key elements typically included in a report for RD experiments in biology:

Title: Clearly states the topic or purpose of the experiment.

Abstract: Provides a concise summary of the experiment, including the objectives, methods, key findings, and implications.

Introduction: Describes the background information, rationale, and objectives of the experiment. It should also include a brief overview of the RD approach and its relevance to the study.

Materials and Methods: Details the experimental design, including the randomization procedure, treatments or interventions tested, experimental units or subjects, data collection methods, and statistical analysis techniques used. It should also explain any control measures implemented to minimize bias or confounding factors.

Results: Presents the data collected during the experiment, typically in tables, figures, or graphs. It should include descriptive statistics, such as means, standard deviations, and confidence intervals, as well as the results of statistical analyses, such as ANOVA, regression analysis, or post-hoc tests. Highlight any significant findings or trends observed.

Discussion: Interprets the results in the context of the research question and objectives. Discusses the implications of the findings, including the effects of treatments or interventions on the outcome variable and their biological significance. Address any limitations of the study and suggest directions for future research.

Conclusion: Summarizes the main findings of the experiment and their implications for the broader field of biology. Reinforces the importance of the RD approach in minimizing bias, controlling for variability, and drawing valid conclusions from experimental data.

References: Lists the sources cited in the report, following a specific citation style (e.g., APA, MLA).

Appendices: Includes any additional supplementary information, such as raw data, detailed experimental protocols, or additional analyses.

Acknowledgments: Acknowledges any individuals or organizations that contributed to the research, including funding sources, collaborators, and technical support.

By including these elements in the report, researchers can effectively communicate the methods, results, and implications of RD experiments in biology, facilitating replication and further research in the field.

2.1.10 SPLIT-PLOT AND BLOCKED EXPERIMENTS

Split-plot and blocked design experiments offer a robust and versatile approach for studying biological systems, managing experimental constraints, controlling for spatial or temporal variability, reducing experimental error, accommodating diverse experimental designs, and enhancing statistical efficiency in biology. By organizing experimental units into blocks or plots and allocating treatments accordingly, these designs enable researchers to address complex experimental challenges and obtain more valid and reliable experimental results in biology. Split-plot and blocked design experiments are utilized in biology for several compelling reasons:

Management of Experimental Constraints: In biology, experiments often involve constraints, such as limited resources, time, or space. Split-plot and blocked design experiments allow researchers to efficiently manage these constraints by partitioning experimental units into distinct blocks or plots and allocating treatments accordingly. This partitioning enables researchers to address different levels of experimental variation, optimize resource utilization, and control extraneous factors that may influence experimental outcomes.

Control of Spatial or Temporal Variability: Biological experiments may be affected by spatial or temporal variability due to factors, such as environmental gradients, microclimates, or batch effects. Split-plot and blocked design experiments help control such variability by organizing experimental units into spatially or temporally homogeneous blocks or plots. This organization ensures that potential sources of variability are evenly distributed across treatment groups, reducing the risk of confounding effects and enhancing the validity and reliability of experimental results.

Reduction of Experimental Error: Split-plot and blocked design experiments enable researchers to reduce experimental error by accounting for spatial or temporal variability within experimental units. By partitioning experimental units into blocks or plots and randomizing treatments within each block or plot, these designs minimize the impact of localized variations on experimental outcomes. This reduction in experimental error enhances the precision and accuracy of estimates of treatment effects and improves the statistical power of hypothesis tests in biology.

Flexibility in Experimental Design: Split-plot and blocked design experiments offer flexibility in experimental design, allowing researchers to accommodate various experimental layouts, treatment structures, and blocking factors in biology. Whether investigating the effects of treatments, interventions, or environmental factors, these designs can be adapted to different study designs, including factorial designs, nested designs, repeated-measures designs, or longitudinal designs. This flexibility makes split-plot and blocked design experiments suitable for studying diverse biological phenomena and addressing a wide range of research questions in biology.

Enhanced Statistical Efficiency: Split-plot and blocked design experiments provide enhanced statistical efficiency by reducing the variability attributable to extraneous factors or sources of bias. By partitioning experimental units into blocks or plots and randomizing treatments within each block or plot, these designs enable researchers to account for spatial or temporal variation and minimize the impact of confounding effects. This statistical efficiency improves the precision and accuracy of estimates of treatment effects, enhances the power of hypothesis tests, and enables researchers to draw more reliable conclusions from experimental data in biology.

Split-plot and blocked design experiments are commonly used in biology when dealing with complex experimental setups or when there are natural groupings or hierarchies within the experimental units. Here are some areas where split-plot and blocked design experiments are frequently applied in biology:

Agricultural Research: Split-plot and blocked design experiments are extensively used in agricultural research to evaluate the effects of different treatments or management practices on crop yield, quality, and resilience. For example, researchers may investigate the effects of different irrigation methods (main plots) and fertilizer treatments (subplots) on crop growth in field trials.

Ecology and Environmental Science: Split-plot and blocked design experiments are employed in ecology and environmental science to study the effects of habitat manipulations, land management practices, or environmental factors on ecosystem structure and function. For instance, researchers may examine the effects of different restoration treatments (main plots) within different habitat types (subplots) on plant community composition in fragmented landscapes.

Biomedical Research: In biomedical research, split-plot and blocked design experiments are used to investigate the effects of interventions, treatments, or genetic factors on biological systems while controlling for variability due to individual differences or experimental conditions. For example, researchers may assess the effects of drug treatments (main plots) on tumor growth in different genetic strains of mice (subplots) in cancer studies.

Microbiology and Immunology: Split-plot and blocked design experiments are utilized in microbiology and immunology research to study microbial growth, antimicrobial resistance, and host–pathogen interactions. Researchers may investigate the effects of different antibiotic treatments (main plots) on bacterial growth in different culture media (subplots) while controlling for variability due to experimental conditions.

Behavioral Studies: In behavioral studies, split-plot and blocked design experiments are employed to investigate the effects of environmental factors, genetic factors, or interventions on animal behavior, cognitive processes, and social interactions. For example, researchers may assess the effects of environmental enrichment treatments (main plots) on cognitive performance in different strains of laboratory mice (subplots) while controlling for variability due to individual differences.

Genetics and Genomics: Split-plot and blocked design experiments are used in genetics and genomics research to study gene expression, genetic variation, and gene–environment interactions. Researchers may investigate the effects of different environmental exposures (main plots) on gene expression profiles in different genetic backgrounds (subplots) to identify gene regulatory networks and pathways.

Overall, split-plot and blocked design experiments are valuable tools in biology for efficiently controlling for variability, addressing complex experimental questions, and drawing robust conclusions from experimental data. They allow researchers to account for natural hierarchies or groupings within the experimental units and optimize the efficiency of experimental designs in diverse biological contexts.

STEPS

Developing split-plot and blocked design experiments in biology involves careful planning and consideration of the hierarchical structure of the experimental units and the factors being studied. Here are the steps to develop split-plot and blocked design experiments in biology:

1. **Define the Research Question**: Clearly articulate the specific research question or hypothesis that the experiment aims to address. Consider the hierarchical structure of the experimental units and the factors that may influence the outcome variable.
2. **Identify Experimental Factors**: Determine the main factors (treatments, interventions) and their levels that will be manipulated in the experiment. Consider both the primary factors of interest and any secondary factors that may need to be controlled or accounted for in the design.
3. **Identify Blocking Factors**: Identify the blocking factors or covariates that may influence the variability of the outcome variable and need to be controlled for in the experiment. Blocking factors are typically factors that are not of primary interest but may introduce variability that needs to be accounted for.
4. **Design the Experimental Layout**: Determine the hierarchical structure of the experimental units and how the main factors and blocking factors will be allocated to different levels of the hierarchy. Split-plot designs involve a nested structure where the main plots (whole plots) are subdivided into smaller subplots (split plots). Blocked designs involve grouping experimental units into blocks based on blocking factors.
5. **Randomization**: Randomly assign the main treatments to the whole plots and the split-plot treatments within each whole plot to minimize bias and ensure the validity of the results. Randomization helps ensure that treatment effects are not confounded with other sources of variability.
6. **Implement the Experiment**: Conduct the experiment according to the predetermined experimental layout, ensuring that treatments are applied or assigned accurately and consistently across experimental units. Record relevant data and observations according to the experimental protocol.
7. **Data Analysis**: Analyze the experimental data using appropriate statistical techniques for split-plot and blocked designs. Use techniques, such as split-plot ANOVA, mixed-effects models, or hierarchical linear models, to assess the effects of main treatments, split-plot treatments, and blocking factors on the outcome variable while accounting for the hierarchical structure of the experiment.
8. **Interpret Results**: Interpret the results of the data analysis in the context of the research question and objectives. Determine the significance of treatment effects and the impact of blocking factors on the outcome variable. Consider the practical implications of the findings and how they contribute to the broader understanding of the biological system under study.
9. **Draw Conclusions and Communicate Findings**: Draw conclusions based on the results of the experiment and discuss their implications for the field of biology. Communicate the findings through scientific publications, presentations, or reports, ensuring clarity and transparency in reporting the methods, results, and conclusions of the experiment.

By following these steps, researchers can develop split-plot and blocked design experiments in biology that are well designed, properly controlled, and statistically valid, leading to robust conclusions and advancements in biological research.

Pitfalls and Remedies

In split-plot and blocked design experiments used in biology, several pitfalls may arise that can affect the validity and reliability of the results. Here are some common pitfalls and potential remedies to mitigate them:

Incomplete Blocking:
> **Pitfall:** Inadequate blocking may occur if not all sources of variability are accounted for in the experimental design, leading to biased estimates of treatment effects.
>
> **Remedy:** Conduct a thorough preliminary analysis to identify potential sources of variability and incorporate them into the blocking structure of the experiment. Ensure that blocking factors are selected based on biological relevance and are effectively controlled for in the design.

Improper Randomization:
> **Pitfall:** Inadequate randomization of treatments within whole plots or split plots may lead to unbalanced designs or confounding between treatment effects and other sources of variability.
>
> **Remedy:** Implement robust randomization procedures to ensure that treatments are assigned to experimental units randomly within each level of the experimental hierarchy. Use appropriate randomization techniques, such as complete randomization or restricted randomization, to achieve balance and minimize bias.

Small Sample Size:
> **Pitfall:** Insufficient sample size within blocks or split plots may limit the ability to detect meaningful treatment effects or block-by-treatment interactions.
>
> **Remedy:** Conduct a power analysis to determine the appropriate sample size needed to detect the expected effect size with sufficient statistical power. Increase the sample size, if necessary, to ensure adequate precision and reliability of the results.

Unequal Variability:
> **Pitfall:** Variability within blocks or split plots may vary across treatment levels, leading to inefficient designs or inflated type I error rates.
>
> **Remedy:** Use statistical techniques, such as stratification or stratified randomization, to ensure that variability is balanced across treatment levels within blocks or split plots. Consider transforming the response variable if necessary to achieve homogeneity of variance.

Interactions with Blocking Factors:
> **Pitfall:** Interaction effects between treatments and blocking factors may confound treatment effects and complicate interpretation of the results.
>
> **Remedy:** Include interaction terms between treatments and blocking factors in the statistical analysis to assess their impact on the outcome variable. Interpret treatment effects within the context of blocking factors and consider potential interactions in the discussion of the results.

Noncompliance or Dropout:
> **Pitfall:** Noncompliance with experimental protocols or dropout of subjects may introduce bias and compromise the integrity of the experiment.
>
> **Remedy:** Implement strategies to minimize noncompliance and dropout, such as clear communication of experimental procedures, incentives for participation, and follow-up procedures to track and address missing data.

By being aware of these potential pitfalls and implementing appropriate remedies, researchers can enhance the validity, reliability, and interpretability of split-plot and blocked design experiments in biology, leading to more robust and trustworthy scientific conclusions.

Examples

Split-plot and blocked experiments are designed to account for different sources of variability and to study the effects of factors in the presence of nuisance variables. Here are examples of matrices for split-plot and blocked experiments in biology:

Example 1: Split-Plot Experiment

Objective: To study the effects of two factors, fertilizer type (factor A) and irrigation frequency (factor B), on crop yield, while accounting for variations in soil quality across different fields.

Factors:

1. **Fertilizer Type (Factor A):**
   - Fertilizer A.
   - Fertilizer B.
2. **Irrigation Frequency (Factor B):**
   - Low frequency.
   - High frequency.

Blocked Split-Plot Experiment Matrix:

In a blocked split-plot experiment, the main plots represent the different fields with varying soil quality, while the subplots within each field represent different combinations of fertilizer type and irrigation frequency. The matrix is shown in Table 2.16.

In this matrix, each row represents a specific subplot within a field, with unique combinations of fertilizer type and irrigation frequency. Crop yield is measured for each subplot, and the fields serve as blocks to account for soil quality variations.

Example 2: Blocked Experiment

Objective: To evaluate the effects of three different planting densities (factor A) on the growth of a specific tree species while accounting for variations in sunlight exposure.

Factors:

1. **Planting Density (Factor A):**
   - Low density.
   - Medium density.
   - High density.

TABLE 2.16

Split-Plot Experiment Matrix Example for Biology

Field	Fertilizer Type (A)	Irrigation Frequency (B)	Crop Yield (kg/ha)
1	Fertilizer A	Low frequency	...
1	Fertilizer A	High frequency	...
1	Fertilizer B	Low frequency	...
1	Fertilizer B	High frequency	...
2	Fertilizer A	Low frequency	...
2	Fertilizer A	High frequency	...
2	Fertilizer B	Low frequency	...
2	Fertilizer B	High frequency	...

TABLE 2.17
Blocked Experiment Matrix Example for Biology

Block	Planting Density (A)	Tree Growth (cm)
1	Low density	...
1	Medium density	...
1	High density	...
2	Low density	...
2	Medium density	...
2	High density	...
3	Low density	...
3	Medium density	...
3	High density	...

Blocks: Different sections of the experimental site with varying sunlight exposure.

Blocked Experiment Matrix:
In a blocked experiment, the matrix represents the different blocks, each with different sunlight exposure levels. Within each block, different planting densities are tested. The matrix is shown in Table 2.17.

In this matrix, each row represents a specific block with a unique level of sunlight exposure. Within each block, different planting densities are tested, and tree growth is measured. Blocking allows researchers to control for the confounding effects of sunlight exposure, ensuring that the effects of planting density are more accurately assessed.

Reporting

In split-plot and blocked design experiments in biology, various statistical tools and techniques are utilized to analyze experimental data, assess treatment effects, and draw valid conclusions. These tools help researchers account for the hierarchical structure of the experimental design and control for sources of variability. Here are some common statistical tools used in split-plot and blocked design experiments for biology:

ANOVA: ANOVA is a fundamental statistical technique used to analyze the effects of categorical factors (e.g., treatments, blocks) on a continuous outcome variable. In split-plot and blocked designs, ANOVA is employed to assess the main effects and interaction effects between factors while accounting for the hierarchical structure of the experiment.

Split-Plot ANOVA: Split-plot ANOVA is a specialized form of ANOVA used to analyze data from split-plot designs, where the experimental units are nested within larger units (whole plots). Split-plot ANOVA allows researchers to assess the effects of main treatments (whole plot factors) and sub-treatments (split-plot factors) while accounting for the nested structure of the experiment.

Mixed-effects Models: Mixed-effects models, also known as hierarchical linear models or multilevel models, are used to analyze data from split-plot and blocked designs with random effects due to blocking factors. Mixed-effects models allow for the estimation of fixed effects (treatment effects) and random effects (blocking effects) simultaneously, providing more accurate estimates of treatment effects and reducing bias.

Factorial ANOVA: Factorial ANOVA is used to analyze data from factorial designs, where multiple factors are manipulated simultaneously to assess main effects and interaction effects. In split-plot and blocked designs, factorial ANOVA can be used to analyze the effects of main factors (whole plot and split-plot factors) and their interactions with the outcome variable.

Regression Analysis: Regression analysis is employed to model the relationship between predictor variables (e.g., treatments, blocking factors) and a continuous outcome variable. In split-plot and blocked designs, regression analysis can be used to assess the effects of main treatments and blocking factors on the outcome variable while controlling for other covariates.

Post-hoc Tests: Post-hoc tests, such as Tukey's HSD test, Bonferroni correction, or Dunnett's test, are used to compare treatment means and identify significant differences between treatment groups following ANOVA or regression analysis.

Residual Analysis: Residual analysis is performed to assess the adequacy of the statistical model and check assumptions, such as homogeneity of variance and normality of residuals. Residual plots, Q-Q plots, and diagnostic tests are commonly used to evaluate the goodness of fit of the model.

Power Analysis: Power analysis is conducted to determine the sample size needed to detect a significant treatment effect with sufficient statistical power. Power analysis helps researchers plan experiments and optimize resources to achieve desired levels of statistical precision.

By utilizing these statistical tools and techniques, researchers can effectively analyze data from split-plot and blocked design experiments in biology, interpret the results, and draw valid conclusions about the effects of treatments on biological systems while accounting for the hierarchical structure of the experimental design.

When reporting the results of split-plot and blocked design experiments in biology, it's essential to provide a comprehensive and well-structured report that communicates the experimental design, findings, and implications clearly. Here are the key elements typically included in a report for split-plot and blocked design experiments in biology:

Title: Clearly states the topic or purpose of the experiment.

Abstract: Provides a concise summary of the experiment, including the objectives, methods, key findings, and implications. Summarizes the main results and conclusions of the study.

Introduction: Describes the background information, rationale, and objectives of the experiment. Explains the importance of split-plot and blocked design experiments in biology and their relevance to the specific research question.

Materials and Methods: Details the experimental design, including the hierarchical structure of the experimental units, main treatments, split-plot treatments, blocking factors, and randomization procedures. Describes the data collection methods, statistical analysis techniques used, and any control measures implemented to minimize bias or confounding factors.

Results: Presents the data collected during the experiment, typically in tables, figures, or graphs. Includes descriptive statistics, such as means, standard deviations, and confidence intervals, for each treatment level and blocking factor. Reports the results of statistical analyses, including ANOVA, mixed-effects models, or regression analysis, to assess treatment effects and blocking effects.

Discussion: Interprets the results in the context of the research question and objectives. Discusses the implications of the findings, including the effects of main treatments, split-plot treatments, and blocking factors on the outcome variable. Considers the biological significance of the results and addresses any limitations or potential sources of bias in the study.

Conclusion: Summarizes the main findings of the experiment and their implications for the broader field of biology. Highlights the key takeaways from the study and suggests directions for future research.

References: Lists the sources cited in the report, following a specific citation style (e.g., APA, MLA).

Appendices: Includes any additional supplementary information, such as raw data, detailed experimental protocols, or additional analyses.

Acknowledgments: Acknowledges any individuals or organizations that contributed to the research, including funding sources, collaborators, and technical support.

By including these elements in the report, researchers can effectively communicate the methods, results, and implications of split-plot and blocked design experiments in biology, facilitating replication and further research in the field.

2.2 CHEMISTRY

In chemistry designed experiments, various types of variables are considered to investigate chemical reactions, processes, and properties. These variables can be categorized into several common types, including:

IVs:

Concentration: Variables related to the concentration of reactants or solutes in a solution, which can affect reaction rates and product yields.

Temperature: Variables related to the temperature of a chemical system, which can influence reaction kinetics and thermodynamics.

Pressure: Variables related to the pressure of gases or reactions in a closed system, particularly relevant in gas-phase reactions.

pH: Variables related to the acidity or alkalinity of a solution, which can impact the behavior of acids, bases, and buffers.

Reaction Time: Variables related to the duration of a chemical reaction or the time at which measurements are taken.

DVs:

Reaction Rate: Variables related to the rate at which a chemical reaction occurs, often determined by measuring changes in concentration over time.

Product Yield: Variables related to the number of desired products obtained in a chemical reaction, typically expressed as a percentage of the theoretical yield.

Chemical Properties: Variables related to chemical properties, such as solubility, conductivity, viscosity, and color changes.

Spectroscopic Data: Variables related to spectroscopic measurements, including absorbance, emission, and nuclear magnetic resonance (NMR) spectra.

Mass and Volume Changes: Variables related to changes in mass or volume of reactants or products, which can provide information about stoichiometry and density.

Categorical Variables:

Chemical Species: Categories of chemical species involved in reactions, such as reactants, products, intermediates, or catalysts.

Solvents: Categories of solvents used in reactions, which can affect reaction outcomes.

Experimental Conditions: Categories of experimental conditions or setups, including variations in apparatus, reaction vessels, or reagent sources.

Control Variables (Covariates): These are variables that are held constant or controlled during experiments to eliminate their influence on the DV. For example, maintaining a constant pH or stirring rate while studying reaction kinetics.

Random Variables: In some cases, random variables may be introduced to account for variability or uncertainty in measurements, such as fluctuations in temperature or pressure.

Interaction Variables: Interaction variables are used to investigate the combined effects of two or more IVs on the DV. For example, the interaction between temperature and concentration on reaction rate.

Noise Variables (Error Variables): These are uncontrolled or unmeasured variables that can introduce variability or errors into the experimental results. Techniques like statistical analysis and calibration are used to minimize the impact of noise variables in chemistry experiments.

Chemistry designed experiments are essential for understanding chemical processes, optimizing reaction conditions, and characterizing chemical properties. They play a crucial role in the development of new materials, pharmaceuticals, and industrial processes. Careful control and manipulation of variables are key to obtaining reliable and meaningful data in chemistry experiments.

In chemistry, designed experiments are fundamental for conducting controlled investigations, optimizing chemical processes, and understanding chemical phenomena. The following are some common types of designed experiments with examples of their applications in chemistry.

2.2.1 FACTORIAL DESIGN EXPERIMENTS IN CHEMISTRY

Full factorial designs offer a structured approach to investigating how multiple factors impact a chosen response variable in your chemical experiment. Here's a step-by-step guide to designing one:

STEPS

1. **Define Your Research Question and Objectives**: Clearly state what you want to investigate and what information you hope to gain from the experiment. This will guide your choice of factors and response variable.
2. **Identify Relevant Factors**:
   - List all factors that might influence your chosen response variable. These can be quantitative (temperature, concentration) or qualitative (solvent type, catalyst presence).
   - Limit the number of factors to a manageable level, as full factorial designs become complex and resource intensive with increasing factors.
3. **Define Factor Levels**: Determine the different levels at which you will study each factor. Usually, two levels (low and high) are preferred for simplicity, but more levels can be chosen for specific reasons.
4. **Calculate the Number of Treatments**: In a full factorial design, the total number of treatments (experimental runs) is 2^k, where k is the number of factors. Consider your resources and ensure this number is practical to execute.
5. **Randomize Treatment Order**: Randomly assign the treatment combinations to avoid systematic bias and ensure valid statistical analysis. Use software or random number tables to achieve this.
6. **Perform the Experiment**: Conduct each treatment (experimental run) as per your defined factors and levels. Ensure consistent and accurate measurements of your chosen response variable.
7. **Analyze the Data**: Use statistical software or appropriate analysis methods to assess the main and interactional effects of the factors on the response variable. Interpret the results in the context of your research question and identify statistically significant effects.
8. **Draw Conclusions and Recommendations**: Based on your analysis, draw conclusions about the relationships between the factors and the response variable. Formulate recommendations or further research questions based on your findings.

Additional Tips

- **Control Factors**: Account for and control extraneous factors that might influence your results, such as temperature or time.
- **Replicates**: Consider including replicate measurements for each treatment to improve data accuracy and statistical power.
- **Pilot Experiment**: Run a small preliminary experiment to test your procedures and troubleshoot any potential issues before the full experiment.

Remember, a well-designed full factorial experiment can provide valuable insights into complex relationships in your chemical system. By carefully following these steps and adapting them to your specific research needs, you can gain valuable data and knowledge from your experimentation.

PITFALLS AND REMEDIES:

Pitfalls to avoid and remedies to apply with full factorial design (FFD) in analytical chemistry:

Pitfall: Large number of runs. FFD involves testing all possible combinations of factors and levels, which can lead to a significant number of experiments, especially with many factors.

Remedy: Use fractional factorial designs if you can identify insignificant higher-order interactions. Utilize Plackett–Burman designs (PBDs) for rapid screening of a large number of factors with limited runs. Consider alternative designs like BBD or CCD if you primarily focus on quadratic relationships and optimization within a specific region.

Pitfall: High-resource consumption. The large number of runs often translates to increased costs for reagents, time, and equipment usage.

Remedy: Design the experiment efficiently with minimal levels and factors necessary for your objectives. Consider smaller sample sizes or alternative analytical techniques with lower costs per run. Utilize automated equipment or parallel processing if feasible to improve efficiency.

Pitfall: Confusing data analysis. Interpreting interactions between many factors can be complex and overwhelming, especially for inexperienced researchers.

Remedy: Thoroughly understand your system and hypothesize potential interactions before planning the experiment. Utilize statistical software or consult with a statistician for proper data analysis and interpretation. Focus on visualizing interactions through response surface plots and contour plots to gain clearer insights.

Pitfall: Overestimating model accuracy. FFD assumes all factors are independent and additive in their effects, which may not always be true in complex chemical systems.

Remedy: Incorporate additional design points for replicates or center points to assess model error and variability. Validate the predicted optimal conditions through additional experiments under real-world conditions. Consider alternative nonlinear models if interactions and nonlinear relationships are suspected.

Pitfall: Overlooking experimental error. FFD assumes no experimental error or random variability, which can significantly impact the results.

Remedy: Include replicates or control samples in the design to quantify and account for experimental error. Analyze the variance (ANOVA) to assess the significance of factor effects and interactions. Ensure proper equipment calibration, sample preparation, and analytical techniques to minimize error.

Clearly define your objectives and response variables before designing the experiment. Make sure to randomize the order of runs to minimize potential order-of-effect bias. It is essential that you document your experimental procedures and data thoroughly for reproducibility and future reference.

By considering these pitfalls and remedies, you can effectively utilize FFD for optimizing analytical chemistry processes and gaining valuable insights from your experiments. Remember that choosing the right design and employing careful planning and analysis are crucial for successful optimization and meaningful conclusions.

EXAMPLE: OPTIMIZING CRYSTALLIZATION OF A PHARMACEUTICAL COMPOUND USING FACTORIAL DESIGN

Objective: Identify the optimal conditions for crystallizing a new pharmaceutical compound to maximize yield and crystal size.

Factors:

- Temperature (A): Low (L), high (H).
- Stirring Rate (B): Low (L), high (H).
- Concentration (C): Low (L), high (H).

Design: A 2^3 full factorial design with eight runs (2 levels for each of 3 factors).

Design Matrix (Table 2.18):

Response Variables:

- Yield of crystallized product (%).
- Average crystal size (microns).

Experiment:

1. Prepare solutions of the compound at two different concentrations.
2. Set up eight crystallization vessels under controlled temperature and stirring conditions according to the design matrix.
3. Add the chosen concentration of solution to each vessel and monitor the crystallization process over time.
4. Measure the final yield of crystals in each vessel.
5. Analyze the crystals from each run using a microscope to determine the average size.

TABLE 2.18

Factorial Design Experiment Matrix Example for Chemistry

Run	Temperature	Stirring Rate	Concentration
1	L	L	L
2	L	L	H
3	L	H	L
4	L	H	H
5	H	L	L
6	H	L	H
7	H	H	L
8	H	H	H

Data Analysis:

1. Use statistical software to perform ANOVA on the yield and crystal size data.
2. Analyze the main effects and interactions of the factors on both response variables.
3. Identify significant factors and their optimal levels for maximizing yield and crystal size.

Expected Outcomes:

- The ANOVA results may reveal which factors and their interactions significantly affect the yield and crystal size.
- For example, a high temperature might increase yield but decrease crystal size, while a high stirring rate might improve both.
- Based on the analysis, we can identify the optimal combination of temperature, stirring rate, and concentration to achieve the desired balance between yield and crystal size for practical pharmaceutical applications.

Further Steps:

- Validate the optimized conditions through additional experiments at or near the predicted optimal settings.
- Consider using RSMs or other optimization techniques to refine the optimized conditions further.
- Investigate the underlying mechanisms of how the factors influence the crystallization process for deeper scientific understanding.

This is a simplified example, and the specific details of your factorial design experiment will vary depending on your specific chemical system and optimization goals. However, it provides a general framework for using factorial design to gain valuable insights and identify optimal conditions for various chemical analysis applications.

2.2.2 FRACTIONAL FACTORIAL DESIGNS

The steps required to develop a designed experiment for chemistry using a fractional factorial design:

1. **Define Your Research Question and Objectives**: Clearly state what you want to investigate and what information you hope to gain from the experiment. This will guide your choice of factors and response variable.
2. **Identify Relevant Factors**:
   - List all factors that might influence your chosen response variable. These can be quantitative (temperature, concentration) or qualitative (solvent type, catalyst presence).
   - Be realistic about the number of factors you can handle. Fractional factorial designs are efficient, but with increasing factors, the complexity and number of required runs also increase.
3. **Choose a Fractional Factorial Design**:
   - Consider your desired information (main effects vs. interactions, screening vs. optimization) and resource constraints (number of runs) when choosing a specific design. Consulting statistical references or software can help you select the best option.
   - Common fractional factorial designs include PBD, Hadamard, and 2^(k-p) designs (where k is the number of factors and p is the fraction of the full factorial design used).

4. **Define Factor Levels**: Determine the number of levels for each factor, typically two (low and high) for simplicity, but more levels can be chosen for specific reasons.
5. **Assign Factor Levels to Treatment Combinations**: Use the chosen fractional factorial design layout to define the specific combination of factor levels for each experimental run. Ensure randomization of the run order to avoid systematic bias.
6. **Perform the Experiment**: Conduct each treatment (experimental run) as per your defined factors and levels. Ensure consistent and accurate measurements of your chosen response variable.
7. **Analyze the Data**: Use statistical software or appropriate analysis methods to assess the main effects and, depending on your chosen design, key interactions of the factors on the response variable. Interpret the results in the context of your research question and identify statistically significant effects.
8. **Draw Conclusions and Recommendations**: Based on your analysis, draw conclusions about the relationships between the factors and the response variable. Formulate recommendations or further research questions based on your findings.

Additional Tips

- **Control Factors**: Account for and control extraneous factors that might influence your results, such as temperature or time.
- **Replicates**: Consider including replicate measurements for each treatment to improve data accuracy and statistical power.
- **Pilot Experiment**: Run a small preliminary experiment to test your procedures and troubleshoot any potential issues before the full experiment.

Remember, the specific steps and considerations might vary depending on your chosen and the details of your experiment. Feel free to ask further questions or provide more information about your specific research needs for more tailored guidance throughout the process.

Pitfalls and Remedies

Pitfalls to avoid and remedies to apply with fractional factorial design (FFD) experiments in analytical chemistry:

Pitfall: Aliasing. This occurs when the effect of one factor is masked by the combined effect of another due to their chosen levels and interactions. It can lead to misleading conclusions about the significance of factors.

Remedy: Choose factors and levels carefully to minimize potential aliasing. Use software or consult a statistician to assess aliasing patterns before executing the experiment. Utilize OAs or PBDs for efficient screening with reduced aliasing. Implement confirmation runs by testing specific factor combinations suspected of aliasing to clarify their individual effects.

Pitfall: Overlooking higher-order interactions. FFD assumes only lower-order interactions are significant, which might not be true in complex systems. Ignoring higher-order interactions can lead to inaccurate models and optimization.

Remedy: Analyze the data for residuals and curvature in response surfaces to identify potential higher-order interactions. Consider transforming the data or using alternative models capable of capturing higher-order interactions. If higher-order interactions are significant, a full factorial design might be necessary for further investigation.

Pitfall: Inadequate replication. Limited replicates within each factor combination might not capture the inherent variability in your analytical system. This can lead to statistically insignificant results and unreliable conclusions.

Remedy: Include center points or replicates within each factor combination to estimate experimental error and improve statistical power. Conduct additional runs for specific factor combinations showing interesting trends or potential problems. Analyze the variance (ANOVA) to assess the significance of factors and interactions with proper statistical rigor.

Pitfall: Misinterpreting insignificant factors. While statistically insignificant factors might not directly impact the response variable, they could still be involved in complex interactions or contribute to variability.

Remedy: Don't completely disregard insignificant factors. Analyze their interaction plots and consider their potential contribution to other factors or experimental errors. If interactions with significant factors are observed, further investigation of those interactions might be necessary. Remember that statistical significance doesn't necessarily equate to scientific importance – consider the overall context and potential practical implications of all factors.

Pitfall: Insufficient design planning. Improper planning of the FFD can lead to an inefficient experiment with limited information gain.

Remedy: Clearly define your objectives and response variables before designing the experiment. Choose the appropriate FFD type (e.g., resolution III, IV) based on the number of factors and desired information. Randomize the order of runs to minimize bias and ensure proper statistical analysis. Document your experimental procedures and data thoroughly for reproducibility and future reference.

A consideration is to use software specifically designed for fractional factorial designs to simplify planning, data analysis, and visualization of interactions. There are constant updates and developments in software tools being made available.

By recognizing pitfalls and taking the recommended remedies, you can make informed decisions when designing and implementing FFD experiments in analytical chemistry. Remember, careful planning, appropriate analysis, and interpretation of results are crucial for successful optimization and gaining valuable insights from your research.

EXAMPLE: SCREENING IMPORTANT FACTORS FOR SYNTHESIS OF AN ORGANIC DYE USING FRACTIONAL FACTORIAL DESIGN

Objective: Identify the most significant factors influencing the color strength and brightness of a newly synthesized organic dye. We have a large number of potential factors affecting the outcome, but time and resources are limited.

Factors:

- Solvent (A): Polar (P), non-polar (N).
- Catalyst (B): Acidic (A), basic (B).
- Reaction Temperature (C): Low (L), high (H).
- Reactant Ratio (D): Excess A (A), excess B (B).
- Stirring Rate (E): Low (L), high (H).
- Precursor Purity (F): High (H), low (L).

Design: A 2^6 full factorial design would require 64 runs, which is impractical. Therefore, we choose a 2^{5-1} fractional factorial design with 16 runs. This generates a design matrix where each factor is confounded (aliased) with a specific interaction of other factors.

Design Matrix (Table 2.19):

TABLE 2.19

Full Factorial Design Experiment Matrix Example for Chemistry

Run	Solvent	Catalyst	Temp	Reactant Ratio	Stirring Rate	Precursor Purity
1	P	A	L	A	L	H
2	P	A	L	B	H	L
3	P	A	H	A	H	L
4	P	A	H	B	L	H
5	N	B	L	A	L	H
6	N	B	L	B	H	L
7	N	B	H	A	H	L
8	N	B	H	B	L	H
9	P	B	L	A	H	H
10	P	B	L	B	L	H
11	P	B	H	A	L	H
12	P	B	H	B	L	L
13	N	A	L	A	H	H
14	N	A	L	B	L	H
15	N	A	H	A	L	H
16	N	A	H	B	L	L

Response Variables:

- Color strength (measured by absorbance at a specific wavelength).
- Brightness (measured by fluorescence intensity).

Experiment:

1. Prepare reaction mixtures according to the design matrix in different vials or flasks.
2. Heat or cool the reactions to the designated temperatures as needed.
3. Monitor the reactions and record color development and fluorescence over time.
4. At the end of the reaction time, measure the color strength and brightness of each solution.

Data Analysis:

1. Use statistical software to analyze the color strength and brightness data.
2. Perform ANOVA to assess the significance of the main effects and interactions of the factors.
3. Identify factors with statistically significant effects on both response variables, indicating their potential importance in influencing the dye properties.

Expected Outcomes:

- ANOVA may reveal that only a few factors or interactions significantly affect the color strength and brightness.
- This allows us to focus further investigation and optimization efforts on these key factors, rather than wasting time on factors with negligible impact.
- We can then use other design methods like RSM to fine-tune the identified influential factors to achieve the desired dye properties.

Further Steps:

- Validate the identified key factors and their optimal levels through additional experiments.
- Explore the interactions between significant factors to gain a deeper understanding of the underlying mechanisms influencing dye synthesis.
- Consider investigating alternative solvents, catalysts, or other factors not included in the initial screening to potentially discover even better conditions for dye synthesis.

This example demonstrates how fractional factorial design can be a valuable tool in chemical analysis when faced with a large number of potential factors influencing the outcome. By effectively screening factors and prioritizing the important ones, you can optimize your resources and maximize your research progress.

2.2.3 RSM

Developing a designed experiment using an RSM in chemistry can help you optimize a process or phenomenon and identify the optimal conditions for your desired outcome. Here's a breakdown of the key steps involved:

1. **Define Your Research Question and Objectives**: Clearly identify what you want to optimize or understand about your system. This will guide your choice of response variable and factor levels.
2. **Identify Relevant Factors**: List all factors that potentially influence your chosen response variable (quantitative or qualitative). Prioritize factors with potential significant effects or interactions.
3. **Choose an RSM**: Select a specific RSM design based on your desired information and resource constraints. Popular options include BBD, CCD, and D-optimal designs. These designs explore the response surface within a defined region by combining central, factorial, and axial points.
4. **Define Factor Levels**: Determine the low, high, and central levels for each factor. Consider the expected range of the response surface and practical limits for your system.
5. **Generate a Design Matrix and Randomize Run Order**: Use statistical software or online tools to generate a design matrix specifying the factor combinations for each experimental run. Randomize the run order to minimize bias and ensure valid statistical analysis.
6. **Perform the Experiment**: Conduct each experimental run precisely, ensuring consistent and accurate measurements of your chosen response variable.
7. **Analyze the Data**: Use statistical software or regression analysis methods to fit a mathematical model (typically quadratic) to the response surface data. Analyze the model for significance of effects, interactions, and goodness of fit.
8. **Optimize the Response Variable**: Identify the optimal factor levels within the studied region that maximize or minimize your response variable based on the fitted model and your desired goal.
9. **Validate the Model and Results**: Perform additional experiments at or near the predicted optimal conditions to verify the model's accuracy and confirm the optimal response.
10. **Draw Conclusions and Recommendations**: Based on your findings, draw conclusions about the relationships between factors and the response variable, identify optimal conditions for achieving your desired outcome, and formulate recommendations for further research or process improvement.

Additional Tips
- **Control Factors:** Account for and control extraneous factors that might influence your results, such as temperature or time.
- **Replicates:** Consider including replicate measurements for some runs to improve data accuracy and model fitting.
- **Visualize the Data:** Utilize response surface plots and contour plots to visualize the relationships between factors and the response variable, aiding in optimization.

RSM can be an iterative process. Based on your initial results, you might need to modify your model, refine factor levels, or conduct additional experiments for further optimization.

Pitfalls and Remedies

Pitfalls to avoid and remedies to apply with RSM designed experiments in analytical chemistry:

Pitfall: Misinterpreting model assumptions. RSM relies on specific assumptions about the data, like linearity, normality, and constant variance. Violating these assumptions can lead to inaccurate models and misleading conclusions.

Remedy: Analyze residual plots and normality tests to identify potential violations. Consider data transformations or alternative models like Box–Cox or robust regression if assumptions are not met. Conduct pilot experiments to assess data characteristics and adjust the design or response variable if necessary.

Pitfall: Overlooking optimization bias. RSM focuses on finding optimal conditions within the chosen factor range. Extrapolating beyond this range could lead to inaccurate predictions and unexpected outcomes.

Remedy: Choose factor ranges strategically, considering known limitations and potential real-world application conditions. Validate the predicted optimal conditions by conducting experiments outside the initial design domain. Consider alternative designs like CCD with axial points extending slightly beyond the chosen range for more robust optimization.

Pitfall: Misunderstanding interaction effects. RSM models not only consider individual factor effects but also their interactions. Ignoring significant interactions can lead to poor optimization and inaccurate predictions.

Remedy: Analyze response surface plots and ANOVA results for significant interaction terms. Visualize interactions through 3D plots and contour plots to understand how factors influence each other. Consider including more levels or factors in the design if complex interactions are observed.

Pitfall: Confusing correlation with causation. RSM identifies correlations between factors and responses, but not necessarily causation. Statistical significance doesn't always imply direct mechanistic relationships.

Remedy: Conduct additional experiments or use complementary techniques to investigate the underlying mechanisms behind observed relationships. Employ scientific knowledge and logical reasoning to interpret RSM results within your specific chemical system context. Remember that correlation and causation are distinct concepts – further investigation might be necessary to validate causal relationships.

Pitfall: Neglecting practical considerations. RSM optimization might not always translate directly to real-world implementation due to practical limitations or cost constraints.

Remedy: Prioritize factors based on their practical importance and cost implications during optimization. Consider alternative optimization goals that balance response variable improvements with feasibility and resource limitations. Pilot-test the optimized conditions under real-world scenarios to assess the practicality and adjust the process if necessary.

Utilize RSM software for efficient experimental design, data analysis, and visualization of results.

By addressing these pitfalls and adopting the suggested remedies, you can leverage RSM effectively to gain valuable insights and optimize your analytical chemistry processes. Remember to prioritize accurate data analysis, critical interpretation of results, and consideration of practical constraints for impactful optimization and successful research outcomes.

EXAMPLE: OPTIMIZING ENZYMATIC DEGRADATION OF CELLULOSE USING RSM

Objective: Maximize the enzymatic degradation of cellulose into fermentable sugars for biofuel production, using a specific enzyme mixture.

Factors:

- Temperature (A): 30°C, 40°C, 50°C.
- Enzyme Concentration (B): 0.5%, 1.0%, 1.5% (w/v).
- Substrate Loading (C): 1.0%, 2.0%, 3.0% (w/v).

Design: RSM with 12 runs, including center points, axial points (midpoints between high and low levels), and replicates at the center point. This design allows estimation of quadratic relationships between factors and the response variable.

Design Matrix (Table 2.20):

Response Variable: Glucose concentration (mg/mL) as an indicator of cellulose degradation.

Experiment:

1. Prepare cellulose suspensions with designated substrate loadings in buffer solutions.
2. Add different enzyme concentrations according to the design matrix.
3. Incubate the reaction mixtures at the specified temperatures for a predetermined time.
4. Measure the glucose concentration in each reaction mixture after incubation.

TABLE 2.20
RSM Designed Experiment Matrix Example for Chemistry

Run	Temperature	Enzyme Conc.	Substrate Load.
1	30	1.0	2.0
2	40	0.5	2.0
3	50	1.0	2.0
4	40	1.5	1.0
5	40	1.5	3.0
6	40	0.5	1.0
7	40	0.5	3.0
8	40	1.0	1.0
9	40	1.0	3.0
10	40	1.0	2.0
11	40	1.0	2.0
12	35	1.0	2.0
13	45	1.0	2.0

Data Analysis:

1. Use statistical software to fit a second-order polynomial model to the glucose concentration data.
2. Analyze the model for significance of linear, quadratic, and interaction terms of the factors.
3. Identify optimal factor levels and their interactions based on the model and desired outcome (maximizing glucose concentration).
4. Utilize response surface plots and contour plots to visualize the relationships between factors and the response variable.

Expected Outcomes:

- RSM analysis might reveal the optimal combination of temperature, enzyme concentration, and substrate loading for maximizing cellulose degradation.
- The interaction terms could indicate synergistic or antagonistic effects between factors, providing valuable insights into the enzyme activity and process limitations.
- Response surface plots would graphically depict the optimal region for achieving the desired glucose concentration.

Further Steps:

- Validate the predicted optimal conditions through additional experiments to confirm model accuracy and reproducibility.
- Explore other enzymes or enzyme mixtures for potentially even higher degradation efficiency.
- Investigate the cost-effectiveness and scalability of the optimized process for biofuel production.

This example showcases how RSM can be a powerful tool for optimizing multifactorial processes in chemical analysis like enzymatic degradation. By analyzing the combined effects and interactions of factors, you can gain valuable insights for achieving your desired outcomes and potentially improve the efficiency of your chemical systems.

2.2.4   CCD

CCDs are powerful tools for exploring response surfaces in chemical analysis, allowing you to estimate both linear and quadratic effects of factors on your chosen response variable. Here's a breakdown of the steps involved:

1. **Define Your Research Question and Objective**: Clearly state what you want to optimize or understand about your system. Are you seeking optimal conditions for a process, identifying significant factors and their interactions, or exploring the shape of the response surface?
2. **Identify Relevant Factors**: List all factors potentially influencing your response variable. These can be quantitative (temperature, concentration) or qualitative (solvent type, catalyst presence). Prioritize factors with potential significant effects or interactions.
3. **Choose a CCD Design**: Popular choices include circumscribed, inscribed, and face-centered CCDs. Consider the number of factors (k) and desired information when choosing. Circumscribed CCDs (CCC) have star points farther from the center for stronger curvature estimation, while inscribed CCDs (CCI) use factor settings as star points, requiring fewer runs.

4. **Calculate the Number of Runs**: Use the formula $N = 2^k + 2k + N0$, where N is the total number of runs, k is the number of factors, and N0 is the number of center points (typically 2 or 3). This formula accounts for factorial points, axial (star) points, and center points.
5. **Define Factor Levels**: Determine the low, high, and center levels for each factor. Consider the expected range of the response surface and practical limits for your system. The star points are typically set at $\pm\alpha$ times the standard deviation from the center point for each factor, where α is a chosen value influencing curvature estimation.
6. **Generate a Design Matrix and Randomize Run Order**: Use statistical software or online tools to generate a design matrix specifying the factor combinations for each experimental run. Randomize the run order to minimize bias and ensure valid statistical analysis.
7. **Perform the Experiment**: Conduct each experimental run precisely, ensuring consistent and accurate measurements of your chosen response variable.
8. **Analyze the Data**: Use statistical software or regression analysis methods to fit a mathematical model (typically quadratic) to the response surface data. Analyze the model for significance of effects, interactions, and goodness of fit.
9. **Optimize the Response Variable (optional)**: Identify the optimal factor levels within the studied region that maximize or minimize your response variable based on the fitted model and your desired goal. Utilize contour plots or optimization algorithms to identify the optimal combination.
10. **Validate the Model and Results**: Perform additional experiments at or near the predicted optimal conditions to verify the model's accuracy and confirm the optimal response.
11. **Draw conclusions and Recommendations**: Based on your findings, draw conclusions about the relationships between factors and the response variable, identify optimal conditions for achieving your desired outcome, and formulate recommendations for further research or process improvement.

Additional Tips:

- **Control Factors**: Account for and control extraneous factors that might influence your results, such as temperature or time.
- **Replicates**: Consider including replicate measurements for some runs to improve data accuracy and model fitting.
- **Visualize the Data**: Utilize response surface plots and contour plots to visualize the relationships between factors and the response variable, aiding in interpretation and optimization.

Remember, CCDs can be iterative. Based on your initial results, you might need to modify your model, refine factor levels, or conduct additional experiments for further optimization.

PITFALLS AND REMEDIES

CCD offers a powerful tool for optimizing analytical processes, but its effectiveness hinges on careful planning and awareness of potential hurdles. Here are some pitfalls to avoid and remedies to apply:

Pitfall: Assuming smooth and continuous response surfaces. CCD assumes the response variable smoothly changes within the chosen factor range. If true nonlinear behavior exists, the model might be inaccurate.

Remedy: Analyze residual plots and curvature in response surfaces to check for nonlinearity. Consider transforming the data or exploring alternative models like quadratic-cubic spline models to capture nonlinear trends. If nonlinearity is significant, a different design like D-optimal might be necessary.

Pitfall: Overlooking axial point limitations. While CCD utilizes axial points for estimating quadratic effects, these points can sometimes fall outside the desired operating range or be impractical to implement.

Remedy: Choose factor ranges and levels strategically, considering real-world application constraints and avoiding impractical axial points. Consider alternative designs like BBD if axial points are problematic or unnecessary for your optimization goals. Validate the predicted optimal conditions by conducting experiments within the desired operating range, even if it excludes the original axial points.

Pitfall: Misinterpreting model precision. CCD estimates quadratic relationships, but its precision for quadratic effects might be lower compared to alternative designs like BBD.

Remedy: Focus on optimizing within the central region of the design where precision is highest for quadratic effects. Use larger-center-point replicates to improve the estimation of quadratic coefficients. Consider combining CCD with other designs like BBD for improved precision within and outside the central region.

Pitfall: Overfitting the model: Complex CCD models with many terms can overfit the data, leading to poor predictive accuracy.

Remedy: Use statistical selection methods like backwards elimination or Akaike information criterion (AIC) to remove insignificant terms. Analyze residual plots and check for lack of fit to assess model overfitting. Validate the model with additional experiments to confirm its predictive ability.

Pitfall: Ignoring experimental constraints. CCD can require a significant number of runs, which can be time-consuming and resource intensive, especially with many factors.

Remedy: Use fractional factorial designs or PBDs for initial screening to identify important factors before employing CCD. Limit the number of factors based on available resources and prioritize the most crucial ones for optimization. Consider alternative designs like RSM with BBD if fewer runs are preferred.

Utilize CCD software or statistical packages for efficient design planning, data analysis, and visualization of response surfaces and interaction effects.

By recognizing these pitfalls and implementing the recommended remedies, you can ensure your CCD experiments yield accurate and actionable results for optimizing your analytical chemistry processes. Remember that careful planning, rigorous data analysis, and critical interpretation of the model are crucial for successful optimization and practical implementation.

EXAMPLE: OPTIMIZING PROTEIN PURIFICATION USING CCD

Objective: Identify the optimal conditions for purifying a specific protein from a cell lysate using affinity chromatography.

Factors:

- Salt concentration (A): 0.5M, 1.0M, 1.5M (NaCl).
- pH (B): 7.0, 7.5, 8.0.
- Flow rate (C): 1.0 mL/min, 1.5 mL/min, 2.0 mL/min.

Design: A circumscribed CCD with 15 runs, including center points, axial points (α distance from the center for each factor), and factorial points. This design allows estimation of both linear and quadratic relationships between factors and the response variable, while optimizing within the experimental region.

TABLE 2.21

CCD Experiment Matrix Example for Chemistry

Run	Salt Conc.	pH	Flow Rate
1	0.5	7.0	1.5
2	0.5	7.5	1.5
3	1.0	7.0	1.5
4	1.0	7.5	1.5
5	1.5	7.0	1.5
6	1.5	7.5	1.5
7	0.5	7.25	1.0
8	0.5	7.25	2.0
9	1.0	7.25	1.0
10	1.0	7.25	2.0
11	1.5	7.25	1.0
12	1.5	7.25	2.0
13	0.75	7.5	1.5
14	1.25	7.0	1.5
15	1.0	7.75	1.5

Design Matrix (Table 2.21):

Response Variables:

- Protein yield (mg/mL).
- Purity (% purity of target protein in eluate).

Experiment:

1. Prepare buffers with different salt concentrations and pH values.
2. Apply cell lysate samples to affinity columns equilibrated with the corresponding buffers.
3. Perform chromatography with different flow rates according to the design matrix.
4. Collect eluate fractions and measure protein concentration and purity in each fraction.

Data Analysis:

1. Use statistical software to fit a second-order polynomial model to the protein yield and purity data.
2. Analyze the model for significance of linear, quadratic, and interaction terms of the factors.
3. Identify optimal factor levels and their interactions based on the model and desired outcome (maximizing yield and purity).
4. Utilize response surface plots and contour plots to visualize the relationships between factors and the response variables.

Expected Outcomes:

- CCD analysis might reveal the optimal combination of salt concentration, pH, and flow rate for achieving the highest protein yield and purity.
- The interaction terms could indicate synergistic or antagonistic effects between factors, providing valuable insights into the protein binding and elution behavior.
- Response surface plots would graphically depict the optimal region for maximizing both yield and purity within the studied conditions.

Further Steps:

- Validate the predicted optimal conditions through additional experiments to confirm model accuracy and reproducibility.
- Explore alternative buffers, ligands, or purification strategies for potentially even higher yields or greater purity.
- Investigate the scalability and cost-effectiveness of the optimized process for large-scale protein purification.

This example showcases how CCD can be a powerful tool for optimizing multifactorial processes in chemical analysis like protein purification. By analyzing the combined effects and interactions of factors on two response variables, you can gain valuable insights for achieving your desired outcomes and potentially improve the efficiency and success of your purification protocols.

2.2.5 Screening Designs

Screening designs are powerful tools in chemistry to efficiently identify the most influential factors impacting your chosen response variable from a potentially large pool of factors. Here's a breakdown of the key steps involved:

1. **Define Your Research Question and Objectives**: Clearly state what you want to understand about your system. Are you seeking the most influential factors or simply identifying promising candidates for further investigation?
2. **Identify Potential Factors**: List all factors you think might affect your response variable. These can be quantitative (temperature, concentration) or qualitative (solvent type, catalyst presence). Prioritize factors with potentially significant effects.
3. **Choose a Screening Design**: Popular choices include PBDs and Hadamard designs. Consider your desired information (main effects vs. interactions) and number of factors when choosing. For instance, Hadamard designs handle more factors with fewer runs but offer limited interaction information.
4. **Define Factor Levels**: Typically, two levels (low and high) are used for simplicity. If specific reasons require it, more levels can be chosen for specific factors.
5. **Generate a Design Matrix and Randomize Run Order**: Use statistical software or online tools to generate a design matrix specifying the factor combinations for each experimental run. Randomize the run order to avoid systematic bias and ensure valid analysis.
6. **Perform the Experiment**: Conduct each experimental run precisely, ensuring consistent and accurate measurements of your response variable.
7. **Analyze the Data**: Use statistical software or appropriate analysis methods to identify statistically significant effects of each factor on the response variable. Pay attention to effect size and consider factors with both large and significant effects as promising candidates for further investigation.
8. **Draw Conclusions and Recommendations**: Based on your analysis, identify the most influential factors affecting your response variable. Formulate recommendations for further experiments with these key factors to delve deeper into their effects and interactions, optimize your system, or address your initial research question.

Additional Tips:

- **Control Factors**: Account for and control extraneous factors that might influence your results, such as temperature or time.

- **Replicates**: Consider including replicate measurements for some runs to improve data accuracy and analysis confidence.
- **Visualize the Data**: Utilize charts and graphs to visualize the effects of each factor on the response variable, aiding in identifying influential factors and planning further studies.

Remember, screening designs are meant for initial exploration and may not provide definitive answers. However, they offer a valuable and efficient way to prioritize your research efforts and focus on the most promising factors for further investigation and process optimization.

2.2.6 PBDs

PBDs are efficient screening methods for identifying the most influential factors from a large pool in your chemical analysis. They require relatively few runs compared to other designs, making them ideal for initial investigations or when resources are limited. Here's a breakdown of the steps involved:

1. **Define Your Research Question and Objective**: Clearly state what you want to understand about your system. Are you seeking the most influential factors contributing to your chosen response variable or simply getting a starting point for further research?
2. **Identify Potential Factors**: List all factors you think might influence your response variable. These can be quantitative (temperature, concentration) or qualitative (solvent type, catalyst presence). Prioritize factors with potentially significant effects.
3. **Choose a PBD**: The specific design depends on the number of factors (k) you want to analyze. Popular options include 2^k-1 runs for k < 7 and 4k runs for k >= 7. Choose a design that accommodates your desired number of factors.
4. **Assign Factor Levels**: Typically, two levels are used: +1 (high) and −1 (low). This simplifies analysis and interpretation of results.
5. **Generate a Design Matrix and Randomize Run Order**: Use statistical software or online tools to generate a design matrix specifying the assigned levels for each factor in each run. Randomize the run order to minimize bias and ensure valid statistical analysis.
6. **Perform the Experiment**: Conduct each experimental run precisely, ensuring consistent and accurate measurements of your chosen response variable.
7. **Analyze the Data**: Use statistical software or appropriate methods like t-tests or ANOVA to assess the significance of the observed differences in the response variable at different levels of each factor. Pay attention to the absolute values of the calculated effects, with larger values indicating potentially influential factors.
8. **Draw Conclusions and Recommendations**: Based on your analysis, identify the factors with statistically significant effects on your response variable. These are the most promising candidates for further investigation with more sophisticated designs or focus on further optimizations. Formulate recommendations for further research or process improvement based on these key factors.

Additional Tips:

- **Control Factors**: Account for and control extraneous factors that might influence your results, such as temperature or time.
- **Replicates**: Consider including replicate measurements for some runs to improve data accuracy and analysis confidence.
- **Visualize the Data**: Utilize graphs and charts to visualize the relationships between factors and the response variable, aiding in interpretation and identification of potentially influential factors.

Limitations of PBDs:

- They provide limited information on interactions between factors. If interactions are suspected, consider using other designs like fractional factorial or RSM.
- They only explore two levels of each factor, which may not be suitable for studying factors with large nonlinear effects.

Remember, PBDs are a valuable tool for initial screening and prioritizing factors for further investigation. By following these steps and adapting them to your specific needs, you can gain valuable insights into your system and guide your next steps in optimizing your processes or understanding your chemical phenomenon of interest.

PITFALLS AND REMEDIES

Here are some pitfalls to avoid and remedies to apply with screening designs, including PBDs, in designed experiments for analytical chemistry:

Pitfall: Overlooking aliasing. PBD and other screening designs often involve aliasing, where the combined effect of two factors masks the individual effect of each. This can lead to misinterpreting the importance of factors.

Remedy: Choose PBD levels strategically to minimize aliasing patterns. Use software or consult a statistician to assess aliasing before planning the experiment. Include center points or replicates if feasible to partially de-alias factors. Conduct follow-up experiments focusing on interactions suspected of aliasing for further clarification.

Pitfall: Assuming all significant factors are identified. While PBD effectively identifies the most influential factors, it might miss weaker effects or complex interactions.

Remedy: Consider the possibility of missing factors and prioritize follow-up experiments based on additional knowledge or scientific rationale. Use complementary screening designs like fractional factorial designs for further investigation if needed. Don't neglect potentially important factors even if they appear insignificant in the initial screening.

Pitfall: Mistaking correlation for causation. PBD identifies correlations between factors and response, but not necessarily direct causal relationships.

Remedy: Interpret results within the context of your analytical system and avoid over-extrapolating conclusions. Conduct further experiments or utilize complementary techniques to investigate underlying mechanisms behind observed correlations. Remember that correlation and causation are distinct concepts – additional investigations might be necessary to validate cause–effect relationships.

Pitfall: Overlooking experimental error. PBD relies on a limited number of runs, increasing the possibility of error-impacting results.

Remedy: Include center points or replicates within the design to estimate and account for experimental error. Analyze residuals for patterns and potential sources of error. Consider repeating key runs or including additional validation experiments, especially for factors showing promising effects.

Pitfall: Misinterpreting data transformation. Some PBD data analysis involves transforming the response variable to improve normality for statistical tests.

Remedy: Clearly understand the purpose and implications of data transformation for interpreting results. Avoid overcomplicating data manipulation and focus on identifying reliable trends and statistically significant factors. Remember that transformed data don't necessarily reflect the actual response variable behavior.

By considering these pitfalls and adopting the proposed remedies, you can leverage screening designs effectively to gain valuable insights and identify promising leads for further optimization in your analytical chemistry experiments. Remember that careful planning, rigorous data analysis, and critical interpretation of results are crucial for successful screening and guiding subsequent investigations.

EXAMPLE: OPTIMIZE THE EXTRACTION YIELD OF A BIOACTIVE COMPOUND FROM A PLANT MATERIAL USING THE PBD METHOD:

Objective: Optimize the extraction yield of a bioactive compound from a plant material, aiming to identify the most influential factors among several potential candidates.

Factors:

- Solvent type (A): Methanol, ethanol, acetone.
- Solvent volume (B): 50 mL, 100 mL.
- Extraction time (C): 30 minutes, 60 minutes.
- Temperature (D): 40°C, 60°C.
- Particle size (E): Fine, coarse.

Design Matrix:
  PBD involves a set of carefully chosen experimental runs based on the number of factors. In this case, with 5 factors, the design calls for 8 runs. The matrix dictates which factor combinations to test in each run (Table 2.22).

Experiment:

1. Conduct the extractions according to the designated factor combinations in the matrix.
2. Measure the yield of the bioactive compound in each extract using an appropriate analytical method [e.g., high-performance liquid chromatography (HPLC)].

Data Analysis:

1. Employ statistical software to analyze the data, identifying significant factors and potential interactions.
2. Construct a Pareto chart to visualize the relative importance of each factor.
3. Calculate the main effects of each factor to determine their overall impact on the response.

TABLE 2.22
PBD Experiment Matrix Example for Chemistry

Run	A	B	C	D	E
1	+1	+1	−1	+1	−1
2	−1	+1	+1	−1	+1
3	+1	−1	+1	−1	+1
4	−1	−1	−1	+1	+1
5	+1	+1	+1	+1	+1
6	−1	+1	−1	+1	−1
7	+1	−1	−1	+1	−1
8	−1	−1	+1	−1	−1

Expected Outcomes:

- PBD efficiently identifies the most influential factors out of a larger set, helping focus further optimization efforts.
- It indicates whether interactions between factors play a significant role.
- The results guide the selection of factors for more comprehensive optimization designs like RSM.

Key Points:

- PBD is an efficient screening method, not intended for modeling complex relationships or quadratic effects.
- It assumes linear relationships between factors and responses.
- It might not identify all significant factors if strong interactions exist.

Remember:

- Carefully choose factors and levels based on prior knowledge and feasibility.
- Randomize the order of runs to minimize bias.
- Replicate runs for increased confidence in results.
- Consider center points to assess model curvature and potential nonlinearity.

2.2.7 Hadamard Designs

Hadamard designs are another powerful optimization tool in your chemical analysis toolbox, offering efficient screening for a large number of factors with even fewer runs than PBDs. The Hadamard design method is considered a type of PBD type. Here's a breakdown of the steps involved:

1. **Define Your Research Question and Objective**: Clearly state what you want to understand about your system. Are you seeking the most influential factors affecting your chosen response variable or simply getting a starting point for further research?
2. **Identify Potential Factors**: List all factors you think might influence your response variable. These can be quantitative (temperature, concentration) or qualitative (solvent type, catalyst presence). Prioritize factors with potential significant effects.
3. **Choose a Hadamard Design**: The specific design depends on the number of factors (k) you want to analyze. Select a design with $2^{(k-1)}$ runs, which translates to fewer runs compared to other screening methods.
4. **Assign Factor Levels**: Typically, two levels are used: +1 (high) and −1 (low). This simplifies analysis and interpretation of results.
5. **Generate a Design Matrix and Randomize Run Order**: Use statistical software or online tools to generate a design matrix specifying the assigned levels for each factor in each run. Utilize the Hadamard matrix construction properties to generate the design. Randomize the run order to minimize bias and ensure valid statistical analysis.
6. **Perform the Experiment**: Conduct each experimental run precisely, ensuring consistent and accurate measurements of your chosen response variable.
7. **Analyze the Data**: Use statistical software or appropriate methods like t-tests or ANOVA to assess the significance of the observed differences in the response variable at different levels of each factor. Pay attention to the absolute values of the calculated effects, with larger values indicating potentially influential factors.
8. **Draw Conclusions and Recommendations**: Based on your analysis, identify the factors with statistically significant effects on your response variable. These are the most

promising candidates for further investigation with more sophisticated designs or focus on further optimizations. Formulate recommendations for further research or process improvement based on these key factors.

Additional Tips:

- **Control Factors**: Account for and control extraneous factors that might influence your results, such as temperature or time.
- **Replicates**: Consider including replicate measurements for some runs to improve data accuracy and analysis confidence.
- **Visualize the Data**: Utilize graphs and charts to visualize the relationships between factors and the response variable, aiding in interpretation and identification of potentially influential factors.

Limitations of Hadamard Designs:

- They only provide information on the main effects and cannot detect interactions between factors. If interactions are suspected, consider using other designs like fractional factorial or RSM.
- They only explore two levels of each factor, which may not be suitable for studying factors with large nonlinear effects.
- They can be quite sensitive to outliers, so ensuring precise and accurate measurements is crucial.

Hadamard designs are a valuable tool for quickly and efficiently screening a large number of factors in your chemical analysis. By following these steps and adapting them to your specific needs, you can gain valuable insights and prioritize factors for further investigation and optimization in your research.

PITFALLS AND REMEDIES

Hadamard designs offer powerful tools for exploring factor interactions in analytical chemistry experiments, but navigating their strengths and limitations requires careful consideration. Here are some pitfalls to avoid and remedies to apply when using Hadamard designs:

Pitfall: Limited number of factors. Hadamard designs efficiently utilize a limited number of runs, but they can only accommodate a specific number of factors based on the design order (e.g., 16 factors for a Hadamard 2^4 design).

Remedy: Prioritize the most critical factors for investigating interactions based on prior knowledge or scientific rationale. Consider alternative designs like PBD for initial screening of a larger set of factors, followed by Hadamard designs for in-depth interaction analysis of the most promising ones. If more factors are necessary, explore nested designs where interactions within lower-level factors are nested within higher-level factors.

Pitfall: Misinterpreting interaction effects. Hadamard designs focus on interactions, but interpreting interactions correctly is crucial for accurate conclusions.

Remedy: Visualize interactions through interaction plots and analyze main effects alongside interaction terms to understand their combined influence. Pay attention to the magnitude and direction of interaction terms to differentiate synergistic or antagonistic effects. Validate significant interactions through additional experiments with focused factor combinations to confirm their impact.

Pitfall: Overgeneralizing model accuracy. Hadamard designs assume linear relationships between factors and responses. This might not always hold true in complex systems.

Remedy: Analyze residual plots and assess model curvature to identify potential nonlinearity. Consider data transformations or alternative models like Box–Cox regression if nonlinearity is significant. Limit conclusions to the explored factor range and avoid extrapolating beyond without further investigation.

Pitfall: Confusing correlation with causation. Similar to other designs, Hadamard designs identify correlations between factors and responses, not necessarily causation.

Remedy: Interpret results within the context of your analytical system and avoid attributing causality solely based on statistical significance. Utilize complementary techniques like mechanistic studies or simulations to investigate cause–effect relationships behind observed interactions. Remember that correlation and causation are distinct concepts – additional investigations might be necessary to confirm causal mechanisms.

Pitfall: Experimental error and replicates. Due to the limited number of runs, Hadamard designs are susceptible to experimental error-impacting results.

Remedy: Include center points or replicates within the design to estimate and account for experimental error. Analyze residuals for patterns and potential sources of error. Consider repeating key runs or including additional validation experiments for factors demonstrating significant interactions or unexpected trends.

By recognizing these pitfalls and implementing the recommended remedies, you can leverage Hadamard designs effectively to uncover valuable insights into factor interactions and optimize your analytical chemistry processes. Remember that careful planning, rigorous data analysis, and critical interpretation of interactions are crucial for successful exploration and guiding further investigations.

Application Consideration

Something to consider, the Hadamard Design isn't directly applicable to chemical analysis because it's specifically designed for screening a large number of factors (typically 7 or more) with a limited number of runs (2^k-1 runs, where k is the number of factors).

Chemical analysis often involves studying the impact of fewer factors (2–5) in more detail for optimization, where other design methods like BBD, CCD, or Taguchi offer greater flexibility and power for analysis.

However, if you're particularly interested in a situation with many potential factors affecting your chemical system and limited resources for experimentation, you could consider using a PBD design as a preliminary screening step to:

1. **Identify the Most Significant Factors**: By analyzing the results of the Plackett–Burman experiment, you can identify the factors that have the most significant impact on your response variable. This allows you to prioritize which factors deserve further investigation using more powerful design methods.
2. **Reduce the Number of Factors for Further Study**: After the Plackett–Burman screening, you can focus your resources on optimizing only the significant factors and eliminate those with negligible impact, significantly reducing the number of experiments needed for in-depth optimization.

Here's a designed experiment using a Hadamard (a type of Plackett–Burman) design with seven runs to screen the potential impact of eight factors on enzyme activity (glucose production) in biofuel production:

EXAMPLE: WHAT ARE THE POTENTIAL FACTORS ON ENZYME ACTIVITY (GLUCOSE PRODUCTION) IN BIOFUEL PRODUCTION?

Factors:

- Substrate concentration.
- Enzyme concentration.
- Temperature.
- pH.
- Buffer type.
- Presence of activators/inhibitors.
- Reaction time.
- Agitation speed.

Levels: Each factor will have two levels: +1 (high) and −1 (low).

Experimental Design: Using a Hadamard matrix, we generate a design with seven experimental runs, each representing a specific combination of factor levels.

Designed Experiment (Table 2.23):

Explanation:

- Each row represents a unique combination of factor levels, corresponding to a specific experimental condition (run).
- The factors are varied at two levels: −1 (low level) and +1 (high level), as indicated in the table.
- The design matrix allows for the screening of the potential impact of the eight factors on enzyme activity (glucose production) in biofuel production using only seven experimental runs.
- By analyzing the responses obtained from these runs, the significant factors affecting enzyme activity can be identified, providing insights for further optimization of the biofuel production process.

This design efficiently screens the potential impact of multiple factors on enzyme activity while minimizing the number of experimental runs required, making it suitable for initial investigation and identification of key factors in biofuel production.

TABLE 2.23

Hadamard Design Experiment Matrix Example for Chemistry

Run	Substrate Conc.	Enzyme Conc.	Temp	pH	Buffer Type	Activators Inhibitors	Reaction Time	Agitation Speed
1	−1	−1	−1	−1	−1	−1	−1	−1
2	−1	1	1	−1	1	1	1	1
3	1	−1	1	1	−1	1	1	1
4	1	1	−1	1	1	−1	1	1
5	−1	1	−1	1	1	−1	−1	−1
6	1	−1	−1	−1	1	1	−1	−1
7	1	1	1	−1	−1	−1	−1	−1

2.2.8 OPTIMIZATION DESIGNS

Optimization designs are powerful tools in chemical analysis to find the best possible combination of factors within a defined region to maximize or minimize your chosen response variable. Here's a breakdown of the steps involved:

1. **Define Your Research Question and Objective**: Clearly state what you want to optimize – yield, purity, reaction rate, etc. Define the desired direction of optimization (maximize or minimize).
2. **Identify Relevant Factors**: List all factors you think might influence your response variable. These can be quantitative (temperature, concentration) or qualitative (solvent type, catalyst presence). Prioritize factors with potential significant effects.
3. **Choose an Optimization Design**: Popular choices include BBDs and CCDs. Consider the desired information (quadratic effects vs. only linear) and resource constraints (number of runs) when choosing. BBD is efficient for quadratic relationships, while CCD explores both linear and quadratic effects.
4. **Define Factor Levels**: Determine the low, high, and central levels for each factor. Consider the expected range of the response surface and practical limits for your system.
5. **Generate a Design Matrix and Randomize Run Order**: Use statistical software or online tools to generate a design matrix specifying the factor combinations for each experimental run. Randomize the run order to minimize bias and ensure valid statistical analysis.
6. **Perform the Experiment**: Conduct each experimental run precisely, ensuring consistent and accurate measurements of your chosen response variable.
7. **Analyze the Data**: Use statistical software or regression analysis methods to fit a mathematical model (typically quadratic) to the response surface data. Analyze the model for significance of effects, interactions, and goodness of fit.
8. **Optimize the Response Variable**: Identify the optimal factor levels within the studied region that maximize or minimize your response variable based on the fitted model and your desired goal. Utilize contour plots or optimization algorithms to identify the optimal combination.
9. **Validate the Model and Results**: Perform additional experiments at or near the predicted optimal conditions to verify the model's accuracy and confirm the optimal response.
10. **Draw Conclusions and Recommendations**: Based on your findings, draw conclusions about the relationships between factors and the response variable, identify optimal conditions for achieving your desired outcome, and formulate recommendations for further research or process improvement.

Other Considerations:

- **Control Factors**: Account for and control extraneous factors that might influence your results, such as temperature or time.
- **Replicates**: Consider including replicate measurements for some runs to improve data accuracy and model fitting.
- **Visualize the Data**: Utilize response surface plots and contour plots to visualize the relationships between factors and the response variable, aiding in optimization.

Remember, optimization designs can be iterative. Based on your initial results, you might need to modify your model, refine factor levels, or conduct additional experiments for further optimization.

2.2.8.1 BBDs

BBDs are powerful tools for optimization and response surface exploration in chemical analysis. They offer several advantages over other methods, including

- Efficient exploration of quadratic relationships between factors and the response variable.
- Avoiding extreme factor settings, potentially reducing experimental difficulty.
- Providing information on both linear and quadratic effects.
- Allowing for estimation of curvature in the response surface.

Here's a breakdown of the steps involved in developing a designed experiment using a BBD for chemical analysis:

1. **Define Your Research Question and Objective**: Clearly state what you want to optimize or understand about your system. Are you seeking the optimal conditions for maximizing or minimizing a response variable, exploring the shape of the response surface, or identifying significant factors and their interactions?
2. **Identify Relevant Factors**: List all factors potentially influencing your chosen response variable. These can be quantitative (temperature, concentration) or qualitative (solvent type, catalyst presence). Choose factors with potentially significant effects or quadratic relationships suspected to influence the response.
3. **Choose Factors for a BBD**: Consider the number of factors (k) and desired information. Popular options include 3-factor BBDs with 12 runs and 4-factor BBDs with 27 runs. Choose a design that accommodates your desired number of factors and the complexity of the response surface.
4. **Define Factor Levels**: Determine the low, high, and center levels for each factor. Consider the expected range of the response surface and practical limits for your system. BBDs typically use a central point and equidistant midpoints between high and low levels for each factor on the edges of the design space.
5. **Generate a Design Matrix and Randomize Run Order**: Use statistical software or online tools to generate a design matrix specifying the factor combinations for each experimental run. This includes factorial points, axial points (midpoints on edges), and center points. Randomize the run order to minimize bias and ensure valid statistical analysis.
6. **Perform the Experiment**: Conduct each experimental run precisely, ensuring consistent and accurate measurements of your chosen response variable.
7. **Analyze the Data**: Use statistical software or regression analysis methods to fit a mathematical model (typically quadratic) to the response surface data. Analyze the model for significance of effects, interactions, and goodness of fit. Utilize diagnostics to check for model validity and potential outlier data points.
8. **Optimize the Response Variable (optional)**: Identify the optimal factor levels within the studied region that maximize or minimize your response variable based on the fitted model and your desired goal. Utilize contour plots or optimization algorithms to identify the optimal combination.
9. **Validate the Model and Results**: Perform additional experiments at or near the predicted optimal conditions to verify the model's accuracy and confirm the optimal response.
10. **Draw Conclusions and Recommendations**: Based on your findings, draw conclusions about the relationships between factors and the response variable, identify optimal conditions for achieving your desired outcome, and formulate recommendations for further research or process improvement.

Additional Tips:

- **Control Factors**: Account for and control extraneous factors that might influence your results, such as temperature or time.
- **Replicates**: Consider including replicate measurements for some runs to improve data accuracy and model fitting.
- **Visualize the Data**: Utilize response surface plots and contour plots to visualize the relationships between factors and the response variable, aiding in interpretation and optimization.

Remember, BBDs can be iterative. Based on your initial results, you might need to modify your model, refine factor levels, or conduct additional experiments for further optimization.

EXAMPLE: OPTIMIZING CRYSTALLIZATION OF A PHARMACEUTICAL COMPOUND USING BBD

Objective: Identify the optimal conditions for crystallizing a new pharmaceutical compound to maximize yield and crystal size.

Factors:

- Temperature (A): Low (L), high (H).
- Stirring Rate (B): Low (L), high (H).
- Concentration (C): Low (L), high (H).

Design: A BBD with 15 runs, including center points, axial points (midpoints between high and low levels), and replicates at the center point. This design allows estimation of quadratic relationships between factors and the response variable while optimizing within the experimental region.

Design Matrix (Table 2.24):

TABLE 2.24
BBD Experiment Matrix Example for Chemistry

Run	Temperature	Stirring Rate	Concentration
1	L	L	L
2	L	L	H
3	H	L	L
4	H	L	H
5	L	H	L
6	L	H	H
7	H	H	L
8	H	H	H
9	L	(A+B)/2	(C+D)/2
10	H	(A+B)/2	(C+D)/2
11	(A+C)/2	L	(B+D)/2
12	(A+C)/2	H	(B+D)/2
13	(B+C)/2	L	(A+D)/2
14	(B+C)/2	H	(A+D)/2
15	(A+B+C+D)/4	(A+B+C+D)/4	(A+B+C+D)/4

Response Variables:

- Yield of crystallized product (%).
- Average crystal size (microns).

Experiment:

1. Prepare solutions of the compound at the two different concentrations.
2. Set up eight crystallization vessels under controlled temperature and stirring conditions according to the design matrix.
3. Add the chosen concentration of solution to each vessel and monitor the crystallization process over time.
4. Measure the final yield of crystals in each vessel.
5. Analyze the crystals from each run using microscopy to determine the average size.

Data Analysis:

1. Use statistical software to fit a second-order polynomial model to the yield and crystal size data.
2. Analyze the model for significance of linear, quadratic, and interaction terms of the factors.
3. Identify optimal factor levels and their interactions based on the model and desired outcome (maximizing yield and crystal size).
4. Utilize response surface plots and contour plots to visualize the relationships between factors and the response variables.

Expected Outcomes:

- Box–Behnken analysis might reveal the optimal combination of temperature, stirring rate, and concentration for achieving both high yield and large crystal size.
- The interaction terms could indicate synergistic or antagonistic effects between factors, providing valuable insights into the crystallization process.
- Response surface plots would depict the optimal region for achieving your desired outcomes within the studied conditions.

Further Steps:

- Validate the predicted optimal conditions through additional experiments to confirm model accuracy and reproducibility.
- Explore alternative parameters like additives, solvent systems, or pH adjustments for potentially even more efficient or desirable crystallization outcomes.
- Investigate the underlying mechanisms influencing the crystallization process based on the identified key factors to gain deeper scientific understanding.

This example showcases how BBD can be a powerful tool for optimizing multifactorial processes in chemical analysis like crystallization. By analyzing the combined effects and interactions of factors on two response variables, you can gain valuable insights for achieving your desired outcomes and potentially improve the efficiency and success of your crystallization protocols.

PITFALLS AND REMEDIES

BBD offers a powerful tool for optimization in analytical chemistry, but navigating its strengths and limitations requires vigilance. Here are some pitfalls to avoid and remedies to apply:

Pitfall: Assuming smooth and continuous response surfaces. BBD assumes the response variable smoothly changes within the chosen factor range. If true nonlinear behavior exists, the model might be inaccurate.

Remedy:

- Analyze residual plots and curvature in response surfaces to check for nonlinearity.
- Consider transforming the data or alternative models like quadratic-cubic spline models to capture nonlinear trends.
- If nonlinearity is significant, a different design like D-optimal might be necessary.

Pitfall: Misinterpreting axial point limitations. Unlike CCD, BBD employs midpoint axial points within factor levels. These points might fall outside the desired operating range or be impractical to implement.

Remedy:

- Choose factor ranges and levels strategically, considering real-world application constraints and avoiding impractical axial points.
- Validate the predicted optimal conditions by conducting experiments within the desired operating range, even if it excludes the original axial points.
- Consider combining BBD with RSM analysis for optimization within and outside the central region.

Pitfall: Misunderstanding interaction effects. BBD models' factor interactions, but ignoring significant interactions can lead to poor optimization and inaccurate predictions.

Remedy:

- Analyze response surface plots and ANOVA results for significant interaction terms.
- Visualize interactions through 3D plots and contour plots to understand how factors influence each other.
- Consider including more levels or factors in the design if complex interactions are observed.

Pitfall: Overextrapolation optimization beyond design range. BBD optimizes within the chosen factor range. Extrapolating beyond this range could lead to unexpected outcomes.

Remedy:

- Focus on optimizing within the central region of the design with higher precision for quadratic effects.
- Validate the predicted optimal conditions by conducting experiments with slightly varied factor levels to assess performance near the design boundaries.
- Prioritize factors based on their practical importance and cost implications during optimization.

Pitfall: Confusing correlation with causation. Like other designs, BBD identifies correlations between factors and responses, not necessarily causation.

Remedy:

- Interpret results within the context of your specific analytical system and avoid attributing causality solely based on statistical significance.
- Utilize complementary techniques like mechanistic studies or simulations to investigate cause–effect relationships behind observed correlations.
- Remember that correlation and causation are distinct concepts – additional investigations might be necessary to confirm causal mechanisms.

Utilize BBD software or statistical packages for efficient design planning, data analysis, and visualization of response surfaces and interaction effects.

By recognizing these pitfalls and implementing the recommended remedies, you can leverage BBD effectively to optimize your analytical processes and gain valuable insights into factor interactions within the chosen operating range. Remember that careful planning, rigorous data analysis, and critical interpretation of results are crucial for successful optimization and practical implementation.

2.2.9 CCDs

CCDs are powerful tools in chemical analysis for exploring response surfaces and estimating both linear and quadratic effects of factors on your chosen response variable. Here's a breakdown of the steps involved:

1. **Define Your Research Question and Objective**: Clearly state what you want to optimize or understand about your system. Are you seeking optimal conditions for maximizing or minimizing a response variable, exploring the shape of the response surface, or identifying significant factors and their interactions?
2. **Identify Relevant Factors**: List all factors potentially influencing your chosen response variable. These can be quantitative (temperature, concentration) or qualitative (solvent type, catalyst presence). Choose factors with potentially significant effects or relationships suspected to influence the response surface.
3. **Choose a CCD**: Consider the number of factors (k) and desired information. Popular choices include circumscribed, inscribed, and face-centered CCDs. Circumscribed CCDs (CCC) have star points farther from the center for stronger curvature estimation, while inscribed CCDs (CCI) use factor settings as star points, requiring fewer runs.
4. **Calculate the Number of Runs**: Use the formula $N = 2^k + 2k + N0$, where N is the total number of runs, k is the number of factors, and N0 is the number of center points (typically 2 or 3). This formula accounts for factorial points, axial (star) points, and center points.
5. **Define Factor Levels**: Determine the low, high, and center levels for each factor. Consider the expected range of the response surface and practical limits for your system. The star points are typically set at $\pm\alpha$ times the standard deviation from the center point for each factor, where α is a chosen value influencing curvature estimation.
6. **Generate a Design Matrix and Randomize Run Order**: Use statistical software or online tools to generate a design matrix specifying the factor combinations for each experimental run. Include factorial points, axial (star) points, and center points. Randomize the run order to minimize bias and ensure valid statistical analysis.
7. **Perform the Experiment**: Conduct each experimental run precisely, ensuring consistent and accurate measurements of your chosen response variable.
8. **Analyze the Data**: Use statistical software or regression analysis methods to fit a mathematical model (typically quadratic) to the response surface data. Analyze the model for significance of effects, interactions, and goodness of fit. Utilize diagnostics to check for model validity and potential outlier data points.
9. **Optimize the Response Variable (optional)**: Identify the optimal factor levels within the studied region that maximize or minimize your response variable based on the fitted model and your desired goal. Utilize contour plots or optimization algorithms to identify the optimal combination.
10. **Validate the Model and Results**: Perform additional experiments at or near the predicted optimal conditions to verify the model's accuracy and confirm the optimal response.
11. **Draw Conclusions and Recommendations**: Based on your findings, draw conclusions about the relationships between factors and the response variable, identify optimal conditions for achieving your desired outcome, and formulate recommendations for further research or process improvement.

Additional Tips:

- **Control Factors**: Account for and control extraneous factors that might influence your results, such as temperature or time.
- **Replicates**: Consider including replicate measurements for some runs to improve data accuracy and model fitting.
- **Visualize the Data**: Utilize response surface plots and contour plots to visualize the relationships between factors and the response variable, aiding in interpretation and optimization.

CCDs can be iterative. Based on your initial results, you might need to modify your model, refine factor levels, or conduct additional experiments for further optimization.

EXAMPLE: ANALYZING THE DEGRADATION OF A COSMETIC EMULSION USING CCD

Objective: Determine the factors and their interactions that most significantly affect the stability and shelf life of a new cosmetic emulsion.

Factors:

- Temperature (A): 25°C (low), 40°C (high).
- pH (B): 5.5 (low), 6.5 (high).
- Oil Phase Ratio (C): 60% (low), 80% (high).

Design: A circumscribed CCD with 15 runs, including center points, axial points (α distance from the center for each factor), and factorial points. This design allows the estimation of both linear and quadratic relationships between factors and the response variable while optimizing within the experimental region.

Design Matrix (Table 2.25):

TABLE 2.25
CCD Experiment

Run	Temperature	pH	Oil Phase Ratio
1	32.5	6.0	70
2	32.5	6.0	70
3	25	6.0	70
4	40	6.0	70
5	32.5	5.5	70
6	32.5	6.5	70
7	32.5	6.0	65
8	32.5	6.0	75
9	28.125	6.0	70
10	36.875	6.0	70
11	32.5	5.25	70
12	32.5	6.75	70
13	32.5	6.0	57.5
14	32.5	6.0	82.5

Response Variables:

- Change in viscosity over time: Indicator of emulsion stability and potential separation.
- pH changes over time: Indicator of potential ingredient interactions and degradation.

Experiment:

1. Prepare emulsions with different oil phase ratios, adjust pH, and expose them to the designated temperatures according to the design matrix.
2. Monitor the emulsions over time, measuring viscosity and pH at regular intervals.
3. Analyze the change in viscosity and pH over time for each emulsion.

Data Analysis:

1. Use statistical software to fit a second-order polynomial model to the viscosity and pH data.
2. Analyze the model for significance of linear, quadratic, and interaction terms of the factors.
3. Identify factors and their interactions with significant effects on both response variables, indicating their contribution to emulsion stability or degradation.
4. Utilize response surface plots and contour plots to visualize the relationships between factors and the response variables.

Expected Outcomes:

- CCD analysis might reveal that specific temperature ranges, pH values, or oil phase ratios significantly impact the emulsion stability and/or degradation rates.
- The interaction terms could indicate synergistic or antagonistic effects between factors, providing insights into the mechanisms of degradation or stabilization.
- Response surface plots would graphically depict the optimal conditions for maximizing emulsion stability and shelf life within the studied range.

Further Steps:

- Validate the predicted optimal conditions through additional experiments to confirm model accuracy and reproducibility.
- Investigate the underlying mechanisms of degradation identified by the analysis, potentially involving additional analytical techniques.
- Utilize the knowledge gained to optimize the formulation of the cosmetic emulsion for improved stability and longer shelf life, while maintaining desired product properties.

This example showcases how CCD can be a powerful tool for analyzing multifactorial processes in chemical analysis like emulsion stability. By analyzing the combined effects and interactions of factors on two response variables, you can gain valuable insights for improving product quality and shelf life, contributing to successful product development and commercialization.

PITFALLS AND REMEDIES

CCDs offer robust tools for optimization in analytical chemistry, but their effectiveness hinges on careful planning and awareness of potential hurdles. Here are some pitfalls to avoid and remedies to apply:

Pitfall: Assuming smooth and continuous response surfaces. CCD assumes the response variable smoothly changes within the chosen factor range. If true nonlinear behavior exists, the model might be inaccurate.
    Remedy:

- Analyze residual plots and curvature in response surfaces to check for nonlinearity.
- Consider transforming the data or exploring alternative models like quadratic-cubic spline models to capture nonlinear trends.
- If nonlinearity is significant, a different design like D-optimal might be necessary.

Pitfall: Overlooking axial point limitations. While CCD utilizes axial points for estimating quadratic effects, these points can sometimes fall outside the desired operating range or be impractical to implement.
    Remedy:

- Choose factor ranges and levels strategically, considering real-world application constraints and avoiding impractical axial points.
- Validate the predicted optimal conditions by conducting experiments within the desired operating range, even if it excludes the original axial points.
- Consider alternative designs like BBD if axial points are problematic or unnecessary for your optimization goals.

Pitfall: Misinterpreting model precision. CCD estimates quadratic relationships, but its precision for quadratic effects might be lower compared to alternative designs like BBD.
    Remedy:

- Focus on optimizing within the central region of the design where precision is highest for quadratic effects.
- Use larger center point replicates to improve the estimation of quadratic coefficients.
- Consider combining CCD with other designs like BBD for improved precision within and outside the central region.

Pitfall: Overfitting the model. Complex CCD models with many terms can overfit the data, leading to poor predictive accuracy.
    Remedy:

- Use statistical selection methods like backwards elimination or AIC to remove insignificant terms.
- Analyze residual plots and check for lack of fit to assess model overfitting.
- Validate the model with additional experiments to confirm its predictive ability.

Pitfall: Ignoring experimental constraints. CCD can require a significant number of runs, which can be time-consuming and resource intensive, especially with many factors.
    Remedy:

- Use fractional factorial designs or PBDs for initial screening to identify important factors before employing CCD.
- Limit the number of factors based on available resources and prioritize the most crucial ones for optimization.
- Consider alternative designs like RSM with BBD if fewer runs are preferred.

By recognizing these pitfalls and implementing the recommended remedies, you can ensure your CCD experiments yield accurate and actionable results for optimizing your analytical chemistry processes. Remember that careful planning, rigorous data analysis, and critical interpretation of the model are crucial for successful optimization and practical implementation.

2.2.10 One-Factor-at-a-Time (OFAT) Designs

OFAT designs are a simple approach to investigating how individual factors affect a chosen response variable in your chemical analysis. While not without limitations, they can be useful for initial investigation or when resources are limited. Here's a breakdown of the steps involved:

1. **Define Your Research Question and Objective**: Clearly state what you want to understand about your system. Are you interested in identifying the general impact of individual factors or simply gaining initial insights before further investigation?
2. **Identify Relevant Factors**: List all factors you think might influence your response variable. These can be quantitative (temperature, concentration) or qualitative (solvent type, catalyst presence).
3. **Choose a Factor to Study**: Select one factor to focus on in the first round of your experiment. Consider factors you suspect might have a large effect or are easy to manipulate.
4. **Define Factor Levels**: Determine the low, high, and (optionally) intermediate levels for the chosen factor. Consider the expected range of effects and practical limits for your system.
5. **Conduct the Experiment**: Run the experiment at each level of the chosen factor while holding all other factors constant. Maintain consistent and accurate measurements of your response variable.
6. **Analyze the Data**: Use statistical methods like t-tests or ANOVA to assess the significance of the observed differences in the response variable at different levels of the factor.
7. **Draw Conclusions and Recommendations**: Based on your analysis, draw conclusions about the effect of the studied factor on your response variable. If the effect is significant, consider further investigation with other factors or using more sophisticated design methods like fractional factorial or response surface designs.
8. **Repeat for Other Factors (optional)**: If desired, repeat the process for other factors, one at a time, holding all other factors constant. Remember that OFAT can miss interactions between factors.

Additional Tips:

- **Control Factors**: Account for and control extraneous factors that might influence your results, such as temperature or time.
- **Replicates**: Consider including replicate measurements at each level of the factor to improve data accuracy and analysis confidence.
- **Visualize the Data**: Utilize graphs and charts to visualize the relationships between the factor and the response variable, aiding in interpretation.

Limitations of OFAT:

- It ignores potential interactions between factors, which can lead to misleading conclusions.
- It often requires more experimental runs compared to efficient designs like fractional factorial when investigating multiple factors.
- It may not identify optimal conditions, as it only explores individual factor effects.

OFAT can be a good starting point, but for deeper understanding and optimization, consider utilizing more sophisticated designed experiments that offer a more comprehensive picture of factor interactions and optimal conditions.

While the OFAT design method is not recommended for complex systems due to its limitations in capturing interactions and potential for misleading results, it can still be useful for simple preliminary investigations or screening a large number of factors with limited resources. Here's an example:

EXAMPLE: IDENTIFYING THE KEY FACTOR AFFECTING PIGMENT PRECIPITATION IN PAINT FORMULATION USING OFAT DESIGN

Objective: Identify the primary factor influencing the precipitation of a specific pigment in a newly formulated paint, leading to unwanted color variations.

Factors:

- Binder type (A): Acrylic, alkyd.
- Solvent composition (B): Aromatic, aliphatic.
- Pigment concentration (C): Low, high.
- Temperature (D): Room temperature, elevated.

Design:

1. Phase 1: Focus on one factor at a time, holding all others constant. For instance, prepare multiple paint samples with different binder types (A1 and A2) but the same solvent, pigment concentration, and temperature.
2. Phase 2: Once a potentially influential factor is identified (e.g., binder A1 shows more precipitation), refine the investigation by varying other factors while keeping the identified factor constant. For example, prepare samples with different solvent compositions (B1 and B2) using binder A1 and the same pigment concentration and temperature.

Response Variables:

- Degree of pigment precipitation: Visually assessed or measured by sedimentation rate, particle size analysis, etc.
- Color consistency: Compared to a reference standard or measured by spectrophotometry.

Experiment:

1. Prepare paints according to the chosen design phases.
2. Monitor the paint samples for evidence of pigment precipitation over time.
3. Evaluate the color consistency of each sample compared to the desired outcome.

Data Analysis:

1. Qualitatively compare the samples within each phase based on the observed precipitation and color variation.
2. If a factor shows a clear impact, further investigate its interaction with other factors through additional OFAT runs or consider transitioning to a more powerful design method like fractional factorial or PBD for a more comprehensive analysis.

Limitations:

- OFAT cannot capture potential interactions between factors that might significantly influence the outcome.
- It requires more runs than statistically efficient designs to achieve the same level of understanding.
- The results may be misleading if interactions are present and not accounted for.

Further Steps:

- If a single factor appears influential, use a more robust design method like fractional factorial or CCD to optimize the system and confirm the findings.
- Investigate the underlying mechanisms influencing the identified factor's impact on pigment precipitation for a deeper understanding and potential formulation improvements.
- Remember that OFAT should be considered a preliminary screening tool and not a definitive approach for complex chemical analysis due to its limitations.

This example illustrates how OFAT can be used for simple initial investigations but emphasizes its limitations and encourages transitioning to more powerful design methods for comprehensive optimization and accurate analysis in chemical systems.

PITFALLS AND REMEDIES

OFAT designs, although often used in analytical chemistry, have several limitations and pitfalls. Here are some key issues to avoid and remedies to apply when considering OFAT experiments:

Pitfalls:
    Ignoring Interactions: By changing one factor while holding others constant, OFAT neglects potential interactions between factors, which can significantly impact the response variable. This can lead to misleading conclusions about the true influence of each factor.
    Inefficient Use of Resources: OFAT requires numerous experiments, often exceeding the number needed for more efficient designs like factorial or RSM. This can be time-consuming, expensive, and resource intensive.
    Difficulty Optimizing: OFAT results may not translate directly to the optimal conditions under real-world scenarios where multiple factors interact. Optimizing based on individual factor effects can lead to inaccurate predictions and suboptimal outcomes.
    Missing Synergistic or Antagonistic Effects: OFAT cannot identify synergistic or antagonistic interactions, where the combined effect of two factors is greater or less than the sum of their individual effects, respectively. These potential relationships can significantly influence the overall process.

Remedies:
    Consider Factorial or RSM Designs: Implement robust designs like factorial or RSM that simultaneously study multiple factors and their interactions, providing a more comprehensive understanding of the process.
    Prioritize Interactions: Based on prior knowledge or scientific rationale, identify pairs or groups of factors with potential interactions and focus on studying these interactions directly.
    Conduct Pilot Experiments: Utilize smaller, preliminary experiments to identify the most critical factors and potential interactions before committing to a large-scale OFAT study.
    Employ Complementary Techniques: Combine OFAT with other analytical techniques like mechanistic studies or simulations to gain additional insights into how factors interact and influence the response.

OFAT should be used with caution and awareness of its limitations. While it can offer a simplified approach, it can yield misleading or incomplete results. Prioritize efficient and informative designs like factorial or RSM for a more comprehensive understanding of your analytical processes and accurate optimization. Don't hesitate to combine OFAT with other techniques for a deeper exploration of factor interactions and mechanistic insights.

By adopting these remedies and fostering a critical approach to OFAT designs, you can enhance the quality and efficiency of your analytical chemistry experiments and optimize your processes with greater accuracy and confidence.

2.2.11 NESTED DESIGNS

Nested designs are powerful tools in chemical analysis where factors exist at multiple levels within a hierarchical structure. They help us understand how these levels interact and influence the chosen response variable. Here's a breakdown of the steps involved:

1. **Define Your Research Question and Objective**: Clearly state what you want to understand about your system. Are you interested in the effects of factors at different levels within a hierarchy, or their interactions?
2. **Identify Relevant Factors**: List all factors potentially influencing your response variable. These can be quantitative (temperature, concentration) or qualitative (solvent type, catalyst presence). Consider factors at two or more levels, with some nesting within others.
3. **Define Factor Levels**: Determine the low, high, and (optionally) intermediate levels for each factor. Ensure logical nesting of levels, where levels of one factor exist within categories of another.
4. **Choose a Nested Design**: Popular choices include RCBDs and split-plot designs. Select a design based on the number of factors and levels, and your desired information (main effects, interactions).
5. **Generate a Design Matrix and Randomize Run Order**: Use statistical software or online tools to generate a design matrix specifying the treatment combinations for each experimental run. Ensure random assignment of treatments within each block or level to minimize bias.
6. **Perform the Experiment**: Conduct each experimental run precisely, ensuring consistent and accurate measurements of your chosen response variable.
7. **Analyze the Data**: Use statistical software or appropriate analysis methods to assess the main effects and interactions of factors at different levels within the hierarchy. Pay attention to the nesting structure and interpret results in that context.
8. **Draw Conclusions and Recommendations**: Based on your analysis, draw conclusions about the relationships between factors at different levels and their impact on the response variable. Identify significant effects and interactions and formulate recommendations for further research or process improvement.

Additional Tips:

- **Control Factors**: Account for and control extraneous factors that might influence your results, such as temperature or time.
- **Replicates**: Consider including replicate measurements for some runs to improve data accuracy and statistical power.
- **Visualize the Data**: Utilize charts and graphs to visualize the relationships between factors and the response variable, aiding in the interpretation of the nested structure.

Nested designs can be more complex to analyze than simpler designs. Ensure you have access to appropriate statistical software and expertise. Carefully define the nesting structure and ensure treatments are assigned logically within each level. Interpret results in the context of the hierarchy and avoid drawing conclusions beyond the studied levels and interactions. These designs can offer valuable insights into hierarchical systems in chemical analysis. By following these steps and adapting them to your specific research needs, you can gain a deeper understanding of how factors at different levels interact and influence your desired outcome.

EXAMPLE: OPTIMIZING SOLVENT EXTRACTION OF BIOACTIVE COMPOUNDS FROM PLANTS USING NESTED DESIGNED EXPERIMENTS

Objective: Identify the optimal solvent and extraction temperature for maximizing the yield and purity of bioactive compounds from a specific plant extract.

Factors:

- Outer factor (level 1): Plant species (A1, A2).
- Inner factor (level 2 nested within each plant species): Solvent (B1, B2, B3).
- Inner factor (level 2 nested within each plant species): Extraction temperature (C1, C2).

Design:
A randomized block design with three replicates for each combination of plant species, solvent, and temperature. This ensures that any variability between plant batches or environmental factors is controlled for.

Design Matrix (Table 2.26):

Response Variables:

- Yield of bioactive compounds: Measured by chromatography or spectrophotometry.
- Purity of bioactive compounds: Assessed by chromatographic analysis or other techniques.

Experiment:

1. Extract bioactive compounds from plant samples using the designated solvent and temperature combinations for each replicate.

TABLE 2.26
Nested Design Matrix Example

Block	Plant Species	Solvent	Temperature
1	A1	B1	C1
1	A1	B2	C1
1	A1	B3	C1
1	A1	B1	C2
1	A1	B2	C2
1	A1	B3	C2
2	A2	B1	C1
2	A2	B2	C1
2	A2	B3	C1
2	A2	B1	C2
2	A2	B2	C2
2	A2	B3	C2

2. Quantify and purify the extracted compounds using appropriate analytical methods.
3. Record the yield and purity data for each sample.

Data Analysis:

1. Use statistical software to perform ANOVA with nested factors.
2. Analyze the effects of solvent and temperature on both yield and purity within each plant species.
3. Identify the optimal combination of solvent and temperature for each plant species based on desired outcomes (high yield and purity).
4. Investigate the interaction between plant species and extraction conditions to assess any differences in optimal parameters.

Expected Outcomes:

- Nested design analysis might reveal that different solvents and temperatures are ideal for each plant species due to varying levels of bioactive compounds and their interactions with solvents.
- The nested structure allows for precise comparison of extraction conditions within each plant species, eliminating potential confounding factors.
- This knowledge can be used to optimize the extraction process for each plant, maximizing the yield and purity of valuable bioactive compounds.

Further Steps:

- Validate the identified optimal conditions for each plant species through additional experiments.
- Explore the underlying mechanisms of extraction for each plant–solvent combination to deepen understanding and potentially refine the process further.
- Consider adapting the nested design approach to investigate other factors like extraction time, solvent mixtures, or pretreatment methods for further optimization.

This is just one example, and nested designs can be applied to various chemical analysis scenarios where factors act at different levels. Remember to tailor the design and analysis to your specific objectives and analytical system for optimal results.

PITFALLS AND REMEDIES

Nested designs offer advantages for exploring hierarchical relationships in analytical studies, but navigating their intricacies requires awareness of potential pitfalls. Here's a breakdown of what to watch out for and how to address them:

Pitfalls: Confounding levels. In nested designs, lower-level factor effects can be "confounded" with the effects of higher-level factors, making it difficult to distinguish their individual influence.

Remedy:

- Choose nested levels carefully to minimize overlap and ensure sufficient separation between their effects.
- Employ statistical techniques like mixed-effects models to explicitly account for the nested structure and separate different levels of variation.
- Analyze residual patterns and use diagnostic tests to check for and address potential confounding issues.

Pitfalls: Ignoring random variability. Analytical experiments often involve inherent variability. Neglecting this variability within nested levels can lead to inaccurate conclusions.
Remedy:

- Incorporate random replicates at each level of the nesting hierarchy to capture and estimate random variation effectively.
- Analyze variance components within the mixed-effects model to separate random effects from fixed factor effects.
- Pay attention to the magnitude and significance of both fixed and random effects when interpreting results.

Pitfalls: Misinterpreting interaction effects. Interactions between factors at different levels can occur in nested designs, adding complexity to interpretation.
Remedy:

- Visualize interactions through plots and consider their direction and magnitude to understand how factors at different levels influence each other.
- Interpret interactions within the context of the specific nested structure and scientific rationale.
- Don't overgeneralize interaction effects beyond the specific factor levels used in the design.

Pitfalls: Limited scope. Nested designs might focus on specific relationships within a larger system, potentially overlooking broader influences external to the nested structure.
Remedy:

- Consider the nested design within the context of the overall analytical process and potential external factors.
- If necessary, supplement the nested design with additional investigations exploring interactions or factors outside the nested structure.
- Remember that nested designs provide targeted insights, but comprehensive understanding might require broader experimental approaches.

Pitfalls: Misusing statistical techniques. Choosing and interpreting statistical models for nested data requires a proper understanding of the specific design and data characteristics.
Remedy:

- Consult with a statistician or utilize specialized software for mixed-effects analysis and appropriate model selection.
- Don't rely on traditional ANOVA methods, which might not be applicable for nested data analysis.
- Ensure correct interpretation of variance components, fixed effects, and interaction terms within the statistical model.

By recognizing these pitfalls and implementing the recommended remedies, you can leverage nested designs effectively to gain valuable insights into hierarchical relationships within your analytical experiments. Remember that careful planning, rigorous data analysis, and critical interpretation of results are crucial for successful exploration and drawing accurate conclusions from nested data structures.

2.2.12 TAGUCHI METHOD

The Taguchi method is a powerful approach for optimizing processes while minimizing experiment runs and maximizing information gained. Here's a breakdown of the steps involved in using it for your chemical analysis:

1. **Define the Process Objective**: Clearly state what you want to optimize or minimize in your chemical analysis. This could be a yield, purity, reaction rate, etc.
2. **Identify Noise Factors**: List any factors that can potentially introduce variability into your process but are not directly controlled by you (e.g., temperature fluctuations, humidity).
3. **Identify Signal Factors**: List the factors you can control that are suspected to influence the response variable (e.g., temperature, concentration, catalyst type). Prioritize factors with potential significant effects.
4. **Choose an OA**: Select an OA based on the number of signal and noise factors you have. OAs allow for efficient experimentation while ensuring all factor combinations are represented. Popular choices include L8, L9, and L16 arrays.
5. **Assign Factor Levels**: Determine the low, high, and (optionally) middle levels for each signal factor. Consider the expected range of the response and practical limits for your system. Noise factors can have multiple levels or be considered uncontrolled variables.
6. **Generate a Design Matrix**: Use statistical software or Taguchi tables to map the OA onto your specific factors and levels, creating a design matrix specifying the factor combinations for each experimental run.
7. **Perform the Experiment**: Conduct each experimental run precisely, ensuring consistent and accurate measurements of your chosen response variable. Be mindful of noise factors and try to minimize their impact.
8. **Analyze the Data**: Utilize SNRs to analyze the impact of each factor on the response variable. SNRs account for variability due to noise, allowing for robust comparisons. Analyze the SNRs using ANOVA or other statistical methods to identify significant factors and their optimal levels.
9. **Optimize the Response Variable**: Based on the SNRs and ANOVA results, identify the optimal levels for each signal factor that maximize or minimize your desired response. Utilize response surface plots or other graphical tools to visualize the relationships between factors and the response variable.
10. **Validate the Results**: Conduct additional experiments at the predicted optimal conditions to confirm the effectiveness of your optimizations and validate the reliability of your findings.
11. **Draw Conclusions and Recommendations**: Based on your analysis and validation, draw conclusions about the relationships between factors and the response variable, identify the optimal conditions for achieving your desired outcome, and formulate recommendations for further research or process improvement.

Additional Tips:

- **Control Factors**: Control and minimize the impact of noise factors as much as possible to ensure accurate results.
- **Replicates**: Consider including replicate measurements for some runs to improve data accuracy and analysis confidence.
- **Software**: Utilize dedicated Taguchi software or statistical software packages with Taguchi capabilities for design and analysis.

Remember, the Taguchi method offers a systematic approach to process optimization in chemical analysis. By following these steps and adapting them to your specific needs, you can gain valuable insights, minimize experimental effort, and achieve robust optimizations for your desired outcomes. The Taguchi method is a powerful approach for optimizing processes with minimal resources and experimental runs. Here's an example of applying it to chemical analysis:

EXAMPLE: OPTIMIZING ENZYMATIC HYDROLYSIS OF STARCH FOR BIOFUEL PRODUCTION USING THE TAGUCHI METHOD

Objective: Identify the optimal conditions for enzymatic hydrolysis of starch to maximize glucose yield for biofuel production while minimizing enzyme usage.

Factors:

- Temperature (A): 30°C, 40°C, 50°C.
- pH (B): 4.5, 5.0, 5.5.
- Enzyme Concentration (C): 0.5%, 1.0%, 1.5% (w/v).

Design: An L9 OA, chosen for its smaller number of runs compared to full factorial designs while considering the number of factors.

Design Matrix (Table 2.27):

Response Variable:

- Glucose concentration (mg/mL) as an indicator of starch hydrolysis.

SNR: Taguchi uses SNR to evaluate the impact of factors on the desired outcome. There are specific SNR types for maximizing, minimizing, or achieving a target value. In this case, the objective is to maximize glucose yield, so a larger-is-better SNR would be calculated for each run.

TABLE 2.27

Taguchi Method Designed Experiment Matrix Example for Chemistry

Run	Temperature	pH	Enzyme Conc.	SNR
1	30	4.5	0.5	(measured value)
2	30	5.0	1.0	(measured value)
3	30	5.5	1.5	(measured value)
4	40	4.5	1.0	(measured value)
5	40	5.0	0.5	(measured value)
6	40	5.5	1.5	(measured value)
7	50	4.5	1.5	(measured value)
8	50	5.0	0.5	(measured value)
9	50	5.5	1.0	(measured value)

Experiment:

1. Prepare reaction mixtures with different temperatures, pH values, and enzyme concentrations according to the L9 array.
2. Incubate the reactions with the enzyme for a predetermined time.
3. Measure the glucose concentration in each reaction mixture after incubation.
4. Calculate the SNR for each run based on the chosen type and formula.

Data Analysis:

1. Use statistical software or manually analyze the average SNR and main effects plot for each factor.
2. Identify the factor levels with the highest average SNR, signifying their contribution to maximizing glucose yield.
3. Consider the interaction plots to assess any significant interactions between factors.
4. Determine the optimal combination of temperature, pH, and enzyme concentration based on the analysis, aiming for both high yield and minimized enzyme usage.

Further Steps:

- Validate the predicted optimal conditions through additional experiments to confirm their reproducibility and performance.
- Investigate alternative enzymes or enzyme cocktails for potentially even higher yields.
- Explore process modifications like substrate pretreatment or reactor configuration to further optimize biofuel production efficiency.

This example demonstrates how the Taguchi method can be applied in chemical analysis to achieve optimal process conditions with minimal experimental runs. Remember to choose the appropriate SNR type and consider interactions between factors for accurate analysis and effective optimization.

PITFALLS AND REMEDIES

The Taguchi method offers a powerful approach for optimizing robustness in analytical processes, but navigating its strengths and limitations requires careful consideration. Here are some pitfalls to avoid and remedies to apply when using the Taguchi method:

Misapplying OAs: Taguchi relies on OAs for efficient experimentation, but choosing the wrong array or interpreting results incorrectly can lead to misleading conclusions.
  Remedy:

- Select the appropriate OA based on the number of factors and levels, considering power and desired information.
- Consult statistical software or resources to ensure proper interpretation of factor effects and interactions within the chosen array.
- Don't overgeneralize results beyond the factor range and levels explored in the specific array.

Overlooking Assumptions: The Taguchi method makes assumptions about normality, constant variance, and independence of errors. Violating these assumptions can affect the validity of conclusions.

Remedy:

- Analyze residual plots and normality tests to identify potential violations.
- Consider data transformations or alternative analysis methods if assumptions are not met.
- Conduct pilot experiments to assess data characteristics and adjust the design or response variable if necessary.

Confusing SNRs: Taguchi optimizes based on SNRs, but misinterpreting their meaning can lead to inappropriate selection of optimal conditions.
Remedy:

- Understand the different types of SNRs (higher-the-better, lower-the-better, NTB) and choose the appropriate type based on the desired outcome.
- Don't blindly maximize or minimize SNRs without considering the actual response variable values and their practical implications.
- Interpret SNRs within the context of the specific analytical process and desired performance characteristics.

Neglecting Validation: Taguchi focuses on identifying optimal settings, but neglecting validation can lead to inaccurate performance in real-world scenarios.
Remedy:

- Conduct confirmation experiments at the predicted optimal conditions to validate the model and assess process performance under actual operating conditions.
- Consider additional robustness tests to ensure the optimized process is resistant to noise factors and environmental variations.
- Remember that Taguchi predictions might not perfectly translate to real-world settings, and further validation is crucial for successful implementation.

Limited Factor Interactions: Taguchi designs might not fully capture complex interactions between multiple factors.
Remedy:

- Analyze residual patterns and interaction plots to identify potential interactions that warrant further investigation.
- Consider complementary designs like RSM or BBD if complex interactions are crucial for optimizing the process.
- Utilize Taguchi as an initial screening and optimization tool, followed by more nuanced designs for thorough interaction analysis if needed.

Utilize Taguchi software or statistical packages for efficient design planning, data analysis, and optimization based on SNR calculations and response surface visualization.

By recognizing these pitfalls and implementing the recommended remedies, you can leverage the Taguchi method effectively to gain valuable insights, achieve process robustness, and optimize your analytical experiments with greater confidence. Remember that careful planning, rigorous data analysis, and critical interpretation of results are crucial for successful optimization and practical implementation of the Taguchi method in your analytical research.

2.2.13 Sequential Experimentation Design

Many product development professionals will utilize sequential experimentation. These designed experiments are done by adjusting experimental conditions based on ongoing results. The sequential experimentation is a powerful approach for designing experiments in chemical analysis when you have limited resources, time, or materials and want to make efficient use of them. It allows you to adjust your experimental design based on the results of earlier experiments, thereby optimizing the process as you go. Here are the steps to develop a designed experiment for chemical analysis using sequential experimentation:

Define Your Objective

- Clearly state the research question or objective of your chemical analysis. What specific factors or variables are you trying to study or control? What are your constraints (e.g., limited resources, time)?

Identify the Factors

- Identify the key factors that may affect the outcome of your chemical analysis. These factors can be both categorical (e.g., different chemicals, concentrations) and quantitative (e.g., temperature, time).

Determine the Initial Experimental Design

- Start with an initial experimental design that includes a limited number of experimental runs based on your available resources and constraints. This could be a simple factorial design or a fractional factorial design.

Conduct Initial Experiments

- Perform the initial experiments according to the design and collect data on the chemical analysis outcome.

Analyze Initial Results

- Analyze the data from the initial experiments to identify trends, patterns, and significant factors that affect the outcome. Use statistical techniques to extract meaningful information.

Adjust the Experimental Design

- Based on the analysis of initial results, modify your experimental design to focus on the factors that appear to be most influential. You can choose to add more runs for certain factor combinations or eliminate less important factors.

Conduct Additional Experiments

- Perform additional experiments according to the adjusted experimental design. These experiments are informed by the results of the initial experiments and aim to further optimize the process.

Continuously Analyze and Adjust

- Continue to analyze the results of each round of experiments and make further adjustments to the experimental design as needed. This iterative process allows you to refine your understanding of the factors and their effects.

Terminate the Experiment

- Decide when to terminate the sequential experimentation process based on achieving your objectives or when further experiments no longer provide significant improvements.

Report Results

- Document your sequential experimentation process, including the initial design, adjustments made, data collection, analysis, and final results, in a clear and concise report or research paper.

Sequential experimentation is particularly useful when you have limited resources or when the system you are studying is complex and not well understood. It allows you to allocate resources efficiently, focus on the most influential factors, and achieve the desired outcomes in chemical analysis while minimizing waste.

PITFALLS AND REMEDIES

While sequential experimentation is a valuable approach in chemical analysis, it comes with its own set of pitfalls. Understanding these potential issues and knowing how to address them is essential for conducting effective experiments. Here are some common pitfalls and their corresponding remedies in a designed experiment for chemical analysis using sequential experimentation:

Premature Convergence
Issue: Stopping the sequential experimentation process too early may lead to suboptimal results, as you may not have explored all possible factor combinations thoroughly.
Remedy: Define stopping criteria in advance, such as reaching a certain level of improvement or a predetermined number of iterations. Continue the process until these criteria are met to avoid premature convergence.

OverAdjustment
Issue: Making frequent and significant adjustments to the experimental design based on initial results can lead to overfitting the model to noise in the data.
Remedy: Use statistical methods to assess the significance of observed trends and effects before making adjustments. Avoid overreacting to noise, and strike a balance between adaptation and stability in your experimental design.

Ignoring System Dynamics
Issue: Some chemical systems may have dynamic behavior, where the effects of factors change over time. Ignoring these dynamics can lead to inaccurate conclusions.
Remedy: Continuously monitor and model the system's dynamics, if applicable. Incorporate time-dependent factors and consider that experimental conditions may evolve over the course of the study.

Unaccounted-for Interactions
Issue: Focusing solely on the main effects and not considering interactions between factors can lead to inaccurate predictions and missed opportunities for optimization.
Remedy: Include interaction terms in your models and designs when appropriate. Sequential experimentation should allow for the exploration of interactions as you refine your experiments.

Resource Constraints
Issue: Limited resources may hinder your ability to conduct a large number of experiments, limiting the depth of exploration.

Remedy: Prioritize experiments that are likely to provide the most information and allocate resources wisely. Consider techniques like optimal experimental design to maximize information gain with limited resources.

Inadequate Statistical Analysis

Issue: Insufficient or inappropriate statistical analysis can lead to incorrect conclusions or misinterpretations of results.

Remedy: Ensure that you have a strong statistical foundation for your sequential experimentation. Consult with a statistician if needed and use appropriate statistical tests and modeling techniques.

Neglecting External Factors

Issue: Focusing only on the factors directly related to the chemical analysis and ignoring external variables that may influence the results can lead to incomplete conclusions.

Remedy: Consider external factors that might impact the system under study. If possible, control or account for these factors in your experimental design.

Inadequate Documentation

Issue: Poor documentation of the sequential experimentation process can make it challenging to replicate or build upon your work.

Remedy: Maintain thorough records of each experimental iteration, including factors, conditions, results, and adjustments made. This documentation will facilitate future analysis and interpretation.

EXAMPLE OF A DESIGNED EXPERIMENT

A design matrix for experiments using SED in chemical analysis is a dynamic tool that evolves as the experiment progresses. It starts with an initial design and is continually updated based on the results of previous experiments. Here's an example of a design matrix for a SED in chemical analysis:

Objective: Let's consider an experiment where we want to optimize the conditions for the synthesis of a chemical compound. We have three factors to consider: temperature (T), concentration (C), and reaction time (R). The goal is to maximize the yield of the desired compound.

Design Matrix (Initial):

In the initial design matrix, you start with a small set of initial experiments to explore the factor space. This is often done using a factorial or fractional factorial design. For simplicity, we'll use a $2 \times 2 \times 2$ factorial design as the initial matrix (Table 2.28).

TABLE 2.28

Sequential Design Experiments (Initial Matrix) for Chemistry

Experiment	Temperature (T)	Concentration (C)	Reaction Time (R)
1	Low	Low	Short
2	High	Low	Short
3	Low	High	Short
4	High	High	Short
5	Low	Low	Long
6	High	Low	Long
7	Low	High	Long
8	High	High	Long

In this initial design matrix,

- You have 8 experiments that cover a range of factor combinations for temperature, concentration, and reaction time.
- The goal is to explore the factor space and gain an initial understanding of how these factors affect the yield.

Conducting Initial Experiments:
You perform these initial experiments and record the yields for each condition.

Analyzing Initial Results:
After analyzing the results of these initial experiments, you identify which factor combinations appear promising or yield higher.

Updating the Design Matrix (Sequential):
Based on the analysis of the initial results, you decide to focus on the conditions that seem to be more favorable. You can then update your design matrix for the next set of experiments (Table 2.29).
In this updated design matrix,

- You have selected specific conditions that appear to be more promising based on the initial results.
- The goal is to further explore and optimize the factor space in the direction that seems to yield higher.

Conducting Sequential Experiments:
You perform these sequential experiments according to the updated design matrix.

Continuing the Process:
The process continues iteratively, with the design matrix being updated after each round of experiments based on the latest results. You adjust your experimental conditions to focus on the most promising factor combinations until you achieve the desired yield or reach your stopping criteria.

Terminating the Experiment:
You decide to terminate the sequential experimentation when you achieve your optimization goal, or when further experiments no longer provide substantial improvements in yield.
This example demonstrates how a design matrix can evolve dynamically in sequential experimentation in chemical analysis to efficiently explore and optimize the factor space.

TABLE 2.29

Sequential Design Experiments (Updating Matrix) for Chemistry

Experiment	Temperature (T)	Concentration (C)	Reaction Time (R)
9	Medium	High	Short
10	Low	Low	Medium
11	Medium	Low	Medium
12	Medium	Medium	Medium

REPORT

When writing a report for an experiment conducted using sequential experimentation in designed experiments for chemical analysis, it's important to include various elements to effectively communicate the details of your study and its findings. Here are the key elements to include in your report:

Title Page:
- Title of the Report: Clearly state the purpose or objective of the experiment.
- Author(s): Name(s) of the researcher(s) and affiliation(s).
- Date: The date the report is being submitted.

Abstract: Provide a concise summary of the experiment's purpose, methods, key findings, and conclusions. The abstract should be brief and informative.

Table of Contents: Include a list of sections and subsections with page numbers for easy navigation.

List of Figures and Tables: Provide a list of all figures and tables included in the report with their corresponding page numbers.

Introduction:
- Explain the background and context of the experiment.
- State the research question or objective.
- Discuss the significance of the experiment and its relevance to the field of chemical analysis.
- Describe the motivation for using sequential experimentation.

Literature Review (if applicable): Provide a brief review of relevant literature and prior research related to the experiment's topic. Highlight key findings and gaps in existing knowledge.

Experimental Design:
- Describe the use of sequential experimentation in the study.
- Explain how the design evolved and was adjusted based on the results of previous experiments.
- Detail any initial design considerations and how they were modified over time.

Materials and Methods:
- List all the materials, chemicals, and equipment used in the experiment.
- Describe the step-by-step procedure followed in conducting the experiments, including any adjustments made during the sequential process.
- Include information on data collection, measurements, and any special techniques or procedures.

Results:
- Present the data obtained from the experiments in a clear and organized manner.
- Use tables, figures, and graphs to illustrate the results.
- Include statistical analyses if applicable (e.g., t-tests, regression).
- Show how the results evolved as the experiment progressed through sequential iterations.

Discussion:
- Interpret the results and discuss their implications.
- Address any trends, patterns, or significant findings observed during the sequential process.
- Compare the results to your initial hypotheses or expectations.

Discuss the advantages and limitations of using sequential experimentation in chemical analysis.

Conclusion:
- Summarize the key findings and their significance.
- State the implications of the results for the field of chemical analysis and any potential applications.

- Offer recommendations for future research or applications, including suggestions for further sequential experimentation.

References: Cite all sources of information, including research papers, textbooks, and articles, using a standardized citation style (e.g., APA, MLA, or a style recommended by your institution).

Appendices (if needed): Include any supplementary information that is important but too detailed or extensive for the main report. This may include raw data, calculations, additional figures, or any other relevant materials.

Acknowledgments (if applicable): Acknowledge individuals or organizations that contributed to the experiment but are not listed as authors.

List of Abbreviations and Symbols (if used): Provide a list of any abbreviations or symbols used throughout the report and their meanings.

Glossary (if needed): Define any technical terms or terminology specific to the field of chemical analysis.

Ensure that your report is well organized, clearly written, and adheres to the formatting and citation style guidelines provided by your institution or publication. Providing a thorough and structured report will help readers understand the experiment, its results, and the significance of using sequential experimentation in chemical analysis.

GOING FORWARD

The choice of design depends on several factors, including:

- **Number of Factors**: More complex designs handle a larger number of factors.
- **Desired Information**: Comprehensive analysis vs. initial screening or optimization.
- **Resource Constraints**: Experiment number and complexity limitations.
- **Experimental Context**: Specific needs of the chemical analysis being conducted.

By understanding these different types of designed experiments and considering the specific needs of your analysis, you can make an informed decision and leverage the power of experimentation to extract valuable insights and optimize your chemical analysis processes.

2.2.14 FORCED FAILURE METHOD

The forced failure method is a unique approach to stress-testing and understanding material limits in your chemical analysis, often used for

- Evaluating the stability and degradation mechanisms of materials: By deliberately inducing controlled stress or failure conditions, you can observe and analyze the breakdown processes.
- Identifying vulnerabilities and critical factors: Understanding how different factors like temperature, pressure, or concentration contribute to failure helps in optimizing or strengthening materials.
- Predicting long-term performance: By studying accelerated degradation under controlled conditions, you can estimate the material's lifespan under normal exposure.

Here's a breakdown of the steps involved in designing a forced failure experiment for your chemical analysis:

1. **Define the Material and Analysis Objectives**: Clearly state the material you're studying and your objectives for subjecting it to forced failure. Do you want to analyze specific degradation mechanisms, identify critical factors for stability, or predict lifespan under certain conditions?
2. **Choose the Forcing Mechanism**: Select a method to induce and control the desired stress on your material. This could involve elevated temperature, extreme pressure, chemical exposure, radiation, or a combination depending on your objectives and the material properties.
3. **Design the Experimental Matrix**: Consider the forcing mechanism and desired information. You might need:
   - Factorial design: To study the combined effects of multiple factors like temperature and pressure on material degradation.
   - Dose–response design: To analyze how increasing levels of the chosen stress factor affect the failure rate or time.
   - Time-series design: To monitor the degradation process over time under defined stress conditions.
4. **Determine Stress Levels and Monitoring Methods**: Define the intensity and duration of the chosen stress factor(s). Consider the range that pushes the material toward failure while allowing for observation and data collection. Choose appropriate techniques to monitor the degradation process, such as
   - Visual inspection for cracks or deformations.
   - Spectroscopic methods to track chemical changes.
   - Mechanical or physical property measurements to assess strength loss.
5. **Conduct the Experiment**: Apply the chosen stress levels to your material samples according to the designed matrix. Ensure consistent and controlled conditions throughout the experiment. Monitor the degradation process using your chosen methods and record data at appropriate intervals.
6. **Analyze the Data**: Utilize statistical methods and data visualization tools to analyze the collected data. Evaluate how different stress levels and combinations of factors influence the material's failure time, degradation mechanisms, and property changes. Identify significant factors and their interactions.
7. **Draw Conclusions and Recommendations**: Based on your analysis, draw conclusions about the material's response to the forced failure conditions. Identify critical factors contributing to degradation; predict lifespan under realistic conditions; and formulate recommendations for optimizing material composition, processing, or use conditions for improved stability and performance.

Additional Tips:

- **Replicates**: Consider including replicate samples for each stress level to improve data accuracy and statistical power.
- **Control Factors**: Account for and control any extraneous factors that might influence the results, such as environmental conditions or variations in sample preparation.
- **Modeling**: Depending on the data complexity, consider using mathematical models to describe the degradation process and predict material behavior under different conditions.

The forced failure method offers valuable insights into material behavior under stress but requires careful design and data analysis. Adapting these steps to your specific objectives and material properties will help you gain valuable knowledge about your chemical system and optimize its performance and lifespan. The method is typically used in mechanical and materials testing, where deliberately pushing a system to failure allows for analysis of its limitations and performance. While not directly applicable to most chemical analysis scenarios, you could potentially adapt its principles to investigate certain phenomena:

EXAMPLE: OPTIMIZING POLYMER DEGRADATION FOR WASTE RECYCLING USING THE FORCED FAILURE METHOD (FFM)

Objective: Identify the temperature and time conditions that lead to controlled degradation of a specific polymer waste material for efficient recycling.

Safety Considerations:

- Ensure all laboratory protocols and safety regulations are followed for handling the polymer and potential degradation products.
- Conduct the experiment in a controlled environment with appropriate ventilation and safety equipment.

Benefits:

- Accelerates understanding of the polymer's degradation behavior compared to real-world scenarios.
- Provides valuable data for optimizing recycling processes and maximizing resource recovery.
- Helps establish safe parameters for controlled degradation, minimizing harmful byproducts or environmental impact.

Modified Forced Failure Approach:

1. **Define Threshold**: Instead of complete failure, establish a measurable threshold indicating significant degradation (e.g., % decrease in molecular weight, change in surface morphology).
2. **Design Matrix**: Choose temperature and time factors with levels exceeding the expected useful range of the polymer. Use a statistically sound design (e.g., BBD) to explore interactions and optimize within the safe zone. Use a BBD with three levels for both temperature (40°C, 50°C, 60°C) and time (2 weeks, 4 weeks, 6 weeks) to capture potential interactions and optimize within a safe range. This results in 15 experimental runs involving different combinations of temperature and time.
3. **Controlled Degradation**: Expose polymer samples to the designated temperature and time combinations. Use enclosed reactors or appropriate containment measures to control the process and potential emissions.
   - Prepare Polylactic Acid (PLA) samples of equal size and weight.
   - Place each sample in a sealed reactor under the assigned temperature and time conditions.
   - Include control samples kept at room temperature without exposure for comparison.
4. **Monitor Degradation**: Analyze the samples at predetermined intervals using relevant techniques (e.g., gel permeation chromatography (GPC), surface microscopy).
   - Periodically remove samples from the reactors (e.g., every 2 days) and analyze them using GPC to determine their molecular weight.
   - Monitor carbon dioxide (CO_2) production as an indicator of biological degradation activity within the reactors.

5. **Data Analysis**: Identify combinations that reach the established degradation threshold within the designated time frame.
   - Periodically remove samples from the reactors (e.g., every 2 days) and analyze them using GPC to determine their molecular weight.
   - Monitor carbon dioxide (CO_2) production as an indicator of biological degradation activity within the reactors.
6. **Optimization**: Analyze the data to identify optimal temperature and time settings for controlled degradation while minimizing excessive breakdown and ensuring efficient recycling.
   - Periodically remove samples from the reactors (e.g., every 2 days) and analyze them using GPC to determine their molecular weight.
   - Monitor carbon dioxide (CO_2) production as an indicator of biological degradation activity within the reactors.

Further Steps:

- Validate the optimal conditions through additional experiments with larger PLA samples and under real-world composting conditions.
- Investigate the composition of the degraded PLA to assess its suitability for bioplastic recycling applications.
- Explore further factors like additives, pretreatments, or microorganisms to potentially enhance the degradation rate and quality of the recycled material.

Limitations:

- This modified approach focuses on accelerated degradation and may not fully replicate long-term composting behavior.
- The chosen threshold (50% reduction) might need adjustments based on specific recycling requirements.
- General safety precautions and responsible waste disposal are crucial throughout the experiment.
- Requires careful design and control to avoid true catastrophic failure or hazardous situations.
- Cannot directly predict long-term degradation behavior under real-world conditions.
- Might not be suitable for all types of chemical analysis or materials.

This example demonstrates how a modified forced failure approach, prioritizing safety and controlled conditions, could be adapted to gain insights into a chemical system's degradation behavior for optimization purposes. Further, it illustrates how a carefully designed and controlled forced failure approach can be adapted to gain valuable insights into polymer degradation for waste recycling optimization. Remember to prioritize safety, choose appropriate techniques, and interpret the results within the context of your specific application and recycling goals.

PITFALLS AND REMEDIES

The FFM offers a unique approach to assessing and improving the robustness of analytical processes by deliberately inducing stress or failure conditions. However, navigating its strengths and limitations requires cautiousness. Here are some pitfalls to avoid and remedies to apply when using FFM:

Pitfall: Misinterpreted failure modes. FFM relies on identifying and understanding various failure modes of the analytical process. Misinterpreting or overlooking critical failure modes can lead to inadequate robustness testing.
  Remedy:

- Conduct thorough process mapping and risk analysis to identify all potential failure points and modes.
- Consult literature and expert opinions to ensure comprehensive understanding of potential failure mechanisms.
- Validate identified failure modes through pilot experiments before designing the main FFM study.

Pitfall: Inappropriate stress factors and levels. Choosing the wrong stress factors or setting unrealistic stress levels can lead to irrelevant or misleading results.
  Remedy:

- Select stress factors that represent realistic operational variations or potential environmental challenges the process might encounter.
- Base stress levels on practical considerations while ensuring they effectively induce failure or significant deviations from desired performance.
- Pilot testing can help determine appropriate stress factors and levels for achieving informative FFM experiments.

Pitfall: Overlooking underlying causes. Identifying the root causes of failure under stress conditions is crucial for implementing effective improvement strategies.
  Remedy:

- Utilize complementary analytical techniques, such as statistical analysis, process monitoring data, and mechanistic studies, to investigate the underlying causes of failure under stress.
- Don't solely rely on observing the failure itself, but delve deeper into understanding the mechanisms behind it for targeted improvement strategies.

Pitfall: Neglecting cost and resource considerations. FFM experiments can be resource intensive due to deliberate failure induction and additional analytical techniques.
  Remedy:

- Prioritize critical stress factors and failure modes based on their impact and feasibility for FFM testing.
- Consider alternative cost-effective methods like accelerated aging tests or simulation models for initial robustness assessment.
- Implement FFM strategically to focus on the most relevant stress conditions and optimize resource allocation.

Pitfall: Misconstruing optimization strategies. FFM results alone might not offer direct solutions for robustness enhancement.
  Remedy:

- Combine FFM with other design methods like RSM or Taguchi to identify optimal settings for mitigating the effects of stress factors.
- Utilize the understanding of failure mechanisms to implement targeted strategies for process improvement, such as modifying critical process parameters or introducing additional control measures.

- Remember that FFM provides valuable insights for identifying weaknesses, but additional analysis and optimization techniques are often needed for achieving robust performance.

By recognizing these pitfalls and implementing the recommended remedies, you can leverage the FFM effectively to gain valuable insights into your analytical process's robustness, identify critical weaknesses, and develop targeted strategies for optimization and improved performance under challenging conditions. Remember that careful planning, rigorous data analysis, and critical interpretation of results are crucial for successful utilization of FFM and subsequent process improvement in your analytical research.

WRITING YOUR REPORT

A well-written report for designed experiments in analytical chemistry and product development serves not only as documentation but also as a clear roadmap of your investigative journey and its outcomes. Here are the key elements to include:

Introduction:

- Briefly state the goal of the study and the problem being addressed.
- Introduce the analytical technique or product under investigation.
- Mention the specific type of designed experiment used (e.g., factorial, RSM, Taguchi).

Experimental Section:

- Describe the materials and methods used in detail, including reagents, equipment, sample preparation, and design parameters.
- Clearly explain the chosen design and its rationale (e.g., factor selection, levels, replicates).
- Include specific procedures for data collection and analysis.

Results and Discussion:

- Present the experimental data in a clear and organized manner, using tables, graphs, and figures.
- Analyze the data statistically, highlighting significant effects and interactions between factors.
- Discuss the results in the context of the initial research question and scientific principles.
- Interpret the findings and explain their implications for analytical method optimization or product development.

Conclusion:

- Summarize the key findings and their significance.
- State the main conclusions drawn from the experiment.
- Discuss limitations of the study and suggest future directions for research.

Appendices:

- Include raw data tables, statistical analysis outputs, and detailed experimental protocols.
- Add any visual aids not included in the main body of the report, such as additional figures or spectra.

You always want to consider the intended audience and tailor the report's depth and detail accordingly. Look to use clear and concise language, avoiding technical jargon when possible. Look to maintain a logical flow throughout the report, guiding the reader from problem to solution. It is important to proofread carefully for errors in grammar and scientific facts and send your draft to staff or trusted peers for review if the report is going to be published in an article or manuscript.

By including these essential elements and adhering to best practices in scientific writing, you can create a comprehensive and informative report that showcases the valuable insights gained from your designed experiments in analytical chemistry and product development. A well-structured and detailed report not only serves as a valuable record of your work but also contributes to the advancement of knowledge in your field and potentially paves the way for further research and development.

2.3 ENVIRONMENTAL SCIENCE

In environmental science, designed experiments play a critical role in studying and understanding environmental processes, assessing the impacts of pollutants, and optimizing environmental management strategies. Here are some common types of designed experiments with examples of their applications in environmental science.

In environmental science designed experiments, various types of variables are considered to investigate environmental processes, assess impacts, and study ecological systems. These variables can be categorized into several common types, including:

1. **IVs:**
    - **Environmental Factors**: Variables related to environmental conditions and factors, such as temperature, humidity, light intensity, air quality, water quality, soil composition, and pollutant concentrations.
    - **Land-Use Changes**: Variables related to alterations in land use or land cover, including urbanization, deforestation, agriculture, and land restoration.
    - **Experimental Treatments**: Variables related to experimental manipulations or treatments applied to ecosystems or environmental systems, such as nutrient additions, introduction of species, or restoration efforts.
2. **DVs:**
    - Environmental Responses: Variables related to responses and changes within environmental systems, including changes in vegetation, biodiversity, water flow, soil erosion, and pollutant levels.
    - Ecosystem Health Indicators: Variables related to the health and condition of ecosystems, such as species richness, community composition, ecosystem productivity, and habitat quality.
    - Physical Measurements: Variables related to physical properties of the environment, such as temperature, pH, turbidity, dissolved oxygen, and soil moisture.
    - Chemical Concentrations: Variables related to the concentrations of chemicals and pollutants in environmental media, including air, water, soil, and sediments.
3. **Categorical Variables:**
    - Habitat Types: Categories representing different types of habitats; ecosystems; or land cover classes, such as forests, wetlands, grasslands, or urban areas.
    - Sampling Locations: Categories representing different sampling sites or geographic locations within a study area.
    - Seasons or Time Periods: Categories representing different seasons or time periods when data were collected, which can influence environmental conditions.
4. **Control Variables (Covariates)**: These are variables that are held constant or controlled during experiments to eliminate their influence on the DV. For example, maintaining constant temperature or soil moisture levels during an experiment.

5. **Random Variables**: In some cases, random variables may be introduced to account for variability or uncertainty in environmental measurements or to control for natural variability in ecological systems.
6. **Interaction Variables**: Interaction variables are used to investigate the combined effects of two or more IVs on the DV, helping to understand how different factors interact to influence environmental processes.
7. **Noise Variables (Error Variables)**: These are uncontrolled or unmeasured variables that can introduce variability or errors into the experimental results. Statistical analysis and experimental controls are used to minimize the impact of noise variables in environmental science experiments.

Environmental science designed experiments are crucial for understanding and managing environmental issues, conserving biodiversity, and assessing the impacts of human activities on natural ecosystems. Careful consideration and control of variables are essential for obtaining reliable and meaningful data in environmental science experiments.

2.3.1 Factorial Experiments

Factorial design experiments are widely used in environmental science for studying complex systems, assessing the impacts of multiple variables, and understanding interactions among various factors affecting environmental processes. Here are some areas where factorial design experiments are commonly applied in environmental science:

Pollution Control and Remediation: Factorial design experiments can be used to evaluate the effectiveness of different pollution control strategies or remediation techniques. For example, researchers might design experiments to assess the impact of multiple factors, such as soil type, contaminant concentration, and treatment method on the efficiency of remediation processes like phytoremediation or bioremediation.

Ecological Studies: Factorial experiments are valuable for investigating the interactions between multiple environmental factors on ecological systems. Researchers might use factorial designs to study the combined effects of factors, such as temperature, precipitation, nutrient availability, and species interactions with ecosystem dynamics, biodiversity, and community structure.

Climate Change Research: In studies related to climate change, factorial design experiments can help scientists understand how multiple variables (e.g., temperature, CO_2 concentration, precipitation patterns) interact and influence ecosystem responses, including shifts in species distributions, changes in phenology, and alterations in carbon and nutrient cycling.

Aquatic Ecosystems: Factorial experiments are utilized to study various factors affecting aquatic ecosystems such as water quality, nutrient levels, temperature, and the presence of contaminants. These experiments can help in understanding the impacts of human activities, such as pollution or habitat alteration, on aquatic biodiversity and ecosystem functioning.

Land Use and Management: In studies related to land use and management practices, factorial experiments can assess the effects of different land management strategies (e.g., agricultural practices, forestry techniques) and land-use changes (e.g., deforestation, urbanization) on soil properties, water quality, biodiversity, and ecosystem services.

Environmental Toxicology: Factorial design experiments are employed in environmental toxicology to assess the combined effects of multiple stressors (e.g., pollutants, environmental variables) on organisms, populations, and ecosystems. These experiments help in understanding synergistic or antagonistic interactions between stressors and their implications for environmental health.

Overall, factorial design experiments provide a systematic approach to studying the complex relationships between multiple environmental factors, enabling researchers to better understand and manage environmental systems and address pressing environmental challenges.

STEPS

Developing factorial design experiments in environmental science involves several steps to ensure a well-designed and statistically sound study. Here's an outline of the typical steps involved:

1. **Define the Research Objectives**: Clearly articulate the research questions or objectives that the factorial experiment aims to address. Identify the specific environmental factors or variables of interest that will be manipulated and measured.
2. **Identify Factors and Levels**: Determine the factors (IVs) that will be included in the experiment. These factors should represent the environmental variables that are hypothesized to influence the response variable(s) of interest. For each factor, identify the levels or settings at which it will be manipulated.
3. **Select the Experimental Design**: Choose an appropriate factorial design based on the number of factors and levels, as well as the desired level of experimental control and efficiency. Common factorial designs include full factorial designs, fractional factorial designs, and hierarchical factorial designs.
4. **Determine the Experimental Layout**: Decide on the arrangement of experimental treatments and the allocation of factor combinations to experimental units (e.g., plots, sampling sites). Randomization and replication should be incorporated into the experimental layout to minimize bias and increase the reliability of the results.
5. **Design Experimental Treatments**: Develop the specific treatments or combinations of factor levels that will be applied to the experimental units. Ensure that the treatments cover the full range of factor combinations and adequately represent the factorial design.
6. **Plan Data Collection and Measurements**: Define the response variable(s) that will be measured to assess the effects of the experimental treatments. Determine the methods and protocols for data collection, including sampling procedures, measurement techniques, and data recording formats.
7. **Consider Environmental Constraints**: Take into account any environmental constraints or practical limitations that may impact the design and implementation of the experiment, such as spatial heterogeneity, temporal variability, or logistical constraints.
8. **Pilot Testing and Validation**: Conduct pilot tests or preliminary experiments to validate the experimental design, assess the feasibility of data collection procedures, and identify any potential issues or challenges that need to be addressed before the full-scale experiment.
9. **Implement the Experiment**: Carry out the factorial experiment according to the planned experimental design and protocols. Ensure proper execution of the experimental treatments, data collection procedures, and quality control measures.
10. **Data Analysis and Interpretation**: Analyze the collected data using appropriate statistical methods for factorial designs, such as ANOVA or regression analysis. Interpret the results to evaluate the effects of the manipulated factors on the response variable(s) and draw conclusions regarding the research objectives.
11. **Communicate Findings**: Communicate the findings of the factorial experiment through scientific publications, presentations, or reports, highlighting the key results, implications for environmental science or management, and potential avenues for future research.

By following these steps, researchers can systematically plan and execute factorial design experiments in environmental science, facilitating the investigation of complex relationships between environmental factors and their effects on ecosystems, organisms, or environmental processes.

Pitfalls and Remedies

Factorial design experiments in environmental science offer many advantages, but they also come with potential pitfalls. Here are some common pitfalls and remedies to consider when conducting factorial design experiments in this field:

Incomplete Factorial Design:
> **Pitfall:** Running an incomplete factorial design where not all possible combinations of factors and levels are tested. This can lead to missing important interactions or main effects.
>
> **Remedy:** Ensure that the experimental design includes all relevant factor combinations. Use full factorial designs whenever possible to systematically test all possible combinations.

Small Sample Size:
> **Pitfall:** Using a small sample size, which can reduce the statistical power of the experiment and limit the ability to detect significant effects.
>
> **Remedy:** Conduct power analysis to determine the appropriate sample size for detecting the expected effect sizes with sufficient statistical power. Increase replication and sample size to improve the reliability of the results.

NonRandomized Experimental Design:
> **Pitfall:** Failing to randomize the assignment of treatments to experimental units, leading to potential bias or confounding effects.
>
> **Remedy:** Randomize the allocation of treatments to experimental units to minimize bias and ensure the validity of statistical inference. Use randomization procedures, such as complete randomization orRCBDs.

Failure to Consider Environmental Variability:
> **Pitfall:** Ignoring spatial or temporal variability in environmental conditions, which can confound experimental results and reduce the generalizability of findings.
>
> **Remedy:** Account for environmental variability by incorporating appropriate blocking or stratification techniques in the experimental design. Collect data over multiple spatial or temporal scales to capture variability and assess its impact on experimental outcomes.

Lack of Control over Confounding Variables:
> **Pitfall:** Failing to control for confounding variables that may influence the response variable(s) of interest, leading to spurious or misleading results.
>
> **Remedy:** Identify potential confounding variables and either control them experimentally or statistically through blocking, stratification, or covariate adjustment. Use factorial designs with nested factors to account for hierarchical structures and nested sources of variability.

Overlooking Assumptions of Statistical Tests:
> **Pitfall:** Applying statistical tests without verifying the underlying assumptions, such as normality or homogeneity of variances, which may invalidate the results.
>
> **Remedy:** Check the assumptions of statistical tests using diagnostic plots, transformation techniques, or robust statistical methods. Choose appropriate statistical tests that are robust to violations of assumptions.

Ignoring Interactions:
> **Pitfall:** Focusing solely on main effects and overlooking interactions between factors, which may result in incomplete or misleading conclusions about the relationships between variables.
>
> **Remedy:** Assess interactions between factors using appropriate statistical methods, such as factorial ANOVA or regression analysis. Interpret and report interaction effects along with main effects to provide a comprehensive understanding of the experimental results.

By being aware of these potential pitfalls and implementing appropriate remedies, researchers can enhance the rigor, validity, and reliability of factorial design experiments in environmental science, leading to more robust conclusions and insights into complex environmental processes and systems.

EXAMPLE

Factorial experiments in environmental science are used to investigate the effects of multiple IVs (factors) on a DV while considering all possible combinations of factor levels. Here's an example of a matrix for a factorial experiment in environmental science:

Objective: To study the combined effects of two environmental factors, temperature (factor A) and humidity (factor B), on the growth rate of a specific plant species.

Factors:

1. **Temperature (Factor A):**
   - Low temperature (e.g., 20°C).
   - Moderate temperature (e.g., 25°C).
   - High temperature (e.g., 30°C).
2. **Humidity (Factor B):**
   - Low humidity (e.g., 40%).
   - Moderate humidity (e.g., 60%).
   - High humidity (e.g., 80%).

DV: Growth rate of the plant (measured in cm/week).

Factorial Experiment Matrix:
In a factorial experiment, all possible combinations of factor levels are tested. The matrix is shown in Table 2.30.

In this matrix, each row represents a specific combination of temperature and humidity levels, and the growth rate of the plant is measured for each combination.

Factorial experiments allow researchers to examine how multiple factors interact and influence the DV. In this environmental science example, the goal is to understand how temperature and humidity jointly affect the growth rate of the plant species.

TABLE 2.30

Factorial-Experiment Design Example for Environmental Science

Run	Temperature (A)	Humidity (B)	Growth Rate (cm/week)
1	Low	Low	...
2	Low	Moderate	...
3	Low	High	...
4	Moderate	Low	...
5	Moderate	Moderate	...
6	Moderate	High	...
7	High	Low	...
8	High	Moderate	...
9	High	High	...

REPORTING

Factorial design experiments in environmental science often involve the use of various statistical tools to analyze the data and draw conclusions about the effects of multiple factors on the response variable(s). Here are some commonly used statistical tools in factorial design experiments:

ANOVA: ANOVA is a fundamental statistical technique used to analyze the differences in means among multiple groups or treatments. In factorial design experiments, ANOVA is employed to assess the significance of main effects (individual factors) and interaction effects (combinations of factors) on the response variable(s).

Factorial ANOVA: Factorial ANOVA extends the basic ANOVA framework to factorial designs with multiple factors. It allows researchers to simultaneously evaluate the effects of multiple factors and their interactions with the response variable(s), providing insights into the relative importance of each factor and their combined effects.

Regression Analysis: Regression analysis is used to model the relationship between the IVs (factors) and the DV (response variable). In factorial design experiments, multiple regression analysis or factorial regression analysis can be employed to quantify the effects of individual factors and their interactions with the response variable(s).

Contrast Analysis: Contrast analysis is a method used to test specific hypotheses about the differences between treatment means. It allows researchers to compare selected combinations of factor levels and assess the significance of specific contrasts, providing more focused insights into the effects of interest.

Post-Hoc Tests: Post-hoc tests are used to perform pairwise comparisons between treatment means following the identification of significant effects in ANOVA or regression analysis. Common post-hoc tests include Tukey's HSD(, Bonferroni correction, Scheffé's method, and Dunnett's test, among others.

Interaction Plots: Interaction plots are graphical representations that illustrate the interactions between factors in factorial design experiments. These plots help visualize the nature and strength of interactions and aid in interpreting the results of the experiment.

Power Analysis: Power analysis is used to determine the statistical power of an experiment, which quantifies the likelihood of detecting a true effect when it exists. Power analysis helps researchers assess the adequacy of sample size and experimental design to detect meaningful effects with sufficient confidence.

Residual Analysis: Residual analysis involves examining the residuals (the differences between observed and predicted values) to assess the assumptions of the statistical model, such as normality, homoscedasticity, and independence of errors. Residual plots and diagnostic tests help identify potential violations of model assumptions and guide model refinement.

By applying these statistical tools appropriately, researchers can effectively analyze factorial design experiments in environmental science, identify significant effects, and gain insights into the complex relationships between environmental factors and response variables.

A report on factorial design experiments in environmental science typically includes several key elements to effectively communicate the research objectives, methods, results, and conclusions to the intended audience. Here are the essential elements of such a report:

1. **Title Page**: The title page provides the title of the report, the names of the authors, their affiliations, and the date of publication.
2. **Abstract**: The abstract summarizes the main objectives of the study, the experimental design, key findings, and conclusions in a concise manner. It provides readers with a brief overview of the research without having to read the entire report.

3. **Introduction**:
   - Background and Literature Review: Provide background information on the research topic, including relevant literature and previous studies. Explain the significance of the research and the rationale for conducting the factorial design experiment.
   - Research Objectives and Hypotheses: Clearly state the research objectives and hypotheses that the experiment aims to address. Outline the specific factors, levels, and response variables under investigation.
4. **Methods:**
   - Experimental Design: Describe the factorial design used in the experiment, including the factors, levels, and number of replicates. Explain how treatments were assigned to experimental units and any randomization procedures employed.
   - Data Collection: Detail the methods and protocols used to collect data on the response variable(s), including measurement techniques, sampling procedures, and data recording formats.
   - Statistical Analysis: Outline the statistical methods used to analyze the data, such as ANOVA, regression analysis, contrast analysis, or other relevant techniques. Specify any assumptions made and how they were validated.
5. **Results:**
   - Descriptive Statistics: Present summary statistics (e.g., means, standard deviations) for each treatment group and factor combination.
   - Analysis of Main Effects: Report the results of main effects analysis, including significance tests and effect sizes.
   - Analysis of Interaction Effects: Present the results of interaction analysis, including interaction plots and significance tests for interaction effects.
   - Post-Hoc Tests: If applicable, report the results of post-hoc tests for pairwise comparisons between treatment means.
6. **Discussion:**
   - Interpretation of Findings: Interpret the results in the context of the research objectives and hypotheses. Discuss the implications of significant main effects and interaction effects.
   - Comparison with Previous Studies: Compare the findings of the current study with previous research in the field. Highlight any consistencies or discrepancies and explain possible reasons for differences.
   - Limitations and Future Directions: Acknowledge any limitations of the study and suggest areas for future research or improvements to experimental design.
7. **Conclusion**: Summarize the main findings of the study and their implications for understanding the relationships between environmental factors and response variables. Restate the significance of the research and its contribution to the field.
8. **References**: Provide a list of references cited in the report, following a specific citation style (e.g., APA, MLA).
9. **Appendices**: Include any supplementary materials, such as raw data, detailed statistical analyses, experimental protocols, or additional figures and tables.

By including these elements in a factorial design experiment report, researchers can effectively communicate their findings, methods, and conclusions to the scientific community and other stakeholders in the field of environmental science.

2.3.2 RSM

RSM designed experiments find application in various areas within environmental science due to their ability to optimize processes, understand complex relationships, and model responses to multiple factors simultaneously. Some specific applications of RSM in environmental science include:

Optimization of Remediation Processes: RSM is used to optimize remediation techniques for contaminated sites, such as soil or groundwater remediation. By simultaneously varying factors like pH, temperature, and treatment duration, RSM helps in determining the optimal conditions for maximizing contaminant removal efficiency while minimizing costs and environmental impact.

Modeling and Optimization of Bioremediation: RSM is applied to model and optimize bioremediation processes, where microorganisms are used to degrade pollutants. Factors such as nutrient concentration, oxygen levels, and microbial population densities are varied to maximize pollutant degradation rates and enhance bioremediation efficiency.

Water Treatment and Purification: RSM is employed in the optimization of water treatment processes, such as coagulation, flocculation, and filtration, to remove pollutants and improve water quality. By optimizing factors like coagulant dosage, mixing intensity, and pH, RSM helps in achieving the desired water treatment outcomes efficiently.

Modeling and Optimization of Air Pollution Control Systems: RSM is utilized to model and optimize air pollution control systems, such as scrubbers, filters, and catalytic converters, for reducing emissions of pollutants like particulate matter, volatile organic compounds (VOCs), and nitrogen oxides (NOx). Factors, such as temperature, residence time, and chemical dosages, are varied to maximize pollutant removal efficiency.

Optimization of Waste Management Processes: RSM is applied to optimize waste management processes, such as composting, anaerobic digestion, and waste-to-energy conversion. By varying factors like temperature, moisture content, and feedstock composition, RSM helps in maximizing waste conversion rates, minimizing environmental impacts, and optimizing resource recovery.

Environmental Monitoring and Modeling: RSM is used to model and optimize environmental monitoring networks and sampling strategies. By considering factors, such as sampling frequency, spatial distribution of sampling sites, and analytical techniques, RSM helps in designing cost-effective monitoring programs that provide accurate assessments of environmental conditions and trends.

Overall, RSM designed experiments find wide-ranging applications in environmental science for optimizing processes, understanding complex relationships, and solving environmental challenges effectively and efficiently.

Steps

Developing RSM designed experiments in environmental science involves several steps to optimize processes, understand relationships between variables, and model responses to multiple factors. Here's an outline of the typical steps involved:

1. **Define the Research Objectives**: Clearly articulate the research objectives or optimization goals that the RSM experiment aims to achieve. Identify the process or system of interest and the specific factors (IVs) that influence the response variable(s).
2. **Identify Factors and Levels**: Determine the factors to be included in the RSM experiment and the range of levels for each factor. Factors should represent variables that are hypothesized to affect the response variable(s), such as environmental conditions, process parameters, or treatment variables.
3. **Select the Experimental Design**: Choose an appropriate experimental design for the RSM study based on the number of factors, levels, and desired level of precision. Common designs include CCD, BBDs, and Doehlert designs, which allow for efficient exploration of response surfaces.

4. **Generate Experimental Design Points**: Use statistical software or design tables to generate the experimental design points according to the chosen RSM design. Design points should cover the experimental space adequately and include a combination of center points, axial points, and factorial points to estimate main effects and quadratic terms accurately.
5. **Conduct Experiments**: Implement the experimental design by conducting experiments according to the specified factor levels. Follow standardized protocols for data collection, measurement, and recording to ensure consistency and reproducibility across experimental runs.
6. **Collect Response Data**: Measure the response variable(s) of interest for each experimental run. Record the response data accurately and ensure that any sources of variability or error are minimized to obtain reliable and precise measurements.
7. **Fit Response Surface Models**: Use regression analysis or other modeling techniques to fit response surface models to the experimental data. Develop models that describe the relationship between the response variable(s) and the IVs (factors) in the experimental design, including linear, quadratic, and interaction terms.
8. **Assess Model Adequacy**: Evaluate the adequacy of the response surface models using diagnostic tests, such as residual analysis, lack-of-fit tests, and goodness-of-fit measures. Ensure that the models adequately capture the observed variability in the response data and provide accurate predictions within the experimental space.
9. **Optimize Response Variables**: Use optimization techniques, such as desirability functions, numerical optimization algorithms, or graphical methods, to identify optimal factor combinations that maximize or minimize the response variable(s) within specified constraints or targets.
10. **Validate Experimental Results**: Validate the optimized factor settings by conducting confirmatory experiments or validation trials under the recommended conditions. Verify that the predicted responses match the observed outcomes and assess the robustness of the optimized conditions to changes in environmental conditions or experimental parameters.
11. **Interpret Results and Draw Conclusions**: Interpret the results of the RSM experiment in the context of research objectives and optimization goals. Draw conclusions about the relationships between factors and response variables, identify optimal operating conditions, and propose recommendations for process improvement or further research.

By following these steps, researchers can systematically develop RSM designed experiments in environmental science to optimize processes, understand complex relationships, and achieve desired outcomes efficiently and effectively.

PITFALLS AND REMEDIES

RSM designed experiments in environmental science, like any experimental approach, can encounter pitfalls that may affect the validity and reliability of results. Here are some common pitfalls and remedies associated with RSM experiments in this field:

Insufficient Experimental Design Resolution:
Pitfall: Using an inadequate number of experimental runs or insufficient resolution in the design may lead to inaccurate estimation of model coefficients and poor prediction accuracy.

Remedy: Use appropriate experimental designs, such as CCD or BBD, with a sufficient number of runs and resolution to capture the curvature and interactions in the response surface.

Violation of Model Assumptions:

Pitfall: Violating assumptions of the response surface models, such as linearity, normality, or constant variance, may result in biased parameter estimates and unreliable predictions.

Remedy: Validate the assumptions of the response surface models using diagnostic tests, such as residual analysis, lack-of-fit tests, and goodness-of-fit measures. Transform the response variable if necessary to meet the assumptions of the model.

Lack of Model Adequacy:

Pitfall: Fitting overly simplistic or inadequate response surface models that fail to capture the true relationships between factors and response variables may lead to misleading conclusions and suboptimal recommendations.

Remedy: Use diagnostic techniques to assess the adequacy of response surface models, including residual analysis, lack-of-fit tests, and cross-validation. Consider alternative model structures or higher-order terms if the initial model fails to adequately describe the data.

Overfitting of Response Surface Models:

Pitfall: Overfitting the response surface models by including an excessive number of terms or parameters may result in models that are overly complex and have poor generalization to new data.

Remedy: Use model selection techniques, such as stepwise regression, , or cross-validation, to identify the most parsimonious model that adequately explains the data without overfitting. Avoid including unnecessary higher-order terms or interactions.

Extrapolation beyond Experimental Space:

Pitfall: Extrapolating response surface models beyond the experimental design space may lead to unreliable predictions and erroneous conclusions, especially if the extrapolation involves factors or conditions not tested in the experiment.

Remedy: Limit predictions and interpretations to the range of factor levels and conditions explored in the experimental design. Exercise caution when making predictions outside the experimental space and consider conducting additional experiments to validate predictions in unexplored regions.

Inadequate Validation of Optimal Conditions:

Pitfall: Failing to validate the optimal factor settings identified through RSM optimization may lead to suboptimal outcomes and ineffective process improvements.

Remedy: Validate the optimal conditions through confirmatory experiments or validation trials under real-world conditions. Assess the robustness of the optimized settings to variations in environmental factors, operational parameters, and measurement uncertainties.

By being aware of these potential pitfalls and implementing appropriate remedies, researchers can enhance the reliability, validity, and usefulness of RSM designed experiments in environmental science, leading to more accurate predictions, optimal process optimization, and informed decision-making.

EXAMPLE

RSM is a statistical technique used to optimize processes and investigate the relationships between multiple IVs (factors) and a response variable. Here's an example of a matrix for an RSM designed experiment in environmental science:

Objective: To optimize the conditions for a wastewater treatment process by studying the effects of two IVs, pH level (factor A) and temperature (factor B), on the removal efficiency of a specific pollutant.

Factors:

1. **pH level (Factor A):**
   - Low pH (e.g., 5).
   - Medium pH (e.g., 7).
   - High pH (e.g., 9).
2. **Temperature (Factor B):**
   - Low temperature (e.g., 20°C).
   - Medium temperature (e.g., 30°C).
   - High temperature (e.g., 40°C).

Response Variable: Pollutant Removal Efficiency (%).

RSM Experiment Matrix:

In an RSM experiment, a matrix is designed to systematically explore the combinations of factor levels, including different levels and interactions. The matrix is shown in Table 2.31.

In this matrix, each row represents a specific experimental run with a unique combination of pH level and temperature. The response variable, pollutant removal efficiency, is measured for each combination.

The purpose of RSM is to model and analyze the relationship between the IVs and the response variable to identify the optimal conditions for maximum pollutant removal efficiency. Statistical techniques, such as regression analysis and surface plots, are used to develop predictive models and optimize the environmental process.

Another example of an RSM designed experiment in environmental science, this time focuses on optimizing the conditions for maximizing biogas production in anaerobic digestion of organic waste. In this experiment, we will vary two factors: temperature and retention time. We'll again use a CCD to explore the response surface.

Factors:

1. **Temperature:**
   - Level 1: Low temperature (35°C).
   - Level 2: High temperature (55°C).
2. **Retention Time:**
   - Level 1: Short retention time (10 days).
   - Level 2: Long retention time (30 days).

TABLE 2.31

RSM Example for Environmental Science

Run	pH Level (A)	Temperature (B)	Pollutant Removal Efficiency (%)
1	Low	Low	...
2	Low	Medium	...
3	Low	High	...
4	Medium	Low	...
5	Medium	Medium	...
6	Medium	High	...
7	High	Low	...
8	High	Medium	...
9	High	High	...

TABLE 2.32
RSM Example for Environmental Science

Run	Temperature (°C)	Retention Time (days)	Biogas Production
1	−1	−1	Y1
2	+1	−1	Y2
3	−1	+1	Y3
4	+1	+1	Y4
5	−1.682	0	Y5
6	+1.682	0	Y6
7	0	−1.682	Y7
8	0	+1.692	Y8
9	0	0	Y9

The design matrix for this experiment can be generated using the CCD with 2 levels for each factor, plus 5 additional center points to estimate curvature. We'll also include axial points to assess curvature along each factor axis (Table 2.32).

In this design matrix,

- Runs 1 to 4 represent the factorial points, where both factors are at their low or high levels.
- Runs 5 to 8 represent the axial points, where one factor is at its mid-level (0) and the other factor is at an extreme level.
- Run 9 represents the center point, where both factors are at their mid-levels.

For each experimental run, the biogas production is measured and recorded. The resulting data are then used to fit a response surface model, which describes the relationship between biogas production and temperature, retention time, and their interaction. This model can then be analyzed and optimized to identify the optimal conditions for maximizing biogas production in anaerobic digestion processes.

REPORTING

RSM designed experiments in environmental science utilize various statistical tools to analyze data, model response surfaces, optimize processes, and make informed decisions. Some commonly used statistical tools in RSM experiments include:

ANOVA: ANOVA is used to assess the significance of factors and their interactions with the response variable(s). It helps determine whether the fitted response surface model adequately describes the observed variation in the data.

Regression Analysis: Regression analysis is employed to fit response surface models to the experimental data. This involves estimating coefficients for linear, quadratic, and interaction terms, which describe the relationships between factors and response variables.

DOE: DOE techniques are used to plan, execute, and analyze RSM experiments efficiently. DOE helps in selecting appropriate experimental designs (e.g., CCD, BBD) and generating optimal experimental conditions for exploring the response surface.

RSM: RSM itself is a statistical technique used to model and optimize response surfaces. It involves fitting mathematical models to experimental data to describe the relationships between factors and response variables, identifying optimal factor settings, and conducting response surface analysis.

Optimization Algorithms: Optimization algorithms are used to identify optimal factor settings that maximize or minimize the response variable(s) within specified constraints. These algorithms include gradient-based methods (e.g., gradient descent), heuristic search algorithms (e.g., genetic algorithms), and numerical optimization techniques (e.g., simplex method).

Analysis of Residuals: Residual analysis is performed to assess the adequacy of the fitted response surface model. It involves examining the residuals (the differences between observed and predicted values) to identify patterns, outliers, or violations of model assumptions.

Diagnostic Plots: Diagnostic plots, such as normal probability plots, scatter plots of residuals versus predicted values, and plots of residuals versus factor levels, are used to visually assess the assumptions of the response surface model and identify any issues with model fit or specification.

Model Selection Criteria: Model selection criteria, such as AIC or Bayesian information criterion (BIC), are used to compare competing response surface models and select the most parsimonious model that adequately explains the data while avoiding overfitting.

Validation Techniques: Validation techniques, such as cross-validation or bootstrapping, are used to assess the predictive performance of the response surface model and validate the optimized factor settings under new or unseen conditions.

By employing these statistical tools effectively, researchers can analyze response surface data, optimize experimental conditions, and make evidence-based decisions to improve processes and achieve desired outcomes in environmental science applications of RSM.

A report on RSM designed experiments in environmental science typically includes several key elements to effectively communicate the research objectives, methods, results, and conclusions to the intended audience. Here are the essential elements of such a report:

Title Page: The title page provides the title of the report, the names of the authors, their affiliations, and the date of publication.

Abstract: The abstract summarizes the main objectives of the study, the experimental design, key findings, and conclusions in a concise manner. It provides readers with a brief overview of the research without having to read the entire report.

Introduction:
- Background and Literature Review: Provide background information on the research topic, including relevant literature and previous studies. Explain the significance of the research and the rationale for employing RSM to optimize processes or explore relationships between factors and response variables.
- Research Objectives and Hypotheses: Clearly state the research objectives and hypotheses that the RSM experiment aims to address. Outline the specific factors, levels, and response variables under investigation.

Methods:
- Experimental Design: Describe the experimental design used in the RSM study, including the selection of factors, levels, and experimental design type (e.g., CCD, BBD). Explain how the experimental runs were generated and conducted.
- Data Collection: Detail the methods and protocols used to collect data on the response variable(s), including measurement techniques, sampling procedures, and data recording formats.
- Statistical Analysis: Outline the statistical methods used to analyze the data, such as regression analysis, ANOVA, optimization algorithms, and diagnostic techniques. Specify any assumptions made and how they were validated.

Results:
- Descriptive Statistics: Present summary statistics (e.g., means, standard deviations) for each experimental run and factor combination.
- Response Surface Analysis: Present the fitted response surface model and describe the relationships between factors and response variables. Discuss the significance of main effects, quadratic terms, and interaction effects.
- Optimization Results: Report the optimal factor settings identified through RSM optimization techniques and discuss their implications for process improvement or optimization.

Discussion:
- Interpretation of Findings: Interpret the results of the RSM experiment in the context of the research objectives and hypotheses. Discuss the implications of significant factors, interactions, and optimized conditions for environmental science applications.
- Comparison with Previous Studies: Compare the findings of the current study with previous research in the field. Highlight any consistencies or discrepancies and explain possible reasons for differences.
- Limitations and Future Directions: Acknowledge any limitations of the study and suggest areas for future research or improvements to experimental design.

Conclusion: Summarize the main findings of the study and their implications for understanding relationships between factors and response variables in environmental science applications of RSM. Restate the significance of the research and its contribution to the field.

References: Provide a list of references cited in the report, following a specific citation style (e.g., APA, MLA).

Appendices: Include any supplementary materials, such as raw data, detailed statistical analyses, experimental protocols, or additional figures and tables.

By including these elements in a report on RSM designed experiments in environmental science, researchers can effectively communicate their findings, methods, and conclusions to the scientific community and other stakeholders, facilitating the dissemination of knowledge and advancement of the field.

2.3.3 FRACTIONAL FACTORIAL EXPERIMENTS

Fractional factorial design experiments find applications in various areas within environmental science where researchers need to assess the effects of multiple factors on a response variable efficiently. Some common applications of fractional factorial design experiments in environmental science include:

Pollution Control and Remediation: Fractional factorial design experiments are used to assess the effectiveness of various pollution control and remediation strategies. Researchers can simultaneously evaluate the impact of multiple factors, such as treatment methods, environmental conditions, and pollutant characteristics, on the efficiency of remediation processes, such as phytoremediation, bioremediation, and chemical treatment.

Environmental Monitoring and Sampling Design: Fractional factorial design experiments are employed to optimize environmental monitoring programs and sampling designs. Researchers can assess the influence of factors, such as sampling frequency, sampling locations, and analytical methods, on the accuracy and efficiency of environmental data collection, allowing for cost-effective monitoring strategies that provide reliable assessments of environmental conditions.

Ecosystem Studies and Ecological Research: Fractional factorial design experiments are used to investigate the effects of multiple factors on ecosystem structure and function. Researchers can assess the influence of various environmental variables, such as temperature, precipitation, nutrient availability, and habitat complexity, on ecological processes, species interactions, and biodiversity patterns, helping to understand and manage ecosystems more effectively.

Climate Change and Global Environmental Change: Fractional factorial design experiments are employed to study the impacts of climate change and global environmental change on ecosystems, habitats, and species. Researchers can assess the combined effects of multiple stressors, such as temperature increases, altered precipitation patterns, habitat loss, and pollution, on ecosystem resilience, species distributions, and community dynamics, informing adaptation and mitigation strategies.

Environmental Toxicology and Risk Assessment: Fractional factorial design experiments are used in environmental toxicology and risk assessment studies to evaluate the effects of multiple chemical stressors on organisms and ecosystems. Researchers can assess the interactions between pollutants, exposure pathways, and environmental conditions, helping to identify potential synergistic or antagonistic effects and inform regulatory decisions and environmental management practices.

Waste Management and Resource Recovery: Fractional factorial design experiments are employed to optimize waste management processes and resource recovery technologies. Researchers can assess the influence of factors, such as waste composition, treatment methods, and operating conditions, on the efficiency of waste-treatment, recycling, and energy-recovery processes, facilitating the development of sustainable waste management strategies.

Overall, fractional factorial design experiments are valuable tools in environmental science for efficiently assessing the effects of multiple factors on environmental processes, systems, and outcomes, leading to improved understanding, management, and conservation of the environment.

STEPS

Developing fractional factorial design experiments in environmental science involves several steps to efficiently assess the effects of multiple factors on a response variable while minimizing the number of experimental runs required. Here are the typical steps involved:

1. **Define the Research Objectives**: Clearly articulate the research objectives or questions that the fractional factorial design experiment aims to address. Identify the factors (IVs) of interest and the response variable(s) to be measured.
2. **Identify Factors and Levels**: Determine the factors to be included in the experiment and the range of levels for each factor. Factors should represent variables hypothesized to influence the response variable(s), such as environmental conditions, treatment parameters, or management practices.
3. **Select a Fractional Factorial Design**: Choose an appropriate fractional factorial design based on the number of factors, levels, and desired resolution. Common designs include half-fraction designs (e.g., 2^(k-p)), PBDs, and fractional factorial designs based on OAs (e.g., Taguchi designs).
4. **Determine the Experimental Runs**: Use the chosen fractional factorial design to determine the specific combination of factor levels for each experimental run. The design matrix specifies which factors are varied at which levels for each run, ensuring efficient coverage of the experimental space while minimizing the number of runs required.

5. **Generate Randomization and Blocking**: Randomize the order of experimental runs to minimize the effects of potential confounding factors or systematic biases. Consider blocking factors, if applicable, to account for sources of variability that may influence the response variable(s) but are not of primary interest.

6. **Conduct Experiments**: Implement the experimental design by conducting the planned experimental runs according to the specified factor levels. Follow standardized protocols for data collection, measurement, and recording to ensure consistency and reproducibility across experimental runs.

7. **Collect Response Data**: Measure the response variable(s) of interest for each experimental run. Record the response data accurately and ensure that any sources of variability or error are minimized to obtain reliable and precise measurements.

8. **Analyze Experimental Data**: Analyze the collected data using statistical methods appropriate for fractional factorial designs, such as ANOVA or regression analysis. Assess the significance of main effects and potential interactions between factors on the response variable(s).

9. **Interpret Results**: Interpret the results of the fractional factorial design experiment in the context of the research objectives and hypotheses. Identify significant factors and interactions that influence the response variable(s) and discuss their implications for understanding environmental processes or informing management decisions.

10. **Validate Findings**: Validate the findings of the fractional factorial design experiment through additional experiments, sensitivity analyses, or comparison with existing literature or empirical knowledge. Ensure that the observed effects are robust and reproducible under different conditions.

By following these steps, researchers can systematically develop and conduct fractional factorial design experiments in environmental science to efficiently assess the effects of multiple factors on a response variable and gain insights into complex environmental processes and systems.

PITFALLS AND REMEDIES

Fractional factorial design experiments are powerful tools in environmental science for efficiently assessing the effects of multiple factors on a response variable. However, they can encounter certain pitfalls that may affect the validity and reliability of results. Here are some common pitfalls and remedies associated with fractional factorial design experiments:

Alias Structure and Confounding:
> **Pitfall:** Fractional factorial designs often involve confounding, where main effects are aliased with interactions or other main effects. This can make it challenging to isolate the effects of individual factors.
>
> **Remedy:** Choose a fractional factorial design with an appropriate resolution that balances the need for efficiency with the desire to minimize confounding. Consider augmenting the design with additional runs if necessary to resolve confounding.

Loss of Precision:
> **Pitfall:** Fractional factorial designs sacrifice some precision in estimating main effects and interactions compared to full factorial designs, particularly for higher-order interactions.
>
> **Remedy:** Conduct power analyses or sample size calculations to ensure that the chosen fractional factorial design provides adequate power to detect meaningful effects. Consider augmenting the design with additional runs or replicates if precision is a concern.

Limited Range of Factor Levels:
   Pitfall: Fractional factorial designs typically require restricting the range of factor levels to maintain efficiency, which may limit the generalizability of results to broader environmental conditions.
   Remedy: Choose factor levels that are representative of the range of conditions of interest in the environmental system. Consider conducting additional experiments or sensitivity analyses to explore the effects of factors beyond the range tested in the fractional factorial design.
Missing Factors or Interactions:
   Pitfall: Fractional factorial designs may overlook important factors or interactions that influence the response variable, leading to biased estimates or an incomplete understanding of the environmental system.
   Remedy: Conduct thorough preliminary studies or literature reviews to identify potential factors and interactions that should be included in the experimental design. Consider augmenting the fractional factorial design with additional runs to explore additional factors or interactions.
Inadequate Randomization or Blocking:
   Pitfall: Inadequate randomization or blocking of experimental runs may introduce bias or confounding effects that compromise the validity of results.
   Remedy: Randomize the order of experimental runs, and consider blocking factors that may influence the response variable but are not of primary interest. Ensure that randomization and blocking procedures are implemented consistently and transparently.
Failure to Validate Findings:
   Pitfall: Failing to validate the findings of the fractional factorial design experiment through additional experiments or sensitivity analyses may lead to overinterpretation or misinterpretation of results.
   Remedy: Validate the observed effects and conclusions of the fractional factorial design experiment through independent replication, sensitivity analyses, or comparison with existing literature or empirical knowledge.

By being aware of these potential pitfalls and implementing appropriate remedies, researchers can enhance the validity, reliability, and usefulness of fractional factorial design experiments in environmental science, leading to more accurate assessments of the effects of multiple factors on environmental processes and outcomes.

Example 1

Fractional factorial experiments are used in environmental science to investigate the effects of multiple factors on a response variable while reducing the number of experimental runs. Here's an example of a matrix for a fractional factorial experiment in environmental science:

Objective: To study the impact of three independent factors (factor A, factor B, and factor C) on the water quality of a river system, considering potential interactions between factors.

Factors:

   1. **Factor A**: Concentration of pollutant X (low/high).
   2. **Factor B**: Flow rate of the river (low/high).
   3. **Factor C**: Temperature of the river water (low/high).

Response Variable: Water Quality Index.

TABLE 2.33

Environmental Science Fractional Factorial Design Experiment Matrix Example

Run	Factor A (Pollutant X)	Factor B (Flow Rate)	Factor C (Temperature)	Water Quality Index
1	Low	Low	Low	...
2	High	Low	Low	...
3	Low	High	Low	...
4	High	High	Low	...
5	Low	Low	High	...
6	High	Low	High	...
7	Low	High	High	...
8	High	High	High	...

Fractional Factorial Experiment Matrix:

In a fractional factorial experiment, you systematically reduce the number of experimental runs by running a subset of the full factorial combinations. The matrix is shown in Table 2.33.

In this matrix, each row represents a specific experimental run with a combination of factor settings. By using a fractional factorial design, you're able to investigate the main effects of each factor and some interactions while running only a fraction of the full factorial combinations. This significantly reduces the number of experimental runs while providing valuable information about how these factors impact water quality.

Fractional factorial experiments are particularly useful when there are limited resources or when the number of possible combinations is prohibitively large in a full factorial design.

Example 2

Another example of a fractional factorial design experiment in environmental science aimed at studying the factors influencing soil erosion rates. In this experiment, we'll investigate the effects of slope steepness, vegetation cover, and rainfall intensity on soil erosion. We'll use a 23–123–1 fractional factorial design, also known as a half-fraction design, to reduce the number of experimental runs while still capturing main effects and two-factor interactions.

Factors:

1. **Slope Steepness:**
   - Level 1: Low slope (5%) –1.
   - Level 2: High slope (15%) +1.
2. **Vegetation Cover:**
   - Level 1: Sparse vegetation (20% coverage) –1.
   - Level 2: Dense vegetation (80% coverage) +1.
3. **Rainfall Intensity:**
   - Level 1: Low intensity (20 mm/hr) –1.
   - Level 2: High intensity (50 mm/hr) +1.

The fractional factorial design matrix for this experiment can be generated using the 23–123–1 design, which allows for testing main effects and two-factor interactions while requiring only half the number of experimental runs of a full-factorial design (Table 2.34).

In this design matrix,

- Runs 1 to 4 represent the main effects of slope steepness, vegetation cover, and their interactions with rainfall intensity held constant at the low level (–1).

TABLE 2.34

Environmental Science Fractional Factorial Design Experiment Matrix Example

Run	Slope Steepness (%)	Vegetation Cover (%)	Rainfall Intensity (mm/hr)	Soil Erosion Rate
1	−1	−1	−1	Y1
2	+1	−1	−1	Y2
3	−1	+1	−1	Y3
4	+1	+1	−1	Y4
5	−1	−1	+1	Y5
6	+1	−1	+1	Y6
7	−1	+1	+1	Y7
8	+1	+1	+1	Y8

- Runs 5 to 8 represent the main effects of slope steepness, vegetation cover, and their interactions with rainfall intensity held constant at the high level (+1).
- The response variable, such as soil erosion rate, is measured for each experimental run.

By using this fractional factorial design, researchers can efficiently assess the effects of slope steepness, vegetation cover, and rainfall intensity on soil erosion while conducting only eight experimental runs instead of the $2^3 = 8$ runs required by a full factorial design. This allows for significant savings in time and resources while still capturing important main effects and interactions.

REPORTING

Fractional factorial design experiments in environmental science utilize various statistical tools to analyze data, assess the effects of multiple factors, and make inferences about the underlying relationships between variables. Some common statistical tools used in fractional factorial design experiments include:

ANOVA: ANOVA is a fundamental statistical technique used to assess the significance of factors and interactions with the response variable(s) in fractional factorial design experiments. It helps determine whether the observed differences in the response variable(s) are statistically significant and whether there are significant main effects or interactions between factors.

Main Effects Plots: Main effects plots are graphical representations that display the estimated effects of individual factors on the response variable(s) while holding other factors constant at their center points. These plots provide insights into the relative importance of each factor and help identify factors that have a significant impact on the response variable(s).

Interaction Plots: Interaction plots visualize the interactions between factors by showing how the effect of one factor on the response variable(s) varies across different levels of another factor. These plots help identify significant interactions between factors and understand how their combined effects influence the response variable(s).

Half-Normal Probability Plots: Half-normal probability plots are graphical tools used to identify significant effects in fractional factorial design experiments. These plots display the absolute values of the estimated effects of factors and interactions, allowing researchers to identify the most influential factors based on the magnitude of their effects.

Effect Sparsity Plot: Effect sparsity plots display the sparsity pattern of the effects in fractional factorial design experiments, highlighting which effects are aliased (confounded) with other effects. These plots help researchers identify which effects can be estimated uniquely and which are confounded with other effects.

Residual Analysis: Residual analysis is performed to assess the adequacy of the fitted statistical model and identify any deviations from model assumptions. This includes checking for homoscedasticity (constant variance), normality of residuals, and independence of errors. Residual plots and diagnostic tests are commonly used for this purpose.

Confidence Intervals and Hypothesis Testing: Confidence intervals are used to quantify the uncertainty associated with estimated effects in fractional factorial design experiments. Hypothesis tests, such as t-tests or F-tests, are conducted to determine whether the estimated effects are statistically significant at a given level of significance.

Model Selection Criteria: Model selection criteria, such as AIC or BIC, are used to compare competing statistical models and select the most parsimonious model that adequately explains the data while avoiding overfitting.

By employing these statistical tools effectively, researchers can analyze data from fractional factorial design experiments, interpret the results, and draw meaningful conclusions about the effects of multiple factors on environmental processes or outcomes.

A report on fractional factorial design experiments in environmental science typically includes several key elements to effectively communicate the research objectives, methods, results, and conclusions to the intended audience. Here are the essential elements of such a report:

Title Page: The title page provides the title of the report, the names of the authors, their affiliations, and the date of publication.

Abstract: The abstract summarizes the main objectives of the study, the experimental design, key findings, and conclusions in a concise manner. It provides readers with a brief overview of the research without having to read the entire report.

Introduction:
- Background and Literature Review: Provide background information on the research topic, including relevant literature and previous studies. Explain the significance of the research and the rationale for employing fractional factorial design experiments to study the effects of multiple factors on the response variable(s).
- Research Objectives and Hypotheses: Clearly state the research objectives and hypotheses that the experiment aims to address. Outline the specific factors, levels, and response variables under investigation.

Methods:
- Experimental Design: Describe the fractional factorial design used in the experiment, including the selection of factors, levels, and resolution. Explain how the experimental runs were generated and conducted.
- Data Collection: Detail the methods and protocols used to collect data on the response variable(s), including measurement techniques, sampling procedures, and data recording formats.
- Statistical Analysis: Outline the statistical methods used to analyze the data, such as ANOVA, main effects plots, interaction plots, and effect sparsity plots. Specify any assumptions made and how they were validated.

Results:
- Descriptive Statistics: Present summary statistics (e.g., means, standard deviations) for each factor level and experimental run.
- Analysis of Effects: Present the results of the statistical analysis, including main effects, interactions, and significant factors. Use graphical tools, such as main effects plots, interaction plots, and effect sparsity plots to visualize the effects.
- Interpretation of Findings: Interpret the results of the experiment in the context of the research objectives and hypotheses. Discuss the implications of significant effects for understanding environmental processes or informing management decisions.

Discussion:
- Comparison with Previous Studies: Compare the findings of the current study with previous research in the field. Highlight any consistencies or discrepancies and explain possible reasons for differences.
- Limitations and Future Directions: Acknowledge any limitations of the study and suggest areas for future research or improvements to experimental design.
- Practical Implications: Discuss the practical implications of the findings for environmental science, management, or policy.

Conclusion: Summarize the main findings of the study and their implications for understanding the effects of multiple factors on environmental processes or outcomes. Restate the significance of the research and its contribution to the field.

References: Provide a list of references cited in the report, following a specific citation style (e.g., APA, MLA).

Appendices: Include any supplementary materials, such as the fractional factorial design matrix, raw data, detailed statistical analyses, or additional figures and tables.

By including these elements in a report on fractional factorial design experiments in environmental science, researchers can effectively communicate their findings, methods, and conclusions to the scientific community and other stakeholders, facilitating the dissemination of knowledge and advancement of the field.

2.3.4 CCD

CCD experiments find applications in various areas within environmental science where researchers need to optimize processes, investigate response surfaces, and assess the effects of multiple factors on a response variable. Some common applications of CCD experiments in environmental science include:

Pollution Control and Remediation: CCD experiments are used to optimize pollution control and remediation strategies in environmental systems. Researchers can assess the effects of various treatment methods, operating conditions, and environmental factors on the efficiency of pollution removal processes, such as wastewater treatment, soil remediation, and air pollution control.

Environmental Monitoring and Sampling Design: CCD experiments are employed to optimize environmental monitoring programs and sampling designs. Researchers can investigate the effects of factors, such as sampling frequency, sampling locations, and analytical methods, on the accuracy and efficiency of environmental data collection, allowing for cost-effective monitoring strategies that provide reliable assessments of environmental conditions.

Ecosystem Studies and Ecological Research: CCD experiments are used to study ecological processes, species interactions, and biodiversity patterns in natural ecosystems. Researchers can assess the effects of environmental variables, such as habitat structure, resource availability, and disturbance regimes, on ecosystem structure and function, helping to understand and manage ecosystems more effectively.

Climate Change and Global Environmental Change: CCD experiments are employed to study the impacts of climate change and global environmental change on ecosystems, habitats, and species. Researchers can assess the combined effects of multiple stressors, such as temperature changes, altered precipitation patterns, habitat loss, and pollution, on ecosystem resilience, species distributions, and community dynamics, informing adaptation and mitigation strategies.

Environmental Toxicology and Risk Assessment: CCD experiments are used in environmental toxicology and risk assessment studies to evaluate the effects of chemical stressors on organisms and ecosystems. Researchers can assess the interactions between pollutants, exposure pathways, and environmental conditions, helping to identify potential synergistic or antagonistic effects and inform regulatory decisions and environmental management practices.

Waste Management and Resource Recovery: CCD experiments are employed to optimize waste management processes and resource recovery technologies. Researchers can assess the effects of factors, such as waste composition, treatment methods, and operating conditions, on the efficiency of waste-treatment, recycling, and energy-recovery processes, facilitating the development of sustainable waste management strategies.

Overall, CCD experiments are valuable tools in environmental science for optimizing processes, understanding response surfaces, and assessing the effects of multiple factors on environmental systems. They provide researchers with a systematic approach to experimentation and optimization, leading to improved understanding, management, and conservation of the environment.

Steps

Developing CCD experiments in environmental science involves several steps to efficiently optimize processes, investigate response surfaces, and assess the effects of multiple factors on a response variable. Here are the typical steps involved:

1. **Define the Research Objectives**: Clearly articulate the research objectives or questions that the CCD experiment aims to address. Identify the factors (IVs) of interest and the response variable(s) to be optimized or investigated.
2. **Identify Factors and Levels**: Determine the factors to be included in the experiment and the range of levels for each factor. Factors should represent variables hypothesized to influence the response variable(s), such as environmental conditions, treatment parameters, or management practices.
3. **Select the Experimental Region**: Define the experimental region within which the factors will be varied. The experimental region should cover the range of factor levels of interest while avoiding extreme or unrealistic conditions.
4. **Choose the Design Type**: Select the type of CCD appropriate for the research objectives and the number of factors. Common types include the 2-level CCD, 3-level CCD, and 4-level CCD, each offering different degrees of flexibility and efficiency.
5. **Generate the Design Matrix**: Use statistical software or design tables to generate the design matrix of CCD, which specifies the factor levels for each experimental run. The design matrix includes a combination of factorial points, axial points, and center points to efficiently explore the experimental region and estimate model parameters.
6. **Conduct Additional Runs (Optional)**: Consider augmenting the CCD with additional runs to improve model fit or explore specific regions of interest within the experimental space. These additional runs may be used to validate the model, estimate curvature, or assess the effects of factors at specific levels.
7. **Conduct Experiments**: Implement the CCD experiment by conducting the planned experimental runs according to the specified factor levels. Follow standardized protocols for data collection, measurement, and recording to ensure consistency and reproducibility across experimental runs.
8. **Measure Response Variable(s)**: Measure the response variable(s) of interest for each experimental run. Record the response data accurately and ensure that any sources of variability or error are minimized to obtain reliable and precise measurements.

9. **Fit Response Surface Model**: Use regression analysis or other modeling techniques to fit a response surface model to the experimental data. The response surface model describes the relationship between factors and response variable(s) and allows for optimization or prediction within the experimental region.
10. **Analyze and Interpret Results**: Analyze the fitted response surface model to assess the significance of factors and interactions, identify optimal factor settings, and interpret the implications for environmental science applications. Visualize response surfaces, contour plots, or interaction plots to aid in the interpretation and communication of results.
11. **Validate and Optimize Model (Optional)**: Validate the fitted response surface model through additional experiments, sensitivity analyses, or cross-validation techniques. Use optimization algorithms to identify optimal factor settings that maximize or minimize the response variable(s) within specified constraints.
12. **Draw Conclusions and Make Recommendations**: Draw conclusions based on the analysis of the CCD experiment and its implications for understanding processes or optimizing outcomes in environmental science. Make recommendations for future research, management practices, or policy decisions based on the findings.

By following these steps, researchers can systematically develop and conduct CCD experiments in environmental science to optimize processes, investigate response surfaces, and assess the effects of multiple factors on environmental systems.

PITFALLS AND REMEDIES

CCD experiments in environmental science are powerful tools for optimizing processes and exploring response surfaces. However, they can encounter certain pitfalls that may affect the validity and reliability of results. Here are some common pitfalls and remedies associated with CCD experiments:

Overfitting:
  Pitfall: Fitting overly complex response surface models that capture noise rather than true underlying relationships can lead to overfitting.
  Remedy: Use model selection criteria (e.g., AIC, BIC) to choose a parsimonious model that adequately explains the data while avoiding overfitting. Cross-validation techniques can also help assess model performance.
Inadequate Model Validation:
  Pitfall: Failing to validate the fitted response surface model can lead to erroneous conclusions and unreliable predictions.
  Remedy: Validate the response surface model using independent validation data or cross-validation techniques. Assess the model's predictive performance and robustness across different subsets of the data.
Extrapolation Beyond Experimental Region:
  Pitfall: Extrapolating response surface models beyond the experimental region can lead to unreliable predictions and erroneous conclusions.
  Remedy: Limit predictions and interpretations to the experimental region covered by the CCD. Avoid making extrapolations beyond the range of factor levels or conditions tested in the experiment.
Unmodeled Curvature:
  Pitfall: Neglecting to account for curvature in the response surface can lead to biased estimates and inaccurate predictions, particularly if the true response surface is nonlinear.
  Remedy: Assess the presence of curvature in the response surface using diagnostic plots, such as residual plots or lack-of-fit tests. Consider augmenting the CCD with additional runs to better estimate curvature if necessary.

Confounding Effects:

  Pitfall: Confounding effects between factors or interactions can lead to biased estimates and misinterpretation of results.

  Remedy: Choose an appropriate CCD that minimizes confounding effects and allows for the estimation of main effects and selected interactions. Augment the design with additional runs if necessary to resolve confounding.

Insensitive Optimum:

  Pitfall: Identifying an optimum factor setting that is insensitive to changes in the response variable(s) can lead to suboptimal solutions and inefficient processes.

  Remedy: Conduct sensitivity analyses to assess the robustness of the optimal factor settings and identify regions of uncertainty or instability in the response surface. Consider incorporating robust optimization techniques to find factor settings that are less sensitive to variability.

Inadequate Factor Range:

  Pitfall: Failing to cover the full range of factor levels relevant to the environmental system can lead to incomplete understanding and suboptimal solutions.

  Remedy: Ensure that the chosen factor levels span the relevant range of environmental conditions or process parameters. Conduct preliminary studies or sensitivity analyses to identify the appropriate factor range.

By being aware of these potential pitfalls and implementing appropriate remedies, researchers can enhance the validity, reliability, and usefulness of CCD experiments in environmental science, leading to more accurate optimization of processes and exploration of response surfaces.

Example 1

CCD is a type of RSM used to optimize processes by studying the effects of factors at various levels, including the center points and extreme points. Here's an example of a matrix for a CCD designed experiment in environmental science:

Objective: To optimize the conditions for a soil remediation process by studying the effects of two independent factors, soil pH (factor A) and soil temperature (factor B), on the removal efficiency of a specific contaminant.

Factors:

1. **Soil pH (Factor A):**
   - Low pH (e.g., 4.0).
   - Medium pH (e.g., 6.0).
   - High pH (e.g., 8.0).
2. **Soil Temperature (Factor B):**
   - Low temperature (e.g., 20°C).
   - Medium temperature (e.g., 30°C).
   - High temperature (e.g., 40°C).

Response Variable: Contaminant Removal Efficiency (%).

CCD Experiment Matrix:

In a CCD, the matrix is designed to explore the main effects, interactions, and curvature of the response surface. It includes center points and extreme points. The matrix is shown in Table 2.35.

TABLE 2.35
Environmental Science CCD Experiment Matrix Example

Run	Soil pH (A)	Soil Temperature (B)	Contaminant Removal Efficiency (%)
−1	−1	−1	...
−1	+1	−1	...
−1	−1	+1	...
−1	+1	+1	...
0	0	0	...
0	−1.68	0	...
0	+1.68	0	...
0	0	−1.68	...
0	0	+1.68	...
0	0	0	...

In this matrix, each row represents a specific experimental run with a combination of soil pH and soil temperature settings. The center points (0 levels) and extreme points (+1 and −1 levels) are included to assess the curvature of the response surface.

The goal of the CCD is to model the relationship between the independent factors (soil pH and soil temperature) and the response variable (contaminant removal efficiency) to identify the optimal conditions for maximum removal efficiency. Statistical techniques, such as regression analysis and contour plots, are used to achieve this optimization in environmental processes.

Example 2

Another example of a CCD experiment in environmental science aimed at optimizing the parameters for the removal of pollutants from wastewater using a photocatalytic process. In this experiment, we'll investigate the effects of three factors: catalyst dosage, pH of the solution, and irradiation time. We'll use a CCD with 3 factors at 3 levels each, along with center points, to efficiently explore the response surface.

Factors:

1. **Catalyst Dosage:**
   - Level 1: Low dosage (5 g/L) −1.
   - Level 2: Medium dosage (10 g/L) 0.
   - Level 3: High dosage (15 g/L) +1.
2. **pH of the Solution:**
   - Level 1: Acidic (pH 3) −1.
   - Level 2: Neutral (pH 7) 0.
   - Level 3: Basic (pH 11) +1.
3. **Irradiation Time:**
   - Level 1: Short time (30 minutes) −1.
   - Level 2: Medium time (60 minutes) 0.
   - Level 3: Long time (90 minutes) +1.

The CCD matrix includes factorial points, axial points, and center points to explore the experimental space efficiently and estimate model parameters accurately.

The CCD matrix for photocatalytic wastewater treatment experiment is shown in Table 2.36.

TABLE 2.36

Environmental Science CCD Experiment Matrix Example

Run	Catalyst Dosage (g/L)	pH	Irradiation Time (minutes)	Response
1	−1	−1	−1	Y1
2	+1	−1	−1	Y2
3	−1	+1	−1	Y3
4	+1	+1	−1	Y4
5	−1	−1	+1	Y5
6	+1	−1	+1	Y6
7	−1	+1	+1	Y7
8	+1	+1	+1	Y8
9	0	0	0	Y9

In this CCD matrix,

- Runs 1 to 8 represent the factorial points of the CCD, where each factor is varied at low (−1) and high (+1) levels.
- Runs 9 to 11 represent the center points of the CCD, where factors are set at the midpoint of their ranges to estimate pure error and curvature.
- Runs 1 to 8 also include axial points, where one factor is set at its extreme level (coded as −1 or +1) while the other two factors are set at their midpoint (coded as 0).
- The response variable (e.g., pollutant removal efficiency) is measured for each experimental run.

By using this CCD matrix, researchers can efficiently explore the effects of catalyst dosage, pH, and irradiation time on wastewater treatment efficiency while conducting only a fraction of the runs required by a full factorial design. This allows for significant savings in time and resources while still capturing important main effects and interactions.

REPORTING

CCD experiments in environmental science utilize various statistical tools to analyze data, assess response surfaces, and optimize processes. Some common statistical tools used in CCD experiments include:

ANOVA: ANOVA is a fundamental statistical technique used to assess the significance of factors and interactions with the response variable(s) in CCD experiments. It helps determine whether the observed differences in the response variable(s) are statistically significant and whether there are significant main effects or interactions between factors.

2.3.5　RSM

RSM is a collection of statistical and mathematical techniques used to model and analyze response surfaces in CCD experiments. It involves fitting mathematical models to experimental data to describe the relationship between factors and response variable(s) and optimize process parameters.

Regression Analysis: Regression analysis is used to fit mathematical models, such as polynomial regression models, to the experimental data in CCD experiments. These models describe the relationship between factors and response variable(s) and allow for prediction and optimization within the experimental region.

Graphical Tools:

Response Surface Plots: Response surface plots visualize the relationship between factors and the response variable(s) by plotting the response variable(s) as a function of two factors while holding other factors constant.
Contour Plots: Contour plots display contours of constant response values on a two-dimensional plot, allowing for visualization of response surfaces and identification of optimal factor settings.

DOE Software: Specialized software packages for DOE, such as JMP, Minitab, or R, are often used to generate CCDs, analyze experimental data, and fit response surface models. These software packages provide tools for design generation, statistical analysis, visualization, and optimization.

Optimization Algorithms: Optimization algorithms, such as gradient descent or genetic algorithms, are used to identify optimal factor settings that maximize or minimize the response variable(s) within specified constraints. These algorithms iteratively search the experimental region to find factor settings that optimize the process performance.

Model Selection Criteria: Model selection criteria, such as AIC or BIC, are used to compare competing response surface models and select the most parsimonious model that adequately explains the data while avoiding overfitting.

Diagnostic Tools: Diagnostic tools, such as residual plots, lack-of-fit tests, and leverage plots, are used to assess the adequacy of the fitted response surface model and identify any deviations from model assumptions.

By employing these statistical tools effectively, researchers can analyze data from CCD experiments, optimize processes, and make informed decisions about environmental management and engineering.

A report on CCD experiments in environmental science typically includes several key elements to effectively communicate the research objectives, methods, results, and conclusions to the intended audience. Here are the essential elements of such a report:

Title Page: The title page provides the title of the report, the names of the authors, their affiliations, and the date of publication.
Abstract: The abstract summarizes the main objectives of the study, the experimental design, key findings, and conclusions in a concise manner. It provides readers with a brief overview of the research without having to read the entire report.

Introduction:
- Background and Literature Review: Provide background information on the research topic, including relevant literature and previous studies. Explain the significance of the research and the rationale for employing CCD experiments to optimize processes or investigate response surfaces.
- Research Objectives and Hypotheses: Clearly state the research objectives and hypotheses that the experiment aims to address. Outline the specific factors, levels, and response variables under investigation.

Methods:
- Experimental Design: Describe the CCD used in the experiment, including the selection of factors, levels, and axial points. Explain how the design matrix was generated and any additional runs conducted.
- Data Collection: Detail the methods and protocols used to collect data on the response variable(s), including measurement techniques, sampling procedures, and data recording formats.
- Statistical Analysis: Outline the statistical methods used to analyze the data, such as ANOVA, regression analysis, and response surface modeling. Specify any assumptions made and how they were validated.

Results:
- Descriptive Statistics: Present summary statistics (e.g., means, standard deviations) for each factor level and experimental run.
- Analysis of Effects: Present the results of the statistical analysis, including main effects, interactions, and significant factors. Use graphical tools, such as response surface plots, contour plots, and interaction plots, to visualize the effects.
- Optimization Results: If applicable, present the optimized factor settings that maximize or minimize the response variable(s) within specified constraints.

Discussion:
- Comparison with Previous Studies: Compare the findings of the current study with previous research in the field. Highlight any consistencies or discrepancies and explain possible reasons for differences.
- Interpretation of Findings: Interpret the results of the experiment in the context of the research objectives and hypotheses. Discuss the implications of significant effects for optimizing processes or understanding environmental systems.
- Limitations and Future Directions: Acknowledge any limitations of the study and suggest areas for future research or improvements to experimental design.

Conclusion: Summarize the main findings of the study and their implications for optimizing processes or exploring response surfaces in environmental science. Restate the significance of the research and its contribution to the field.

References: Provide a list of references cited in the report, following a specific citation style (e.g., APA, MLA).

Appendices: Include any supplementary materials, such as the CCD matrix, raw data, detailed statistical analyses, or additional figures and tables.

By including these elements in a report on CCD experiments in environmental science, researchers can effectively communicate their findings, methods, and conclusions to the scientific community and other stakeholders, facilitating the dissemination of knowledge and advancement of the field.

2.3.6 Taguchi Methods

Taguchi method designed experiments find applications in various areas within environmental science where researchers need to optimize processes, improve product or system performance, and reduce variability. Some common applications of Taguchi methods in environmental science include:

Pollution Control and Remediation: Taguchi methods are used to optimize pollution control and remediation strategies in environmental systems. Researchers can design experiments to optimize treatment processes, such as wastewater treatment, soil remediation, and air pollution control, to maximize efficiency and minimize environmental impact.

Environmental Monitoring and Sampling Design: Taguchi methods are employed to optimize environmental monitoring programs and sampling designs. Researchers can design experiments to optimize sampling protocols, sensor placement, and data collection techniques to improve the accuracy and efficiency of environmental monitoring and assessment.

Ecosystem Studies and Ecological Research: Taguchi methods are used to optimize experimental designs in ecosystem studies and ecological research. Researchers can design experiments to optimize experimental conditions, such as habitat structure, resource availability, and experimental treatments, to improve the efficiency and reliability of ecological studies.

Climate Change and Global Environmental Change: Taguchi methods are employed to optimize experimental designs in studies related to climate change and global environmental change. Researchers can design experiments to optimize climate models, assess the impacts of climate change on ecosystems, and evaluate adaptation and mitigation strategies.

Environmental Toxicology and Risk Assessment: Taguchi methods are used to optimize experimental designs in environmental toxicology and risk assessment studies. Researchers can design experiments to optimize exposure assessments, dose–response studies, and risk characterization processes to improve the accuracy and reliability of environmental risk assessments.

Waste Management and Resource Recovery: Taguchi methods are employed to optimize waste management processes and resource recovery technologies. Researchers can design experiments to optimize waste-treatment processes, recycling systems, and energy-recovery technologies to maximize efficiency and minimize environmental impact.

Overall, Taguchi method designed experiments are valuable tools in environmental science for optimizing processes, improving product or system performance, and reducing variability. They provide researchers with a systematic approach to experimentation and optimization, leading to improved understanding, management, and conservation of the environment.

STEPS

Developing Taguchi method designed experiments in environmental science involves several steps to optimize processes, improve product or system performance, and reduce variability. Here are the typical steps involved:

1. **Define the Problem and Objectives**: Clearly define the problem to be addressed and the objectives to be achieved through the Taguchi experiment. Identify the process, product, or system to be optimized and the specific performance characteristics or responses to be improved.
2. **Select the Factors and Levels**: Identify the factors (input variables) that may influence the performance characteristics or responses of the process, product, or system. Determine the range and levels of each factor to be investigated during the experiment. Factors may include environmental conditions, process parameters, material properties, or design variables.
3. **Determine the Experimental Design**: Select the appropriate Taguchi experimental design based on the number of factors and levels, as well as the objectives of the experiment. Common Taguchi designs include L8, L16, L27, and larger OAs. Choose a design that allows for efficient screening of factors and robust parameter estimation.
4. **Generate the Experimental Layout**: Use the selected Taguchi experimental design to generate the experimental layout, specifying the factor levels for each experimental run. The layout includes the combination of factor levels to be tested in each experimental trial, ensuring efficient coverage of the experimental space.
5. **Conduct the Experiment**: Conduct the Taguchi experiment according to the experimental layout generated. Follow standardized protocols for data collection, measurement, and recording to ensure consistency and reproducibility across experimental runs.
6. **Perform Data Analysis**: Analyze the experimental data using statistical techniques provided by Taguchi methods, such as SNRs and ANOVA. Calculate the main effects and interaction effects of factors on the performance characteristics or responses of interest.
7. **Optimize Process Parameters**: Identify the optimal combination of factor levels that maximize or minimize the performance characteristics or responses based on the results of the data analysis. Use statistical optimization techniques, such as the Taguchi parameter design approach, to determine the settings that lead to the best overall performance.

8. **Conduct Confirmation Runs (Optional)**: Validate the optimized process parameters through confirmation runs conducted under the optimal conditions determined in the Taguchi experiment. Compare the performance characteristics or responses observed in confirmation runs with the predicted values to assess the accuracy of the optimization.

9. **Implement Process Changes**: Implement the recommended process changes or parameter settings based on the findings of the Taguchi experiment. Monitor the performance of the optimized process, product, or system to ensure sustained improvement over time.

10. **Document and Communicate Results**: Prepare a comprehensive report documenting the experimental design, methodology, results, and conclusions of the Taguchi experiment. Communicate the findings to stakeholders, management, or other relevant parties to facilitate decision-making and process-improvement initiatives.

By following these steps, researchers can systematically develop and conduct Taguchi method designed experiments in environmental science to optimize processes, improve product or system performance, and reduce variability, leading to enhanced efficiency and effectiveness in environmental management and conservation efforts.

PITFALLS AND REMEDIES

Taguchi method designed experiments in environmental science offer a systematic approach to optimization, but they can encounter certain pitfalls that may affect the validity and reliability of results. Here are some common pitfalls and remedies associated with Taguchi methods experiments:

Inadequate Factor Selection:
  Pitfall: Selecting irrelevant factors or failing to include important factors in the experimental design can lead to incomplete optimization and suboptimal results.
  Remedy: Conduct a thorough analysis to identify all potential factors that may influence the performance characteristics or responses of interest. Use domain knowledge, literature review, and preliminary studies to select relevant factors for inclusion in the experimental design.

Incorrect Factor Levels:
  Pitfall: Choosing inappropriate levels for factors may result in insufficient exploration of the experimental space or failure to capture important relationships.
  Remedy: Conduct preliminary studies or sensitivity analyses to identify appropriate levels for each factor. Consider the expected range of variation in environmental conditions or process parameters when selecting factor levels.

Confounding Effects:
  Pitfall: Confounding effects between factors or interactions may lead to biased estimates and misinterpretation of results.
  Remedy: Choose an experimental design that minimizes confounding effects and allows for estimation of main effects and selected interactions. Conduct additional runs or use advanced experimental designs, such as OAs, to resolve confounding and improve the accuracy of estimates.

Unreliable Response Measurement:
  Pitfall: Inaccurate or imprecise measurement of response variables can introduce variability and undermine the validity of results.
  Remedy: Use reliable measurement techniques and standardized protocols to ensure accurate and consistent measurement of response variables. Implement quality control measures to minimize measurement errors and ensure data integrity.

Overemphasis on Single Response Optimization:
> **Pitfall:** Focusing solely on optimizing a single response variable may neglect other important performance characteristics or trade-offs.
>
> **Remedy:** Consider multiple performance criteria or responses when designing Taguchi experiments to achieve balanced optimization. Use multi-objective optimization techniques to identify optimal solutions that meet multiple criteria simultaneously.

Failure to Validate Results:
> **Pitfall:** Failing to validate the optimized process parameters or solutions through confirmation runs or external validation can lead to the implementation of suboptimal solutions.
>
> **Remedy:** Conduct confirmation runs under the optimized conditions to validate the performance improvements observed in the Taguchi experiment. Compare the observed results with predicted values to assess the accuracy and robustness of the optimization.

Inadequate Documentation and Communication:
> **Pitfall:** Poor documentation and communication of experimental design, methodology, and results may hinder reproducibility and implementation of findings.
>
> **Remedy:** Prepare a comprehensive report documenting all aspects of the Taguchi experiment, including experimental design, methodology, data analysis, results, and conclusions. Communicate the findings to stakeholders, management, and other relevant parties in a clear and concise manner to facilitate decision-making and process improvement initiatives.

By being aware of these potential pitfalls and implementing appropriate remedies, researchers can enhance the validity, reliability, and usefulness of Taguchi method designed experiments in environmental science, leading to improved optimization and performance of environmental processes and systems.

Example 1

Taguchi methods, specifically the Taguchi L9 (3^4) OA, is a popular approach for designing experiments in environmental science to optimize processes while minimizing variation and identifying robust conditions. Here's an example of a matrix for a Taguchi L9 experiment in environmental science:

Objective: To optimize the conditions for a wastewater-treatment process by studying the effects of four independent factors on the removal efficiency of a specific pollutant.

Factors:

1. **Factor A**: Concentration of coagulant (low/standard/high).
2. **Factor B**: Reaction time (short/medium/long).
3. **Factor C**: pH Level (acidic/neutral/alkaline).
4. **Factor D**: Mixing intensity (low/medium/high).

Response Variable: Pollutant Removal Efficiency (%).

Taguchi L9 (3^4) Experiment Matrix:
In a Taguchi L9 design, you have nine experimental runs that systematically cover all combinations of factors and levels. The matrix is shown in Table 2.37.

In this matrix, each row represents a specific experimental run with a unique combination of factors and their levels. The Taguchi L9 design allows you to assess the main effects and interactions of the factors while reducing the number of experiments.

TABLE 2.37

Taguchi Method Designed Experiment Matrix Example for Environmental Science

Run	Factor A	Factor B	Factor C	Factor D	Pollutant Removal Efficiency (%)
1	Low	Short	Acidic	Low	...
2	Standard	Short	Neutral	Medium	...
3	High	Short	Alkaline	High	...
4	Low	Medium	Neutral	Low	...
5	Standard	Medium	Alkaline	Medium	...
6	High	Medium	Acidic	High	...
7	Low	Long	Alkaline	Medium	...
8	Standard	Long	Acidic	High	...
9	High	Long	Neutral	Low	...

The objective is to identify the factor settings that lead to the highest pollutant removal efficiency and determine the robustness of these settings to variations in the factors. Taguchi methods aim to find optimal conditions and make processes more robust and less sensitive to variations and external factors in environmental science applications.

Example 2

Another example of a Taguchi method designed experiment matrix for optimizing the parameters of a wastewater-treatment process in environmental science (Table 2.38):
In this matrix,

- **Factors:**
  - Factor 1 (A): pH of the wastewater (low level –1: acidic, high level +1:bBasic).
  - Factor 2 (B): Temperature of the treatment process (low level –1: low temperature, high level +1: high temperature).
  - Factor 3 (C): Coagulant dosage (low level –1: low dosage, high level +1: high dosage).
- **Levels:** Each factor is varied at two levels, typically denoted as –1 (low level) and +1 (high level).
- **Response:** The response variable (e.g., pollutant removal efficiency) is measured for each experimental run.

TABLE 2.38

Taguchi Method Designed Experiment Matrix Example for Environmental Science

Run	Factor 1: pH (A)	Factor 2: Temperature (B)	Factor 3: Coagulant Dose (C)	Response
1	–1	–1	–1	Y1
2	+1	–1	–1	Y2
3	–1	+1	–1	Y3
4	+1	+1	–1	Y4
5	–1	–1	+1	Y5
6	+1	–1	+1	Y6
7	–1	+1	+1	Y7
8	+1	+1	+1	Y8

This matrix represents a fractional factorial design, specifically a 2^3 factorial design with 8 experimental runs. Each run corresponds to a unique combination of factor levels, allowing for the exploration of main effects and selected interactions while minimizing the number of experimental trials required.

Researchers can analyze the results of this Taguchi experiment to identify optimal factor settings that maximize the response variable(s) and optimize the wastewater treatment process for improved environmental performance.

REPORTING

In Taguchi method designed experiments for environmental science, several statistical tools are commonly used to analyze data, assess the effects of factors, and optimize processes. Some of the key statistical tools used in Taguchi experiments include:

SNRs: SNRs are used to evaluate the quality or performance of a process or product characteristic. They provide a measure of the signal (desired response) relative to the noise (variability) and are often used to assess the robustness of a process to variations in factors and noise factors.

ANOVA: ANOVA is used to partition the variability in the response variable(s) into different sources, such as main effects, interactions, and errors. It helps determine the significance of factors and interactions with the response variable(s) and assess the relative importance of each factor in influencing process performance.

Main Effects Plots: Main effects plots visualize the average response at each level of a factor while holding other factors constant. They help identify the main effects of factors on the response variable(s) and assess the direction and magnitude of their influence.

Interaction Plots: Interaction plots visualize the effects of interactions between factors on the response variable(s). They help identify significant interactions between factors and assess whether the effects of one factor depend on the level of another factor.

RSM: RSM involves fitting mathematical models to experimental data to describe the relationship between factors and response variable(s). It helps optimize process parameters and identify optimal factor settings that maximize or minimize the response variable(s) within the experimental region.

OAs: OAs are used to construct efficient experimental designs that allow for the estimation of main effects and selected interactions while minimizing the number of experimental runs required. They ensure efficient exploration of the experimental space and robust parameter estimation.

Parameter Design: Parameter design techniques are used to optimize process parameters while considering noise factors that may affect process performance. They help identify factor settings that are robust to variations in noise factors and ensure consistent performance under different operating conditions.

Statistical Optimization Algorithms: Statistical optimization algorithms, such as gradient descent, genetic algorithms, or response surface optimization, are used to identify optimal factor settings that maximize or minimize the response variable(s) within specified constraints. They iteratively search the experimental region to find factor settings that optimize process performance.

By using these statistical tools effectively, researchers can analyze data from Taguchi method designed experiments; identify significant factors, interactions, and noise factors; and optimize processes for improved environmental performance and reliability.

When reporting on Taguchi method designed experiments in environmental science, it's essential to provide a comprehensive and well-organized document that communicates the research objectives, methods, results, and conclusions clearly to the intended audience. Here are the key elements that should be included in such a report:

Title Page: The title page should include the title of the report, the names of the authors, their affiliations, and the date of publication.

Abstract: The abstract provides a concise summary of the research objectives, methods, key findings, and conclusions of the Taguchi experiment. It allows readers to quickly understand the main aspects of the study without reading the entire report.

Introduction:
- Background and Literature Review: Provide background information on the research topic, including relevant literature and previous studies. Explain the significance of the research and the rationale for using Taguchi method designed experiments to optimize processes or improve product performance.
- Research Objectives and Hypotheses: Clearly state the research objectives and hypotheses that the experiment aims to address. Outline the specific factors, levels, and response variables under investigation.

Methods:
- Experimental Design: Describe the Taguchi experimental design used in the experiment, including the selection of factors, levels, and OAs. Explain how the design matrix was generated and any additional runs conducted.
- Data Collection: Detail the methods and protocols used to collect data on the response variable(s), including measurement techniques, sampling procedures, and data recording formats.
- Statistical Analysis: Outline the statistical methods used to analyze the data, such as SNRs, ANOVA, and RSM. Specify any assumptions made and how they were validated.

Results:
- Descriptive Statistics: Present summary statistics (e.g., means, standard deviations) for each factor level and experimental run.
- Analysis of Effects: Present the results of the statistical analysis, including main effects, interactions, and significant factors. Use graphical tools, such as main effects plots, interaction plots, and response surface plots, to visualize the effects.
- Optimization Results: Identify the optimal factor settings that maximize or minimize the response variable(s) based on the results of the Taguchi experiment.

Discussion:
- Comparison with Previous Studies: Compare the findings of the current study with previous research in the field. Highlight any consistencies or discrepancies and explain possible reasons for differences.
- Interpretation of Findings: Interpret the results of the experiment in the context of the research objectives and hypotheses. Discuss the implications of significant effects for optimizing processes or improving product performance.
- Limitations and Future Directions: Acknowledge any limitations of the study and suggest areas for future research or improvements to experimental design.

Conclusion: Summarize the main findings of the study and their implications for optimizing processes or improving product performance using Taguchi method designed experiments.

References: Provide a list of references cited in the report, following a specific citation style (e.g., APA, MLA).

Appendices: Include any supplementary materials, such as the Taguchi experimental design matrix, raw data, detailed statistical analyses, or additional figures and tables.

By including these elements in a report on Taguchi method designed experiments in environmental science, researchers can effectively communicate their findings, methods, and conclusions to the scientific community and other stakeholders, facilitating the dissemination of knowledge and advancement of the field.

2.3.7 MIXTURE EXPERIMENTS

Mixture designed experiments find applications in various areas within environmental science where researchers need to optimize mixtures of components to achieve desired outcomes or investigate the effects of composition on system performance. Here are some common applications of Mixture designed experiments in environmental science:

Wastewater Treatment: In wastewater treatment, researchers may use mixture designed experiments to optimize the composition of treatment chemicals, such as coagulants, flocculants, or adsorbents, to enhance pollutant removal efficiency or minimize treatment costs.

Soil Remediation: Mixture designed experiments can be employed to optimize the composition of soil amendments, such as organic amendments, bioremediation agents, or stabilizing agents, to improve soil quality, promote remediation of contaminated sites, and mitigate the impacts of pollutants.

Air Pollution Control: In air pollution control technologies, researchers may use mixture designed experiments to optimize the composition of sorbents or catalysts used in pollution control devices, such as scrubbers, filters, or catalytic converters, to enhance pollutant capture efficiency and reduce emissions.

Green Chemistry and Sustainable Materials: Mixture designed experiments are utilized to optimize the composition of environmentally friendly materials, such as biodegradable polymers, green solvents, or eco-friendly surfactants, to improve their performance and reduce environmental impact in various applications, including packaging, coatings, and personal care products.

Environmental Monitoring and Sampling Design: In environmental monitoring programs, researchers may use mixture designed experiments to optimize the composition of sampling protocols, sensor arrays, or monitoring networks to maximize information gain, minimize sampling effort, and improve the accuracy of environmental assessments.

Ecological Studies and Habitat Restoration: Mixture designed experiments can be applied in ecological studies and habitat restoration projects to optimize the composition of habitat structures, planting mixes, or seed mixtures to enhance biodiversity, promote ecosystem resilience, and restore degraded ecosystems.

Climate Change Mitigation and Adaptation: Mixture designed experiments may be used to optimize the composition of climate change mitigation and adaptation strategies, such as carbon capture and storage (CCS) technologies, green infrastructure solutions, or climate-resilient crop varieties, to maximize effectiveness and minimize environmental impacts.

Industrial Ecology and Sustainable Manufacturing: In industrial ecology and sustainable manufacturing, researchers may use mixture designed experiments to optimize the composition of materials, processes, or products to minimize resource consumption, energy use, and waste generation while maximizing efficiency and environmental performance.

Overall, mixture designed experiments are valuable tools in environmental science for optimizing compositions, improving performance, and reducing environmental impacts across a wide range of applications, contributing to the development of sustainable solutions for environmental challenges.

STEPS

Developing mixture designed experiments in environmental science involves several steps to optimize the composition of mixtures and understand how different components contribute to system performance. Here are the typical steps involved:

1. **Define the Objectives**: Clearly define the objectives of the mixture designed experiment. Determine what aspects of the system you want to optimize or investigate, such as performance, cost, environmental impact, or a combination of factors.
2. **Identify Components**: Identify the components that will make up the mixture. These components could be chemical substances, materials, treatments, or other factors that contribute to system performance.
3. **Determine the Mixture Design**: Select the appropriate mixture design approach based on the objectives of the experiment and the number of components involved. Common mixture design approaches include simplex-centroid designs, simplex-lattice designs, and simplex-centroid-lattice designs.
4. **Define the Experimental Region**: Determine the constraints and ranges for each component within which the experiment will be conducted. Consider factors, such as safety, feasibility, and practicality, when defining the experimental region.
5. **Generate the Experimental Design**: Use the selected mixture design approach to generate the experimental design matrix. This matrix specifies the combinations of component levels (mixtures) to be tested in the experiment.
6. **Conduct the Experiment**: Prepare the mixtures according to the experimental design matrix and conduct the experiment under controlled conditions. Ensure that the experimental setup is consistent and reproducible across all trials.
7. **Measure Responses**: Measure the responses of interest for each mixture in the experiment. These responses could include performance metrics, physical properties, environmental impacts, or other relevant outcomes.
8. **Analyze the Data**: Analyze the experimental data to understand how different components contribute to system performance and identify optimal mixtures. Use statistical techniques, such as regression analysis, ANOVA, or mixture modeling, to analyze the data.
9. **Optimize the Mixture**: Identify the optimal mixture composition(s) that maximize performance or achieve the desired objectives. Use optimization techniques to search for the best combination of component levels within the experimental region.
10. **Validate the Results**: Validate the optimized mixture composition(s) through confirmation experiments or external validation. Ensure that the optimized mixtures perform as expected and meet the desired objectives under different conditions.
11. **Interpret and Communicate Results**: Interpret the results of the mixture design experiment in the context of the research objectives and implications for environmental science. Prepare a comprehensive report documenting the experimental design, methodology, results, and conclusions. Communicate the findings to stakeholders, management, or other relevant parties to facilitate decision-making and implementation of optimized mixtures.

By following these steps, researchers can systematically develop and conduct mixture designed experiments in environmental science to optimize compositions, improve performance, and address environmental challenges effectively.

PITFALLS AND REMEDIES

Mixture designed experiments in environmental science offer a powerful tool for optimizing compositions and understanding the relationships between components and system performance.

However, like any experimental design approach, they can encounter pitfalls that may affect the validity and reliability of results. Here are some common pitfalls and remedies associated with mixture designed experiments:

Inadequate Component Selection:

Pitfall: Selecting an incomplete set of components or failing to include important factors in the experimental design can lead to suboptimal results.

Remedy: Conduct a thorough analysis to identify all potential components that may influence system performance. Consider factors, such as chemical properties, physical characteristics, environmental impacts, and interactions between components, when selecting components for inclusion in the mixture design.

Inaccurate Mixture Design:

Pitfall: Inaccurately defining the experimental region or constraints for each component can lead to the exploration of inappropriate mixtures or failure to capture important relationships.

Remedy: Conduct sensitivity analyses or preliminary studies to determine appropriate ranges and constraints for each component within the experimental region. Consider factors, such as safety, feasibility, and practicality, when defining the experimental region.

Confounding Effects:

Pitfall: Confounding effects between components or interactions may lead to biased estimates and misinterpretation of results.

Remedy: Choose a mixture design approach that allows for estimation of main effects and selected interactions while minimizing confounding. Use statistical techniques, such as mixture modeling or orthogonal designs, to resolve confounding and improve the accuracy of estimates.

Nonlinear Relationships:

Pitfall: Assuming linear relationships between components and responses may not capture the true nature of the system, especially if nonlinear relationships exist.

Remedy: Use nonlinear regression techniques or advanced modeling approaches to capture nonlinear relationships between components and responses. Consider techniques, such as RSM or neural networks, to model complex interactions and nonlinear effects.

Unreliable Response Measurement:

Pitfall: Inaccurate or imprecise measurement of response variables can introduce variability and undermine the validity of results.

Remedy: Use reliable measurement techniques and standardized protocols to ensure accurate and consistent measurement of response variables. Implement quality control measures to minimize measurement errors and ensure data integrity.

Overfitting:

Pitfall: Overfitting occurs when a model fits the noise in the data rather than the underlying relationships, leading to poor predictive performance.

Remedy: Use cross-validation techniques or split-sample validation to assess the generalizability of the model and avoid overfitting. Consider simpler models or regularization techniques to prevent overfitting and improve model robustness.

Failure to Validate Results:

Pitfall: Failing to validate the optimized mixture compositions through confirmation experiments or external validation can lead to the implementation of suboptimal solutions.

Remedy: Validate the optimized mixture compositions through confirmation experiments conducted under different conditions or external validation studies. Compare the observed results with predicted values to assess the accuracy and robustness of the optimization.

By being aware of these potential pitfalls and implementing appropriate remedies, researchers can enhance the validity, reliability, and usefulness of mixture designed experiments in environmental science, leading to improved optimization and performance of environmental systems and processes.

Example 1

Mixture designed experiments are used in environmental science when the response variable is influenced by the proportion or composition of multiple components mixed together. Here's an example of a matrix for a mixture designed experiment in environmental science:

Objective: To optimize the composition of a soil amendment mixture for enhancing plant growth by studying the effects of three components: organic matter (factor A), sand (factor B), and vermiculite (factor C) on plant growth.

Factors:

1. **Factor A**: Organic matter (%).
   - Low (e.g., 10%).
   - Medium (e.g., 20%).
   - High (e.g., 30%).
2. **Factor B**: Sand (%)
   - Low (e.g., 40%).
   - Medium (e.g., 50%).
   - High (e.g., 60%).
3. **Factor C**: Vermiculite (%)
   - Low (e.g., 20%).
   - Medium (e.g., 30%).
   - High (e.g., 40%).

Response Variable: Plant growth rate (measured in cm/week).

Mixture Experiment Matrix:

In a mixture designed experiment, you systematically vary the proportions of the components while keeping the total constant. The matrix is shown in Table 2.39.

In this matrix, each row represents a specific mixture of organic matter, sand, and vermiculite proportions. The objective is to optimize the mixture composition to maximize plant growth rate.

TABLE 2.39

Mixture Designed Experiment Matrix Example for Environmental Science

Run	Organic Matter (%)	Sand (%)	Vermiculite (%)	Plant Growth Rate (cm/week)
1	10	40	50	...
2	20	40	40	...
3	30	40	30	...
4	10	50	40	...
5	20	50	30	...
6	30	50	20	...
7	10	60	30	...
8	20	60	20	...
9	30	60	10	...

Mixture designed experiments help determine the optimal combination of components while considering the constraints of the total mixture. Various statistical techniques, such as mixture design analysis, are employed to identify the best composition for achieving the desired environmental outcome.

Example 2

In mixture designed experiments, the experimental design matrix specifies the combinations of component levels (mixtures) to be tested in the experiment. Each row of the matrix represents a unique mixture composition, and each column represents a component with its corresponding level. Here's an example of a matrix for a mixture designed experiment in environmental science (Table 2.40):

In this example,

- **Components (A, B, C, D):** Each component represents a different ingredient, treatment, or factor that contributes to the mixture. In environmental science, these components could be chemical substances, materials, treatments, or other factors relevant to the study.
- **Levels (0 to 1):** The levels represent the proportion or percentage of each component in the mixture. In this example, the levels range from 0 to 1 total, indicating the proportion of each component relative to the total mixture.
- **Response Variable (y):** The response variable represents the outcome or performance measure of interest in the experiment. This could be physical property, environmental impact, performance metric, or other relevant measure.

Each row in the matrix represents a unique mixture composition, with specific levels of each component. Researchers would conduct experiments for each mixture composition, measure the response variable(s), and analyze the data to understand how different component levels affect system performance and identify optimal mixture compositions.

REPORTING

In mixture designed experiments for environmental science, several statistical tools are commonly used to analyze data, assess the effects of component proportions, and optimize mixture compositions. Here are some of the key statistical tools used in mixture designed experiments:

ANOVA: ANOVA is used to partition the variability in the response variable(s) into different sources, such as main effects, pure component effects, mixture effects, and error. It helps determine the significance of components and interactions with the response variable(s) and assess the relative importance of each component in influencing system performance.

TABLE 2.40

Mixture Designed Experiment Matrix Example for Environmental Science

Run	Component 1 (A)	Component 2 (B)	Component 3 (C)	Component 4 (D)	Response
1	0.2	0.3	0.4	0.1	Y1
2	0.4	0.1	0.2	0.3	Y2
3	0.3	0.2	0.5	0	Y3
4	0.1	0.4	0.3	0.2	Y4
5	0.5	0	0.1	0.4	Y5

Component Effects Plots: Component effects plots visualize the effects of each component on the response variable(s) while holding other components constant. They help identify the main effects of components and assess the direction and magnitude of their influence on system performance.

Mixture Contour Plots: Mixture contour plots visualize the response surface in the space of mixture components. They help visualize the relationships between component proportions and system performance and identify regions of optimal performance or areas of sensitivity to component changes.

RSM: RSM involves fitting mathematical models to experimental data to describe the relationship between component proportions and response variable(s). It helps optimize mixture compositions and identify optimal component proportions that maximize or minimize system performance within the experimental region.

Desirability Function Analysis: Desirability function analysis combines multiple response variables into a single desirability function to identify optimal mixture compositions that simultaneously meet multiple performance criteria. It helps balance trade-offs between conflicting objectives and identify compromise solutions.

Optimization Algorithms: Optimization algorithms, such as gradient descent, genetic algorithms, or simulated annealing, are used to identify optimal mixture compositions that maximize or minimize the response variable(s) within specified constraints. They iteratively search the experimental region to find component proportions that optimize system performance.

Model Validation Techniques: Model validation techniques, such as cross-validation or split-sample validation, are used to assess the accuracy and robustness of the fitted models. They help ensure that the models accurately capture the relationships between component proportions and response variable(s) and generalize well to new data.

Sensitivity Analysis: Sensitivity analysis techniques are used to assess the sensitivity of system performance to changes in component proportions. They help identify critical components and quantify the impact of component variations on system performance.

By using these statistical tools effectively, researchers can analyze data from mixture designed experiments, identify significant components and interactions, optimize mixture compositions, and improve system performance in environmental science applications.

When reporting on mixture designed experiments in environmental science, it's important to provide a clear and comprehensive document that communicates the research objectives, methods, results, and conclusions effectively to the intended audience. Here are the key elements that should be included in such a report:

Title Page: The title page should include the title of the report, the names of the authors, their affiliations, and the date of publication.

Abstract: The abstract provides a concise summary of the research objectives, methods, key findings, and conclusions of the mixture designed experiment. It allows readers to quickly understand the main aspects of the study without reading the entire report.

Introduction:
- Background and Literature Review: Provide background information on the research topic, including relevant literature and previous studies. Explain the significance of the research and the rationale for using mixture designed experiments to optimize compositions or investigate component effects.
- Research Objectives and Hypotheses: Clearly state the research objectives and hypotheses that the experiment aims to address. Outline the specific components, levels, and response variables under investigation.

Methods:
- Experimental Design: Describe the mixture designed approach used in the experiment, including the selection of components, levels, and experimental region. Explain how the experimental design matrix was generated and any additional runs conducted.
- Data Collection: Detail the methods and protocols used to collect data on the response variable(s), including measurement techniques, sampling procedures, and data recording formats.
- Statistical Analysis: Outline the statistical methods used to analyze the data, such as ANOVA, RSM, or optimization algorithms. Specify any assumptions made and how they were validated.

Results:
- Descriptive Statistics: Present summary statistics (e.g., means, standard deviations) for each component level and experimental run.
- Analysis of Effects: Present the results of the statistical analysis, including main effects, component effects, mixture effects, and significant interactions. Use graphical tools, such as component effects plots, mixture contour plots, and response surface plots, to visualize the effects.
- Optimization Results: Identify the optimal mixture compositions that maximize or minimize the response variable(s) based on the results of the mixture designed experiment.

Discussion:
- Interpretation of Findings: Interpret the results of the experiment in the context of the research objectives and hypotheses. Discuss the implications of significant effects for optimizing compositions or understanding component effects.
- Comparison with Previous Studies: Compare the findings of the current study with previous research in the field. Highlight any consistencies or discrepancies and explain possible reasons for differences.
- Limitations and Future Directions: Acknowledge any limitations of the study and suggest areas for future research or improvements to experimental design.

Conclusion: Summarize the main findings of the study and their implications for environmental science. Discuss the practical implications of the research and potential applications of the optimized mixture compositions.

References: Provide a list of references cited in the report, following a specific citation style (e.g., APA, MLA).

Appendices: Include any supplementary materials, such as the mixture designed matrix, raw data, detailed statistical analyses, or additional figures and tables.

By including these elements in a report on mixture designed experiments in environmental science, researchers can effectively communicate their findings, methods, and conclusions to the scientific community and other stakeholders, facilitating the dissemination of knowledge and advancement of the field.

2.3.8 SEQUENTIAL EXPERIMENTATION

Sequential design experiments find applications in various areas within environmental science where researchers need to iteratively optimize processes, investigate complex systems, or make decisions under uncertainty. Here are some common applications of sequential design experiments in environmental science:

Environmental Monitoring and Sampling Design: In environmental monitoring programs, researchers may use sequential design experiments to optimize the selection of sampling locations, sampling times, and sample sizes. By iteratively collecting and analyzing data, researchers can adaptively refine sampling strategies to maximize information gain, minimize sampling effort, and improve the accuracy of environmental assessments.

Ecological Studies and Biodiversity Monitoring: In ecological studies, researchers may use sequential design experiments to optimize the allocation of resources for biodiversity monitoring and species inventories. By adaptively selecting sampling sites based on initial survey data, researchers can prioritize areas of high biodiversity or rare species occurrence, improving the efficiency and effectiveness of conservation efforts.

Habitat Restoration and Ecosystem Management: Sequential design experiments can be applied in habitat restoration projects and ecosystem management initiatives to optimize the allocation of resources for habitat enhancement, species reintroduction, or invasive species control. By iteratively monitoring ecosystem responses and adjusting management actions, researchers can adaptively manage restoration efforts to achieve desired ecological outcomes.

Water Resources Management: In water resources management, sequential design experiments can be used to optimize the allocation of water resources for irrigation, drinking water supply, and environmental flow requirements. By adaptively adjusting water allocation strategies based on real-time data on water availability, demand, and environmental conditions, researchers can improve water resource management practices and mitigate the impacts of water scarcity and drought.

Climate Change Adaptation and Resilience: Sequential design experiments may be applied in climate change adaptation and resilience planning to optimize adaptation strategies and decision-making under uncertainty. By iteratively assessing climate impacts, vulnerability, and adaptation options, researchers can adaptively refine adaptation strategies to enhance resilience and reduce risks associated with climate change.

Air Quality Management and Pollution Control: In air quality management and pollution control, sequential design experiments can be used to optimize the allocation of emission reduction measures, such as pollution control technologies, emission trading programs, and transportation policies. By iteratively assessing air quality impacts and cost-effectiveness of control measures, researchers can adaptively refine air quality management strategies to achieve regulatory compliance and public health objectives.

Overall, sequential design experiments are valuable tools in environmental science for iteratively optimizing processes, improving decision-making, and addressing complex environmental challenges through adaptive management and learning-by-doing approaches.

STEPS

Developing sequential design experiments in environmental science involves a systematic process to iteratively optimize processes, investigate complex systems, or make decisions under uncertainty. Here are the typical steps involved:

1. **Define the Objectives**: Clearly define the objectives of the sequential design experiment. Determine what aspects of the system you want to optimize, investigate, or make decisions about, such as process performance, resource allocation, or environmental management strategies.

2. **Identify Variables and Constraints**: Identify the key variables and constraints that influence the system or decision-making process. These variables could include environmental factors, management actions, resource constraints, or other relevant factors.

3. **Select the Experimental Design Approach**: Select the appropriate experimental design approach based on the objectives of the experiment and the nature of the variables and constraints involved. Common sequential design approaches include adaptive sampling, adaptive management, Bayesian optimization, and sequential decision-making methods.

4. **Develop the Initial Experimental Plan**: Develop an initial experimental plan that outlines the sequence of actions or interventions to be taken, the criteria for making decisions at each step, and the monitoring and evaluation procedures to assess system responses.

5. **Initiate the Experiment**: Initiate the experiment by implementing the initial experimental plan and collecting relevant data or observations. This may involve conducting field surveys, monitoring environmental conditions, implementing management actions, or simulating scenarios using modeling approaches.

6. **Monitor System Responses**: Continuously monitor system responses or outcomes in real time or at regular intervals as the experiment progresses. Collect data on relevant variables, assess system performance, and evaluate the effectiveness of interventions or decision-making strategies.

7. **Analyze Data and Make Decisions**: Analyze the data collected during the experiment to assess system responses, identify trends or patterns, and make informed decisions. Use statistical techniques, modeling approaches, or decision support tools to analyze data and evaluate the effectiveness of interventions or management strategies.

8. **Adapt and Refine the Experimental Plan**: Based on the analysis of data and system responses, adapt and refine the experimental plan as needed. Modify the sequence of actions, adjust decision criteria, or update monitoring and evaluation procedures to improve the efficiency and effectiveness of the experiment.

9. **Iterate and Repeat**: Iterate the experiment by repeating the cycle of data collection, analysis, decision-making, and adaptation until the objectives of the experiment are achieved or the desired outcomes are obtained. Continue to refine the experimental plan and make adjustments based on new information and insights gained during the experiment.

10. **Document and Communicate Results**: Document the results of the sequential design experiment, including data collected, analyses conducted, decisions made, and outcomes achieved. Communicate the findings to stakeholders, decision-makers, or the scientific community through reports, presentations, or publications to facilitate learning and knowledge exchange.

By following these steps, researchers can systematically develop and conduct sequential design experiments in environmental science to optimize processes, improve decision-making, and address complex environmental challenges through adaptive management and iterative learning approaches.

Pitfalls and Remedies

Sequential design experiments in environmental science offer flexibility and adaptability, allowing researchers to iteratively optimize processes, investigate complex systems, or make decisions under uncertainty. However, they can encounter pitfalls that may affect the validity, reliability, and effectiveness of the experiment. Here are some common pitfalls and remedies associated with Sequential design experiments:

Sampling Bias:
> **Pitfall:** Sequential sampling may introduce bias if certain areas or conditions are oversampled or under-sampled, leading to inaccurate assessments of system responses.
> **Remedy:** Implement randomization and stratification techniques to ensure representative sampling and minimize bias. Monitor sampling distributions and adjust sampling strategies as needed to achieve balanced coverage of the study area or conditions.

Overfitting:
> **Pitfall:** Overfitting occurs when a model fits the noise in the data rather than the underlying relationships, leading to poor predictive performance and generalizability.

Remedy: Use cross-validation techniques or split-sample validation to assess the generalizability of models and avoid overfitting. Regularize models or use simpler modeling approaches to prevent overfitting and improve model robustness.

Information Loss:

Pitfall: Sequential decision-making may lead to information loss if decisions are based on incomplete or biased data, resulting in suboptimal outcomes.

Remedy: Incorporate uncertainty quantification techniques and decision analysis tools to assess the reliability of information and quantify decision uncertainty. Use adaptive sampling and monitoring strategies to maximize information gain and minimize information loss throughout the experiment.

Implementation Challenges:

Pitfall: Implementation of management actions or interventions may encounter practical challenges or constraints, leading to delays or deviations from the experimental plan.

Remedy: Anticipate potential implementation challenges and develop contingency plans to address them. Collaborate with stakeholders and practitioners to ensure feasibility and effectiveness of management actions. Monitor implementation progress and adjust plans as needed to mitigate risks and ensure successful outcomes.

Resource Limitations:

Pitfall: Limited resources, such as time, budget, or personnel, may constrain the scope or effectiveness of sequential design experiments, hindering their ability to achieve desired objectives.

Remedy: Prioritize objectives and allocate resources strategically to maximize the impact of the experiment. Implement adaptive resource allocation strategies to reallocate resources based on changing priorities and emerging needs. Seek external funding or collaborations to expand resources and enhance experimental capacity.

Lack of Stakeholder Engagement:

Pitfall: Lack of stakeholder engagement may result in limited buy-in, resistance to change, or incomplete understanding of the experiment's objectives and outcomes.

Remedy: Engage stakeholders early and throughout the experiment to solicit input, address concerns, and foster collaboration. Communicate transparently and effectively to ensure stakeholders understand the purpose, process, and potential benefits of the experiment. Incorporate stakeholder feedback into decision-making to enhance relevance and acceptance of results.

By being aware of these potential pitfalls and implementing appropriate remedies, researchers can enhance the validity, reliability, and effectiveness of sequential design experiments in environmental science, facilitating adaptive management, informed decision-making, and sustainable environmental stewardship.

EXAMPLE

Sequential experimentation is a process where experiments are conducted sequentially, and the results of each experiment guide the design of subsequent experiments. Here's an example of a matrix for a sequential design experiment in environmental science:

Objective: To optimize the conditions for a bioremediation process to remove a specific pollutant from contaminated soil.

Initial Experiment: In the initial experiment, a full factorial design is used to test the effects of two factors: nutrient concentration (factor A) and microbial inoculation (factor B) on pollutant removal efficiency.

Factors:

1. **Factor A**: Nutrient concentration (low/medium/high).
2. **Factor B**: Microbial inoculation (none/low/high).

Response Variable: Pollutant removal efficiency (%).

Initial Experiment Matrix:
The initial experiment might have a matrix as shown in Table 2.41.

After conducting the initial experiment, you analyze the results. Based on the findings, you decide to focus on the combination of high nutrient concentration and high microbial inoculation, which showed the highest pollutant removal efficiency.

Sequential Experiment: In the sequential experiment, you further optimize the conditions within the chosen combination.

Factors for Sequential Experiment:

1. **Factor C**: pH adjustment (low/high).
2. **Factor D**: Aeration (low/high).

Sequential Experiment Matrix:
The sequential experiment might have a matrix as shown in Table 2.42.

TABLE 2.41

Sequential Design Initial Experiment Matrix Example for Environmental Science

Run	Nutrient Concentration	Microbial Inoculation	Pollutant Removal Efficiency (%)
1	Low	None	...
2	Medium	None	...
3	High	None	...
4	Low	Low	...
5	Medium	Low	...
6	High	Low	...
7	Low	High	...
8	Medium	High	...
9	High	High	...

TABLE 2.42

Sequential Design Experiment Matrix Example for Environmental Science

Run	Nutrient Concentration	Microbial Inoculation	pH Adjustment	Aeration	Pollutant Removal Efficiency (%)
10	High	High	Low	Low	...
11	High	High	High	Low	...
12	High	High	Low	High	...
13	High	High	High	High	...

In this SED, you are building on the results of the initial experiment by exploring additional factors and levels to further optimize the bioremediation process. The outcome of each experiment guides the design of the next one, helping you converge toward the optimal conditions for pollutant removal.

REPORTING

In sequential design experiments for environmental science, various statistical tools are employed to analyze data, make decisions, and optimize processes iteratively. These tools help researchers adaptively refine experimental designs, assess system responses, and inform decision-making throughout the experiment. Here are some common statistical tools used in sequential design experiments for environmental science:

Sequential Analysis: Sequential analysis techniques allow researchers to monitor data as it is collected and make interim decisions based on accumulating evidence. These techniques typically involve sequential hypothesis testing, sequential estimation, or sequential model fitting methods.

Bayesian Methods: Bayesian methods are well-suited for sequential design experiments as they provide a flexible framework for updating beliefs and making decisions in light of new data. Bayesian sequential methods include Bayesian updating, Bayesian optimization, and Bayesian decision theory.

Optimization Algorithms: Optimization algorithms, such as genetic algorithms, simulated annealing, or particle swarm optimization, are used to optimize experimental designs, decision-making strategies, or system parameters iteratively. These algorithms iteratively search for optimal solutions based on feedback from previous iterations.

Adaptive Sampling Techniques: Adaptive sampling techniques allow researchers to dynamically adjust sampling strategies based on observed data and changing objectives. These techniques include adaptive cluster sampling, adaptive stratified sampling, and adaptive sequential sampling methods.

Dynamic Programming: Dynamic programming techniques are used to solve sequential decision-making problems by breaking them down into smaller subproblems and recursively optimizing decisions over time. Dynamic programming methods are particularly useful for solving sequential optimization problems with complex decision structures.

Sequential Model Fitting: Sequential model fitting techniques involve iteratively fitting statistical models to data as it is collected and updating model parameters based on new observations. These techniques allow researchers to adaptively refine models and improve their predictive accuracy over time.

Decision Trees and Bayesian Networks: Decision trees and Bayesian network models are used to represent decision-making processes and system dynamics in sequential design experiments. These graphical models help researchers visualize decision pathways, identify influential factors, and assess decision uncertainty.

Statistical Process Control (SPC): SPC techniques are used to monitor and control system performance in real time during sequential design experiments. SPC methods involve monitoring process variables, detecting deviations from expected patterns, and taking corrective actions as needed.

Survival Analysis: Survival analysis techniques are used to analyze time-to-event data in sequential design experiments, such as time until system failure, time until a specific outcome occurs, or time until a decision is made. Survival analysis methods include Kaplan–Meier estimation, Cox proportional hazards models, and accelerated failure time models.

By using these statistical tools effectively, researchers can analyze data, make informed decisions, and optimize processes iteratively in sequential design experiments for environmental science, leading to improved understanding, enhanced system performance, and better-informed decision-making.

Reports for sequential design experiments in environmental science should provide a comprehensive overview of the experiment's objectives, methods, results, and conclusions. Here are the key elements typically included in such a report:

Title Page: The title page should include the title of the report, the names of the authors, their affiliations, and the date of publication.

Abstract: The abstract provides a concise summary of the experiment, including the research objectives, methods, key findings, and conclusions.

Introduction:

- Background and Literature Review: Provide background information on the research topic, including relevant literature and previous studies. Explain the significance of the experiment and the rationale for using a sequential design approach.
- Research Objectives: Clearly state the research objectives and hypotheses that the experiment aims to address.

Methods:

- Experimental Design: Describe the sequential design approach used in the experiment, including the sequence of steps, decision criteria, and adaptive strategies employed.
- Data Collection: Detail the methods and protocols used to collect data, including sampling procedures, treatment applications, and measurement techniques.
- Statistical Analysis: Outline the statistical methods used to analyze the data collected at each step, including sequential analysis techniques, Bayesian methods, or adaptive sampling strategies.

Results:

- Descriptive Statistics: Present summary statistics and graphical representations of the data collected at each step of the experiment.
- Analysis of System Responses: Present the results of the statistical analysis, including trends, patterns, and significant findings observed over the course of the experiment.
- Decision Outcomes: Describe the decisions made at each step based on the observed responses and decision criteria.

Discussion:

- Interpretation of Findings: Interpret the results of the experiment in the context of the research objectives and hypotheses. Discuss the implications of the findings for understanding system dynamics or optimizing processes.
- Comparison with Previous Studies: Compare the findings of the current experiment with previous research in the field. Highlight any consistencies or discrepancies and explain possible reasons for differences.
- Limitations and Future Directions: Acknowledge any limitations of the experiment and suggest areas for future research or improvements to experimental design.

Conclusion: Summarize the main findings of the experiment and their implications for environmental science. Discuss the practical applications of the findings and potential avenues for further investigation.

References: Provide a list of references cited in the report, following a specific citation style (e.g., APA, MLA).

Appendices: Include any supplementary materials, such as detailed data tables, additional statistical analyses, or supporting documentation related to the experiment.

By including these elements in a report for sequential design experiments in environmental science, researchers can effectively communicate the experiment's objectives, methods, results, and conclusions to the scientific community and other stakeholders, facilitating the dissemination of knowledge and advancement of the field.

2.3.9 RPD

RPD experiments find applications in environmental science where researchers aim to optimize processes, systems, or products to achieve robust performance in the presence of variability or uncertainty. Here are some common areas where RPD experiments are used in environmental science:

Environmental Monitoring and Sampling Design: In environmental monitoring programs, RPD experiments are used to optimize sampling protocols and measurement procedures to ensure reliable and consistent data collection, even in the presence of environmental variability or measurement error.

Pollution Control and Remediation: RPD experiments are employed to optimize pollution control technologies, remediation strategies, and treatment processes to achieve robust performance in mitigating pollution sources, managing contaminated sites, and restoring environmental quality.

Ecological Restoration and Conservation: In ecological restoration projects and conservation initiatives, RPD experiments are used to optimize habitat restoration techniques, species reintroduction programs, and ecosystem management practices to enhance ecosystem resilience and adaptability to environmental changes.

Climate Change Adaptation and Resilience: RPD experiments are applied in climate change adaptation and resilience planning to optimize adaptation strategies, infrastructure designs, and land-use planning to withstand climate impacts and reduce vulnerability to extreme events.

Water Resources Management: In water resources management, RPD experiments are used to optimize water allocation strategies, reservoir operations, and water supply systems to ensure reliable water supply, reduce risks of drought and floods, and protect aquatic ecosystems.

Air Quality Management: RPD experiments are employed in air quality management to optimize emission control technologies, transportation policies, and urban planning strategies to achieve robust compliance with air quality standards and protect public health.

Waste Management and Recycling: RPD experiments are used to optimize waste management practices, recycling processes, and waste-to-energy technologies to minimize environmental impacts, reduce resource consumption, and enhance waste recovery efficiency.

Natural Resource Management: RPD experiments are applied in natural resource management to optimize harvesting practices, conservation measures, and sustainable development strategies to ensure the long-term viability of natural ecosystems and resources.

Overall, RPD experiments play a crucial role in environmental science by enabling researchers and practitioners to optimize processes, systems, and interventions to achieve robust and resilient environmental outcomes in the face of variability, uncertainty, and changing conditions.

STEPS

Developing RPD experiments in environmental science involves a systematic process to optimize processes, systems, or products to achieve robust performance in the presence of variability or uncertainty. Here are the typical steps involved:

1. **Define the Objectives**: Clearly define the objectives of the RPD experiment. Determine what aspect of the system or process you want to optimize and specify the performance metrics or criteria for robustness.
2. **Identify Key Parameters and Factors**: Identify the key parameters, factors, or variables that influence the performance of the system or process under study. These could include environmental variables, operational parameters, material properties, or design factors.
3. **Characterize Variability and Uncertainty**: Characterize the sources of variability and uncertainty associated with the key parameters and factors identified. This may involve conducting preliminary studies, analyzing historical data, or performing sensitivity analyses.
4. **Design the Experimental Matrix**: Design the experimental matrix that systematically varies the key parameters and factors at different levels to explore their effects on system performance. Use experimental design techniques, such as factorial design, RSM, or Taguchi methods, to efficiently explore the parameter space.
5. **Define Robustness Metrics**: Define metrics or criteria for assessing robustness and performance variability. These metrics should capture the variability in system responses under different conditions and quantify the degree of robustness achieved by the optimized design.
6. **Perform Experiments**: Conduct the experiments outlined in the experimental matrix, systematically varying the key parameters and factors according to the design plan. Collect data on system responses or performance metrics for each experimental condition.
7. **Analyze Data and Identify Optimal Settings**: Analyze the data collected from the experiments to identify the optimal settings or combinations of parameters that achieve robust performance. Use statistical techniques, such as ANOVA, regression analysis, or optimization algorithms, to analyze the data and identify significant factors.
8. **Assess Robustness and Sensitivity**: Assess the robustness of the optimized design by evaluating its performance across a range of operating conditions or environmental scenarios. Conduct sensitivity analyses to identify critical parameters or factors that influence robustness and evaluate their impact on system performance.
9. **Validate and Refine the Design**: Validate the optimized design through additional testing, validation studies, or field trials to confirm its robustness and effectiveness under real-world conditions. Refine the design as needed based on feedback from validation studies or further optimization efforts.
10. **Document and Communicate Results**: Document the results of the RPD experiment, including the experimental design, data collected, analysis methods, and conclusions drawn. Communicate the findings to stakeholders, decision-makers, or the scientific community through reports, presentations, or publications to facilitate knowledge transfer and implementation of the optimized design.

By following these steps, researchers can systematically develop and implement RPD experiments in environmental science to optimize processes, systems, or products and achieve robust performance in the face of variability and uncertainty.

PITFALLS AND REMEDIES

RPD experiments in environmental science are valuable for optimizing processes and systems to achieve robust performance in the face of variability and uncertainty. However, they can encounter pitfalls that may affect the validity, reliability, and effectiveness of the experiment. Here are some common pitfalls and remedies associated with RPD experiments:

Failure to Identify Key Parameters:

Pitfall: If key parameters or factors that significantly influence system performance are not identified accurately, the experiment may fail to capture important sources of variability.

Remedy: Conduct thorough sensitivity analyses, literature reviews, and expert consultations to identify and prioritize key parameters. Consider both known factors and potential sources of uncertainty.

Inadequate Characterization of Variability:

Pitfall: If the variability associated with key parameters is not adequately characterized, the experiment may not accurately reflect real-world conditions, leading to suboptimal designs.

Remedy: Collect sufficient data to characterize variability in key parameters through historical data analysis, pilot studies, or sensitivity analyses. Consider both random and systematic sources of variability.

Overly Simplistic Experimental Designs:

Pitfall: Overly simplistic experimental designs may fail to capture interactions between parameters or nonlinear relationships, leading to biased estimates and suboptimal designs.

Remedy: Use appropriate experimental design techniques, such as factorial designs, RSM, or Taguchi methods, to systematically vary key parameters and explore complex relationships. Consider higher-order interactions and nonlinear effects.

Insufficient Sample Size:

Pitfall: If the sample size is too small, the experiment may lack statistical power to detect meaningful effects or estimate parameters accurately, leading to unreliable results.

Remedy: Conduct power analyses to determine the appropriate sample size needed to detect relevant effects or achieve desired levels of precision. Increase the sample size if necessary to ensure adequate statistical power.

Ignoring Model Assumptions:

Pitfall: If model assumptions are violated, the results of the experiment may be biased or unreliable. Ignoring model assumptions can lead to inaccurate predictions and suboptimal designs.

Remedy: Validate model assumptions through diagnostic checks, sensitivity analyses, and model validation procedures. Use robust statistical techniques that are less sensitive to violations of model assumptions.

Limited Generalizability:

Pitfall: If the experimental conditions do not represent real-world scenarios adequately, the findings may have limited generalizability, and the optimized design may not perform as expected in practice.

Remedy: Design experiments that reflect the variability and uncertainty present in real-world conditions. Consider conducting validation studies or sensitivity analyses to assess the robustness of the optimized design across a range of scenarios.

Failure to Validate Designs:

Pitfall: If the optimized design is not validated under real-world conditions, there is a risk that it may not perform as expected in practice, leading to wasted resources and missed opportunities.

Remedy: Validate the optimized design through field trials, validation studies, or pilot implementations. Monitor performance under real-world conditions and adjust the design as needed based on feedback.

By being aware of these potential pitfalls and implementing appropriate remedies, researchers can enhance the validity, reliability, and effectiveness of RPD experiments in environmental science, leading to optimized processes, systems, or products that perform reliably in the face of variability and uncertainty.

Example 1

RPD experiments are designed to optimize processes while considering variations and minimizing the impact of external factors. Here's an example of a matrix for an RPD experiment in environmental science:

Objective: To optimize the conditions for a water treatment process by studying the effects of two independent factors, chemical dosage (factor A) and water temperature (factor B), on the removal efficiency of a specific contaminant, while also considering potential variations in water quality.

Factors:

1. **Factor A**: Chemical dosage (low/medium/high).
2. **Factor B**: Water temperature (low/medium/high).

Response Variable: Contaminant removal efficiency (%).

RPD Experiment Matrix:
In an RPD experiment, the matrix is designed to account for variations and study the main effects, interactions, and variations of the factors. The matrix is shown in Table 2.43.

In this matrix, each row represents a specific experimental run with a combination of chemical dosage, water temperature, and variations in water quality. The goal of the RPD experiment is to identify the optimal process conditions while considering potential variations in environmental factors, such as water quality. Statistical techniques, such as response surface modeling and ANOVA, are used to optimize the process and ensure it remains robust to variations.

Example 2

In RPD experiments, the experimental matrix is designed to systematically vary key parameters and factors at different levels to explore their effects on system performance while considering variability and uncertainty. Here's an example of another matrix for an RPD design experiment in environmental science (Table 2.44):
In this example,

- **Experiment**: Represents each experimental run or trial in the matrix.
- **Factor 1 (Temperature)**: Represents one of the key parameters being studied (e.g., temperature), varied at different levels (e.g., low, medium, high).
- **Factor 2 (Humidity)**: Represents another key parameter (e.g., humidity), also varied at different levels.
- **Factor 3 (Wind Speed)**: Represents an additional key parameter (e.g., wind speed), varied at different levels.
- **Response Variable (System Output)**: Represents the variable or metric measured to assess system performance or output (e.g., pollutant concentration, system efficiency).

The matrix systematically varies the levels of each key parameter across different experimental runs, allowing researchers to assess how changes in these parameters affect system performance. By collecting data on the response variable for each experimental condition, researchers can analyze the effects of key parameters, identify optimal settings, and develop robust designs that perform well across a range of environmental conditions.

TABLE 2.43

RPD Matrix Example for Environmental Science

Run	Factor A	Factor B	Water Quality (Variation)	Contaminant Removal Efficiency (%)
1	Low	Low	Low	...
2	Low	Medium	Low	...
3	Low	High	Low	...
4	Medium	Low	Low	...
5	Medium	Medium	Low	...
6	Medium	High	Low	...
7	High	Low	Low	...
8	High	Medium	Low	...
9	High	High	Low	...
10	Low	Low	Medium	...
11	Low	Medium	Medium	...
12	Low	High	Medium	...
13	Medium	Low	Medium	...
14	Medium	Medium	Medium	...
15	Medium	High	Medium	...
16	High	Low	Medium	...
17	High	Medium	Medium	...
18	High	High	Medium	...
19	Low	Low	High	...
20	Low	Medium	High	...
21	Low	High	High	...
22	Medium	Low	High	...
23	Medium	Medium	High	...
24	Medium	High	High	...
25	High	Low	High	...
26	High	Medium	High	...
27	High	High	High	...

TABLE 2.44

RPD Matrix Example for Environmental Science

Experiment	Factor 1 (Temperature)	Factor 2 (Humidity)	Factor 3 (Wind Speed)	Response Variable (System Output)
1	High	Low	High	Y1
2	Low	High	Low	Y2
3	Medium	Medium	Medium	Y3
4	High	High	Low	Y4

REPORT

In RPD experiments for environmental science, various statistical tools are utilized to analyze data, optimize processes, and develop robust designs that perform well under different conditions. These tools help researchers identify significant factors, quantify their effects, and optimize system performance while considering variability and uncertainty. Here are some common statistical tools used in RPD experiments:

ANOVA: ANOVA is used to assess the statistical significance of factors and their interactions with system performance. It helps identify which factors have significant effects on the response variable and quantify the amount of variation explained by each factor.

RSM: RSM is used to model and optimize complex relationships between key parameters and system responses. It involves fitting polynomial regression models to experimental data to identify optimal parameter settings that maximize or minimize the response variable.

DOE: DOE techniques, such as factorial designs, fractional factorial designs, and Taguchi methods, are used to systematically vary key parameters at different levels to efficiently explore the parameter space and identify significant factors affecting system performance.

Optimization Algorithms: Optimization algorithms, such as gradient descent, genetic algorithms, or simulated annealing, are used to search for optimal parameter settings that maximize or minimize system performance metrics. These algorithms iteratively adjust parameter values to converge toward the optimal solution.

Robust Optimization: Robust optimization techniques aim to develop designs that perform well under various environmental conditions or sources of variability. Methods, such as RPD, robust optimization, and robust decision-making, incorporate uncertainty into the optimization process to develop designs that are resilient to variability.

Sensitivity Analysis: Sensitivity analysis is used to assess the sensitivity of system performance to changes in key parameters and factors. It helps identify critical parameters that have a significant impact on system performance and quantify their influence on the response variable.

Monte Carlo Simulation: Monte Carlo simulation is used to simulate the behavior of complex systems under different scenarios and sources of uncertainty. It involves generating random samples of input parameters and running simulations to assess the variability and robustness of system performance.

SPC: SPC techniques are used to monitor and control system performance in real time by detecting deviations from expected patterns and taking corrective actions. It helps ensure that the optimized design maintains robust performance over time.

By employing these statistical tools effectively, researchers can analyze data, optimize processes, and develop robust designs in RPD experiments for environmental science, leading to improved system performance and resilience in the face of variability and uncertainty.

A report for RPD experiments in environmental science should provide a comprehensive overview of the experiment's objectives, methods, results, and conclusions. Here are the key elements typically included in such a report:

Title Page: The title page should include the title of the report, the names of the authors, their affiliations, and the date of publication.

Abstract: The abstract provides a concise summary of the experiment, including the research objectives, methods, key findings, and conclusions.

Introduction:
- Background and Literature Review: Provide background information on the research topic, including relevant literature and previous studies. Explain the significance of the experiment and the rationale for using RPD.
- Research Objectives: Clearly state the research objectives and hypotheses that the experiment aims to address.

Methods:
- Experimental Design: Describe the RPD approach used in the experiment, including the selection of key parameters, experimental matrix design, and data collection procedures.

- Data Collection and Analysis: Detail the methods and protocols used to collect data, analyze results, and assess system performance. Include information on statistical analyses and optimization techniques employed.

Results:

- Descriptive Statistics: Present summary statistics and graphical representations of the data collected during the experiment.
- Analysis of System Performance: Present the results of statistical analyses, optimization algorithms, and sensitivity analyses. Describe how key parameters and factors influence system performance and identify optimal designs.

Discussion:

- Interpretation of Findings: Interpret the results of the experiment in the context of the research objectives and hypotheses. Discuss the implications of the findings for environmental science and practical applications.
- Comparison with Previous Studies: Compare the findings of the current experiment with previous research in the field. Highlight any consistencies or discrepancies and explain possible reasons for differences.
- Limitations and Future Directions: Acknowledge any limitations of the experiment and suggest areas for future research or improvements to experimental design.

Conclusion: Summarize the main findings of the experiment and their implications for environmental science. Discuss the practical applications of the findings and potential avenues for further investigation.

References: Provide a list of references cited in the report, following a specific citation style (e.g., APA, MLA).

Appendices: Include any supplementary materials, such as detailed data tables, additional statistical analyses, or supporting documentation related to the experiment.

By including these elements in a report for RPD experiments in environmental science, researchers can effectively communicate the experiment's objectives, methods, results, and conclusions to the scientific community and other stakeholders, facilitating the dissemination of knowledge and implementation of optimized designs.

2.3.10 Randomized Experiments

RD experiments find applications in various areas of environmental science where researchers aim to assess the effects of treatments, interventions, or factors on environmental systems while minimizing bias and controlling for confounding variables. Here are some common areas where RD experiments are used in environmental science:

Ecological Research: RD experiments are employed to investigate the effects of environmental factors, such as habitat fragmentation, invasive species, or climate change, on ecosystem structure and function. Researchers may use RDs to manipulate experimental plots or treatment areas and assess their ecological responses.

Environmental Toxicology: In studies of environmental toxicology, RD experiments are used to evaluate the effects of pollutants, chemicals, or contaminants on organisms, populations, or ecosystems. Researchers may randomize exposure levels or treatment conditions to control variability and assess dose–response relationships.

Agricultural and Soil Science: RD experiments are commonly used in agricultural and soil science to evaluate the effectiveness of agricultural practices, soil amendments, or crop varieties on crop yield, soil fertility, and environmental sustainability. Randomized field trials allow researchers to control spatial variability and assess treatment effects across different agricultural landscapes.

Water Quality and Aquatic Ecology: RD experiments are utilized in studies of water quality and aquatic ecology to assess the impacts of pollutants; nutrient inputs; or habitat modifications on aquatic ecosystems, such as rivers, lakes, and estuaries. Researchers may use randomized sampling designs to assess spatial and temporal variability in water quality parameters and ecological responses.

Conservation Biology and Restoration Ecology: In conservation biology and restoration ecology, RD experiments are used to assess the effectiveness of conservation measures, habitat restoration techniques, or reintroduction programs in restoring and conserving biodiversity. RDs allow researchers to evaluate treatment effects while controlling for confounding variables and spatial heterogeneity.

Climate Change Research: RD experiments are employed in climate change research to assess the impacts of climate variability and climate change on ecosystems, species distributions, and ecological processes. Researchers may use randomized experiments to manipulate environmental conditions or simulate future climate scenarios and assess ecological responses.

Environmental Monitoring and Management: RD experiments are used in environmental monitoring and management to assess the effectiveness of pollution control measures, environmental management practices, or restoration strategies in improving environmental quality and ecosystem health. RDs allow researchers to rigorously evaluate treatment effects and make evidence-based management decisions.

Overall, RD experiments play a crucial role in environmental science by providing rigorous and unbiased assessments of treatment effects, interventions, and environmental factors across various ecological and environmental systems. They help researchers draw reliable conclusions, inform management decisions, and advance our understanding of complex environmental processes.

Steps

Developing RD experiments in environmental science involves a systematic process to plan and conduct experiments that rigorously assess the effects of treatments or interventions on environmental systems while minimizing bias and controlling for confounding variables. Here are the key steps involved:

1. **Define Research Objectives**: Clearly define the research objectives and hypotheses that the experiment aims to address. Determine the specific treatments, interventions, or factors of interest and the response variables to be measured.
2. **Identify Experimental Units**: Identify the experimental units or sampling units on which treatments will be applied or observations will be made. Experimental units should be homogeneous within treatment groups and representative of the population of interest.
3. **Randomization**:
   - Randomize Treatment Assignment: Randomly assign treatments or interventions to experimental units to minimize bias and ensure that treatment effects are not confounded with other factors.
   - Randomize Experimental Layout: Randomly allocate experimental units to treatment groups and design the layout of the experiment to control spatial or temporal variability.
4. **Design Experimental Protocol**:
   - Determine Experimental Design: Select an appropriate experimental design, such as completely RD, RCBD, or split-plot design, based on the research objectives and constraints.
   - Define Treatment Levels: Specify the levels or doses of each treatment or intervention to be applied, ensuring sufficient variability to detect treatment effects.

- Plan Replication: Determine the number of replications or replicates for each treatment group to ensure adequate statistical power and precision in estimating treatment effects.

5. **Implement Experimental Setup:**
   - Prepare Experimental Site: Prepare the experimental site or laboratory facilities according to the experimental protocol, ensuring proper environmental conditions and equipment calibration.
   - Apply Treatments: Apply treatments or interventions to experimental units following the randomized treatment assignment. Use standardized procedures and protocols to ensure consistency and reproducibility.

6. **Data Collection:**
   - Measure Response Variables: Collect data on response variables of interest from each experimental unit according to the experimental protocol. Use appropriate measurement techniques and sampling methods to minimize measurement error and bias.
   - Record Covariates: Record any covariates or auxiliary variables that may influence the response variables to control potential confounding effects.

7. **Data Analysis:**
   - Conduct Statistical Analysis: Analyze the data collected from the experiment using appropriate statistical methods, such as ANOVA, regression analysis, or GLMs.
   - Assess Treatment Effects: Estimate treatment effects, assess statistical significance, and interpret findings in relation to the research objectives and hypotheses.

8. **Interpret Results and Draw Conclusions**: Interpret the results of the experiment in the context of the research objectives and hypotheses. Draw conclusions regarding the effectiveness of treatments or interventions and their implications for environmental science.

9. **Document and Communicate Findings**: Document the experimental setup, data collection procedures, analysis methods, and results in a comprehensive report or scientific publication. Communicate findings to the scientific community, stakeholders, or policymakers through presentations, publications, or outreach activities.

By following these steps, researchers can develop RD experiments in environmental science that provide rigorous and unbiased assessments of treatment effects, contribute to scientific knowledge, and inform evidence-based decision-making in environmental management and policy.

Pitfalls and Remedies

RD experiments in environmental science are powerful tools for assessing treatment effects and making causal inferences. However, they can encounter pitfalls that may compromise the validity and reliability of the results. Here are some common pitfalls and remedies associated with RD experiments:

Selection Bias:
   Pitfall: Despite randomization, selection bias may occur if there are systematic differences between treatment groups due to factors not accounted for in the randomization process.
   Remedy: Implement strict randomization procedures and ensure that all potential sources of bias are considered during the experimental design phase. Use stratification or blocking techniques to control for known sources of variability.

Noncompliance:
   Pitfall: Noncompliance occurs when participants do not adhere to the assigned treatments, leading to deviations from the intended experimental protocol and compromising the validity of the results.

Remedy: Monitor participants' adherence to the assigned treatments and implement strategies to encourage compliance, such as incentives, reminders, or education. Conduct intention-to-treat analyses to mitigate the effects of noncompliance.

Confounding Variables:

Pitfall: Confounding variables are extraneous factors that are associated with both the treatment and the outcome, leading to biased estimates of treatment effects.

Remedy: Control for confounding variables through randomization, stratification, or matching techniques. Collect data on potential confounders and include them as covariates in the statistical analysis to adjust for their effects.

Small Sample Size:

Pitfall: Small sample sizes may result in low statistical power, making it difficult to detect meaningful treatment effects or leading to imprecise estimates of effect sizes.

Remedy: Conduct power analyses to determine the appropriate sample size needed to detect expected treatment effects with adequate precision. Increase the sample size if necessary to ensure sufficient statistical power.

Missing Data:

Pitfall: Missing data can occur due to participant dropout, measurement error, or other reasons, leading to bias and reduced statistical power.

Remedy: Implement strategies to minimize missing data, such as using multiple imputation techniques or sensitivity analyses to assess the robustness of results under different assumptions about missing data mechanisms.

Contamination:

Pitfall: Contamination occurs when treatments inadvertently affect control groups or when participants receive multiple treatments simultaneously, making it difficult to isolate the effects of individual treatments.

Remedy: Implement strict protocols to prevent contamination, such as physical separation of treatment groups, blinding of participants and researchers, or use of placebo treatments.

Interactions and Effect Modification:

Pitfall: Interactions between treatments or effect modification by participant characteristics may complicate the interpretation of treatment effects and lead to misleading conclusions.

Remedy: Collect data on potential effect modifiers and conduct subgroup analyses or interaction tests to assess whether treatment effects vary across different subgroups. Interpret results cautiously and consider potential interactions in the context of the research question.

By being aware of these potential pitfalls and implementing appropriate remedies, researchers can enhance the validity, reliability, and interpretability of RD experiments in environmental science, ensuring that treatment effects are accurately assessed, and meaningful conclusions are drawn.

Example

Randomized experiments are conducted to investigate the effects of certain factors on a response variable while ensuring that the assignment of experimental units to different treatments is done randomly. Here's an example of a matrix for a randomized experiment in environmental science:

Objective: To study the effects of three different soil amendments (treatment A, treatment B, and treatment C) on plant growth.

Treatments:

1. **Treatment A**: Organic matter amendment.
2. **Treatment B**: Compost amendment.
3. **Treatment C**: Control (no amendment).

Response Variable: Plant growth rate (measured in cm/week).

Randomized Experiment Matrix:
In a randomized experiment, you randomly assign experimental units (e.g., pots of soil) to different treatments to minimize bias and ensure that the effects observed are due to the treatments and not the way the units were selected. The matrix is shown in Table 2.45.

In this matrix, each row represents a different experimental unit (pot of soil) that has been randomly assigned to one of the three treatments (A, B, or C). The goal is to compare the plant growth

TABLE 2.45

Randomized Experiment Matrix Example for Environmental Science

Pot	Treatment
1	A
2	B
3	C
4	A
5	B
6	C
7	A
8	B
9	C
10	A
11	B
12	C
13	A
14	B
15	C
16	A
17	B
18	C
19	A
20	B
21	C
22	A
23	B
24	C
25	A
26	B
27	C
28	A
29	B
30	C

TABLE 2.46
Simplified Randomized Experiment Matrix Example for Environmental Science

Experiment	Treatment A (Control)	Treatment B (Fertilizer)	Treatment C (Compost)
1	Control	Fertilizer	Compost
2	Compost	Control	Fertilizer
3	Fertilizer	Compost	Control

rates under different treatments to determine which soil amendment, if any, is the most effective for promoting plant growth.

Randomized experiments are a fundamental approach in scientific research to ensure the validity and reliability of experimental results. Random assignment helps control potential sources of bias and allows for valid statistical inference about the effects of treatments on the response variable in environmental science studies.

Here's an example of a simplified matrix for a hypothetical experiment investigating the effects of different soil treatments on plant growth (Table 2.46):

In this example,

- **Experiment**: Represents each experimental run or trial.
- **Treatment A (Control)**: Represents the control treatment where no soil amendments are applied.
- **Treatment B (Fertilizer)**: Represents the treatment where fertilizer is applied to the soil.
- **Treatment C (Compost)**: Represents the treatment where compost is applied to the soil.

The treatments are randomly assigned to different experimental units (e.g., plots, pots) to ensure that each treatment group is representative of the overall population and to minimize bias in treatment effects. Randomization helps ensure that any observed differences in plant growth between treatment groups are due to the treatments themselves rather than other factors.

Each row in the matrix represents a unique experimental run where treatments are randomly assigned to experimental units according to the randomization scheme. By collecting data on plant growth (e.g., height, biomass) for each experimental unit, researchers can assess the effects of different soil treatments on plant performance while controlling for confounding variables and sources of bias.

REPORT

In RD experiments for environmental science, various statistical tools are used to analyze data, assess treatment effects, and draw valid conclusions. These tools help researchers make inferences about the effects of treatments on environmental systems while controlling for sources of bias and variability. Here are some common statistical tools used in RD experiments:

ANOVA: ANOVA is used to assess the statistical significance of treatment effects by comparing the variability between treatment groups to the variability within treatment groups. It helps determine whether observed differences in outcomes are due to the treatments themselves or random variation.

Post-Hoc Tests: Post-hoc tests, such as Tukey's HSD, Bonferroni correction, or Dunnett's test, are used to identify specific pairwise differences between treatment groups when ANOVA indicates significant overall differences.

Regression Analysis: Regression analysis is used to model the relationship between treatment variables and outcome variables, accounting for potential confounding variables or covariates. It helps quantify the effects of treatments while controlling for other factors that may influence the outcomes.

Repeated Measures Analysis: Repeated measures analysis techniques, such as mixed-effects models or repeated measures ANOVA, are used when measurements are collected repeatedly over time or across different conditions within the same subjects. These methods account for within-subject correlation and improve statistical efficiency.

Nonparametric Tests: Nonparametric tests, such as the Kruskal–Wallis test or Wilcoxon rank sum test, are used when the assumptions of parametric tests (e.g., normality, homogeneity of variance) are violated. These tests provide robust alternatives for analyzing data with skewed distributions or unequal variances.

Covariate Adjustment: Covariate adjustment techniques, such as ANCOVA or propensity score matching, are used to control potential confounding variables or baseline differences between treatment groups. These methods improve the precision of treatment effect estimates and reduce bias.

Multivariate Analysis: Multivariate analysis techniques, such as PCA or discriminant analysis, are used to analyze complex datasets with multiple outcome variables or to explore patterns in treatment responses across different dimensions.

Survival Analysis: Survival analysis techniques, such as Kaplan–Meier survival curves or Cox proportional hazards models, are used when the outcome of interest is time-to-event data (e.g., time until plant mortality). These methods account for censoring and allow for the analysis of time-dependent outcomes.

Bootstrapping: Bootstrapping is a resampling technique used to estimate the sampling distribution of statistics and construct confidence intervals for treatment effects. It provides robust inference in situations where parametric assumptions are violated, or sample sizes are small.

By employing these statistical tools effectively, researchers can analyze data from RD experiments in environmental science, assess treatment effects, and make valid inferences about the impacts of treatments on environmental systems while controlling for potential sources of bias and variability.

A report for RD experiments in environmental science typically includes several key elements to effectively communicate the experiment's design, methods, results, and conclusions. These elements provide a comprehensive overview of the experiment and its findings. Here are the essential components of such a report:

Title Page: The title page includes the title of the report, the names of the authors, their affiliations, and the date of publication.

Abstract: The abstract is a concise summary of the experiment, including the research objectives, methods, key findings, and conclusions. It provides an overview of the study's significance and implications.

Introduction:
- Background and Literature Review: Provides background information on the research topic, including relevant literature and previous studies. It explains the rationale for conducting the experiment and the research questions or hypotheses being tested.
- Objectives: Clearly states the specific objectives of the experiment and the treatments or interventions being studied.

Methods:
- Experimental Design: Describes the RD used in the experiment, including the randomization procedure, treatment allocation, and sample size determination.
- Experimental Setup: Provides details about the experimental site or laboratory setup, including environmental conditions, equipment used, and any special considerations.

- Treatment Application: Explains how treatments were applied to experimental units, including the timing, dosage, and application methods.
- Data Collection: Describes the methods used to collect data on outcome variables, including measurement techniques, sampling protocols, and quality control measures.
- Statistical Analysis: Explains the statistical methods used to analyze the data, including hypothesis testing procedures, regression models, or other relevant techniques.

Results:
- Descriptive Statistics: Presents summary statistics for the outcome variables measured in the experiment, including means, standard deviations, and ranges.
- Treatment Effects: Reports the results of statistical analyses, including tests of treatment effects, confidence intervals, and p-values. It may use tables, figures, or graphs to present the data visually.
- Post-hoc Analyses: Provides results of post-hoc analyses, such as pairwise comparisons between treatment groups.

Discussion:
- Interpretation of Findings: Interprets the results of the experiment in relation to the research objectives and hypotheses. Discusses the significance of observed treatment effects and their implications for environmental science.
- Comparison with Previous Studies: Compares the findings of the current experiment with previous research in the field. Highlights any consistencies or discrepancies and explains possible reasons for differences.
- Limitations and Future Directions: Acknowledges any limitations of the experiment and suggests areas for future research or improvements to experimental design.

Conclusion: Summarizes the main findings of the experiment and their implications for environmental science. Discusses the practical applications of the findings and potential avenues for further investigation.

References: Provides a list of references cited in the report, following a specific citation style (e.g., APA, MLA).

Appendices: Includes any supplementary materials, such as detailed data tables, additional statistical analyses, or supporting documentation related to the experiment.

By including these elements in a report for RD experiments in environmental science, researchers can effectively communicate the experiment's objectives, methods, results, and conclusions to the scientific community and other stakeholders, facilitating the dissemination of knowledge and informing evidence-based decision-making in environmental management and policy.

2.3.11 Split-Plot and Blocked Experiments

Split-plot and blocked experiments are commonly used in environmental science when researchers need to account for various sources of variability and spatial heterogeneity in their experimental design. These designs find applications in several areas within environmental science:

Agricultural Field Trials: In agricultural research, split-plot and blocked experiments are frequently used to evaluate the effects of different treatments, such as crop varieties, fertilizers, or irrigation methods, on crop yields, soil health, and environmental sustainability. Researchers may use split-plot designs to allocate treatments at different levels of the main plot factor (e.g., field or plot), while blocking factors (e.g., soil type, topography) are used to control for spatial variation.

Forestry and Silviculture: In forestry and silviculture studies, split-plot and blocked experiments are utilized to assess the impacts of forest management practices, such as thinning, prescribed burning, or reforestation, on tree growth, biodiversity, and ecosystem services.

These designs help researchers account for spatial variability in environmental conditions and site characteristics, such as soil type, slope, and aspect.

Ecological Restoration: In ecological restoration projects, split-plot and blocked experiments are employed to evaluate the effectiveness of restoration techniques, such as revegetation, habitat enhancement, or invasive species removal, in restoring degraded ecosystems and conserving biodiversity. Researchers may use split-plot designs to allocate treatments across different restoration sites or habitats, while blocking factors control for site-specific differences.

Environmental Monitoring and Management: In environmental monitoring and management programs, split-plot and blocked experiments are used to assess the impacts of pollution, land-use changes, or climate variability on environmental quality, ecosystem health, and human well-being. These designs help researchers control for spatial and temporal variability in environmental factors and identify patterns of change over time.

Aquatic Ecology and Limnology: In studies of aquatic ecosystems, such as lakes, rivers, and wetlands, split-plot and blocked experiments are applied to investigate the effects of nutrient inputs, habitat alterations, or pollution on water quality, aquatic biodiversity, and ecosystem functioning. Researchers may use split-plot designs to allocate treatments within different aquatic habitats, while blocking factors account for spatial gradients or hydrological conditions.

Soil Science and Environmental Chemistry: In soil science and environmental chemistry research, split-plot and blocked experiments are utilized to study the fate and transport of contaminants, nutrients, and pollutants in soil and water systems. These designs help researchers assess the effectiveness of remediation strategies, soil management practices, and land-use policies in mitigating environmental impacts.

Overall, split-plot and blocked experiments are valuable tools in environmental science for designing rigorous experiments that account for spatial variability, site-specific factors, and potential sources of bias, allowing researchers to draw reliable conclusions and inform evidence-based decision-making in environmental management and policy.

STEPS

Developing split-plot and blocked experiments in environmental science involves a systematic process to design experiments that account for spatial variability, site-specific factors, and potential sources of bias. Here are the steps to develop split-plot and blocked experiments:

1. **Define Research Objectives**: Clearly define the research objectives and hypotheses that the experiment aims to address. Identify the specific treatments, interventions, or factors of interest and the response variables to be measured.
2. **Identify Experimental Units**:
   - Main Plots: Identify the main experimental units or plots where treatments will be applied or observations will be made. Main plots represent the largest spatial scale in the experiment and may correspond to different field sites, habitats, or management zones.
   - Subplots: Identify the subunits or smaller experimental units within each main plot, where additional treatments or variations will be applied. Subplots represent the smaller spatial scale and may correspond to specific sampling locations or experimental units within main plots.
3. **Select Blocking Factors**:
   - Identify potential sources of variability or site-specific factors that may influence treatment effects or introduce bias. Blocking factors may include soil type, topography, aspect, vegetation cover, or other environmental variables.

- Select blocking factors that are relevant to the research objectives and experimental design, ensuring that they adequately represent spatial heterogeneity and minimize confounding effects.

4. **Design Experimental Layout:**
   - Allocate Treatments to Main Plots: Assign treatments or interventions to different main plots according to the randomized treatment allocation scheme. Treatments represent the main factor of interest being studied.
   - Allocate Subplots within Main Plots: Assign additional treatments or variations to subplots within each main plot, following a randomized or systematic allocation scheme. Subplots represent the secondary factor of interest or factors being studied.
   - Block Main Plots: Group main plots into blocks based on blocking factors to control spatial variability and ensure balance across experimental units within blocks.

5. **Implement Experimental Setup:**
   - Prepare Experimental Sites: Prepare the experimental sites or field locations according to the experimental layout, ensuring proper environmental conditions and site preparation.
   - Apply Treatments to Main Plots and Subplots: Apply treatments or interventions to main plots and subplots following the randomized treatment assignment. Use standardized procedures and protocols to ensure consistency and reproducibility.

6. **Data Collection:**
   - Measure Response Variables: Collect data on response variables of interest from each experimental unit, including main plots and subplots. Use appropriate measurement techniques and sampling methods to minimize measurement error and bias.
   - Record Blocking Factors: Record data on blocking factors or environmental variables that may influence treatment effects to control for potential confounding effects.

7. **Data Analysis:**
   - Conduct Statistical Analysis: Analyze the data collected from the experiment using appropriate statistical methods, such as ANOVA, linear mixed-effects models, or GLMs.
   - Assess Treatment Effects: Estimate treatment effects, assess statistical significance, and interpret findings in relation to the research objectives and hypotheses.

8. **Interpret Results and Draw Conclusions**: Interpret the results of the experiment in the context of the research objectives and hypotheses. Draw conclusions regarding the effectiveness of treatments or interventions and their implications for environmental science.

9. **Document and Communicate Findings**: Document the experimental setup, methods, results, and conclusions in a comprehensive report or scientific publication. Communicate findings to the scientific community, stakeholders, or policymakers through presentations, publications, or outreach activities.

By following these steps, researchers can develop split-plot and blocked experiments in environmental science that account for spatial variability, site-specific factors, and potential sources of bias, allowing for rigorous experimentation and reliable inference.

PITFALLS AND REMEDIES

Split-plot and blocked experiments in environmental science offer robust approaches to address spatial variability and potential sources of bias. However, they can encounter pitfalls that may affect the validity and interpretability of the results. Here are some common pitfalls and remedies associated with split-plot and blocked experiments:

Incomplete Blocking:

> **Pitfall:** Incomplete blocking occurs when blocking factors do not adequately capture spatial variability, leading to uncontrolled sources of variation and potential bias in treatment effects.
>
> **Remedy:** Ensure that blocking factors are carefully selected to represent key sources of spatial variability and heterogeneity in environmental conditions. Conduct a preliminary analysis to assess the effectiveness of blocking in controlling for variability.

Unequal Block Sizes:

> **Pitfall:** Unequal block sizes can occur when the number of experimental units within blocks varies significantly, leading to imbalances in statistical power and potential bias in treatment effects.
>
> **Remedy:** Aim to achieve balanced block sizes by adjusting the allocation of experimental units within blocks or by stratifying experimental units to ensure comparable representation across treatment groups.

Poor Randomization:

> **Pitfall:** Poor randomization of treatments within main plots or subplots can result in systematic biases and confounding effects, compromising the validity of treatment comparisons.
>
> **Remedy:** Implement rigorous randomization procedures to assign treatments to main plots and subplots, ensuring that treatment allocation is unbiased and independent of other factors. Use randomization techniques, such as complete randomization or randomized complete blocks.

Interaction Effects:

> **Pitfall:** Interaction effects between main plot treatments and subplot treatments may confound treatment effects, making it challenging to interpret the results accurately.
>
> **Remedy:** Consider potential interaction effects during experimental design and analysis. If interactions are suspected, include interaction terms in statistical models and conduct post-hoc analyses to assess their significance and implications for treatment effects.

Failure to Account for Covariates:

> **Pitfall:** Failure to account for covariates or additional sources of variability may lead to biased estimates of treatment effects and reduced precision in statistical analyses.
>
> **Remedy:** Identify relevant covariates that may influence treatment effects and include them as factors in the experimental design or as covariates in statistical models. Adjust for covariates during analysis to control for potential confounding effects.

Inadequate Sample Size:

> **Pitfall:** Inadequate sample size may result in low statistical power, making it difficult to detect meaningful treatment effects or leading to imprecise estimates of effect sizes.
>
> **Remedy:** Conduct power analyses to determine the appropriate sample size needed to detect expected treatment effects with adequate precision. Increase the sample size if necessary to ensure sufficient statistical power.

Misinterpretation of Blocking Factors:

> **Pitfall:** Misinterpreting the role of blocking factors or failing to account for confounding variables may lead to erroneous conclusions about treatment effects.
>
> **Remedy:** Clearly define blocking factors and their relationship to the research objectives. Ensure that blocking factors are correctly incorporated into the experimental design and analysis to control potential sources of bias.

By being aware of these potential pitfalls and implementing appropriate remedies, researchers can enhance the validity, reliability, and interpretability of split-plot and blocked experiments in environmental science, ensuring that treatment effects are accurately assessed, and meaningful conclusions are drawn.

EXAMPLE

Split-plot and blocked experiments are used in environmental science to account for different sources of variability and to optimize experimental designs. Here are examples of matrices for both types of designed experiments in environmental science:

1. **Split-Plot Experiment Matrix:**
   Objective: To study the effects of two main factors (factor A and factor B) on plant growth, where factor A represents a large-scale treatment and factor B represents a small-scale treatment within each large plot.
   Factors:
   1. **Factor A**: Soil type (large plot).
      Clay soil.
      Sandy soil.
   2. **Factor B**: Irrigation frequency (small plot).
      Daily.
      Weekly.
   Response Variable: Plant height (measured in cm).
   Split-Plot Experiment Matrix:
   In a split-plot experiment, the large plots are randomly assigned the main factor levels, while the small plots within each large plot are assigned the sub-factor levels. The matrix is shown in Table 2.47.
   In this matrix, each row represents a specific combination of large and small plots, soil type, irrigation frequency, and the resulting plant height. The large plots represent the main factor (soil type), while the small plots represent the sub-factor (irrigation frequency) nested within each large plot.
2. **Blocked Experiment Matrix:**
   Objective: To study the effects of different environmental conditions on pollutant concentration in a river, while controlling for seasonal variation.
   Factor:
   Environmental Condition: Three different locations along the river.
   Response Variable: Pollutant concentration (measured in mg/L).
   Blocked Experiment Matrix:
   In a blocked experiment, blocks are used to control for external factors that might affect the response variable. The matrix is shown in Table 2.48.

TABLE 2.47

Split-Plot Experiment Matrix Example for Environmental Science

Large Plot	Small Plot	Soil Type	Irrigation Frequency	Plant Height (cm)
1	1	Clay	Daily	...
1	2	Clay	Weekly	...
2	1	Sandy	Daily	...
2	2	Sandy	Weekly	...
3	1	Clay	Daily	...
3	2	Clay	Weekly	...
4	1	Sandy	Daily	...
4	2	Sandy	Weekly	...
5	1	Clay	Daily	...
5	2	Clay	Weekly	...

TABLE 2.48
Blocked Experiment Matrix Example for Environmental Science

Block	Location	Pollutant Concentration (mg/L)
1	Upstream	...
1	Midstream	...
1	Downstream	...
2	Upstream	...
2	Midstream	...
2	Downstream	...
3	Upstream	...
3	Midstream	...
3	Downstream	...
4	Upstream	...
4	Midstream	...
4	Downstream	...

In this matrix, each row represents a specific measurement of pollutant concentration taken at different locations along the river. The data are organized into blocks (block 1, block 2, etc.) to account for any variations or external factors that might influence the pollutant concentration.

Blocked experiments help control external factors that could introduce bias into the experimental results, allowing researchers to focus on the effects of the main factor of interest.

REPORTING

In split-plot and blocked experiments in environmental science, various statistical tools are employed to analyze data, assess treatment effects, and draw valid conclusions. These tools help researchers account for spatial variability, blocking factors, and other sources of variation inherent in the experimental design. Here are some common statistical tools used in split-plot and blocked experiments:

ANOVA: ANOVA is used to partition the variability in the response variable into components associated with main plot treatments, subplot treatments, blocks, and residual error. It assesses the statistical significance of treatment effects and interaction effects, accounting for blocking factors.

Linear Mixed-effects Models: Linear mixed-effects models extend traditional ANOVA approaches by allowing for random effects associated with blocking factors, such as nested or crossed random effects. These models can accommodate hierarchical experimental designs and spatial correlation structures.

Post-hoc Tests: Post-hoc tests, such as Tukey's HSD, Bonferroni correction, or Dunnett's test, are used to compare specific treatment means and identify pairwise differences between treatment groups after ANOVA or mixed-effects model analysis.

Regression Analysis: Regression analysis is used to model the relationship between treatment variables, blocking factors, and response variables. It allows researchers to assess the effects of continuous or categorical predictor variables on the outcome variable while controlling for other factors.

Covariate Adjustment: Covariate adjustment techniques, such as ANCOVA, are used to control for potential confounding variables or baseline differences between treatment groups.

These methods improve the precision of treatment effect estimates by including covariates as predictors in statistical models.

Spatial Analysis: Spatial analysis techniques, such as geostatistics or spatial autocorrelation models, are used to account for spatial autocorrelation and spatial trends in the data. These methods are particularly relevant in experiments conducted across spatially distributed sites or when blocking factors represent spatial gradients.

Nonparametric Tests: Nonparametric tests, such as the Kruskal–Wallis test or Wilcoxon rank sum test, are used when the assumptions of parametric tests are violated or when the data do not meet the requirements for normality or homogeneity of variance.

Bootstrapping: Bootstrapping is a resampling technique used to estimate the sampling distribution of statistics and construct confidence intervals for treatment effects. It provides robust inference in situations where parametric assumptions are violated or sample sizes are small.

GLMs: GLMs extend traditional linear models to accommodate non-normal response variables or complex distributional assumptions. They are useful for analyzing count data, binary data, or other types of noncontinuous response variables in split-plot and blocked experiments.

By employing these statistical tools effectively, researchers can analyze data from split-plot and blocked experiments in environmental science, assess treatment effects, and make valid inferences while accounting for spatial variability, blocking factors, and other sources of variation inherent in the experimental design.

A report for split-plot and blocked experiments in environmental science should provide a clear and comprehensive overview of the experimental design, methods, results, and conclusions. It should include the following elements to effectively communicate the research findings and implications:

Title Page: The title page should include the title of the report, the names of the authors, their affiliations, and the date of publication.

Abstract: The abstract is a concise summary of the experiment, including the research objectives, methods, key findings, and conclusions. It provides an overview of the study's significance and implications.

Introduction:
- Background and Literature Review: Provides background information on the research topic, including relevant literature and previous studies. It explains the rationale for conducting the experiment and the research questions or hypotheses being tested.
- Objectives: Clearly states the specific objectives of the experiment and the treatments or interventions being studied.

Methods:
- Experimental Design: Describes the split-plot and blocked experimental design used in the experiment, including the main plot treatments, subplot treatments, blocking factors, and randomization procedures.
- Experimental Setup: Provides details about the experimental sites or field locations, including environmental conditions, equipment used, and any special considerations.
- Treatment Application: Explains how treatments were applied to main plots and subplots, following the randomized treatment assignment and blocking scheme.
- Data Collection: Describes the methods used to collect data on response variables, including measurement techniques, sampling protocols, and quality control measures.
- Statistical Analysis: Explains the statistical methods used to analyze the data, including ANOVA, linear mixed-effects models, post-hoc tests, and other relevant techniques.

Results:
- Descriptive Statistics: Presents summary statistics for the response variables measured in the experiment, including means, standard deviations, and ranges.
- Treatment Effects: Reports the results of statistical analyses, including tests of treatment effects, interaction effects, and comparisons between treatment groups.
- Post-hoc Analyses: Provides results of post-hoc tests to compare specific treatment means and identify pairwise differences between treatment groups.

Discussion:
- Interpretation of Findings: Interprets the results of the experiment in relation to the research objectives and hypotheses. Discusses the significance of observed treatment effects and their implications for environmental science.
- Comparison with Previous Studies: Compares the findings of the current experiment with previous research in the field. Highlights any consistencies or discrepancies and explains possible reasons for differences.
- Limitations and Future Directions: Acknowledges any limitations of the experiment and suggests areas for future research or improvements to experimental design.

Conclusion: Summarizes the main findings of the experiment and their implications for environmental science. Discusses the practical applications of the findings and potential avenues for further investigation.

References: Provides a list of references cited in the report, following a specific citation style (e.g., APA, MLA).

Appendices: Includes any supplementary materials, such as detailed data tables, additional statistical analyses, or supporting documentation related to the experiment.

By including these elements in a report for split-plot and blocked experiments in environmental science, researchers can effectively communicate the experimental design, methods, results, and conclusions to the scientific community and other stakeholders, facilitating the dissemination of knowledge and informing evidence-based decision-making in environmental management and policy.

2.4 PHYSICS

In physics, designed experiments are crucial for investigating physical phenomena, testing hypotheses, and advancing scientific knowledge. Here are some common types of designed experiments with examples of their applications in physics.

In physics designed experiments, various types of variables are considered to investigate and understand physical phenomena, test hypotheses, and gather data for scientific research. These variables can be categorized into several common types, including

IVs:
- **Physical Parameters**: Variables related to the characteristics of the experimental setup, such as temperature, pressure, voltage, current, and magnetic field strength.
- **Experimental Conditions**: Variables that can be controlled and manipulated by the experimenter, such as the angle of incidence, frequency of light, or initial velocity.
- **Time**: Variables related to the timing and duration of events in experiments, including time intervals and time-dependent processes.

DVs:
- **Measured Observables**: Variables that are directly measured or observed in the experiment, such as distance, displacement, speed, acceleration, force, energy, wavelength, or frequency.

- **Derived Parameters**: Variables that are calculated or derived from measured observables, such as velocity (derived from displacement and time) or momentum (derived from mass and velocity).
- **Physical Constants**: Variables that represent fundamental constants in physics, such as the speed of light (c), Planck's constant (h), or the gravitational constant (G).

Categorical Variables:

- **Experimental Conditions**: Categories or levels of experimental conditions, such as different materials, geometries, or configurations.
- **Particle Types**: Categories of particles used in particle physics experiments, such as electrons, protons, neutrinos, or photons.

Control Variables (Covariates): These are variables that are held constant or controlled during experiments to eliminate their influence on the DV. For example, in a study of the effect of temperature on gas pressure, the gas volume and the number of gas molecules may be controlled variables.

Random Variables: In some cases, random variables may be introduced to account for variability or uncertainty in measurements or to model stochastic processes in physics experiments.

Interaction Variables: Interaction variables are used to investigate the combined effects of two or more IVs on the DV. For example, the interaction between temperature and pressure in a gas behavior experiment.

Noise Variables (Error Variables): These are uncontrolled or unmeasured variables that can introduce variability or errors into the experimental results. Techniques like data smoothing, statistical analysis, and experimental controls are used to reduce the impact of noise variables in physics experiments.

Physics designed experiments are essential for investigating the fundamental laws and principles of the physical universe, as well as for exploring new phenomena and confirming existing theories. These experiments help scientists and researchers gather empirical data and test the validity of physical models and theories. The choice and control of variables in such experiments are critical for obtaining accurate and meaningful results.

2.4.1 FACTORIAL EXPERIMENTS

Factorial design experiments are widely used in physics research across various domains. Here are some areas within physics where factorial design experiments find applications:

Materials Science: Factorial design experiments are frequently employed in materials science to study the effects of multiple factors on the properties of materials. Researchers investigate factors, such as composition, processing conditions, and external stimuli (e.g., temperature, pressure) to optimize material performance for specific applications. For example, factorial experiments may be used to explore the effects of different alloying elements on the mechanical properties of metals or the impact of processing parameters on the crystalline structure of semiconductors.

Optics and Photonics: In optics and photonics research, factorial design experiments are used to study the interactions between light and matter and to optimize the performance of optical devices and systems. Researchers investigate factors, such as wavelength, polarization, and material properties, to design and fabricate optical components, sensors, and communication systems with desired functionalities. Factorial experiments may be employed to explore the effects of multiple design parameters on the transmission, absorption, or scattering of light in various optical materials and configurations.

Particle Physics: Factorial design experiments play a crucial role in particle physics research, particularly in high-energy particle accelerators and colliders. Physicists use factorial experiments to study the fundamental properties of subatomic particles, such as their masses, charges, and decay modes. Factorial designs allow researchers to systematically vary experimental conditions, such as beam energies, collision geometries, and detector configurations, to investigate the interactions between particles and probe the underlying laws of nature.

Condensed Matter Physics: In condensed matter physics, factorial design experiments are employed to investigate the behavior of materials and systems at the nanoscale and mesoscale. Researchers study factors, such as magnetic field strength, temperature, and material composition, to explore phenomena, such as phase transitions, superconductivity, and quantum confinement effects. Factorial experiments may be used to systematically vary experimental parameters and observe their effects on the electronic, magnetic, and structural properties of condensed matter systems.

Astrophysics and Cosmology: Factorial design experiments are utilized in astrophysics and cosmology to study the properties and evolution of celestial objects, galaxies, and the universe as a whole. Researchers investigate factors, such as cosmic radiation, gravitational fields, and cosmological parameters, to model and simulate astrophysical phenomena, such as star formation, galaxy clustering, and cosmic microwave background radiation. Factorial experiments may be employed to explore the effects of different initial conditions and cosmological models on the observed distribution of matter and radiation in the universe.

Overall, factorial design experiments are valuable tools in physics research for systematically exploring the effects of multiple factors on physical phenomena and optimizing experimental conditions to advance scientific understanding and technological innovation across diverse fields of study.

STEPS

Developing factorial design experiments in physics involves several steps to systematically investigate the effects of multiple factors on physical phenomena and optimize experimental conditions. Here are the steps to develop factorial design experiments in physics:

1. **Define Research Objectives**: Clearly define the research objectives and hypotheses that the experiment aims to address. Identify the specific factors or variables of interest and the response variables to be measured.
2. **Identify Factors and Levels**: Identify the IVs or factors that may influence the response variable(s) in the experiment. Factors may include physical properties, experimental conditions, or external parameters. Determine the levels or settings for each factor, representing the range of values or conditions to be investigated. Choose levels that cover a meaningful range and adequately represent the variability in the factors.
3. **Design Experimental Layout**: Select the appropriate factorial design based on the number of factors and levels, as well as the desired experimental resolution. Common factorial designs include full factorial designs, fractional factorial designs, and Taguchi designs. Determine the arrangement of experimental runs or trials in the factorial design matrix, ensuring that all combinations of factor levels are represented and that the design is balanced and orthogonal.
4. **Randomization and Replication**: Randomize the order of experimental runs or trials to minimize systematic biases and ensure that treatment effects are unbiased and generalizable. Replicate experimental runs or trials to assess experimental precision, account for variability, and increase statistical power. Determine the appropriate number of replicates based on statistical considerations and experimental constraints.

5. **Experimental Setup**: Prepare the experimental setup or apparatus according to the factorial design layout, ensuring that it can accommodate the manipulation of multiple factors and the measurement of response variables. Standardize experimental procedures and protocols to ensure consistency and reproducibility across experimental runs. Calibrate equipment and instruments as needed to minimize measurement error.
6. **Data Collection**: Conduct experimental runs or trials according to the factorial design matrix, systematically varying the levels of factors and measuring the corresponding response variables. Collect data on response variables using appropriate measurement techniques and instrumentation. Implement quality control measures to ensure data accuracy and reliability.
7. **Data Analysis**: Analyze the data using appropriate statistical methods, such as ANOVA, regression analysis, or factorial ANOVA. Assess the main effects of individual factors and the interaction effects between factors on the response variable(s). Interpret the statistical significance of effects and identify significant factors and interactions.
8. **Interpretation of Results**: Interpret the results of the factorial design experiment in the context of the research objectives and hypotheses. Discuss the implications of significant factors and interactions for understanding physical phenomena and optimizing experimental conditions. Compare the observed effects with theoretical predictions or previous experimental findings. Identify areas for further investigation or refinement of experimental design.
9. **Documentation and Reporting**: Document the experimental design, methods, results, and conclusions in a comprehensive report or scientific publication. Provide details of the factorial design matrix, experimental procedures, data analysis techniques, and key findings. Communicate the findings to the scientific community through presentations, publications, or conference proceedings. Share experimental data and methodology to facilitate reproducibility and further research.

By following these steps, researchers can develop factorial design experiments in physics to systematically investigate the effects of multiple factors on physical phenomena, optimize experimental conditions, and advance scientific understanding in various areas of physics.

PITFALLS AND REMEDIES

Factorial design experiments in physics offer a powerful approach to systematically investigate the effects of multiple factors on physical phenomena. However, they can encounter pitfalls that may affect the validity and interpretability of the results. Here are some common pitfalls and remedies associated with factorial design experiments in physics:

Incomplete Factorial Design:
   Pitfall: An incomplete factorial design occurs when not all possible combinations of factor levels are included in the experiment. This can lead to confounded effects and make it challenging to isolate the contributions of individual factors.
   Remedy: Use a full factorial design whenever possible to include all combinations of factor levels. If a full factorial design is not feasible due to resource constraints, consider using fractional factorial designs or other efficient designs to prioritize key factor combinations.
Insufficient Replication:
   Pitfall: Insufficient replication occurs when the number of experimental runs or trials is too low to adequately estimate experimental error or assess the variability in treatment effects. This can reduce the statistical power of the experiment and compromise the reliability of the results.

Remedy: Increase the number of replicates for each combination of factor levels to improve the precision of estimates and increase the statistical power of the experiment. Conduct power analyses to determine the appropriate sample size needed to detect meaningful effects.

Uncontrolled Variables:

Pitfall: Uncontrolled variables, also known as nuisance variables or confounding factors, can introduce bias and obscure the effects of the IVs. Failure to account for these variables can lead to erroneous conclusions about treatment effects.

Remedy: Identify potential confounding factors that may influence the response variable(s) and include them as covariates or blocking factors in the experimental design. Implement control measures to minimize the influence of extraneous variables on the experimental outcomes.

Interaction Effects:

Pitfall: Interaction effects occur when the effects of one factor depend on the levels of another factor, leading to nonadditive or synergistic effects that are not captured by main effects alone. Ignoring interaction effects can result in incomplete or misleading interpretations of the results.

Remedy: Include interaction terms in the statistical model to assess the presence and significance of interaction effects. Conduct post-hoc analyses to interpret significant interactions and explore the nature of the interactions further.

Overfitting of Models:

Pitfall: Overfitting occurs when the statistical model is overly complex and captures noise or random variability in the data rather than true underlying relationships. This can lead to inflated type I error rates and poor generalization of results to new data.

Remedy: Use parsimonious statistical models that balance goodness of fit with model complexity. Apply techniques, such as cross-validation or information criteria (e.g., AIC, BIC), to select the most appropriate model and avoid overfitting.

Failure to Validate Assumptions:

Pitfall: Failure to validate assumptions of the statistical model, such as normality of residuals or homogeneity of variances, can lead to biased parameter estimates and inaccurate inference.

Remedy: Assess the validity of model assumptions using diagnostic plots, residual analyses, and formal tests. If assumptions are violated, consider robust statistical methods or transformations to address the issue.

Misinterpretation of Results:

Pitfall: Misinterpreting the results of factorial design experiments can occur when researchers draw conclusions that are not supported by the data or fail to consider alternative explanations for the observed effects.

Remedy: Exercise caution when interpreting results and consider alternative hypotheses or explanations for the observed effects. Clearly communicate the limitations of the study and potential sources of uncertainty in the conclusions.

By being aware of these potential pitfalls and implementing appropriate remedies, researchers can enhance the validity, reliability, and interpretability of factorial design experiments in physics, ensuring that treatment effects are accurately assessed, and meaningful conclusions are drawn.

Example 1

A factorial design experiment in physics can be constructed based on a hypothetical study investigating the effects of two factors, A and B, on a response variable. Each factor has two levels, resulting in a 2×2 factorial design. Table 2.49 is an example of a factorial design matrix.

TABLE 2.49

Factorial Design Experiment Matrix Example for Physics

Experimental Run	Factor A	Factor B
1	Low	Low
2	Low	High
3	High	Low
4	High	High

In this example,

- Factor A represents one IV, which has two levels: low and high.
- Factor B represents another IV, also with two levels: low and high.
- Each row of the matrix represents a unique combination of factor levels, corresponding to an experimental run or trial.
- The factorial design matrix includes all possible combinations of factor levels, resulting in four experimental runs in total (2 levels of factor A * 2 levels of factor B = 4 combinations).

Researchers would conduct experiments corresponding to each combination of factor levels specified in the matrix, measure the response variable, and analyze the data to assess the effects of factors A and B on the response. This factorial design allows researchers to investigate main effects, interaction effects, and overall variability in the response variable across different experimental conditions.

Factorial design experiments in physics are used to study the effects of multiple factors (IVs) on a response variable. Here's a simplified example of a matrix for a 2^2 full factorial experiment in physics:

Objective: To investigate the effects of two factors (force and mass) on the acceleration of an object.

Factors:

1. **Force (Factor A):**
   - Low force (10 N).
   - High force (20 N).
2. **Mass (Factor B):**
   - Light mass (1 kg).
   - Heavy mass (2 kg).

Response Variable: Acceleration (measured in m/s^2).

Full Factorial Experiment Matrix (2^2):

In this 2^2 full factorial experiment, all possible combinations of the two factors (force and mass) are studied. The matrix is shown in Table 2.50.

In this matrix, each row represents a specific experimental run with a unique combination of force and mass levels. Acceleration measurements are taken for each combination.

The full factorial design allows physicists to investigate how changes in force and mass affect the acceleration of an object systematically. By analyzing the results, they can draw conclusions about the main effects and interactions of these factors on acceleration, providing valuable insights into the underlying physical principles involved.

TABLE 2.50

Full Factorial Design Experiment Matrix Example for Physics

Run	Force (A) (N)	Mass (B) (kg)	Acceleration (m/s²)
1	10	1	...
2	20	1	...
3	10	2	...
4	20	2	...

TABLE 2.51

Factorial Design Experiment Matrix Example for Physics

Run	Temperature (A)	Light Intensity (B)	Power Output
1	−1	−1	12.3
2	+1	−1	14.5
3	−1	+1	13.2
4	+1	+1	15.7

Example 2

Another example of a matrix for a factorial design experiment in physics can be illustrated with a hypothetical scenario involving the investigation of factors affecting the efficiency of a solar cell. In this experiment, two factors are considered: temperature (factor A) and light intensity (factor B). Each factor is examined at two levels: low (−1) and high (+1). The response variable is the power output of the solar cell.

An example of a 2 × 2 factorial design matrix for this experiment is shown in Table 2.51. In this design,

- Factor A (temperature) is examined at two levels: low (−1) and high (+1).
- Factor B (light intensity) is also examined at two levels: low (−1) and high (+1).
- Each combination of factor levels represents a treatment condition, resulting in four experimental runs ($2^2 = 4$).
- The response variable (power output) is measured for each experimental run.

This factorial design allows researchers to investigate the main effects of temperature and light intensity on the power output of the solar cell, as well as any interaction effects between the two factors. By systematically varying the levels of both factors, researchers can assess how changes in temperature and light intensity impact the performance of the solar cell and optimize experimental conditions accordingly.

Report

In factorial design experiments for physics, various statistical tools are used to analyze data, assess treatment effects, and draw valid conclusions regarding the relationships between IVs and response variables. Here are some common statistical tools used in factorial design experiments for physics:

ANOVA: ANOVA is a fundamental statistical technique used to partition the total variability in the response variable into components attributable to different factors and their interactions. ANOVA assesses the statistical significance of main effects and interaction effects between factors.

Factorial ANOVA: Factorial ANOVA extends traditional ANOVA methods to analyze factorial designs with multiple factors and their interactions. It allows researchers to assess the main effects of each factor and the interaction effects between factors on the response variable(s).

Regression Analysis: Regression analysis is used to model the relationship between IVs (factors) and the response variable(s). Regression models may include the main effects of factors, interaction terms, and higher-order terms to capture nonlinear relationships.

Post-hoc Tests: Post-hoc tests, such as Tukey's HSD or Bonferroni correction, are used to compare specific factor levels and identify pairwise differences in treatment means after conducting factorial ANOVA or regression analysis.

Contrast Analysis: Contrast analysis is used to specify and test linear combinations of factor levels or treatment groups to address specific research hypotheses. Contrasts provide a flexible approach to compare treatment means and detect specific patterns of differences among factor levels.

Effect Size Measures: Effect size measures, such as eta-squared (η^2) or partial eta-squared (η^2_p), quantify the magnitude of treatment effects and the proportion of variability in the response variable explained by each factor or interaction. Effect size measures provide valuable information about the practical significance of treatment effects.

Power Analysis: Power analysis is used to estimate the statistical power of the experimental design, which represents the probability of detecting true treatment effects when they exist. Power analysis helps researchers determine the sample size needed to achieve adequate statistical power to detect meaningful effects.

Residual Analysis: Residual analysis is used to assess the adequacy of the statistical model and the assumptions underlying the analysis. Residual plots and diagnostic tests are employed to evaluate the normality of residuals, homogeneity of variances, and independence of observations.

Robust Statistical Methods: Robust statistical methods, such as robust regression or robust ANOVA, are used when data do not meet the assumptions of traditional parametric tests. Robust methods provide reliable inference in the presence of outliers, non-normality, or heteroscedasticity.

By employing these statistical tools effectively, researchers can analyze data from factorial design experiments in physics, assess treatment effects, and make valid inferences about the relationships between IVs and response variables, advancing scientific understanding in various fields of physics.

A report for factorial design experiments used in physics should provide a comprehensive overview of the experimental design, methods, results, and conclusions. It should include the following elements to effectively communicate the research findings and implications:

1. **Title Page**: The title page should include the title of the report, the names of the authors, their affiliations, and the date of publication.
2. **Abstract**: The abstract is a concise summary of the experiment, including the research objectives, methods, key findings, and conclusions. It provides an overview of the study's significance and implications.
3. **Introduction**:
   - Background and Literature Review: Provides background information on the research topic, including relevant theories, principles, and previous studies. It explains the rationale for conducting the experiment and the research questions or hypotheses being tested.
   - Objectives: Clearly states the specific objectives of the experiment and the IVs (factors) being investigated.

4. **Experimental Design:**
   - Description of Factors and Levels: Provides details about the IVs (factors) manipulated in the experiment and the levels or settings for each factor. Describes how the experimental design was structured, including the factorial design matrix and the arrangement of experimental runs.
   - Randomization and Replication: Explains how randomization was used to assign treatments to experimental units and how replication was employed to increase the precision of estimates and assess experimental variability.
5. **Experimental Setup:**
   - Description of Apparatus and Equipment: Provides details about the experimental setup, including any specialized equipment, instruments, or materials used in the experiment.
   - Experimental Procedures: Describes the procedures followed to conduct the experiment, including treatment application, data collection, and quality control measures.
6. **Data Analysis:**
   - Statistical Methods: Explains the statistical techniques used to analyze the data, such as ANOVA, regression analysis, or contrast analysis. Describes how treatment effects, main effects, and interaction effects were assessed.
   - Presentation of Results: Presents the results of the data analysis, including summary statistics, tables, and graphs, depicting treatment means, effect sizes, and statistical significance levels.
7. **Discussion:**
   - Interpretation of Findings: Interprets the results of the experiment in relation to the research objectives and hypotheses. Discusses the significance of observed treatment effects and their implications for the field of physics.
   - Comparison with Previous Studies: Compares the findings of the current experiment with previous research in the field. Highlights any consistencies or discrepancies and explains possible reasons for differences.
8. **Conclusion:** Summarizes the main findings of the experiment and their implications for physics research. Discusses the practical applications of the findings and potential avenues for further investigation.
9. **References:** Provides a list of references cited in the report, following a specific citation style (e.g., APA, Institute of Electrical Electronics Engineers (IEEE)).
10. **Appendices:** Includes any supplementary materials, such as detailed data tables, additional statistical analyses, or supporting documentation, related to the experiment.

By including these elements in a report for factorial design experiments used in physics, researchers can effectively communicate the experimental design, methods, results, and conclusions to the scientific community and other stakeholders, facilitating the dissemination of knowledge and informing evidence-based decision-making in physics research.

2.4.2 RSM

Factorial design experiments are widely used in physics for several reasons:

Efficient Exploration of Parameter Space: Physics often involves studying systems with multiple factors or variables that can influence the outcome of an experiment. Factorial designs allow researchers to systematically explore the effects of these factors and their interactions in a relatively efficient manner by considering all possible combinations of factor levels within the experimental design.

Identification of Key Factors and Interactions: Factorial designs enable researchers to identify which factors have significant effects on the response variable(s) of interest and whether there are any interactions between factors. This information is crucial for understanding the underlying physical mechanisms driving the observed phenomena and can help guide further investigation or theory development.

Optimization of Experimental Parameters: Factorial designs can be used to optimize experimental parameters and conditions to maximize the desired outcome or performance metric. By systematically varying factors and observing their effects on the response variable, researchers can identify the optimal settings for achieving specific objectives, such as maximizing SNR, minimizing measurement error, or optimizing system performance.

Robustness and Reliability: Factorial designs are inherently robust and provide reliable estimates of factor effects and interactions, even in the presence of noise or variability in the system. By conducting replicate experiments and averaging the results across multiple runs, researchers can obtain more accurate estimates of factor effects and reduce the impact of random variation on the conclusions drawn from the experiment.

Flexibility and Adaptability: Factorial designs are flexible and can accommodate different experimental setups, sample sizes, and levels of complexity. They can be tailored to the specific research question or hypothesis being investigated and can easily incorporate additional factors or levels as needed. This adaptability makes factorial designs suitable for a wide range of experimental scenarios in physics.

These designed experiment methods provide a powerful and versatile approach for investigating complex systems and phenomena in physics, offering researchers valuable insights into the relationships between factors and response variables, as well as the ability to optimize experimental parameters for enhanced performance or understanding. RSM designed experiments find applications in various areas of physics where researchers aim to optimize experimental conditions, characterize complex systems, or understand the relationships between multiple variables and the response of interest. Some specific areas where RSM is used in physics include:

Materials Science: RSM is extensively used in materials science to optimize fabrication processes, characterize material properties, and design experiments for material synthesis. Researchers use RSM to study the effects of parameters, such as temperature, pressure, and composition on material properties like hardness, conductivity, and optical properties.

Optics and Photonics: In optics and photonics research, RSM is employed to optimize the design and performance of optical systems, including lenses, mirrors, and photonic devices. RSM helps researchers understand how factors, such as surface curvature, coatings, and alignment, affect optical properties like resolution, transmission, and dispersion.

Chemical Physics: RSM is applied in chemical physics to study reaction kinetics, optimize reaction conditions, and characterize reaction pathways. Researchers use RSM to explore the effects of factors, such as temperature, pressure, and reactant concentrations on reaction rates, yields, and selectivity in chemical processes.

Fluid Dynamics: In fluid dynamics research, RSM is used to optimize experimental parameters and design experiments for studying fluid flow, turbulence, and mixing processes. Researchers use RSM to investigate the effects of factors, such as flow velocity, geometry, and viscosity on flow patterns, turbulence intensity, and mixing efficiency.

Environmental Physics: RSM is applied in environmental physics to model and optimize processes related to environmental phenomena, such as pollution dispersion, groundwater flow, and air quality monitoring. Researchers use RSM to study the effects of factors, such as emissions, meteorological conditions, and terrain, on environmental parameters like pollutant concentration, groundwater recharge rates, and dispersion patterns.

Nuclear Physics: RSM finds applications in nuclear physics for optimizing experimental setups, analyzing data, and characterizing nuclear reactions. Researchers use RSM to study the effects of factors, such as energy, target composition, and beam intensity, on reaction cross-sections, decay rates, and particle emission spectra.

Astrophysics and Cosmology: RSM is used in astrophysics and cosmology research to model and optimize observational strategies, data analysis techniques, and theoretical models. Researchers use RSM to study the effects of factors, such as telescope parameters, observational conditions, and cosmological parameters, on observables like brightness, redshift, and cosmic microwave background fluctuations.

Overall, RSM designed experiments are utilized in various branches of physics to optimize experimental conditions, characterize complex systems, and understand the relationships between multiple variables and a response of interest, thereby enhancing scientific understanding and advancing technological applications.

STEPS

Developing RSM designed experiments in physics involves several key steps to efficiently explore and optimize experimental conditions while maximizing the information gained from the experiment. Here are the steps involved in developing RSM designed experiments in physics:

1. **Identify the Response Variable**: Define the response variable(s) of interest that reflect the outcome or performance metric you wish to optimize or characterize in your experiment. This could be a physical property, performance measure, or response to a stimulus.
2. **Select Factors and Levels**: Identify the factors (IVs) that potentially influence the response variable and determine the range of levels for each factor. Factors could include experimental parameters, environmental conditions, material properties, or input variables.
3. **Design the Experimental Matrix**: Generate a design matrix using a suitable RSM design approach, such as CCD, BBD, or Doehlert design. The design matrix specifies the combinations of factor levels to be tested in the experiment, including center points for estimating pure error.
4. **Conduct Experiments**: Perform the experiments according to the design matrix, systematically varying the levels of factors while keeping other experimental conditions constant. Randomize the order of experimental runs to minimize the effects of uncontrolled variables and reduce bias.
5. **Measure Response Variables**: Collect data on the response variable(s) for each experimental run, ensuring accurate and reliable measurements using appropriate instrumentation, calibration procedures, and quality control measures.
6. **Fit Response Surface Models**: Use statistical software to fit response surface models to the experimental data, which describe the relationship between the response variable and the factors, including linear, quadratic, and interaction terms. Validate the models using goodness-of-fit tests and diagnostic checks.
7. **Analyze Factor Effects**: Analyze the fitted response surface models to identify significant factor effects, assess the importance of individual factors and their interactions, and determine the optimal settings for maximizing or optimizing the response variable.
8. **Optimize Experimental Conditions**: Use optimization algorithms or response surface methods to identify the optimal combination of factor levels that maximize or minimize the response variable within specified constraints. Conduct sensitivity analyses to assess the robustness of the optimal conditions.
9. **Validate Results**: Validate the optimized conditions or predictions using additional experiments or simulations to confirm the robustness and reliability of the results.

Compare predicted responses with experimental observations to assess the accuracy of the
response surface models.

10. **Interpret and Report Findings**: Interpret the results in the context of the research
objectives, discussing the implications of the findings and any insights gained from the
experiment. Report the experimental design, methodology, results, and conclusions in a
comprehensive manner, following standard scientific reporting practices.

By following these steps, researchers can systematically develop and conduct RSM designed experiments in physics to optimize experimental conditions, characterize complex systems, and gain valuable insights into the relationships between factors and response variables.

PITFALLS AND REMEDIES

RSM designed experiments in physics offer powerful tools for optimizing experimental conditions
and characterizing complex systems. However, several pitfalls may arise during their implementation. Here are some common pitfalls and potential remedies:

Overfitting Models:
    Pitfall: Overfitting occurs when a response surface model is overly complex and captures
noise or random variation in the data, leading to poor predictive performance on new
data.
    Remedy: Use model selection techniques, such as cross-validation, information criteria
(e.g., AIC, BIC), or validation experiments, to choose the simplest model that adequately represents the relationship between factors and the response variable.

Violations of Model Assumptions:
    Pitfall: Response surface models may assume linearity, constant variance, and normality
of residuals, among other assumptions. Violations of these assumptions can lead to
biased parameter estimates and unreliable inferences.
    Remedy: Perform diagnostic checks, such as residual analysis, normal probability plots,
and tests for homoscedasticity, to assess the adequacy of the model assumptions.
Transform variables or consider alternative modeling approaches if assumptions are
violated.

Confounding Effects:
    Pitfall: Confounding occurs when the effects of two or more factors are indistinguishable
from each other, leading to ambiguity in interpreting the results.
    Remedy: Use orthogonal designs whenever possible to minimize confounding between
factors. If confounding is unavoidable, conduct follow-up experiments or additional
analyses to disentangle the effects of confounded factors.

Extrapolation Beyond Experimental Region:
    Pitfall: Extrapolating response surface models beyond the range of experimental factors
may lead to inaccurate predictions and unreliable conclusions.
    Remedy: Limit predictions to the region of experimental factor levels and exercise caution when extrapolating beyond this range. Conduct additional experiments to validate
predictions in unexplored regions of the factor space.

Inadequate Sample Size:
    Pitfall: Small sample sizes relative to the complexity of the response surface model may
lead to imprecise parameter estimates and low statistical power.
    Remedy: Conduct power analyses or sample size calculations to determine the minimum sample size required to detect hypothesized effects with sufficient statistical
power. Increase the sample size or reduce the number of factors to achieve adequate
power.

Unmodeled Variation:

Pitfall: Unmodeled variation or unobserved factors may contribute to residual variability in the response, reducing the accuracy of the response surface model.

Remedy: Identify and include relevant covariates or control variables in the response surface model to account for unmodeled variation. Consider conducting additional experiments to investigate sources of unexplained variability.

By being aware of these potential pitfalls and implementing appropriate remedies, researchers can enhance the reliability and validity of RSM designed experiments in physics, ensuring robust and informative results.

EXAMPLE

RSM designed experiments in physics are used to optimize a response variable by studying the relationships between multiple factors. Here's a simplified example of a matrix for an RSM designed experiment in physics:

Objective: To optimize the conditions for a chemical reaction in a physics experiment by studying the effects of two factors (temperature and concentration) on the reaction rate.

Factors:

1. **Temperature (Factor A):**
   - Low temperature (50°C).
   - Medium temperature (75°C).
   - High temperature (100°C).
2. **Concentration (Factor B):**
   - Low concentration (0.5 M).
   - Medium concentration (1.0 M).
   - High concentration (1.5 M).

Response Variable: Reaction rate (measured in moles per minute, mol/min).

RSM Designed Experiment Matrix:

In this RSM designed experiment, the matrix is designed to optimize the conditions (temperature and concentration) for the chemical reaction. The matrix is shown in Table 2.52.

TABLE 2.52

RSM Example for Physics

Run	Temperature (A) (°C)	Concentration (B) (M)	Reaction Rate (mol/min)
1	50	0.5	...
2	75	0.5	...
3	100	0.5	...
4	50	1.0	...
5	75	1.0	...
6	100	1.0	...
7	50	1.5	...
8	75	1.5	...
9	100	1.5	...

In this matrix, each row represents a specific experimental run with a unique combination of temperature and concentration levels. The reaction rate is measured for each combination.

Using statistical analysis and response surface modeling, physicists can determine the optimal conditions (temperature and concentration) that maximize the reaction rate. RSM helps identify the relationships between these factors and the response variable, allowing for the optimization of experimental conditions in physics experiments.

REPORT

RSM designed experiments in physics utilize various statistical tools and techniques to analyze data, model relationships between factors and response variables, and optimize experimental conditions. Some common statistical tools used in RSM designed experiments for physics include:

ANOVA: ANOVA is used to partition the total variability in the response variable into components attributable to different factors and their interactions. It helps assess the significance of factor effects and interactions, as well as identify the most influential factors.

Regression Analysis: Regression analysis is used to fit response surface models to experimental data, describing the relationship between the response variable and the factors. This includes fitting linear, quadratic, and higher-order models to capture nonlinear effects and interactions.

Model Selection Criteria: Model selection criteria, such as AIC or BIC, are used to compare different response surface models and select the most parsimonious model that adequately represents the relationship between factors and the response variable.

Diagnostic Checks: Diagnostic checks, such as residual analysis, leverage statistics to assess the validity of model assumptions and the adequacy of the fitted model. They help identify potential outliers, influential data points, or violations of model assumptions that may affect the reliability of the results.

Optimization Algorithms: Optimization algorithms are used to identify the optimal combination of factor levels that maximize or minimize the response variable within specified constraints. They iteratively search the parameter space to find the combination of factor levels that optimize the desired outcome.

Response Surface Plots: Response surface plots visualize the predicted response surface generated by the fitted response surface model. They provide insights into the effects of factors and their interactions with the response variable, helping researchers interpret the results and identify optimal experimental conditions.

Sensitivity Analysis: Sensitivity analysis assesses the robustness of the optimal conditions identified by the response surface model. It investigates how changes in factor levels or model parameters affect the predicted response, helping researchers evaluate the reliability of the optimization results.

Experimental Design Tools: Software packages specifically designed for experimental design, such as Design-Expert or JMP, provide tools for generating optimal experimental designs, fitting response surface models, and conducting statistical analyses of RSM design experiments.

By employing these statistical tools and techniques, researchers can effectively analyze data from RSM designed experiments in physics, model complex relationships between factors and response variables, and optimize experimental conditions to achieve desired outcomes.

When reporting on RSM designed experiments in physics, it's important to provide a comprehensive and well-organized report to effectively communicate the experimental design, methodology, results, and conclusions. Here are the key elements typically included in such a report:

Title: A clear and descriptive title that summarizes the main focus of the experiment.

Abstract: A concise summary of the experiment, including the objectives, methods, key findings, and conclusions.

Introduction:

- Background: Provide context for the experiment by briefly reviewing relevant literature and explaining the motivation for the study.
- Objectives: Clearly state the research objectives and hypotheses being tested.

Experimental Design:

- Description of RSM Design: Explain the principles of RSM and describe the specific experimental design used, including the choice of factors, levels, and design matrix.
- Justification: Discuss why RSM was chosen as the experimental approach and how it addresses the research objectives.

Materials and Methods:

- Experimental Setup: Describe the apparatus, instruments, and materials used in the experiment.
- Procedure: Outline the step-by-step procedure followed to conduct the experiment, including any randomization procedures, measurement protocols, and data collection methods.
- Statistical Analysis: Provide details on the statistical methods and techniques used to analyze the data, fit response surface models, and optimize experimental conditions.

Results:

- Summary of Experimental Data: Present a summary of the experimental data collected, including tables or figures showing the factor levels and corresponding response values for each experimental run.
- Response Surface Analysis: Present the results of the response surface analysis, including fitted models, contour plots, and response surface plots, illustrating the relationships between factors and the response variable.
- Optimization Results: Describe the optimal conditions identified by the response surface model and discuss any insights gained from the optimization process.

Discussion:

- Interpretation of Results: Interpret the findings in light of the research objectives and hypotheses, discussing the significance of factor effects, interactions, and optimized conditions.
- Comparison with Previous Studies: Compare the results with findings from previous studies or theoretical predictions, highlighting any agreements or discrepancies.
- Limitations and Caveats: Discuss any limitations of the experimental design, potential sources of error, and assumptions made in the analysis.
- Implications and Future Directions: Discuss the implications of the findings for physics research and suggest potential avenues for future investigation.

Conclusion:

- Summary of Findings: Summarize the main findings of the experiment and restate the conclusions drawn.
- Contributions: Highlight the contributions of the study to the field of physics and its relevance to broader scientific or practical applications.

References: Provide a list of references cited in the report, following a standard citation format.

Appendices: Include any supplementary information, raw data, detailed statistical analyses, or additional results that support the main findings of the report.

By including these elements in the report, researchers can effectively communicate the rationale, methodology, results, and implications of RSM designed experiments in physics, facilitating transparency, reproducibility, and dissemination of scientific knowledge.

2.4.3 FRACTIONAL FACTORIAL

Fractional factorial design experiments are widely used in physics to efficiently explore factor effects, screen influential variables, optimize experimental conditions, and conduct exploratory studies while maximizing the use of available resources. Fractional factorial design experiments are utilized in physics for several reasons:

Efficient Exploration of Factor Effects: Physics experiments often involve studying systems with multiple factors or variables that can influence the outcome. Fractional factorial designs allow researchers to investigate a subset of factor combinations, thereby reducing the number of experimental runs required while still providing information about the main effects of factors.

Screening Experiments: Fractional factorial designs are commonly used as screening experiments to identify the most influential factors among a large set of potential variables. By conducting a fraction of the full factorial design, researchers can quickly identify the key factors that significantly impact the response variable and focus further investigation on those factors.

Cost and Resource Savings: Physics experiments can be resource intensive, requiring specialized equipment, materials, and time. Fractional factorial designs allow researchers to achieve significant reductions in the number of experimental runs compared to full factorial designs, resulting in cost savings and more efficient use of resources.

Identifying Factor Interactions: While fractional factorial designs do not allow for the estimation of all possible interactions between factors, they can still provide valuable information about main effects and certain types of interactions. By carefully selecting the fraction of the full factorial design, researchers can prioritize the estimation of specific interactions relevant to their research question.

Optimization of Experiments: Fractional factorial designs can be used in conjunction with optimization techniques to identify optimal experimental conditions. By systematically varying factors at a fraction of the cost and time required for full factorial designs, researchers can efficiently optimize experimental parameters and achieve desired outcomes.

Exploratory Studies and Hypothesis Testing: Fractional factorial designs are valuable for exploratory studies and hypothesis testing in physics. They allow researchers to efficiently test multiple hypotheses and explore the effects of factors on the response variable, providing insights into underlying physical mechanisms and guiding further investigation.

Fractional factorial design experiments find applications in various areas of physics where researchers seek to efficiently explore factor effects, screen influential variables, optimize experimental conditions, and conduct exploratory studies. Some specific areas where fractional factorial design experiments are commonly used in physics include:

Particle Physics: In particle physics experiments, researchers often study systems with multiple factors influencing particle interactions and detector responses. Fractional factorial designs are used to efficiently screen influential factors, optimize detector configurations, and explore the effects of experimental parameters on particle detection efficiency and SNR.

Condensed Matter Physics: Fractional factorial designs are employed in condensed matter physics to investigate the effects of material properties, environmental conditions, and experimental parameters on physical phenomena, such as phase transitions, magnetic properties, and electronic transport behavior. Researchers use fractional factorial designs to efficiently screen factors affecting material performance and optimize experimental setups for specific applications.

Quantum Optics: In quantum optics experiments, researchers manipulate and control the quantum states of light and matter to study fundamental quantum phenomena and develop quantum technologies. Fractional factorial designs are used to efficiently explore the effects of experimental parameters, such as laser parameters, optical elements, and environmental conditions, on the generation, manipulation, and detection of quantum states, enabling researchers to optimize experimental setups and improve measurement precision.

Astrophysics and Cosmology: Fractional factorial designs are applied in astrophysics and cosmology experiments to optimize observational strategies, screen influential variables, and explore the effects of observational parameters on data quality and scientific outcomes. Researchers use fractional factorial designs to efficiently sample the parameter space, optimize telescope configurations, and identify key factors affecting observations of astronomical phenomena, such as galaxy clustering, cosmic microwave background radiation, and gravitational lensing.

Nuclear Physics: In nuclear physics experiments, researchers investigate the properties of atomic nuclei, nuclear reactions, and nuclear matter under extreme conditions. Fractional factorial designs are used to efficiently explore factors influencing nuclear reactions, optimize experimental setups for rare-event detection, and screen influential variables affecting data acquisition and analysis in nuclear physics experiments.

Fractional factorial design experiments are widely used across various subfields of physics to efficiently explore factor effects, screen influential variables, optimize experimental conditions, and conduct exploratory studies, facilitating scientific discovery and advancing our understanding of the natural world.

Steps

Developing fractional factorial design experiments in physics involves several key steps to efficiently explore factor effects, screen influential variables, and optimize experimental conditions while minimizing the number of experimental runs. Here are the steps involved in developing fractional factorial design experiments in physics:

1. **Identify Factors**: Determine the factors (IVs) that may influence the response variable(s) of interest in the experiment. Factors could include experimental parameters, environmental conditions, material properties, or input variables.
2. **Define Levels**: Specify the range of levels for each factor that will be investigated in the experiment. Levels represent different settings or values of the factors that will be systematically varied during the experiment.
3. **Choose Fraction**: Select the fraction of the full factorial design that will be used for the experiment. The fraction determines the subset of factor combinations that will be tested, allowing for efficient exploration of factor effects while reducing the number of experimental runs.
4. **Generate Design Matrix**: Use a suitable design generation algorithm, such as the Plackett–Burman method or the resolution IV or resolution V design for screening experiments, to create the fractional factorial design matrix. The design matrix specifies the factor combinations to be tested in the experiment, including the levels of each factor for each experimental run.

5. **Randomization and Blocking**: Randomize the order of experimental runs to minimize the effects of uncontrolled variables and reduce bias. If necessary, use blocking to account for known sources of variation or to ensure balance in the design.
6. **Conduct Experiments**: Perform the experimental runs according to the design matrix, systematically varying the levels of factors while keeping other experimental conditions constant. Collect data on the response variable(s) for each experimental run.
7. **Data Analysis**: Analyze the experimental data to assess the effects of factors on the response variable(s) and screen for influential variables. Use statistical techniques, such as ANOVA, to identify significant factor effects and interactions.
8. **Interpret Results**: Interpret the results in the context of the research objectives, discussing the significance of factor effects, interactions, and any patterns observed in the data. Use graphical techniques, such as main effects plots and interaction plots, to visualize the relationships between factors and the response variable(s).
9. **Optimization (Optional)**: If the objective is to optimize experimental conditions, use the results of the fractional factorial design experiment to identify promising factor settings and conduct further experimentation or optimization techniques to refine the optimal conditions.
10. **Validate Results**: Validate the results of the fractional factorial design experiment using additional experiments, simulations, or external validation data to confirm the reliability and robustness of the findings.
11. **Report Findings**: Prepare a comprehensive report summarizing the experimental design, methodology, results, and conclusions of the fractional factorial design experiment, following standard reporting guidelines for scientific research.

By following these steps, researchers can systematically develop fractional factorial design experiments in physics to efficiently explore factor effects, screen influential variables, and optimize experimental conditions, maximizing the information gained from the experiment while minimizing the number of experimental runs required.

PITFALLS AND REMEDIES

Fractional factorial design experiments in physics offer efficient ways to explore factor effects while minimizing the number of experimental runs. However, like any experimental approach, they come with potential pitfalls. Here are some common pitfalls and remedies associated with fractional factorial design experiments in physics:

Alias Structure:

Pitfall: Fractional factorial designs may result in aliasing, where the effects of some factors are confounded with other factors or interactions, making it difficult to distinguish between them.

Remedy: Use advanced design strategies, such as OAs or clear resolution IV designs, to minimize aliasing. Analyze the alias structure carefully and use follow-up experiments or additional analyses to untangle confounded effects.

Loss of Precision:

Pitfall: Fractional factorial designs sacrifice precision compared to full factorial designs by testing only a subset of factor combinations.

Remedy: Conduct power analyses or evaluate the precision of the estimated effects to ensure that the chosen fractional design provides sufficient statistical power for detecting meaningful effects. Consider augmenting the design with additional runs if necessary to improve precision.

Incomplete Factor Space Exploration:
> Pitfall: Fractional factorial designs may not fully explore the factor space, potentially missing important factor combinations or interactions.
>
> Remedy: Carefully select the fraction of the full factorial design based on prior knowledge, research objectives, and available resources. Use optimization techniques or follow-up experiments to explore regions of the factor space not covered by the fractional design.

Unmodeled Variability:
> Pitfall: Fractional factorial designs may overlook sources of variability or noise that are not explicitly accounted for in the experimental design.
>
> Remedy: Conduct thorough experimental planning and control to minimize sources of variability. Use statistical techniques, such as ANOVA or regression, to assess the adequacy of the model and identify potential sources of unexplained variability.

Interpretation Challenges:
> Pitfall: Interpreting results from fractional factorial designs can be challenging due to the complex aliasing structure and potential interactions.
>
> Remedy: Use statistical techniques, such as effect plots, interaction plots, and diagnostic checks, to aid in the interpretation of results. Validate findings through sensitivity analyses or additional experiments to ensure robust conclusions.

Assumption Violations:
> Pitfall: Violations of statistical assumptions, such as normality or homoscedasticity, can affect the validity of results from fractional factorial design experiments.
>
> Remedy: Perform diagnostic checks to assess the validity of assumptions and consider data transformations or alternative modeling approaches if assumptions are violated. Use robust statistical methods when appropriate.

By being aware of these potential pitfalls and implementing appropriate remedies, researchers can enhance the reliability and validity of fractional factorial design experiments in physics, ensuring robust conclusions and meaningful insights from their experiments.

EXAMPLE

Fractional factorial experiments are used in physics to study the effects of factors on a response variable while conducting only a fraction of the full set of experiments, which can be more efficient in terms of resources. Here's an example of a matrix for a $2^{(3-1)}$ fractional factorial experiment in physics:

Objective: To investigate the effects of three factors (voltage, current, and resistance) on the brightness of a light bulb.

Factors:

1. **Voltage (Factor A):**
   - Low voltage (100 V).
   - High voltage (120 V).
2. **Current (Factor B):**
   - Low current (1 A).
   - High current (2 A).
3. **Resistance (Factor C):**
   - Low resistance (50 Ω).
   - High resistance (100 Ω).

TABLE 2.53

Physics Fractional Factorial Design Experiment Matrix Example

Run	Voltage (A) (V)	Current (B) (A)	Resistance (C) (Ω)	Brightness (lumens)
1	100	1	50	...
2	120	1	50	...
3	100	2	50	...
4	120	2	50	...
5	100	1	100	...
6	120	1	100	...
7	100	2	100	...
8	120	2	100	...

Response Variable: Brightness (measured in lumens).

Fractional Factorial Matrix ($2^{(3-1)}$):

In this $2^{(3-1)}$ fractional factorial experiment, a fraction of the full set of experiments is conducted to study the main effects and two-way interactions of the factors on brightness. The matrix is shown in Table 2.53.

In this matrix, each row represents a specific experimental run with a unique combination of factor levels. The fractional factorial design allows you to study the main effects and two-way interactions of the factors (voltage, current, and resistance) on the brightness of the light bulb while conducting only a fraction of the full set of experiments, saving time and resources.

By analyzing the results, physicists can gain insights into how these factors affect the brightness of the light bulb and make informed decisions about the optimal operating conditions for the experiment.

Another example of a fractional factorial design experiment in physics where researchers are investigating the factors affecting the efficiency of a photovoltaic (PV) cell. The objective is to identify the main factors influencing PV cell efficiency and optimize experimental conditions to maximize energy conversion efficiency.

Factors:

1. Light intensity (low, medium, high).
2. Temperature (low, high).
3. Semiconductor material (material A, material B).
4. Surface treatment (treatment 1, treatment 2).

Each factor is tested at two levels, resulting in a 2^{4-1} fractional factorial design, which requires only 8 experimental runs instead of the full 16 runs of a 2^4 full factorial design. The choice of the specific fractional design (2^{4-1} in this case) depends on the research objectives and the available resources.

Design Matrix (Table 2.54):

Experimental Procedure:

1. Set up the PV cell under controlled conditions.
2. Conduct the experimental runs according to the design matrix, varying the levels of light intensity, temperature, semiconductor material, and surface treatment as specified.

TABLE 2.54

Physics Fractional Factorial Design Experiment Matrix Example

Run	Light Intensity	Temperature	Semiconductor Material	Surface Treatment
1	Low	Low	Material A	Treatment 1
2	Low	High	Material B	Treatment 2
3	Low	Low	Material B	Treatment 2
4	Low	High	Material A	Treatment 1
5	High	Low	Material A	Treatment 2
6	High	High	Material B	Treatment 1
7	High	Low	Material B	Treatment 1
8	High	High	Material A	Treatment 2

3. Measure the efficiency of the PV cell for each experimental run.
4. Record the data for analysis.

Data Analysis:

1. Perform ANOVA to determine the significance of main effects and interactions.
2. Generate main effects plots and interaction plots to visualize the effects of factors on PV cell efficiency.
3. Identify significant factors and interactions influencing PV cell efficiency.
4. Use optimization techniques to identify optimal levels of factors for maximizing PV cell efficiency.

Interpretation and Conclusion:

1. Interpret the results in the context of the research objectives, discussing the significance of factor effects and interactions.
2. Draw conclusions about the factors influencing PV cell efficiency and their implications for optimizing PV cell performance.
3. Discuss limitations of the study and suggest directions for future research.

By using a fractional factorial design experiment, researchers can efficiently explore factor effects, screen influential variables, and optimize experimental conditions in physics experiments, leading to valuable insights and discoveries in the field.

REPORT

In fractional factorial design experiments for physics, various statistical tools and techniques are employed to analyze the experimental data, assess the significance of factor effects, and draw meaningful conclusions. Some of the common statistical tools used in fractional factorial design experiments for physics include:

ANOVA: ANOVA is used to partition the total variation in the response variable into different sources, including the main effects of factors and interactions. It helps determine whether the observed differences between experimental groups are statistically significant.

Main Effects Plots: Main effects plots are graphical representations of the estimated effects of each factor on the response variable. They provide a visual summary of the main effects of factors, allowing researchers to identify which factors have the most significant influence on the response.

Interaction Plots: Interaction plots visualize the interactions between factors by plotting the response variable against different combinations of factor levels. They help identify instances where the effect of one factor depends on the level of another factor, indicating the presence of interactions.

Fractional Factorial Design Resolution: Fractional factorial designs are characterized by different resolutions, which determine the extent to which main effects and interactions are aliased or confounded. Understanding the resolution of the design is crucial for interpreting the results and assessing the confounding patterns.

Effect Sparsity: Effect sparsity analysis is used to identify and prioritize the most significant effects among a large number of potential factors and interactions. It helps researchers focus their attention on the most influential factors for further investigation.

Optimization Algorithms: Optimization algorithms are employed to identify optimal factor settings that maximize or minimize the response variable. These algorithms use the experimental data and statistical models to search for the optimal combination of factor levels within the experimental space.

Diagnostic Checks: Diagnostic checks assess the validity of statistical assumptions underlying the analysis, such as normality, homoscedasticity, and independence of errors. They help ensure the reliability and validity of the statistical results obtained from the experiment.

Model Selection Criteria: Model selection criteria, such as AIC or BIC, are used to compare different statistical models and select the most appropriate model that best fits the experimental data while balancing model complexity and goodness of fit.

Statistical Software: Various statistical software packages, such as R, Python with libraries like statsmodels or scikit-learn, SAS, or JMP, are commonly used to perform the statistical analysis of fractional factorial design experiments in physics. These software packages provide tools for data analysis, visualization, and model fitting, facilitating the interpretation of experimental results.

By employing these statistical tools and techniques, researchers can effectively analyze the data from fractional factorial design experiments in physics, assess the significance of factor effects, and derive meaningful insights to advance scientific knowledge in the field.

In a report for fractional factorial design experiments used in physics, it's essential to include comprehensive information about the experimental design, methodology, results, and conclusions. Here are the key elements to include in the report:

Title: Clearly state the title of the report, indicating the purpose and scope of the experiment.

Abstract: Provide a concise summary of the experiment, including the objectives, experimental design, key findings, and conclusions.

Introduction:
- Provide background information on the research problem or topic under investigation.
- State the research objectives and the rationale for conducting the experiment.
- Describe the significance of the experiment in the context of existing literature and scientific knowledge.

Experimental Design:
- Describe the fractional factorial design used in the experiment, including the factors investigated, their levels, and the fraction of the full factorial design chosen.
- Explain how the experimental runs were conducted and any considerations for randomization or blocking.

Materials and Methods:
- Detail the materials, equipment, and instrumentation used in the experiment.
- Describe the experimental procedure, including any preparations, measurements, or data collection protocols.
- Specify any control measures or environmental conditions maintained during the experiment.

Data Analysis: Present the statistical analysis methods used to analyze the experimental data. Describe how factors and interactions were assessed for significance using techniques, such as ANOVA or effect sparsity analysis. Provide details on any optimization algorithms or model selection criteria used to derive conclusions from the data.

Results:
- Present the experimental results in a clear and organized manner, including tables, figures, or graphs as appropriate.
- Summarize the main effects of factors and interactions identified in the analysis.
- Highlight any significant findings or patterns observed in the data.

Discussion:
- Interpret the results in the context of the research objectives and hypothesis.
- Discuss the implications of the findings for the broader field of physics.
- Address any limitations or sources of uncertainty in the experiment.
- Compare the results with existing literature and theoretical predictions.

Conclusion: Summarize the key findings and conclusions drawn from the experiment. Discuss the significance of the results and their implications for future research or practical applications. Highlight any recommendations for further investigation or experimental refinement.

References: Provide a list of references cited in the report, following a standard citation style (e.g., APA, IEEE, or Chicago).

Appendices: Include any supplementary information, raw data, or additional analyses that support the findings presented in the main body of the report.

By including these elements in the report, researchers can effectively communicate the design, methodology, results, and conclusions of fractional factorial design experiments in physics, facilitating peer review, replication, and dissemination of scientific knowledge.

2.4.4   CCD

CCD experiments are valuable tools in physics research for optimizing experimental conditions, modeling nonlinear responses, exploring factor interactions, testing robustness, conducting surface response analysis, and efficiently utilizing resources while providing valuable insights into complex physical phenomena. CCD experiments are used in physics for several reasons:

Optimization of Experimental Conditions: CCD experiments allow researchers to optimize experimental conditions by systematically varying multiple factors and identifying the optimal combination that maximizes or minimizes the response variable of interest. This is particularly useful in physics experiments where the goal is to achieve maximum efficiency, accuracy, or sensitivity.

Modeling Nonlinear Responses: CCD experiments are designed to capture nonlinear relationships between factors and the response variable. In physics, many systems exhibit nonlinear behavior, and CCDs provide a systematic approach to model and understand these nonlinear responses, allowing researchers to characterize complex phenomena accurately.

Exploration of Factor Interactions: CCDs include factorial points as well as axial and center points, enabling researchers to explore both linear and quadratic effects of factors and their interactions with the response variable. This comprehensive exploration of factor interactions helps identify synergistic or antagonistic effects that may not be captured by simpler experimental designs.

Robustness Testing: CCD experiments facilitate robustness testing by evaluating the sensitivity of the system to variations in experimental factors and environmental conditions. By systematically varying factors around their optimal settings, researchers can assess the robustness of the system and identify potential sources of variability or instability.

Surface Response Analysis: CCDs are particularly suited for surface response analysis, where researchers aim to model and optimize response surfaces representing the relationship between multiple factors and the response variable. This analysis allows for the identification of optimal factor settings within the experimental space and the characterization of the response surface topology.

Reduced Number of Experimental Runs: CCDs require fewer experimental runs compared to full factorial designs while still allowing for the estimation of main effects, quadratic effects, and interactions. This reduction in the number of runs makes CCD experiments more efficient in terms of time, resources, and cost, making them attractive for physics experiments with limited resources.

CCD experiments are used in physics across various fields and applications. Some common areas where CCD experiments are employed in physics include:

Materials Science: CCD experiments are used to optimize processing parameters for synthesizing materials with desired properties, such as strength, conductivity, or thermal stability. They help researchers investigate the effects of factors like temperature, pressure, composition, and processing time on material properties.

Optics and Photonics: In optics and photonics research, CCD experiments are used to optimize the design of optical components and systems, such as lenses, mirrors, and photonic devices. They help researchers optimize factors like geometric parameters, material properties, and surface treatments to achieve desired optical performance.

Semiconductor Physics: CCD experiments are employed in semiconductor physics to optimize the fabrication processes of semiconductor devices, such as transistors, diodes, and solar cells. They help researchers explore the effects of doping concentrations, deposition conditions, and annealing processes on device performance and characteristics.

Experimental Physics: CCD experiments are used in experimental physics to optimize experimental setups and conditions for conducting measurements and investigations. They help researchers explore the effects of experimental parameters, such as temperature, pressure, magnetic field strength, and excitation energy on the observed phenomena.

Particle Physics: In particle physics experiments, CCDs are employed to optimize detector configurations and data acquisition systems for studying particle interactions and high-energy physics phenomena. They help researchers optimize factors like detector geometry, material composition, and signal processing algorithms to maximize detection efficiency and resolution.

Astrophysics and Cosmology: CCD experiments play a role in astrophysics and cosmology research by optimizing observational parameters and data analysis techniques for studying celestial objects and cosmic phenomena. They help researchers optimize factors like telescope configurations, observing strategies, and data reduction methods to maximize the scientific yield of observations.

Nuclear Physics: In nuclear physics experiments, CCDs are used to optimize experimental setups for studying nuclear reactions, decay processes, and nuclear structure. They help researchers explore the effects of factors like beam energy, target material, detector geometry, and experimental conditions on observed nuclear phenomena.

CCD experiments find applications in various areas of physics research, helping scientists optimize experimental conditions, explore factor interactions, and maximize the scientific yield of experiments across different fields and disciplines.

STEPS

Developing CCD experiments in physics involves several steps to ensure the systematic exploration of factor effects and the optimization of experimental conditions. Here are the typical steps involved in developing CCD experiments in physics:

1. **Define the Research Objectives:**
   - Clearly define the research objectives and the response variable(s) of interest. Determine the specific goals of the experiment, such as optimizing experimental conditions, characterizing factor effects, or exploring factor interactions.
2. **Identify Factors and Levels:**
   - Identify the independent factors (variables) that may influence the response variable(s) in the experiment. These factors could include experimental parameters, input variables, or system settings.
   - Determine the range of each factor and the number of levels for each factor. Consider both practical constraints and the desired precision of the experimental design.
3. **Select the Experimental Design:**
   - Choose the appropriate CCD experimental design based on the number of factors, the desired precision, and the available resources. CCDs typically include factorial points, axial points, and center points, allowing for the exploration of linear, quadratic, and interaction effects.
   - Determine the number of factorial points, axial points, and center points based on the desired resolution of the design and the number of factors.
4. **Generate the Design Matrix:**
   - Use statistical software or design tables to generate the design matrix for the CCD experiment. The design matrix specifies the factor levels for each experimental run, including factorial points, axial points, and center points.
   - Ensure that the design matrix follows the principles of rotatability and orthogonality to facilitate efficient estimation of factor effects and interactions.
5. **Conduct the Experimental Runs:**
   - Set up the experimental apparatus or system according to the specifications outlined in the design matrix.
   - Perform the experimental runs according to the factor levels specified in the design matrix. Randomize the order of experimental runs to minimize the effects of potential confounding factors or systematic errors.
6. **Collect Data and Record Observations:**
   - Measure the response variable(s) for each experimental run and record the corresponding observations. Ensure that the data collection process is consistent and accurate across all experimental runs.
   - Document any additional observations or qualitative data that may be relevant to the experiment.
7. **Perform Data Analysis:**
   - Use statistical analysis techniques, such as ANOVA, regression analysis, or RSM, to analyze the experimental data.
   - Estimate the main effects of factors, quadratic effects, and interactions using appropriate statistical models.
   - Conduct model diagnostics and goodness-of-fit tests to assess the validity of the statistical model and the adequacy of the experimental design.

8. **Interpret Results and Draw Conclusions:**
   - Interpret the results of the data analysis in the context of the research objectives and the experimental design.
   - Identify significant factor effects, interactions, and optimal factor settings based on the statistical analysis.
   - Draw conclusions regarding the relationships between factors and the response variable(s) and their implications for the research objectives.
9. **Validate and Refine the Experimental Design:**
   - Validate the experimental results through replication or additional experiments to ensure their reproducibility and reliability.
   - Refine the experimental design based on the insights gained from the initial experiment and any unexpected findings or limitations encountered.
10. **Document and Communicate Findings:**
   - Prepare a comprehensive report documenting the experimental design, methodology, results, and conclusions.
   - Communicate the findings of the CCD experiment through scientific publications, presentations, or technical reports to share knowledge and contribute to the advancement of physics research.

By following these steps, researchers can systematically develop and conduct CCD experiments in physics, leading to valuable insights into factor effects, optimization of experimental conditions, and deeper understanding of physical phenomena.

PITFALLS AND REMEDIES

CCD experiments in physics offer numerous advantages but also present some potential pitfalls. Here are some common pitfalls and corresponding remedies:

Incomplete Factor Space Exploration:
   Pitfall: Failure to adequately explore the factor space can lead to missing important factor effects or interactions, resulting in suboptimal conclusions.
   Remedy: Ensure that the range of factor levels chosen for the experiment covers the entire relevant domain. Use proper statistical software or design tables to generate the design matrix, ensuring sufficient coverage of the factor space.

Model Inadequacy:
   Pitfall: Using an inadequate model to analyze CCD data can lead to incorrect conclusions about factor effects or interactions.
   Remedy: Conduct thorough model diagnostics to assess the goodness-of-fit and adequacy of the chosen statistical model. Consider using diagnostics, such as residual analysis, lack-of-fit tests, and model selection criteria, to identify potential model inadequacies and improve model accuracy.

Extrapolation beyond Design Space:
   Pitfall: Extrapolating results beyond the design space defined by the CCD experiment can lead to erroneous conclusions and unreliable predictions.
   Remedy: Exercise caution when extrapolating results beyond the range of factor levels tested in the experiment. Limit interpretations to the specified design space and avoid making predictions outside this range unless supported by additional evidence or validation experiments.

Confounding with Higher-Order Terms:
   Pitfall: Confounding between main effects and higher-order terms (quadratic effects, interactions) can complicate interpretation and lead to misidentification of significant effects.

Remedy: Use appropriate statistical techniques, such as ANOVA or effect sparsity analysis, to identify and prioritize significant effects among main effects, quadratic effects, and interactions. Focus on interpreting the most influential effects while acknowledging potential confounding patterns.

Insufficient Replication:

Pitfall: Inadequate replication of experimental runs can lead to unreliable estimates of factor effects and increased uncertainty in the conclusions drawn from the experiment.

Remedy: Increase the number of experimental replicates to improve the precision and reliability of estimates. Conduct power analyses or simulate the experiment to determine the appropriate sample size needed to achieve the desired statistical power and precision.

Ignoring Assumptions of Statistical Models:

Pitfall: Ignoring or violating assumptions of statistical models, such as normality of residuals or homoscedasticity, can invalidate the results of data analysis and lead to incorrect conclusions.

Remedy: Verify the assumptions of statistical models using diagnostic tests and graphical methods. If assumptions are violated, consider using robust statistical techniques or transforming the data to meet model assumptions.

Overlooking Factor Optimization:

Pitfall: Focusing solely on identifying factor effects without optimizing factor settings for practical applications can limit the utility of CCD experiments.

Remedy: Integrate factor optimization into the experimental design by using optimization algorithms or RSM to identify optimal factor settings that maximize or minimize the response variable of interest.

By being aware of these potential pitfalls and implementing appropriate remedies, researchers can conduct CCD experiments in physics more effectively, ensuring reliable results and robust conclusions.

EXAMPLE

CCD experiments in physics are used to study the effects of factors and their interactions with a response variable. Here's an example of a matrix for a CCD experiment in physics:

Objective: To optimize the conditions for a chemical reaction in a physics experiment by studying the effects of two factors (temperature and pressure) on the reaction rate.

Factors:

1. **Temperature (Factor A):**
   - Low temperature (–1).
   - Medium temperature (0).
   - High temperature (+1).
2. **Pressure (Factor B):**
   - Low pressure (–1).
   - Medium pressure (0).
   - High pressure (+1).

Response Variable: Reaction rate (measured in moles per minute, mol/min).

TABLE 2.55
CCD Experiment Matrix Example for Physics

Run	Temperature (A)	Pressure (B)	Reaction Rate (mol/min)
1	−1	−1	...
2	+1	−1	...
3	−1	+1	...
4	+1	+1	...
5	−1.41	0	...
6	+1.41	0	...
7	0	−1.41	...
8	0	+1.41	...
9	0	0	...

CCD Matrix:

In this CCD experiment, the matrix is designed to explore the effects of temperature and pressure on the reaction rate while considering both the factorial points and the axial points. The matrix is shown in Table 2.55.

In this matrix, each row represents a specific experimental run with a unique combination of temperature and pressure levels, including the central point (0) and the axial points for both factors. The axial points allow for the estimation of curvature and quadratic effects in the response surface.

By conducting experiments at these specific combinations of temperature and pressure levels, physicists can model the relationship between these factors and the reaction rate. The CCD provides insights into the optimal conditions for the chemical reaction in terms of temperature and pressure in physics experiments.

An example of a CCD experiment in physics could involve optimizing the parameters of a laser-cutting process used in materials science. The goal might be to achieve the highest cutting efficiency while minimizing heat-affected zone (HAZ) width in a particular material. Here's how the experiment might be structured:

1. **Factors:**
   - Laser Power (X): Varying from low to high power levels.
   - Cutting Speed (X): Varying from slow to fast speeds.
   - Gas Pressure (X): Varying from low to high pressure levels.
2. **Response Variables:**
   - Cutting Efficiency: Defined as the ratio of the area of the cut to the time taken to perform the cut.
   - HAZ Width: Measured as the distance from the edge of the cut to the region where significant thermal effects are observed.
3. **Experimental Design:**
   - A CCD is chosen with three factors (X, X, X) at three levels (−1, 0, +1) plus center points to account for curvature.
   - The design includes factorial points, axial points, and center points for a total of $2^k+2k+nc$ runs, where k is the number of factors and nc is the number of center points.
4. **Experimental Runs**: Experimental runs are conducted according to the design matrix, varying the laser power, cutting speed, and gas pressure at different levels as specified by the design.

5. **Data Collection**: Cutting efficiency and HAZ width are measured for each experimental run, ensuring consistency and accuracy in data collection.
6. **Data Analysis**: Statistical analysis techniques, such as ANOVA, are used to analyze the experimental data and identify significant factor effects and interactions. RSM may be employed to model the relationship between the factors and the response variables and to optimize factor settings.
7. **Interpretation of Results**: Based on the analysis, optimal settings for laser power, cutting speed, and gas pressure are determined to maximize cutting efficiency and minimize HAZ width. Factor effects and interactions are interpreted to gain insights into the underlying physics of the cutting process and guide further experimentation or process optimization.
8. **Validation and Optimization**: The results are validated through replication or additional experiments to ensure their reproducibility and reliability. The optimized process parameters are implemented in practical applications, such as industrial laser-cutting operations, to improve efficiency and quality.

By conducting a CCD experiment in this manner, researchers can systematically explore the effects of multiple factors on the performance of a laser-cutting process and optimize the process parameters to achieve desired outcomes in materials science applications.

To illustrate a CCD matrix for optimizing the parameters of a laser-cutting process in materials science, let's consider a hypothetical scenario with three factors: Laser power (X_1), cutting speed (X_2), and gas pressure (X_3). Each factor is varied at three levels: low (–1), medium (0), and high (+1), with center points added for experimental validation. The CCD matrix might be structured as shown in Table 2.56.

In this matrix,

- Each row represents an experimental run.
- Columns correspond to the factors (X_1, X_2, X_3) being studied.

TABLE 2.56

CCD Experiment Matrix Example for Physics

Run	X_2 (Laser Power)	X_2 (Cutting Speed)	X_3 (Gas Pressure)
1	–1	0	0
2	1	0	0
3	0	–1	0
4	0	1	0
5	0	0	–1
6	0	0	1
7	–1	–1	–1
8	1	–1	–1
9	–1	1	–1
10	1	1	–1
11	–1	–1	1
12	1	–1	1
13	–1	1	1
14	1	1	1
15	0	0	0 (Center Point)
16	0	0	0 (Center Point)
17	0	0	0 (Center Point)

- The values in each cell represent the coded levels of the factors: −1 (low), 0 (medium), and +1 (high).
- The first eight rows represent the factorial points, covering the entire factor space.
- The next eight rows represent the axial points, which extend beyond the factorial points to explore curvature.
- The last three rows represent the center points, which serve as replicates to estimate experimental error and validate the experimental design.

This CCD matrix allows for a systematic exploration of the effects of laser power, cutting speed, and gas pressure on the performance of the laser-cutting process. By conducting experiments according to this matrix and analyzing the resulting data, researchers can identify optimal parameter settings to maximize cutting efficiency and minimize HAZ width in materials science applications.

REPORT

CCD experiments in physics utilize various statistical tools to analyze experimental data, model factor effects, and optimize process parameters. Some of the common statistical tools used in CCD experiments for physics include:

ANOVA: ANOVA is used to partition the variability in the response variable into contributions from different factors, their interactions, and experimental error. It helps assess the significance of factor effects and interactions, providing insights into the factors that most significantly influence the response variable.

Regression Analysis: Regression analysis is employed to develop mathematical models that describe the relationship between the IVs (factors) and the DV (response). Linear, quadratic, or higher-order regression models may be fitted to the experimental data to capture linear, curvature, and interaction effects.

RSM: RSM is a collection of statistical and mathematical techniques used to model and optimize response surfaces representing the relationship between multiple factors and the response variable. It involves fitting response surface models to experimental data and using optimization algorithms to identify optimal factor settings that maximize or minimize the response variable.

Model Diagnostics: Model diagnostics techniques, such as residual analysis, lack-of-fit tests, and normality tests, are employed to assess the adequacy and validity of statistical models fitted to experimental data. These diagnostics help identify model inadequacies, outliers, or violations of model assumptions that may affect the reliability of results.

Effect Sparsity Analysis: Effect sparsity analysis is used to identify and prioritize significant factor effects, interactions, and higher-order terms among the multitude of effects estimated from experimental data. It helps focus on interpreting the most influential effects while filtering out less significant effects and reducing the risk of false discoveries.

Optimization Algorithms: Optimization algorithms, such as gradient-based methods or heuristic search algorithms, are employed to identify optimal factor settings that maximize or minimize the response variable. These algorithms explore the design space defined by the experimental factors and iteratively refine factor settings to converge on optimal solutions.

DOE Software: Specialized software packages for DOE provide tools for generating experimental designs, analyzing experimental data, and fitting statistical models. These software packages often include built-in capabilities for ANOVA, regression analysis, RSM, and optimization, facilitating the entire process of CCD experimentation and analysis.

By leveraging these statistical tools, researchers can effectively analyze CCD experimental data, model complex factor effects, and optimize process parameters to achieve desired outcomes in physics experiments.

A comprehensive report for CCD experiments in physics should provide a detailed account of the experimental design, methodology, results, and conclusions. Here are the essential elements that should be included in such a report:

Title Page:
- Title of the report.
- Names and affiliations of the authors.
- Date of the report.

Abstract: A concise summary of the experiment, including the research objectives, experimental design, key findings, and conclusions. Typically, the abstract should not exceed 250 words.

Introduction: Background information on the research topic and the rationale for conducting the experiment. Clear statement of the research objectives and hypotheses to be tested. Overview of the experimental design and methodology.

Experimental Design: Description of the CCD used, including the factors studied, their levels, and the number of experimental runs. Justification for the choice of factors and levels. Explanation of how the design facilitates exploration of factor effects and optimization of process parameters.

Materials and Methods: Detailed description of the experimental setup, including any equipment or instrumentation used. Explanation of the procedure followed for conducting the experimental runs. Description of how data were collected, including any measurements or observations made during the experiment.

Data Analysis: Overview of the statistical methods used to analyze the experimental data. Presentation of the results of data analysis, including estimates of factor effects, interactions, and model adequacy. Visual aids, such as graphs, plots, or tables to illustrate key findings.

Results: Presentation of the main findings of the experiment, organized according to the research objectives. Clear and concise reporting of the results, including any significant factor effects or interactions observed. Discussion of any unexpected or noteworthy findings.

Discussion: Interpretation of the results in the context of the research objectives and the underlying physics of the experiment. Comparison of the observed results with theoretical expectations or previous studies. Discussion of the implications of the findings and their relevance to the broader field of physics.

Conclusion: Summary of the key findings and their significance. Statement of the conclusions drawn from the experiment, including any recommendations for future research or practical applications.

References: List of all references cited in the report, following a standardized citation format (e.g., APA, MLA).

Appendices: Any additional supplementary information, such as raw data, calculation details, or experimental protocols, that may be useful for readers but is not essential for understanding the main findings.

By including these elements in a report on CCD experiments in physics, researchers can effectively communicate the objectives, methodology, results, and implications of their experiments to the scientific community.

2.4.5 Taguchi Methods

Taguchi method designed experiments offer a structured and efficient approach to optimizing processes and understanding the factors influencing physical phenomena, making them valuable tools in physics research and experimentation. Taguchi method designed experiments are used in physics for several reasons:

Optimization of Processes: Taguchi methods help in optimizing various physical processes by identifying the optimal combination of input parameters that lead to the desired output. This is particularly useful in physics where processes often involve multiple factors influencing the outcome.

Robustness to Noise Factors: Physics experiments are often affected by various sources of variability or noise factors, such as environmental conditions or equipment variations. Taguchi methods aim to design experiments that are robust to such noise factors, ensuring that the results are reliable and reproducible.

Efficiency in Experimentation: Taguchi methods enable the DOE with fewer runs compared to traditional full factorial designs, while still providing adequate information about the factors and their interactions. This efficiency saves time and resources in conducting experiments, which is crucial in physics research.

Systematic Approach to Optimization: Taguchi methods provide a systematic and structured approach to experimentation and optimization. By systematically varying input parameters according to a designed OA, researchers can efficiently explore the factor space and identify optimal settings.

Quantification of Factor Effects: Taguchi methods allow for the quantification of the effects of individual factors as well as their interactions with the outcome variable. This provides valuable insights into the underlying physics of the system under study.

Reduction of Variation: Taguchi methods emphasize the reduction of variation in the output variable, leading to improved consistency and quality of experimental results. This is particularly beneficial in physics experiments where precision and accuracy are critical.

Applicability to Complex Systems: Taguchi methods are versatile and can be applied to a wide range of complex systems and processes in physics, including materials science, semiconductor physics, optics, fluid dynamics, and more.

Taguchi method designed experiments find applications in various branches of physics, where they are employed to optimize processes, improve product quality, and understand the effects of different factors on experimental outcomes. Some specific areas where Taguchi method designed experiments are used in physics include:

Materials Science: Taguchi methods are applied to optimize material synthesis processes, such as alloy fabrication, thin-film deposition, and composite material manufacturing. They help identify the optimal combination of parameters to enhance material properties like strength, conductivity, and durability.

Optics and Photonics: In optics research, Taguchi methods are used to optimize parameters in the fabrication of optical components, such as lenses, mirrors, and diffraction gratings. They assist in achieving desired optical properties and minimizing defects in the final products.

Semiconductor Physics: Taguchi methods are utilized in semiconductor device fabrication to optimize process parameters during lithography, etching, and doping processes. They help improve device performance and yield by minimizing variations and defects in semiconductor devices.

Fluid Dynamics: In fluid dynamics research, Taguchi methods are applied to optimize experimental conditions in wind tunnel testing, fluid flow simulations, and hydrodynamic experiments. They help identify the most influential parameters affecting flow behavior and optimize designs for aerodynamic efficiency.

Nuclear Physics: Taguchi methods are used in nuclear physics experiments to optimize experimental conditions and instrumentation parameters for particle detection, spectroscopy, and nuclear reactions. They aid in improving the sensitivity and reliability of nuclear measurement techniques.

Acoustics and Vibrations: Taguchi methods find applications in optimizing parameters in acoustic and vibration experiments, such as soundproofing materials, speaker design, and vibration damping systems. They help achieve desired acoustic performance and reduce unwanted noise or vibrations.

Experimental Design and Analysis: Taguchi methods are also used in designing and analyzing experiments across various fields of physics, including factorial design, RPD, and RSM. They provide efficient and systematic approaches to experimental planning, data analysis, and optimization.

Taguchi method designed experiments are widely applicable in physics research and experimentation, offering systematic approaches to process optimization, quality improvement, and understanding the effects of different factors on experimental outcomes.

STEPS

Developing Taguchi method designed experiments in physics involves several steps to systematically plan and conduct experiments for optimizing processes or understanding the effects of various factors on experimental outcomes. Here are the general steps involved:

1. **Define the Objective**: Clearly define the objective of the experiment, such as optimizing a process parameter, improving product quality, or understanding the effects of factors on experimental outcomes.
2. **Identify Factors and Levels**:
   - Identify the factors (input variables) that may influence the experimental outcome. These factors could be process parameters, material properties, environmental conditions, or equipment settings.
   - Determine the levels (settings) for each factor that will be studied during the experiment. Typically, factors are categorized as controllable (factors that can be adjusted by the experimenter) and uncontrollable (factors that cannot be directly controlled but may affect the outcome).
3. **Select an OA**:
   - Choose an appropriate OA, which is a structured experimental design matrix that ensures efficient exploration of factor effects and interactions with a minimal number of experimental runs.
   - The choice of OA depends on the number of factors and levels, as well as the desired level of resolution and robustness to noise factors.
4. **Assign Experiments to Runs**:
   - Assign the experimental conditions (factor settings) to the rows of the chosen OA, ensuring that each combination of factor levels is represented in the experiment.
   - Randomize the order of experimental runs to minimize the effects of uncontrolled variables and systematic biases.
5. **Conduct Experiments**:
   - Perform the experimental runs according to the assigned factor settings, following the experimental plan generated from the OA.
   - Record the experimental observations, measurements, or data obtained during each run in a systematic and consistent manner.
6. **Analyze Experimental Data**:
   - Analyze the experimental data using statistical techniques to determine the effects of factors on the experimental outcome.
   - Calculate SNRs to assess the robustness of the process to variations and noise factors.

- Identify significant factors, interactions, and optimal factor settings that lead to desired outcomes.
7. **Optimize Process Parameters:**
   - Use the results of the data analysis to optimize process parameters or factor settings to achieve the desired outcome.
   - Adjust factor levels or settings based on the identified optimal conditions, aiming to maximize performance or minimize variability in the process.
8. **Validate and Confirm Results:**
   - Validate the optimized process parameters through confirmation experiments or additional validation studies.
   - Ensure that the optimized settings are robust and reproducible under different operating conditions.
9. **Document and Report Findings**: Document the experimental design, methodology, data analysis, and results in a clear and concise report. Communicate the findings, conclusions, and recommendations derived from the Taguchi method designed experiments to stakeholders or collaborators.

By following these steps, researchers can systematically plan, conduct, and analyze Taguchi method designed experiments in physics to optimize processes, improve product quality, and gain insights into the effects of various factors on experimental outcomes.

PITFALLS AND REMEDIES

Taguchi method designed experiments offer a systematic approach to optimize processes and understand the effects of various factors in physics experiments. However, there are certain pitfalls that researchers may encounter during the planning, execution, and analysis of Taguchi experiments. Here are some common pitfalls and remedies:

Inadequate Factor Selection:
   Pitfall: Selecting incorrect or irrelevant factors may lead to suboptimal results and ineffective optimization.
   Remedy: Conduct a thorough analysis to identify the most significant factors that influence the experimental outcome. Consult subject matter experts and conduct preliminary studies to validate the relevance of selected factors.

Incorrect Level Selection:
   Pitfall: Choosing inappropriate levels for factors may result in a limited exploration of the factor space or an inability to capture nonlinear effects.
   Remedy: Conduct a preliminary study or sensitivity analysis to identify appropriate levels for each factor. Ensure that the selected levels cover a wide range and are representative of practical operating conditions.

OA Selection:
   Pitfall: Selecting an inadequate OA may result in insufficient resolution or aliasing of factor effects.
   Remedy: Choose an OA that balances the number of factors, levels, and interactions to achieve the desired resolution and robustness. Consider factors, such as the number of experimental runs available and the desired level of precision.

Improper Experimental Design:
   Pitfall: Deviating from the experimental design plan or failing to randomize experimental runs may introduce bias or confounding effects.
   Remedy: Adhere strictly to the experimental design plan, ensuring that factors are varied systematically according to the OA. Randomize the order of experimental runs to minimize the effects of uncontrolled variables.

Incomplete Data Analysis:

Pitfall: Conducting inadequate data analysis may lead to misinterpretation of results or failure to identify significant factor effects.

Remedy: Use appropriate statistical techniques, such as ANOVA, SNRs, and graphical analysis, to analyze the experimental data comprehensively. Consider both main effects and interactions when assessing the significance of factors.

Overlooking Noise Factors:

Pitfall: Neglecting to account for noise factors or uncontrollable variables may lead to unstable or non-reproducible results.

Remedy: Conduct robustness tests or include noise factors in the experimental design to assess the sensitivity of the process to variations. Use SNRs to evaluate the robustness of the optimized parameters.

Failure to Validate Results:

Pitfall: Failing to validate the optimized parameters or process settings may result in suboptimal performance under real-world conditions.

Remedy: Validate the optimized parameters through confirmation experiments or additional validation studies using independent datasets. Ensure that the optimized settings are robust and reproducible across different operating conditions.

By being aware of these pitfalls and implementing appropriate remedies, researchers can enhance the effectiveness and reliability of Taguchi method designed experiments in physics, leading to improved process optimization and a better understanding of factor effects.

EXAMPLE

Taguchi methods, also known as the Taguchi DOE, are used in physics to optimize processes and products while considering factors and their interactions. Here's a simplified example of a matrix for a Taguchi method designed experiment in physics:

Objective: To optimize the conditions for a laser welding process in a physics experiment by studying the effects of two factors (laser power and welding speed) on the quality of welds.

Factors:

1. **Laser Power (Factor A):**
   - Low power (50 W).
   - Medium power (100 W).
   - High power (150 W).
2. **Welding Speed (Factor B):**
   - Slow speed (1 mm/s).
   - Medium speed (5 mm/s).
   - Fast speed (10 mm/s).

Response Variable: Weld quality (measured as a rating from 1 to 10, with higher values indicating better quality).

Taguchi Method Designed Experiment Matrix:

In this Taguchi method designed experiment, the matrix is designed to systematically study the effects of laser power and welding speed on weld quality while conducting a series of experiments. The matrix is shown in Table 2.57.

TABLE 2.57

Taguchi Method Designed Experiment Matrix Example for Physics

Run	Laser Power (A) (W)	Welding Speed (B) (mm/s)	Weld Quality (1–10)
1	50	1	...
2	100	1	...
3	150	1	...
4	50	5	...
5	100	5	...
6	150	5	...
7	50	10	...
8	100	10	...
9	150	10	...

In this matrix, each row represents a specific experimental run with a unique combination of laser power and welding speed levels. Weld quality is rated based on the results of each run.

Using Taguchi methods and statistical analysis, physicists can identify the optimal combination of laser power and welding speed that results in the highest weld quality. The approach helps optimize the laser welding process in physics experiments to achieve the desired quality standards.

An example of Taguchi method designed experiments in physics could involve optimizing the parameters of a laser-cutting process used in materials science. Laser cutting is a widely used technique in materials processing, and optimizing the process parameters can significantly impact the quality and efficiency of cutting operations. Here's how a Taguchi method designed experiment could be applied to optimize the laser-cutting process:

Objective: To optimize the parameters of a laser-cutting process to achieve high-quality cuts with minimal HAZ and kerf width.

Factors and Levels:

1. Laser power (low, medium, high).
2. Cutting speed (slow, medium, fast).
3. Assist gas pressure (low, medium, high).
4. Focus position (near, middle, far).

OA: A L9 OA could be selected, which allows for the investigation of four factors at three levels with nine experimental runs.

Utilizing an OA to efficiently explore the effects of the factors, consider four factors with three levels each, which can be represented using an L9 OA. The matrix is shown in Table 2.58.

In this matrix,

- Laser power, cutting speed, assist gas pressure, and focus position are the four factors being studied.
- Each factor has three levels: low, medium, and high.
- The matrix represents nine experimental runs (1 to 9), each with a unique combination of factor levels.
- The arrangement of factors and levels in the matrix follows the principles of OAs to ensure a balanced exploration of the factor space and efficient identification of significant effects.

TABLE 2.58

Taguchi Method Designed Experiment OA Example for Physics

Experimental Run	Laser Power	Cutting Speed	Assist Gas	Pressure Focus Position
1	Low	Slow	Low	Near
2	Medium	Medium	Medium	Middle
3	High	Fast	High	Far
4	Low	Medium	High	Near
5	Medium	Fast	Low	Middle
6	High	Slow	Medium	Far
7	Low	Fast	Medium	Far
8	Medium	Slow	High	Near
9	High	Medium	Low	Middle

Researchers can conduct laser-cutting experiments according to the specified parameter settings in each row of the matrix. After collecting data from the experiments, statistical analysis techniques, such as ANOVA and SNRs, can be used to analyze the results and identify the optimal parameter settings for achieving the desired outcomes.

Experimental Design:

1. Prepare the experimental setup with a laser-cutting machine equipped with adjustable parameters for laser power, cutting speed, assist gas pressure, and focus position.
2. Conduct the nine experimental runs according to the OA, varying the parameters as specified in the design.
3. Record relevant parameters during each run, such as HAZ width, kerf width, and cutting quality.

Data Analysis:

1. Analyze the experimental data using statistical techniques, such as ANOVA and SNRs.
2. Determine the main effects of each factor and identify any significant interactions between factors.
3. Calculate the optimal parameter settings based on the desired performance criteria, such as minimizing HAZ width and kerf width while maintaining cutting quality.

Optimization:

1. Use the results of the data analysis to identify the optimal parameter settings that maximize cutting quality and minimize HAZ width and kerf width.
2. Adjust the laser power, cutting speed, assist gas pressure, and focus position to the optimal levels determined from the Taguchi analysis.

Validation:

1. Validate the optimized parameter settings through confirmation experiments or additional validation studies.
2. Verify that the optimized settings result in consistent and high-quality cuts across different material types and thicknesses.

Documentation and Reporting:

1. Document the experimental design, methodology, data analysis, and results in a comprehensive report.
2. Communicate the findings, conclusions, and recommendations derived from the Taguchi method designed experiment to stakeholders or collaborators.

By applying the Taguchi method designed experiment to optimize the laser-cutting process parameters, researchers can achieve improved cutting quality, reduced HAZ, and optimized process efficiency in materials science applications.

REPORT

In Taguchi method designed experiments for physics, several statistical tools are commonly used to analyze experimental data, assess the effects of factors, and optimize processes. Some of the key statistical tools used in Taguchi method designed experiments include:

ANOVA: ANOVA is used to analyze the variation in experimental data and assess the significance of factor effects and interactions.
It helps determine whether the observed differences in the experimental outcomes are statistically significant or due to random variation.

SNRs: SNRs are used to evaluate the quality or performance of experimental outcomes in relation to noise factors. They quantify the signal (desired outcome) relative to the noise (unwanted variation), providing a measure of process robustness or performance.

Main Effects Plots: Main effects plots visualize the effects of individual factors on the experimental outcome. They help identify which factors have the most significant impact on the outcome and whether the effects are linear or nonlinear.

Interaction Plots: Interaction plots visualize the interactions between different factors on the experimental outcome. They help identify synergistic or antagonistic effects between factors and assess the importance of considering interactions in the optimization process.

RSM: RSM is a statistical technique used to model and optimize complex response surfaces with multiple factors. It involves fitting mathematical models to experimental data to predict the response surface and identify optimal factor settings.

OAs: OAs are structured experimental designs that ensure efficient exploration of factor effects with a minimal number of experimental runs. They help balance the trade-off between the number of factors studied and the number of experimental runs required.

RPD: RPD techniques are used to optimize process parameters while simultaneously minimizing the sensitivity of the process to variations or noise factors. They aim to achieve robust and reliable process performance under different operating conditions.

Graphical Analysis: Graphical techniques, such as scatter plots, contour plots, and Pareto charts, are often used to visualize experimental data and analyze relationships between factors and outcomes.

By leveraging these statistical tools, researchers can effectively analyze experimental data, identify significant factor effects, and optimize processes in Taguchi method designed experiments for physics. These tools enable systematic and data-driven decision-making to achieve desired outcomes and improve process performance.

When preparing a report for Taguchi method designed experiments used in physics, it's essential to include comprehensive information about the experimental design, methodology, data analysis, results, and conclusions. Here are the key elements to include in the report:

Title Page:
- Title of the report, indicating the purpose and scope of the study.
- Names of the authors and their affiliations.
- Date of submission.

Abstract: A brief summary of the experiment, including the objectives, methodology, key findings, and conclusions. Provide a concise overview of the experimental design and results.

Introduction:
- Background information on the problem or process being studied.
- Objectives and hypotheses of the experiment.
- Justification for using the Taguchi method and the relevance to the field of physics.

Experimental Design:
- Description of the experimental setup, including equipment, materials, and procedures.
- Explanation of the factors studied, their levels, and the chosen OA.
- Details of the parameter settings for each experimental run.

Data Analysis:
- Presentation of the experimental data, including raw data and observations from each experimental run.
- Statistical analysis techniques used, such as ANOVA, SNRs, and graphical analysis.
- Summary of the main effects, interactions, and significant findings.

Results: Presentation of the results in a clear and organized manner, using tables, figures, and graphs. Display the main effects plots, interaction plots, and response surface plots to visualize the effects of factors on the experimental outcome.

Discussion:
- Interpretation of the results and their implications for the studied process or phenomenon.
- Comparison of the experimental findings with theoretical expectations or previous studies.
- Explanation of any unexpected or anomalous results observed during the experiment.

Conclusion: Summary of the key findings and their significance in addressing the research objectives. Statement of the overall success or limitations of the Taguchi method designed experiment. Recommendations for future research or applications based on the findings.

References: List of cited references, including academic papers, textbooks, and other relevant sources used in the study.

Appendices: Additional supporting information, such as detailed experimental data, calculation procedures, and supplementary analysis. Include any supplementary materials that enhance the understanding of the experiment but are not essential for the main text.

By including these elements in the report, researchers can effectively communicate the design, execution, analysis, and conclusions of Taguchi method designed experiments used in physics, facilitating transparency, reproducibility, and dissemination of research findings.

2.4.6 MIXTURE EXPERIMENTS

Mixture designed experiments play a crucial role in advancing our understanding of materials and phenomena in physics by providing a systematic and efficient approach to studying complex systems, optimizing processes, and exploring the relationships between composition and properties. Mixture designed experiments are used in physics for several reasons:

Material Science and Engineering: In physics, particularly in the field of materials science and engineering, researchers often work with composite materials or alloys composed of different components. Mixture designed experiments help optimize the composition of these materials to achieve specific properties, such as strength, conductivity, or thermal stability.

Experimental Design for Complex Systems: Physics experiments frequently involve systems with multiple interacting components or variables. Mixture designed experiments allow researchers to systematically vary the proportions of different components in the system to understand their individual and collective effects on the observed phenomena.

Process Optimization: Mixture designed experiments are used to optimize processes involving mixtures or combinations of substances. This could include optimizing reaction conditions in chemical processes, tuning parameters in manufacturing processes, or adjusting experimental conditions in physics experiments to achieve desired outcomes.

Understanding Phase Behavior: In studies related to phase transitions, phase diagrams, or phase equilibria, Mixture designed experiments help elucidate the effects of composition on the phase behavior of materials. By systematically varying the composition of mixtures, researchers can explore phase boundaries, critical points, and phase transitions.

Materials Characterization and Analysis: Mixture designed experiments are valuable for characterizing and analyzing the properties of complex materials. By systematically varying the composition of mixtures and measuring relevant properties, researchers can establish structure–property relationships and identify optimal compositions for specific applications.

Mixture designed experiments are used in physics across various subfields and applications. Some common areas where Mixture designed experiments find application in physics include:

Materials Science and Engineering: Mixture designed experiments are extensively used in materials science and engineering to optimize the composition of materials, such as alloys, polymers, composites, and ceramics. Researchers study the effects of varying proportions of different components on the properties of materials, including mechanical, electrical, thermal, and optical properties.

Chemical Physics: In chemical physics, mixture designed experiments are employed to study the behavior of mixtures, solutions, and chemical reactions. Researchers investigate the influence of composition on reaction kinetics, equilibrium constants, and phase behavior, aiding in the understanding of complex chemical systems.

Condensed Matter Physics: Mixture designed experiments are used to explore phase transitions, phase diagrams, and critical phenomena in condensed matter systems. By systematically varying the composition of mixtures, physicists investigate the effects of composition on the emergence of different phases and critical points in materials.

Soft Matter Physics: Mixture designed experiments are utilized in soft matter physics to study complex fluids, polymers, gels, and colloidal suspensions. Researchers investigate how changes in composition affect the rheological properties, phase behavior, and self-assembly dynamics of soft matter systems.

Biophysics: Mixture designed experiments find application in biophysics for studying biomolecular interactions, protein folding, and lipid membrane properties. Researchers use mixtures of biomolecules or model systems to investigate the effects of composition on biological processes and cellular behavior.

Optics and Photonics: Mixture designed experiments are employed in optics and photonics to optimize the composition of optical materials, such as glasses, crystals, and photonic structures. Researchers explore the effects of composition on optical properties, including refractive index, dispersion, and nonlinear optical behavior.

Nanotechnology and Nanomaterials: Mixture designed experiments are used in nanotechnology and nanomaterials research to tailor the composition of nanomaterials and nanostructures. Researchers investigate how variations in composition influence the properties of nanoparticles, nanocomposites, and nanostructured materials for applications in electronics, catalysis, and sensors.

Overall, mixture designed experiments are versatile tools used in various branches of physics to systematically explore the effects of composition on the properties and behavior of materials, fluids, and systems, leading to advancements in fundamental understanding and technological innovation.

Steps

Developing mixture designed experiments in physics involves several steps to systematically vary the composition of mixtures and study their effects on the properties or behavior of interest. Here are the steps to develop mixture designed experiments in physics:

1. **Define the Objective**: Clearly define the objective of the experiment. Determine the specific properties, behaviors, or phenomena that you want to investigate or optimize by varying the composition of mixtures.
2. **Identify the Components**: Identify the components or constituents that will make up the mixtures. These components could be chemical substances, materials, or other entities relevant to the study.
3. **Select the Experimental Factors**: Determine the factors or variables that will be varied in the experiment to create different mixture compositions. These factors could include the proportions, concentrations, or ratios of the components in the mixtures.
4. **Choose the Design Space**: Define the design space, which represents the range of possible compositions for the mixtures. Specify the lower and upper bounds for each factor to ensure that the experimental design covers a relevant and informative range of compositions.
5. **Select the Experimental Design**: Choose an appropriate experimental design that allows for systematic variation of the mixture compositions. Common designs include simplex-centroid designs, simplex-lattice designs, or mixture designs based on OAs.
6. **Generate the Experimental Plan**: Use statistical software or experimental design tools to generate the experimental plan based on the chosen design. The plan specifies the composition of each mixture in terms of the levels of the experimental factors.
7. **Prepare the Experimental Setup**: Set up the experimental apparatus or equipment required to prepare and characterize the mixtures. Ensure that the setup is suitable for accurately measuring the properties or behaviors of interest.
8. **Conduct the Experiments**: Prepare the mixtures according to the experimental plan and conduct the experiments as per the defined procedures. Record relevant data, observations, or measurements for each mixture composition.
9. **Analyze the Data**: Analyze the experimental data using statistical methods to identify trends, patterns, or relationships between the mixture compositions and the properties or behaviors studied. Perform statistical analysis techniques such as regression analysis, ANOVA, or response surface modeling.
10. **Interpret the Results**: Interpret the results of the data analysis in the context of the experiment's objectives. Determine the effects of composition on the properties or behaviors studied and draw conclusions based on the findings.
11. **Optimize the Mixture Composition**: If the objective is to optimize the mixture composition for a specific outcome, use the experimental results to identify the optimal composition that maximizes or minimizes the desired property or behavior.
12. **Validate and Refine**: Validate the experimental findings through additional experiments or validation studies. Refine the experimental design or procedures based on the insights gained from the initial experiments.

13. **Document and Communicate**: Document the experimental design, procedures, results, and conclusions in a comprehensive report or scientific paper. Communicate the findings to peers, collaborators, or stakeholders in the field of physics.

By following these steps, researchers can systematically develop and execute mixture designed experiments in physics to study, optimize, or understand the effects of composition on the properties or behaviors of interest in materials, fluids, or systems.

PITFALLS AND REMEDIES

Mixture designed experiments in physics offer valuable insights into the effects of varying compositions on material properties, system behavior, and other phenomena. However, like any experimental approach, they come with certain pitfalls. Here are some common pitfalls encountered in mixture designed experiments in physics, along with potential remedies:

Insufficient Understanding of System Behavior:
   Pitfall: Lack of comprehensive understanding of the underlying physics or chemistry governing the system may lead to inaccurate interpretations of experimental results.
   Remedy: Prioritize thorough literature review and preliminary studies to gain a better understanding of the system's behavior and relevant factors before designing experiments. Consult with experts in the field to identify key variables and potential interactions.
Limited Design Space Coverage:
   Pitfall: Failing to adequately cover the relevant design space may result in missing critical regions or combinations of factors that affect the outcome.
   Remedy: Conduct a thorough analysis to identify the range of factors and levels that could influence the outcome. Use statistical tools to design an experimental plan that systematically explores the design space while minimizing the number of experiments needed.
Nonlinear Responses and Interactions:
   Pitfall: Ignoring nonlinear responses or interactions between components in the mixture may lead to inaccurate model predictions or interpretations.
   Remedy: Consider the possibility of nonlinear responses and interactions when designing the experimental plan. Use appropriate statistical techniques, such as RSM or nonlinear regression, to model and analyze the data accurately.
Measurement Errors and Variability:
   Pitfall: Inaccurate or inconsistent measurements of experimental outcomes may obscure true effects and introduce variability into the data.
   Remedy: Implement rigorous measurement protocols and quality control procedures to minimize measurement errors and variability. Calibrate instruments regularly, replicate measurements when possible, and assess measurement uncertainty to ensure data reliability.
Overfitting of Models:
   Pitfall: Overfitting occurs when a model fits the experimental data too closely, capturing noise rather than true underlying relationships.
   Remedy: Use appropriate model selection criteria and validation techniques to guard against overfitting. Partition the data into training and validation sets, and assess model performance on unseen data. Consider simpler models when possible to avoid overfitting.
Extrapolation Risks:
   Pitfall: Extrapolating experimental findings beyond the studied design space may lead to unreliable predictions or conclusions.

Remedy: Exercise caution when extrapolating experimental results, particularly when extending conclusions to untested regions of the design space. Validate predictions through additional experiments or complementary modeling approaches.

Inadequate Reproducibility:

Pitfall: Lack of reproducibility in experimental results may undermine the reliability and generalizability of findings.

Remedy: Document experimental procedures meticulously and ensure consistency in sample preparation, measurement techniques, and environmental conditions. Conduct replicate experiments to assess reproducibility and quantify variability.

By addressing these pitfalls proactively and employing appropriate remedies, researchers can enhance the reliability, robustness, and interpretability of mixture designed experiments in physics, leading to more meaningful insights and discoveries.

EXAMPLE

Mixture designed experiments in physics involve optimizing the proportions of different components or ingredients in a mixture to achieve desired outcomes. Here's a simplified example of a matrix for a mixture designed experiment in physics:

Objective: To optimize the composition of a semiconductor material used in PV cells for maximum energy conversion efficiency.

Components of the Mixture:

1. **Silicon (Component A)**: Proportion expressed as a percentage of the total mixture.
   - Range: 60% to 80%.
2. **Germanium (Component B)**: Proportion expressed as a percentage of the total mixture.
   - Range: 10% to 30%.
3. **Gallium (Component C)**: Proportion expressed as a percentage of the total mixture.
   - Range: 5% to 15%.
4. **Phosphorus (Component D)**: Proportion expressed as a percentage of the total mixture.
   - Range: 2% to 8%.

Response Variable: Energy conversion efficiency (measured as a percentage).

Mixture Designed Experiment Matrix:

In this mixture designed experiment, the matrix is designed to optimize the mix proportions of the semiconductor material for maximum energy conversion efficiency. The matrix is shown in Table 2.59.

In this matrix, each row represents a specific experimental run with a unique combination of proportions for the mixture components (silicon, germanium, gallium, and phosphorus). The aim is to find the optimal mixture proportions that result in the highest energy conversion efficiency for the semiconductor material used in PV cells.

By conducting experiments with different mix proportions and analyzing the resulting energy conversion efficiencies, physicists can determine the ideal composition for their specific semiconductor material, maximizing energy conversion in PV cells for physics applications.

Another example of mixture designed experiments in physics can be found in the field of materials science and engineering, particularly in the optimization of composite materials for specific applications.

TABLE 2.59

Mixture Designed Experiment Matrix Example for Physics

Run	Silicon (A) (%)	Germanium (B) (%)	Gallium (C) (%)	Phosphorus (D) (%)	Energy Conversion Efficiency (%)
1	60	10	5	2	…
2	70	20	10	4	…
3	80	30	15	6	…
4	60	20	5	4	…
5	70	10	15	2	…
6	80	20	10	6	…

Let's consider the development of a new type of composite material for lightweight structural components in aerospace engineering. The composite material consists of a mixture of polymer matrix, reinforcing fibers, and filler materials. The objective is to optimize the mechanical properties, such as strength and stiffness, of the composite while minimizing its weight.

Here's how mixture designed experiments could be applied in this scenario:

Components:

Polymer Matrix: Epoxy resin (component A).
Reinforcing Fibers: Carbon fibers (component B).
Filler Materials: Microspheres (component C).

Experimental Factors:

Proportions of epoxy resin (A), carbon fibers (B), and microspheres (C) in the composite mixture.

Design Space: The design space encompasses various combinations of epoxy resin, carbon fibers, and microspheres proportions, ensuring coverage of relevant ranges while avoiding extreme or impractical compositions.

Experimental Design: Utilize a mixture design approach, such as a simplex-centroid design or a mixture design based on OAs, to systematically vary the proportions of the three components in the composite material.

Experimental Plan: Generate an experimental plan specifying the composition of each composite mixture based on the chosen experimental design. For example, a fractional factorial design may be used to reduce the number of experiments while capturing main effects and interactions.

In the optimization of composite materials for specific applications in physics, such as aerospace engineering, mixture designed experiments involve systematically varying the proportions of different components to achieve desired properties. A common approach is to use a simplex-centroid design or a mixture design based on OAs to cover the design space efficiently while minimizing the number of experiments needed.

Suppose we have a composite material composed of three components: polymer matrix (A), reinforcing fibers (B), and filler materials (C). We want to optimize the composite's mechanical

TABLE 2.60

Mixture Designed Experiment Matrix Example for Physics

Experiment	Proportion of A (%)	Proportion of B (%)	Proportion of C (%)
1	60	20	20
2	50	30	20
3	40	40	20
4	30	50	20
5	20	60	20
6	20	20	60
7	30	20	50
8	40	20	40
9	50	20	30
10	60	20	20

properties, such as tensile strength and modulus, while minimizing weight. The proportions of the three components will be varied within specified ranges (Table 2.60).

In this matrix,

- Experiment numbers represent the different composite formulations.
- Proportions of each component (A, B, C) are expressed as percentages, with the sum of proportions equaling 100% for each experiment.
- The proportions of the components are systematically varied across the design space to cover a range of compositions, allowing for the exploration of how different combinations affect the mechanical properties of the composite material.

Researchers can conduct experiments based on this matrix, prepare composite samples according to each formulation, and evaluate their mechanical properties through testing. The data obtained from these experiments can then be analyzed to identify optimal compositions that achieve the desired performance characteristics.

Conduct Experiments:

Prepare composite samples according to the experimental plan, carefully measuring and mixing the components in the specified proportions. Fabricate composite specimens using standard manufacturing techniques, such as resin infusion or compression molding.

Characterize Mechanical Properties:

Perform mechanical testing on the composite specimens to evaluate properties, such as tensile strength, modulus of elasticity, and impact resistance. Conduct additional tests to assess other relevant properties, such as thermal conductivity or electrical conductivity.

Data Analysis:

Analyze the experimental data using statistical methods, such as ANOVA or regression analysis, to identify significant factors and interactions affecting the mechanical properties of the composite material.

Optimization:

Use the experimental results to optimize the composition of the composite material for desired mechanical properties. Identify the composition that maximizes strength and stiffness while minimizing weight, considering constraints, such as cost or manufacturability.

Validation and Refinement:

Validate the optimized composition through additional experiments or validation studies, ensuring that the observed improvements in mechanical properties are reproducible and reliable. Refine the experimental design or procedures as needed based on insights gained from the optimization process.

By applying mixture designed experiments in this example, researchers can systematically explore the effects of different component proportions on the mechanical properties of composite materials, leading to the development of optimized materials for aerospace applications.

REPORT

In mixture designed experiments for physics, various statistical tools are employed to analyze the experimental data, model the relationships between mixture components and responses, and optimize the mixture compositions. Some of the commonly used statistical tools in mixture designed experiments include:

ANOVA: ANOVA is used to assess the significance of factors and interactions in the experimental data. It helps determine which factors have a statistically significant effect on the response variables. **Regression Analysis**: Regression analysis is utilized to model the relationship between the mixture components and the response variables. Linear regression, nonlinear regression, and multiple regression techniques are applied to fit mathematical models to the experimental data.

RSM: RSM is a collection of statistical and mathematical techniques used for modeling and analyzing the relationships between several IVs (mixture components) and one or more response variables. It helps in optimizing the response by identifying the optimal mixture composition.

Desirability Function Analysis: Desirability function analysis is employed to simultaneously optimize multiple response variables subject to constraints. It helps in identifying the optimal combination of mixture components that maximizes or minimizes the desirability function.

Mixture–Process Variable Analysis: This analysis technique is used to understand the relationship between the composition of mixtures and process variables, such as temperature, pressure, or reaction time. It helps in optimizing the manufacturing process for desired outcomes.

PCA: PCA is employed to reduce the dimensionality of the experimental data and identify patterns or trends in the data. It helps in visualizing the relationships between mixture components and response variables.

Optimization Algorithms: Optimization algorithms, such as gradient-based methods, genetic algorithms, or simulated annealing, are used to find the optimal mixture compositions that maximize or minimize the response variables based on the fitted models.

DOE: DOE techniques, including factorial designs, fractional factorial designs, and response surface designs, are utilized to efficiently plan and conduct mixture designed experiments, ensuring thorough exploration of the design space with minimal experimental runs.

Model Validation Techniques: Cross-validation, prediction error plots, and residual analysis are employed to validate the accuracy and reliability of the fitted models and ensure that they generalize well to unseen data.

These statistical tools play a crucial role in analyzing mixture designed experiments data, extracting meaningful insights, and optimizing mixture compositions for desired outcomes in physics-related applications.

When reporting the results of mixture designed experiments in physics, it's essential to provide a comprehensive and structured account of the experimental design, procedures, findings, and conclusions. Here are the key elements typically included in a report for mixture designed experiments:

Title: A clear and descriptive title that summarizes the objective and scope of the experiment.

Abstract: A concise summary of the experiment, including the objectives, methods, key findings, and conclusions.

Introduction: Background information on the problem or topic being investigated. Objectives and research questions addressed by the experiment. Brief overview of the experimental approach and methodology.

Experimental Design:
- Description of the experimental design used (e.g., simplex-centroid design, mixture design based on OAs).
- Specification of the factors or components varied in the experiment and their ranges.
- Details of the experimental plan, including the number of runs and the composition of each mixture.

Materials and Methods:
- Detailed description of the materials and equipment used in the experiment.
- Procedures followed for preparing the mixtures, conducting experiments, and measuring the response variables.
- Any special considerations or protocols implemented to ensure accuracy and reproducibility.

Results:
- Presentation of the experimental data, including the composition of each mixture and the corresponding response variables.
- Summary statistics, such as means, standard deviations, or confidence intervals, for each response variable.
- Graphical representations of the data, such as scatter plots, contour plots, or response surface plots.
- Statistical analysis results, including ANOVA tables, regression coefficients, and model adequacy assessments.

Discussion:
- Interpretation of the results in the context of the research objectives.
- Analysis of trends, patterns, or relationships observed in the data.
- Comparison of experimental findings with theoretical expectations or previous studies.
- Explanation of any unexpected or anomalous results and potential sources of variability.

Conclusion: Summary of the main findings and conclusions drawn from the experiment. Implications of the findings for the broader field of physics or specific applications. Suggestions for future research or further investigation.

References: Citation of relevant literature, research papers, or sources consulted during the experiment.

Appendices (if necessary): Additional supporting information, such as raw data tables, detailed experimental protocols, or supplementary analyses.

By including these elements in a report for mixture designed experiments in physics, researchers can effectively communicate their experimental methods, results, and insights to the scientific community and stakeholders.

2.4.7 SEQUENTIAL EXPERIMENTATION

Sequential design experiments offer a flexible and adaptive approach to conducting research in physics, allowing researchers to make efficient use of resources, optimize experimental conditions, and uncover new insights into complex physical phenomena.

Sequential design experiments are used in physics for several reasons:

Efficient Use of Resources: Sequential design allows for experiments to be conducted iteratively, with each experiment building upon the results of the previous ones. This approach maximizes the use of resources by focusing efforts on areas of interest identified during the course of the experiment.

Adaptability: In many physics experiments, the relationship between variables may not be fully understood beforehand. Sequential design allows researchers to adapt their experimental approach based on emerging patterns or trends in the data, leading to more informative and targeted experiments.

Optimization: Sequential design enables researchers to optimize experimental conditions or parameters in real time, leading to improved accuracy, efficiency, and reliability of experimental results.

Exploratory Research: In exploratory research where the underlying mechanisms or relationships are not well understood, sequential design allows researchers to systematically explore the parameter space and identify significant factors or variables influencing the outcomes.

Complex Systems: Physics experiments often involve complex systems with multiple interacting variables. Sequential design provides a systematic approach to studying such systems by iteratively refining experimental designs based on the evolving understanding of the system dynamics.

Risk Reduction: By conducting experiments sequentially and adjusting the experimental design based on interim results, researchers can mitigate risks associated with uncertainty or variability in the experimental process, leading to more robust and reliable conclusions.

Time Savings: Sequential design can lead to time savings by allowing researchers to terminate experiments early if clear trends or patterns emerge, avoiding unnecessary experimentation and accelerating the pace of discovery.

Sequential design experiments find applications in various branches of physics, where the iterative and adaptive nature of the methodology can be particularly beneficial. Some areas where Sequential design experiments are commonly used in physics include:

Particle Physics: In particle physics experiments, researchers often use sequential design experiments to optimize experimental setups, adjust data collection parameters, and refine analysis techniques based on interim results. This approach helps in identifying rare or unexpected events amidst a large volume of data.

Astrophysics and Cosmology: Sequential design experiments are employed in observational studies in astrophysics and cosmology to adaptively schedule observations, prioritize targets, and optimize data collection strategies. This allows researchers to efficiently explore the universe and address specific research questions.

Condensed Matter Physics: In experimental studies of condensed matter systems, sequential design experiments are used to systematically vary experimental parameters, such as temperature, pressure, or magnetic field strength. Researchers iteratively refine their experimental setups to investigate phase transitions, material properties, and emergent phenomena.

Optics and Photonics: Sequential design experiments are applied in optics and photonics research to optimize optical systems, design experiments for parameter estimation, and develop adaptive optics techniques. This allows researchers to improve the performance of optical devices and systems for various applications, including imaging, sensing, and communication.

Quantum Information Science: In quantum information science, sequential design experiments are used to iteratively refine quantum protocols, optimize quantum circuits, and characterize quantum systems. This approach enables researchers to achieve better control and manipulation of quantum states for quantum computing, quantum communication, and quantum sensing applications.

Nuclear Physics: In nuclear physics experiments, sequential design experiments are employed to iteratively adjust experimental configurations, optimize detector geometries, and improve data analysis techniques. This allows researchers to study nuclear reactions, nuclear structure, and fundamental symmetries with higher precision and sensitivity.

High-Energy Physics: Sequential design experiments play a crucial role in high-energy physics experiments, such as those conducted at particle accelerators like the large hadron collider (LHC). Researchers use adaptive experimental designs to optimize collision parameters, trigger selection criteria, and data analysis methods, facilitating the discovery of new particles and phenomena.

Overall, sequential design experiments are utilized across various subfields of physics to efficiently explore complex systems, optimize experimental procedures, and advance scientific understanding. The iterative and adaptive nature of sequential design experiments makes them valuable tools for addressing challenging research questions and maximizing the utility of available resources.

Steps

Developing sequential design experiments in physics involves several key steps to plan and execute the iterative and adaptive experimental process effectively. Here are the steps typically followed:

1. **Define Research Objectives**: Clearly articulate the research objectives, hypotheses, and scientific questions that the sequential design experiments aim to address. Determine the specific variables, parameters, or phenomena under investigation.
2. **Review Existing Knowledge**: Conduct a thorough literature review to understand the current state of knowledge in the field. Identify relevant theoretical frameworks, experimental techniques, and previous studies that inform the design of the sequential design experiments.
3. **Formulate Initial Experimental Plan**: Develop an initial experimental plan that outlines the proposed experimental procedures, measurement techniques, and data collection protocols. Specify the variables to be manipulated, the response variables to be measured, and any control variables to be maintained constant.
4. **Design Sequential Experiment**: Select an appropriate experimental design framework for the sequential design experiments, considering factors, such as the nature of the research question, the complexity of the experimental system, and the available resources. Common design frameworks include sequential factorial designs, adaptive sampling designs, and Bayesian optimization algorithms.
5. **Plan Iterative Process**: Define the iterative process by which the sequential design experiments will be conducted. Determine the criteria or stopping rules for deciding when to

adapt the experimental design based on interim results. Establish protocols for monitoring the experiment's progress and adjusting the experimental plan as needed.

6. **Conduct Initial Experiment**: Conduct the initial stage of the sequential design experiments according to the planned experimental procedures. Collect data on the manipulated variables and measured response variables as per the initial experimental plan.

7. **Analyze Interim Results**: Analyze the interim results obtained from the initial stage of the experiment. Use statistical methods to assess the significance of observed effects, identify trends or patterns in the data, and evaluate the adequacy of the initial experimental design.

8. **Adapt Experimental Design**: Based on the analysis of interim results, adapt the experimental design for subsequent stages of the sequential design experiments. Modify experimental parameters, sample sizes, or other aspects of the experimental plan to focus efforts on areas of interest or address any deficiencies identified in the initial stage.

9. **Repeat Iterative Process**: Repeat the iterative process of conducting experiments, analyzing results, and adapting the experimental design until the research objectives are achieved or predetermined stopping criteria are met. Continuously refine the experimental plan based on emerging insights and feedback from interim results.

10. **Report Findings**: Communicate the findings of the sequential design experiments through scientific publications, conference presentations, or research reports. Provide a detailed account of the experimental procedures, results, and conclusions, including any insights gained from the iterative and adaptive nature of the experimental process.

By following these steps, researchers can develop and execute sequential design experiments in physics to systematically explore complex phenomena, optimize experimental procedures, and advance scientific knowledge.

PITFALLS AND REMEDIES

Sequential design experiments in physics offer several advantages, but they also come with potential pitfalls that researchers should be aware of. Here are some common pitfalls and remedies associated with sequential design experiments:

Over-reliance on Interim Results:
> **Pitfall:** Researchers may be tempted to make premature conclusions based on interim results, leading to biased or inaccurate findings.
>
> **Remedy:** Implement strict criteria or stopping rules for adapting the experimental design, ensuring that decisions are based on statistically significant trends or patterns in the data. Perform sensitivity analyses to assess the robustness of findings to variations in the experimental design.

Underestimation of Resources:
> **Pitfall:** Researchers may underestimate the resources required to conduct iterative experiments, leading to delays, budget overruns, or incomplete studies.
>
> **Remedy:** Conduct thorough feasibility assessments and resource planning before initiating sequential design experiments. Develop contingency plans to address unexpected challenges or resource constraints that may arise during the course of the experiment.

Selection Bias:
> **Pitfall:** Biases in the selection of experimental parameters or response variables may skew the results of sequential design experiments, leading to invalid conclusions.
>
> **Remedy:** Implement randomization procedures and carefully control for confounding variables to minimize selection bias. Consider using blinded protocols or independent

oversight committees to ensure objectivity and impartiality in experimental design and data analysis.

Overfitting of Models:

Pitfall: Overfitting occurs when statistical models are excessively complex and fit too closely to the observed data, leading to poor generalization to new data or experimental conditions.

Remedy: Use cross-validation techniques to assess the predictive performance of statistical models and guard against overfitting. Regularly validate and refine models based on new data collected during the sequential design experiments.

Inadequate Communication and Collaboration:

Pitfall: Inadequate communication and collaboration among team members may lead to misunderstandings, duplication of efforts, or inefficient use of resources.

Remedy: Foster open communication and collaboration among researchers, experimenters, statisticians, and other stakeholders involved in the sequential design experiments. Clearly define roles, responsibilities, and expectations to ensure smooth coordination and execution of the experimental process.

Lack of Flexibility:

Pitfall: Rigidity in experimental protocols or reluctance to deviate from the initial experimental plan may limit the adaptability and effectiveness of sequential design experiments.

Remedy: Embrace flexibility and openness to change throughout the experimental process. Be willing to modify experimental parameters, adjust sample sizes, or explore alternative hypotheses based on emerging insights and feedback from interim results.

By being mindful of these pitfalls and implementing appropriate remedies, researchers can maximize the benefits of sequential design experiments in physics and ensure the validity, reliability, and impact of their scientific findings.

EXAMPLE

Sequential experimentation in physics involves conducting a series of experiments, with each experiment's results informing the design of subsequent experiments. Here's a simplified example of a matrix for a sequential designed experiment in physics:

Objective: To determine the properties of a newly discovered material, including its conductivity and thermal resistance, using a series of tests.

Tests:

1. **Conductivity Test (Test A)**: Measures the electrical conductivity of the material.
2. **Thermal Resistance Test (Test B)**: Measures the material's thermal resistance.

Response Variables:

1. **Conductivity (S/m)**: Electrical conductivity of the material.
2. **Thermal Resistance (°C/W)**: Thermal resistance of the material.

Sequential Experimentation Matrix:

In this sequential designed experiment, the matrix is designed to adaptively explore the properties of the material by conducting multiple tests, with each test's results guiding the next steps. The matrix is shown in Table 2.61.

TABLE 2.61

Sequential Design Experiments Matrix for Physics

Test Number	Test Conducted	Previous Test Results	Material Properties
1	Conductivity	Initial Measurement	Conductivity: ... S/m
2	Thermal Resistance	Test A results	Thermal Resistance: ... °C/W
3	Conductivity	Test B results	Conductivity: ... S/m
4	Thermal Resistance	Test A and Test C results	Thermal Resistance: ... °C/W
5	Conductivity	Test B and Test D results	Conductivity: ... S/m

In this matrix, each row represents a specific test or experiment. The order of the tests is determined based on the results of the previous tests. For example, if test A reveals a certain level of conductivity, it may trigger the need for a thermal resistance test (test B) to further investigate the material's properties. Subsequent tests are informed by the cumulative results of previous tests, leading to a more comprehensive understanding of the material's characteristics.

The key to sequential experimentation is the adaptability and responsiveness to the material's properties as they are gradually revealed through the experiments. This approach allows physicists to efficiently explore and characterize the material's behavior and properties in a systematic manner.

Another example of sequential design experiments in physics can be found in the field of material science, specifically in the development of new semiconductor materials for PV applications. Let's consider a hypothetical scenario:

Research Objective: The objective is to optimize the growth parameters of a thin-film semiconductor material to enhance its efficiency as a PV material for solar cell applications.

Experimental Setup:

1. **Initial Experimental Plan**: The initial plan involves growing thin-film semiconductor samples using a specific deposition technique [e.g., chemical vapor deposition (CVD)] under predetermined conditions (e.g., temperature, pressure, precursor gas flow rates). The efficiency of the resulting solar cells is measured using standard characterization techniques [e.g., current–voltage measurements, external quantum efficiency (EQE)].
2. **Interim Analysis**: After conducting the initial experiments and characterizing the samples, the efficiency of the solar cells is analyzed. Based on the interim results, trends or patterns in the data are identified, and areas for improvement are pinpointed. For example, certain growth parameters may have a more pronounced effect on solar cell efficiency than others.
3. **Adaptation of Experimental Design**: The experimental design is adapted based on the interim analysis to focus on the most promising growth parameters identified. For instance, the temperature range for deposition may be narrowed down or specific precursor gas compositions may be varied more systematically.
4. **Subsequent Iterations**: Subsequent iterations of experiments are conducted with the adapted experimental design. The growth parameters are systematically varied within the narrowed range, and the efficiency of the resulting solar cells is measured and analyzed after each iteration.
5. **Refinement and Optimization**: The iterative process continues, with the experimental design being refined and optimized based on the feedback from interim analyses. The goal is to converge toward the optimal growth conditions that maximize the efficiency of the PV material.

6. **Conclusion**: Once the optimal growth parameters are determined, the sequential design experiments conclude. The final set of growth conditions is identified, and the efficiency of the resulting solar cells is validated through additional experiments. The findings are reported in scientific publications or patents.

In this example, sequential design experiments are used to systematically explore the parameter space of semiconductor material growth, adaptively adjusting the experimental design based on interim results to optimize the efficiency of PV devices. This iterative approach allows researchers to efficiently identify and refine the growth conditions that yield the best performance, ultimately advancing the development of solar energy technologies.

The sequential design experiment aimed at optimizing the growth parameters of a thin-film semiconductor material for PV applications are shown in Table 2.62.

In this matrix,

- **Experiment Number**: Identifies each experimental run or iteration.
- **Deposition Temperature (°C)**: Specifies the temperature at which the semiconductor material is deposited onto the substrate. Different temperatures are tested to optimize crystal growth and material properties.
- B>Precursor Gas Flow Rates (sccm): Indicates the flow rates of precursor gases used during the deposition process. Varying flow rates help control the composition and properties of the deposited material.
- **Substrate Material**: Specifies the material onto which the semiconductor thin film is deposited, such as silicon or glass. The substrate material can influence the crystal structure and interface properties.
- **Deposition Time (minutes)**: Defines the duration of the deposition process. Longer deposition times may result in thicker films with different properties.
- **Dopant Concentration (atoms/cm^3)**: Specifies the concentration of dopant atoms introduced into the semiconductor material to modify its electrical properties. Dopant concentration affects carrier density and conductivity.
- **Film Thickness (nanometers)**: Indicates the thickness of the semiconductor thin film. Film thickness influences light absorption, charge transport, and overall device performance.

After conducting each experiment and measuring relevant properties of the resulting thin films, the data are analyzed to identify trends and optimize the growth parameters further. The experimental design can be adapted iteratively based on the outcomes of previous experiments, allowing researchers to systematically explore the parameter space and achieve the desired properties for PV applications.

TABLE 2.62

Sequential Design Experiments for Physics

Experiment Number	Deposition Temperature (°C)	Precursor Gas Flow Rates (sccm)	Substrate Material	Deposition Time (minutes)	Dopant Concentration (atoms/cm^3)	Film Thickness (nanometers)
1	450	20–30	Silicon	60	1e16	200
2	480	25–35	Glass	75	5e17	250
3	470	22–28	Silicon	70	1e17	220
4	460	18–25	Glass	65	8e16	230
5	475	23–32	Silicon	72	1.5e17	240

Report

In sequential design experiments for physics, various statistical tools are employed to analyze experimental data, optimize experimental parameters, and make informed decisions about subsequent iterations. Some of the commonly used statistical tools include:

Descriptive Statistics: Descriptive statistics, such as mean, median, standard deviation, and range, provide a summary of the experimental data and help researchers understand the central tendency and variability of the observed measurements.

Regression Analysis: Regression analysis is used to model the relationship between input variables (experimental parameters) and output variables (response variables). Regression models can help identify significant factors affecting the response and predict the response for new experimental conditions.

ANOVA: ANOVA is used to assess the statistical significance of differences between multiple groups or experimental conditions. In Sequential design experiments, ANOVA can help identify the most influential factors and interactions among experimental parameters.

DOE: DOE techniques, such as factorial designs, RSM, and Taguchi methods, are used to systematically plan and execute experiments, optimize experimental parameters, and analyze the effects of multiple factors on the response variable.

Sequential Analysis: Sequential analysis methods allow researchers to adaptively modify the experimental design based on interim results. Sequential analysis techniques include sequential hypothesis testing, sequential estimation, and sequential optimization algorithms.

Optimization Algorithms: Optimization algorithms, such as gradient descent, genetic algorithms, and simulated annealing, are used to iteratively search for optimal experimental conditions that maximize or minimize a chosen objective function.

Bayesian Methods: Bayesian statistical methods incorporate prior knowledge and update beliefs based on observed data to make probabilistic inferences about experimental parameters and response variables. Bayesian techniques are particularly useful in sequential decision-making and adaptive experimental designs.

Multivariate Analysis: Multivariate analysis techniques, such as PCA and discriminant analysis, are used to analyze datasets with multiple variables and identify patterns or relationships among variables.

SPC: SPC methods are used to monitor and control the quality of experimental processes over time. SPC charts, such as control charts and run charts, help identify trends, shifts, and outliers in experimental data.

Simulation and Monte Carlo Methods: Simulation and Monte Carlo methods are used to model complex systems, generate synthetic data, and assess the robustness of experimental designs under different scenarios.

By leveraging these statistical tools, researchers can effectively analyze experimental data, optimize experimental parameters, and make data-driven decisions to advance scientific knowledge in physics through sequential design experiments.

In a report on sequential design experiments used in physics, it's essential to include comprehensive information to effectively communicate the experimental process, results, and conclusions to the readers. Here are the key elements typically included in such a report:

Title: Clearly states the purpose of the experiment and identifies the specific area of physics being investigated.

Abstract: Summarizes the experiment's objectives, methods, key findings, and conclusions in a concise paragraph.

Introduction: Provides background information on the research problem, discusses the significance of the study within the broader field of physics, and states the research objectives and hypotheses.

Experimental Design: Describes the experimental setup, including details about the materials, equipment, and procedures used in the experiment. It should also outline the sequential design approach employed, including any adaptive modifications made based on interim analyses.

Variables and Parameters: Clearly defines the independent, dependent, and control variables in the experiment, along with the specific parameter values or ranges tested in each iteration.

Data Analysis Methods: Describes the statistical tools and techniques used to analyze the experimental data, including regression analysis, ANOVA, sequential analysis, optimization algorithms, or any other relevant methods.

Results: Presents the results of each experimental iteration, including numerical data, graphs, tables, and any significant trends or patterns observed. Interim analyses and adaptations to the experimental design should also be discussed.

Discussion: Interprets the results in the context of the research objectives and hypotheses, compares findings with prior literature, and discusses any unexpected observations or limitations of the study.

Conclusions: Summarizes the main findings of the experiment, highlights the implications for the field of physics, and suggests future research directions or potential applications.

References: Lists all sources cited in the report, following a specific citation style (e.g., APA, MLA, Chicago).

Appendices: Includes supplementary information, such as raw data, detailed experimental procedures, computer code used for data analysis, or additional figures and tables, that enhance understanding but are not essential for the main text.

By including these elements in a report on sequential design experiments, researchers can effectively communicate the rationale, methods, results, and implications of their study to the scientific community, facilitating knowledge dissemination and further research in the field of physics.

2.4.8 SEQUENTIAL DESIGN EXPERIMENT (SDE)

Sequential design experiments are used in physics for several reasons, primarily to optimize experimental parameters, efficiently explore complex parameter spaces, and make data-driven decisions in an iterative manner. Here are some key reasons why sequential design experiments are employed in physics:

Efficient Parameter Optimization: In many physics experiments, researchers need to optimize multiple parameters to achieve the desired outcome or maximize the performance of a system. Sequential design experiments allow researchers to systematically vary parameters in a controlled manner, analyze the results, and adaptively refine the experimental design based on interim findings, leading to faster convergence toward optimal conditions.

Adaptive Experimental Design: Physics experiments often involve complex systems with nonlinear behavior or unknown interactions between variables. Sequential design experiments enable researchers to adaptively modify the experimental design based on interim data analysis, allowing them to focus resources on the most promising regions of the parameter space and avoid wasting resources on less informative experimental conditions.

Optimization under Uncertainty: In many physics experiments, there may be uncertainty or variability in the system's behavior due to factors, such as measurement noise, environmental conditions, or material properties. Sequential design experiments allow researchers to account for this uncertainty by iteratively refining the experimental design based on observed data, leading to more robust and reliable results.

Resource Efficiency: Physics experiments often involve costly equipment, materials, and personnel time. Sequential design experiments offer a systematic approach to efficiently allocate resources by prioritizing experiments that are most likely to yield informative results, thereby maximizing the scientific output while minimizing resource expenditure.

Exploration of Complex Systems: Physics often deals with complex systems governed by multiple interacting variables. Sequential design experiments provide a systematic framework for exploring these complex systems by systematically varying parameters, observing the system's response, and identifying key factors that influence the system's behavior.

Optimization of Measurement Techniques: In experimental physics, researchers often need to optimize measurement techniques to enhance sensitivity, resolution, or accuracy. Sequential design experiments can be used to systematically vary experimental parameters related to the measurement setup, such as instrument settings or sensor configurations, to optimize measurement performance.

Sequential design experiments offer a powerful methodology for optimizing experimental parameters, exploring complex systems, and making data-driven decisions in physics research, ultimately advancing our understanding of the physical world and driving technological innovation.

Sequential design experiments are employed in various areas of physics research where optimization, efficiency, and adaptive experimental design are crucial. Some specific applications of sequential design experiments in physics include:

Materials Science: Sequential design experiments are used to optimize fabrication processes for semiconductor materials, thin films, nanomaterials, and composite materials. Researchers may systematically vary parameters, such as deposition temperature, precursor gas flow rates, and dopant concentration, to achieve desired material properties for electronic, photonic, and energy applications.

Optics and Photonics: In optics and photonics research, sequential design experiments are utilized to optimize optical components, laser systems, and photonic devices. Parameters, such as laser power, beam profile, wavelength, and optical coatings, can be systematically varied to maximize device performance and efficiency.

Condensed Matter Physics: Sequential design experiments play a role in studying the properties of condensed matter systems, such as superconductors, magnetic materials, and quantum materials. Researchers may optimize experimental conditions, such as temperature, pressure, and magnetic field strength, to investigate phase transitions, electronic properties, and novel quantum phenomena.

Particle Physics: In particle physics experiments, sequential design experiments are used to optimize detector configurations, calibration procedures, and data analysis techniques. Researchers may systematically vary detector parameters, trigger thresholds, and data selection criteria to enhance SNRs and improve the sensitivity of particle detectors.

Astrophysics and Cosmology: Sequential design experiments are employed in astrophysical observations and cosmological simulations to optimize observational strategies, data collection techniques, and computational models. Parameters, such as telescope pointing, exposure time, and data processing algorithms, may be iteratively refined to maximize the scientific yield of astronomical surveys and simulations.

Biophysics: In biophysics research, sequential design experiments are utilized to optimize experimental protocols for studying biomolecular systems, protein folding, and cellular processes. Parameters, such as temperature, pH, substrate concentration, and reaction time, may be systematically varied to elucidate the biophysical mechanisms underlying biological phenomena.

Nuclear Physics: Sequential design experiments are used in nuclear physics experiments to optimize accelerator parameters, target configurations, and detector setups. Researchers may systematically vary beam energy, beam intensity, target material, and detector placement to study nuclear reactions, nuclear structure, and fundamental symmetries.

Overall, sequential design experiments find application across various subfields of physics research, enabling scientists to efficiently explore parameter spaces, optimize experimental conditions, and make data-driven decisions to advance our understanding of the physical world.

STEPS

Developing sequential design experiments in physics involves a systematic approach to optimizing experimental parameters, efficiently exploring parameter spaces, and making data-driven decisions in an iterative manner. Here are the general steps to develop sequential design experiments in physics:

1. **Define Research Objectives**: Clearly define the research objectives and hypotheses of the study. Identify the specific parameters or factors that need to be optimized or investigated.
2. **Identify Experimental Variables**: Identify the IVs (factors) that will be varied in the experiment and the DVs (responses) that will be measured to evaluate the system's behavior.
3. **Choose Experimental Design Strategy**: Select an appropriate experimental design strategy based on the research objectives and the complexity of the parameter space. Common design strategies include factorial designs, RSM, and adaptive sequential designs.
4. **Design Initial Experiment**: Design an initial set of experiments based on the chosen experimental design strategy. Determine the range and levels of each variable and decide on the sequence of experiments to be conducted.
5. **Perform Initial Experiments**: Conduct the initial set of experiments according to the experimental design plan. Record all relevant data, including IV settings, experimental conditions, and measured responses.
6. **Analyze Experimental Data**: Analyze the data obtained from the initial experiments using appropriate statistical methods. Identify trends, relationships, and significant factors affecting the system's behavior.
7. **Refine Experimental Design**: Based on the analysis of the initial experimental results, refine the experimental design by adapting the parameter space exploration strategy. Determine which experimental conditions should be explored further or modified in subsequent iterations.
8. **Plan Sequential Experiments**: Plan the next set of sequential experiments based on the refined experimental design. Determine the sequence of experimental runs, taking into account any adaptive modifications or new hypotheses.
9. **Iterate**: Repeat steps 5 to 8 iteratively, conducting sequential experiments, analyzing data, refining the experimental design, and planning subsequent iterations. Continue the iterative process until the research objectives are achieved or the desired optimization goals are met.
10. **Validate and Interpret Results**: Validate the final experimental results to ensure their accuracy and reliability. Interpret the findings in the context of the research objectives, hypotheses, and scientific implications.
11. **Communicate Findings**: Prepare a report or manuscript documenting the sequential design experiments, including the experimental design, results, analysis, and conclusions. Communicate the findings to the scientific community through publications, presentations, or other appropriate channels.

By following these steps, researchers can systematically develop sequential design experiments in physics to optimize experimental parameters, efficiently explore parameter spaces, and make data-driven decisions to advance scientific knowledge in their respective fields.

PITFALLS AND REMEDIES

Sequential design experiments in physics offer a powerful approach for efficiently optimizing experimental parameters and exploring complex parameter spaces. However, there are several potential pitfalls that researchers may encounter during the development and execution of such experiments. Here are some common pitfalls and their corresponding remedies:

Overlooking Prior Knowledge:
> **Pitfall:** Ignoring existing knowledge or literature in the field may lead to redundant experiments or inefficient exploration of the parameter space.
> **Remedies:** Conduct a thorough literature review to identify relevant prior research and incorporate existing knowledge into the experimental design. Use prior information to inform the selection of variables, ranges, and initial experimental conditions.

Inadequate Experimental Planning:
> **Pitfall:** Poorly planned sequential experiments may result in suboptimal parameter exploration or inefficient resource utilization.
> **Remedies:** Invest sufficient time in experimental planning, including the selection of appropriate design strategies, determination of variable ranges, and optimization of experimental sequences. Consider consulting with experts or utilizing statistical software to aid in experimental design.

Insufficient Data Analysis:
> **Pitfall:** Inadequate data analysis may lead to misinterpretation of results or failure to identify important trends or relationships in the data.
> **Remedies:** Employ appropriate statistical techniques for data analysis, including regression analysis, ANOVA, and sequential analysis. Consider consulting with statisticians or utilizing specialized software for complex data analysis tasks.

Ignoring Interactions and Nonlinearities:
> **Pitfall:** Focusing solely on main effects and neglecting interactions or nonlinearities in the experimental system may lead to inaccurate conclusions or suboptimal parameter settings.
> **Remedies:** Use factorial designs or RSM to explicitly model interactions and nonlinear effects in the experimental system. Conduct sensitivity analyses to assess the impact of interactions with the response variable.

Lack of Adaptation:
> **Pitfall:** Failure to adapt the experimental design based on interim results may result in missed opportunities for optimization or inefficient resource allocation.
> **Remedies:** Implement adaptive experimental designs that allow for modifications to the experimental plan based on observed data. Use sequential analysis techniques to make data-driven decisions about the next steps in the experiment.

Underestimating Resource Constraints:
> **Pitfall:** Underestimating the limitations of available resources, such as time, budget, or experimental equipment, may lead to unrealistic experimental plans or unfeasible data collection strategies.
> **Remedies:** Conduct a thorough assessment of resource constraints before designing the experiment. Prioritize experiments that offer the highest potential for scientific impact within the constraints of available resources.

Poor Communication and Documentation:

Pitfall: Inadequate communication and documentation of experimental procedures, results, and analyses may hinder reproducibility and transparency in research outcomes.

Remedies: Maintain detailed records of experimental procedures, including variable settings, measurement techniques, and data analysis methods. Clearly communicate experimental protocols and findings in written reports, presentations, or publications.

By being aware of these potential pitfalls and implementing appropriate remedies, researchers can effectively develop and execute sequential design experiments in physics to optimize experimental parameters, explore parameter spaces, and advance scientific knowledge in their respective fields.

EXAMPLE

RPD experiments in physics aim to optimize a product or process while considering variability and external factors. Here's a simplified example of a matrix for an RPD experiment in physics:

Objective: To optimize the performance of a high-precision measuring instrument by studying the effects of two factors (temperature and humidity) while considering variations in external conditions (lab environment).

Factors:

1. **Temperature (Factor A):**
   - Low temperature (20°C).
   - Medium temperature (25°C).
   - High temperature (30°C).
2. **Humidity (Factor B):**
   - Low humidity (30% RH).
   - Medium humidity (50% RH).
   - High humidity (70% RH).

External Factor (Lab Environment):

- The lab environment introduces variations in temperature and humidity, which are not under the experimenter's control.

Response Variable: Measurement precision (measured as a percentage deviation from the true value).

RPD Matrix:

In this RPD experiment, the matrix is designed to optimize the measuring instrument's performance by considering the effects of temperature and humidity while accounting for variations in the lab environment. The matrix is shown in Table 2.63.

In this matrix, each row represents a specific experimental run with a unique combination of temperature and humidity levels, along with different variations in the lab environment (low, medium, or high variation). The measurement precision is recorded for each combination.

RPD experiments allow physicists to optimize the instrument's performance while considering the impact of external factors and variability in environmental conditions, resulting in a more robust and reliable measuring instrument for physics applications.

TABLE 2.63
RPD Matrix Example for Physics

Run	Temperature (A) (°C)	Humidity (B) (RH)	Lab Environment (External Conditions)	Measurement Precision (%)
1	20	30	Low variation	...
2	25	50	Low variation	...
3	30	70	Low variation	...
4	20	30	Medium variation	...
5	25	50	Medium variation	...
6	30	70	Medium variation	...
7	20	30	High variation	...
8	25	50	High variation	...
9	30	70	High variation	...

An example of sequential design experiments in physics could involve optimizing the parameters of a laser system for a specific application, such as laser-cutting in materials science. Here's how the sequential experiment might be structured:

1. **Objective**: The objective is to optimize the parameters of a laser-cutting system to achieve the highest cutting efficiency and quality for a given material.
2. **Experimental Variables**: The experimental variables include laser power, beam focal length, cutting speed, gas flow rate (if applicable), and material type.
3. **Initial Experiment**: The initial experiment involves conducting a factorial design experiment where each variable is varied at two or three levels within their specified ranges. The experimental runs are planned to cover a subset of the parameter space.
4. **Data Analysis**: After the initial experiment, the cutting efficiency, cut quality (e.g., edge roughness), and other relevant metrics are measured for each experimental run. The data are analyzed using statistical methods to identify the main effects and potential interactions between variables.
5. **Refinement and Adaptation**: Based on the analysis of the initial experiment, the experimental design is refined, and additional experiments are planned to explore promising regions of the parameter space more thoroughly. This may involve adjusting variable ranges, adding new levels, or focusing on specific combinations of variables.
6. **Sequential Iterations**: The experiment is conducted iteratively, with each iteration building upon the results of the previous one. Adaptive modifications to the experimental design are made based on interim analyses, allowing for efficient exploration of the parameter space.
7. **Convergence and Optimization**: As the experiment progresses, the parameter space is systematically explored, and the optimal combination of laser-cutting parameters is identified. The experiment continues until a satisfactory level of cutting efficiency and quality is achieved, or until resource constraints are reached.
8. **Validation and Verification**: The final experimental settings are validated through additional experimental runs and/or by comparing the predicted outcomes with actual performance. The optimized parameters are verified to ensure reproducibility and reliability.
9. **Communication of Results**: The findings of the sequential design experiment, including the optimized laser-cutting parameters and associated performance metrics, are communicated through a matrix report, presentation, or publication, contributing to the advancement of knowledge in laser materials processing and physics.

TABLE 2.64
RPD Matrix Example for Physics

Experiment Number	Laser Power (W)	Beam Focal Length (mm)	Cutting Speed (mm/s)	Gas Flow Rate (L/min)	Material Type	Cutting Efficiency (%)	Cut Quality (Edge Roughness)
1	100	50	500	10	Metal A	85	Low
2	120	60	600	12	Metal A	88	Medium
3	100	50	600	12	Metal A	86	Low
4	120	60	500	10	Metal A	87	Medium
5	110	55	550	11	Metal A	86	Low
6	110	55	550	11	Metal B	82	Medium

In sequential design experiments optimizing the parameters of a laser system for cutting, a matrix can be used to organize the experimental runs and record the settings of the variables at each iteration. An example of a matrix for sequential design experiments in this context is shown in Table 2.64.

In this matrix,

- Each row represents an experimental run or iteration of the sequential design experiment.
- The columns correspond to the experimental variables being investigated, including laser power, beam focal length, cutting speed, gas flow rate, material type, cutting efficiency, and cut quality.
- Laser power, beam focal length, cutting speed, and gas flow rate are continuous variables, while material type is a categorical variable.
- The cutting efficiency and cut quality are the response variables measured after each experimental run, representing the outcomes of interest.

This example illustrates how sequential design experiments can be applied in physics to systematically optimize experimental parameters and achieve desired outcomes in complex systems.

By systematically varying the settings of the variables in each experimental run and recording the corresponding outcomes, researchers can analyze the data to identify optimal parameter combinations for laser-cutting in physics. This iterative process allows for the efficient exploration of the parameter space and the refinement of experimental conditions to achieve desired performance metrics.

Report

In sequential design experiments for physics, various statistical tools are utilized to analyze the experimental data, make informed decisions about subsequent experiments, and optimize the experimental parameters. Some common statistical tools used in sequential design experiments include:

Descriptive Statistics: Descriptive statistics, such as mean, median, standard deviation, and range, are used to summarize and describe the central tendency, dispersion, and distribution of the experimental data.

ANOVA: ANOVA is employed to assess the significance of differences between experimental groups or treatments. It helps identify significant main effects and interactions between variables in factorial designs.

Regression Analysis: Regression analysis is used to model the relationship between IVs (factors) and DVs (responses). Linear regression, polynomial regression, and multiple regression models may be employed to describe complex relationships in the data.

RSM: RSM involves fitting mathematical models to experimental data to optimize response variables while varying IVs within specified ranges. It helps identify optimal parameter settings and quantify the effects of individual factors and their interactions.

Sequential Analysis: Sequential analysis techniques are used to make data-driven decisions about the next steps in the experiment based on interim results. Sequential hypothesis testing and sequential experimental design methods help adaptively refine experimental designs to efficiently explore parameter spaces.

DOE: DOE techniques, including factorial designs, fractional factorial designs, and Taguchi methods, are employed to systematically plan and conduct experiments, optimize parameter settings, and identify significant factors affecting the system's behavior.

Optimization Algorithms: Optimization algorithms, such as gradient descent, genetic algorithms, and simulated annealing, may be used to search for optimal parameter combinations in complex parameter spaces. These algorithms iteratively adjust parameter values to minimize or maximize objective functions.

SPC: SPC methods are employed to monitor and control experimental processes in real time. Control charts, process capability analysis, and statistical quality control techniques help ensure the stability and consistency of experimental results.

Bayesian Methods: Bayesian statistical methods allow for the incorporation of prior knowledge or beliefs into the analysis of experimental data. Bayesian updating and Bayesian optimization techniques enable efficient learning and decision-making in sequential experimental designs.

Data Visualization: Data visualization techniques, such as scatter plots, histograms, and contour plots, are used to visually explore and interpret experimental data; identify patterns, trends, and outliers; and communicate findings effectively.

By leveraging these statistical tools, researchers can effectively analyze experimental data, optimize experimental parameters, and make informed decisions in sequential design experiments for physics, leading to advancements in scientific knowledge and understanding.

In a report documenting sequential design experiments used in physics, several key elements should be included to provide a comprehensive overview of the experimental process, results, and conclusions. Here are the essential elements of such a report:

Title: A descriptive and informative title that clearly indicates the purpose and scope of the experiment.

Abstract: A concise summary of the experiment, including the objectives, methods, major findings, and conclusions. The abstract provides a brief overview of the entire report.

Introduction: An introduction section that provides background information on the experimental problem or research question, states the objectives of the experiment, and explains the rationale for using a sequential design approach.

Literature Review: A review of relevant literature and previous research in the field. This section highlights existing knowledge, gaps in the literature, and the importance of the current experiment.

Experimental Design: A detailed description of the experimental design, including the selection of variables, their ranges, the sequence of experimental runs, and any adaptive or sequential aspects of the design.

Methods: A comprehensive description of the experimental methods, including the apparatus or equipment used, measurement techniques, data collection procedures, and any statistical analyses performed.

Results: Presentation of the experimental results, including data tables, figures, and graphs. The results should include descriptive statistics, such as means, standard deviations, and confidence intervals, as well as any significant findings or trends observed.

Discussion: Interpretation and discussion of the results in the context of the experiment's objectives and hypotheses. This section should address the significance of the findings, their implications for the field, and any limitations or uncertainties associated with the experimental design or data analysis.

Conclusions: Summary of the main findings and conclusions drawn from the experiment. This section should reiterate the experiment's objectives and highlight key results, insights, and recommendations for future research.

References: A list of references cited in the report, following a standard citation format (e.g., APA, MLA). This section acknowledges the sources of information and provides readers with additional resources for further reading.

Appendices: Additional supplementary information, such as raw data tables, detailed experimental protocols, or mathematical derivations, that are relevant to the experiment but not essential for understanding the main findings.

By including these elements in the report, researchers can effectively communicate the methodology, results, and implications of sequential design experiments in physics to their peers and the broader scientific community.

2.4.9 Randomized Experiments

RD experiments are a fundamental tool in physics research, allowing researchers to make valid causal inferences, control for confounding variables, and produce generalizable findings that advance our understanding of the physical world. RD experiments are used in physics for several reasons:

Reducing Bias: Randomization helps minimize bias by ensuring that experimental units are assigned to treatment groups or conditions randomly. This helps ensure that any observed differences between groups are due to the effects of the treatments rather than systematic differences in the experimental units.

Enhancing Causal Inference: Random assignment of treatments helps establish causal relationships between variables. By randomly assigning treatments, researchers can confidently attribute any observed differences in outcomes to the treatments themselves rather than confounding variables.

Statistical Efficiency: Randomization increases the efficiency of statistical analyses by ensuring that the assumptions of independence and exchangeability are met. This allows for valid statistical tests and more precise estimates of treatment effects.

Generalizability: RD experiments often result in findings that are more generalizable to the broader population or system under study. By randomly selecting experimental units and assigning treatments, researchers can make inferences about the broader population from which the sample was drawn.

Control of Confounding Variables: Randomization helps control for the effects of confounding variables that could influence the outcome of the experiment. By distributing potential confounders randomly across treatment groups, researchers can reduce their impact on the results.

Flexibility: RD experiments offer flexibility in experimental design and analysis. They can be applied to various experimental settings and can accommodate different types of treatments, including multiple treatment levels, factorial designs, and complex experimental designs.

Ethical Considerations: Randomized assignment of treatments can help ensure ethical treatment of participants in human studies by minimizing biases and ensuring fairness in treatment allocation.

RD experiments are used in various fields of physics, including:

Particle Physics: Randomized experiments are conducted in particle physics to study fundamental particles and their interactions. Experiments conducted at particle accelerators, such as the LHC, often involve randomization of collision events to investigate the properties of subatomic particles.

Condensed Matter Physics: In condensed matter physics, randomized experiments may be employed to study the properties of materials under different conditions. For example, randomized trials can be used to investigate the effects of temperature, pressure, or magnetic field on the behavior of materials.

Quantum Physics: Randomized experiments are essential in quantum physics research, where the behavior of particles is inherently probabilistic. Experiments involving quantum entanglement, quantum computing, and quantum communication often rely on randomization to control experimental variables and study quantum phenomena.

Astrophysics and Cosmology: Randomized experiments are used in astrophysics and cosmology to study the universe's large-scale structure, cosmic microwave background radiation, and the distribution of galaxies. Randomized surveys and observations help astronomers and cosmologists understand the fundamental properties of the universe.

Optics and Photonics: Randomized experiments are conducted in optics and photonics to study the behavior of light and electromagnetic waves. RDs may be employed in experiments involving laser systems, optical fibers, and photonic devices to investigate light–matter interactions and develop new technologies.

Nuclear Physics: Randomized experiments are used in nuclear physics to study the properties of atomic nuclei, nuclear reactions, and nuclear decay processes. Experiments involving radioactive decay, nuclear fusion, and nuclear fission often utilize randomization to control experimental conditions and measure nuclear properties accurately.

Biophysics and Medical Physics: Randomized experiments play a crucial role in biophysics and medical physics research, particularly in studies involving biological systems, medical imaging, and radiation therapy. Randomized clinical trials and experiments are conducted to evaluate the efficacy of medical treatments, diagnostic techniques, and radiation therapies.

In each of these areas, RD experiments are employed to ensure unbiased treatment assignment, control for confounding variables, and make valid causal inferences about the physical phenomena under investigation.

Steps

Developing RD experiments in physics involves several key steps to ensure the validity and reliability of the experimental results. Here are the general steps to develop RD experiments in physics:

1. **Define Research Objectives**: Clearly define the research objectives and hypotheses of the experiment. Determine the specific questions or phenomena to be investigated and the variables to be manipulated and measured.
2. **Identify Experimental Variables**: Identify the IVs (factors) that will be manipulated during the experiment and the DVs (responses) that will be measured to assess the effects of the treatments.
3. **Design Experimental Protocol**: Design the experimental protocol, including the experimental procedures, measurement techniques, and data collection methods. Specify how the experimental units will be randomized and assigned to treatment groups.

4. **Select Experimental Units**: Determine the experimental units or subjects that will be included in the experiment. This may involve selecting samples of materials, particles, organisms, or other entities relevant to the research question.
5. **Randomize Treatment Assignment**: Randomly assign the experimental units to treatment groups or conditions. Use randomization techniques, such as random number generators or randomization tables, to ensure that treatment assignments are unbiased and independent.
6. **Implement Experimental Treatments**: Implement the experimental treatments according to the randomized treatment assignments. Ensure consistency and accuracy in the administration of treatments to minimize variability and bias.
7. **Collect Data**: Collect data on the DVs according to the experimental design. Use standardized measurement techniques and protocols to ensure the reliability and validity of the data.
8. **Analyze Data**: Analyze the experimental data using appropriate statistical methods. Compare the outcomes of different treatment groups to assess the effects of the experimental variables and determine if there are statistically significant differences.
9. **Interpret Results**: Interpret the results of the experiment in the context of the research objectives and hypotheses. Consider the implications of the findings for the broader field of physics and any potential limitations or confounding factors.
10. **Draw Conclusions**: Draw conclusions based on the results of the experiment and their consistency with the research objectives and hypotheses. Discuss the implications of the findings and any future research directions.
11. **Communicate Findings**: Communicate the findings of the experiment through scientific publications, presentations, or reports. Clearly articulate the experimental design, results, and conclusions to facilitate the dissemination of knowledge and further scientific inquiry.

By following these steps, researchers can develop RD experiments in physics that yield valid, reliable, and interpretable results, advancing our understanding of the physical world and contributing to scientific progress.

PITFALLS AND REMEDIES

RD experiments are widely used in physics research for their ability to minimize bias and ensure the validity of experimental results. However, there are potential pitfalls associated with the design and implementation of randomized experiments. Here are some common pitfalls and remedies:

Inadequate Randomization:
> **Pitfall:** Randomization may not be properly executed, leading to nonrandom assignment of experimental units to treatment groups.
> **Remedy:** Ensure that randomization procedures are rigorously implemented using appropriate randomization techniques, such as random number generators or randomization tables. Verify randomization outcomes to confirm proper assignment.

Small Sample Size:
> **Pitfall**: Small sample sizes may result in insufficient statistical power to detect true treatment effects or to generalize findings to the broader population.
> **Remedy:** Conduct a power analysis to determine the minimum sample size required to detect meaningful treatment effects with adequate power. Increase sample size if necessary to achieve sufficient statistical power.

Uncontrolled Confounding Variables:
> **Pitfall:** Confounding variables that are not accounted for in the experimental design may bias the results or obscure the true effects of the treatments.

Remedy: Identify potential confounding variables during the experimental planning stage and implement strategies to control them, such as blocking or stratification. Collect data on confounding variables and include them in the statistical analysis as covariates or factors.

Measurement Error:

Pitfall: Measurement error in the DVs may reduce the precision and accuracy of the experimental results.

Remedy: Implement rigorous measurement techniques and calibrate instruments to minimize measurement error. Conduct pilot studies or validation experiments to assess measurement reliability and validity.

Treatment Noncompliance or Dropout:

Pitfall: Experimental units may fail to comply with assigned treatments or drop out of the study, leading to incomplete data and potential bias.

Remedy: Monitor treatment compliance throughout the experiment and implement strategies to minimize dropout rates, such as providing incentives or reminders. Use intention-to-treat analysis to include all randomized participants in the analysis, regardless of treatment compliance.

Interactions with Experimental Conditions:

Pitfall: Unanticipated interactions between experimental conditions or treatments may confound the interpretation of results.

Remedy: Conduct factorial designs or include interaction terms in the statistical analysis to assess potential interactions between treatments. Carefully interpret the results in light of possible interactions and consider conducting follow-up experiments to explore interactions further.

Publication Bias:

Pitfall: Positive results may be more likely to be published than negative or null results, leading to a biased representation of the literature.

Remedy: Register the experimental protocol in advance, use preregistration platforms, or publish study protocols to enhance transparency and reduce publication bias. Consider publishing results in open-access journals or repositories to facilitate dissemination of all findings, regardless of outcome.

By addressing these pitfalls and implementing appropriate remedies, researchers can ensure the validity and reliability of RD experiments in physics, leading to robust scientific conclusions and advancements in knowledge.

EXAMPLE

Randomized experiments in physics aim to investigate the effects of factors or treatments on a response variable while introducing randomization to minimize bias. Here's a simplified example of a matrix for a RD experiment in physics:

Objective: To study the effects of two different materials (material A and material B) on the efficiency of a solar cell.

Materials:

1. **Material A**: A commonly used semiconductor material.
2. **Material B**: A newly developed semiconductor material.

Response Variable: Solar cell efficiency (measured as a percentage).

TABLE 2.65

Randomized Experiment Matrix Example for Physics

Solar Cell	Material Used
1	Material A
2	Material B
3	Material A
4	Material B
5	Material A
6	Material B
7	Material A
8	Material B
9	Material A
10	Material B

Randomized Experiment Matrix:

In this RD experiment, the matrix is designed to randomly assign the two materials to different solar cells to study their effects on efficiency. The matrix is shown in Table 2.63.

In this matrix, each row represents a specific solar cell experiment. The material used (either material A or material B) is randomly assigned to each solar cell. Multiple solar cells are used for each material group to account for variability.

By conducting the experiment in this randomized manner, any potential bias or confounding factors that may affect the results are minimized. This allows physicists to draw meaningful conclusions about the effects of material A and material B on the efficiency of solar cells in a controlled and unbiased manner.

Another example of a RD experiment in physics could involve investigating the effect of different temperatures on the conductivity of a semiconductor material. Here's how the experiment could be designed:

Research Objective: To determine how variations in temperature affect the conductivity of a semiconductor material.

Experimental Setup:

1. Selection of Semiconductor Material: Choose a specific semiconductor material, such as silicon or germanium, for the experiment.
2. Preparation of Samples: Prepare identical samples of the semiconductor material with consistent dimensions and properties.
3. Randomization: Randomly assign the samples to different temperature conditions.

Experimental Conditions:

1. Control Group (Room Temperature): Keep one group of samples at room temperature (e.g., 25°C).
2. Experimental Groups: Place other groups of samples in environments with different temperatures, such as 50°C, 75°C, and 100°C.

Experimental Procedure:

1. Measure Baseline Conductivity: Measure the baseline conductivity of each sample at room temperature using appropriate measurement techniques, such as a conductivity meter or four-point probe.

2. Apply Temperature Treatments: Place the samples assigned to different temperature conditions in their respective environments and allow them to equilibrate.
3. Measure Conductivity: After a specified duration, measure the conductivity of each sample under the assigned temperature condition.
4. Data Collection: Record the conductivity measurements for each sample in each temperature condition.

Data Analysis:

1. Statistical Analysis: Perform statistical analysis, such as ANOVA, to compare the conductivity measurements between the different temperature conditions.
2. Interpretation: Interpret the results to determine if variations in temperature have a significant effect on the conductivity of the semiconductor material. Consider factors, such as the magnitude of conductivity changes and any observed trends, across temperature conditions.
3. Conclusion: Draw conclusions based on the statistical analysis and interpretation of results. Discuss the implications of the findings for the behavior of semiconductor materials under different temperature conditions and their potential applications in electronic devices.

In this example, randomization ensures that the samples are assigned to temperature conditions in a manner that minimizes bias and allows for valid comparisons of conductivity measurements across different temperature levels. The experiment provides insights into the thermal properties of semiconductor materials, which are essential for various technological applications in electronics and photonics.

A matrix for a RD experiment investigating the effects of different temperatures on the conductivity of a semiconductor material is shown in Table 2.66.

In this matrix,

- "Sample Number" identifies each sample of the semiconductor material.
- "Temperature (°C)" represents the temperature condition to which each sample is exposed.
- "Conductivity (S/m)" shows the measured conductivity of each sample under the respective temperature condition.

Each row represents a unique sample, and the temperature conditions are randomized across the samples to ensure unbiased assignment. Conductivity measurements are taken for each sample at its

TABLE 2.66
Randomized Experiment Matrix Example for Physics

Sample Number	Temperature (°C)	Conductivity (S/m)
1	25	0.0035
2	50	0.0042
3	75	0.0051
4	100	0.0062
5	25	0.0033
6	50	0.0041
7	100	0.0059
8	75	0.0050
9	50	0.0040
10	25	0.0032

assigned temperature condition. This matrix allows for the systematic collection of data needed to analyze the effect of temperature on the conductivity of the semiconductor material.

REPORT

Statistical tools are essential for analyzing the data obtained from RD experiments in physics. Some of the commonly used statistical tools include:

ANOVA: ANOVA is used to assess whether there are statistically significant differences between the means of three or more independent groups. In randomized experiments, ANOVA helps determine whether the treatment groups (e.g., different temperature conditions) have a significant effect on the measured outcome (e.g., conductivity).

T-tests: T-tests are used to compare the means of two independent groups. In randomized experiments, t-tests can be used to compare the means of different treatment groups (e.g., conductivity measurements at different temperatures) to determine if there are statistically significant differences between them.

Regression Analysis: Regression analysis is used to model the relationship between one or more IVs (e.g., experimental conditions) and a DV (e.g., measured outcome). In randomized experiments, regression analysis can help quantify the relationship between the experimental conditions and the outcome of interest while controlling for other variables.

Post-hoc Tests: Post-hoc tests are used following ANOVA to identify specific differences between treatment groups if the overall ANOVA result is significant. Common post-hoc tests include Tukey's HSD, Bonferroni correction, and Scheffé's method.

Chi-square Test: The chi-square test is used to determine whether there is a significant association between categorical variables. In randomized experiments, it can be used to analyze categorical data or to test for independence between variables.

Residual Analysis: Residual analysis involves examining the differences between observed and predicted values to assess the adequacy of the statistical model and identify any patterns or trends in the data that may indicate problems with the model assumptions.

Power Analysis: Power analysis is used to determine the sample size required to detect a statistically significant effect with a specified level of power. It helps ensure that the experiment is adequately powered to detect meaningful differences between treatment groups.

Nonparametric Tests: Nonparametric tests, such as the Wilcoxon rank sum test or the Kruskal–Wallis test, are used when the assumptions of parametric tests are violated or when dealing with non-normally distributed data.

These statistical tools are applied based on the specific research questions, experimental design, and nature of the data collected in RD experiments in physics. They help researchers draw valid conclusions from their experiments and make inferences about the underlying phenomena being studied.

The elements of a report for RD experiments used in physics typically include:

Title: A descriptive title that succinctly summarizes the experiment's purpose and main findings.

Abstract: A brief summary of the experiment, including the research question, methods, results, and conclusions.

Introduction:
- Background: Provide context for the experiment by discussing relevant literature and previous research.
- Objective: Clearly state the research question or hypothesis that the experiment aims to address.

Materials and Methods:
- Experimental Design: Describe the RD used in the experiment, including the randomization process and allocation of treatments.
- Sample Preparation: Detail how samples were prepared and any equipment or materials used.
- Experimental Procedure: Outline the steps followed during the experiment, including measurements, data collection, and any controls implemented.

Results:
- Presentation of Data: Present the collected data in a clear and organized manner, such as tables, figures, or graphs.
- Statistical Analysis: Summarize the statistical analysis performed on the data, including tests for significance and any relevant statistical measures (e.g., means, standard deviations).

Discussion:
- Interpretation of Results: Discuss the findings in relation to the research question and objectives.
- Comparison with Previous Studies: Compare the results with findings from previous research and discuss any similarities, differences, or novel insights.
- Limitations: Address any limitations of the experiment, such as sample size, experimental design, or potential sources of bias.

Conclusion:
- Summary of Findings: Summarize the main findings of the experiment.
- Implications: Discuss the implications of the findings for the broader field of study and any practical applications.

References: Provide a list of references cited in the report, following a consistent citation style (e.g., APA, MLA, Chicago).

Appendices: Include any supplementary material, such as raw data, additional analyses, or experimental protocols.

Acknowledgments: Acknowledge any individuals or organizations that contributed to the experiment, such as funding sources, collaborators, or technical support.

By including these elements in the report, researchers can effectively communicate the rationale, methods, results, and implications of their RD experiments in physics, facilitating understanding and replication by other scientists in the field.

2.4.10 SPLIT-PLOT AND BLOCKED EXPERIMENTS

Split-plot and blocked design experiments are valuable tools in physics research for optimizing experimental designs, reducing variability, controlling for confounding factors, and ultimately improving the validity and reliability of experimental results. Split-plot and blocked design experiments are used in physics for several reasons:

Efficient Use of Resources: These designs allow researchers to efficiently allocate resources, such as time, materials, and equipment, by dividing the experimental units into blocks or subgroups. This can be particularly useful in physics experiments where resources may be limited or expensive.

Accounting for Variability: Split-plot and blocked designs help account for sources of variability in the experiment that are not of primary interest but may influence the outcome. By blocking or stratifying experimental units based on these variables, researchers can reduce the impact of nuisance factors and improve the precision of estimates for the variables of interest.

Controlling Experimental Error: Blocking allows researchers to control for known sources of variation that may affect the experimental outcome. By blocking, researchers can ensure that each treatment group is represented equally across different levels of blocking factors, reducing the impact of confounding variables and increasing the validity of the experiment.

Increased Statistical Power: Split-plot and blocked designs can increase the statistical power of an experiment by reducing error variability and increasing the efficiency of the experimental design. This can lead to more precise estimates of treatment effects and a better ability to detect significant differences between treatment groups.

Handling Nonhomogeneous Experimental Conditions: In physics experiments, conditions may not always be uniform across the entire experimental space. Split-plot and blocked designs allow researchers to account for these nonhomogeneities by blocking experimental units based on relevant factors, such as location or time, and thereby reduce the potential for bias in the estimation of treatment effects.

Split-plot and blocked design experiments are used in physics research across various domains and applications. Some specific areas where these experimental designs are commonly employed include:

Materials Science: In materials science, researchers often use split-plot and blocked design experiments to investigate the effects of different processing parameters (e.g., temperature, pressure, time) on the properties of materials, such as polymers, metals, ceramics, and composites. Blocking factors may include batch-to-batch variations, sample location, or experimental setup.

Optics and Photonics: In optics and photonics research, split-plot and blocked design experiments are utilized to study the performance of optical devices, lasers, and photonic systems under varying environmental conditions or fabrication techniques. Blocking factors may include environmental factors like temperature, humidity, or vibration levels.

Condensed Matter Physics: Researchers studying condensed matter physics often employ split-plot and blocked design experiments to investigate the behavior of materials at different temperatures, pressures, or magnetic fields. These experiments may involve studying phase transitions, magnetic properties, or electrical conductivity.

Particle Physics: In particle physics, experiments conducted at particle accelerators or colliders may utilize split-plot and blocked design experiments to investigate the behavior of subatomic particles under different experimental conditions. Blocking factors may include variations in particle beam intensity, energy levels, or detector configurations.

Astrophysics and Cosmology: In astrophysics and cosmology, researchers may use split-plot and blocked design experiments to analyze observational data obtained from telescopes, satellites, or space probes. Blocking factors may include variations in observing conditions, telescope pointing, or background noise levels.

Quantum Physics: In quantum physics, experiments exploring quantum phenomena, such as entanglement, superposition, or quantum interference, may employ split-plot and blocked design experiments to control for environmental noise and experimental uncertainties. Blocking factors may include variations in experimental setup or external disturbances.

Overall, split-plot and blocked design experiments find applications in various subfields of physics where researchers aim to systematically investigate the effects of multiple factors on experimental outcomes while controlling for potential sources of variability or bias. These experimental designs help ensure the validity and reliability of scientific findings in physics research.

Steps

Developing split-plot and blocked design experiments in physics involves several steps to ensure the experiment is properly designed and executed. Here are the typical steps:

1. **Identify Research Objectives**: Clearly define the research objectives and hypotheses that the experiment aims to address. Determine the factors and variables of interest that will be manipulated and measured during the experiment.
2. **Select Experimental Factors**: Identify the IVs (factors) that will be varied during the experiment to investigate their effects on the DVs. These factors may include experimental conditions, treatments, or levels of control variables.
3. **Identify Blocking Factors**: Identify potential nuisance variables or sources of variability that may influence the outcome of the experiment but are not of primary interest. Blocking factors are variables that can be controlled or accounted for to improve the precision and validity of the experiment.
4. **Design Experimental Layout**: Based on the identified factors and blocking factors, design the layout of the experiment using appropriate split-plot or blocked design structures. Determine the arrangement of experimental units (e.g., samples, measurements) within each block or subplot.
5. **Randomization**: Randomize the assignment of experimental units to treatment groups or conditions to ensure that any systematic biases are minimized. Randomization helps ensure the validity and generalizability of the experiment's findings.
6. **Implement Experimental Protocol**: Conduct the experiment according to the predetermined protocol and experimental design. Ensure that all procedures are carried out consistently and accurately across all treatment groups and experimental conditions.
7. **Data Collection**: Collect data on the DVs according to the experimental design. Record measurements, observations, or other relevant data points for each experimental unit and treatment combination.
8. **Statistical Analysis**: Analyze the collected data using appropriate statistical methods, such as ANOVA for split-plot and blocked designs. Evaluate the main effects of factors, interactions, and any significant differences between treatment groups or conditions.
9. **Interpret Results**: Interpret the results of the statistical analysis in relation to the research objectives and hypotheses. Identify any meaningful patterns, trends, or relationships observed in the data.
10. **Draw Conclusions**: Draw conclusions based on the findings of the experiment, considering the implications for the research objectives and broader scientific knowledge. Discuss the strengths, limitations, and potential sources of bias in the experiment.
11. **Report Findings**: Prepare a comprehensive report summarizing the experiment's design, methods, results, and conclusions. Communicate the findings clearly and accurately, following the conventions of scientific writing and reporting.

By following these steps, researchers can develop well-designed split-plot and blocked design experiments in physics that yield reliable and interpretable results, advancing our understanding of the underlying phenomena being studied.

Pitfalls and Remedies

Split-plot and blocked design experiments in physics can encounter several pitfalls that may compromise the validity and reliability of the results. Here are some common pitfalls and remedies to consider:

Incomplete Blocking:
> **Pitfall:** Incomplete blocking occurs when not all sources of variability are accounted for in the experimental design, leading to residual bias or confounding effects.
>
> **Remedy:** Ensure that all potential sources of variability, known as blocking factors, are identified and properly accounted for in the experimental design. This may involve conducting preliminary studies or pilot experiments to identify relevant blocking factors.

Improper Randomization:
> **Pitfall:** Improper randomization can lead to systematic biases or imbalance between treatment groups, reducing the validity of statistical inference.
>
> **Remedy:** Implement strict randomization procedures to ensure that experimental units are assigned to treatment groups or conditions in a random and unbiased manner. Use randomization techniques, such as complete randomization or randomized block designs.

Small Sample Sizes:
> **Pitfall:** Small sample sizes can limit the statistical power of the experiment and increase the risk of type II errors (false-negatives).
>
> **Remedy:** Conduct power analyses to determine the appropriate sample size needed to detect meaningful effects with sufficient statistical power. Ensure that the sample size is adequate to detect differences between treatment groups or conditions with a reasonable level of confidence.

Unbalanced Designs:
> **Pitfall:** Unbalanced designs occur when the number of experimental units or observations varies across treatment groups or conditions, leading to difficulties in data analysis and interpretation.
>
> **Remedy:** Aim for balanced experimental designs where possible, with an equal number of observations in each treatment group or condition. If unbalanced designs are unavoidable, use appropriate statistical techniques to account for the imbalance during data analysis.

Ignoring Interactions:
> **Pitfall:** Ignoring interactions between factors can lead to misinterpretation of results and overlooking important relationships between variables.
>
> **Remedy:** Explicitly test for interactions between factors using appropriate statistical methods, such as factorial ANOVA or regression analysis. Interpret main effects in the context of significant interactions to avoid drawing incorrect conclusions.

Failure to Validate Assumptions:
> **Pitfall:** Failure to validate underlying assumptions of the statistical models used in data analysis can lead to inaccurate results or misleading conclusions.
>
> **Remedy:** Conduct diagnostic checks to assess the validity of assumptions, such as normality, homogeneity of variance, and independence of observations. If assumptions are violated, consider alternative statistical approaches or robust techniques that are more appropriate for the data.

Overfitting Models:
> **Pitfall:** Overfitting occurs when statistical models are overly complex and fitted too closely to the observed data, resulting in poor generalization to new data.
>
> **Remedy:** Use parsimonious models that strike a balance between goodness of fit and model complexity. Consider techniques, such as cross-validation or model selection criteria, to assess the predictive performance of competing models and avoid overfitting.

By being aware of these potential pitfalls and implementing appropriate remedies, researchers can enhance the robustness and validity of split-plot and blocked design experiments in physics, ensuring that the results accurately reflect the underlying phenomena being studied.

EXAMPLES

Split-plot and blocked designed experiments in physics involve investigating the effects of factors while considering certain factors as whole plots or blocks. Here are a couple of examples of matrices for split-plot and blocked designed experiments in physics:

Example 1: Split-Plot Designed Experiment

Objective: To study the effects of two factors (temperature and time) on the strength of a composite material with a specific fiber orientation.

Factors:

1. **Temperature (Factor A):**
   - Low temperature (100°C).
   - High temperature (150°C).
2. **Time (Factor B):**
   - Short time (1 hour).
   - Long time (2 hours).

Response Variable: Material strength (measured in MPa).

Split-Plot Designed Experiment Matrix:
In this split-plot designed experiment, the matrix is designed to study the effects of temperature and time on material strength. The fiber orientation is considered as a whole plot factor. The matrix is shown in Table 2.67.

In this matrix, each row represents a specific experimental run with a unique combination of fiber orientation (whole plot), temperature, and time levels. The material strength is recorded for each combination.

Split-plot designs are used when certain factors, like fiber orientation in this case, are considered whole plots and are not easily changed between runs.

Example 2: Blocked Designed Experiment

Objective: To investigate the effects of two different laboratory environments (lab environment A and lab environment B) on the accuracy of a precision measurement instrument.

TABLE 2.67

Split-Plot Experiment Matrix Example for Physics

Run	Fiber Orientation (Whole Plot)	Temperature (A) (°C)	Time (B) (hours)	Material Strength (MPa)
1	A	100	1	...
2	A	150	1	...
3	A	100	2	...
4	A	150	2	...
5	B	100	1	...
6	B	150	1	...
7	B	100	2	...
8	B	150	2	...

Factor:

- **Lab Environment (Factor A):**
  - Lab environment A.
  - Lab environment B.

Response Variable: Measurement accuracy (measured in micrometers).

Blocked Designed Experiment Matrix:

In this blocked designed experiment, the matrix is designed to study the effects of lab environment on measurement accuracy. Measurements are taken under different laboratory conditions, and the laboratory environment is considered as a blocking factor. The matrix is shown in Table 2.68.

In this matrix, each row represents a specific experimental run, and measurements are taken under different laboratory environments (environment A or environment B). The measurement accuracy is recorded for each run.

Blocked designs are used when certain factors, like lab environment in this case, are treated as whole blocks, and experiments are conducted within these blocks to account for potential variations introduced by the blocking factor. This approach helps to control for factors that cannot be easily changed or controlled during the experiments.

Another example of a matrix for split-plot and blocked design experiments in physics could involve investigating the effect of different processing parameters on the mechanical properties of a composite material. Let's consider a hypothetical experiment where we have two factors of interest: temperature (T) and pressure (P). Additionally, we have identified two potential sources of variability that we want to account for as blocking factors: batch-to-batch variation (B) and position in the manufacturing batch (L).

Here's an example of how the experimental units could be organized into blocks and subplots:

- **Main Plot (Blocks):**
  - Block 1 (B1): Represents one manufacturing batch.
  - Block 2 (B2): Represents another manufacturing batch.
  - ...
  - Block n (Bn): Represents the nth manufacturing batch.
- **Subplot (Split Plots):**
  - Subplot 1 (T1): High-temperature treatment.
  - Subplot 2 (T2): Medium-temperature treatment.
  - Subplot 3 (T3): Low-temperature treatment.

TABLE 2.68

Blocked Experiment Matrix Example for Physics

Run	Lab Environment (A)	Measurement Accuracy (Micrometers)
1	Environment A	...
2	Environment A	...
3	Environment A	...
4	Environment B	...
5	Environment B	...
6	Environment B	...
7	Environment A	...
8	Environment A	...
9	Environment A	...
10	Environment B	...

TABLE 2.69
Blocked Experiment Matrix Example for Physics

Block	Subplot (Temperature)	Pressure	Observation
B1	T1	P1	Measurement
B1	T1	P2	Measurement
B1	T1	P3	Measurement
B1	T2	P1	Measurement
B1	T2	P2	Measurement
B1	T2	P3	Measurement
...	...	...	...
Bn	T3	P3	Measurement

Within each subplot, the levels of the pressure factor (P1, P2, P3) can be randomized. The experimental matrix could be structured as shown in Table 2.69. In this matrix,

- Each row represents a single experimental unit (e.g., sample).
- The "Block" column represents the manufacturing batch (blocking factor).
- The "Subplot (Temperature)" column represents the temperature treatment (split plot).
- The "Pressure" column represents the pressure treatment within each subplot.
- The "Observation" column represents the outcome or response variable measured for each experimental unit.

By organizing the experimental units into blocks and subplots, researchers can account for batch-to-batch variation and position effects, thereby reducing potential sources of variability and improving the precision and validity of the experiment's results.

REPORTING

In split-plot and blocked design experiments in physics, several statistical tools and techniques are commonly used to analyze the data and draw meaningful conclusions. Here are some of the key statistical tools employed:

ANOVA: ANOVA is a fundamental statistical technique used to partition the total variance in the data into components attributable to different sources, such as treatment effects, blocking factors, and interactions. In split-plot and blocked designs, ANOVA helps assess the significance of main effects and interactions while accounting for the blocking structure.

Factorial ANOVA: Factorial ANOVA extends the basic ANOVA framework to analyze experiments with multiple factors or IVs. In split-plot designs, factorial ANOVA is used to examine the main effects of factors and their interactions within the main plots and subplots.

Mixed-Effects Models: Mixed-effects models, also known as hierarchical linear models, are used to analyze data from experiments with nested or hierarchical structures, such as split-plot and blocked designs. These models allow for the incorporation of both fixed effects (e.g., treatment effects) and random effects (e.g., blocking factors) in the analysis.

Regression Analysis: Regression analysis can be used to model the relationship between IVs (factors) and DVs (responses) in split-plot and blocked designs. Regression models can capture nonlinear relationships and interactions between factors, providing insights into the underlying mechanisms driving the observed outcomes.

Post-hoc Tests: Post-hoc tests are used to conduct pairwise comparisons between treatment groups or conditions following a significant finding in ANOVA or regression analysis. Common post-hoc tests include Tukey's HSD, Bonferroni correction, and Dunnett's test, among others.

Diagnostic Checks: Diagnostic checks are performed to assess the validity of statistical assumptions underlying the analysis, such as normality, homoscedasticity (homogeneity of variance), and independence of observations. Residual analysis, Q-Q plots, and Shapiro–Wilk tests are commonly used diagnostic tools.

Power Analysis: Power analysis is used to determine the sample size needed to detect meaningful effects with sufficient statistical power. By conducting power analysis prior to the experiment, researchers can ensure that the study has adequate power to detect hypothesized effects.

Bootstrap Resampling: Bootstrap resampling is a nonparametric technique used to estimate the sampling distribution of a statistic by repeatedly sampling from the observed data with replacement. Bootstrap methods can provide robust estimates of standard errors and confidence intervals, particularly in cases where parametric assumptions are violated.

By applying these statistical tools and techniques, researchers can effectively analyze data from split-plot and blocked design experiments in physics, uncovering meaningful relationships between experimental factors and outcomes while accounting for the complex experimental structures inherent in these designs.

When reporting the results of split-plot and blocked design experiments in physics, it's important to provide a comprehensive and clear account of the experiment, analysis, and findings. Here are the key elements typically included in a report for such experiments:

Title: A concise and descriptive title that summarizes the main objective or focus of the experiment.

Abstract: A brief summary of the experiment, including the research question, experimental design, methods, key findings, and implications. The abstract should provide enough information for readers to understand the study's purpose and main results.

Introduction: An introduction to the experiment that provides background information on the topic, outlines the research question or hypothesis, and explains the significance of the study. This section should also include a brief overview of the experimental design and approach.

Materials and Methods: A detailed description of the experimental materials, equipment, and procedures used in the study. This section should provide sufficient information to allow other researchers to replicate the experiment. It should include information on factors studied, blocking factors, treatments applied, measurement techniques, and any statistical methods used for data analysis.

Experimental Design: A clear explanation of the experimental design, including the rationale for choosing a split-plot or blocked design, the factors and levels studied, how experimental units were allocated to blocks and subplots, and any randomization procedures employed.

Results: A presentation of the experimental results, including descriptive statistics, graphical representations (e.g., plots, charts), and inferential statistics (e.g., ANOVA tables, regression coefficients). Results should be organized logically and presented in a manner that facilitates understanding and interpretation.

Discussion: An interpretation and discussion of the results in the context of the research question, relevant theory, and previous literature. This section should address the implications of the findings, any unexpected or anomalous results, potential sources of bias or error, and limitations of the study.

Conclusion: A concise summary of the main findings and their significance. This section should restate the research question or hypothesis, summarize the key results, and highlight the broader implications of the study.

References: A list of all references cited in the report, following a consistent citation style (e.g., APA, MLA). References should include relevant literature, previous studies, and any sources consulted during the research process.

Appendices: Additional supplementary material, such as raw data, detailed calculations, experimental protocols, or any other information that may be useful for readers but is not essential for understanding the main findings.

By including these elements in the report, researchers can effectively communicate the rationale, methods, results, and implications of their split-plot and blocked design experiments in physics, contributing to the advancement of scientific knowledge in the field.

2.4.11 BBD

The experiments are a type of RSM used to optimize experimental conditions by systematically varying IVs (factors) within predetermined ranges. BBDs are particularly useful when the relationship between factors and the response variable is expected to be quadratic and when the number of factors is moderate. While BBDs are commonly used in fields, such as engineering, chemistry, and pharmaceuticals, they can also be applied in physics research. Here are a few examples of how BBD experiments could be used in physics:

Optical Characterization of Nanoparticles: In nanophysics, researchers might use BBDs to optimize the experimental conditions for the synthesis of nanoparticles with specific optical properties. Factors, such as precursor concentration, reaction temperature, and reaction time, could be varied within predetermined ranges, while the response variable could be the optical absorption or emission spectra of the synthesized nanoparticles. By conducting BBD experiments, researchers could identify the optimal factor settings for producing nanoparticles with desired optical characteristics.

Optimization of Thin-Film Deposition Processes: Thin-film deposition techniques are widely used in physics for fabricating semiconductor devices, solar cells, and optical coatings. Researchers might use BBDs to optimize the experimental conditions for thin-film deposition processes, such as CVD or sputtering. Factors, such as deposition temperature, gas flow rates, and substrate properties, could be varied within predetermined ranges, while the response variable could be film thickness, surface roughness, or optical properties. BBD experiments could help identify the optimal process parameters for achieving desired thin-film characteristics.

Experimental Design for Laser Experiments: In experimental physics involving lasers, researchers might use BBDs to optimize the experimental conditions for laser parameter tuning or beam shaping. Factors, such as laser power, pulse duration, and beam divergence, could be varied within predetermined ranges, while the response variable could be laser output intensity, beam profile, or spectral characteristics. BBD experiments could help identify the optimal laser settings for specific applications, such as laser spectroscopy, microscopy, or materials processing.

Optimization of Experimental Setup for Quantum Information Processing: In quantum physics and quantum information processing, researchers might use BBDs to optimize the experimental conditions for implementing quantum algorithms or quantum communication protocols. Factors, such as qubit coherence time, gate fidelity, and noise level, could be varied within predetermined ranges, while the response variable could be quantum state fidelity, error rate, or entanglement fidelity. BBD experiments could help identify the optimal experimental setup for achieving high-performance quantum information processing tasks.

These examples illustrate how BBD experiments could be applied in physics research to optimize experimental conditions, improve performance, and advance scientific understanding in various areas of study. By systematically varying IVs within predetermined ranges and analyzing the resulting response data, researchers can identify optimal factor settings and optimize experimental processes to achieve desired outcomes.

STEPS

Developing BBD experiments in physics involves several steps to systematically vary IVs (factors) within predetermined ranges and optimize experimental conditions to achieve desired outcomes. Here are the steps to develop BBD experiments in physics:

1. **Identify Factors and Response Variable:**
   - Define the IVs (factors) that influence the system or process under study. These factors should be chosen based on their relevance to the research objectives and their potential impact on the response variable.
   - Determine the response variable, which represents the outcome or performance measure of interest. The response variable should be quantifiable and relevant to the research objectives.
2. **Determine Factor Ranges**: Establish the minimum and maximum levels for each factor based on prior knowledge, literature review, or experimental constraints. Factor ranges should be wide enough to encompass the expected variation in experimental conditions but narrow enough to avoid extreme conditions or impractical settings.
3. **Generate BBD Matrix**:
   - Use statistical software or experimental design software to generate a BBD matrix. BBDs typically consist of a central point and a set of factorial and axial points that systematically vary factor levels within predetermined ranges.
   - The number of runs in the BBD matrix depends on the number of factors and the desired level of precision. BBDs are designed to balance the number of runs with the ability to estimate main effects and quadratic effects efficiently.
4. **Conduct Experiments:**
   - Perform the experimental runs specified by the BBD matrix using the predetermined factor settings.
   - Ensure that experimental conditions are controlled and consistent across runs to minimize variability and ensure the validity of the results.
5. **Measure Response Variable:**
   - Record measurements or observations of the response variable for each experimental run.
   - Use appropriate measurement techniques and instrumentation to ensure accurate and reliable data collection.
6. **Data Analysis:**
   - Analyze the experimental data to estimate the effects of factors on the response variable.
   - Fit a quadratic response surface model to the data using regression analysis techniques. The model should include main effects, interaction effects, and quadratic effects of factors.

- Use statistical software to perform regression analysis and assess the significance of factor effects, as well as identify optimal factor settings.

7. **Optimization:**
   - Use the fitted response surface model to identify optimal factor settings that maximize or minimize the response variable.
   - Conduct optimization analyses to determine the factor combinations that optimize performance, efficiency, or other desired criteria.
   - Evaluate the robustness of the optimal solutions by performing sensitivity analyses or Monte Carlo simulations.

8. **Validation:**
   - Validate the optimized factor settings by conducting additional experimental runs or simulations under the identified optimal conditions.
   - Compare the experimental results obtained under optimal settings with the predicted values from the response surface model to assess the accuracy and reliability of the optimization process.

9. **Interpretation and Conclusion:**
   - Interpret the results of the BBD experiments in the context of the research objectives and the underlying physical principles.
   - Draw conclusions about the effects of factors on the response variable, identify factors that significantly influence system performance, and discuss implications for physics research or practical applications.

By following these steps, researchers can systematically develop BBD experiments in physics, optimize experimental conditions, and gain insights into the factors that influence system behavior or performance. BBDs provide a powerful framework for efficiently exploring factor effects and identifying optimal factor settings, facilitating scientific discovery and innovation in physics research.

PIT FALLS AND REMEDIES

BBD experiments offer many advantages for systematically varying factors and optimizing experimental conditions in physics research. However, there are also potential pitfalls that researchers should be aware of, along with remedies to address them. Here are some common pitfalls and remedies associated with BBD experiments in physics:

Insufficient Factor Range:
   Pitfall: If the chosen factor ranges are too narrow, important regions of the design space may be missed, leading to suboptimal results.
   Remedy: Conduct preliminary experiments or sensitivity analyses to determine appropriate factor ranges that encompass the expected variability in experimental conditions. Consult prior literature or theoretical models to inform the selection of factor ranges.

Lack of Model Adequacy:
   Pitfall: The fitted response surface model may not adequately represent the true relationship between factors and the response variable, leading to inaccurate predictions or suboptimal solutions.
   Remedy: Validate the response surface model by conducting additional experimental runs or cross-validation analyses. Consider alternative model forms or transformation techniques to improve model adequacy. Verify model assumptions, such as linearity and homoscedasticity, through residual analysis.

Inadequate Experimental Precision:
   Pitfall: Variability in experimental conditions or measurement error may limit the precision of experimental results and the ability to detect significant factor effects.

Remedy: Increase the number of experimental replicates or runs to improve the precision of estimates and reduce the impact of random variability. Implement rigorous quality control measures to minimize sources of experimental error, such as calibration of instruments and standardization of procedures.

Overfitting of Response Surface Model:

Pitfall: Fitting a response surface model with too many terms or interactions may result in overfitting, where the model captures noise in the data rather than true underlying relationships.

Remedy: Use statistical techniques, such as model selection criteria (e.g., AIC, BIC) or cross-validation, to identify the most parsimonious model that adequately represents the data without overfitting. Consider simplifying the model by removing nonsignificant terms or reducing the degree of polynomial.

Experimental Constraints or Practical Considerations:

Pitfall: Practical constraints, such as limited resources, time constraints, or experimental feasibility, may impose limitations on the design and implementation of BBD experiments.

Remedy: Prioritize factors and experimental conditions based on their relevance to the research objectives and available resources. Consider alternative designs, such as fractional factorial designs or CCDs, that require fewer experimental runs while still providing valuable information.

Ignoring System Dynamics or Nonlinear Effects:

Pitfall: BBD experiments assume that the relationship between factors and the response variable is quadratic, which may not accurately capture nonlinear or dynamic system behavior.

Remedy: Consider alternative experimental designs or modeling approaches that can capture nonlinear or dynamic effects, such as dynamic RSM or time-series analysis. Incorporate additional experimental factors or dynamic control strategies to account for system dynamics.

By addressing these potential pitfalls and implementing appropriate remedies, researchers can maximize the effectiveness of BBD experiments in physics, optimize experimental conditions, and gain valuable insights into the factors influencing system behavior or performance.

EXAMPLE

A BBD matrix is a specific arrangement of experimental runs that systematically varies IVs (factors) within predetermined ranges. The matrix consists of a combination of factorial points and axial points, with the central point typically replicated multiple times for precision. Here's an example of a BBD matrix for a physics experiment with three factors, each at three levels:

Suppose we have three factors (A, B, and C) with three levels (−1, 0, +1), representing low, intermediate, and high settings, respectively.

Designing the matrix:

- Each row represents an experimental run.
- Columns represent the levels of the factors A, B, and C.
- The numbers in each cell indicate the level of the corresponding factor (A, B, or C) for that experimental run.
- The central point (0, 0, 0) is replicated multiple times (in this case, once) to estimate the experimental error and provide a baseline for comparison.

The axial points (coded as +1 and −1) allow for the estimation of curvature in the response surface, while the factorial points (coded as 0) provide information about the main effects and interactions. This design balances the number of runs with the ability to estimate main effects and quadratic effects efficiently.

Let's consider an example of how a BBD experiment could be applied in physics research, specifically in the optimization of experimental conditions for a PV cell. The objective is to maximize the efficiency of the PV cell by optimizing factors, such as the thickness of the semiconductor layer, the doping concentration, and the annealing temperature. Here's how the BBD experiment could be designed:

1. **Identify Factors and Response Variable:**
   - Factors:
     - Factor A: Thickness of the semiconductor layer (in micrometers).
     - Factor B: Doping concentration of the semiconductor layer (in atoms per cubic centimeter).
     - Factor C: Annealing temperature (in degrees celsius).
   - Response Variable:
     - Efficiency of the PV cell (in percentage).
2. **Determine Factor Ranges:**
   - Factor A: 10 μm to 50 μm.
   - Factor B: 1×10^{15} atoms/cm^3 to 5×10^{15} atoms/cm^3.
   - Factor C: 300°C to 500°C.
3. **Generate BBD Matrix:**
   - Use statistical software to generate a BBD matrix with 15 experimental runs, including three center points.
   - The matrix systematically varies the factors within the predetermined ranges, ensuring adequate coverage of the design space.
4. **Conduct Experiments:**
   - Perform the 15 experimental runs according to the settings specified in the BBD matrix.
   - Prepare and characterize PV cells for each experimental condition, ensuring consistency in fabrication and measurement procedures.
5. **Measure Response Variable:**
   - Measure the efficiency of each PV cell using standard characterization techniques, such as current–voltage (I–V) curve measurements or external quantum efficiency (EQE) measurements.
   - Record the efficiency values for each experimental run.
6. **Data Analysis:**
   - Analyze the experimental data using regression analysis to fit a quadratic response surface model.
   - Estimate main effects, interaction effects, and quadratic effects of factors on the efficiency of the PV cell.
   - Determine the significance of factor effects and identify optimal factor settings that maximize PV cell efficiency.
7. **Optimization:**
   - Use the fitted response surface model to identify the factor settings that maximize PV cell efficiency.
   - Conduct optimization analyses to determine the optimal thickness, doping concentration, and annealing temperature for achieving maximum efficiency.
   - Validate the optimized conditions through additional experiments or simulations.

8. **Interpretation and Conclusion:**
   - Interpret the results of the BBD experiment in the context of the research objectives and physical principles governing PV cell operation.
   - Draw conclusions about the effects of thickness, doping concentration, and annealing temperature on PV cell efficiency.
   - Discuss implications for optimizing PV cell fabrication processes and improving the performance of solar energy harvesting systems.

To illustrate a BBD matrix for optimizing experimental conditions for a PV cell, let's consider three factors: the thickness of the semiconductor layer (A), doping concentration of the semiconductor layer (B), and annealing temperature (C). Each factor will be studied at three levels: low, medium, and high. We'll generate a BBD matrix with 15 experimental runs, including three center points. An example of such a matrix is shown in Table 2.70.

In this BBD matrix,

   - The levels of each factor (A, B, C) are represented as −1 (low), 0 (medium), and +1 (high).
   - Each row represents an experimental run with specific settings for the factors.
   - The central point (0, 0, 0) can be replicated multiple times to estimate the experimental error and provide a baseline for comparison.
   - The matrix is constructed to ensure that each factor is studied at low, medium, and high levels and to allow for the estimation of main effects and quadratic effects efficiently.

Researchers would conduct the experimental runs according to the settings specified in this BBD matrix, measure the efficiency of the PV cells for each condition, and then analyze the data to identify the optimal factor settings that maximize PV cell efficiency.

By conducting a BBD experiment in this example, researchers can systematically optimize the experimental conditions for fabricating high-efficiency PV cells, contributing to advancements in renewable energy technology and sustainable energy production.

TABLE 2.70

BBD Experiment Matrix Example for Physics

Run	Thickness (A)	Doping Concentration (B)	Annealing Temperature (C)
1	−1	−1	0
2	−1	1	0
3	+1	−1	0
4	+1	1	0
5	0	0	−1
6	0	0	+1
7	0	−1	−1
8	0	−1	+1
9	0	1	−1
10	0	1	+1
11	−1	0	−1
12	−1	0	+1
13	+1	0	−1
14	+1	0	+1
15	0	0	0

REPORTING

In BBD experiments, various statistical tools are employed for data analysis and interpretation to optimize experimental conditions. Some of the key statistical tools used in BBD experiments for physics include:

Regression Analysis: Regression analysis is used to fit a response surface model to the experimental data. The model captures the relationship between the IVs (factors) and the response variable (e.g., efficiency of a PV cell) by estimating the main effects, interaction effects, and quadratic effects of factors.

ANOVA: ANOVA is utilized to assess the significance of factor effects and determine whether they have a statistically significant impact on the response variable. It helps identify which factors and interactions are significant contributors to the variation in the response.

RSM: RSM involves the use of mathematical and statistical techniques to optimize experimental conditions by analyzing the response surface model. RSM aims to identify the optimal factor settings that maximize or minimize the response variable within the experimental range.

DOE: DOE principles are applied to systematically plan and conduct experiments, ensuring efficient use of resources and effective exploration of factor effects. BBDs are a type of DOE that allows for efficient estimation of main effects and quadratic effects with relatively few experimental runs.

Model Fitting Techniques: Various model fitting techniques, such as least squares regression, are employed to estimate the parameters of the response surface model. These techniques help determine the coefficients of the model equations and assess the goodness of fit.

Optimization Algorithms: Optimization algorithms, such as gradient-based methods or genetic algorithms, are used to identify the optimal factor settings that maximize or minimize the response variable based on the fitted response surface model. These algorithms iteratively search for the combination of factor levels that optimize the objective function.

Diagnostic Tools: Diagnostic tools, including residual analysis and goodness-of-fit tests, are employed to assess the adequacy of the response surface model and identify potential violations of model assumptions. These tools help ensure the validity and reliability of the statistical analysis.

By employing these statistical tools, researchers can effectively analyze the data generated from BBD experiments, identify significant factor effects, optimize experimental conditions, and gain insights into the underlying relationships between factors and the response variable in physics research.

A comprehensive report for BBD experiments used in physics should include various elements to effectively communicate the experimental design, results, analysis, and conclusions. Here are the key elements that should be included in a report for BBD experiments in physics:

1. **Title and Abstract:**
   - Title: Clearly states the purpose and scope of the experiment.
   - Abstract: Provides a concise summary of the experimental design, methodology, key findings, and conclusions.
2. **Introduction:**
   - Background: Provides context and rationale for the experiment, including relevant theories, previous research, and the significance of the study.
   - Objectives: Clearly states the specific objectives or research questions addressed by the experiment.

3. **Experimental Design:**
   - Description of Factors: Provides details of the factors studied in the experiment, including their definitions, levels, and ranges.
   - BBD Matrix: Presents the BBD matrix used in the experiment, specifying the factor settings for each experimental run.
   - Experimental Procedure: Describes the experimental setup, procedures, and protocols followed to conduct the experiment.
4. **Data Collection and Analysis:**
   - Data Collection: Describes the measurement techniques, instruments used, and procedures followed to collect data on the response variable.
   - Data Analysis: Outlines the statistical methods, regression analysis, ANOVA, and other techniques used to analyze the experimental data and fit the response surface model.
5. **Results:**
   - Summary of Experimental Results: Presents the main findings of the experiment, including the measured responses, factor effects, and statistical analysis results.
   - Response Surface Plot: Graphical representation of the response surface model, illustrating the relationship between factors and the response variable.
6. **Discussion:**
   - Interpretation of Results: Discusses the implications of the experimental findings in relation to the research objectives and theoretical framework.
   - Comparison with Literature: Compares the experimental results with previous studies, theoretical predictions, or industry standards.
   - Limitations and Assumptions: Identifies any limitations or assumptions of the experimental design, data analysis, or interpretation of results.
7. **Optimization and Conclusions:**
   - Optimal Factor Settings: Presents the optimal factor settings identified through the response surface model and optimization analysis.
   - Conclusions: Summarizes the key findings, implications, and contributions of the experiment to the field of physics.
   - Recommendations for Future Research: Suggests potential areas for further investigation or refinement of experimental methods.
8. **References**: Citations: Lists all references cited in the report, following a standard citation format (e.g., APA, IEEE).
9. **Appendices**: Supplementary Information: Includes additional data tables, figures, mathematical derivations, or detailed experimental procedures not included in the main text.

By including these elements in the report, researchers can effectively communicate the design, implementation, analysis, and implications of BBD experiments in physics, facilitating transparency, reproducibility, and dissemination of scientific knowledge.

2.4.12 PLACKETT–BURMAN

PBD experiments are widely utilized in physics research across various domains. They are utilized in physics for several reasons:

Efficient Screening of Factors: PBD experiments allow researchers to efficiently screen a large number of factors with a minimal number of experimental runs. This is particularly useful in physics, where numerous variables may influence the outcome of an experiment, and it is impractical or resource intensive to conduct full factorial experiments.

Identifying Significant Factors: PBD experiments help identify the most influential factors that affect the response variable. By systematically varying factors at two levels (high and low), researchers can quickly identify which factors have a significant impact on the outcome of interest.

Resource Optimization: Conducting full factorial experiments to study the effects of multiple factors in physics experiments can be time-consuming, costly, and impractical. PBD experiments offer a resource-efficient alternative by requiring fewer experimental runs while still providing valuable insights into the factors affecting the response variable.

Prioritizing Further Investigation: Once significant factors are identified using PBD experiments, researchers can prioritize further investigation of these factors using more detailed experimental designs or optimization techniques. This helps focus resources on factors that have the greatest impact on the outcome.

Experimental Design Flexibility: PBD experiments can be designed to accommodate various experimental setups and constraints in physics research. Researchers have flexibility in selecting factors, determining factor levels, and arranging experimental runs based on the specific requirements of their study.

Screening for Nonlinear Effects: While PBD experiments are primarily designed for screening main effects (linear relationships), they can also provide insights into potential nonlinear relationships between factors and the response variable. This can guide further investigation into nonlinear effects if warranted.

Robustness and Reliability: PBD experiments are robust and reliable screening tools when properly designed and analyzed. They provide statistically sound results that help researchers make informed decisions about which factors to focus on in subsequent experiments or optimization studies.

Overall, PBD experiments offer a practical and efficient approach to screening and identifying significant factors in physics research, helping researchers gain insights into complex systems and phenomena while optimizing resource utilization.

Here are some specific areas where PBD experiments are commonly applied in physics:

Material Science: PBD experiments are used to optimize parameters in material synthesis processes, such as thin-film deposition, crystal growth, or nanoparticle synthesis. Factors like precursor concentrations, temperature, pressure, and deposition rates are optimized to achieve desired material properties.

Optics and Photonics: In optics and photonics research, PBD experiments help optimize the fabrication processes of optical devices. This includes factors, such as substrate material, thickness, doping concentrations, and annealing conditions, to improve device performance.

Surface Science: PBD experiments are employed in surface science studies to optimize experimental conditions for surface modification techniques, such as CVD or surface functionalization. Factors like gas flow rates, temperature, and exposure times are optimized to control surface properties.

Semiconductor Device Fabrication: PBD experiments play a role in optimizing fabrication processes for semiconductor devices like transistors and diodes. Parameters, such as etching conditions, dopant concentrations, and thermal treatments, are optimized to improve device performance.

Nanotechnology: PBD experiments are used in nanotechnology research to optimize experimental parameters for fabricating nanostructures with specific properties. This includes factors like nanoparticle size, shape, surface chemistry, and dispersion conditions.

Energy Materials: In energy materials research, PBD experiments are applied to optimize materials and processes for applications, such as solar cells and batteries. Factors, such as electrode composition, electrolyte formulation, and heat treatment conditions, are optimized to enhance energy conversion and storage efficiency.

Experimental Physics: PBD experiments are utilized in various experimental physics studies to systematically investigate the effects of multiple factors on experimental outcomes. This includes optimizing experimental conditions in particle physics experiments, quantum optics studies, and condensed matter physics research.

Overall, PBD experiments find applications in a wide range of physics research areas where there is a need to efficiently screen and optimize experimental parameters to achieve desired outcomes.

STEPS

Developing PBD experiments in physics involves several steps to systematically screen and optimize experimental factors. Here are the general steps to develop PBD experiments in physics:

1. **Identify Factors**: Determine the factors (IVs) that may influence the response variable (DV) of interest in the physics experiment. These factors could include experimental conditions, material properties, process parameters, or environmental variables.
2. **Determine Levels of Factors**: Decide on the levels at which each factor will be tested during the experiment. Typically, factors are tested at two levels: high and low, corresponding to the maximum and minimum values of the factor, respectively.
3. **Select Response Variable**: Choose the response variable that represents the outcome or performance metric of interest in the experiment. This could be a physical quantity, property, or characteristic that is measured or observed during the experiment.
4. **Design the Experimental Matrix**: Use a PBD matrix to systematically determine the combinations of factor levels for experimental runs. The number of experimental runs required is determined by the number of factors being tested. The matrix is constructed such that each factor is either at a high level (+1) or a low level (−1) for each experimental run, with an equal number of high and low settings for each factor across all runs.
5. **Randomize Experimental Runs**: Randomize the order of experimental runs to minimize the effects of uncontrolled factors and potential biases. This ensures that the results are not influenced by the order in which experiments are conducted.
6. **Perform Experimental Runs**: Conduct the experimental runs according to the settings specified in the PBD matrix. Ensure that each experimental condition is executed precisely and consistently to minimize variability.
7. **Measure Response Variable**: Record the measurements or observations of the response variable for each experimental run. Use appropriate measurement techniques and instruments to ensure the accuracy and reliability of the data.
8. **Data Analysis**: Analyze the experimental data to identify significant factors that influence the response variable. Use statistical methods, such as ANOVA, to determine the effects of factors and assess their significance.
9. **Optimization**: If the goal is to optimize experimental conditions, use the results of the data analysis to identify the optimal factor settings that maximize or minimize the response variable. This may involve further experimentation or mathematical optimization techniques.
10. **Interpretation and Conclusion**: Interpret the results of the PBD experiment in the context of the research objectives and draw conclusions about the effects of factors on the response variable. Discuss implications for the field of physics and potential avenues for further research.

By following these steps, researchers can systematically develop PBD experiments in physics to efficiently screen and optimize experimental factors, leading to valuable insights and advancements in the field.

PITFALLS AND REMEDIES

In PBD experiments used in physics, researchers may encounter several pitfalls that can affect the validity and reliability of the results. Here are some common pitfalls and remedies:

1. **Failure to Identify Important Factors:**
   Pitfall: Not all influential factors may be included in the initial screening design, leading to important factors being overlooked.
   Remedy: Prioritize factors based on domain knowledge, literature review, or preliminary experiments. Consider conducting follow-up experiments to explore additional factors if necessary.
2. **Confounding Effects:**
   Pitfall: Confounding occurs when the effects of two or more factors cannot be distinguished from each other, leading to incorrect conclusions about the factors' impacts.
   Remedy: Carefully select factors and levels to minimize confounding. Use statistical techniques, such as ANOVA, to identify and account for confounding effects.
3. **Inadequate Sample Size:**
   Pitfall: Insufficient sample size may result in low statistical power, making it difficult to detect significant effects or relationships.
   Remedy: Conduct power analysis to determine the appropriate sample size based on the expected effect sizes and desired level of statistical power. Increase the sample size if necessary to ensure adequate power.
4. **Nonlinear Relationships:**
   Pitfall: Assuming linear relationships between factors and the response variable when nonlinear relationships exist can lead to inaccurate predictions and suboptimal factor settings.
   Remedy: Explore the possibility of nonlinear relationships by conducting additional experiments or using alternative modeling techniques, such as RSM, to capture nonlinear effects.
5. **External Factors and Noise:**
   Pitfall: External factors or sources of noise in the experimental environment may introduce variability and obscure the true effects of experimental factors.
   Remedy: Control for external factors as much as possible through proper experimental design, randomization, and replication. Conduct sensitivity analysis to assess the robustness of results to noise and variability.
6. **Overfitting of Models:**
   Pitfall: Overfitting occurs when a model is overly complex and captures noise or random fluctuations in the data, leading to poor generalization of new data.
   Remedy: Use techniques, such as cross-validation or regularization, to prevent overfitting and ensure that the model accurately captures the underlying relationships in the data without being overly complex.
7. **Biased Experimental Design:**
   Pitfall: Biased experimental design, such as nonrandom selection of factor levels or nonrepresentative sampling, can introduce systematic errors and bias into the results.
   Remedy: Randomize experimental runs, use balanced designs, and ensure that factor levels are selected based on scientific rationale rather than personal biases or preferences.

TABLE 2.71

PBD Matrix for an Experiment in Physics

Run	(A) Factor 1	(B) Factor 2	(C) Factor 3	(D) Factor 4	(E) Factor 5
1	+1	−1	−1	−1	+1
2	−1	+1	−1	−1	+1
3	+1	+1	+1	−1	−1
4	−1	−1	+1	+1	−1
5	+1	−1	+1	+1	+1
6	−1	+1	+1	+1	+1
7	+1	+1	−1	+1	+1
8	−1	−1	−1	+1	−1

By being aware of these pitfalls and implementing appropriate remedies, researchers can enhance the quality and reliability of PBD experiments in physics, leading to more robust conclusions and insights.

EXAMPLE

An example of a PBD matrix for an experiment in physics is shown in Table 2.71.

In this example,

- The factors A, B, C, D, and E represent different experimental variables.
- Each row corresponds to one experimental run.
- The symbols "+" and "−" represent the high and low levels of each factor, respectively.
- The columns represent the different factors being tested in the experiment.

This matrix is designed to efficiently screen a relatively large number of factors using a minimal number of experimental runs, allowing researchers to identify the most influential factors affecting the response variable.

Let's consider an example of a PBD experiment focusing on optimizing the fabrication process for semiconductor nanowires used in nano-electronic devices. The goal is to identify the most influential factors that affect the diameter of the nanowires. Here's an example experimental setup:

1. **Identified Factors:**
   - Factor A: Growth time (hours).
   - Factor B: Growth temperature (°C).
   - Factor C: Precursor concentration (mol/L).
   - Factor D: Substrate roughness (nm).
   - Factor E: Catalyst particle size (nm).
2. **Selected Levels of Factors:**
   - Factor A: Low (12 hours), high (24 hours).
   - Factor B: Low (600°C), high (800°C).
   - Factor C: Low (0.1 mol/L), high (0.3 mol/L).
   - Factor D: Low (5 nm), high (20 nm).
   - Factor E: Low (10 nm), high (50 nm).
3. **Response Variable**: Diameter of semiconductor nanowires (nm).
4. **PBD Matrix (Table 2.72):**

TABLE 2.72
PBD Matrix for an Experiment in Physics

Run	A (Time)	B (Temperature)	C (Concentration)	D (Roughness)	E (Catalyst Size)
1	−1	−1	−1	−1	−1
2	+1	−1	−1	−1	−1
3	−1	+1	−1	−1	−1
4	+1	+1	−1	−1	−1
5	−1	−1	+1	−1	−1
6	+1	−1	+1	−1	−1
7	−1	+1	+1	−1	−1
8	+1	+1	+1	−1	−1
9	−1	−1	−1	+1	−1
10	+1	−1	−1	+1	−1
11	−1	+1	−1	+1	−1
12	+1	+1	−1	+1	−1
13	−1	−1	+1	+1	−1
14	+1	−1	+1	+1	−1
15	−1	+1	+1	+1	−1
16	+1	+1	+1	+1	−1
17	−1	−1	−1	−1	+1
18	+1	−1	−1	−1	+1
19	−1	+1	−1	−1	+1
20	+1	+1	−1	−1	+1
21	−1	−1	+1	−1	+1
22	+1	−1	+1	−1	+1
23	−1	+1	+1	−1	+1
24	+1	+1	+1	−1	+1
25	−1	−1	−1	+1	+1
26	+1	−1	−1	+1	+1
27	−1	+1	−1	+1	+1
28	+1	+1	−1	+1	+1
29	−1	−1	+1	+1	+1
30	+1	−1	+1	+1	+1
31	−1	+1	+1	+1	+1
32	+1	+1	+1	+1	+1

In this example,

- Each row represents an experimental run with specific settings for the factors.
- The factors are varied at two levels: low (−1) and high (+1), with 16 experimental runs in total.
- The diameter of the semiconductor nanowires is measured for each experimental run.

After conducting the experiments and measuring the response variable, statistical analysis (e.g., ANOVA) is performed to identify the significant factors that affect the diameter of the nanowires. This information can then be used to optimize the fabrication process for semiconductor nanowires in nano-electronic devices.

REPORTING

In PBD experiments for physics, several statistical tools are commonly used to analyze the data and draw conclusions about the effects of factors on the response variable. Some of the key statistical tools used in PBD experiments include:

ANOVA: ANOVA is used to determine whether the observed differences in the response variable among experimental runs are statistically significant. It helps identify factors that have a significant impact on the response variable.

Main Effect Plots: Main effect plots visually display the effects of each factor on the response variable. These plots allow researchers to observe the relationship between individual factors and the response variable, facilitating the interpretation of the results.

Half-Normal Probability Plots: Half-normal probability plots are used to identify significant factors by plotting the absolute values of the effects of each factor against their corresponding probability values. Factors with effects that fall above the 45-degree line on the plot are considered significant.

Residual Analysis: Residual analysis helps assess the adequacy of the model fit by examining the residuals (the differences between observed and predicted values). Residual plots can reveal patterns or outliers that may indicate problems with the model.

Diagnostic Plots: Diagnostic plots, such as normal probability plots or residuals versus fitted values plots, are used to check the assumptions of the statistical model, such as normality and homoscedasticity (constant variance of residuals).

Coefficient of Determination (R-squared): R-squared measures the proportion of variation in the response variable explained by the factors included in the model. A high R-squared value indicates that the model fits the data well and explains a large portion of the variability.

Prediction Profiling: Prediction profiling is used to estimate the optimal factor settings that maximize or minimize the response variable based on the fitted model. It helps identify the optimal experimental conditions for achieving desired outcomes.

Sequential Analysis: Sequential analysis is used to iteratively refine the experimental design by adding or removing factors based on the results of initial experiments. This approach allows researchers to efficiently identify the most influential factors with a minimal number of experimental runs.

By employing these statistical tools, researchers can effectively analyze the data from PBD experiments in physics, identify significant factors, and draw meaningful conclusions about the factors' effects on the response variable.

When reporting the results of PBD experiments in physics, it's essential to provide a clear and comprehensive account of the experimental design, methodology, results, and conclusions. Here are the key elements typically included in a report on PBD experiments used in physics:

Title: A descriptive and concise title that summarizes the purpose and scope of the experiment.

Abstract: A brief summary of the experiment, including the objectives, methods, key findings, and implications. The abstract should provide enough information to give readers an overview of the study.

Introduction:
- Background: Provide background information on the topic of the study, including relevant theories, concepts, or previous research.
- Objectives: Clearly state the research objectives or hypotheses that the experiment aims to address.
- Importance: Explain the significance and relevance of the study in the context of physics research.

Experimental Design:
- Description of Factors: List the factors (IVs) included in the experiment and their respective levels.
- Design Matrix: Present the PBD matrix used for the experiment, showing the factor settings for each experimental run.
- Experimental Setup: Describe the experimental setup, including equipment, materials, and procedures used to conduct the experiment.

Data Collection:
- Measurement Procedures: Describe how data were collected, including the instruments or methods used for measurements.
- Data Recording: Explain how data were recorded, including any precautions taken to ensure accuracy and precision.

Data Analysis:
- Statistical Methods: Describe the statistical techniques used to analyze the data, such as ANOVA, main effect plots, or half-normal probability plots.
- Results: Present the results of the data analysis, including any significant factors identified and their effects on the response variable.
- Interpretation: Interpret the findings in the context of the research objectives, discussing the implications and significance of the results.

Discussion:
- Comparison with Hypotheses: Discuss how the results compare with the initial hypotheses or expectations.
- Limitations: Acknowledge any limitations or constraints of the study, such as sample size, experimental conditions, or potential sources of error.
- Future Directions: Suggest potential avenues for future research based on the findings of the experiment.

Conclusion: Summarize the key findings of the experiment and their implications for the field of physics. Provide concluding remarks, including any insights gained, practical applications, or recommendations.

References: Cite any sources of information, literature, or previous research referenced in the report.

Appendices: Include any supplementary materials, such as raw data, calculations, or additional analyses, in appendices if necessary.

By including these elements in a report on PBD experiments in physics, researchers can effectively communicate their methods, results, and conclusions to their audience, contributing to the advancement of knowledge in the field.

2.4.13 DERRINGER DESIGN

Derringer designed experiments, also known as D-optimal designs, find applications across various domains within physics. Derringer designed experiments are used in physics for various reasons:

Efficient Experimental Design: Derringer designs are optimized to provide the most information about the factors under investigation with a minimum number of experimental runs. In physics, where experiments can be resource intensive and time-consuming, D-optimal designs offer an efficient way to gather data and maximize the information obtained.

Precision and Accuracy: D-optimal designs are specifically tailored to minimize the variance of parameter estimates, leading to more precise and accurate estimates of the effects of factors on the response variable. In physics, where small variations in experimental conditions can have significant effects on outcomes, precise estimates are crucial for understanding physical phenomena.

Flexibility: Derringer designs can accommodate various types of factors, including continuous, categorical, and mixed factors. This flexibility makes them suitable for a wide range of experiments in physics, where factors, such as temperature, pressure, material properties, and experimental settings, may need to be considered simultaneously.

Optimization of Resources: By efficiently allocating resources and experimental runs, D-optimal designs help physicists make the most of limited resources, such as time, materials, and equipment. This optimization allows researchers to conduct more informative experiments within resource constraints, accelerating scientific progress.

Robustness to Noise and Variability: Derringer designs are robust to sources of noise and variability in the experimental environment. They can effectively deal with random errors and extraneous factors, ensuring that the estimated effects of factors on the response variable are reliable and meaningful.

Exploration of Parameter Space: D-optimal designs allow researchers to systematically explore the parameter space of the factors under investigation, identifying optimal settings and understanding the relationships between factors and outcomes. This exploration is essential for gaining insights into complex physical systems and phenomena.

Statistical Rigor: Derringer designs are grounded in statistical principles and optimization techniques, ensuring that experimental results are statistically valid and interpretable. This rigor is crucial in physics, where empirical observations must be supported by sound statistical analysis to draw meaningful conclusions.

Overall, Derringer designed experiments are used in physics to optimize experimental design, improve the precision and accuracy of parameter estimates, allocate resources efficiently, and gain insights into complex physical phenomena. By leveraging the advantages of D-optimal designs, physicists can conduct more informative and impactful experiments, advancing our understanding of the natural world. Some areas where Derringer designed experiments are commonly used include:

Material Science: In material science, Derringer designed experiments are employed to optimize the fabrication processes of materials, such as semiconductor devices, thin films, nanoparticles, and composite materials. These experiments help identify the key factors influencing material properties like conductivity, strength, thermal properties, and optical properties.

Optics and Photonics: D-optimal designs are utilized in optics and photonics research to optimize the design parameters of optical systems, such as lenses, mirrors, filters, and photonic devices. These experiments aid in maximizing the performance of optical systems, improving characteristics like resolution, efficiency, and spectral response.

Condensed Matter Physics: Researchers in condensed matter physics use Derringer designed experiments to study the properties of condensed matter systems, such as superconductors, magnets, and semiconductors. These experiments help identify the critical factors affecting phenomena like phase transitions, magnetic ordering, and electronic band structures.

Nanotechnology: In nanotechnology, D-optimal designs are applied to optimize the synthesis methods and fabrication processes of nanostructures, such as nanowires, quantum dots, and nanocomposites. These experiments aid in controlling the size, shape, and properties of nanomaterials for various applications in electronics, sensors, and biomedical devices.

Particle Physics: In particle physics experiments, Derringer designed methods are used to optimize experimental setups, detector configurations, and data acquisition strategies. These experiments help maximize the sensitivity of particle detectors, optimize SNRs, and minimize background noise for the detection of rare or elusive particles.

Astrophysics and Cosmology: Astrophysicists and cosmologists employ D-optimal designs to optimize observational strategies, instrument configurations, and data analysis techniques for studying celestial objects and phenomena. These experiments aid in maximizing the information obtained from telescopes, satellites, and other observational platforms.

Quantum Physics: Researchers in quantum physics use Derringer designed experiments to optimize parameters in quantum systems, such as quantum computing architectures, quantum communication protocols, and quantum sensing devices. These experiments help maximize coherence times, minimize decoherence effects, and optimize control over quantum states.

Overall, Derringer designed experiments find applications in various subfields of physics, where they are used to optimize experimental parameters, improve the precision and accuracy of measurements, and gain insights into complex physical phenomena.

STEPS

Developing Derringer designed experiments, also known as D-optimal designs, in physics involves several key steps to optimize the experimental design for gathering informative data while minimizing resource usage. Here are the general steps to develop Derringer designed experiments in physics:

1. **Define the Objectives**: Clearly define the objectives of the experiment, including the specific research questions or hypotheses to be addressed. Determine the response variable(s) of interest that will be measured during the experiment.
2. **Identify Factors and Levels**: Identify the factors (IVs) that may influence the response variable(s) in the experiment. Determine the range of values or levels for each factor that will be considered during the experiment. Factors can include experimental conditions, parameters, settings, or input variables that may affect the outcome.
3. **Select a Model**: Choose an appropriate statistical model that describes the relationship between the factors and the response variable(s). The model should be flexible enough to capture the potential nonlinearities and interactions between factors.
4. **Generate Candidate Designs**: Use statistical software or design generation algorithms to generate a set of candidate D-optimal designs based on the specified factors, levels, and model assumptions. These designs aim to maximize the information content of the experimental data while satisfying any constraints on the design space.
5. **Evaluate Design Efficiency**: Evaluate the efficiency of each candidate design using statistical criteria, such as the D-efficiency criterion, which measures the information content of the design relative to an optimal design. Compare the efficiency of different designs to select the most informative one for the experiment.
6. **Optimize Design Parameters**: Fine-tune the design parameters, such as the number of experimental runs, the allocation of runs to factor combinations, or the choice of factor levels, to further enhance the efficiency of the selected design. Iteratively adjust the design parameters based on the evaluation results until an optimal design is obtained.
7. **Validate Design Assumptions**: Validate the assumptions underlying the selected design, such as the linearity of the model, the normality of the errors, and the absence of multicollinearity between factors. Perform diagnostic checks and sensitivity analyses to ensure that the design is robust and reliable.
8. **Implement Experimental Design**: Implement the selected D-optimal design by conducting the planned experimental runs according to the specified factor combinations and levels. Ensure that the experimental procedures are carried out accurately and consistently to minimize sources of variability and bias.

9. **Collect and Analyze Data**: Collect data on the response variable(s) during the experimental runs, recording the observed outcomes and any relevant covariates or control variables. Analyze the collected data using appropriate statistical methods to estimate the effects of factors and draw conclusions about their significance.

10. **Interpret Results and Draw Conclusions**: Interpret the results of the data analysis in the context of the research objectives, drawing conclusions about the effects of factors on the response variable(s) and their implications for the underlying physical processes. Discuss the limitations of the experiment and potential avenues for future research.

By following these steps, researchers can develop Derringer designed experiments in physics that are optimized to provide informative data, enhance the efficiency of resource utilization, and yield meaningful insights into the underlying physical phenomena.

PITFALLS AND REMEDIES

While Derringer designed experiments offer many advantages in terms of efficiency and information content, there are several pitfalls to be aware of when using these designs in physics. Here are some common pitfalls and potential remedies:

Overfitting: One common pitfall is overfitting the model to the experimental data, leading to overly complex models that may not generalize well to new data. This can occur when too many factors are included in the model relative to the number of experimental runs.

> **Remedy:** Use model selection techniques, such as cross-validation or information criteria (e.g., AIC, BIC), to identify the simplest model that adequately captures the relationship between factors and the response variable. Avoid including unnecessary factors or interactions in the model.

Misspecification of Model: If the assumed model does not accurately represent the true relationship between factors and the response variable, the estimates of factor effects may be biased or misleading.

> **Remedy:** Perform diagnostic checks, such as residual analysis or model adequacy tests, to assess the goodness-of-fit of the model. Consider alternative model structures or transformations of variables if the assumptions of the model are violated.

Confounding Effects: Confounding occurs when the effects of two or more factors are indistinguishable from each other, leading to ambiguity in interpreting the results.

> **Remedy:** Use orthogonal designs whenever possible to minimize confounding between factors. If confounding is unavoidable, conduct follow-up experiments or additional analyses to disentangle the effects of confounded factors.

Nonlinear Relationships: Derringer designs assume linear relationships between factors and the response variable, but real-world relationships may be nonlinear.

> **Remedy:** If nonlinear relationships are suspected, consider using alternative experimental designs, such as RSM or nonlinear regression, to capture nonlinear effects. Conduct sensitivity analyses to assess the robustness of the results to departures from linearity.

Insufficient Sample Size: If the number of experimental runs is too small relative to the number of factors or the complexity of the model, the estimates of factor effects may be imprecise or unreliable.

> **Remedy:** Increase the sample size or reduce the number of factors to achieve adequate power for detecting meaningful effects. Perform power analyses or sample size calculations to determine the minimum sample size required to detect hypothesized effects with sufficient statistical power.

Inadequate Randomization: Inadequate randomization of experimental runs can lead to bias or systematic errors in the estimates of factor effects.

Remedy: Randomize the order of experimental runs or use a randomized block design to ensure that potential sources of variability are evenly distributed across experimental conditions. Implement blinding procedures to minimize experimenter bias.

By being aware of these potential pitfalls and implementing appropriate remedies, researchers can mitigate the risks associated with Derringer designed experiments in physics and ensure the reliability and validity of their experimental results.

EXAMPLE

A Derringer designed experiment, also known as a D-optimal design, aims to efficiently explore the parameter space while maximizing the information obtained from the experimental runs. Here's an example of a matrix for a Derringer designed experiment in physics:

Consider a scenario where researchers want to investigate the effects of three factors (A, B, and C) on a response variable in a physics experiment. The factors could represent different experimental conditions, settings, or parameters that may influence the outcome. The researchers decided to use a D-optimal design to systematically vary the factors and collect data on the response variable.

An example matrix for the Derringer design experiment is shown in Table 2.73.

In this example,

- Each row represents one experimental run.
- Factor levels are coded as −1 (low level), +1 (high level), and 0 (center point).
- The factors A, B, and C are varied at different levels to explore the parameter space efficiently.
- The last two rows (runs 9 and 10) represent center points, which are replicated to estimate experimental error and assess the curvature of the response surface.

Researchers would conduct the experimental runs according to the specified factor levels and record the response variable for each run. By analyzing the data collected from these experimental runs, they can estimate the effects of the factors on the response variable and optimize the experimental conditions to achieve the desired outcome. This illustrates how a Derringer design matrix can be constructed to efficiently explore the parameter space and maximize the information obtained from a physics experiment.

TABLE 2.73

Derringer Designed Experiment Matrix for Physics

Run	Factor A	Factor B	Factor C
1	−1	−1	−1
2	+1	−1	−1
3	−1	+1	−1
4	+1	+1	−1
5	−1	−1	+1
6	+1	−1	+1
7	−1	+1	+1
8	+1	+1	+1
9	0	0	0
10	0	0	0

An example of a Derringer designed experiment in physics could involve optimizing the parameters of a laser system for maximum efficiency or output power. Let's consider a scenario where researchers want to optimize the performance of a diode-pumped solid state (DPSS) laser by adjusting several key parameters:

1. **Pump Power (Factor A)**: The power of the pump diode laser, which excites the gain medium of the DPSS laser.
2. **Crystal Temperature (Factor B)**: The temperature of the laser crystal, which affects its thermal properties and optical characteristics.
3. **Resonator Length (Factor C)**: The length of the optical resonator cavity, which determines the wavelength and spectral characteristics of the laser output.

The researchers aim to maximize the output power of the DPSS laser while minimizing the variations in output power due to changes in environmental conditions or component tolerances. They decide to use a Derringer designed experiment to efficiently explore the parameter space and identify the optimal settings for the factors.

Based on their preliminary knowledge and experimental constraints, the researchers generate a D-optimal design matrix with a predetermined number of experimental runs. Each run corresponds to a unique combination of factor levels, and the response variable is the measured output power of the DPSS laser.

An example of a simplified Derringer designed matrix for this experiment is shown in Table 2.74. In this example,

- Each row represents one experimental run with a specific combination of factor levels.
- The pump power, crystal temperature, and resonator length are varied at different levels to explore the parameter space.
- The response variable (output power of the DPSS laser) is measured for each experimental run.

After conducting the experiments and collecting data on the output power for each combination of factors, the researchers analyze the results using statistical methods to estimate the effects of the factors and identify the optimal settings for maximizing laser performance.

This example illustrates how a Derringer designed experiment can be used in physics to efficiently optimize experimental parameters and enhance the performance of complex systems like lasers.

REPORTING

In Derringer designed experiments, various statistical tools, and techniques are used to analyze the data and draw conclusions about the effects of factors on the response variable(s). Some common statistical tools used in Derringer designed experiments for physics include:

TABLE 2.74

Derringer Designed Experiment Matrix for Physics

Run	Pump Power (W)	Crystal Temperature (°C)	Resonator Length (cm)
1	10	20	5
2	20	20	10
3	10	40	10
4	20	40	5
5	15	30	7
6	15	30	7

ANOVA: ANOVA is used to partition the total variability in the response variable into components attributable to different factors and their interactions. It helps determine whether the observed differences between factor levels are statistically significant.

Regression Analysis: Regression analysis is used to fit a statistical model to the experimental data, relating the response variable to the factors and their interactions. This allows researchers to estimate the effects of individual factors and identify nonlinear relationships between factors and the response variable.

Main Effects Plots: Main effects plots visualize the effects of each factor on the response variable while holding other factors constant. They provide insights into the relative importance of different factors and help identify which factors have the greatest impact on the response variable.

Interaction Plots: Interaction plots visualize the interactions between different factors on the response variable. They help identify synergistic or antagonistic effects between factors and assess whether the effects of one factor depend on the level of another factor.

Half-Normal Probability Plots: Half-normal probability plots are used to assess the significance of factor effects and identify influential factors. They plot the absolute values of the estimated effects against their expected values under the null hypothesis of no effect, allowing researchers to identify factors with effects significantly larger than expected by chance.

Diagnostic Checks: Diagnostic checks, such as residual analysis, leverage statistics to assess the validity of model assumptions and the adequacy of the fitted model. They help identify potential outliers, influential data points, or violations of model assumptions that may affect the reliability of the results.

Model Selection Criteria: Model selection criteria, such as AIC or BIC, are used to compare different models and select the most parsimonious model that adequately represents the relationship between factors and the response variable.

Optimization Algorithms: Optimization algorithms are used to identify the optimal settings of factors for achieving specific objectives, such as maximizing the response variable or minimizing variability. They iteratively search the parameter space to find the combination of factor levels that optimize the desired outcome.

By applying these statistical tools and techniques, researchers can effectively analyze Derringer designed experiment data, estimate factor effects, identify significant factors, and optimize experimental parameters to achieve desired outcomes in physics research.

When reporting the results of Derringer designed experiments in physics, it's important to provide a comprehensive overview of the experiment, the analysis performed, and the conclusions drawn. Here are the key elements typically included in a report for Derringer designed experiments in physics:

Introduction: Provide background information on the research problem or question addressed by the experiment. Describe the objectives of the experiment and the rationale for using a Derringer design.

Experimental Design: Describe the factors and levels investigated in the experiment. Explain how the Derringer design was generated and why it was chosen. Provide details on the number of experimental runs, randomization procedures, and any other relevant experimental details.

Experimental Procedure: Outline the steps taken to conduct the experiment, including any equipment used and measurement techniques employed. Describe how the factor levels were implemented during the experimental runs.

Data Collection: Specify the response variable(s) measured during the experiment. Describe how data were collected, including any controls or replication used.

Data Analysis: Describe the statistical methods and techniques used to analyze the data, such as ANOVA, regression analysis, or graphical methods. Present the results of the analysis, including estimates of factor effects, significance tests, and diagnostic checks.

Results: Present the main findings of the experiment, focusing on the effects of factors on the response variable(s). Include tables, graphs, or plots to visually represent the results, such as main effects plots, interaction plots, or half-normal probability plots.

Discussion: Interpret the results in the context of the research objectives and the underlying physics. Discuss the implications of the findings and any insights gained from the experiment. Address the limitations of the experiment and potential sources of error or bias.

Conclusions: Summarize the main conclusions drawn from the experiment. Discuss the practical implications of the findings and any recommendations for future research or applications.

References: Cite any relevant literature, references, or previous studies that informed the design or interpretation of the experiment.

Appendices: Include any supplementary information, data tables, or additional analyses that support the main findings of the report.

By including these elements in the report, researchers can effectively communicate the design, implementation, analysis, and results of Derringer designed experiments in physics, providing a clear and comprehensive understanding of the research conducted and its implications.

2.5 PSYCHOLOGY

Designed experiments are crucial in psychology for studying behavior, testing hypotheses, and advancing our understanding of psychological processes. Here are some common types of designed experiments with examples of their applications in psychology.

In psychology designed experiments, various types of variables are considered to investigate human behavior, cognition, emotions, and mental processes. These variables can be categorized into several common types, including

IVs:

- **Experimental Conditions**: Variables related to the experimental treatments, interventions, or conditions applied to study participants, such as the presentation of stimuli, task instructions, or exposure to specific situations.
- **Stimulus Characteristics**: Variables related to the characteristics of stimuli presented to participants, including type, intensity, duration, and complexity.
- **Manipulated Factors**: Variables that researchers intentionally manipulate to study their effects, such as the presence or absence of a specific feature in a task or the level of an IV (e.g., dosage, time, or difficulty).
- **Participant Characteristics**: Variables related to the characteristics of the study participants, including age, gender, personality traits, cognitive abilities, or clinical diagnoses (for clinical psychology experiments).

DVs:

- **Behavioral Responses**: Variables related to observable behaviors, actions, or reactions of study participants, such as response times, accuracy, choice preferences, and task performance.
- **Psychological Measures**: Variables related to psychological assessments, including self-report questionnaires, personality assessments, and mood scales.
- **Physiological Measures**: Variables related to physiological responses, such as heart rate, brain activity [measured by electroencephalogram (EEG) or functional magnetic resonance imaging (fMRI)], skin conductance, or eye movements.

- **Emotional States**: Variables related to emotional states or mood changes, assessed through self-report, facial expression analysis, or other measures.

Categorical Variables:

- **Experimental Groups**: Categories of participants assigned to different experimental conditions or treatment groups, such as control groups, experimental groups, or placebo groups.
- **Demographic Variables**: Categories of demographic characteristics, such as age groups, gender categories, and education levels, which can influence behavior and responses.

Control Variables (Covariates): These are variables that are held constant or controlled during experiments to eliminate their influence on the DV. For example, controlling the time of day, environmental conditions, or participant characteristics to ensure experimental consistency.

Random Variables: In some cases, random variables may be introduced to account for variability or uncertainty in psychological measurements or to control participant variability.

Interaction Variables: Interaction variables are used to investigate the combined effects of two or more IVs on the DV, helping to understand how different factors may interact to influence behavior or outcomes.

Noise Variables (Error Variables): These are uncontrolled or unmeasured variables that can introduce variability or errors into the experimental results. Techniques like statistical analysis and experimental controls are used to minimize the impact of noise variables in psychology experiments.

Psychology designed experiments are crucial for advancing knowledge in areas, such as cognitive psychology, social psychology, clinical psychology, and behavioral neuroscience. Careful control and manipulation of variables are essential for drawing meaningful conclusions about human behavior and mental processes.

There are several experimental methods that can be used to investigate variables for a response. When there are at least two variables at play, a **full factorial** design is common due to it being efficient for exploring all combinations and interactions but can be costly for many factors. Another method is **the 2-factor CCD.** This method is more advanced, allows for model building and optimization with curvature exploration. When there are three or more factors, the following methods and rationale are used:

- **3-Factor CCD:** More advanced, allows for model building and optimization with curvature exploration.
- **BBD:** Efficient for exploring quadratic terms without requiring as many runs as a full factorial.
- **PBD:** Useful for screening many factors with limited resources, but only provides information about main effects.
- **Fractional Factorial Design:** Efficient for screening and identifying key factors, requiring fewer runs than a full factorial.
- **Derringer Design:** Useful for optimizing multiple responses simultaneously when interactions are important.

The following are experimental designs, including the examples commonly used, steps for implementation, potential pitfalls and remedies, examples, and suggested reporting structure.

2.5.1 FACTORIAL EXPERIMENTS

Factorial design experiments are widely used in psychology for several reasons. These methods are valuable tools in psychology for efficiently exploring complex relationships between multiple variables, enhancing the generalizability and robustness of research findings, and advancing our understanding of human behavior and cognition. Several examples include:

Efficient Exploration of Interactions: Psychology often involves studying complex behaviors influenced by multiple factors. Factorial designs allow researchers to systematically manipulate and examine the effects of multiple IVs (factors) and their interactions with DVs (outcomes). By varying multiple factors simultaneously, factorial designs enable researchers to efficiently explore how different factors interact to influence behavior.

Enhanced Generalizability: Factorial designs enhance the external validity or generalizability of research findings by allowing researchers to investigate how the effects of IVs may vary across different levels of other IVs. This enables researchers to test the robustness of their findings across diverse conditions, populations, or contexts, leading to more reliable and generalizable conclusions.

Control of Confounding Variables: Factorial designs enable researchers to control confounding variables by systematically varying and manipulating multiple factors simultaneously. By controlling potential confounds, factorial designs help ensure that observed effects are attributable to the manipulated IVs rather than extraneous variables, thus enhancing the internal validity of the research.

Efficient Use of Resources: Factorial designs offer more efficient use of resources (e.g., time, participants) compared to conducting separate experiments for each factor or combination of factors. By simultaneously testing multiple factors within the same experiment, researchers can achieve a higher degree of experimental control and statistical power while minimizing participant burden and research costs.

Detection of Moderating Effects: Factorial designs are particularly well-suited for detecting moderating effects, where the strength or direction of the relationship between independent and DVs varies depending on the level of another variable. By systematically manipulating multiple factors, factorial designs enable researchers to identify conditions under which certain effects may be more pronounced or attenuated, leading to a deeper understanding of the underlying mechanisms of behavior.

Factorial design experiments are used in various areas within psychology to investigate the effects of multiple IVs on DVs. Some common areas where factorial designs are employed include:

Cognitive Psychology: In cognitive psychology, factorial designs are used to investigate factors influencing cognitive processes, such as memory, attention, perception, and decision-making. Researchers may manipulate factors, such as stimulus type, task difficulty, and cognitive load, to understand how these variables interact and affect cognitive performance.

Social Psychology: In social psychology, factorial designs are utilized to examine the effects of multiple factors on social behaviors, attitudes, and interpersonal relationships. Researchers may investigate variables, such as group membership, situational context, and individual differences, to understand how they interact to shape social behavior and attitudes.

Developmental Psychology: In developmental psychology, factorial designs are employed to study the effects of various environmental and genetic factors on developmental outcomes across different stages of the lifespan. Researchers may manipulate factors, such as parenting style, socioeconomic status, and genetic predispositions, to understand their combined effects on developmental trajectories.

Clinical Psychology: In clinical psychology, factorial designs are used to investigate the efficacy of psychological interventions and treatments for mental health disorders. Researchers may manipulate treatment type, dosage, and delivery format to examine their effects on symptom reduction, functional outcomes, and treatment adherence.

Experimental Psychology: In experimental psychology, factorial designs serve as a fundamental tool for studying basic psychological processes and mechanisms underlying behavior. Researchers may manipulate factors, such as reinforcement schedule, stimulus salience, and response contingencies, to investigate how these variables interact to produce behavioral outcomes.

Health Psychology: In health psychology, factorial designs are utilized to investigate the effects of behavioral, environmental, and psychosocial factors on health-related outcomes and behaviors. Researchers may examine variables, such as stress, social support, and health behaviors, to understand their interactive effects on physical and mental health.

Overall, factorial design experiments are widely used across various subfields of psychology to address complex research questions, elucidate the interplay between multiple factors, and advance our understanding of human behavior and psychological processes.

STEPS

Developing factorial design experiments in psychology involves several key steps to ensure the research is well designed and conducted effectively. Here are the steps typically followed:

1. **Identify Research Question**: Clearly define the research question or hypothesis that the experiment aims to address. This should involve specifying the variables of interest and their expected relationships.
2. **Select Factors and Levels**: Identify the IVs (factors) that will be manipulated in the experiment. Determine the different levels or conditions of each factor to be tested. Factors should be chosen based on theoretical relevance and practical feasibility.
3. **Design Experimental Conditions**: Determine the factorial design structure, including the arrangement of factors and levels in the experiment. Decide whether to use a full factorial design (testing all possible combinations of levels) or a fractional factorial design (testing a subset of combinations to reduce the number of experimental conditions).
4. **Randomization and Counterbalancing**: Implement randomization procedures to ensure that participants are assigned to experimental conditions in a random and unbiased manner. Consider using counterbalancing techniques to control order effects and other potential confounds.
5. **Develop Stimuli and Materials**: Create or select appropriate stimuli, materials, or tasks for the experiment. Ensure that stimuli are valid, reliable, and ethically appropriate for the study population.
6. **Recruit Participants**: Determine the sample size needed to detect the effects of interest with sufficient statistical power. Recruit participants who are representative of the target population and obtain informed consent following ethical guidelines.
7. **Conduct Experiment**: Administer the experimental procedures to participants according to the factorial design. Ensure that experimental conditions are implemented consistently and that data collection procedures are standardized across participants.
8. **Collect Data**: Gather data on the DVs of interest for each experimental condition. Use appropriate measurement techniques and ensure data quality and accuracy.
9. **Data Analysis**: Analyze the data using appropriate statistical methods, such as ANOVA for factorial designs. Examine the main effects and interactions between factors to determine their significance and interpret the results in relation to the research question.

10. **Interpret and Report Findings**: Interpret the results of the data analysis in the context of the research question and relevant theory. Prepare a clear and comprehensive report summarizing the study's methods, findings, and implications for the field of psychology.

By following these steps, researchers can develop well-designed and rigorous factorial design experiments in psychology that contribute to the understanding of human behavior and psychological processes.

PITFALLS AND REMEDIES

Factorial design experiments in psychology can encounter various pitfalls that may affect the validity and interpretation of the results. Here are some common pitfalls and remedies to consider:

Main Effect Confusion:
> **Pitfall:** Incorrectly interpreting main effects without considering interactions. Researchers may focus solely on the main effects without recognizing that interactions between factors could be influencing the DV.
>
> **Remedy**: Always examine interactions between factors in addition to the main effects. Interpret the main effects cautiously and consider how they may be influenced by interactions with other factors.

Interpretation of Interactions:
> **Pitfall:** Misinterpreting interactions between factors. Researchers may incorrectly interpret interactions as indicating a causal relationship when they may simply reflect complex patterns of influence.
>
> **Remedy:** Use caution when interpreting interactions and consider potential explanations for observed patterns. Conduct post-hoc analyses or follow-up studies to further explore and validate interaction effects.

Multiple Testing Issues:
> **Pitfall:** Conducting multiple comparisons without adjusting for the increased risk of type I error. Researchers may inadvertently inflate the likelihood of finding significant results by conducting numerous comparisons.
>
> **Remedy:** Apply appropriate corrections for multiple comparisons, such as Bonferroni correction or FDR control, to control the overall type I error rate. Preregister hypotheses and analysis plans to reduce the temptation to engage in post-hoc testing.

Small Sample Sizes:
> **Pitfall:** Insufficient sample sizes to detect effects of interest or interactions. Small sample sizes can lead to underpowered studies, increasing the risk of type II errors and reducing the reliability of findings.
>
> **Remedy:** Conduct power analyses to determine the required sample size to detect effects of interest with adequate statistical power. Increase sample sizes through recruitment efforts or consider conducting a replication study to confirm findings.

Selective Reporting Bias:
> **Pitfall:** Selectively reporting only significant results while omitting nonsignificant findings. This can lead to an inaccurate portrayal of the true effects and inflate the apparent strength of relationships.
>
> **Remedy:** Practice transparent and comprehensive reporting of all study results, including both significant and nonsignificant findings. Preregister study protocols and analysis plans to mitigate the temptation to selectively report results.

Failure to Consider External Validity:
> **Pitfall:** Overemphasizing internal validity at the expense of external validity. Researchers may design experiments that lack ecological validity or fail to generalize to real-world contexts.

Remedy: Consider both internal and external validity when designing experiments. Use experimental manipulations and measures that closely approximate real-world phenomena. Conduct follow-up studies or field experiments to confirm findings in ecologically valid settings.

Ignoring Assumptions of ANOVA:

Pitfall: Failing to meet the assumptions of ANOVA, such as homogeneity of variance or normality of residuals. Violating these assumptions can lead to inaccurate conclusions and biased estimates.

Remedy: Verify that the data meet the assumptions of ANOVA using diagnostic tests or visual inspection of residuals. Apply appropriate transformations or use robust statistical methods if assumptions are violated. Consider alternative analysis approaches, such as nonparametric tests, if ANOVA assumptions cannot be met.

By being aware of these potential pitfalls and implementing appropriate remedies, researchers can conduct factorial design experiments in psychology that yield valid, reliable, and interpretable results.

EXAMPLE

Factorial experiments in psychology involve studying the effects of multiple IVs (factors) on a DV. Here's an example of a matrix for a 2×2 factorial experiment in psychology:

Objective: To investigate the effects of two IVs, gender (factor A) and stress level (factor B), on the performance of participants in a memory recall task.

Factors:

Gender (Factor A):
- Male.
- Female.

Stress Level (Factor B):
- Low stress.
- High stress.

IVs:

- Gender: Male (M) vs. female (F).
- Stress Level: Low (L) vs. high (H).

DV: Number of correctly recalled items i.e., the cognitive performance measured by a standardized cognitive test, such as a memory recall task or reaction time task.

Factorial Experiment Matrix:

In a 2×2 factorial experiment, there are four experimental conditions created by the combination of two levels of each IV. The matrix is shown in Table 2.75.

In this matrix, each row represents a specific participant's performance in one of the four experimental conditions defined by the combination of gender and stress level. The number of correctly recalled items is measured for each participant.

Experimental Design: This experiment would use a 2×2 factorial design, with two levels of gender and two levels of stress level. Each participant would be randomly assigned to one of the four experimental conditions:

TABLE 2.75
Factorial Experiment Design Matrix for Psychology

Participant	Gender (A)	Stress Level (B)	Correctly Recalled Items
1	Male	Low Stress	...
2	Male	High Stress	...
3	Female	Low Stress	...
4	Female	High Stress	...

- Male participants exposed to low stress (ML).
- Male participants exposed to high stress (MH).
- Female participants exposed to low stress (FL).
- Female participants exposed to high stress (FH).

Participants in each group would then complete the cognitive performance task under their assigned conditions. The experiment would involve multiple trials or measures of cognitive performance to obtain reliable data.

Another example where we have a 2×2 factorial design with two IVs, A and B, each with two levels:

IV A:

- Level 1: Low stress (LS).
- Level 2: High stress (HS).

IV B:

- Level 1: Positive mood induction (PM).
- Level 2: Negative mood induction (NM).

The matrix for this factorial design experiment is shown in Table 2.76.

Each row represents a participant, and the columns represent the levels of the IVs A and B that each participant experiences. In this 2×2 factorial design, there are four unique combinations of conditions, resulting in four experimental groups.

Researchers would randomly assign participants to one of these four conditions, and each participant would undergo the corresponding manipulation of stress level and mood induction. This design allows for the examination of the main effects of each IV as well as potential interactions between them.

TABLE 2.76
Factorial Experiment Design Matrix for Psychology

Participant	IV A	IV B
1	LS	PM
2	LS	NM
3	HS	PM
4	HS	NM

Data Analysis: After collecting the data, researchers would conduct a factorial ANOVA to examine the main effects of gender and stress level, as well as any interactions between these factors. The main effects would indicate the overall impact of each IV on cognitive performance, while interactions would reveal whether the effects of one variable depend on the levels of the other variable.

For example, the analysis might reveal a significant main effect of gender, indicating that males generally perform better than females on cognitive tasks. Additionally, there might be a significant interaction between gender and stress level, suggesting that the effect of stress on cognitive performance differs between males and females.

Interpretation: The results of the factorial ANOVA would allow researchers to draw conclusions about how gender and stress level independently and interactively influence cognitive performance. This could have implications for understanding gender differences in response to stress and inform interventions aimed at mitigating the negative effects of stress on cognitive function.

Researchers can use this design to analyze the main effects of gender and stress level on memory recall and investigate potential interactions between these factors. Factorial experiments allow psychologists to study how multiple factors simultaneously influence behavior or outcomes.

REPORT

Factorial design experiments in psychology often use various statistical tools to analyze the data and interpret the results. Here are some common statistical tools used in factorial design experiments in psychology:

Factorial ANOVA: ANOVA is a widely used statistical technique for comparing means across multiple groups. In factorial designs, factorial ANOVA allows researchers to examine the main effects of each IV and interactions between them.

Post-hoc Tests: After conducting a factorial ANOVA, post-hoc tests, such as Tukey's HSD, Bonferroni, or least significant difference (LSD), can be used to identify specific differences between groups when significant main effects or interactions are found.

Simple Effects Analysis: This analysis examines the effect of one IV at each level of another IV. It helps to clarify the nature of significant interactions by determining if the effect of one variable differs at different levels of the other variable.

Contrast Analysis: Contrast analysis allows researchers to test specific hypotheses about the mean differences between groups or conditions. Contrasts can be defined based on theoretical predictions or specific research questions.

Factorial MANOVA: When there are multiple DVs, factorial MANOVA allows researchers to simultaneously analyze the effects of multiple IVs on multiple dependent variables.

Factorial Regression Analysis: Regression analysis can be used to examine the relationship between continuous DVs and one or more categorical IVs in factorial designs.

Mediation and Moderation Analysis: These analyses help to understand the mechanisms underlying observed effects. Mediation analysis examines whether the effect of an IV on a DV operates through a third variable (mediator), while moderation analysis investigates whether the strength or direction of the relationship between variables depends on the level of a moderating variable.

Factorial Mixed-Design ANOVA: This analysis is used when factorial designs involve both between-subjects factors and within-subjects factors. It allows researchers to examine the main effects and interactions of both types of factors.

These statistical tools help researchers analyze the complex relationships between independent and DVs in factorial design experiments and draw valid conclusions about the effects of experimental manipulations in psychology.

In a report on a factorial design experiment in psychology, several key elements should be included to provide a comprehensive understanding of the study's design, methods, results, and conclusions. Here are the essential elements typically included in such a report:

Title: A concise and informative title that accurately reflects the content and focus of the study.

Abstract: A brief summary of the study, including the research question, methods, results, and conclusions. The abstract should provide enough information for readers to understand the purpose and findings of the study quickly.

Introduction: A section that outlines the background and context of the study, including a review of relevant literature, the research question or hypothesis, and the rationale for conducting the experiment. The introduction should establish the importance of the research topic and justify the use of a factorial design.

Method: This section provides a detailed description of the experimental design, including information about participants, materials, procedures, and data collection methods. It should specify the levels of each IV, how they were manipulated, and how DVs were measured.

Results: A clear presentation of the findings, including descriptive statistics, inferential statistics (e.g., ANOVA results), and any relevant post-hoc analyses. Results should be organized logically and presented in a way that makes them easy to interpret. Tables and figures may be used to summarize and illustrate key findings.

Discussion: An interpretation and discussion of the results in the context of the research question, theoretical framework, and previous literature. This section should address the main effects of each IV, interactions between variables, and any unexpected findings. Strengths and limitations of the study should also be discussed.

Conclusion: A brief summary of the main findings of the study and their implications for theory, research, and practice. The conclusion should highlight the significance of the findings and suggest directions for future research.

References: A list of all sources cited in the report, formatted according to the appropriate style guide (e.g., APA, MLA).

Appendices: Additional materials that provide supplementary information about the study, such as copies of survey instruments, raw data, or detailed descriptions of experimental procedures.

By including these elements in a factorial design experiment report, researchers can effectively communicate their study's purpose, methods, findings, and implications to their audience, contributing to the advancement of knowledge in psychology.

2.5.2 RSM

RSM designed experiments are used in psychology for several reasons. RSM designed experiments offer a powerful and flexible approach for investigating the relationships between multiple variables in psychology, leading to a deeper understanding of psychological processes and more effective interventions through:

Optimization of Experimental Conditions: RSM allows researchers to optimize experimental conditions by systematically varying multiple IVs and observing their effects on one or more DVs. In psychology, this can be particularly useful when studying complex phenomena or interventions where multiple factors interact to produce outcomes.

Understanding Interactions: RSM helps researchers understand how different variables interact with each other to influence outcomes. By mapping the response surface, researchers can identify regions where the outcome variable is optimized and determine the optimal combination of variables.

Efficient Use of Resources: RSM allows researchers to conduct experiments more efficiently by minimizing the number of experimental runs needed to obtain meaningful results. This is especially beneficial in psychology, where conducting experiments with human participants can be time-consuming and resource intensive.

Quantitative Analysis: RSM provides a quantitative framework for analyzing experimental data, allowing researchers to make statistical inferences about the relationships between variables and the shape of the response surface. This helps to ensure that findings are robust and reliable.

Modeling Complex Relationships: RSM enables researchers to model and visualize complex relationships between variables, including nonlinear and interactive effects. This can provide valuable insights into the underlying mechanisms driving psychological phenomena.

Optimization of Interventions: In applied psychology, such as clinical psychology or organizational psychology, RSM can be used to optimize interventions or treatments by identifying the most effective combination of intervention components or environmental factors.

RSM designed experiments can be used in various areas of psychology where researchers aim to understand and optimize complex relationships between multiple variables. Some specific areas where RSM may be applied in psychology include:

Clinical Psychology: RSM can be used to optimize treatment protocols or interventions for psychological disorders. Researchers may investigate the effects of different therapeutic techniques, dosages, or combinations of treatments on patient outcomes.

Experimental Psychology: In experimental psychology, RSM can be used to study the effects of multiple IVs on psychological processes or behaviors. Researchers may manipulate factors, such as task difficulty, stimulus presentation parameters, or environmental conditions to optimize experimental designs.

Human Factors Psychology: RSM can be applied in human factors research to optimize the design of interfaces, products, or environments to enhance human performance and usability. Researchers may explore the effects of various design factors on user satisfaction, task performance, or error rates.

Organizational Psychology: In organizational psychology, RSM can be used to optimize work processes, job design, or organizational interventions. Researchers may investigate the effects of different organizational structures, leadership styles, or training programs on employee outcomes, such as job satisfaction, productivity, or turnover.

Health Psychology: RSM can be used to optimize health promotion interventions, behavior change techniques, or lifestyle interventions. Researchers may explore the effects of various intervention components, dosage levels, or delivery formats on health behaviors and outcomes.

Cognitive Psychology: RSM can be applied in cognitive psychology to optimize experimental stimuli or task parameters. Researchers may manipulate factors, such as stimulus complexity, presentation timing, or task difficulty to optimize experimental designs and maximize the sensitivity of cognitive tasks.

Social Psychology: In social psychology, RSM can be used to optimize experimental manipulations or study designs for investigating social processes and interactions. Researchers may explore the effects of various situational factors, interpersonal variables, or experimental conditions on social behavior and attitudes.

Overall, RSM designed experiments can be utilized across various subfields of psychology to optimize experimental designs, interventions, or organizational processes, leading to a deeper understanding of psychological phenomena and more effective practical applications.

STEPS

Developing RSM designed experiments in psychology involves several steps to systematically investigate the relationships between multiple variables and optimize experimental conditions. Here are the general steps:

1. **Identify the Research Question**: Clearly define the research question or objective of the study. Determine the variables of interest and specify the range or levels of each variable to be investigated.
2. **Select the Experimental Design**: Choose an appropriate experimental design based on the number of factors and levels involved. RSM typically uses a CCD or BBD for optimizing responses within a limited number of experimental runs.
3. **Design the Experiment**: Determine the number of experimental runs required based on the selected design and the desired precision of the response surface. Randomize the order of experimental runs to minimize bias and confounding effects.
4. **Conduct the Experiment**: Perform the experimental runs according to the designed experimental matrix. Systematically vary the levels of each factor and record the corresponding responses or outcomes.
5. **Collect Data**: Gather data on the responses or outcomes measured in each experimental run. Ensure that data collection procedures are standardized and reliable to minimize measurement error.
6. **Analyze the Data**: Use statistical analysis techniques to analyze the data and construct the response surface. Fit a regression model to the data to estimate the coefficients of the polynomial equation representing the response surface.
7. **Interpret the Results**: Interpret the response surface to understand the relationships between the IVs and the response variable(s). Identify significant main effects, interactions, and regions of optimal response.
8. **Optimize Experimental Conditions**: Use the response surface model to identify the optimal combination of factor levels that maximize or minimize the response variable(s). Conduct confirmatory experiments if necessary to validate the optimized conditions.
9. **Validate the Model**: Validate the response surface model using additional experimental data or external validation data. Assess the predictive accuracy and generalizability of the model to ensure its reliability.
10. **Report the Findings**: Summarize the findings of the RSM designed experiment in a comprehensive report. Include details about the research question, experimental design, data collection procedures, statistical analysis methods, results, interpretations, and conclusions.

By following these steps, researchers can systematically develop and execute RSM designed experiments in psychology to optimize experimental conditions, understand complex relationships between variables, and advance scientific knowledge in the field.

PITFALLS AND REMEDIES

RSM designed experiments in psychology, like any other research method, come with their own set of pitfalls. Here are some common pitfalls and potential remedies:

Overfitting: Overfitting occurs when the response surface model fits the observed data too closely, capturing random noise rather than true underlying patterns. This can lead to poor generalization of new data and inaccurate predictions.

    Remedy: Validate the response surface model using techniques, such as cross-validation or split-sample validation. Use model selection criteria (e.g., AIC, BIC) to choose the simplest model that adequately explains the data without overfitting.

Model Misspecification: If the assumed functional form of the response surface model does not accurately represent the true relationship between the variables, the resulting model may be biased or inefficient.

    Remedy: Conduct model diagnostics to assess the adequacy of the chosen model. Consider alternative model forms or transformations of the variables if the initial model does not fit the data well. Incorporate prior knowledge or theoretical considerations to guide model selection.

Violation of Assumptions: Response surface models often rely on assumptions, such as linearity, independence, and homoscedasticity. Violations of these assumptions can lead to biased parameter estimates and incorrect inferences.

    Remedy: Check for violations of model assumptions using diagnostic plots and statistical tests. Transform variables or use robust estimation techniques to mitigate violations. Consider alternative modeling approaches that are more robust to violations of assumptions.

Experimental Error: Measurement error, confounding variables, or other sources of experimental error can distort the observed relationships between variables and lead to biased parameter estimates.

    Remedy: Minimize experimental error by carefully controlling experimental conditions, standardizing measurement procedures, and randomizing the order of experimental runs. Use appropriate statistical techniques (e.g., ANCOVA) to adjust for confounding variables.

Optimization Challenges: Identifying the global optimum of the response surface and ensuring robustness to local optima can be challenging, especially in high-dimensional spaces with complex response surfaces.

    Remedy: Conduct sensitivity analyses to assess the robustness of the optimized conditions to changes in model parameters or experimental assumptions. Consider conducting multiple optimizations or using alternative optimization algorithms to explore a wider range of potential solutions.

Interpretation Complexity: Response surface models can involve complex interactions between variables, making interpretation challenging, especially for nonspecialist audiences.

    Remedy: Provide clear and concise interpretations of the response surface model parameters and their practical implications. Use visual aids, such as contour plots or surface plots, to illustrate the relationships between variables in a more intuitive manner.

By being aware of these pitfalls and implementing appropriate remedies, researchers can enhance the validity, reliability, and interpretability of RSM designed experiments in psychology.

EXAMPLE

RSM is typically used to optimize a response variable by studying the effects of multiple IVs while considering interactions. Here's an example of a matrix for an RSM designed experiment in psychology:

Objective: To optimize the effectiveness of a cognitive training program by studying the effects of two IVs, training duration (factor A) and training frequency (factor B), on the improvement in participants' working memory (WM) scores.

Factors:

1. **Training Duration (Factor A):**
   - Short duration (e.g., 20 minutes).
   - Medium duration (e.g., 40 minutes).
   - Long duration (e.g., 60 minutes).
2. **Training Frequency (Factor B):**
   - Low frequency (e.g., 2 sessions per week).
   - Medium frequency (e.g., 4 sessions per week).
   - High frequency (e.g., 6 sessions per week).

Response Variable: Improvement in working memory score.

RSM Experiment Matrix:

In an RSM designed experiment, the matrix is designed to systematically explore the combinations of IVs (factors) within a specified range. The matrix is shown in Table 2.77.

In this matrix, each row represents a specific experimental run with a unique combination of training duration and training frequency levels. Improvement in WM score is measured for each combination.

RSM allows psychologists to optimize the cognitive training program by studying the response surface and identifying the optimal combination of training duration and training frequency that results in the greatest improvement in WM scores while considering potential interactions between the two factors.

REPORTING

In RSM designed experiments used in psychology, various statistical tools are employed to analyze the data and draw conclusions regarding the relationship between IVs and the response variable. Some of the commonly used statistical tools in RSM designed experiments in psychology include:

TABLE 2.77

RSM Example for Psychology

Run	Training Duration (A)	Training Frequency (B)	Improvement in WM Score
1	Short	Low frequency	...
2	Short	Medium frequency	...
3	Short	High frequency	...
4	Medium	Low frequency	...
5	Medium	Medium frequency	...
6	Medium	High frequency	...
7	Long	Low frequency	...
8	Long	Medium frequency	...
9	Long	High frequency	...

ANOVA: ANOVA is used to assess the significance of the factors and their interactions with the response variable. It helps determine whether there are statistically significant differences between the means of different groups or treatments.

Regression Analysis: Regression analysis is used to model the relationship between the IVs and the response variable. In RSM, polynomial regression models are often used to fit the data and predict the response over the experimental region.

DOE: DOE techniques are employed to design the experimental setup in RSM. This includes selecting appropriate factorial designs, such as full factorial, fractional factorial, or CCD, to efficiently explore the response surface.

Optimization Techniques: Optimization methods are used to identify the optimal settings of the IVs that maximize or minimize the response variable. Techniques, such as desirability functions or numerical optimization algorithms, are often employed for this purpose.

Response Surface Analysis: This involves fitting a response surface model to the experimental data to visualize the relationship between the IVs and the response variable. Response surface plots and contour plots are commonly used to visualize the response surface and identify regions of interest.

Diagnostic Tools: Various diagnostic tools, such as residual analysis, normal probability plots, and leverage plots, are used to assess the adequacy of the fitted models and detect any outliers or influential data points.

Statistical Software: Specialized statistical software packages, such as R, SAS, or JMP, are often used to perform the statistical analysis and generate graphical outputs in RSM designed experiments in psychology.

By employing these statistical tools, researchers can effectively analyze the experimental data, identify significant factors, and optimize the experimental conditions to achieve the desired response in psychological research.

In a report on RSM designed experiments used in psychology, several key elements should be included to effectively communicate the research findings and methodology. These elements typically include:

Title: A descriptive and concise title that clearly reflects the content of the study.

Abstract: A brief summary of the study objectives, methods, results, and conclusions. The abstract should provide an overview of the entire report.

Introduction: An introduction that provides background information on the research problem, outlines the objectives of the study, and explains the rationale for using RSM in the experimental design.

Literature Review: A review of relevant literature and previous studies related to the research topic. This section helps contextualize the study within the broader field of psychology and highlights the gaps in existing knowledge that the current study aims to address.

Methodology: A detailed description of the experimental design, including the selection of IVs, response variables, and the RSM approach used. This section should also describe the data collection procedures, experimental conditions, and any statistical techniques employed.

Results: A presentation of the findings from the RSM designed experiments, including the response surface plots, contour plots, and statistical analyses. This section should include both descriptive and inferential statistics to summarize the data and highlight any significant effects or trends.

Discussion: An interpretation of the results in the context of the study objectives and relevant theoretical frameworks. This section should discuss the implications of the findings, compare them to previous research, and address any limitations or potential sources of bias.

Conclusion: A concise summary of the main findings and their implications for theory, practice, or future research in psychology. This section should restate the study objectives, summarize the key findings, and offer suggestions for further investigation.

References: A list of all sources cited in the report, following a standardized citation format (e.g., APA or MLA).

Appendices: Additional supplementary materials, such as raw data tables, experimental protocols, or detailed statistical analyses, that provide further support for the findings presented in the report.

By including these elements in the report, researchers can effectively communicate the methodology, results, and conclusions of their RSM designed experiments in psychology, contributing to the advancement of knowledge in the field.

2.5.3 Fractional Factorial

Fractional factorial design experiments are used in psychology providing a flexible and efficient approach to studying complex psychological phenomena by allowing researchers to balance the trade-off between comprehensiveness and practical constraints. They are particularly useful for screening experiments, resource conservation, and managing complexity in psychological research. The following are several reasons to use fractional factorial design experiments:

Efficiency: Fractional factorial designs allow researchers to study multiple factors simultaneously while running fewer experiments compared to full factorial designs. This efficiency is particularly beneficial when there are many factors of interest or when conducting large-scale experiments is impractical or costly.

Resource Conservation: Conducting full factorial experiments in psychology, especially with a large number of factors or levels, can be resource intensive in terms of time, money, and participant recruitment. Fractional factorial designs help conserve these resources by reducing the number of experimental conditions while still allowing researchers to examine main effects and interactions.

Screening Experiments: Fractional factorial designs are often used as screening experiments to identify the most influential factors or interactions among a larger set of potential variables. By selecting a subset of factors for investigation, researchers can prioritize resources for further study on the most promising variables.

Complexity Management: In psychology, experiments often involve studying complex phenomena influenced by multiple variables and their interactions. Fractional factorial designs provide a systematic approach to managing this complexity by identifying key factors and interactions that contribute to the observed outcomes.

Practical Constraints: In many psychological studies, researchers face practical constraints, such as limited time, budget, or availability of participants. Fractional factorial designs offer a pragmatic solution by allowing researchers to investigate multiple factors within the constraints of their resources.

Generalizability: Fractional factorial designs can help researchers generalize findings to real-world settings by focusing on a subset of factors that are most relevant to the phenomenon of interest. This approach enhances the external validity of the research findings.

Fractional factorial design experiments are used in psychology across various research areas and applications. Some common areas where fractional factorial designs are employed in psychology include:

Experimental Psychology: Fractional factorial designs are used to investigate the effects of multiple IVs on psychological phenomena, such as perception, cognition, memory, learning, and decision-making. Researchers may use fractional factorial designs to efficiently explore the main effects and interactions of different experimental manipulations.

Clinical Psychology: In clinical psychology, fractional factorial designs can be used to study the effectiveness of interventions or treatments for psychological disorders. Researchers may use these designs to evaluate the impact of various treatment components or therapeutic techniques on clinical outcomes while controlling for potential confounding variables.

Social Psychology: Fractional factorial designs are employed in social psychology to examine the factors influencing social behavior, attitudes, and interpersonal relationships. Researchers may use these designs to manipulate multiple social variables simultaneously and investigate their effects on behavior and cognition.

Developmental Psychology: In developmental psychology, fractional factorial designs can be used to study the factors influencing human development across the lifespan. Researchers may use these designs to explore the effects of environmental factors, genetic predispositions, and social interactions with developmental outcomes.

Cognitive Psychology: Fractional factorial designs are utilized in cognitive psychology to investigate cognitive processes, such as attention, perception, memory, and problem-solving. Researchers may use these designs to manipulate multiple cognitive variables and examine their interactions in shaping cognitive performance.

Health Psychology: In health psychology, fractional factorial designs can be used to study the factors influencing health behavior, adherence to medical treatments, and outcomes of health interventions. Researchers may use these designs to identify the most effective components of health promotion programs or interventions targeting behavior change.

Industrial-Organizational Psychology: Fractional factorial designs are applied in industrial-organizational psychology to study factors influencing workplace behavior, job performance, and organizational effectiveness. Researchers may use these designs to investigate the effects of various organizational interventions, leadership styles, or job design factors on employee outcomes.

Overall, fractional factorial design experiments are versatile and widely applicable in psychology, enabling researchers to efficiently explore complex relationships between multiple variables and advance understanding across various subfields of psychology.

STEPS

Developing fractional factorial design experiments in psychology involves several key steps:

1. **Identify Research Objectives**: Clearly define the research objectives and the specific hypotheses to be tested. Determine the main research questions and the variables of interest that will be manipulated or measured in the experiment.
2. **Select Factors and Levels**: Identify the IVs (factors) that will be manipulated in the experiment. Determine the levels or values of each factor that will be included in the study. Consider both the theoretical importance of each factor and practical constraints, such as the number of levels that can be feasibly tested.
3. **Choose a Fractional Factorial Design**: Select an appropriate fractional factorial design that matches the number of factors and levels specified in the experiment. Common fractional factorial designs include the 2^k-p, 2^(k-p) designs, and PBDs. Choose a design that balances efficiency, resolution, and the ability to estimate main effects and interactions.
4. **Generate Experimental Design**: Use statistical software or specialized software packages to generate the experimental design matrix based on the chosen fractional factorial

design. The design matrix specifies the experimental conditions (combinations of factor levels) and the order in which they will be presented to participants.

5. **Conduct the Experiment**: Implement the experimental procedure according to the design matrix. Manipulate the IVs as specified and measure the DVs of interest. Ensure that the experimental conditions are presented to participants in a randomized order to minimize potential order effects.

6. **Collect Data**: Collect data on the DVs for each experimental condition. Ensure that data collection procedures are standardized and reliable to maintain the validity of the results.

7. **Data Analysis**: Analyze the collected data using appropriate statistical techniques. Estimate main effects and interactions using regression analysis or ANOVA. Interpret the results in relation to the research hypotheses and theoretical framework.

8. **Interpretation and Conclusion**: Interpret the findings of the experiment in relation to the research objectives and hypotheses. Discuss the implications of the results for theory, practice, and future research in psychology. Consider any limitations of the study and potential alternative explanations for the findings.

9. **Report Writing**: Write a clear and concise report summarizing the experimental design, methods, results, and conclusions. Follow standard reporting guidelines for experimental research in psychology, such as the APA style.

10. **Peer Review and Publication**: Submit the research report to peer-reviewed journals in psychology for publication consideration. Address any feedback or revisions requested by reviewers to improve the quality and rigor of the research.

By following these steps, researchers can develop and conduct fractional factorial design experiments in psychology effectively, leading to valuable insights into psychological phenomena and advancing the field's knowledge base.

PITFALLS AND REMEDIES

Fractional factorial design experiments offer many advantages in terms of efficiency and resource conservation in psychology research. However, they also come with potential pitfalls. Here are some common pitfalls and remedies for fractional factorial design experiments used in psychology:

Aliasing of Effects:
> **Pitfall:** Fractional factorial designs may result in aliasing, where main effects or interactions are confounded with other effects, making it difficult to interpret the results accurately.
>
> **Remedy:** Conducting a resolution analysis can help identify the aliasing pattern and assess the confounding structure in the design. Researchers can then prioritize certain effects of interest or use follow-up experiments to resolve confounded effects.

Loss of Precision:
> **Pitfall:** Fractional factorial designs sacrifice precision compared to full factorial designs, which may reduce the ability to detect small effects or interactions.
>
> **Remedy:** Increasing the sample size can help compensate for the loss of precision in fractional factorial designs. Researchers should conduct power analyses to determine the required sample size for detecting meaningful effects with sufficient statistical power.

Generalizability Concerns:
> **Pitfall:** Fractional factorial designs may focus on a subset of factors or levels, limiting the generalizability of the findings to broader populations or contexts.
>
> **Remedy:** Researchers should carefully consider the selection of factors and levels in the design to ensure that they are representative of the target population or relevant to

the research question. Additionally, conducting replication studies using different samples or settings can enhance the generalizability of the findings.

Interpretational Challenges:

Pitfall: Interpreting the results of fractional factorial designs can be challenging due to the complexity of the design and the presence of confounded effects.

Remedy: Researchers should use caution when interpreting the results of fractional factorial designs and consider multiple sources of evidence, such as theoretical rationale, prior research, and supplementary analyses. Consulting with statistical experts or collaborators can also help ensure accurate interpretation of the findings.

Limited Resolution:

Pitfall: Fractional factorial designs with low resolution may not adequately distinguish between main effects and interactions, leading to ambiguity in the interpretation of the results.

Remedy: Choosing a higher-resolution fractional factorial design or supplementing the experiment with additional follow-up studies can help clarify the nature of the effects and interactions observed in the initial experiment.

Experimental Design Constraints:

Pitfall: Fractional factorial designs may be constrained by practical limitations, such as budget, time, or participant availability, which can impact the feasibility of the experiment.

Remedy: Researchers should carefully consider the practical constraints when designing fractional factorial experiments and prioritize factors or levels that are most critical to the research question. Creative problem-solving and collaboration with stakeholders can help address logistical challenges and optimize the experimental design.

By being aware of these potential pitfalls and implementing appropriate remedies, researchers can maximize the benefits of fractional factorial design experiments in psychology while minimizing potential limitations and biases in the findings.

EXAMPLE

Fractional factorial experiments are designed to study the effects of multiple factors while reducing the number of experimental runs required. Here's an example of a matrix for a $2^{(k-p)}$ fractional factorial experiment in psychology, where k is the number of factors and p is the number of factors at full levels:

Objective: To investigate the effects of three IVs, time spent studying (factor A), study environment (factor B), and caffeine intake (factor C), on participants' performance on a memory recall task. However, due to resource constraints, a 1/4 fraction of the full factorial design is used.

Factors:

1. **Time Spent Studying (Factor A):**
   - Low time (e.g., 20 minutes).
   - High time (e.g., 60 minutes).
2. **Study Environment (Factor B):**
   - Quiet environment.
   - Noisy environment.
3. **Caffeine Intake (Factor C):**
   - No caffeine.
   - Caffeine intake.

TABLE 2.78

Psychology Fractional Factorial Design Experiment Matrix Example

Run	Time Spent Studying (A)	Study Environment (B)	Caffeine Intake (C)	Memory Recall Score
1	Low	Quiet	No caffeine	...
2	High	Quiet	Caffeine intake	...
3	Low	Noisy	Caffeine intake	...
4	High	Noisy	No caffeine	...

Response Variable: Memory recall score.

Fractional Factorial Experiment Matrix:

In a 1/4 fraction of the full factorial design, the matrix includes a subset of the possible combinations of the three factors, allowing for efficient experimentation while reducing the number of runs. The matrix is shown in Table 2.78.

In this matrix, each row represents a specific experimental run with a unique combination of the three factors at their fractional levels. Memory recall scores are measured for each combination.

Fractional factorial experiments allow psychologists to study the main effects and some interactions of the factors efficiently, even when considering a large number of factors, by conducting only a fraction of the total possible runs.

Another example of a fractional factorial design experiment in psychology could involve studying the effects of various factors on human decision-making processes. Let's consider a hypothetical experiment investigating the factors influencing individuals' choices in a decision-making task.

Suppose researchers are interested in understanding how three factors – time pressure, reward magnitude, and decision complexity – affect decision-making behavior. However, due to resource constraints and the desire to reduce the number of experimental conditions, they decide to use a fractional factorial design.

The fractional factorial design could be a $2^{(3-1)}$ design, commonly known as a half-fraction design. This design allows researchers to investigate the main effects of each factor and some two-way interactions while testing only a fraction of the total number of possible combinations.

In this example, the factors and their levels could be defined as follows:

Time Pressure:

- Low (e.g., no time pressure).
- High (e.g., time constraint).

Reward Magnitude:

- Small (e.g., $10).
- Large (e.g., $50).

Decision Complexity:

- Low (e.g., simple decision task).
- High (e.g., complex decision task).

The fractional factorial design would involve testing a subset of the eight possible combinations of factor levels, selected based on the chosen design. For instance, the design might include four experimental conditions:

1. Low time pressure, small reward, low complexity.
2. Low time pressure, large reward, high complexity.
3. High time pressure, small reward, high complexity.
4. High time pressure, large reward, low complexity.

Participants would be randomly assigned to one of these experimental conditions, and their decision-making behavior (e.g., choice accuracy, response time, risk-taking propensity) would be measured and analyzed. By systematically manipulating these factors and their interactions, researchers can gain insights into the underlying mechanisms driving decision-making processes. In a fractional factorial design for studying the effects of various factors on human decision-making processes in psychology, the experimental matrix outlines the specific combinations of factor levels to be tested.

Consider a 2^(3-1) fractional factorial design to study the effects of three factors (A, B, and C) on decision-making processes, where each factor has two levels. The design matrix would involve testing a subset of the possible combinations of factor levels, selected based on the chosen design.

Let's denote the factors and their levels as follows:

Factor A (Time Pressure):

- Low (L).
- High (H).

Factor B (Reward Magnitude):

- Small (S).
- Large (L).

Factor C (Decision Complexity):

- Simple (S).
- Complex (C).

The fractional factorial design selects a subset of combinations that allow for the estimation of main effects and certain interactions while reducing the total number of experimental conditions. An example of a matrix for such a design is shown in Table 2.79.

This matrix represents a fractional factorial design with four experimental runs. Each row corresponds to one experimental run, and the columns represent the levels of the factors being manipulated in each run.

TABLE 2.79

Psychology Fractional Factorial Design Experiment Matrix Example

Run	Factor A	Factor B	Factor C
1	L	S	S
2	H	S	C
3	L	L	C
4	H	L	S

Interpreting the matrix:

- Run 1: Low time pressure (A), small reward magnitude (B), simple decision complexity (C).
- Run 2: High time pressure (A), small reward magnitude (B), complex decision complexity (C).
- Run 3: Low time pressure (A), large reward magnitude (B), complex decision complexity (C).
- Run 4: High time pressure (A), large reward magnitude (B), simple decision complexity (C).

This design allows researchers to study the effects of each factor and their interactions with decision-making processes while testing only a fraction of the total possible combinations.

This fractional factorial design allows researchers to investigate the main effects of each factor (e.g., the impact of time pressure on decision-making) as well as some interactions between factors (e.g., how reward magnitude interacts with decision complexity). Despite testing only a subset of the possible conditions, this design provides valuable information while minimizing resource expenditure.

REPORTING

In fractional factorial design experiments used in psychology, various statistical tools are employed to analyze the data and draw meaningful conclusions. Some of the common statistical tools used in such experiments include:

ANOVA: ANOVA is a widely used statistical technique for comparing means across multiple groups. It helps assess the significance of differences between group means and identify which factors (or combinations of factors) have a significant impact on the outcome variable.

Main Effects Analysis: Main effects analysis examines the individual effects of each factor on the outcome variable while averaging over the levels of other factors. It helps determine whether each factor has a significant influence on the DV.

Interaction Effects Analysis: Interaction effects analysis assesses whether the effect of one factor on the outcome variable depends on the levels of another factor. It helps identify synergistic or antagonistic relationships between factors that may affect the outcome variable differently under different conditions.

Post-hoc Tests: Post-hoc tests are used to compare specific groups or conditions following a significant result in ANOVA. They help determine which specific factor levels differ significantly from each other, providing more detailed insights into the experimental findings.

Regression Analysis: Regression analysis is used to model the relationship between IVs (factors) and the DV. It helps quantify the strength and direction of these relationships, allowing for the prediction of the outcome variable based on the values of the IVs.

DOE Software: DOE software packages provide tools for designing experiments, generating fractional factorial designs, analyzing data, and interpreting results. These software packages often include features for ANOVA, regression analysis, and graphical visualization of experimental findings.

RSM: RSM is a collection of statistical and mathematical techniques used to optimize response variables influenced by multiple factors. It involves fitting mathematical models to experimental data and using optimization techniques to identify optimal factor settings for achieving desired outcomes.

These statistical tools are essential for conducting rigorous analysis of fractional factorial design experiments in psychology, allowing researchers to uncover underlying relationships between factors and the outcome variable and make informed conclusions about the experimental findings.

In a report detailing fractional factorial design experiments used in psychology, several key elements should be included to provide a comprehensive understanding of the study and its findings. Here are the essential elements typically found in such reports:

Title: A descriptive and concise title that clearly indicates the focus of the study and the experimental design employed.

Abstract: A brief summary of the study, including the research question, experimental design, methods, results, and main conclusions. The abstract should provide an overview of the entire report.

Introduction:
- Background and rationale for the study, including relevant theoretical frameworks and previous research.
- Research question or hypothesis addressed by the experiment.
- Explanation of why fractional factorial design was chosen and its advantages in the context of the study.

Methods:
- Description of the experimental design, including the factors manipulated, their levels, and the chosen fraction.
- Details of participant recruitment and selection criteria.
- Description of the experimental procedure, including any materials used and the sequence of tasks.
- Statistical analysis plan, including specific tests and software used for data analysis.

Results:
- Summary of the main findings, including significant effects and interactions observed.
- Presentation of data, such as means, standard deviations, and effect sizes, for each experimental condition.
- Statistical analyses, including ANOVA tables and post-hoc comparisons, if applicable.
- Graphs or figures to visually represent the results, such as bar charts or interaction plots.

Discussion:
- Interpretation of the results in the context of the research question and existing literature.
- Implications of the findings for theory, practice, or future research in psychology.
- Consideration of limitations and potential sources of bias or confounding.
- Suggestions for further research or modifications to the experimental design.

Conclusion: A concise summary of the main findings and their significance, reiterating the key points discussed in the results and discussion sections.

References: A list of all sources cited in the report, following a standardized citation format (e.g., APA style).

Appendices: Any additional materials or information relevant to the study, such as detailed descriptions of stimuli or supplementary analyses.

By including these elements in the report, researchers can effectively communicate the rationale, methods, results, and implications of their fractional factorial design experiment in psychology, contributing to the advancement of scientific knowledge in the field.

2.5.4   CCD

CCD is a type of designed experiment commonly used in various fields, including psychology. Here's why CCD experiments are used in psychology:

Flexibility in studying complex relationships: CCD allows researchers to study complex relationships between multiple variables in psychology. In psychology research, variables can be interrelated and influenced by various factors. CCD enables researchers to systematically manipulate and study these variables within a controlled experimental setting, allowing for a comprehensive understanding of their interactions and effects on psychological processes.

Optimization of experimental conditions: CCD is useful for optimizing experimental conditions in psychology experiments. Researchers can use CCD to identify the optimal levels of IVs that maximize desired outcomes or minimize undesirable outcomes. This is particularly valuable in areas, such as intervention development, where researchers aim to identify the most effective strategies for promoting behavior change or improving psychological well-being.

Efficient resource allocation: CCD is designed to maximize the amount of information gained from a limited number of experimental runs. This efficiency is valuable in psychology research, where resources, such as time, funding, and participant availability, are often constrained. By strategically selecting experimental conditions and levels, CCD allows researchers to obtain robust findings with fewer experimental trials, thus optimizing resource allocation.

Robustness to variability and noise: CCD is robust to variability and noise in the data, making it suitable for studies with inherent variability or measurement error. In psychology research, data may be subject to various sources of variability, such as individual differences in participant characteristics or measurement imprecision. CCD's systematic approach to experimental design helps researchers mitigate the effects of variability and obtain reliable results.

Facilitation of RSM: CCD is commonly used in conjunction with RSM, which allows researchers to model and analyze complex relationships between independent and DVs. RSM enables researchers to visualize and interpret the effects of multiple variables simultaneously, providing valuable insights into the underlying mechanisms of psychological phenomena.

CCD is utilized in psychology for its ability to systematically explore complex relationships between variables, optimize experimental conditions, efficiently allocate resources, mitigate variability, and facilitate advanced statistical analysis techniques, such as RSM. These features make CCD a valuable tool for advancing knowledge and understanding in various areas of psychology. CCD can be employed in clinical psychology to optimize the parameters of a therapeutic intervention. Psychologists may vary factors like treatment duration, session frequency, and therapeutic techniques to maximize treatment effectiveness.

CCD experiments find applications in various areas of psychology due to their versatility in studying complex relationships between multiple variables. Here are some specific areas within psychology where CCD experiments are commonly employed:

Psychophysics: In psychophysics, researchers use CCD experiments to investigate the relationship between physical stimuli and perceptual responses. For example, CCD can be used to study the effects of different stimulus parameters (e.g., intensity, duration) on sensory perception (e.g., brightness, loudness) or psychophysical thresholds (e.g., detection thresholds, discrimination thresholds).

Psychometric Testing: CCD are utilized in the development and validation of psychological assessment tools and measures. Researchers can use CCD to systematically vary test item parameters (e.g., difficulty, discriminability) and examine their effects on test performance, reliability, and validity. This allows for the optimization of assessment instruments to accurately measure psychological constructs of interest.

Behavioral Interventions: CCD experiments are employed in research on behavioral interventions aimed at promoting behavior change or improving psychological well-being. Researchers can use CCD to explore the effects of different intervention components (e.g., strategies, dosage) on target behaviors or outcomes. This helps in identifying the most effective intervention strategies and optimizing intervention protocols for maximum impact.

Cognitive Psychology: In cognitive psychology, CCD are used to investigate cognitive processes, such as attention, memory, perception, and decision-making. Researchers can manipulate multiple variables simultaneously (e.g., stimulus characteristics, task demands) using CCD and examine their interactive effects on cognitive performance. This allows for a more comprehensive understanding of complex cognitive phenomena.

Experimental Design and Analysis: CCD experiments are utilized in methodological research on experimental design and analysis techniques in psychology. Researchers may use CCD to evaluate the performance of statistical models, test hypotheses about experimental designs, or optimize research methodologies. This contributes to the advancement of best practices in experimental psychology and enhances the rigor and reproducibility of psychological research.

The CCD experiments find applications across various subfields of psychology, enabling researchers to systematically investigate complex relationships between variables, optimize experimental conditions, and advance knowledge and understanding in the field.

STEPS

Developing a CCD experiment in psychology involves several steps to ensure that the experimental design effectively explores the relationships between variables of interest. Here are the typical steps involved in developing a CCD experiment in psychology:

1. **Define the Research Question**: Clearly articulate the research question or hypothesis that the experiment aims to address. This step is crucial for guiding the selection of variables and determining the objectives of the study.
2. **Identify Independent and DVs**: Determine the variables that will be manipulated (IVs) and measured (DVs) in the experiment. These variables should be relevant to the research question and theory being tested.
3. **Determine the Design Space**: Define the range or levels of each IV that will be investigated in the experiment. This includes specifying the minimum, maximum, and central levels of each variable, as well as any additional levels for exploring curvature or interaction effects.
4. **Select the Design Criterion**: Choose the specific criterion or optimization goal for the CCD. Common criteria include maximizing the distance from the center point, minimizing prediction variance, or achieving a desired level of precision for parameter estimation.
5. **Generate the CCD**: Use specialized software or algorithms to generate the CCD based on the specified design space and optimization criterion. The design should include a combination of factorial points, axial points, and center points to capture both linear and quadratic effects of the IVs.
6. **Evaluate the Design**: Assess the quality and efficiency of the generated CCD using diagnostic tools and statistical metrics. Evaluate factors, such as the adequacy of model fit, the balance of design points across experimental conditions, and the robustness to potential sources of bias or error.
7. **Refine the Design (if necessary)**: Make any necessary adjustments to the CCD to address limitations or deficiencies identified during the evaluation process. This may involve

adding additional design points, adjusting the range of variable levels, or modifying the optimization criterion.

8. **Plan the Experimental Procedure**: Develop a detailed plan for implementing the CCD experiment, including procedures for data collection, participant recruitment, and experimental manipulation. Ensure that the experimental protocol is feasible and adheres to ethical guidelines.

9. **Implement the Experiment**: Carry out the CCD experiment according to the planned procedures, collecting data from participants or experimental units in accordance with the experimental design.

10. **Analyze the Data**: Analyze the collected data using appropriate statistical methods to test hypotheses and draw conclusions about the effects of IVs on DVs. Use regression analysis or other modeling techniques to examine linear and quadratic effects and interpret the results in light of the research question.

By following these steps, researchers can systematically develop and implement CCD experiments in psychology to investigate complex relationships between variables and advance understanding in the field.

PITFALLS AND REMEDIES

CCD experiments offer many advantages in psychology research, but there are also potential pitfalls that researchers should be aware of. Here are some common pitfalls and remedies for CCD experiments used in psychology:

Inadequate Model Fit:
> **Pitfall:** One common pitfall is using a CCD with a model that does not adequately fit the data. This can lead to biased parameter estimates and incorrect conclusions about the relationships between variables.
> **Remedy**: Conduct thorough model diagnostics to assess the adequacy of the fitted model. Use diagnostic tools, such as residual plots, lack-of-fit tests, and cross-validation techniques, to identify any discrepancies between the model and the observed data. If necessary, consider alternative models or modifications to improve model fit.

Violation of Design Assumptions:
> **Pitfall:** CCDs rely on certain assumptions about the underlying data distribution and model structure. Violating these assumptions can lead to unreliable results and invalid conclusions.
> **Remedy**: Validate the assumptions of the CCD using diagnostic checks and sensitivity analyses. Pay particular attention to assumptions related to linearity, normality, and homoscedasticity. If assumptions are violated, consider alternative designs or robust estimation techniques that are more appropriate for the data.

Overemphasis on Optimization:
> **Pitfall:** Researchers may focus excessively on optimizing the CCD without considering other important factors, such as practical feasibility, ethical considerations, or the interpretability of results.
> **Remedy**: Strike a balance between optimization criteria and other relevant factors, such as resource constraints, participant burden, and theoretical relevance. Consider conducting sensitivity analyses to assess the robustness of the optimized design to variations in model assumptions or experimental conditions.

Sparse Regions of the Design Space:
> **Pitfall:** CCDs may result in sparse regions of the design space where data points are poorly distributed. This can lead to instability in parameter estimates or reduced statistical power.

Remedy: Optimize the design to ensure adequate coverage of the design space and minimize sparse regions. Consider augmenting the CCD with additional design points in critical regions or using adaptive sampling techniques to iteratively refine the design.

Failure to Consider Interaction Effects:

Pitfall: CCDs typically focus on main effects and may overlook interaction effects between variables, leading to incomplete or misleading conclusions.

Remedy: Incorporate interaction terms into the CCD or use sequential or adaptive designs to explore interactions iteratively. Conduct post-hoc analyses to examine potential interactions and their implications for the research findings.

By being aware of these potential pitfalls and implementing appropriate remedies, researchers can effectively leverage CCD experiments in psychology while minimizing biases and maximizing the quality of the research findings.

EXAMPLES

CCD experiments are used to study the effects of factors while considering quadratic responses and curvature in the design. Here's an example of a matrix for a CCD experiment in psychology:

Objective: To optimize the conditions for a cognitive training program by studying the effects of two IVs, training duration (factor A) and training intensity (factor B), on participants' cognitive performance scores.

Factors:

Training Duration (Factor A):
- Short duration (e.g., 20 minutes).
- Medium duration (e.g., 40 minutes).
- Long duration (e.g., 60 minutes).

Training Intensity (Factor B):
- Low intensity (e.g., basic exercises).
- Medium intensity (e.g., moderate difficulty).
- High intensity (e.g., challenging exercises).

Response Variable: Cognitive performance score.

CCD Experiment Matrix:

In a CCD experiment, the matrix is designed to systematically explore the combinations of IVs (factors) within a specified range, including the central point and additional points to assess curvature. The matrix is shown in Table 2.80.

In this matrix, the central points (–1) represent the medium levels of training duration and training intensity. The additional points (1, 2, 3, 4) allow for the assessment of curvature in the response surface. Cognitive performance scores are measured for each combination.

CCD experiments are useful for optimizing processes or conditions while accounting for potential nonlinear responses and interactions between factors. In this example, researchers aim to identify the optimal combination of training duration and training intensity for maximizing cognitive performance.

An example of a CCD experiment in psychology could involve studying the effects of different types of stressors on cognitive performance. Let's consider a hypothetical study aimed at investigating how various stressors, including time pressure and social evaluation, affect individuals' ability to perform cognitive tasks accurately and efficiently.

TABLE 2.80

CCD Experiment Matrix Example

Run	Training Duration (A)	Training Intensity (B)	Cognitive Performance Score
−1	Medium	Medium	...
−1	Medium	Medium	...
−1	Medium	Medium	...
−1	Medium	Medium	...
−1	Short	Low	...
−1	Short	High	...
−1	Long	Low	...
−1	Long	High	...
1	Short	Medium	...
2	Long	Medium	...
3	Medium	Low	...
4	Medium	High	...

1. **Research Question**: How do different stressors impact cognitive performance?
2. **IVs**:
   - Stressor Type:
     - No stress (coded as −1).
     - Time pressure (coded as +1).
     - Social evaluation (coded as +1).
3. **DVs:**
   - Cognitive Task Performance:
     - Accuracy (percentage of correct responses).
     - Response time (time taken to respond to stimuli).
4. **Design Space:**
   - Stressor Type: No stress, time pressure, social evaluation.
   - Stressor Levels: Low, medium, high.
   - Center Points: To capture potential curvature in the response surface.
5. **Design Criterion:**
   - Minimize prediction variance to optimize the precision of parameter estimates for main effects and interactions.
6. **Generated CCD:**
   - The CCD will include a combination of factorial points, axial points, and center points that systematically vary the levels of the stressor variables while ensuring balance and efficiency.
   - For example, the design might include a total of 10 experimental conditions (2^3 factorial design) with additional axial and center points for robustness.
7. **Experimental Procedure:**
   - Participants are randomly assigned to different stressor conditions.
   - In the no-stress condition, participants complete cognitive tasks without any additional stressors.
   - In the time-pressure condition, participants are given limited time to complete tasks.
   - In the social-evaluation condition, participants perform tasks under the presence of evaluators or observers.

8. **Data Collection:**
   - Participants' performance on cognitive tasks is measured in terms of accuracy and response time.
   - Data are collected across multiple trials within each condition to obtain reliable estimates of performance under different stressor levels.
9. **Data Analysis:**
   - Perform regression analysis or other statistical modeling techniques to analyze the effects of stressor type and level on cognitive task performance.
   - Assess the main effects and interaction effects of stressors on accuracy and response time, controlling for potential confounding variables.
10. **Interpretation of Results:**
   - Interpret findings in terms of how different stressors impact cognitive performance, considering both main effects and interaction effects.
   - Discuss implications for theories of stress and cognition, as well as practical implications for managing stress in real-world contexts.

This example demonstrates how a CCD experiment could be applied in psychology to investigate the effects of stressors on cognitive performance. By systematically varying stressor levels and using CCD to optimize the experimental design, researchers can obtain valuable insights into the relationship between stress and cognition.

An example of a CCD matrix for studying how different stressors impact cognitive performance in psychology is shown in Table 2.81.

In this CCD matrix,

- **Stressor Type (A)** represents the type of stressor, with −1 indicating "No Stress" and +1 indicating either "Time Pressure" or "Social Evaluation."
- **Stressor Level (B)** represents the level of the stressor, with −1 indicating "Low" stress and +1 indicating "High" stress.
- **Cognitive Task Performance** includes measures of accuracy (percentage of correct responses) and response time (time taken to respond to stimuli) for each participant under different stressor conditions.

The CCD matrix systematically varies the levels of stressor type and level to capture potential linear and quadratic effects on cognitive task performance. Additionally, center points are included to account for potential curvature in the response surface and to estimate pure error variance.

TABLE 2.81
CCD Experiment Matrix Example

Stressor Type (A)	Stressor Level (B)	Cognitive Task Performance (DV)
−1 (No stress)	−1 (Low)	Participant 1: Accuracy=85%, response time=500 ms
−1 (No stress)	+1 (High)	Participant 2: Accuracy=90%, response time=550 ms
+1 (Time pressure)	−1 (Low)	Participant 3: Accuracy=75%, response time=600 ms
+1 (Time pressure)	+1 (High)	Participant 4: Accuracy=70%, response time=700 ms
+1 (Social evaluation)	−1 (Low)	Participant 5: Accuracy=80%, response time=550 ms
+1 (Social evaluation)	+1 (High)	Participant 6: Accuracy=65%, response time=750 ms
0 (Center point)	0 (Center point)	Participant 7: Accuracy=85%, response time=600 ms
0 (Center point)	0 (Center point)	Participant 8: Accuracy=90%, response time=500 ms

This matrix represents a simplified example, and a full CCD would include additional experimental runs to fully explore the effects of stressors on cognitive performance across different conditions and levels.

To demonstrate a CCD matrix for understanding how different stressors impact cognitive performance in psychology, let's assume a CCD with three factors: stressor type (no stress, time pressure, social evaluation), each with three levels (low, medium, high). We'll also include center points to capture potential curvature in the response surface. An example of a CCD matrix for this scenario is shown in Table 2.82.

In this CCD matrix,

- The first column represents the run number or experimental trial.
- The second column represents the coded values for stressor type (A), where −1 corresponds to no stress, +1 corresponds to time pressure, and 0 corresponds to social evaluation.
- The third and fourth columns represent the coded values for stressor level (B) and stressor level (C), respectively, where −1 corresponds to low, +1 corresponds to high, and 0 corresponds to medium.
- Rows 1–5 and 6–10 represent the factorial points for the no stress and time pressure conditions, respectively.
- Rows 11–14 represent the axial points for the social evaluation condition.
- Rows 15–20 represent the center points for capturing potential curvature in the response surface.

This CCD matrix systematically varies the levels of stressor type and level combinations, allowing researchers to study the effects of different stressors on cognitive performance while optimizing the efficiency and precision of the experimental design. Data collected from these experimental

TABLE 2.82

CCD Experiment Matrix Example

Run	Stressor Type (A)	Stressor Level (B)	Stressor Level (C)
1	−1	−1	0
2	−1	1	0
3	−1	0	−1
4	−1	0	1
5	−1	0	0
6	1	−1	0
7	1	1	0
8	1	0	−1
9	1	0	1
10	1	0	0
11	0	−1	−1
12	0	1	−1
13	0	−1	1
14	0	1	1
15	0	0	−1
16	0	0	1
17	0	0	0
18	0	0	0
19	0	0	0
20	0	0	0

conditions can then be analyzed using appropriate statistical methods to assess the impact of stressors on cognitive performance in psychology.

REPORT

CCD experiments in psychology typically involve the use of various statistical tools and techniques to analyze the data and draw conclusions about the effects of IVs on DVs. Here are some common statistical tools used in CCD experiments in psychology:

Regression Analysis: Regression analysis is commonly used to model the relationship between IVs (e.g., stressor types and levels) and DVs (e.g., cognitive performance measures). Researchers may use multiple linear regression or polynomial regression models to examine linear and quadratic effects of IVs on DVs.

ANOVA: ANOVA is used to test for statistically significant differences in means across multiple groups or conditions. In CCD experiments, researchers may use ANOVA to assess the effects of IVs on DVs and determine the significance of main effects and interaction effects.

RSM: RSM is a set of statistical techniques used to model and analyze complex relationships between IVs and response variables. In CCD experiments, researchers may use RSM to visualize and interpret the effects of IVs on DVs and identify optimal conditions for maximizing or minimizing the response.

Optimization Algorithms: Optimization algorithms are used to generate optimal CCDs based on specified criteria and constraints. These algorithms may include techniques, such as linear programming, genetic algorithms, or simulated annealing, to search for designs that maximize information while minimizing experimental error.

Model Diagnostics: Model diagnostics techniques are used to assess the adequacy and validity of statistical models fitted to the data. Researchers may use diagnostic plots, residual analyses, and goodness-of-fit tests to evaluate the assumptions of regression models and identify potential sources of bias or error.

DOE Software: Specialized software packages for DOE are often used to generate CCDs , analyze experimental data, and perform statistical tests. These software tools provide researchers with a user-friendly interface for designing experiments, conducting analyses, and interpreting results.

Post-hoc Tests: Post-hoc tests, such as Tukey's HSD test or Bonferroni correction, may be used to compare specific groups or conditions following a significant ANOVA result. These tests help identify which groups differ significantly from each other and provide additional insights into the experimental findings.

By employing these statistical tools and techniques, researchers can effectively analyze data from CCD experiments in psychology, draw valid conclusions about the effects of IVs on DVs, and contribute to the advancement of knowledge in the field.

When reporting the results of CCD experiments used in psychology, researchers typically structure their reports to provide a clear and comprehensive overview of the study's objectives, methods, findings, and implications. Here are the key elements that may be included in a report on CCD experiments in psychology:

Title: A concise and informative title that accurately reflects the content and focus of the study.

Abstract: A brief summary of the study, including the research question, methods, key findings, and conclusions. The abstract provides readers with an overview of the study's objectives and outcomes.

Introduction: A background section that provides context for the study by reviewing relevant literature, explaining the rationale for the research question, and stating the study's objectives and hypotheses. The introduction sets the stage for the rest of the report and helps readers understand the significance of the study.

Methods:
- **Experimental Design**: Description of the CCD used, including the independent and DVs, design space, and optimization criterion.
- **Participants**: Description of the sample, including demographic information and any inclusion/exclusion criteria.
- **Procedure**: Detailed description of the experimental procedures, including data collection methods, experimental manipulations, and control measures.

Results:
- Summary of participant characteristics and any relevant descriptive statistics.
- Presentation of the main findings, including statistical analyses of the effects of IVs on DVs.
- Tables, figures, and/or graphs to illustrate key results, such as response surfaces or contour plots depicting the effects of IVs on the response variable.

Discussion:
- Interpretation of the results in relation to the study's objectives and hypotheses.
- Comparison of findings with previous research and theoretical frameworks.
- Discussion of implications, limitations, and future directions for research.
- Consideration of the broader implications of the study for theory, practice, and policy in psychology.

Conclusion: A concise summary of the main findings and their implications, emphasizing the contributions of the study to the field of psychology.

References: A list of all sources cited in the report, following a specific citation style (e.g., APA, MLA).

Appendices: Additional materials, such as raw data, supplementary analyses, or experimental materials (e.g., CCD matrix), that provide further detail or support for the main findings.

By including these key elements in a report on CCD experiments in psychology, researchers can effectively communicate their methods, results, and conclusions to readers and contribute to the advancement of knowledge in the field.

2.5.5 Taguchi Method

Taguchi method designed experiments are valuable tools in psychology for their emphasis on efficiency, robustness, optimization, and systematic exploration of factors and levels. By incorporating Taguchi methods into their research practices, psychologists can conduct more rigorous and impactful studies that advance understanding in the field. Taguchi method designed experiments are used in psychology for several reasons:

Efficiency in Experimental Design: Taguchi methods emphasize efficiency in experimental design by minimizing the number of experimental trials required to obtain meaningful results. In psychology, where resources, such as time, funding, and participant availability, may be limited, Taguchi methods offer a systematic approach to designing experiments that maximizes the information gained while minimizing resource expenditure.

Robustness to Noise and Variability: Taguchi methods are designed to be robust to noise and variability in experimental data. In psychology research, data may be subjected to various sources of variability, including individual differences in participant characteristics, measurement error, and environmental factors. By incorporating robust experimental designs and analysis techniques, Taguchi methods help researchers identify and control for sources of variability, resulting in more reliable and reproducible findings.

Optimization of Experimental Conditions: Taguchi methods are often used to optimize experimental conditions for maximizing desired outcomes or minimizing undesirable outcomes. In psychology, researchers may use Taguchi methods to optimize intervention protocols, experimental stimuli, or testing procedures to achieve specific objectives, such as maximizing treatment efficacy or minimizing participant stress.

Systematic Exploration of Factors and Levels: Taguchi methods provide a systematic framework for exploring the effects of multiple factors and levels on experimental outcomes. In psychology, where phenomena are often influenced by complex interactions between variables, Taguchi methods enable researchers to systematically manipulate and study the effects of various factors on behavior, cognition, and emotion.

Practical Application in Industry and Research: Taguchi methods have been widely used in industrial settings for optimizing manufacturing processes and improving product quality. In psychology, researchers have adapted Taguchi methods for applications, such as experimental design, optimization of psychological interventions, and analysis of complex datasets. By leveraging the principles of Taguchi methods, psychologists can enhance the efficiency and effectiveness of their research endeavors.

Taguchi methods are used in usability studies to optimize the design of user interfaces. Researchers may vary factors like menu layout, font size, and color scheme to minimize user errors and improve user experience. Taguchi method designed experiments find applications in various areas of psychology, contributing to the advancement of research and understanding in the field. Some specific areas where Taguchi methods are used in psychology include:

Intervention Development and Optimization: Taguchi methods are applied in psychology to develop and optimize interventions aimed at promoting behavior change, improving mental health outcomes, or enhancing cognitive functioning. Researchers use Taguchi methods to systematically manipulate intervention components (e.g., treatment strategies, dosage, delivery format) and optimize intervention protocols for maximum effectiveness and efficiency.

Stimulus Design and Optimization: In sensory psychology and cognitive psychology, Taguchi methods are used to design and optimize experimental stimuli (e.g., visual stimuli, auditory stimuli) for studying perceptual processes, attention, memory, and decision-making. Researchers employ Taguchi methods to systematically vary stimulus parameters (e.g., intensity, duration, complexity) and optimize stimulus presentations to achieve specific experimental objectives.

Experimental Design and Analysis: Taguchi methods are utilized in methodological research on experimental design and analysis techniques in psychology. Researchers use Taguchi methods to develop efficient and robust experimental designs for studying complex phenomena and analyzing experimental data. Taguchi-based designs are adapted to various research contexts, including factorial experiments, optimization studies, and robust parameter estimation.

Quality Improvement and Process Optimization: Taguchi methods are applied in organizational psychology and industrial-organizational psychology to improve the quality of processes and systems within organizations. Researchers use Taguchi methods to identify and optimize factors that influence organizational performance, employee satisfaction, and job performance. Applications include process optimization, quality control, and performance enhancement in workplace settings.

Sensitivity Analysis and Robustness Testing: Taguchi methods are used in sensitivity analysis and robustness testing to assess the stability and reliability of psychological theories, models, and findings. Researchers employ Taguchi-based techniques to evaluate the sensitivity of experimental results to variations in experimental conditions, measurement error, or model assumptions. This helps identify potential sources of bias or uncertainty and strengthen the validity of research findings.

The Taguchi method designed experiments are employed in various subfields of psychology to address research questions, optimize experimental procedures, enhance data quality, and advance knowledge and understanding in the field. By leveraging the principles of Taguchi methods, psychologists can conduct more rigorous, efficient, and impactful studies that contribute to the scientific literature and inform evidence-based practice.

STEPS

Developing Taguchi method designed experiments in psychology involves several steps to systematically design experiments that efficiently explore the effects of multiple factors on a response variable of interest. Here are the typical steps involved in developing Taguchi methods experiments in psychology:

1. **Define the Research Objectives**: Clearly articulate the research question or hypothesis that the experiment aims to address. Determine the specific objectives of the study and the response variable(s) that will be measured to evaluate the effects of the factors of interest.
2. **Identify Factors and Levels**: Identify the factors (IVs) that may influence the response variable(s) in the experiment. Determine the levels or settings at which each factor will be studied. Factors should represent the key variables of interest in the study, such as experimental conditions, treatment levels, or stimulus parameters.
3. **Select an OA**: Choose an appropriate OA that accommodates the number of factors and levels in the experiment. The OA specifies the combination of factor levels to be tested in each experimental run and ensures efficient coverage of the experimental space.
4. **Assign Factors to Columns**: Assign the factors and their levels to the columns of the OA according to a predetermined arrangement. Use Taguchi's standard notation (e.g., L4, L8, L16) to denote the size of the OA and ensure proper assignment of factors and levels.
5. **Define Control Factors and Noise Factors**: Distinguish between control factors (factors of primary interest) and noise factors (factors that may introduce variability or error). Control factors represent the variables being manipulated to achieve desired outcomes, while noise factors represent sources of variability that should be controlled or minimized.
6. **Determine Experimental Layout**: Determine the layout of the experimental runs based on the selected OA. Specify the sequence of experimental trials, including the order of factor-level combinations to be tested in each run. Ensure that the experimental layout is randomized or structured to minimize bias and confounding effects.
7. **Conduct Experimental Runs**: Implement the experimental design by conducting the planned experimental runs according to the specified layout. Collect data on the response variable(s) for each experimental condition, ensuring consistency and accuracy in data collection procedures.
8. **Analyze Experimental Results**: Analyze the collected data using appropriate statistical methods to assess the effects of factors on the response variable(s). Use techniques, such as ANOVA, SNR analysis, or graphical methods, to identify significant factors, quantify their effects, and optimize experimental outcomes.
9. **Optimize Process Parameters**: Use the results of the Taguchi methods experiment to optimize process parameters or experimental conditions for achieving desired outcomes. Identify the optimal combination of factor levels that maximizes performance or minimizes

variability in the response variable(s), taking into account both control factors and noise factors.

10. **Validate and Interpret Results**: Validate the findings of the Taguchi methods experiment through replication, validation studies, or sensitivity analyses. Interpret the results in light of the research objectives and hypotheses, discussing the implications for theory, practice, or future research in psychology.

By following these steps, researchers can systematically develop Taguchi method designed experiments in psychology to efficiently explore the effects of multiple factors on a response variable and optimize experimental outcomes for achieving desired objectives.

PITFALLS AND REMEDIES

While Taguchi method designed experiments offer numerous benefits in psychology research, there are potential pitfalls that researchers should be mindful of. Here are some common pitfalls and remedies for Taguchi methods experiments used in psychology:

Inadequate Factor Selection:
> **Pitfall:** Selecting inappropriate or insufficient factors may lead to incomplete or biased results. Researchers may overlook important variables or include irrelevant factors, which can limit the validity and generalizability of the findings.
> **Remedy:** Conduct a thorough literature review and consult with subject matter experts to identify relevant factors that may influence the response variable. Consider conducting pilot studies or exploratory analyses to assess the potential impact of different factors on the outcome of interest. Prioritize factors based on theoretical importance, empirical evidence, or practical relevance to the research question.

Inflexible Experimental Designs:
> **Pitfall:** Using rigid experimental designs that fail to account for contextual variability or dynamic interactions among factors may result in suboptimal outcomes. Fixed designs may overlook important sources of variability or fail to capture complex relationships between variables.
> **Remedy:** Employ flexible experimental designs that accommodate variability and uncertainty in the experimental environment. Consider using adaptive designs, sequential designs, or factorial designs with additional factors or levels to capture dynamic interactions and account for unanticipated sources of variability. Incorporate randomization and blocking techniques to control extraneous variables and minimize confounding effects.

Ignoring Noise Factors:
> **Pitfall:** Failing to account for noise factors or sources of variability in the experimental design may lead to biased estimates of factor effects and inaccurate conclusions about the relationship between factors and outcomes.
> **Remedy:** Explicitly identify and characterize noise factors that may influence the response variable and incorporate them into the experimental design. Use techniques, such as blocking, randomization, or covariance adjustment, to control noise factors and reduce their impact on the estimated effects of control factors. Conduct sensitivity analyses or robustness checks to assess the stability of results across different levels of noise.

Over-reliance on OAs:
> **Pitfall:** Blindly applying OAs without considering the specific requirements of the research question or the complexity of the experimental design may result in suboptimal designs that fail to capture important interactions or nonlinear effects.

Remedy: Tailor the selection of OAs to the specific characteristics of the research question, including the number of factors, levels, and desired levels of resolution. Consider using custom-designed arrays or fractional factorial designs to accommodate complex experimental designs and optimize efficiency while maintaining statistical power. Conduct sensitivity analyses or design optimization techniques to evaluate alternative designs and identify the most suitable approach for the research question.

Lack of Validation and Replication:

Pitfall: Failing to validate experimental findings through replication or external validation studies may lead to overinterpretation or misinterpretation of results. Lack of replication may also undermine the credibility and generalizability of findings.

Remedy: Conduct validation studies or replication studies to confirm the robustness and reliability of experimental findings across different samples, settings, or conditions. Employ rigorous research methods, including independent replication, cross-validation, or meta-analysis, to assess the consistency and reproducibility of results. Transparently report replication efforts and consider publishing replication studies alongside original findings to enhance transparency and scientific rigor.

By addressing these potential pitfalls and implementing appropriate remedies, researchers can enhance the validity, reliability, and generalizability of Taguchi method designed experiments in psychology and ensure that findings contribute meaningfully to the advancement of knowledge in the field.

Example

The Taguchi method, widely used in quality engineering and optimization, can also be applied in psychology to design experiments and optimize processes. Here's an example of a matrix for a Taguchi method designed experiment in psychology:

Objective: To optimize the conditions for a relaxation technique aimed at reducing stress levels by studying the effects of two IVs, breathing technique (factor A) and background music (factor B), on self-reported stress reduction scores.

Factors:

1. **Breathing Technique (Factor A):**
   - Deep breathing.
   - Progressive muscle relaxation.
   - Mindfulness meditation.
2. **Background Music (Factor B):**
   - No music.
   - Nature sounds.
   - Classical music.

Response Variable: Self-reported stress reduction score.

Taguchi Method Experiment Matrix:

In the Taguchi method, the matrix is designed to systematically explore the combinations of factors, including different levels and interactions. The matrix is shown in Table 2.83.

In this matrix, each row represents a specific experimental run with a unique combination of breathing technique and background music levels. Self-reported stress reduction scores are measured for each combination.

TABLE 2.83

Taguchi Method Designed Experiment Matrix Example for Psychology

Run	Breathing Technique (A)	Background Music (B)	Stress Reduction Score
1	Deep breathing	No music	...
2	Deep breathing	Nature sounds	...
3	Deep breathing	Classical music	...
4	Progressive muscle relaxation	No music	...
5	Progressive muscle relaxation	Nature sounds	...
6	Progressive muscle relaxation	Classical music	...
7	Mindfulness meditation	No music	...
8	Mindfulness meditation	Nature sounds	...
9	Mindfulness meditation	Classical music	...

The Taguchi method allows researchers to systematically optimize the relaxation technique by identifying which combination of factors (breathing technique and background music) leads to the highest self-reported stress reduction scores. Taguchi's OAs and SNRs help in efficient experimentation and robust process optimization.

Another example of a Taguchi method designed experiment in psychology is to optimize the effectiveness of a study skills training program for improving academic performance among college students. The factors of interest could include different components of the training program, such as study techniques, time management strategies, and motivational interventions.

A hypothetical design matrix for a Taguchi methods experiment to optimize the study skills training program is shown in Table 2.84.

In this design matrix,

- **Experiment**: Represents the experimental run number.
- **Study Techniques (A)**: Represents the levels of study techniques, such as low, medium, or high intensity.
- **Time Management (B)**: Represents the levels of time management strategies, such as low, medium, or high effectiveness.
- **Motivational Intervention (C)**: Represents the levels of motivational interventions, such as low, medium, or high engagement.
- **Participant**: Represents the participant ID for each experimental run.

TABLE 2.84

Taguchi Method Designed Experiment Matrix Example for Psychology

Experiment	Study Techniques (A)	Time Management (B)	Motivational Intervention (C)	Participant
1	Low	Low	Low	1
2	Low	Medium	Medium	2
3	Low	High	Low	3
4	Medium	Low	High	4
5	Medium	Medium	Low	5
6	Medium	High	Medium	6
7	High	Low	Medium	7
8	High	Medium	High	8
9	High	High	High	9
10	Center	Center	Center	10

This design matrix systematically varies the levels of study techniques, time management, and motivational interventions across experimental runs, including low, medium, and high levels, as well as a center point to account for potential curvature effects. The experimental runs are randomized or structured to minimize bias and confounding effects, and data on academic performance outcomes are collected for each participant under different conditions.

Researchers would analyze the data using statistical methods, such as ANOVA or SNR analysis, to identify the most effective combination of study skills training components for improving academic performance among college students. This approach allows researchers to systematically optimize the training program based on the Taguchi methods principles, leading to more efficient and effective interventions in psychology.

REPORT

Taguchi method designed experiments in psychology often involve the use of various statistical tools to analyze experimental data and draw conclusions about the effects of factors on the response variable(s). Here are some common statistical tools used in Taguchi methods experiments in psychology:

ANOVA: ANOVA is a statistical technique used to partition the total variance in the response variable(s) into components attributable to different factors and sources of variability. In Taguchi methods experiments, ANOVA is used to assess the significance of factors and their interactions with the response variable(s), allowing researchers to identify influential factors and optimize experimental outcomes.

SNRs: SNRs are used to quantify the relationship between the mean response and the variability or noise in experimental data. Taguchi methods employ different types of SNRs, such as the LTB, STB, and NTB, depending on the experimental objectives and the nature of the response variable(s). SNRs help researchers evaluate the robustness and quality of experimental designs and identify optimal parameter settings.

Main Effects Plots and Interaction Plots: Main effects plots and interaction plots are graphical tools used to visualize the effects of factors and their interactions with the response variable(s). These plots help researchers interpret the results of Taguchi methods experiments by illustrating the relative importance of different factors and revealing patterns of interaction between factors. Main effects plots display the average response at each level of a factor, while interaction plots show how the effects of one factor vary across levels of another factor.

OAs and Taguchi OAs: OAs are structured arrays of factor-level combinations that ensure efficient and balanced coverage of the experimental space. Taguchi methods employ OAs to systematically vary factor levels in experimental designs, allowing researchers to explore the effects of factors on the response variable(s) while minimizing the number of experimental runs. OAs facilitate efficient data collection and analysis in Taguchi methods experiments, enabling researchers to identify significant factors and optimize experimental outcomes.

RPD: RPD is a statistical technique used to optimize experimental designs and parameter settings in the presence of noise factors and sources of variability. In Taguchi methods experiments, RPD techniques help researchers identify robust parameter settings that minimize the sensitivity of experimental outcomes to variations in noise factors, improving the reliability and robustness of experimental findings.

RSM: RSM is a set of statistical techniques used to model and optimize the relationship between factors and response variable(s) in multifactor experiments. In Taguchi methods experiments, RSM is used to develop predictive models of experimental outcomes, identify optimal parameter settings, and explore complex relationships between factors. RSM techniques complement Taguchi methods by providing detailed insights into the effects of factors and facilitating optimization of experimental designs.

By employing these statistical tools and techniques, researchers can effectively analyze data from Taguchi method designed experiments in psychology, identify influential factors, optimize experimental outcomes, and advance understanding in the field.

When reporting the results of Taguchi method designed experiments in psychology, researchers typically include several key elements to provide a comprehensive overview of the study's objectives, methods, findings, and implications. Here are the elements commonly included in a report on Taguchi methods experiments in psychology:

Title: A clear and informative title that accurately reflects the content and focus of the study.

Abstract: A concise summary of the study, including the research question, experimental design, key findings, and conclusions. The abstract provides readers with an overview of the study's objectives and outcomes.

Introduction:
- Background: Review of relevant literature and theoretical frameworks related to the research question.
- Research Objectives: Statement of the research question or hypothesis and the specific objectives of the study.
- Importance of the Study: Explanation of the significance and relevance of the research question to the field of psychology.

Methods:
- Experimental Design: Description of the Taguchi methods experimental design, including factors, levels, OAs, and response variable(s).
- Participants: Description of the sample, including demographic characteristics and any inclusion/exclusion criteria.
- Procedure: Detailed description of the experimental procedures, including data collection methods, experimental manipulations, and control measures.
- Data Analysis: Explanation of the statistical methods used to analyze the experimental data, including ANOVA, SNRs, and other relevant techniques.

Results:
- Description of Experimental Runs: Presentation of the experimental design matrix and summary of the experimental runs conducted.
- Data Analysis: Presentation of the statistical analyses conducted to assess the effects of factors on the response variable(s), including ANOVA results, SNRs, and main effects/ interaction plots.
- Findings: Summary of the main findings, including significant factors, optimal parameter settings, and implications for the research question.

Discussion:
- Interpretation of Results: Discussion of the implications of the findings in relation to the research question and theoretical frameworks.
- Comparison with Previous Research: Comparison of the current findings with previous research in the field, highlighting similarities, differences, and contributions to existing knowledge.
- Limitations: Identification of potential limitations of the study, including sources of bias, confounding factors, and constraints of the experimental design.
- Future Directions: Suggestions for future research directions, including areas for further investigation, methodological improvements, and theoretical developments.

Conclusion: A concise summary of the main findings and their implications, emphasizing the contributions of the study to the field of psychology.

References: A list of all sources cited in the report, following a specific citation style (e.g., APA, MLA).

Appendices: Additional materials, such as the experimental design matrix, raw data, statistical analyses, or supplementary results that provide further detail or support for the main findings.

By including these key elements in a report on Taguchi method designed experiments in psychology, researchers can effectively communicate their methods, results, and conclusions to readers and contribute to the advancement of knowledge in the field.

2.5.6 SEQUENTIAL EXPERIMENTATION

Sequential design experiments offer several advantages in psychology research, including efficiency, flexibility, statistical power, optimization, adaptive allocation, early stopping rules, and real-time monitoring. By leveraging these benefits, researchers can conduct more rigorous, informative, and impactful studies that advance understanding in the field of psychology. Sequential design experiments are used in psychology for several reasons:

Efficiency: Sequential design experiments allow researchers to iteratively refine their experimental design based on accumulating data. This iterative approach can lead to more efficient use of resources, as researchers can adaptively allocate resources to the most informative experimental conditions.

Flexibility: Sequential design experiments offer flexibility in exploring complex research questions that may involve multiple factors or levels. Researchers can modify the experimental design in response to emerging trends or unexpected findings, allowing for greater flexibility in hypothesis testing and exploration of alternative hypotheses.

Statistical Power: By continuously monitoring the data as they are collected, sequential design experiments can achieve higher statistical power compared to traditional fixed-sample designs. This increased power enables researchers to detect smaller effect sizes or differences between experimental conditions, leading to more reliable and robust conclusions.

Optimization: Sequential design experiments can be used to optimize experimental parameters or intervention protocols in psychology. Researchers can iteratively refine experimental conditions or treatment strategies based on accumulating data, leading to improved outcomes and more effective interventions.

Adaptive Allocation: Sequential design experiments allow for an adaptive allocation of resources to experimental conditions based on interim data analysis. Researchers can allocate more resources to promising experimental conditions or discontinue unpromising conditions, maximizing the efficiency of data collection and hypothesis testing.

Early Stopping Rules: Sequential design experiments can incorporate early stopping rules based on predefined criteria for statistical significance or effect size. Researchers can terminate the experiment early if clear and significant results are observed, reducing the time and resources required to complete the study.

Real-time Monitoring: Sequential design experiments enable real-time monitoring of data quality and experimental progress. Researchers can identify and address issues, such as data collection errors, participant dropout, or unforeseen confounding variables, as they arise, ensuring the integrity and validity of the study.

Psychologists may use sequential experimentation to optimize the parameters of a psychological assessment tool. They adjust factors like item difficulty, response format, and scoring algorithm based on real-time data and participant performance.

Sequential design experiments find applications in various areas of psychology, contributing to the advancement of research and understanding in the field. Here are some specific areas where sequential design experiments are used in psychology:

Clinical Trials: Sequential design experiments are commonly used in clinical psychology and behavioral medicine to evaluate the effectiveness of psychological interventions, pharmacological treatments, or medical procedures. Researchers may use sequential trial designs, such as sequential parallel comparison design (SPCD) or group sequential design, to monitor treatment efficacy, adjust sample sizes, or terminate trials early based on interim analyses of treatment effects.

Neuroscience Research: In cognitive neuroscience and neuropsychology, sequential design experiments are employed to study brain–behavior relationships, neural mechanisms underlying cognitive processes, or effects of brain stimulation techniques. Researchers may use sequential experimental designs to adaptively allocate stimuli, tasks, or conditions based on real-time neural activity or behavioral responses, allowing for more efficient hypothesis testing and data collection.

Intervention Development and Optimization: Sequential design experiments are used in intervention research and program evaluation to develop and optimize psychological interventions, prevention programs, or educational interventions. Researchers may use sequential trial designs to iteratively refine intervention components, assess treatment fidelity, or evaluate the effectiveness of adaptive interventions tailored to individual needs or preferences.

Psychological Assessment: In psychometric research and psychological assessment, sequential design experiments are employed to develop and validate assessment tools, tests, or measures of psychological constructs. Researchers may use sequential experimental designs to iteratively refine item pools, assess measurement reliability and validity, or identify optimal item selection algorithms based on accumulating data from test-takers.

Behavioral Economics: In experimental psychology and behavioral economics, sequential design experiments are used to study decision-making processes, economic behaviors, or social preferences. Researchers may use sequential trial designs to adaptively adjust experimental conditions, incentives, or task parameters based on participants' responses or choices, allowing for more precise estimation of economic parameters and behavioral effects.

Longitudinal Studies: In developmental psychology and longitudinal research, sequential design experiments are employed to study developmental trajectories, individual differences, or the effects of early interventions over time. Researchers may use sequential trial designs to track participants' progress, adjust intervention strategies, or assess long-term outcomes based on repeated measurements collected at multiple time points.

Sequential design experiments are used in various subfields of psychology to address research questions, optimize experimental procedures, enhance data quality, and advance knowledge and understanding in the field. By leveraging the flexibility and efficiency of sequential trial designs, psychologists can conduct more rigorous, informative, and impactful studies that contribute to theory development, evidence-based practice, and policy formulation.

STEPS

Developing sequential design experiments in psychology involves several steps to systematically plan and execute experiments in an adaptive and iterative manner. Here are the typical steps involved in developing sequential design experiments in psychology:

1. **Define Research Objectives**: Clearly articulate the research question or hypothesis that the experiment aims to address. Determine the specific objectives of the study and the response variable(s) that will be measured to evaluate the effects of experimental manipulations or interventions.

2. **Identify Factors and Levels**: Identify the factors (IVs) that may influence the response variable(s) in the experiment. Determine the levels or settings at which each factor will be studied. Factors should represent the key variables of interest in the study, such as experimental conditions, treatment strategies, or participant characteristics.

3. **Select Experimental Design**: Choose an appropriate experimental design that allows for adaptive allocation of resources and sequential analysis of data. Common designs used in sequential design experiments include group sequential design, adaptive allocation design, or Bayesian adaptive design. Select a design that suits the specific objectives and constraints of the research question.

4. **Determine Stopping Rules**: Specify stopping rules or criteria for terminating the experiment based on interim analyses of accumulating data. Stopping rules may be based on statistical significance, effect size, futility bounds, or ethical considerations. Determine when and under what conditions the experiment will be stopped to ensure valid and interpretable results.

5. **Plan Data Collection Procedures**: Develop data collection procedures and protocols for collecting data on the response variable(s) and relevant covariates. Determine how data will be collected, recorded, and monitored throughout the experiment. Establish quality control measures to ensure the integrity and reliability of the collected data.

6. **Conduct Pilot Studies**: Conduct pilot studies or simulation studies to assess the feasibility and performance of the planned experimental design. Evaluate the adequacy of the stopping rules, the efficiency of the adaptive allocation procedures, and the robustness of the data collection protocols. Use pilot data to refine the experimental design and identify potential challenges or limitations.

7. **Implement Adaptive Allocation**: Implement adaptive allocation procedures to allocate resources (e.g., participants, treatments, conditions) based on interim analyses of accumulating data. Adaptively adjust sample sizes, treatment assignments, or experimental conditions to maximize the efficiency and informativeness of the experiment while maintaining statistical validity and integrity.

8. **Monitor Data and Conduct Interim Analyses**: Continuously monitor data as they are collected and conduct interim analyses to assess the effects of experimental manipulations or interventions. Evaluate the accumulating evidence against predefined stopping rules and determine whether to continue, modify, or terminate the experiment based on interim results.

9. **Analyze Final Results**: Analyze the final results of the experiment using appropriate statistical methods to assess the effects of factors on the response variable(s). Conduct post-hoc analyses to explore subgroup effects, interaction effects, or potential confounding variables. Interpret the results in light of the research objectives and hypotheses, discussing implications for theory, practice, or future research.

10. **Report Findings**: Prepare a comprehensive report summarizing the experimental design, methods, results, and conclusions. Clearly communicate the rationale, procedures, and outcomes of the experiment, including any adaptive modifications made during the course of the study. Provide sufficient detail and transparency to enable replication and validation of the findings by other researchers.

By following these steps, researchers can systematically develop sequential design experiments in psychology to efficiently address research questions, adaptively allocate resources, and generate valid and interpretable results that contribute to the advancement of knowledge in the field.

PITFALLS AND REMEDIES

Sequential design experiments offer numerous advantages in psychology research, but they also come with potential pitfalls that researchers should be aware of. Here are some common pitfalls and remedies for sequential design experiments used in psychology:

Interim Analysis Bias:

    Pitfall: Conducting multiple interim analyses may increase the likelihood of observing false-positive results due to chance alone, leading to biased estimates of treatment effects.

    Remedy: Implement rigorous procedures to control for multiple comparisons and maintain the overall type I error rate at the desired level. Use appropriate statistical adjustments, such as Bonferroni correction or sequential testing procedures, to account for the increased risk of type I error in interim analyses.

Sample Size Determination:

    Pitfall: Inadequate sample size determination or incorrect assumptions about effect sizes may lead to underpowered studies or misleading conclusions about treatment effects.

    Remedy: Use robust statistical methods to calculate sample size requirements based on realistic effect sizes, power levels, and significance levels. Consider conducting pilot studies or simulations to estimate parameters and inform sample size calculations. Continuously monitor sample size and adaptively adjust recruitment strategies to ensure adequate power to detect meaningful effects.

Adaptive Allocation Biases:

    Pitfall: Adaptive allocation procedures may introduce biases if not properly implemented or if decisions are made based on unblinded interim data.

    Remedy: Implement blinded or masked procedures to minimize biases in adaptive allocation decisions. Use independent data monitoring committees or external statisticians to oversee interim analyses and decision-making processes. Employ predefined algorithms or protocols for adaptive allocation that are transparent, reproducible, and free from researcher discretion.

Stopping Rule Ambiguity:

    Pitfall: Ambiguous or poorly defined stopping rules may lead to premature termination of the experiment, false-positive findings, or inconclusive results.

    Remedy: Clearly define stopping rules and criteria for terminating the experiment based on predetermined thresholds for statistical significance, effect size, or futility. Specify the conditions under which the experiment will be stopped for efficacy, safety, or ethical reasons. Validate stopping rules through simulation studies or sensitivity analyses to ensure their appropriateness and robustness.

Loss of Statistical Power:

    Pitfall: Sequential analyses may lead to loss of statistical power if not properly planned or if adaptive modifications reduce the effective sample size.

    Remedy: Use statistical methods, such as group sequential design or Bayesian adaptive design, to optimize the allocation of resources and maintain statistical power throughout the experiment. Employ flexible stopping boundaries that balance the trade-off between statistical power and type I error rate. Conduct sensitivity analyses to assess the impact of adaptive modifications on statistical power and validity of results.

Ethical Considerations:

    Pitfall: Ethical concerns may arise if adaptive modifications affect participant welfare or introduce undue risks or burdens.

    Remedy: Conduct thorough ethical review and obtain approval from institutional review boards (IRBs) or ethics committees before initiating the experiment. Clearly communicate the rationale, procedures, and potential risks of adaptive modifications to participants and stakeholders. Monitor participant safety and well-being throughout the experiment and intervene if necessary to mitigate risks or address ethical concerns.

By addressing these potential pitfalls and implementing appropriate remedies, researchers can mitigate risks and enhance the validity, reliability, and ethical integrity of sequential design experiments in psychology.

EXAMPLE

Sequential experimentation in psychology involves a series of experiments where the results of each experiment inform the design of subsequent experiments. Here's a simplified example of a matrix for a sequential design experiment in psychology:

Objective: To develop an effective intervention program to improve the attention span of children with attention deficit hyperactivity disorder (ADHD) by studying the effects of different intervention components.

Components to be Tested Sequentially:

1. **Exercise**: A physical activity program.
2. **Diet**: A specific diet plan.
3. **Meditation**: Mindfulness meditation sessions.
4. **Behavioral Therapy**: A structured behavioral therapy program.

Response Variable: Improvement in attention span score (measured using a standardized assessment).

Sequential Experimentation Matrix:

In sequential experimentation, the matrix is designed to test the intervention components sequentially, with each stage building upon the results of the previous stage. The matrix might look like Tables 2.85–2.88:

Stage 1: Testing Exercise:

Stage 2: Testing Diet:

Based on the results of stage 1, you decide to test the diet component:

Stage 3: Testing Meditation:

Based on the results of stage 2, you decide to test the meditation component:

Stage 4: Testing Behavioral Therapy:

Based on the results of stage 3, you decide to test the behavioral therapy component:

TABLE 2.85

Sequential Design Experiments Stage 1 for Psychology

Run	Exercise Component	Attention Span Score
1	None	...
2	Low intensity	...
3	Moderate intensity	...
4	High intensity	...

TABLE 2.86

Sequential Design Experiments Stage 2 for Psychology

Run	Diet Component	Attention Span Score
5	No special diet	...
6	Low sugar diet	...
7	Gluten-free diet	...
8	Balanced diet	...

TABLE 2.87

Sequential Design Experiments Stage 3 for Psychology

Run	Meditation Component	Attention Span Score
9	No meditation	...
10	Guided meditation	...
11	Mindfulness meditation	...
12	Yoga meditation	...

TABLE 2.88

Sequential Design Experiments Stage 4 for Psychology

Run	Behavioral Therapy Component	Attention Span Score
13	No behavioral therapy	...
14	Structured therapy	...
15	Cognitive-behavioral therapy	...
16	Parent training program	...

In this example, each stage represents a set of experimental runs with specific combinations of intervention components. The results of each stage guide the adjustments made in subsequent stages to systematically develop an effective intervention program for children with ADHD. Sequential experimentation allows psychologists to efficiently refine their intervention approach based on the evolving understanding of the treatment's effectiveness.

REPORTING

Sequential design experiments in psychology employ various statistical tools to analyze data, monitor progress, and make adaptive decisions throughout the experiment. Here are some common statistical tools used in sequential design experiments:

ANOVA: ANOVA is used to compare means across multiple groups or conditions to determine whether there are statistically significant differences in outcomes. In sequential design experiments, ANOVA may be used to analyze data at interim points to assess treatment effects or differences between experimental conditions.

Sequential Analysis: Sequential analysis methods are used to monitor accumulating data throughout the experiment and make adaptive decisions based on interim analyses. Sequential methods allow for early stopping, sample size adjustment, or modification of experimental conditions based on predefined criteria for statistical significance or effect size.

Bayesian Methods: Bayesian statistical methods are used to update beliefs or probabilities about treatment effects or experimental outcomes based on accumulating data. Bayesian approaches allow for flexible modeling of uncertainty, incorporation of prior information, and adaptive decision-making in sequential design experiments.

Signal Detection Theory: Signal detection theory is used to analyze data from tasks involving signal detection or discrimination, such as perceptual tasks or diagnostic testing. Signal detection theory allows for the estimation of sensitivity (d') and response bias (criterion) and can be used to evaluate performance in sequential design experiments.

Survival Analysis: Survival analysis methods are used to analyze time-to-event data, such as time until a particular outcome or event occurs. Survival analysis allows for the estimation of survival curves, hazard ratios, and other measures of time-to-event outcomes in sequential design experiments with longitudinal or survival data.

Group Sequential Design: Group sequential design methods allow for adaptive monitoring of accumulating data in clinical trials and other longitudinal studies. Group sequential designs incorporate predefined stopping rules and interim analyses to assess treatment effects, adjust sample size, or terminate the experiment early based on predefined criteria for statistical significance.

RSM: RSM is used to model and optimize the relationship between experimental factors and response variables in multifactor experiments. RSM allows for the exploration of complex relationships between factors and the prediction of optimal parameter settings in sequential design experiments.

Multilevel Modeling: Multilevel modeling techniques are used to analyze hierarchical or nested data structures, such as repeated measures or longitudinal data collected from individuals nested within groups. Multilevel models allow for the estimation of fixed and random effects and can be used to account for within-subject and between-subject variability in sequential design experiments.

By employing these statistical tools and methods, researchers can effectively analyze data, monitor progress, and make adaptive decisions in sequential design experiments in psychology, leading to more efficient and informative studies.

When reporting the results of sequential design experiments in psychology, it's essential to provide a comprehensive overview of the study's objectives, methods, findings, and implications. Here are the key elements typically included in a report on sequential design experiments:

Title: A concise and informative title that accurately reflects the content and focus of the study.

Abstract: A brief summary of the study, including the research question, experimental design, key findings, and conclusions. The abstract should provide readers with a clear understanding of the study's objectives and outcomes.

Introduction:
- Background: Review of relevant literature and theoretical frameworks related to the research question.
- Research Objectives: Statement of the research question or hypothesis and the specific aims of the study.
- Importance of the Study: Explanation of the significance and relevance of the research question to the field of psychology.

Methods:
- Experimental Design: Description of the sequential experimental design, including adaptive allocation procedures, stopping rules, and interim analyses.
- Participants: Description of the sample, including demographic characteristics, recruitment procedures, and inclusion/exclusion criteria.
- Procedure: Detailed explanation of the experimental procedures, data collection methods, and any adaptive modifications made during the course of the study.
- Statistical Analysis: Description of the statistical methods used to analyze the data, including sequential analysis techniques, Bayesian methods, and adjustments for multiple comparisons.

Results:
- Description of Experimental Runs: Presentation of the sequential experimental design matrix and summary of participant recruitment, data collection, and adaptive modifications made during the study.

- Data Analysis: Presentation of the main findings, including results of interim analyses, treatment effects, and differences between experimental conditions.
- Visual Representation: Graphs, tables, or charts illustrating key findings, such as sequential plots of treatment effects or cumulative sample sizes over time.

Discussion:
- Interpretation of Results: Discussion of the implications of the findings in relation to the research question and theoretical frameworks.
- Comparison with Previous Research: Comparison of the current findings with previous research in the field, highlighting similarities, differences, and contributions to existing knowledge.
- Limitations: Identification of potential limitations of the study, such as biases, confounding factors, or constraints of the experimental design.
- Future Directions: Suggestions for future research directions, including areas for further investigation, methodological improvements, and theoretical developments.

Conclusion: A concise summary of the main findings and their implications, emphasizing the contributions of the study to the field of psychology.

References: A list of all sources cited in the report, following a specific citation style (e.g., APA, MLA).

Appendices: Additional materials, such as detailed descriptions of adaptive allocation procedures, interim analysis plans, or supplementary results that provide further detail or support for the main findings.

By including these key elements in a report on sequential design experiments in psychology, researchers can effectively communicate the methods, results, and conclusions of their study and contribute to the advancement of knowledge in the field.

2.5.7 RPD

RPD- experiments offer a systematic and rigorous approach to addressing variability, optimizing interventions, understanding interaction effects, enhancing reproducibility, reducing bias and error, and improving decision-making in psychology research and practice. By leveraging the principles of robust parameter design, researchers can generate more reliable, generalizable, and impactful findings that contribute to the advancement of knowledge and the improvement of human well-being. RPD experiments are used in psychology for several reasons:

Accounting for Variability: Psychology research often deals with complex and multifaceted phenomena, where variability is inherent. RPD experiments allow researchers to systematically account for variability in factors, such as participant characteristics, environmental conditions, or measurement error. By identifying robust solutions that are less sensitive to variability, researchers can ensure the reliability and generalizability of their findings.

Optimizing Interventions: In clinical psychology and behavioral interventions, RPD experiments can be used to optimize intervention protocols or treatment strategies. By systematically varying key parameters and assessing their effects on outcomes, researchers can identify robust intervention strategies that are effective across diverse populations or contexts. This approach helps improve the efficacy and effectiveness of psychological interventions.

Understanding Interaction Effects: RPD experiments enable researchers to study interaction effects between multiple factors or variables. In psychology, interactions between individual differences, situational factors, and treatment modalities are common. RPD allows researchers to systematically explore these interactions and identify robust patterns of effects, enhancing understanding of complex psychological processes.

Enhancing Reproducibility: Reproducibility and replicability are critical aspects of scientific research. RPD experiments facilitate the identification of robust findings that can be replicated across different studies, settings, or samples. By focusing on robust parameter settings and reducing the influence of extraneous factors, RPD enhances the reproducibility of psychological research findings.

Reducing Bias and Error: RPD experiments help mitigate biases and errors that may arise in experimental designs or data collection procedures. By systematically controlling for sources of bias and variability, researchers can improve the internal validity and reliability of their studies, leading to more accurate and trustworthy conclusions.

Improving Decision-Making: RPD experiments provide valuable insights for decision-making in applied psychology settings. Whether in clinical practice, organizational psychology, or educational interventions, RPD helps identify robust solutions that can inform evidence-based decision-making and optimize outcomes for individuals, groups, or organizations.

In clinical trials, RPD may be applied to ensure the robustness of a psychological treatment against variations in patient characteristics and treatment adherence. RPD experiments find applications in various areas of psychology contributing to the understanding and improvement of psychological phenomena and interventions. Here are some specific areas where RPD dexperiments are used in psychology:

Clinical Psychology: RPD experiments are employed in clinical psychology to optimize treatment protocols and interventions for various psychological disorders, such as depression, anxiety, and post-traumatic stress disorder (PTSD). Researchers use RPD to systematically vary treatment parameters, such as dosage, frequency, or delivery format, to identify robust treatment strategies that are effective across diverse populations and contexts.

Behavioral Interventions: In behavioral interventions and behavior change programs, RPD experiments are used to optimize intervention components and strategies for promoting healthy behaviors and reducing risk factors for psychological disorders. Researchers systematically vary intervention parameters, such as incentives, prompts, or feedback mechanisms, to identify robust behavior change techniques that produce sustained outcomes over time.

Educational Psychology: RPD experiments are employed in educational psychology to optimize instructional methods, curriculum design, and learning environments. Researchers use RPD to systematically vary instructional parameters, such as teaching strategies, feedback mechanisms, or assessment formats, to identify robust approaches that enhance student learning outcomes and academic achievement.

Organizational Psychology: In organizational psychology and human resource management, RPD experiments are used to optimize organizational processes, workplace interventions, and leadership strategies. Researchers systematically vary organizational parameters, such as communication protocols, performance incentives, or team structures, to identify robust approaches that improve employee well-being, job satisfaction, and organizational performance.

Health Psychology: RPD experiments are employed in health psychology to optimize health promotion programs, patient interventions, and healthcare delivery systems. Researchers systematically vary intervention parameters, such as communication strategies, behavioral incentives, or treatment modalities, to identify robust approaches that enhance health outcomes, patient adherence, and healthcare efficiency.

Cognitive Psychology: In cognitive psychology and neuroscience, RPD experiments are used to optimize experimental paradigms, stimuli presentations, and data analysis techniques.

Researchers systematically vary experimental parameters, such as task difficulty, stimulus timing, or response modalities, to identify robust methods that minimize confounding effects and enhance the reliability and validity of cognitive assessments.

RPD experiments are used in various subfields of psychology to optimize interventions, enhance learning outcomes, improve organizational processes, promote health behaviors, and advance understanding of cognitive processes. By systematically varying parameters and identifying robust solutions, RPD contributes to the development of evidence-based practices and the improvement of human well-being in diverse psychological contexts.

STEPS

Developing RPD) experiments in psychology involves several steps to systematically optimize interventions, understand psychological phenomena, and improve outcomes. Here are the steps to develop RPD experiments in psychology:

1. **Define Research Objectives**: Clearly articulate the research question or hypothesis that the experiment aims to address. Determine the specific objectives of the study and the outcome variables that will be measured to evaluate the effects of intervention parameters or psychological factors.
2. **Identify Factors and Levels**: Identify the factors (IVs) that may influence the outcome variables in the experiment. Determine the levels or settings at which each factor will be studied. Factors may include intervention components, participant characteristics, environmental conditions, or task parameters.
3. **Determine Key Response Variables**: Identify the primary and secondary outcome variables that will be used to assess the effects of intervention parameters or psychological factors. Choose outcome measures that are sensitive, reliable, and relevant to the research question and objectives.
4. **Design Experimental Matrix**: Develop a matrix that systematically varies the levels of factors and combinations of parameter settings to be tested in the experiment. Use techniques, such as OAs or Latin squares, to efficiently explore the parameter space and identify robust solutions.
5. **Conduct Pilot Studies**: Conduct pilot studies or simulation studies to assess the feasibility and performance of the experimental design. Evaluate the sensitivity of outcome measures, the robustness of parameter settings, and the efficiency of data collection procedures.
6. **Perform Sensitivity Analysis**: Conduct sensitivity analyses to assess the impact of individual factors or parameter settings on the outcome variables. Identify factors that have the greatest influence on outcomes and prioritize them for further investigation or optimization.
7. **Optimize Parameter Settings**: Use statistical techniques, such as RSM, DOE, or optimization algorithms, to identify optimal parameter settings that maximize desired outcomes or minimize variability. Iteratively refine the experimental design based on insights gained from pilot studies and sensitivity analyses.
8. **Validate Robust Solutions**: Validate the robustness of optimized parameter settings through additional experimental trials or validation studies. Assess the stability and reproducibility of outcomes across different samples, settings, or conditions to ensure the generalizability of findings.
9. **Implement Adaptive Strategies**: Implement adaptive strategies to monitor and adjust intervention parameters or experimental conditions based on real-time data analysis and interim results. Continuously evaluate the effectiveness of interventions and make adaptive modifications to enhance outcomes.

10. **Document Findings and Recommendations**: Document the findings of the RPD experiment, including optimized parameter settings, robust solutions, and recommendations for practice or further research. Prepare a comprehensive report summarizing the experimental design, methods, results, and conclusions.

By following these steps, researchers can systematically develop RPD experiments in psychology to optimize interventions, understand psychological processes, and improve outcomes in diverse contexts.

PITFALLS AND REMEDIES

RPD experiments offer numerous advantages in psychology research, but they also come with potential pitfalls that researchers should be aware of. Here are some common pitfalls and remedies for RPD experiments used in psychology:

Overemphasis on Optimization:
  Pitfall: Researchers may focus excessively on optimizing parameter settings without adequately considering the robustness of solutions across different conditions or populations.
  Remedy: Balance optimization objectives with the need for robustness and generalizability. Incorporate sensitivity analyses and validation studies to assess the stability and reproducibility of optimized solutions across diverse samples, settings, or contexts.
Failure to Consider Interactions:
  Pitfall: RPD experiments may overlook interactions between factors or parameter settings, leading to suboptimal solutions or unexpected outcomes.
  Remedy: Systematically explore interactions between factors using factorial designs or RSMs. Analyze interaction effects to identify synergistic or antagonistic relationships between parameters and optimize solutions accordingly.
Limited External Validity:
  Pitfall: Optimized parameter settings may lack external validity or generalizability to real-world settings or populations beyond the study sample.
  Remedy: Incorporate diverse samples, settings, and conditions in the experimental design to enhance external validity. Conduct validation studies or replication experiments in different contexts to assess the robustness and generalizability of optimized solutions.
Inadequate Control of Confounding Factors:
  Pitfall: Confounding factors or uncontrolled variables may influence outcomes, leading to biased or misleading results.
  Remedy: Implement rigorous control procedures to minimize the influence of confounding factors. Use randomization, counterbalancing, or matched sampling techniques to ensure comparability between experimental conditions and control for extraneous variables.
Limited Flexibility in Adaptive Strategies:
  Pitfall: Adaptive strategies may lack flexibility or adaptability to changing conditions or unforeseen circumstances.
  Remedy: Design adaptive strategies that incorporate predefined decision rules or algorithms for modifying parameter settings based on real-time data analysis. Allow for flexibility in adaptive decision-making to accommodate unexpected findings or emergent patterns in the data.
Overfitting of Models:
  Pitfall: Statistical models used to optimize parameter settings may overfit the data, resulting in solutions that are overly tailored to specific samples or conditions.

Remedy: Employ robust statistical techniques and regularization methods to prevent over-fitting and ensure the generalizability of optimized solutions. Validate models using independent datasets or cross-validation techniques to assess their predictive accuracy and robustness.

Ethical Considerations:

Pitfall: Ethical considerations may arise if experimental manipulations pose risks or harm to participants, particularly in clinical or vulnerable populations.

Remedy: Conduct thorough ethical review and obtain approval from IRBs or ethics committees before initiating the experiment. Prioritize participant safety and well-being, and adhere to ethical guidelines and standards throughout the research process.

By addressing these potential pitfalls and implementing appropriate remedies, researchers can enhance the validity, reliability, and ethical integrity of RPD experiments in psychology, leading to more robust and generalizable findings.

EXAMPLES

An RPD experiment in psychology aims to optimize a process or intervention while making it less sensitive to variations and external factors. Here's an example of a matrix for an RPD experiment in psychology:

Objective: To optimize a stress management program by studying the effects of two IVs, mindfulness meditation duration (factor A) and guided relaxation duration (factor B), on reducing participants' self-reported stress levels while making the program robust against variations in instructor expertise.

Factors:

1. **Mindfulness Meditation Duration (Factor A):**
   - Short duration (e.g., 15 minutes).
   - Medium duration (e.g., 30 minutes).
   - Long duration (e.g., 45 minutes).
2. **Guided Relaxation Duration (Factor B):**
   - Short duration (e.g., 10 minutes).
   - Medium duration (e.g., 20 minutes).
   - Long duration (e.g., 30 minutes).

External Factor: Instructor expertise (measured as low, medium, or high expertise).

Response Variable: Self-reported stress reduction score.

RPD Experiment Matrix:

In an RPD experiment, the matrix is designed to optimize the factors while considering robustness against external factors, such as instructor expertise. The matrix is shown in Table 2.89.

(Continuation of the matrix with all possible combinations...)

In this matrix, each row represents a specific experimental run with a unique combination of mindfulness meditation duration, guided relaxation duration, and instructor expertise. The goal is to optimize the stress management program to achieve the greatest reduction in self-reported stress levels while making it robust to variations in instructor expertise.

Statistical tools and analysis in RPD experiments are used to identify factor settings that minimize the sensitivity of the program's effectiveness to external factors, leading to a more reliable and effective stress management program.

TABLE 2.89

RPD Matrix Example for Psychology

Run	Meditation Duration (A)	Relaxation Duration (B)	Instructor Expertise	Self-Reported Stress Reduction
1	Short	Short	Low	...
2	Short	Short	Medium	...
3	Short	Short	High	...
4	Short	Medium	Low	...
5	Short	Medium	Medium	...
6	Short	Medium	High	...
7	Short	Long	Low	...
8	Short	Long	Medium	...
9	Short	Long	High	...

TABLE 2.90

RPD Matrix Example for Psychology

Run	Training	Duration (Minutes)	Cognitive Load (Low/High)	Outcome (Accuracy)
1	20	Low	75%	--
2	20	High	70%	--
3	20	Low	78%	--
4	20	High	72%	--
5	40	Low	80%	--
6	40	High	68%	--
7	40	Low	82%	--
8	40	High	70%	--
9	60	Low	85%	--
10	60	High	65%	--
11	60	Low	88%	--
12	60	High	62%	--

Another example of a matrix for an RPD experiment used in psychology, specifically focusing on optimizing a cognitive training program to improve WM performance in older adults is shown in Table 2.90.

In this example,

- **Training Duration (minutes)**: This factor represents the duration of the cognitive training session, varying from 20 to 60 minutes.
- **Cognitive Load (low/high)**: This factor represents the cognitive load imposed during the training session, with levels of low and high load.
- **Outcome (Accuracy)**: Accuracy in WM tasks is the primary outcome measure, represented as the percentage of correctly completed tasks.

This matrix systematically varies the levels of training duration and cognitive load to evaluate their effects on WM performance in older adults. Researchers can analyze the data collected from these experimental runs to identify robust parameter settings that optimize WM performance across different conditions and durations of cognitive training.

Example of a RPD experiment aimed at systematically varying key parameters of a cognitive-behavioral therapy (CBT) intervention to identify robust strategies for reducing depressive symptoms across diverse populations:

Research Question: How can we optimize the effectiveness of a CBT intervention for reducing symptoms of depression in adults with mild-to-moderate depression?

Objective: To systematically vary key parameters of the CBT intervention to identify robust strategies that maximize reductions in depressive symptoms across diverse adult populations.

Experimental Design:

1. **Factor 1: Session Format**
   - Levels: Individual therapy sessions, group therapy sessions, a combination of individual and group sessions,
2. **Factor 2: Duration of Sessions**
   - Levels: 60 minutes, 90 minutes, 120 minutes,
3. **Factor 3: Homework Assignments**
   - Levels: Regular homework assignments, reduced homework assignments, no homework assignments,

Outcome Measure: Changes in self-reported depressive symptoms assessed using standardized depression scales (e.g., Beck depression inventory-II) administered at baseline, mid-intervention, post-intervention, and follow-up.

Experimental Procedure:

1. Participants are recruited from diverse demographic backgrounds and randomly assigned to different combinations of session format, duration of sessions, and homework assignments.
2. The CBT intervention is delivered over a predetermined period (e.g., 12 weeks), with participants attending sessions according to their assigned format and duration.
3. Participants are instructed to complete homework assignments as prescribed (if applicable) and to attend all scheduled sessions.
4. Assessments of depressive symptoms are conducted at baseline, mid-intervention (e.g., after 6 weeks), post-intervention (e.g., after 12 weeks), and at follow-up (e.g., 3 months after intervention completion).
5. Adaptive modifications may be made to the intervention protocol based on interim analyses of participant responses, such as adjusting session format, duration, or homework assignments.

Data Analysis:

- ANOVA is used to examine the main effects and interaction effects of session format, duration of sessions, and homework assignments on changes in depressive symptoms.
- RSM is employed to identify optimal parameter settings that maximize reductions in depressive symptoms while accounting for interactions between factors.

Results:

- The experiment identifies robust combinations of session format, duration of sessions, and homework assignments that produce significant reductions in depressive symptoms across diverse adult populations.
- Interaction effects between session format, duration of sessions, and homework assignments are explored to understand how different factors interact to influence intervention effectiveness.

Conclusion:

- The study provides evidence-based recommendations for optimizing the design and delivery of CBT interventions to effectively reduce symptoms of depression in adults with mild-to-moderate depression.
- Robust parameter settings identified through the experiment can inform the development of personalized and adaptive CBT interventions tailored to individual needs and preferences.

This example demonstrates how a RPD experiment can be implemented to systematically vary key parameters of a CBT intervention and identify robust strategies for reducing depressive symptoms across diverse adult populations.

Another example of a design matrix used in a RPD experiment to systematically vary key parameters of a CBT intervention and identify robust strategies for reducing depressive symptoms across diverse populations is shown in Table 2.91.

TABLE 2.91

RPD Matrix Example for Psychology

Run	Session Format	Duration of Sessions	Homework Assignments	Outcome (% Reduction in Depressive Symptoms)
1	Individual	60 minutes	Regular	50%
2	Individual	90 minutes	Regular	45%
3	Individual	120 minutes	Regular	40%
4	Individual	60 minutes	Reduced	55%
5	Individual	90 minutes	Reduced	48%
6	Individual	120 minutes	Reduced	42%
7	Individual	60 minutes	None	60%
8	Individual	90 minutes	None	50%
9	Individual	120 minutes	None	45%
10	Group	60 minutes	Regular	55%
11	Group	90 minutes	Regular	52%
12	Group	120 minutes	Regular	48%
13	Group	60 minutes	Reduced	58%
14	Group	90 minutes	Reduced	55%
15	Group	120 minutes	Reduced	50%
16	Group	60 minutes	None	62%
17	Group	90 minutes	None	58%
18	Group	120 minutes	None	55%
19	Combination	60 minutes	Regular	58%
20	Combination	90 minutes	Regular	55%
21	Combination	120 minutes	Regular	52%
22	Combination	60 minutes	Reduced	60%
23	Combination	90 minutes	Reduced	58%
24	Combination	120 minutes	Reduced	55%
25	Combination	60 minutes	None	65%
26	Combination	90 minutes	None	62%
27	Combination	120 minutes	None	60%

In this design matrix,

- **Session Format**: Represents the format of therapy sessions, including individual, group, and a combination of both.
- **Duration of Sessions**: Indicates the length of therapy sessions in minutes, ranging from 60 to 120 minutes.
- **Homework Assignments**: Specifies the type of homework assignments given to participants, such as regular, reduced, or none.
- **Outcome (% Reduction in Depressive Symptoms)**: Reflects the percentage reduction in depressive symptoms observed for each combination of parameters.

This matrix systematically varies the key parameters of the CBT intervention to explore their effects on reducing depressive symptoms across diverse populations. Each row represents a unique combination of session format, duration, and homework assignments, with the corresponding outcome indicating the percentage reduction in depressive symptoms observed.

REPORTING

In RPD experiments in psychology, various statistical tools are employed to analyze data, identify optimal parameter settings, and assess the robustness of solutions. Here are some common statistical tools used in RPD experiments in psychology:

ANOVA: ANOVA is used to analyze the effects of different factors (e.g., intervention components, experimental conditions) on outcome variables (e.g., psychological measures, behavioral outcomes). ANOVA allows researchers to assess the main effects and interaction effects between factors, helping to identify significant sources of variability in the data.

RSM: RSM is a statistical technique used to model and optimize the relationship between input variables (factors) and output variables (responses) in complex systems. RSM allows researchers to identify optimal parameter settings that maximize or minimize desired outcomes while accounting for interactions between factors.

Factorial Design Analysis: Factorial designs are used to systematically vary multiple factors and assess their effects on outcome variables. Researchers analyze factorial designs using statistical methods, such as factorial ANOVA or regression analysis, to identify main effects and interaction effects between factors.

Regression Analysis: Regression analysis is used to model the relationship between predictor variables (e.g., intervention parameters, experimental conditions) and outcome variables (e.g., psychological measures, performance outcomes). Regression models allow researchers to estimate the effects of individual factors while controlling for other variables.

DOE: DOE is a systematic approach to planning and conducting experiments to optimize process parameters and improve product or system performance. In psychology, DOE techniques are used to design experiments that systematically vary intervention components, experimental conditions, or participant characteristics to identify optimal parameter settings.

Sensitivity Analysis: Sensitivity analysis is used to assess the impact of changes in input variables (e.g., intervention parameters, experimental conditions) on output variables (e.g., psychological outcomes, performance measures). Sensitivity analysis helps researchers identify critical factors that have the greatest influence on outcomes and prioritize them for further investigation or optimization.

Monte Carlo Simulation: Monte Carlo simulation techniques are used to model complex systems and evaluate the robustness of solutions under different conditions or scenarios.

Monte Carlo simulations allow researchers to assess the variability and uncertainty in outcomes and make informed decisions about parameter settings.

Bayesian Methods: Bayesian statistical methods are used to update beliefs or probabilities about parameter settings based on observed data. Bayesian approaches allow researchers to incorporate prior knowledge or information into the analysis and make probabilistic predictions about the effects of different factors on outcomes.

By employing these statistical tools and techniques, researchers can effectively analyze data, optimize parameter settings, and assess the robustness of solutions in RPD experiments in psychology, leading to more reliable and generalizable findings.

A report on a RPD experiment used in psychology typically includes several key elements to communicate the research methodology, findings, and implications effectively. Here are the essential elements of such a report:

Title: A concise and informative title that accurately reflects the content and purpose of the study.

Abstract: A brief summary of the research, including the research question, objectives, methodology, key findings, and implications. The abstract provides readers with a quick overview of the study's main points.

Introduction:
- Background: A review of relevant literature and theoretical framework that provides context for the study.
- Research Question/Hypotheses: Clear statement of the research question or hypotheses being tested in the experiment.
- Objectives: Explanation of the specific aims and objectives of the study.

Methodology:
- Experimental Design: Description of the experimental design, including the factors studied, levels of each factor, and rationale for their selection.
- Participants: Details about the study participants, including sample size, demographics, and any inclusion/exclusion criteria.
- Procedure: Overview of the experimental procedure, including how interventions were administered, data collection methods, and any control measures implemented.
- Statistical Analysis: Explanation of the statistical methods used to analyze the data, including any specific techniques for RPD experiments (e.g., ANOVA, RSM).

Results:
- Presentation of Findings: Clear and concise presentation of the study's results, including descriptive statistics, graphical representations (e.g., plots, charts), and inferential statistics.
- Main Effects and Interaction Effects: Discussion of main effects and interaction effects observed in the data, highlighting significant findings and patterns.

Discussion:
- Interpretation of Findings: Analysis and interpretation of the results in relation to the research question and objectives.
- Implications: Discussion of the practical implications of the findings for theory, practice, and future research in psychology.
- Strengths and Limitations: Identification of the strengths and limitations of the study, including potential sources of bias or confounding factors.
- Generalizability: Consideration of the generalizability of the findings to broader populations or contexts.

Conclusion: Summary of the key findings and conclusions drawn from the study, emphasizing their significance and relevance.

References: List of citations for all sources referenced in the report, following a consistent citation style (e.g., APA, MLA).

Appendices: Supplementary materials, such as detailed descriptions of experimental procedures, raw data tables, or additional analyses, that provide further context or support for the findings.

By including these elements in the report, researchers can effectively communicate the rationale, methodology, findings, and implications of their RPD experiment in psychology.

2.5.8 RANDOMIZED EXPERIMENTS

Randomized experiments are valued in psychology for their ability to establish causal relationships, control for confounding variables, enhance internal validity, facilitate replication and generalizability, address ethical considerations, and enable rigorous statistical analyses. These attributes make randomized experiments a cornerstone of empirical research in psychology. They are widely used in clinical psychology to study the effects of psychological interventions on mental health outcomes. Patients are randomly assigned to different treatment groups to assess treatment efficacy. Randomized experiments are widely used in psychology for several reasons:

Causality and Inference: Randomized experiments allow researchers to establish causal relationships between variables. By randomly assigning participants to different conditions, researchers can infer that any observed differences in outcomes are due to the manipulation of the IV, rather than other confounding factors. This strengthens the validity of conclusions drawn from the study.

Control over Confounding Variables: Random assignment helps to control for potential confounding variables that may influence the outcomes of the study. By randomly distributing participants across experimental conditions, researchers can ensure that any individual differences between participants are equally distributed among the groups, reducing the risk of bias.

Internal Validity: Randomized experiments are known for their high internal validity. Internal validity refers to the degree to which a study accurately measures the effect of the IV on the DV, without the influence of extraneous variables. Random assignment enhances internal validity by minimizing the possibility of alternative explanations for observed effects.

Replicability and Generalizability: Randomized experiments enhance the replicability and generalizability of findings. Because random assignment helps to control for confounding variables and biases, the results of randomized experiments are more likely to be replicated across different samples, settings, and conditions, increasing confidence in the generalizability of the findings.

Ethical Considerations: Randomized experiments often involve the manipulation of variables or interventions that could potentially impact participants' well-being. Random assignment helps to ensure that participants have an equal chance of being exposed to different conditions, thereby reducing concerns about unfair treatment or harm.

Statistical Analyses: Randomized experiments are amenable to a wide range of statistical analyses, allowing researchers to assess the effects of interventions or treatments with greater precision and reliability. Random assignment facilitates the use of inferential statistical tests to determine the significance of observed differences between groups.

Randomized experiments are used in various areas of psychology, including but not limited to:

Clinical Psychology: RCTs are commonly used to evaluate the efficacy and effectiveness of psychological interventions for treating mental health disorders, such as depression, anxiety, PTSD, substance abuse, and eating disorders. For example, RCTs may be conducted to compare the effectiveness of different psychotherapeutic approaches (e.g., CBT, mindfulness-based interventions) or pharmacological treatments.

Educational Psychology: Randomized experiments are employed to assess the impact of educational interventions, teaching methods, and classroom strategies on student learning outcomes, academic achievement, and educational attainment. Researchers may conduct RCTs to evaluate the effectiveness of instructional interventions, curriculum modifications, or educational technology tools.

Developmental Psychology: Randomized experiments are used to investigate the effects of various environmental and social factors on child development, such as parenting styles, early childhood interventions, educational programs, and peer interactions. RCTs may be conducted to examine the efficacy of interventions aimed at promoting positive developmental outcomes or mitigating risk factors for developmental delays or behavioral problems.

Social Psychology: Randomized experiments are employed to test hypotheses about social behavior, attitudes, interpersonal relationships, group dynamics, and social influence processes. Researchers may use RCTs to assess the effects of interventions designed to reduce prejudice, increase prosocial behavior, promote healthy relationships, or change attitudes and beliefs.

Cognitive Psychology: Randomized experiments are used to investigate cognitive processes, such as attention, memory, perception, decision-making, and problem-solving. Researchers may conduct RCTs to evaluate the effectiveness of cognitive training programs, memory enhancement techniques, or interventions for cognitive rehabilitation following brain injury or cognitive impairment.

Health Psychology: RCTs are widely used in health psychology to evaluate the effectiveness of interventions aimed at promoting health behavior change, preventing illness, managing chronic conditions, and improving quality of life. RCTs may be conducted to assess the impact of interventions targeting smoking cessation, weight management, physical activity, medication adherence, stress management, and disease self-management.

Industrial-Organizational Psychology: Randomized experiments are used to test interventions and organizational practices aimed at improving workplace productivity, employee satisfaction, leadership effectiveness, team performance, and organizational culture. Researchers may conduct RCTs to evaluate the impact of training programs, leadership development initiatives, performance feedback systems, or organizational interventions aimed at reducing workplace stress and enhancing employee well-being.

RCT designed experiments are employed across various subfields of psychology to investigate causal relationships, evaluate interventions, test theoretical hypotheses, and advance scientific knowledge in the discipline.

STEPS

Developing a RD experiment in psychology involves several key steps. Here's a general outline of the process:

1. **Define Research Question**: Clearly define the research question or hypothesis that the experiment aims to address. The research question should be specific, testable, and relevant to the area of psychology being studied.
2. **Review Literature**: Conduct a comprehensive review of existing literature and theoretical frameworks relevant to the research question. This step helps to identify gaps in knowledge, refine the research question, and inform the design of the experiment.

3. **Identify Variables**: Identify the IV and DV that will be manipulated and measured in the experiment, respectively. The IV is the factor(s) that the researcher manipulates or varies, while the DV is the outcome(s) that are observed and measured.
4. **Design Experimental Conditions**: Determine the experimental conditions or treatment groups that will be compared in the experiment. Randomized experiments typically involve at least two experimental conditions: a treatment group that receives the experimental manipulation and a control group that does not.
5. **Random Assignment**: Randomly assign participants to different experimental conditions to ensure that each participant has an equal chance of being assigned to any condition. Random assignment helps to control potential confounding variables and ensures that any observed differences between groups are due to the experimental manipulation.
6. **Develop Experimental Protocol**: Develop a detailed experimental protocol outlining the procedures and instructions for implementing the experiment. This includes information about participant recruitment, informed consent procedures, experimental manipulations, data collection methods, and any ethical considerations.
7. **Pilot Testing**: Conduct a pilot test or small-scale pilot study to evaluate the feasibility and effectiveness of the experimental protocol. Pilot testing helps to identify and address any potential issues with the procedures or materials before conducting the full experiment.
8. **Data Collection**: Implement the experiment according to the experimental protocol and collect data from participants. Ensure that data collection procedures are standardized and consistent across experimental conditions to minimize bias and ensure reliability.
9. **Data Analysis**: Analyze the collected data using appropriate statistical techniques to test hypotheses and answer the research question. Common statistical analyses for randomized experiments include t-tests, ANOVA, regression analysis, and other inferential statistical tests.
10. **Interpret Results**: Interpret the results of the data analysis in relation to the research question and hypotheses. Determine whether the experimental manipulation had a significant effect on the DV(s) and discuss the implications of the findings.
11. **Report Findings**: Communicate the findings of the experiment through a research report or manuscript. The report should include a detailed description of the research methods, results, interpretations, and conclusions, following the conventions of scientific writing in psychology.
12. **Peer Review and Publication**: Submit the research report to a peer-reviewed journal in the field of psychology for publication. Peer review helps to ensure the quality and validity of the research findings before they are disseminated to the scientific community.

By following these steps, researchers can develop and conduct RD experiments in psychology to investigate causal relationships, test hypotheses, and advance scientific knowledge in the discipline.

PITFALLS AND REMEDIES

RD experiments in psychology are powerful tools for establishing causal relationships and testing hypotheses. However, they can also be susceptible to various pitfalls. Here are some common pitfalls and their corresponding remedies:

Selection Bias:
> **Pitfall**: Despite random assignment, selection bias can occur if there are systematic differences between participants in different experimental conditions.
> **Remedy**: Increase the sample size to enhance the likelihood that randomization will distribute participants evenly across conditions. Additionally, use stratified random sampling to ensure that key demographic variables are balanced across groups.

Attrition Bias:
> **Pitfall**: Attrition bias occurs when participants drop out of the study disproportionately across different experimental conditions, leading to biased results.
>
> **Remedy**: Minimize attrition by implementing strategies to retain participants throughout the study, such as providing incentives, maintaining regular communication, and reducing the burden of participation. Additionally, conduct intention-to-treat analyses to include all randomized participants in the analysis, regardless of whether they completed the intervention.

Experimenter Bias:
> **Pitfall**: Experimenter bias occurs when researchers' expectations or behaviors inadvertently influence participants' responses or the outcome of the study.
>
> **Remedy**: Implement double-blind or triple-blind procedures whenever possible to minimize experimenter bias. In a double-blind study, both participants and experimenters are unaware of the participants' assigned condition, while in a triple-blind study, data analysts are also kept unaware of the conditions. Additionally, standardize procedures and protocols across experimenters to reduce variability in implementation.

Demand Characteristics:
> **Pitfall**: Demand characteristics occur when participants modify their behavior or responses based on their perceptions of the experiment's purpose or the experimenter's expectations.
>
> **Remedy**: Use deception or cover stories to mask the true purpose of the experiment and reduce participants' awareness of the experimental manipulation. Additionally, employ placebo controls or manipulation checks to assess the effectiveness of blinding and detect demand characteristics.

Cross-Contamination:
> **Pitfall**: Cross-contamination occurs when participants in one experimental condition are inadvertently exposed to the intervention or treatment intended for another condition.
>
> **Remedy**: Implement strict protocols to prevent contamination between experimental conditions. This may include physical separation of participants, ensuring confidentiality of intervention materials, and monitoring adherence to the experimental protocol.

Baseline Differences:
> **Pitfall**: Despite random assignment, baseline differences in key variables may still exist between experimental groups, leading to biased results.
>
> **Remedy**: Conduct pretest assessments to measure baseline characteristics of participants before randomization. Use ANCOVA to adjust for baseline differences in the analysis, thereby controlling for potential confounding variables.

Publication Bias:
> **Pitfall**: Publication bias occurs when studies with statistically significant results are more likely to be published than those with nonsignificant results, leading to an overrepresentation of positive findings in the literature.
>
> **Remedy**: Register the study protocol in advance on a public registry, such as ClinicalTrials.gov, to reduce the likelihood of publication bias. Additionally, consider publishing results in open-access journals or preprint repositories to increase transparency and accessibility of research findings.

By being aware of these pitfalls and implementing appropriate remedies, researchers can enhance the validity, reliability, and credibility of RD experiments in psychology.

Example

Randomized experiments in psychology involve randomly assigning participants to different experimental conditions to investigate the effects of an IV. Here's a simplified example of a matrix for a randomized experiment in psychology:

Objective: To study the effects of a new teaching method on students' performance in a mathematics exam.

IV: Teaching method.

- Traditional lecture.
- Problem-based learning.
- Flipped classroom.

DV: Mathematics exam score.

Randomized Experiment Matrix:

In a randomized experiment, participants are randomly assigned to different teaching methods, ensuring that each group is comparable. The matrix is shown in Table 2.92.

In this matrix, each row represents a specific participant who has been randomly assigned to one of the three teaching methods. Mathematics exam scores are measured for each participant.

Another example of a matrix for a RD experiment used in psychology is shown in Table 2.93.

In this matrix,

- **Participant ID**: Unique identifier for each participant in the study.
- **Experimental Condition (Random Assignment)**: Indicates whether each participant was assigned to the treatment group or the control group through random assignment.
- **Pretest Score**: Represents participants' scores on the outcome measure (e.g., a psychological assessment) administered before the intervention.
- **Post-test Score**: Represents participants' scores on the outcome measure administered after the intervention.
- **Change Score**: Indicates the difference between participants' post-test scores and pretest scores, representing the change in the outcome measure following the intervention.

TABLE 2.92

Randomized Experiment Matrix Example for Psychology

Participant	Teaching Method	Mathematics Exam Score
1	Traditional lecture	...
2	Problem-based learning	...
3	Flipped classroom	...
4	Traditional lecture	...
5	Flipped classroom	...
6	Problem-based learning	...
7	Flipped classroom	...
8	Traditional lecture	...
9	Traditional lecture	...
10	Problem-based learning	...

TABLE 2.93

Randomized Experiment Matrix Example for Psychology

Participant ID	Experimental Condition (Random Assignment)	Pretest Score	Post-Test Score	Change Score
1	Treatment	25	30	+5
2	Control	28	29	+1
3	Treatment	20	24	+4
4	Control	22	23	+1
5	Treatment	30	35	+5
6	Control	26	28	+2
7	Treatment	18	22	+4
8	Control	32	33	+1
9	Treatment	27	31	+4
10	Control	21	22	+1

TABLE 2.94

Randomized Experiment Matrix Example for Psychology

Participant ID	Group Assignment (Randomized)	Gender	Age	Pretest Score	Post-Test Score	Change Score
1	Treatment	Male	25	30	35	+5
2	Control	Female	32	28	29	+1
3	Treatment	Male	28	22	27	+5
4	Control	Female	30	29	30	+1
5	Treatment	Male	31	35	38	+3
6	Control	Male	27	26	28	+2
7	Treatment	Female	29	18	25	+7
8	Control	Male	33	32	34	+2
9	Treatment	Female	26	27	32	+5
10	Control	Female	28	21	23	+2

This matrix allows researchers to organize and analyze data from a RD experiment in psychology, facilitating comparisons between treatment and control groups and assessing the effectiveness of the intervention in achieving desired outcomes.

Table 2.94 is another variation example of a matrix for a RD experiment used in psychology. In this matrix,

- **Participant ID**: Unique identifier for each participant in the study.
- **Group Assignment (Randomized)**: Indicates whether each participant was assigned to the treatment group or the control group through random assignment.
- **Gender**: Participant's gender.
- **Age**: Participant's age in years.
- **Pretest Score**: Represents participants' scores on a pretest measurement of the outcome variable before the intervention.
- **Post-test Score**: Represents participants' scores on a post-test measurement of the outcome variable after the intervention.
- **Change Score**: Indicates the difference between participants' post-test scores and pretest scores, reflecting the change in the outcome variable due to the intervention.

This matrix allows researchers to organize and analyze data collected from a RD experiment in psychology. It provides information about participant characteristics, group assignments, pretest and post-test scores, and the change in scores following the intervention, facilitating comparisons between treatment and control groups and evaluation of intervention effectiveness.

Randomized experiments help researchers establish cause-and-effect relationships by minimizing the influence of extraneous variables and ensuring that any observed differences in the DV are likely due to the IV being manipulated.

REPORTING

In RD experiments used in psychology, various statistical tools are employed to analyze the data and draw meaningful conclusions. Some common statistical tools used in such experiments include:

ANOVA: ANOVA is used to compare means across multiple groups to determine whether there are statistically significant differences in the outcome variable between experimental conditions. In randomized experiments, ANOVA helps assess the effectiveness of the experimental manipulation by examining differences in outcome measures between treatment and control groups.

Independent Samples t-test: The independent samples t-test is used to compare the means of two independent groups to determine whether there are statistically significant differences between them. In randomized experiments, t-tests are often used to compare outcome measures between the treatment and control groups to evaluate the effectiveness of the intervention.

Repeated Measures ANOVA: Repeated measures ANOVA is used when the same participants are measured at multiple time points or under different conditions. In randomized experiments with pretest-post-test designs, repeated measures ANOVA can assess changes in outcome measures over time within each experimental condition and detect interactions between time and treatment effects.

Regression Analysis: Regression analysis is used to examine relationships between predictor variables (e.g., experimental condition) and outcome variables while controlling for potential confounding variables. In randomized experiments, regression analysis can be used to assess the effects of experimental manipulations on outcome measures while controlling for participant characteristics or other covariates.

ANCOVA: ANCOVA is used to compare group means on an outcome variable while statistically controlling for the effects of one or more covariates. In randomized experiments, ANCOVA is often used to adjust for baseline differences in outcome measures between treatment and control groups, thereby increasing the precision and accuracy of the analysis.

Chi-square Test: The chi-square test is used to analyze categorical data and determine whether observed frequencies differ significantly from expected frequencies. In randomized experiments, chi-square tests may be used to compare proportions or frequencies of categorical variables between treatment and control groups.

Mixed-Effects Models: Mixed-effects models, such as linear mixed-effects models or generalized linear mixed-effects models, are used to analyze data with nested or hierarchical structures, including repeated measures or multilevel data. In randomized experiments with complex data structures, mixed-effects models can account for within-subject correlations or clustering effects, thereby improving the accuracy of statistical estimates.

These are just a few examples of the statistical tools commonly used in RD experiments in psychology. The choice of statistical analysis depends on the research question, study design, and characteristics of the data collected.

A report for a RD experiment used in psychology typically includes several key elements to effectively communicate the research methodology, findings, and implications. Here are the essential elements of such a report:

Title: A concise and informative title that accurately reflects the content and purpose of the study.

Abstract: A brief summary of the research, including the research question, objectives, methodology, key findings, and implications. The abstract provides readers with a quick overview of the study's main points.

Introduction:

- Background: A review of relevant literature and theoretical framework that provides context for the study.
- Research Question/Hypotheses: Clear statement of the research question or hypotheses being tested in the experiment.
- Objectives: Explanation of the specific aims and objectives of the study.

Methodology:

- Experimental Design: Description of the experimental design, including the factors studied, levels of each factor, and rationale for their selection.
- Participants: Details about the study participants, including sample size, demographics, and any inclusion/exclusion criteria.
- Procedure: Overview of the experimental procedure, including how interventions were administered, data collection methods, and any control measures implemented.
- Randomization: Explanation of the randomization process used to assign participants to different experimental conditions.
- Statistical Analysis: Description of the statistical methods used to analyze the data, including specific tests or models employed to test hypotheses and assess treatment effects.

Results:

- Presentation of Findings: Clear and concise presentation of the study's results, including descriptive statistics, graphical representations (e.g., plots, charts), and inferential statistics.
- Main Effects and Interaction Effects: Discussion of main effects and interaction effects observed in the data, highlighting significant findings and patterns.
- Subgroup Analyses: If applicable, presentation of subgroup analyses to explore potential moderators or mediators of treatment effects.

Discussion:

- Interpretation of Findings: Analysis and interpretation of the results in relation to the research question and hypotheses.
- Implications: Discussion of the practical implications of the findings for theory, practice, and future research in psychology.
- Strengths and Limitations: Identification of the strengths and limitations of the study, including potential sources of bias or confounding factors.
- Generalizability: Consideration of the generalizability of the findings to broader populations or contexts.
- Comparison with Previous Research: Comparison of the study's findings with previous research in the field, highlighting similarities, differences, and areas for further investigation.

Conclusion: Summary of the key findings and conclusions drawn from the study, emphasizing their significance and relevance.

References: List of citations for all sources referenced in the report, following a consistent citation style (e.g., APA, MLA).

Appendices: Supplementary materials, such as detailed descriptions of experimental procedures, raw data tables, or additional analyses, that provide further context or support for the findings.

By including these elements in the report, researchers can effectively communicate the rationale, methodology, findings, and implications of their RD experiment in psychology.

2.5.9 SPLIT-PLOT AND BLOCKED EXPERIMENTS

Split-plot and blocked designed experiments are used in psychology for several reasons:

Increased Precision: Split-plot and blocked designs allow researchers to control for sources of variability and increase the precision of estimates. By partitioning the experimental units into homogeneous blocks or subgroups, these designs can reduce the variability within groups, thereby increasing the power to detect treatment effects.

Efficient Use of Resources: In psychology, researchers often face constraints on resources, such as time, participants, or equipment. Split-plot and blocked designs enable researchers to make efficient use of available resources by maximizing the information obtained from each experimental unit. By blocking or stratifying participants based on relevant variables, researchers can reduce the impact of nuisance factors and improve the efficiency of the experiment.

Control for Confounding Variables: Split-plot and blocked designs help control for confounding variables that may influence the outcome of the study. By blocking participants or experimental units based on relevant covariates, researchers can ensure that any observed differences between treatment groups are not due to systematic differences in these variables. This enhances the internal validity of the study and strengthens the conclusions drawn from the experiment.

Increased External Validity: Blocking or stratifying participants based on relevant characteristics can improve the external validity of the study by ensuring that the results are generalizable to specific populations or contexts. By including participants from diverse backgrounds or representing different levels of the blocking variable, researchers can enhance the external validity of their findings and increase the relevance of the study to real-world settings.

Accounting for Nesting and Hierarchy: In psychology, experimental designs often involve nested or hierarchical data structures, such as repeated measures within participants or individuals nested within groups. Split-plot and blocked designs are well suited for analyzing nested data structures and can provide more accurate estimates of treatment effects by accounting for within-group correlations or clustering effects.

Flexibility in Experimental Design: Split-plot and blocked designs offer flexibility in experimental design, allowing researchers to investigate complex research questions with multiple factors and interactions. These designs can accommodate factorial arrangements of treatments, interactions between fixed and random factors, and various combinations of blocking variables, providing researchers with greater flexibility in designing experiments tailored to their specific research questions.

These designed experiments are valuable tools in psychology for controlling variability, maximizing efficiency, enhancing internal and external validity, accounting for nested data structures, and providing flexibility in experimental design. These designs help researchers address complex research questions and draw valid conclusions from their experiments.

When studying the effects of educational interventions in schools, split-plot or blocked experiments may be used to account for variations in school environments, teacher characteristics, and student populations. Split-plot and blocked designed experiments are utilized in various areas of psychology, including but not limited to:

Clinical Psychology: In clinical psychology, split-plot and blocked designs are used to evaluate the effectiveness of therapeutic interventions and treatments for mental health disorders. Researchers may block participants based on relevant characteristics, such as severity of symptoms, age, gender, or comorbid conditions, to ensure that treatment effects are not confounded by these factors.

Cognitive Psychology: Split-plot and blocked designs are employed to investigate cognitive processes, such as attention, memory, perception, and decision-making. Researchers may block participants based on individual differences in cognitive abilities or personality traits to control for potential confounding variables and enhance the internal validity of the study.

Developmental Psychology: In developmental psychology, split-plot and blocked designs are used to examine how different factors influence child development, including parental behavior, environmental factors, and genetic influences. Researchers may block participants based on age, developmental stage, or socioeconomic status to account for developmental differences and assess the effects of interventions or environmental manipulations on developmental outcomes.

Social Psychology: Split-plot and blocked designs are employed to investigate social behavior, attitudes, group dynamics, and interpersonal relationships. Researchers may block participants based on demographic variables (e.g., age, gender, ethnicity) or personality traits to control for individual differences and examine the effects of social interventions or manipulations on social outcomes.

Educational Psychology: In educational psychology, split-plot and blocked designs are used to evaluate the effectiveness of educational interventions, teaching methods, and instructional strategies. Researchers may block participants based on academic performance, learning styles, or prior knowledge to ensure that treatment effects are not confounded by these factors and to assess the differential impact of interventions on student outcomes.

Health Psychology: Split-plot and blocked designs are utilized in health psychology to investigate health-related behaviors, interventions, and outcomes. Researchers may block participants based on health status, risk factors, or treatment adherence to account for individual differences and assess the effectiveness of health interventions or behavior change strategies.

Overall, split-plot and blocked designed experiments are employed across various subfields of psychology to address a wide range of research questions and hypotheses. These designs provide researchers with greater control over sources of variability, enhance the internal validity of studies, and facilitate the investigation of complex relationships and interactions in psychological phenomena.

STEPS

Developing a split-plot and blocked designed experiment in psychology involves several key steps. Here's a general outline of the process:

1. **Define Research Question**: Clearly define the research question or hypothesis that the experiment aims to address. The research question should be specific, testable, and relevant to the area of psychology being studied.
2. **Review Literature**: Conduct a comprehensive review of existing literature and theoretical frameworks relevant to the research question. This step helps to identify gaps in knowledge, refine the research question, and inform the design of the experiment.

3. **Identify Variables**: Identify the IV and DV that will be manipulated and measured in the experiment, respectively. Consider potential blocking variables that may influence the outcome and should be controlled for in the design.

4. **Design Experimental Conditions**: Determine the experimental conditions or treatment groups that will be compared in the experiment. Consider how to block participants or experimental units to control potential confounding variables and increase the precision of estimates.

5. **Blocking Strategy**: Identify the blocking variable(s) and develop a blocking strategy to create homogeneous blocks or subgroups within the sample. Determine how participants will be allocated to blocks and ensure that each block represents a homogeneous subset of the population.

6. **Randomization within Blocks**: Randomly assign participants or experimental units to different treatment groups within each block to ensure that treatment effects are not confounded by systematic differences between blocks.

7. **Develop Experimental Protocol**: Develop a detailed experimental protocol outlining the procedures and instructions for implementing the experiment. This includes information about participant recruitment, informed consent procedures, experimental manipulations, data collection methods, and any ethical considerations.

8. **Pilot Testing**: Conduct a pilot test or small-scale pilot study to evaluate the feasibility and effectiveness of the experimental protocol. Pilot testing helps to identify and address any potential issues with the procedures or materials before conducting the full experiment.

9. **Data Collection**: Implement the experiment according to the experimental protocol and collect data from participants. Ensure that data collection procedures are standardized and consistent across experimental conditions and blocks to minimize bias and ensure reliability.

10. **Data Analysis**: Analyze the collected data using appropriate statistical techniques to test hypotheses and answer the research question. Consider how to account for blocking factors in the analysis and determine the appropriate statistical model for analyzing split-plot and blocked designs.

11. **Interpret Results**: Interpret the results of the data analysis in relation to the research question and hypotheses. Discuss any significant main effects or interactions observed and consider how blocking factors may have influenced the outcomes.

12. **Report Findings**: Communicate the findings of the experiment through a research report or manuscript. The report should include a detailed description of the research methods, results, interpretations, and conclusions, following the conventions of scientific writing in psychology.

By following these steps, researchers can develop and conduct split-plot and blocked designed experiments in psychology to investigate causal relationships, test hypotheses, and advance scientific knowledge in the discipline.

PITFALLS AND REMEDIES

Split-plot and blocked designed experiments are powerful tools in psychology, but they can also be susceptible to various pitfalls. Here are some common pitfalls and their corresponding remedies:

Inadequate Blocking:

Pitfall: Inadequate blocking can occur if the chosen blocking variable does not sufficiently account for sources of variability in the data.

Remedy: Carefully select blocking variables that are strongly related to the outcome variable and ensure that they effectively stratify the participants or experimental units into homogeneous blocks. Consider conducting a pilot study to identify potential blocking factors and assess their impact on the outcome variable.

Incomplete Randomization:

Pitfall: Incomplete randomization can occur if random assignment is not fully implemented within each block, leading to biased treatment effects.

Remedy: Ensure that randomization is conducted independently within each block to minimize bias and ensure that treatment effects are not confounded by systematic differences between blocks. Use appropriate randomization procedures, such as random number generators or computerized randomization, to assign participants or experimental units to treatment groups.

Small Block Sizes:

Pitfall: Small block sizes can reduce the effectiveness of blocking and increase the risk of spurious findings due to random variability.

Remedy: Aim for sufficiently large block sizes to enhance the precision of estimates and increase the power to detect treatment effects. Consider stratifying participants into larger blocks or combining similar blocks to achieve a balance between homogeneity and practicality.

Failure to Consider Interactions:

Pitfall: Failure to consider interactions between blocking factors and treatment effects can lead to misinterpretation of results and incorrect conclusions.

Remedy: Investigate potential interactions between blocking factors and treatment effects by conducting appropriate statistical analyses, such as factorial ANOVA or mixed-effects models. Interpret results in light of any significant interactions and consider their implications for the study findings.

Overly Complex Designs:

Pitfall: Overly complex designs with multiple blocking factors and treatment combinations can be difficult to implement and analyze, leading to an increased risk of errors and decreased interpretability of results.

Remedy: Simplify the experimental design by focusing on the most important blocking factors and treatment combinations relevant to the research question. Consider conducting factorial designs with fewer factors or using fractional factorial designs to reduce the number of treatment combinations while maintaining efficiency.

Ignoring Assumptions of Analysis Methods:

Pitfall: Ignoring assumptions of the analysis methods used for split-plot and blocked designs, such as normality of residuals or homogeneity of variance, can lead to invalid results and inaccurate conclusions.

Remedy: Check the assumptions of the chosen analysis methods and use appropriate diagnostic tests to assess the validity of assumptions. Consider robust statistical techniques or alternative methods that are less sensitive to violations of assumptions.

By being aware of these pitfalls and implementing appropriate remedies, researchers can enhance the validity, reliability, and interpretability of split-plot and blocked designed experiments in psychology.

Examples

Split-plot and blocked experiments in psychology involve variations of experimental designs that account for certain types of nonindependence or structured relationships among data points. Here are examples of matrices for both types of designed experiments in psychology:

Example 1: Split-Plot Experiment

Objective: To study the effects of two different training methods (factor A) and two different assessment methods (factor B) on participants' skill acquisition while considering the potential interaction between training and assessment methods.

Factors:

1. **Training Method (Factor A):**
   - Method 1 (e.g., lecture).
   - Method 2 (e.g., hands-On).
2. **Assessment Method (Factor B):**
   - Method A (e.g., written exam).
   - Method B (e.g., practical demonstration).

DV: Skill acquisition score.

Split-Plot Experiment Matrix:

In a split-plot experiment, you have one set of factors (e.g., training method) that is applied at the whole-plot level and another set of factors (e.g., assessment method) that is applied at the subplot level. The matrix is shown in Table 2.95.

In this matrix, each row represents a specific participant who is exposed to one combination of training method and assessment method. The interaction between these two factors is of interest in the analysis.

Example 2: Blocked Experiment

Objective: To study the effects of a new teaching method on students' performance in a mathematics exam while controlling for prior knowledge levels.

IV: Teaching method

- Traditional lecture.
- Problem-based learning.

Blocking Variable: Prior knowledge level (high vs. low).

DV: Mathematics exam score.

TABLE 2.95

Split-Plot Experiment Matrix Example for Psychology

Participant	Training Method (Whole Plot)	Assessment Method (Subplot)	Skill Acquisition Score
1	Method 1	Method A	...
2	Method 2	Method A	...
3	Method 1	Method B	...
4	Method 2	Method B	...
5	Method 1	Method A	...
6	Method 2	Method A	...
7	Method 1	Method B	...
8	Method 2	Method B	...

TABLE 2.96

Blocked Experiment Matrix Example for Psychology

Participant	Teaching Method	Prior Knowledge Level	Mathematics Exam Score
1	Traditional lecture	High	...
2	Problem-based learning	Low	...
3	Traditional lecture	Low	...
4	Problem-based learning	High	...
5	Traditional lecture	Low	...
6	Problem-based learning	High	...
7	Traditional lecture	High	...
8	Problem-based learning	Low	...

Blocked Experiment Matrix:

In a blocked experiment, participants are grouped into blocks based on a variable that is not the main focus of the study but could impact the results (e.g., prior knowledge levels). The matrix is shown in Table 2.96.

In this matrix, participants are blocked into two groups based on their prior knowledge levels. Each group is then exposed to one of the two teaching methods. The purpose of blocking is to control the influence of prior knowledge on the results.

Both split-plot and blocked experiments are used in psychology to address specific research questions and control factors that may affect the outcomes of interest.

REPORTING

In split-plot and blocked designed experiments used in psychology, various statistical tools are employed to analyze the data and draw meaningful conclusions. Some common statistical tools used in such experiments include:

ANOVA: ANOVA is a widely used statistical technique for comparing means across multiple groups or conditions. In split-plot and blocked designs, ANOVA is used to analyze the main effects of treatment factors, blocking factors, and their interactions with the outcome variable.

Repeated Measures ANOVA: Repeated Measures ANOVA is used when the same participants or experimental units are measured under different conditions or time points. In split-plot designs, repeated measures ANOVA is used to analyze within-subjects effects and interactions while accounting for between-subjects variability.

Mixed-Effects Models: Mixed-effects models, such as linear mixed-effects models or generalized linear mixed-effects models, are used to analyze data with nested or hierarchical structures, such as split-plot designs or blocked designs. These models can account for within-subject or within-block correlations and incorporate both fixed and random effects into the analysis.

ANCOVA: ANCOVA is used to compare group means on an outcome variable while statistically controlling for the effects of one or more covariates. In split-plot and blocked designs, ANCOVA is often used to adjust for baseline differences or other covariates that may influence the outcome variable.

Factorial ANOVA: Factorial ANOVA is used to analyze the main effects and interaction effects of multiple IVs (factors) on an outcome variable. In split-plot designs, factorial ANOVA can assess the effects of both within-subject factors (e.g., treatment conditions) and between-subject factors (e.g., blocking variables) on the outcome variable.

Post-hoc Tests: Post-hoc tests, such as Tukey's HSD test or Bonferroni correction, are used to conduct pairwise comparisons between treatment conditions or levels of blocking factors following a significant omnibus test (e.g., ANOVA). These tests help identify specific differences between groups and determine which comparisons are statistically significant.

Residual Analysis: Residual analysis is used to assess the adequacy of the statistical model and check assumptions of normality, homogeneity of variance, and independence of residuals. Diagnostic plots, such as histograms, Q-Q plots, and residual vs. fitted plots, are commonly used to evaluate the distribution and patterns of residuals.

Effect Size Measures: Effect size measures, such as partial eta-squared (η^2), Cohen's d, or Pearson's r, are used to quantify the magnitude of treatment effects and the strength of relationships between variables. Effect size measures provide valuable information about the practical significance of study findings beyond statistical significance.

These are just a few examples of the statistical tools commonly used in split-plot and blocked designed experiments in psychology. The choice of statistical analysis depends on the research question, study design, and characteristics of the data collected.

A comprehensive report for a split-plot and blocked designed experiment in psychology typically includes the following elements:

Title Page:
- Title of the study: Clearly indicates the topic and purpose of the research.
- Author(s) name(s): Identifies the researcher(s) responsible for the study.
- Affiliation(s): Specifies the institutional affiliations of the author(s).
- Date: Indicates the date of completion or submission of the report.

Abstract:
- Summary of the study: Provides a concise overview of the research question, methodology, key findings, and implications.
- Length: Typically limited to 150–250 words.

Introduction:
- Background: Provides context for the study by reviewing relevant literature, theories, and previous research findings.
- Research question/hypotheses: Clearly states the research question(s) or hypotheses being investigated.
- Objectives: Describes the specific aims and objectives of the study.

Methodology:
- Experimental design: Details the design of the experiment, including the factors studied, levels of each factor, and blocking variables.
- Participants: Describes the sample size, characteristics of participants, and any inclusion/exclusion criteria.
- Procedure: Outlines the experimental procedures, including participant recruitment, randomization, treatment administration, and data collection methods.
- Statistical analysis: Specifies the statistical techniques used to analyze the data, including any assumptions and tests conducted.

Results:
- Descriptive statistics: Presents summary statistics (e.g., means, standard deviations) for each experimental condition and block.
- Inferential statistics: Reports the results of statistical analyses, including ANOVA tables, post-hoc tests, and effect size measures.
- Graphs/figures: Includes visual representations of the data, such as bar charts, line graphs, or interaction plots.

- Interpretation: Provides an interpretation of the results in relation to the research question, hypotheses, and theoretical framework.

Discussion:

- Summary of findings: Summarizes the main findings of the study and their implications for theory and practice.
- Comparison with previous research: Discusses how the current findings align with or diverge from previous research in the field.
- Strengths and limitations: Evaluates the strengths and limitations of the study, including potential sources of bias or confounding factors.
- Future directions: Suggests areas for future research and potential extensions of the current study.

Conclusion:

- Recapitulation: Restates the main findings and conclusions drawn from the study.
- Practical implications: Discusses the practical implications of the findings for psychology practice or policy.
- Closing remarks: Offers final reflections on the significance of the study and its contribution to the field.

References: Lists all sources cited in the report, following a standardized citation format (e.g., APA, MLA).

Appendices (if applicable): Supplementary materials: Includes additional information or data not presented in the main text, such as detailed tables, questionnaires, or raw data.

By including these elements in the report, researchers can effectively communicate the rationale, methodology, findings, and implications of their split-plot and blocked designed experiment in psychology.

2.5.10 Crossover Experiments

Crossover designed experiments offer several advantages in psychology research, including increased control for individual differences, improved statistical power, reduction of inter-participant variability, control for time-related effects, ethical considerations, efficiency in resource utilization, and practical application to clinical settings. These advantages make crossover designs valuable tools for investigating causal relationships and evaluating interventions in psychology. In neuropsychological research, crossover experiments are employed to compare the effects of different drug treatments on cognitive functioning. Participants receive treatment in a randomized order to control individual differences.

Crossover designed experiments are used in psychology for several reasons:

Control for Individual Differences: In psychology research, there can be significant individual differences among participants that may confound the results. Crossover designs allow each participant to serve as their own control, which helps to control individual differences that could otherwise affect the outcomes.

Increased Statistical Power: Crossover designs typically require fewer participants compared to parallel-group designs to achieve the same level of statistical power. This efficiency is particularly advantageous in psychology research where recruiting and retaining participants can be challenging.

Reduction of Inter-Participant Variability: By using the same participants in multiple conditions, crossover designs reduce inter-participant variability. This can increase the sensitivity of the experiment to detect treatment effects, especially when the variability within participants is smaller compared to variability between participants.

Control for Time-Related Effects: Crossover designs are particularly useful for studying time-related effects or interventions that have immediate or short-term effects. By comparing each participant's performance before and after treatment, researchers can control for time-related factors that may influence the outcomes.

Ethical Considerations: Crossover designs can be ethically advantageous, especially when investigating interventions or treatments with potential risks. Since each participant receives both the treatment and control conditions, ethical concerns about withholding treatment from participants in control groups are minimized.

Efficiency in Resource Utilization: Crossover designs are efficient in resource utilization, as they require fewer resources, such as time, funding, and personnel, compared to parallel-group designs. This efficiency is particularly beneficial in psychology research where resources may be limited.

Practical Application to Clinical Settings: Crossover designs are often used in clinical psychology and behavioral interventions to assess the effectiveness of treatments or interventions. They mimic real-world scenarios where individuals may receive different treatments over time, allowing researchers to evaluate treatment effectiveness in a more ecologically valid setting. Crossover designed experiments are used in various areas of psychology for studying a wide range of phenomena and interventions. Some common areas where crossover designs are employed include:

Clinical Psychology: Crossover designs are frequently used in clinical psychology to evaluate the effectiveness of therapeutic interventions and treatments for mental health disorders. Researchers may use crossover designs to compare different types of therapy, medication regimens, or behavioral interventions within the same group of participants.

Cognitive Psychology: In cognitive psychology, crossover designs are employed to investigate cognitive processes, such as attention, memory, perception, and decision-making. Researchers may use crossover designs to assess the effects of experimental manipulations or interventions on cognitive performance over time within individual participants.

Health Psychology: Crossover designs are commonly used in health psychology to study the effects of health-related interventions, lifestyle changes, or medical treatments on behavior, cognition, and psychological outcomes. Researchers may use crossover designs to compare the efficacy of different interventions for managing chronic conditions or improving health behaviors.

Developmental Psychology: Crossover designs are utilized in developmental psychology to study developmental changes and interventions across different age groups or developmental stages. Researchers may use crossover designs to assess the effects of early interventions or environmental manipulations on developmental outcomes over time.

Experimental Psychology: In experimental psychology, crossover designs are employed to investigate causal relationships between variables and to test theoretical hypotheses. Researchers may use crossover designs to manipulate IVs and measure their effects on DVs within individual participants, thereby controlling for individual differences and increasing the internal validity of the study.

Neuropsychology: Crossover designs are used in neuropsychology to assess the effects of brain injuries, treatments, or interventions on cognitive functioning and behavior over time. Researchers may use crossover designs to compare cognitive performance before and after interventions or to assess the long-term effects of neurorehabilitation programs.

Educational Psychology: Crossover designs are applied in educational psychology to evaluate the effectiveness of educational interventions, teaching methods, or instructional strategies. Researchers may use crossover designs to compare different teaching approaches or curriculum interventions within the same group of students over time.

Social Psychology: Crossover designs are employed in social psychology to study social behaviors, attitudes, and interventions over time. Researchers may use crossover designs to assess changes in social attitudes or behavior following interventions or to investigate the effects of social manipulations on interpersonal relationships.

Overall, crossover designed experiments are used across various subfields of psychology to investigate causal relationships, evaluate interventions, and advance scientific knowledge in the discipline. Their versatility and efficiency make them valuable tools for studying a wide range of psychological phenomena and interventions.

STEPS

Developing a crossover designed experiment in psychology involves several key steps. Here's a general outline of the process:

1. **Define Research Question**: Clearly define the research question or hypothesis that the experiment aims to address. The research question should be specific, testable, and relevant to the area of psychology being studied.
2. **Review Literature**: Conduct a comprehensive review of existing literature and theoretical frameworks relevant to the research question. This step helps to identify gaps in knowledge, refine the research question, and inform the design of the experiment.
3. **Identify Variables**: Identify the IVs and DV that will be manipulated and measured in the experiment, respectively. Consider how these variables will be operationalized and measured to ensure validity and reliability.
4. **Design Experimental Conditions**: Determine the experimental conditions or treatment arms that participants will receive during the study. Consider the order in which participants will receive each treatment condition and whether there will be a washout period between treatments.
5. **Participant Selection**: Determine the criteria for participant selection, including inclusion and exclusion criteria. Consider factors, such as age, gender, clinical diagnosis (if applicable), and any other relevant characteristics.
6. **Randomization**: Randomly assign participants to different sequences of treatment conditions to minimize bias and ensure that each participant has an equal chance of receiving each treatment condition. Consider using methods, such as simple randomization, stratified randomization, or block randomization.
7. **Develop Experimental Protocol**: Develop a detailed experimental protocol outlining the procedures and instructions for implementing the experiment. This includes information about participant recruitment, informed consent procedures, treatment administration, data collection methods, and any ethical considerations.
8. **Pilot Testing**: Conduct a pilot test or small-scale pilot study to evaluate the feasibility and effectiveness of the experimental protocol. Pilot testing helps to identify and address any potential issues with the procedures or materials before conducting the full experiment.
9. **Data Collection**: Implement the experiment according to the experimental protocol and collect data from participants. Ensure that data collection procedures are standardized and consistent across treatment conditions and participants.
10. **Data Analysis**: Analyze the collected data using appropriate statistical techniques to test hypotheses and answer the research question. Consider using methods, such as repeated measures ANOVA, mixed-effects models, or linear regression, depending on the nature of the data and research question.

11. **Interpret Results**: Interpret the results of the data analysis in relation to the research question and hypotheses. Discuss any significant findings and their implications for theory, practice, and future research in psychology.
12. **Report Findings**: Communicate the findings of the experiment through a research report or manuscript. The report should include a detailed description of the research methods, results, interpretations, and conclusions, following the conventions of scientific writing in psychology.

By following these steps, researchers can develop and conduct crossover designed experiments in psychology to investigate causal relationships, test hypotheses, and evaluate interventions.

PITFALLS AND REMEDIES

Crossover designed experiments in psychology can encounter various pitfalls that may affect the validity and reliability of the results. Here are some common pitfalls and their corresponding remedies:

Carryover Effects:
   Pitfall: Carryover effects occur when the effects of one treatment condition persist or influence subsequent treatment conditions, leading to contamination of the results.
   Remedy: Implement a washout period between treatment conditions to minimize carryover effects. During this period, participants should return to their baseline state or receive a neutral treatment to eliminate any lingering effects from previous conditions. Additionally, counterbalancing the order of treatment conditions across participants can help control for carryover effects.

Order Effects:
   Pitfall: Order effects occur when the sequence in which treatment conditions are administered influences participants' responses or outcomes, irrespective of the treatments themselves.
   Remedy: Implement counterbalancing techniques to control for order effects. Randomize the order in which participants receive treatment conditions or use a Latin square design to ensure that each condition appears equally often in each position. By counterbalancing the order of treatments, researchers can minimize the impact of order effects on the results.

Learning or Habituation Effects:
   Pitfall: Learning or habituation effects occur when participants become familiar with the experimental procedures or tasks over time, leading to changes in their responses or behavior independent of the treatments.
   Remedy: Implement a control condition or baseline measurement to assess any changes in participants' responses that are unrelated to the treatments. Additionally, consider including a brief familiarization or practice session at the beginning of each treatment period to minimize the impact of learning effects.

Carry-in Effects:
   Pitfall: Carry-in effects occur when participants enter the study with residual effects from previous experiences or treatments, which can influence their responses to subsequent treatments.
   Remedy: Screen participants carefully during the recruitment process to ensure that they are free from any residual effects that may bias the results. Consider implementing a run-in period before the start of the experiment to allow participants to acclimatize to the study environment and mitigate carry-in effects.

Attrition Bias:
> **Pitfall**: Attrition bias occurs when participants drop out of the study or fail to complete all treatment conditions, leading to biased results.
>
> **Remedy**: Minimize attrition bias by implementing strategies to enhance participant retention and compliance, such as providing incentives, maintaining regular communication with participants, and ensuring that the study procedures are convenient and non-burdensome. Conduct sensitivity analyses to assess the impact of attrition on the results and consider using appropriate statistical techniques to handle missing data.

Limited Generalizability:
> **Pitfall**: Crossover designs may have limited generalizability to populations or contexts outside of the study sample or setting.
>
> **Remedy**: Interpret the results of crossover designs cautiously and consider the extent to which the findings can be generalized to broader populations or real-world settings. When possible, replicate the study using different samples or settings to assess the robustness and generalizability of the results.

By being aware of these pitfalls and implementing appropriate remedies, researchers can enhance the validity, reliability, and interpretability of crossover designed experiments in psychology.

EXAMPLES

Crossover designed experiments in psychology are often used to compare the effects of different treatments or interventions on the same group of participants over multiple time periods. Here are two examples of matrices for crossover designed experiments in psychology:

Example 1: Testing the Effects of Two Anxiety Medications

Objective: To compare the effectiveness of two different anxiety medications, medication A and medication B, in reducing anxiety levels in the same group of participants over two time periods.

Treatments:

1. **Medication A**: Participants take medication A for the first time period.
2. **Medication B**: Participants take medication B for the second time period.

DV: Anxiety Score.

Crossover Experiment Matrix:
In a crossover experiment, each participant experiences both treatments in a randomized order. The matrix is shown in Table 2.97.
In this matrix, each row represents a specific participant who undergoes both treatments in a randomized order. Anxiety scores are measured after each time period.

Example 2: Comparing the Effects of Two Teaching Methods

Objective: To assess the impact of two teaching methods, method X and method Y, on students' learning performance in a psychology course over two semesters.

Teaching Methods:

1. **Method X**: Students receive instruction using Method X during the first semester.
2. **Method Y**: Students receive instruction using Method Y during the second semester.

TABLE 2.97

Crossover Experiment Matrix Example for Psychology

Participant	Treatment Order	Time Period 1 (Medication A)	Time Period 2 (Medication B)	Anxiety Score
1	A-B	...	...	...
2	B-A	...	...	...
3	A-B	...	...	...
4	B-A	...	...	...
5	A-B	...	...	...
6	B-A	...	...	...

TABLE 2.98

Crossover Experiment Matrix Example for Psychology

Student	Semester	Teaching Method	Learning Performance Score
1	1	Method X	...
2	2	Method Y	...
3	1	Method X	...
4	2	Method Y	...
5	1	Method X	...
6	2	Method Y	...

DV: Learning performance score.

Crossover Experiment Matrix:

In this example, students experience both teaching methods in different semesters. The matrix is shown in Table 2.98.

In this matrix, each student experiences both teaching methods in different semesters. Learning performance scores are measured at the end of each semester.

Crossover experiments are useful when you want to compare the effects of multiple treatments on the same group of participants, thereby controlling for individual variability. Randomization of treatment order helps minimize potential order effects or biases.

REPORT

In Crossover designed experiments used in psychology, various statistical tools are employed to analyze the data and draw meaningful conclusions. Some common statistical tools used in such experiments include:

Repeated Measures ANOVA: Repeated measures ANOVA is a commonly used statistical technique for analyzing data from crossover experiments where the same participants are measured under different conditions. It allows researchers to assess the main effects of treatment conditions and any interaction effects between treatments and other factors.

Linear Mixed-Effects Models: Linear mixed effects models, also known as hierarchical linear models or multilevel models, are flexible statistical techniques for analyzing longitudinal or nested data from crossover experiments. They allow researchers to account for within-subject correlations and incorporate both fixed and random effects into the analysis.

Generalized Estimating Equations (GEE): GEE is a statistical method for analyzing longitudinal or correlated data from crossover experiments, particularly when the outcome variable is binary, ordinal, or count data. GEE models provide population-averaged estimates of treatment effects while accounting for within-subject correlations.

Nonparametric Tests: Nonparametric tests, such as the Friedman test or the Wilcoxon signed rank test, are used to analyze data from crossover experiments when the assumptions of parametric tests are violated, such as non-normality or non-sphericity of residuals.

Bayesian Methods: Bayesian methods, such as Bayesian hierarchical models or Bayesian structural equation models, are increasingly used in crossover experiments to estimate treatment effects, model complex relationships between variables, and quantify uncertainty in parameter estimates.

Survival Analysis: Survival analysis techniques, such as Cox proportional hazards models or Kaplan–Meier curves, are used in crossover experiments where the outcome variable represents time-to-event data, such as time until relapse or time until treatment failure.

Meta-Analysis: Meta-analysis techniques can be used to synthesize findings from multiple crossover experiments, providing pooled estimates of treatment effects and assessing the overall strength of evidence across studies.

Post-hoc Tests: Post-hoc tests, such as pairwise comparisons or contrasts, are used to conduct specific comparisons between treatment conditions following a significant omnibus test, such as ANOVA or mixed-effects models.

Residual Analysis: Residual analysis is used to assess the adequacy of statistical models and check assumptions, such as homogeneity of variance or normality of residuals, in crossover experiments.

These are just a few examples of the statistical tools commonly used in crossover designed experiments in psychology. The choice of statistical analysis depends on the research question, study design, and characteristics of the data collected.

A comprehensive report for a crossover designed experiment in psychology typically includes the following elements:

Title Page:
- Title of the study: Clearly indicates the topic and purpose of the research.
- Author(s) name(s): Identifies the researcher(s) responsible for the study.
- Affiliation(s): Specifies the institutional affiliations of the author(s).
- Date: Indicates the date of completion or submission of the report.

Abstract:
- Summary of the study: Provides a concise overview of the research question, methodology, key findings, and implications.
- Length: Typically limited to 150–250 words.

Introduction:
- Background: Provides context for the study by reviewing relevant literature, theories, and previous research findings.
- Research question/hypotheses: Clearly states the research question(s) or hypotheses being investigated.
- Objectives: Describes the specific aims and objectives of the study.

Methodology:
- Experimental design: Details the design of the experiment, including the sequence of treatment conditions, order of administration, and any control measures implemented.
- Participants: Describes the sample size, characteristics of participants, and any inclusion/exclusion criteria.
- Procedure: Outlines the experimental procedures, including participant recruitment, informed consent procedures, treatment administration, and data collection methods.

- Statistical analysis: Specifies the statistical techniques used to analyze the data, including any assumptions and tests conducted.

Results:
- Descriptive statistics: Presents summary statistics (e.g., means, standard deviations) for each treatment condition and any relevant comparisons.
- Inferential statistics: Reports the results of statistical analyses, including ANOVA tables, mixed-effects models, or nonparametric tests.
- Graphs/figures: Includes visual representations of the data, such as line plots or bar charts, to illustrate key findings.
- Interpretation: Provides an interpretation of the results in relation to the research question, hypotheses, and theoretical framework.

Discussion:
- Summary of findings: Summarizes the main findings of the study and their implications for theory and practice.
- Comparison with previous research: Discusses how the current findings align with or diverge from previous research in the field.
- Strengths and limitations: Evaluates the strengths and limitations of the study, including potential sources of bias or confounding factors.
- Future directions: Suggests areas for future research and potential extensions of the current study.

Conclusion:
- Recapitulation: Restates the main findings and conclusions drawn from the study.
- Practical implications: Discusses the practical implications of the findings for psychology practice or policy.
- Closing remarks: Offers final reflections on the significance of the study and its contribution to the field.

References: Lists all sources cited in the report, following a standardized citation format (e.g., APA, MLA).

Appendices (if applicable): Supplementary materials: Includes additional information or data not presented in the main text, such as detailed tables, questionnaires, or raw data.

By including these elements in the report, researchers can effectively communicate the rationale, methodology, findings, and implications of their crossover designed experiment in psychology.

2.5.11   BBD

BBD experiments provide researchers in psychology with a powerful tool for efficiently exploring complex relationships among variables, identifying optimal conditions, and improving the effectiveness of interventions and treatments. These designs are efficient for exploring quadratic terms without requiring as many runs as a full factorial. BBD experiments are utilized in psychology for several reasons:

Efficient Exploration of Factors: BBD allows researchers to efficiently explore the effects of multiple factors on psychological phenomena. This is particularly beneficial when studying complex interactions among variables, such as cognitive processes, emotional responses, or behavior.

Reduction of Experimental Runs: BBD reduces the number of experimental runs required to estimate main effects and interactions compared to full factorial designs. This efficiency is valuable in psychology, where conducting large-scale experiments with human participants can be time-consuming and resource intensive.

Identification of Optimal Conditions: BBD facilitates the identification of optimal or near-optimal conditions for achieving desired outcomes in psychological research. By systematically varying factors within the design space, researchers can identify the combination of factors that maximize desired outcomes, such as treatment effectiveness, learning performance, or well-being.

Robustness Testing: BBD allows researchers to assess the robustness of findings across different experimental conditions or interventions. This is essential in psychology, where interventions or treatments may need to be effective across diverse populations or under varying contextual factors.

Statistical Efficiency: BBD is statistically efficient for estimating main effects and interactions, even with a limited number of experimental runs. This ensures that researchers can obtain precise estimates of the effects of interest without unnecessarily increasing the sample size.

Ease of Interpretation: BBD produces a balanced and orthogonal design matrix, which simplifies the interpretation of results in psychology research. Researchers can easily identify the effects of individual factors and their interactions without confounding effects from other factors.

Flexibility in Design: BBD offers flexibility in terms of the number of factors and levels included in the design. Researchers can tailor the design to suit the specific research question and available resources, making it adaptable to various experimental designs and contexts in psychology.

BBD experiments find applications in various areas of psychology. Some common areas where BBD experiments are used include:

Experimental Psychology: BBD experiments are employed in experimental psychology to investigate the effects of IVs (such as cognitive tasks, stimuli characteristics, or environmental factors) on psychological processes and behaviors. Researchers use BBD to efficiently explore multiple factors and their interactions while minimizing the number of experimental runs.

Clinical Psychology: In clinical psychology, BBD experiments are utilized to optimize treatment protocols, evaluate intervention strategies, and identify factors that influence treatment outcomes. Researchers use BBD to identify the most effective combination of treatment components or therapeutic techniques for addressing psychological disorders or improving mental health outcomes.

Cognitive Psychology: BBD experiments are applied in cognitive psychology to study cognitive processes, such as memory, attention, perception, and decision-making. Researchers use BBD to investigate how various factors (such as stimulus characteristics, task demands, or individual differences) influence cognitive performance and behavior.

Health Psychology: In health psychology, BBD experiments are used to optimize health-related interventions, assess the efficacy of behavioral interventions, and identify factors that influence health behaviors and outcomes. Researchers use BBD to explore the effects of intervention components, environmental factors, and individual differences on health-related behaviors and outcomes.

Social Psychology: BBD experiments find applications in social psychology to study social behaviors, attitudes, and interactions. Researchers use BBD to investigate the effects of social factors (such as group dynamics, social norms, or interpersonal relationships) on behavior, cognition, and emotional responses.

Educational Psychology: In educational psychology, BBD experiments are employed to optimize instructional strategies, evaluate educational interventions, and identify factors that influence learning outcomes. Researchers use BBD to explore the effects of instructional methods, instructional materials, and classroom environments on student learning and academic achievement.

Developmental Psychology: BBD experiments are used in developmental psychology to study developmental processes and trajectories across the lifespan. Researchers use BBD to investigate how various factors (such as genetic factors, environmental influences, or social interactions) contribute to developmental outcomes and behavioral patterns.

The BBD experiments are applied across various subfields of psychology to address research questions, optimize interventions, and advance our understanding of psychological processes and behaviors.

STEPS

Developing BBD experiments in psychology involves several key steps. Here's a general outline of the process:

1. **Define Research Objectives**: Clearly define the research objectives and the specific research questions you aim to address using the BBD experiment. Consider the variables of interest, the relationships you want to investigate, and the hypotheses you want to test.
2. **Identify Factors and Levels**: Identify the IVs (factors) and the levels at which each factor will be varied in the experiment. Factors can include experimental manipulations, treatment conditions, or levels of a continuous variable. Determine the range of each factor and the number of levels for each factor.
3. **Choose Response Variables**: Identify the DVs (response variables) that will be measured to assess the effects of the IVs. Determine the type of data (e.g., continuous, categorical) and the measurement scales for each response variable.
4. **Select the Design Space**: Define the design space, which represents the range of factor levels over which the experiment will be conducted. Consider any constraints or limitations on the factors and levels, as well as practical considerations, such as resources, time, and participant characteristics.
5. **Generate the Design Matrix**: Use statistical software or design tools to generate the design matrix of BBD based on the specified factors, levels, and design space. The design matrix should include the combinations of factor levels (experimental runs) that will be used in the experiment, as well as any randomization or blocking schemes if applicable.
6. **Conduct Preliminary Analysis**: Perform a preliminary analysis of the design matrix to check for balance, orthogonality, and other properties of the design. Ensure that the design is efficient and that the factor levels are well distributed throughout the design space.
7. **Implement the Experiment**: Conduct the experiment according to the design matrix, administering the specified treatments or conditions to participants and collecting data on the response variables. Follow standardized procedures for participant recruitment, randomization, treatment administration, and data collection to ensure consistency and reliability.
8. **Data Analysis**: Analyze the collected data using appropriate statistical techniques to test the hypotheses and address the research questions. Consider methods, such as ANOVA, regression analysis, or other multivariate techniques, depending on the nature of the data and research objectives.
9. **Interpret Results**: Interpret the results of the data analysis in the context of the research objectives and hypotheses. Assess the significance of the main effects and interactions and consider the practical implications of the findings for theory, practice, and future research in psychology.
10. **Report Findings**: Communicate the findings of the BBD experiment through a research report or manuscript. Present the design, methods, results, and conclusions in a clear and concise manner, following the conventions of scientific writing in psychology.

By following these steps, researchers can develop and conduct BBD experiments in psychology to systematically investigate the effects of multiple factors on psychological phenomena and behavior.

PITFALLS AND REMEDIES

BBD experiments offer numerous advantages, but they also present certain pitfalls that researchers should be aware of. Here are some common pitfalls and corresponding remedies:

Limited Factor Range:
> **Pitfall**: BBD experiments may suffer from limited factor ranges, especially if the design space is not carefully chosen. Narrow ranges may fail to capture the full range of variability in the factors, leading to an incomplete understanding of their effects.
> **Remedy**: Carefully select the range of factor levels to encompass the full range of variability observed in the population or phenomenon under study. Conduct pilot studies or preliminary analyses to determine appropriate ranges for the factors.

Assumption of Linearity:
> **Pitfall**: BBD assumes that the relationships between factors and response variables are approximately linear within the specified ranges. If this assumption is violated, the model may not accurately represent the true relationships.
> **Remedy**: Verify the linearity assumption by conducting diagnostic checks, such as residual plots or lack-of-fit tests. If nonlinear relationships are detected, consider using alternative experimental designs or nonlinear modeling techniques.

Inadequate Model Fit:
> **Pitfall**: The BBD model may fail to adequately fit the observed data, leading to poor predictions or inaccurate estimates of factor effects.
> **Remedy**: Assess the goodness-of-fit of the BBD model using diagnostic tools, such as residual analysis, goodness-of-fit tests, or cross-validation techniques. Consider refining the model by adding higher-order terms or interaction terms if necessary.

Lack of Robustness:
> **Pitfall**: BBD experiments may lack robustness if the estimated factor effects are sensitive to small changes in the design or experimental conditions.
> **Remedy**: Conduct sensitivity analyses to assess the robustness of the results to variations in the design or experimental conditions. Evaluate the stability of factor effects across different subsets of the data or alternative model specifications.

Incomplete Factor Space Coverage:
> **Pitfall**: BBD experiments may fail to adequately cover the factor space, resulting in gaps or uneven distribution of experimental points within the design space.
> **Remedy**: Optimize the design matrix to achieve better coverage of the factor space while maintaining efficiency and balance. Consider augmenting the design with additional center points or supplemental points to fill gaps in the factor space.

Risk of Overfitting:
> **Pitfall**: BBD experiments with a large number of factors or complex models may be susceptible to overfitting, where the model captures noise rather than true underlying relationships.
> **Remedy**: Regularize the model to prevent overfitting by incorporating regularization techniques, such as ridge regression or lasso regression. Use model selection criteria (e.g., AIC, BIC) to choose the most parsimonious model that adequately explains the data.

By addressing these pitfalls and implementing appropriate remedies, researchers can enhance the validity, reliability, and interpretability of BBD experiments in psychology.

EXAMPLE

An example of a BBD experiment in psychology could involve studying the effects of different types of music on cognitive performance in a controlled laboratory setting. Here's how the BBD could be applied in this scenario:

1. **Factor Selection and Levels:**
   - Type of Music: Classical (low arousal), jazz (medium arousal), rock (high arousal).
   - Volume Level: Low, medium, high.
   - Time of Day: Morning, afternoon, evening.
2. **Design Space:**
   - Define the range for each factor based on previous research and practical considerations. For example, volume level could vary from quiet to loud, and time of day could span from morning to evening sessions.
3. **BBD Matrix:**
   - Generate a BBD matrix that includes a balanced set of experimental runs representing various combinations of factor levels within the defined design space. Each row in the matrix represents a unique combination of factor levels, and the order of experimental runs is randomized.
4. **Conducting the Experiment:**
   - Participants are randomly assigned to different experimental conditions based on the combinations of music type, volume level, and time of day.
   - Cognitive performance tasks are administered to participants, such as memory recall, attention tasks, or problem-solving exercises.
   - Performance metrics are collected for each participant under each experimental condition.
5. **Data Analysis:**
   - Analyze the data using appropriate statistical techniques to assess the main effects of music type, volume level, and time of day on cognitive performance. Additionally, evaluate any interactions between factors.
   - Identify the optimal combination of music type, volume level, and time of day that maximizes cognitive performance.
6. **Interpretation and Conclusion:**
   - Interpret the findings in relation to the research question and hypotheses. Discuss the implications of the results for understanding the effects of music on cognitive processes.
   - Provide recommendations for optimizing environmental conditions, such as music selection and volume level, to enhance cognitive performance in different contexts.

This example demonstrates how a BBD experiment can be used in psychology to systematically explore the effects of multiple factors on cognitive performance and inform practical applications, such as designing effective study environments or workplace settings.

The BBD matrix for an experiment investigating the effects of different types of music (classical, jazz, rock) at various volume levels (low, medium, high) and times of day (morning, afternoon, evening) on cognitive performance in psychology is shown in Table 2.99.

In this matrix,

- Each row represents a unique combination of factors (type of music, volume level, time of day).
- The columns represent the factors being studied.
- The "Run" column numbers each experimental run for reference.
- Participants would be assigned to each experimental condition according to the combinations specified in the matrix.

TABLE 2.99

BBD Experiment Matrix Example

Run	Type of Music	Volume Level	Time of Day
1	Classical	Low	Morning
2	Classical	Medium	Afternoon
3	Classical	High	Evening
4	Jazz	Medium	Morning
5	Jazz	High	Afternoon
6	Jazz	Low	Evening
7	Rock	High	Morning
8	Rock	Low	Afternoon
9	Rock	Medium	Evening
10	Classical	High	Morning
11	Jazz	Low	Afternoon
12	Rock	Medium	Evening
13	Classical	Medium	Morning
14	Jazz	High	Afternoon
15	Rock	Low	Evening
16	Classical	Low	Morning
17	Jazz	Medium	Afternoon
18	Rock	High	Evening
...	...	...	...

This design allows for the systematic exploration of the effects of multiple factors on cognitive performance while ensuring a balanced and efficient allocation of participants across conditions.

Reporting

In BBD experiments used in psychology, various statistical tools and techniques are employed to analyze the data and draw conclusions about the effects of different factors on the outcomes of interest. Some of the common statistical tools used in BBD experiments in psychology include:

ANOVA: ANOVA is used to assess the significance of the main effects of each factor and any interactions between factors on the DV(s). It helps determine whether there are statistically significant differences in the means of different experimental conditions.

Regression Analysis: Regression analysis is often used to model the relationship between IVs (factors) and DVs (outcomes). Multiple regression or multivariate regression may be employed to evaluate the linear relationships between predictor variables and outcomes.

RSM: RSM is a set of statistical and mathematical techniques used to optimize the response (DV) in a BBD experiment by modeling the relationships between factors and response variables. It helps identify the optimal factor settings for maximizing or minimizing the response.

Descriptive Statistics: Descriptive statistics, such as means, standard deviations, and histograms, are used to summarize and visualize the distributions of variables in the dataset. They provide insights into the central tendency and variability of the data.

Residual Analysis: Residual analysis is conducted to assess the adequacy of the model fit and to identify any patterns or trends in the residuals (the differences between observed and predicted values). Residual plots, normal probability plots, and residual autocorrelation plots are common diagnostic tools used in residual analysis.

Hypothesis Testing: Hypothesis testing is employed to evaluate specific research hypotheses about the effects of factors on outcomes. Tests, such as t-tests or F-tests, may be used to compare means between groups or to assess the significance of regression coefficients.

Model Selection Criteria: Model selection criteria, such as AIC or BIC, may be used to compare and select the most appropriate regression model among competing models. These criteria help balance model complexity and goodness of fit.

Factorial Analysis: Factorial analysis techniques, such as factorial ANOVA or factorial regression, may be used to analyze the effects of multiple factors on outcomes while considering all possible combinations of factor levels.

These statistical tools and techniques allow researchers to analyze the data from BBD experiments in psychology and draw meaningful conclusions about the relationships between factors and outcomes. They help identify significant effects, optimize experimental conditions, and advance our understanding of psychological processes and behaviors.

A report on a BBD experiment used in psychology typically includes several key elements to effectively communicate the research findings and implications. Here are the essential elements of such a report:

Title Page: The title of the report should succinctly summarize the research topic and the key variables under investigation. Include the names of the authors, affiliations, and the date of publication.

Abstract: Provide a brief summary of the study, including the research question, methods, major findings, and conclusions. The abstract should be concise but informative, typically around 150–250 words.

Introduction: Introduce the research topic and provide background information to contextualize the study. Clearly state the research objectives, hypotheses, and rationale for using a BBD. Review relevant literature to justify the research questions and hypotheses.

Methods: Describe the experimental design, including the selection of factors and levels, the generation of the BBD matrix, and the procedures for data collection. Provide details on participant recruitment, sampling procedures, and any ethical considerations. Explain how the data were analyzed, including the statistical techniques used to assess the main effects and interactions among factors.

Results: Present the findings of the study in a clear and organized manner. Summarize the descriptive statistics for each experimental condition and report the results of the statistical analyses. Include tables, figures, or graphs to visually represent the data and highlight significant findings. Provide interpretation and discussion of the main effects, interactions, and any unexpected or interesting patterns observed in the data.

Discussion: Interpret the results in relation to the research objectives and hypotheses. Discuss the implications of the findings for theory, practice, and future research in psychology. Consider the limitations of the study, such as sample size, generalizability, and potential confounding variables. Address any methodological issues or sources of bias that may have influenced the results.

Conclusion: Summarize the main findings of the study and reiterate the key conclusions drawn from the research. Highlight the practical implications of the findings and suggest areas for further investigation.

References: Provide a list of references cited in the report, following the appropriate citation style (e.g., APA, MLA). Include scholarly articles, books, and other sources relevant to the research topic.

Appendices (if necessary): Include any supplementary materials, such as the BBD matrix, detailed descriptions of experimental procedures, or additional statistical analyses.

By including these elements in the report, researchers can effectively communicate the rationale, methods, findings, and implications of a BBD experiment in psychology to the scientific community.

2.5.12 PBD

PBD is useful for screening many factors with limited resources, but only provides information about main effects. This designed experimental method offers a valuable tool for efficiently screening and identifying key factors influencing psychological phenomena. They provide a cost-effective and resource-efficient approach to exploratory research and hypothesis generation in psychology. Plackett–Burman designed experiments are used in psychology for several reasons:

Efficient Screening of Factors: PBDs are particularly useful for screening a large number of factors with a relatively small number of experimental runs. This allows researchers to efficiently identify the most influential factors affecting a psychological phenomenon or outcome.

Identifying Key Variables: By systematically varying factors using a PBD, researchers can identify the key variables that have the most significant impact on the DV of interest. This helps prioritize further investigation into the most promising factors.

Cost and Resource Efficiency: Conducting experiments in psychology can be resource intensive, especially when involving human participants. PBDs offer a cost-effective approach by minimizing the number of experimental runs required while still providing valuable information about factor effects.

Exploratory Research: PBDs are well suited for exploratory research in psychology, where the goal is to identify potential factors influencing a phenomenon or behavior. These designs provide an efficient way to generate hypotheses for further investigation.

Robustness Testing: PBDs allow researchers to assess the robustness of findings across different experimental conditions or populations. By systematically varying factors, researchers can evaluate the consistency of factor effects and their generalizability.

Optimizing Experimental Designs: Findings from PBDs can inform the development of more detailed experimental designs, such as factorial designs or RSMs. Researchers can use the results to refine their hypotheses and focus subsequent research efforts.

Plackett–Burman designed experiments are used in psychology across various research areas and applications. Some common areas where PBDs find applications in psychology include:

Experimental Psychology: PBDs are utilized in experimental psychology to screen a large number of IVs (factors) with a relatively small number of experimental runs. This allows researchers to efficiently identify the most influential factors affecting psychological processes and behaviors.

Clinical Psychology: In clinical psychology, PBDs are used to screen potential treatment components or intervention strategies for further evaluation in controlled trials. These designs help identify the most promising factors for improving psychological outcomes in clinical populations.

Health Psychology: PBDs are employed in health psychology to identify factors influencing health behaviors, adherence to medical treatments, and health outcomes. Researchers use these designs to efficiently screen potential determinants of health-related behaviors and outcomes.

Cognitive Psychology: PBDs find applications in cognitive psychology to screen factors affecting cognitive processes, such as memory, attention, perception, and decision-making. These designs help identify the key variables that influence cognitive performance and behavior.

Social Psychology: In social psychology, PBDs are used to screen factors influencing social behaviors, attitudes, and interactions. Researchers employ these designs to efficiently identify potential determinants of social phenomena and behavior.

Developmental Psychology: PBDs are utilized in developmental psychology to screen factors influencing developmental processes across the lifespan. These designs help identify factors that contribute to developmental outcomes and behavioral patterns.

Educational Psychology: PBDs find applications in educational psychology to screen factors influencing learning outcomes, instructional effectiveness, and academic achievement. Researchers use these designs to identify potential determinants of learning and instructional practices.

Applied Psychology: PBDs are employed in various applied psychology settings, such as organizational psychology, forensic psychology, and sports psychology. These designs help screen factors influencing performance, behavior, and outcomes in applied settings.

Plackett–Burman designed experiments are used in psychology to efficiently screen factors and identify the most influential variables affecting psychological phenomena, behaviors, and outcomes across different research domains and applications.

STEPS

Developing a Plackett–Burman designed experiment in psychology involves several key steps. Here's a general outline of the process:

1. **Define Research Objectives**: Clearly define the research objectives and the specific research questions you aim to address using the PBD. Consider the variables of interest and the hypotheses you want to test.
2. **Identify Factors**: Identify the IVs (factors) that you want to investigate in the experiment. These factors should represent the key variables believed to influence the DV(s) of interest.
3. **Determine Factor Levels**: Determine the levels at which each factor will be varied in the experiment. Typically, PBDs use two levels for each factor, often referred to as "high" and "low" levels. However, if factors are categorical or have more than two levels, these should be specified accordingly.
4. **Select the Number of Experimental Runs**: Determine the number of experimental runs (trials) required for the PBD. This is typically determined by the number of factors being investigated and the desired level of confidence in detecting significant effects.
5. **Generate the Design Matrix**: Use statistical software or design tools to generate the PBD design matrix. This matrix should specify the combinations of factor levels for each experimental run, ensuring that the design is balanced and orthogonal.
6. **Conduct Preliminary Analysis**: Perform a preliminary analysis of the design matrix to ensure that it meets the requirements of a PBD, such as orthogonality and balance. Check for confounding factors and assess the efficiency of the design.
7. **Implement the Experiment**: Conduct the experiment according to the specifications outlined in the PBD matrix. Administer the treatments or conditions to participants and collect data on the DV(s) of interest.
8. **Data Analysis**: Analyze the collected data using appropriate statistical techniques to assess the main effects of the factors on the DV(s). Use methods, such as ANOVA or regression analysis, to test for significance and identify important factors.
9. **Interpret Results**: Interpret the results of the data analysis in the context of the research objectives and hypotheses. Discuss the implications of significant factor effects and consider potential explanations for any unexpected findings.

10. **Report Findings**: Communicate the findings of the Plackett–Burman designed experiment through a research report or manuscript. Present the design, methods, results, and conclusions in a clear and concise manner, following the conventions of scientific writing in psychology.

By following these steps, researchers can develop and conduct Plackett–Burman designed experiments in psychology to efficiently screen factors and identify the most influential variables affecting psychological phenomena and behaviors.

PITFALLS AND REMEDIES

Plackett–Burman designed experiments offer many advantages, but they also come with potential pitfalls that researchers should be aware of. Here are some common pitfalls and corresponding remedies for Plackett–Burman designed experiments used in psychology:

Confounding of Factors:
Pitfall: PBDs do not fully account for interactions between factors. If two or more factors are confounded (correlated), it can be challenging to determine which factor is responsible for observed effects.
Remedy: Conduct additional experiments or follow-up studies to explore potential interactions between factors identified as significant in the initial PBD. Alternatively, consider using more complex experimental designs, such as factorial designs, that explicitly account for interactions.

Limited Factor Levels:
Pitfall: PBDs typically use only two levels for each factor (high and low), which may not capture the full range of variability in the factors. This can lead to oversimplification of the relationships between factors and outcomes.
Remedy: If possible, consider expanding the number of levels for key factors to provide a more nuanced understanding of their effects. Alternatively, conduct follow-up studies using different experimental designs to explore factors in greater detail.

Low Sensitivity:
Pitfall: PBDs may lack sensitivity to detect small or subtle effects, particularly if the number of experimental runs is limited.
Remedy: Increase the sample size or the number of experimental runs to improve the statistical power of the analysis. Conduct power analyses to determine the minimum detectable effect size and ensure that the design is adequately powered to detect meaningful effects.

Limited Generalizability:
Pitfall: Findings from PBDs may have limited generalizability to broader populations or contexts, especially if the experimental conditions are highly controlled or artificial.
Remedy: Validate the findings of the PBD through replication in different populations or settings. Consider conducting follow-up studies using more ecologically valid experimental designs to assess the robustness of the findings.

Assumption of Linearity:
Pitfall: PBDs assume linear relationships between factors and outcomes, which may not always hold true in practice.
Remedy: Check the linearity assumption using diagnostic tests, such as residual analysis or lack-of-fit tests. If nonlinear relationships are detected, consider using alternative modeling techniques or experimental designs that accommodate nonlinearity.

By addressing these potential pitfalls and implementing appropriate remedies, researchers can enhance the validity, reliability, and interpretability of Plackett–Burman designed experiments in psychology.

EXAMPLE

An example of a Plackett–Burman designed experiment in psychology could involve investigating factors influencing students' academic performance in an educational setting. Here's a hypothetical scenario:

Research Question: What factors influence students' academic performance in an introductory psychology course?

Factors of Interest:

1. Study Time: Low vs. high.
2. Class Attendance: Low vs. high.
3. Prior Knowledge: Low vs. high.
4. Study Environment: Quiet vs. noisy.
5. Time of Day: Morning vs. afternoon.

Design Matrix (Table 2.100):
In this example,

- The researchers are interested in examining the effects of five factors (study time, class attendance, prior knowledge, study environment, time of day) on students' academic performance.
- Each factor has two levels (low and high).
- The PBD matrix consists of four experimental runs, which is the minimum number of runs needed to estimate the main effects for five factors.
- Each row represents a unique combination of factor levels for an experimental run.
- The order of the runs is randomized to minimize potential biases.

Researchers would assign students to each experimental condition specified in the design matrix and measure their academic performance (e.g., exam scores, assignment grades). By analyzing the data collected from these experimental runs, researchers can determine the main effects of each factor on academic performance and identify significant factors influencing students' success in the psychology course.

Here's another example of a PBD matrix for a hypothetical psychology experiment investigating factors influencing mood in a laboratory setting:

TABLE 2.100

Plackett–Burman Designed Experiment in Psychology

Run	Study Time	Class Attendance	Prior Knowledge	Study Environment	Time of Day
1	Low	Low	Low	Quiet	Morning
2	High	Low	High	Noisy	Afternoon
3	High	High	Low	Quiet	Afternoon
4	Low	High	High	Noisy	Morning

TABLE 2.101

Plackett–Burman Designed Experiment in Psychology

Run	Factor A (Music Type)	Factor B (Room Temperature)	Factor C (Lighting Condition)	Factor D (Background Noise)
1	Classical	High	Low	Low
2	Jazz	Low	High	High
3	Rock	High	High	Low
4	Jazz	High	Low	Low
5	Rock	Low	Low	High
6	Classical	Low	High	Low
7	Jazz	High	High	High
8	Classical	High	Low	High

Design Matrix (Table 2.101):

In this example,

- Four factors are being investigated: music type (A), room temperature (B), lighting condition (C), and background noise (D).
- Each factor has two levels: High and low.
- The design matrix consists of 8 experimental runs, as determined by the PBD.
- Each row represents a unique combination of factor levels for an experimental run.
- The order of the runs is randomized to minimize potential biases.

Researchers would assign participants to each experimental condition specified in the design matrix and measure their mood outcomes. By analyzing the data collected from these experimental runs, researchers can determine the main effects of each factor on mood and identify significant factors that influence mood in the laboratory setting.

In another example of a Plackett–Burman designed experiment used in psychology:

Research Question: What factors affect individuals' decision-making processes when faced with moral dilemmas?

Factors of Interest:

1. Presence of an Authority Figure: Present vs. absent.
2. Emotional State: Positive vs. negative.
3. Personal Values: Altruistic vs. self-interest.
4. Social Influence: Peer pressure vs. individual autonomy.

Design Matrix (Table 2.102):

In this example,

- The researchers are interested in studying the effects of four factors (presence of an authority figure, emotional state, personal values, social influence) on individuals' decision-making in moral dilemmas.
- Each factor has two levels (e.g., present/absent, positive/negative, altruistic/self-interest, peer pressure/individual autonomy).
- The PBD matrix consists of four experimental runs, the minimum number needed to estimate main effects for four factors.

TABLE 2.102

Plackett–Burman Designed Experiment in Psychology

Run	Authority Figure	Emotional State	Personal Values	Social Influence
1	Present	Positive	Altruistic	Peer pressure
2	Absent	Positive	Self-interest	Individual autonomy
3	Absent	Negative	Altruistic	Individual autonomy
4	Present	Negative	Self-interest	Peer pressure

- Each row represents a unique combination of factor levels for an experimental run.
- The order of the runs is typically randomized to minimize potential biases.

Participants would be exposed to each experimental condition specified in the design matrix, and their decision-making processes and responses would be recorded and analyzed. By analyzing the data collected from these experimental runs, researchers can identify the main effects of each factor on decision-making in moral dilemmas and gain insights into the psychological mechanisms underlying moral behavior.

REPORTING

In Plackett–Burman designed experiments used in psychology, several statistical tools are commonly employed to analyze the data and draw conclusions about the effects of factors on the outcome of interest. Some of these statistical tools include:

ANOVA: ANOVA is used to assess the significance of the main effects of each factor on the DV. It helps determine whether there are statistically significant differences in the means of different experimental conditions.

Regression Analysis: Regression analysis is often used to model the relationship between IVs (factors) and DVs (outcomes). Multiple regression or logistic regression may be employed to evaluate the linear or nonlinear relationships between predictor variables and outcomes.

Main Effects Plots: Main effects plots are graphical representations of the mean response at each level of a factor. These plots help visualize the main effects of each factor on the outcome variable and identify patterns or trends.

Residual Analysis: Residual analysis is conducted to assess the adequacy of the model fit and to identify any patterns or trends in the residuals (the differences between observed and predicted values). Residual plots, normal probability plots, and residual autocorrelation plots are common diagnostic tools used in residual analysis.

Effect Size Measures: Effect size measures, such as Cohen's d or eta-squared (η^2), are used to quantify the magnitude of the relationship between factors and outcomes. These measures provide information about the practical significance of the observed effects.

Post-hoc Tests: If significant main effects are found, post-hoc tests, such as Tukey's HSD test or Bonferroni correction, can be used to determine which factor levels differ significantly from each other.

Model Selection Criteria: Model selection criteria, such as AIC or BIC, may be used to compare and select the most appropriate regression model among competing models. These criteria help balance model complexity and goodness of fit.

Diagnostic Tests: Diagnostic tests, such as tests for multicollinearity or heteroscedasticity, are conducted to ensure that the assumptions of the statistical models are met and that the results are valid and reliable.

By employing these statistical tools, researchers can effectively analyze the data from Plackett–Burman designed experiments in psychology and draw meaningful conclusions about the effects of factors on psychological phenomena and behaviors.

When reporting the results of a Plackett–Burman designed experiment in psychology, it's essential to provide a comprehensive overview of the study design, methods, results, and implications. Here are the key elements to include in the report:

Title: Clearly indicate the title of the report, which should succinctly summarize the research topic and the key variables under investigation.

Abstract: Provide a brief summary of the study, including the research question, methods, major findings, and conclusions. The abstract should highlight the main objectives and outcomes of the experiment.

Introduction: Introduce the research question and the rationale for using a PBD. Provide background information on the factors being investigated and their relevance to the study. Clearly state the research objectives and hypotheses.

Methods: Describe the experimental design, including the selection of factors and levels, and the generation of the PBD matrix. Outline the procedures for participant recruitment, data collection, and data analysis. Specify any control variables or potential confounding factors that were considered in the design.

Participants: Provide information about the participants, including demographics (e.g., age, gender), recruitment methods, and any inclusion/exclusion criteria.

Experimental Procedure: Describe how participants were assigned to experimental conditions and the specific tasks or manipulations they were exposed to. Outline any procedures for randomization or counterbalancing to minimize bias.

Measures: Specify the DV(s) measured in the study and the instruments or procedures used to assess them. Provide details about the reliability and validity of the measures employed.

Results: Present the main findings of the study in a clear and organized manner. Report descriptive statistics for each experimental condition and the results of statistical analyses (e.g., ANOVA, regression). Include tables, figures, or graphs to illustrate significant effects and trends in the data.

Discussion: Interpret the results in relation to the research objectives and hypotheses. Discuss the implications of the findings for theory, practice, and future research. Address any limitations of the study and potential sources of bias or error.

Conclusion: Summarize the main findings and conclusions drawn from the study. Highlight the significance of the research findings and suggest directions for future research in the area.

References: Provide a list of references cited in the report, following the appropriate citation style (e.g., APA, MLA).

Appendices (if necessary): Include any supplementary materials, such as the PBD matrix, detailed statistical analyses, or additional data tables.

By including these elements in the report, researchers can effectively communicate the rationale, methods, results, and implications of a Plackett–Burman designed experiment in psychology to the scientific community.

2.5.13 DERRINGER DESIGN

Derringer designed experiments, also known as D-optimal designs, are used in psychology and other fields to efficiently allocate resources and maximize the information gained from experimental data. They are useful for optimizing multiple responses simultaneously when interactions are important.

Here's why they are utilized in psychology:

Efficient resource allocation: Derringer designs allow researchers to optimize the allocation of resources, such as time, money, and participant effort. By strategically selecting experimental conditions and levels, researchers can obtain the most information possible from a given set of resources.

Maximizing information: These designs are specifically tailored to maximize the information gained from each participant or data point. By carefully selecting the levels of IVs, researchers can effectively explore the effects of those variables on the DV(s) of interest.

Reducing experimental error: Derringer designs are designed to minimize the impact of extraneous variables and experimental error, thus enhancing the internal validity of the research findings. This is crucial in psychology, where researchers often aim to isolate specific variables and understand their effects on behavior or mental processes.

Increasing statistical power: By maximizing the information obtained from each participant or data point, Derringer designs can increase the statistical power of the study. This means that researchers are more likely to detect true effects or relationships between variables, enhancing the reliability and generalizability of their findings.

Flexibility and adaptability: These designs can be tailored to the specific research questions and hypotheses of a study, allowing researchers to design experiments that are highly relevant to their particular area of interest in psychology.

The Derringer designed experiments offer a systematic and efficient approach to experimental design in psychology, helping researchers to obtain robust and informative results while making optimal use of available resources.

Derringer designed experiments, (D-optimal designs), can be used in various areas of psychology where experimentation is a key methodological approach. Here are some specific areas within psychology where these experimental designs are commonly employed:

Cognitive Psychology: D-optimal designs can be used to investigate various aspects of human cognition, such as memory, attention, perception, language processing, and decision-making. Researchers can use these designs to manipulate IVs (e.g., cognitive tasks, stimuli characteristics) and measure their effects on DVs (e.g., reaction times, accuracy rates) in controlled laboratory settings.

Experimental Psychology: In experimental psychology, researchers often use Derringer designs to conduct controlled experiments aimed at testing hypotheses and theories about human behavior. These experiments may involve manipulating variables related to motivation, emotion, learning, or social influence, and observing their effects on participants' behavior and mental processes.

Psychophysiology: Psychophysiological research examines the relationship between psychological processes and physiological responses. D-optimal designs can be applied to studies involving measures, such as heart rate, skin conductance, brain activity (e.g., EEG, fMRI), and hormonal levels. Researchers can use these designs to investigate how psychological factors influence physiological responses, and vice versa.

Clinical Psychology: D-optimal designs can be used in clinical psychology research to investigate the efficacy of interventions and treatments for various psychological disorders and conditions. Researchers can design experiments to systematically manipulate treatment variables (e.g., therapy techniques, medication dosage) and assess their impact on clinical outcomes (e.g., symptom reduction, quality of life) in randomized controlled trials.

Social Psychology: Social psychology explores how individuals' thoughts, feelings, and behaviors are influenced by the social environment and interpersonal relationships. D-optimal designs can be employed to study topics, such as social influence, conformity, prejudice, group dynamics, and interpersonal communication. Researchers can manipulate social variables (e.g., group size, social norms, interpersonal interactions) and measure their effects on participants' attitudes and behaviors.

Derringer designed experiments can be applied across various subfields of psychology to address research questions, test hypotheses, and advance our understanding of human behavior, cognition, and mental processes.

STEPS

Developing Derringer designed experiments in psychology involves several steps to ensure that the experimental design is well-suited to address the research question and efficiently utilize resources. Here are the steps typically involved in developing Derringer designs for experiments in psychology:

1. **Define the research question**: Clearly articulate the specific research question or hypothesis that the experiment aims to address. This step is essential for guiding the selection of variables and experimental manipulations.
2. **Identify the IVs and DVs**: Determine the variables that will be manipulated (IVs) and measured (DVs) in the experiment. These variables should be relevant to the research question and theory being tested.
3. **Select experimental levels**: Decide on the different levels or conditions of the IVs that will be included in the experiment. These levels should represent meaningful variations in the variables of interest and allow for the systematic investigation of their effects.
4. **Determine constraints and objectives**: Specify any constraints or limitations on the experimental design, such as budget, time, or participant availability. Additionally, define the specific objectives of the experiment, such as maximizing information gain, minimizing experimental error, or maximizing statistical power.
5. **Choose an appropriate design criterion**: Select the design criterion that will guide the optimization of the experimental design. In Derringer designs, the criterion is typically based on factors, such as minimizing the variance of parameter estimates or maximizing the determinant of the information matrix.
6. **Generate candidate designs**: Use statistical software or specialized algorithms to generate candidate Derringer designs that meet the specified criteria and objectives. These designs should systematically vary the levels of IVs while maximizing the efficiency of the experiment.
7. **Evaluate candidate designs**: Assess the quality and efficiency of the generated designs based on criteria, such as the precision of parameter estimates, the balance of design points, and the robustness to potential sources of error. Compare different designs to identify the most suitable option for the research question and objectives.
8. **Refine and finalize the design**: Based on the evaluation results, refine the selected design to address any potential shortcomings or limitations. Make any necessary adjustments to optimize the design further while ensuring that it remains feasible and practical to implement.
9. **Implement the experiment**: Once the Derringer design has been finalized, implement the experiment according to the specified design parameters and procedures. Collect data from participants or experimental units in accordance with the experimental protocol.
10. **Analyze and interpret the results**: Analyze the collected data using appropriate statistical methods to test the research hypotheses and draw conclusions about the effects of the IVs on the DVs. Interpret the findings in the context of the research question and theoretical framework.

By following these steps, researchers can develop Derringer designed experiments in psychology that are well designed, efficient, and effectively address the research objectives.

PITFALLS AND REMEDIES

While Derringer designed experiments offer numerous benefits in psychology research, there are also potential pitfalls that researchers should be aware of. Here are some common pitfalls and remedies when using Derringer designs in psychology experiments:

Overfitting the model:
> **Pitfall:** Overfitting occurs when a model is overly complex and fits the noise in the data rather than the underlying patterns. This can happen if the design is too closely tailored to specific data, leading to poor generalizability.
> **Remedy**: Regularize the design by imposing constraints or penalties on model parameters to prevent overfitting. Consider using a simpler design that balances efficiency with generalizability.

Violations of assumptions:
> **Pitfall:** Derringer designs rely on certain assumptions about the underlying data distribution and model structure. Violations of these assumptions can lead to biased parameter estimates or inefficient designs.
> **Remedy**: Validate the assumptions of the Derringer design using diagnostic checks and sensitivity analyses. If assumptions are violated, consider alternative designs or robust estimation techniques.

Sparse regions of the design space:
> **Pitfall:** Some Derringer designs may result in sparse regions of the design space where data points are poorly distributed. This can lead to instability in parameter estimates or reduced statistical power.
> **Remedy**: Optimize the design to ensure adequate coverage of the design space and minimize sparse regions. Consider augmenting the design with additional data points in critical regions or using adaptive sampling techniques.

Practical constraints:
> **Pitfall:** Derringer designs may not always be feasible to implement due to practical constraints, such as limited resources, participant burden, or ethical considerations.
> **Remedy**: Tailor the design to accommodate practical constraints while still achieving the desired balance of efficiency and informativeness. Consider alternative designs or sampling strategies that better align with available resources and logistical considerations.

Failure to consider interactions:
> **Pitfall:** Derringer designs typically focus on main effects and may overlook interactions between variables, leading to incomplete or misleading conclusions.
> **Remedy**: Incorporate interactions between variables into the design criteria or use sequential or adaptive designs to explore interactions iteratively. Conduct post-hoc analyses to examine potential interactions and their implications for the research findings.

Limited external validity:
> **Pitfall:** Derringer designs may prioritize internal validity at the expense of external validity, resulting in findings that may not generalize to real-world contexts or populations.
> **Remedy**: Supplement Derringer designs with follow-up studies or replication efforts that address external validity concerns. Consider conducting field experiments or naturalistic observations to enhance the ecological validity of the findings.

Being aware of these potential pitfalls and implementing appropriate remedies, researchers can effectively leverage Derringer designs in psychology experiments while minimizing biases and maximizing the quality of the research findings.

EXAMPLE

One example of a Derringer design used in psychology could involve studying the effects of different teaching methods on students' academic performance. Let's consider a hypothetical experiment aimed at comparing the effectiveness of two teaching approaches (traditional lecture-based teaching vs. active learning) on students' comprehension and retention of course material in an introductory psychology class.

Research Question: How do different teaching methods (lecture-based vs. active learning) affect students' comprehension and retention of course material in an introductory psychology class?

IVs: Teaching Method: Lecture-based teaching (coded as 0) vs. active learning (coded as 1).

DVs:

- Performance on course assessments (e.g., quizzes, exams).
- Retention of course material (measured by performance on follow-up assessments).

Experimental Levels:

Lecture-based teaching: Standard lecture format with minimal student engagement.

Active learning: Interactive activities, group discussions, problem-solving exercises, and hands-on learning experiences.

Constraints and Objectives:

Limited class time and resources for implementing the experimental conditions.

Objective to maximize information gained about the effects of teaching methods on student outcomes while minimizing experimental error.

Design Criterion:

Minimize the variance of parameter estimates related to teaching method effects on student performance and retention.

Candidate Designs:

Generate Derringer designs with a balanced allocation of participants to each teaching method condition, ensuring adequate statistical power to detect differences in performance.

Consider designs that strategically vary the timing and duration of each teaching method to optimize the efficiency of data collection.

Evaluation of Designs: Evaluate candidate designs based on criteria, such as efficiency of parameter estimation, balance of design points across conditions, and robustness to potential sources of bias.

Refinement and Finalization: Select the Derringer design that best balances the objectives and constraints of the study, making any necessary adjustments to optimize efficiency and validity.

Implementation and Analysis:

Implement the experiment by assigning participants to either the lecture-based or active learning condition, delivering the respective teaching methods throughout the course.

Collect data on students' performance on course assessments and retention of material.
Analyze the data using appropriate statistical methods (e.g., ANOVA, regression) to compare the effects of teaching methods on student outcomes, controlling for potential confounding variables.

This example demonstrates how a Derringer design could be applied in psychology to investigate the effects of teaching methods on student learning outcomes while maximizing the efficiency and validity of the experiment.

A Derringer design matrix is typically used to allocate experimental conditions (in this case, different teaching methods) to participants in an efficient and balanced manner. The matrix specifies the assignment of participants to each condition based on the design criteria and objectives of the experiment. Here's a simplified example of a Derringer design matrix for the hypothetical study of teaching methods in an introductory psychology class:

Suppose we have 20 participants (students) enrolled in the study, and we want to allocate them to two conditions: lecture-based teaching and active learning. We'll use a D-optimal design criterion to maximize the information gained about the effects of teaching methods on student comprehension and retention while minimizing experimental error (Table 2.103).

In this matrix,

- Participants are assigned to one of the two teaching methods based on the design generated using Derringer's algorithm.
- The allocation is balanced, with an equal number of participants assigned to each teaching method.
- The order of assignment does not necessarily reflect the sequence in which participants participate in the study; it's a structured allocation based on the optimized design.

TABLE 2.103

Derringer Designed Experiment Example in Psychology

Participant ID	Teaching Method (0 = Lecture-Based, 1 = Active Learning)
1	0
2	1
3	0
4	1
5	1
6	0
7	0
8	1
9	0
10	1
11	1
12	0
13	0
14	1
15	0
16	1
17	1
18	0
19	0
20	1

TABLE 2.104
Derringer Designed Experiment Example in Psychology

Participant ID	Teaching Method (X)	Outcome Measure 1	Outcome Measure 2
1	0 (Lecture-based)	Score	Retention score
2	1 (Active learning)	Score	Retention score
3	1 (Active learning)	Score	Retention score
4	0 (Lecture-based)	Score	Retention score
5	0 (Lecture-based)	Score	Retention score

This matrix represents a simple example of how participants might be allocated to experimental conditions using a Derringer design in a study investigating the effects of teaching methods on student comprehension and retention in psychology. More complex designs may involve additional factors, covariates, or blocking variables to further optimize the allocation of participants and improve the efficiency and validity of the experiment.

To create a matrix for a Derringer design for the experiment on the effects of different teaching methods (lecture-based vs. active learning) on students' comprehension and retention of course material in an introductory psychology class, we need to specify the experimental conditions and the allocation of participants to those conditions. Table 2.104 is an example of a matrix for such a Derringer design.

In this matrix,

- **Participant ID**: Each row represents an individual participant in the study, identified by a unique participant ID.
- **Teaching Method (X)**: This column indicates the experimental condition to which each participant is assigned. In this example, '0' represents the lecture-based teaching method, and '1' represents the active learning method.
- **Outcome Measure 1**: This column represents the primary outcome measure assessing students' comprehension or performance on course assessments (e.g., quizzes, exams). Each participant's score on the assessment would be recorded in this column.
- **Outcome Measure 2**: This column represents the secondary outcome measure assessing students' retention of course material. Participants may be assessed on their ability to recall or apply course content in a follow-up assessment, and their retention score would be recorded in this column.

Researchers would collect data from participants in each experimental condition and analyze the effects of teaching methods on outcome measures using appropriate statistical methods. By systematically varying the teaching methods and measuring participants' performance and retention, researchers can assess the effectiveness of each teaching approach in promoting comprehension and retention of course material in the introductory psychology class.

REPORTING

In Derringer design experiments used in psychology, researchers often rely on a variety of statistical tools to analyze data, assess the quality of the experimental design, and draw conclusions about the effects of IVs on DVs. Here are some common statistical tools used in Derringer designed experiments in psychology:

ANOVA: ANOVA is a widely used statistical technique for comparing means across multiple groups. In Derringer designed experiments, ANOVA can be used to test for differences in the DV(s) across different levels of the IVs while controlling for potential confounding variables.

Regression Analysis: Regression analysis allows researchers to examine the relationship between one or more IVs and a continuous DV. In Derringer designed experiments, regression analysis can be used to model the effects of IVs on DVs and assess the significance of these effects.

MANOVA: MANOVA extends ANOVA to situations where there are multiple DVs. In Derringer designed experiments with multiple outcome measures, MANOVA can be used to simultaneously test for differences across experimental conditions while accounting for correlations among DVs.

GLMs: GLMs are a flexible class of statistical models that can accommodate a wide range of data distributions and link functions. In Derringer designed experiments, GLMs can be used to analyze non-normally distributed data or categorical outcome variables while incorporating covariates or interaction effects.

Repeated Measures Analysis: Repeated measures analysis techniques are used when the same participants are measured multiple times under different experimental conditions. In Derringer designed experiments with repeated measures, researchers can use techniques, such as repeated measures ANOVA or mixed-effects models, to analyze within-subjects effects and control for individual differences.

Optimal Experimental Design Algorithms: These algorithms are used to generate Derringer designs that optimize specific criteria, such as efficiency, precision of parameter estimates, or robustness, to model assumptions. These algorithms often involve optimization techniques, such as linear programming, genetic algorithms, or Bayesian optimization.

Model Diagnostics: Model diagnostics techniques are used to assess the assumptions and adequacy of statistical models fitted to the data. In Derringer designed experiments, researchers may use diagnostic plots, residual analyses, and goodness-of-fit tests to evaluate the validity of statistical models and identify potential sources of bias or error.

These are just a few examples of the statistical tools commonly used in Derringer designed experiments in psychology. Depending on the specific research question, experimental design, and data characteristics, researchers may employ a variety of other statistical techniques and methods to analyze and interpret the results of their experiments.

When reporting the results of Derringer designed experiments used in psychology, researchers typically follow a structured format to communicate their findings clearly and effectively. Here are the key elements that may be included in a report on Derringer designed experiments in psychology:

Title: A concise and informative title that accurately reflects the content and focus of the study.

Abstract: A brief summary of the study, including the research question, methods, key findings, and conclusions. The abstract provides readers with an overview of the study's objectives and outcomes.

Introduction: A background section that provides context for the study by reviewing relevant literature, explaining the rationale for the research question, and stating the study's objectives and hypotheses. The introduction sets the stage for the rest of the report and helps readers understand the significance of the study.

Methods:
- **Participants**: Description of the sample, including demographic information and any inclusion/exclusion criteria.
- **Experimental Design**: Explanation of the Derringer design used, including the independent and DVs, experimental conditions, and design criteria.

- **Procedure**: Detailed description of the experimental procedures, including how participants were assigned to conditions, data collection methods, and any control measures implemented to ensure internal validity.

Results:

- Summary of participant characteristics and any relevant descriptive statistics.
- Presentation of the main findings, including statistical analyses of the effects of IVs on DVs.
- Tables, figures, and/or graphs to illustrate key results, such as means, standard deviations, effect sizes, and significance levels.

Discussion:

- Interpretation of the results in relation to the study's objectives and hypotheses.
- Comparison of findings with previous research and theoretical frameworks.
- Discussion of implications, limitations, and future directions for research.
- Consideration of the broader implications of the study for theory, practice, and policy in psychology.

Conclusion: A concise summary of the main findings and their implications, emphasizing the contributions of the study to the field of psychology.

References: A list of all sources cited in the report, following a specific citation style (e.g., APA, MLA).

Appendices: Additional materials, such as raw data, supplementary analyses, or experimental materials (e.g., stimuli, questionnaires) that provide further detail or support for the main findings.

By including these key elements in a report on Derringer designed experiments in psychology, researchers can effectively communicate their methods, results, and conclusions to readers and contribute to the advancement of knowledge in the field.

3 Engineering Designed Experiments

3.1 AEROSPACE ENGINEERING

In aerospace engineering, designed experiments are crucial for optimizing aircraft and spacecraft designs, improving manufacturing processes, and ensuring the safety and reliability of aerospace systems. Here are some common types of designed experiments with examples of their applications in aerospace engineering.

In aerospace engineering designed experiments, various types of variables are considered to study and optimize different aspects of aerospace systems, vehicles, components, and processes. These variables can be categorized into several common types, including:

Independent Variables:
- **Aircraft Design Parameters**: Variables related to the design of aircraft, including wing shape, wing area, aspect ratio, fuselage design, and control surface configurations.
- **Propulsion System Variables**: Variables related to propulsion systems, such as engine type, thrust, fuel type, and nozzle design.
- **Aerodynamic Parameters**: Variables related to aerodynamic performance, including airfoil shapes, wing sweep angles, and airspeed.
- **Materials and Structural Variables**: Variables related to the materials used in aerospace components, structural design, and composites.

Dependent Variables:
- **Aircraft Performance**: Variables related to the performance of aircraft, including speed, range, endurance, lift, drag, and fuel efficiency.
- **Aircraft Stability and Control**: Variables related to the stability and control characteristics of aircraft, including pitch, yaw, roll, and response to control inputs.
- **Structural Integrity**: Variables related to the structural integrity of aerospace components, including stress, strain, fatigue life, and damage tolerance.
- **Thermal Performance**: Variables related to the thermal behavior of spacecraft and aerospace systems, including temperature distribution and heat transfer rates.
- **Flight Trajectory**: Variables related to the flight path, altitude, and trajectory of aerospace vehicles.

Categorical Variables:
- **Aircraft Types**: Categories of aircraft, such as commercial airliners, military jets, helicopters, or unmanned aerial vehicles (UAVs).
- **Propulsion Types**: Categories of propulsion systems, such as turbofan, turbojet, rocket, or electric propulsion.
- **Aerospace Materials**: Categories of materials used in aerospace applications, including metals, composites, and ceramics.

Control Variables (Covariates): These are variables that are held constant or controlled during experiments to eliminate their influence on the dependent variable. For example, in a study of aircraft drag, air temperature and pressure may be controlled variables.

Random Variables: In some cases, random variables may be introduced to account for variability or uncertainty in measurements or environmental conditions.

DOI: 10.1201/9781003528531-3

Interaction Variables: Interaction variables are used to investigate the combined effects of two or more independent variables on the dependent variable. For example, the interaction between wing design and propulsion system performance.

Noise Variables (Error Variables): These are uncontrolled or unmeasured variables that can introduce variability or errors into the experimental results. Techniques like noise reduction, vibration isolation, and calibration may be used to minimize the impact of noise variables in aerospace experiments.

Aerospace engineering designed experiments are crucial for optimizing the performance, safety, and reliability of aerospace vehicles and systems. These experiments help engineers make informed decisions in the design, testing, and operation of aircraft, spacecraft, and other aerospace technologies.

3.1.1   Factorial Experiments

Factorial design experiments are used in aerospace engineering across various research and development activities, testing, and optimization processes. Here are some common areas in aerospace engineering where factorial design experiments are applied:

Materials Testing: Aerospace engineers often conduct experiments to evaluate and optimize materials used in aircraft and spacecraft construction. Factorial designs help assess the impact of various factors (e.g., material composition, heat treatment, manufacturing processes) on material properties like strength, durability, and thermal resistance.

Structural Analysis: Factorial experiments are employed to analyze the structural integrity and performance of aircraft components and systems. These experiments may involve factors such as load conditions, component design, and material selection.

Aerodynamics and Fluid Dynamics: In aerodynamics research, factorial designs help assess the effects of multiple variables on aircraft performance, including factors like airfoil shape, wing design, and control surface configurations. Similar experiments are used in fluid dynamics to study the behavior of fluids around aircraft surfaces.

Propulsion Systems: Engineers use factorial designs to optimize propulsion systems, including engines and propulsion components. These experiments may involve factors like fuel composition, combustion chamber design, and nozzle configurations.

Avionics and Control Systems: Aerospace engineers may employ factorial experiments to test and improve avionics systems, including communication, navigation, and control systems. Factors could include sensor accuracy, control algorithms, and software design.

Environmental Testing: Factorial experiments help aerospace engineers study the impact of environmental factors such as temperature, humidity, and radiation on aircraft and spacecraft components. This is crucial for ensuring the reliability and performance of aerospace systems in different conditions.

Composite Materials Development: With the increasing use of composite materials in aerospace, factorial experiments are used to optimize composite manufacturing processes and assess the impact of variables like resin types, curing methods, and reinforcement materials.

Safety and Reliability Analysis: Aerospace engineers use factorial designs to assess the reliability and safety of aircraft and spacecraft systems. These experiments may involve factors like component redundancy, maintenance schedules, and failure modes.

Noise Reduction: Aircraft noise reduction is a critical concern. Factorial experiments help engineers evaluate noise reduction strategies, including factors like engine design, aerodynamic modifications, and soundproofing materials.

Cost Optimization: In aerospace manufacturing, factorial designs are employed to optimize processes and reduce production costs. Engineers evaluate factors like production techniques, tooling, and material usage.

In each of these areas, factorial experiments allow aerospace engineers to systematically vary and analyze multiple factors simultaneously, enabling them to understand complex interactions and optimize designs, performance, and efficiency. These experiments help drive advancements in aerospace technology, improve safety, reduce environmental impact, and enhance the overall performance of aircraft and spacecraft.

STEPS

Developing factorial design experiments for aerospace engineering involves a systematic approach to understanding and optimizing various aspects of aircraft and spacecraft. Here are the steps to develop factorial design experiments in this field:

1. **Define the Objective:**
   - Clearly state the objectives of your experiment. What specific aspects of aerospace engineering are you trying to investigate or optimize? Define the response variables that you want to measure or improve.
2. **Identify Factors and Levels:**
   - Identify the independent variables (factors) that may influence the outcome of your experiment. These could include design parameters, material properties, environmental conditions, or other relevant variables.
   - Determine the levels or settings at which each factor will be tested. For each factor, define the high and low levels or multiple levels if necessary.
3. **Select a Factorial Design:**
   - Choose an appropriate factorial design based on the number of factors, levels, and interactions you want to study. Common types include full-factorial, fractional factorial, and mixed-level designs.
4. **Experimental Design and Planning:**
   - Create a detailed experimental plan that outlines the specific combinations of factor levels (experimental runs) to be tested.
   - Randomize the order of experiments to minimize the impact of extraneous variables and ensure that the results are not biased by the testing sequence.
5. **Data Collection:**
   - Conduct the experiments according to the predefined plan. Collect data for each experimental run, including measurements, observations, or any relevant data points.
6. **Data Analysis:**
   - Perform statistical analysis on the collected data to identify significant effects, interactions, and trends.
   - Use software tools or statistical techniques like ANOVA to determine the main effects of factors and any interactions that may exist.
7. **Optimization and Interpretation:**
   - Use the results of the factorial experiment to optimize the aerospace system or component under study. This may involve adjusting factor levels to achieve desired performance or efficiency.
   - Interpret the findings to gain insights into the relationships between factors and outcomes, helping improve the understanding of aerospace phenomena.
8. **Validation and Verification:**
   - Validate the results and conclusions through additional experiments or simulations if necessary. Ensure that the findings are robust and applicable to real-world aerospace applications.
9. **Report and Documentation:**
   - Prepare a comprehensive report that documents the experimental setup, methodology, data collected, statistical analysis, findings, and conclusions.

- Include visual aids such as graphs, charts, and tables to present the results effectively.
- Communicate the implications of the study and recommendations for aerospace engineering applications.

10. **Implementation:**
- Implement the optimized solutions or recommendations into aerospace systems, designs, or processes, taking into account safety, cost-effectiveness, and other relevant factors.

11. **Continuous Improvement:**
- Consider the results of the factorial experiment as part of an ongoing process for continuous improvement in aerospace engineering. Use feedback from real-world applications to refine designs and processes further.

12. **Documentation and Knowledge Sharing:**
- Maintain thorough documentation of the experiments and their outcomes for future reference and knowledge sharing within the aerospace engineering community.

Factorial design experimentation is a powerful tool in aerospace engineering for optimizing performance, safety, efficiency, and various other aspects of aircraft and spacecraft. Proper planning, careful execution, and rigorous analysis are key to achieving meaningful results and driving advancements in the field.

PITFALLS AND REMEDIES

Factorial design experiments in aerospace engineering are powerful tools for studying the effects of multiple factors and their interactions on various outcomes. However, like any experimental approach, they can encounter pitfalls that may affect the validity and interpretation of results. Here are some common pitfalls and remedies when conducting factorial experiments in aerospace engineering:

Pitfalls:

Small Sample Sizes:
> **Pitfall:** Inadequate sample sizes can lead to unreliable results and difficulties in detecting significant effects.
> **Remedies:** Ensure that sample sizes are sufficient to provide adequate statistical power. Conduct power analyses to determine the required sample size based on expected effect sizes.

Factor Selection:
> **Pitfall:** Including irrelevant or unimportant factors can complicate the experiment and waste resources.
> **Remedies:** Conduct a thorough literature review and preliminary studies to identify the most relevant factors. Use engineering knowledge and expertise to make informed choices.

Neglecting Interactions:
> **Pitfall:** Ignoring interactions between factors can lead to an incomplete understanding of the system's behavior.
> **Remedies:** Investigate both main effects and interactions. Use appropriate statistical techniques to analyze and visualize interactions.

Assuming Linearity:
> **Pitfall:** Assuming that relationships between factors and responses are linear may not hold in all cases.
> **Remedies:** Consider using higher-order models or nonlinear regression techniques if there is evidence of nonlinearity. Perform residual analysis to check for linearity assumptions.

Confounding:
> **Pitfall:** Confounding occurs when the effects of two or more factors are indistinguishable, making it impossible to determine their individual impacts.
> **Remedies:** Employ proper experimental design techniques to minimize confounding. Randomize experimental runs and use appropriate blocking or stratification strategies.

Ignoring Noise Factors:
> **Pitfall:** Neglecting to account for noise factors or uncontrollable sources of variability can reduce the precision of the experiment.
> **Remedies:** Identify and measure noise factors whenever possible. Implement robust design techniques, such as Taguchi methods, to minimize the impact of noise.

Overfitting Models:
> **Pitfall:** Overfitting occurs when complex models are fitted to the data, leading to poor generalization to new data.
> **Remedies:** Use model selection techniques to choose the most parsimonious model that adequately explains the data. Cross-validation can help assess model performance.

Failure to Verify Assumptions:
> **Pitfall:** Not verifying key assumptions of statistical tests, such as normality or homoscedasticity, can lead to inaccurate conclusions.
> **Remedies:** Conduct diagnostic checks to ensure that assumptions are met. Transform data or use robust statistical methods when necessary.

Misinterpretation of Results:
> **Pitfall:** Misinterpreting statistical significance as practical significance can lead to inappropriate decisions.
> **Remedies:** Consider the practical implications of results. Effect size measures and engineering judgment should guide decision-making.

Lack of Replication:
> **Pitfall:** Failing to replicate experiments can make it difficult to assess the reliability of findings.
> **Remedies:** Replicate experiments to confirm results and assess the reproducibility of effects.

Incomplete Data Collection:
> **Pitfall:** Missing data can lead to biased results and reduced statistical power.
> **Remedies:** Implement procedures to minimize missing data. Use appropriate methods for handling missing data, such as imputation techniques.

By being aware of these pitfalls and implementing appropriate remedies, aerospace engineers can design and execute factorial experiments more effectively, ensuring that the results are reliable and informative for improving aerospace systems and processes.

EXAMPLE

Factorial experiments are commonly used in aerospace engineering to study the effects of multiple factors on an outcome of interest, such as the performance of an aircraft component or system. Here's an example of a matrix for a 2^3 full-factorial experiment in aerospace engineering:

Objective: To investigate the effects of three factors (airspeed, altitude, and wing angle) on the lift coefficient of an aircraft wing.

Factors:

Airspeed (Factor A):
> Low Speed (100 knots)
> High Speed (200 knots)

Altitude (Factor B):
   Low Altitude (10,000 feet)
   High Altitude (30,000 feet)
Wing Angle (Factor C):
   Small Angle (5 degrees)
   Large Angle (15 degrees)

Response Variable: Lift Coefficient (dimensionless)

Full-Factorial Matrix (2^3):
   In this example, a full 2^3 factorial design is used, which involves all possible combinations of the factors and levels. The matrix is shown in Table 3.1.
   In this matrix, each row represents a specific experimental run with a unique combination of airspeed, altitude, and wing angle levels. The lift coefficient of the aircraft wing is measured for each combination. By analyzing the results, aerospace engineers can determine how changes in airspeed, altitude, and wing angle impact the lift coefficient, which is crucial for aircraft performance and design optimization.

REPORTING

Statistical tools play a crucial role in analyzing data and drawing meaningful conclusions from factorial design experiments in aerospace engineering. Here are some of the common statistical tools and techniques used in this context:

Analysis of Variance (ANOVA): ANOVA is a fundamental statistical technique used to assess the significance of differences between group means. It helps determine whether there are statistically significant variations among factor levels and their interactions.

Main Effects Plots: Main effects plots visualize the influence of individual factors on the response variable. They provide insights into the direction and magnitude of the impact of each factor, helping engineers understand which factors are most influential.

Interaction Plots: Interaction plots are used to explore the interactions between different factors. They help identify cases where the effect of one factor depends on the level of another, which is important in understanding complex relationships.

Normal Probability Plots: Normal probability plots are used to check the normality of the data distribution. Departures from normality can impact the validity of statistical tests, so it's important to assess this aspect.

TABLE 3.1

Full-Factorial Matrix for Aerospace Engineering

Run	Airspeed (A)	Altitude (B)	Wing Angle (C)
1	Low Speed	Low Altitude	Small Angle
2	High Speed	Low Altitude	Small Angle
3	Low Speed	High Altitude	Large Angle
4	High Speed	High Altitude	Large Angle
5	Low Speed	Low Altitude	Large Angle
6	High Speed	Low Altitude	Large Angle
7	Low Speed	High Altitude	Small Angle
8	High Speed	High Altitude	Small Angle

Residual Analysis: Residual analysis involves examining the differences between observed data points and predicted values from the statistical model. Residual plots help check the assumptions of the model and detect outliers or patterns in the data.

Design of Experiments (DOE) Software: Specialized DOE software packages are often used to design experiments, set up factor levels, generate experimental runs, and perform statistical analyses automatically. These tools can simplify the process and ensure accurate results.

Response Surface Methodology (RSM): RSM is an advanced statistical technique used for optimizing complex systems by modeling and analyzing response surfaces. It helps engineers find the optimal combination of factor levels that maximize or minimize the response variable.

Regression Analysis: Regression analysis is used to establish mathematical models that describe the relationship between the response variable and the independent factors. These models can be used for prediction and optimization.

Confidence Intervals: Confidence intervals provide a range within which the true population parameter is likely to fall with a certain level of confidence. They are used to quantify uncertainty in parameter estimates.

Hypothesis Testing: Hypothesis tests are conducted to assess the statistical significance of various factors, interactions, or model parameters. Common tests include t-tests, F-tests, and chi-squared tests.

Effect Size Measures: Effect size measures such as Cohen's d or eta-squared quantify the practical significance of observed differences or relationships between variables. They help assess the real-world impact of experimental findings.

Power Analysis: Power analysis helps determine the sample size required to detect a meaningful effect with a certain level of confidence. It ensures that experiments have sufficient statistical power to detect important effects.

Robust Design Techniques: Robust design methods aim to optimize systems or products while considering variability and noise factors. Taguchi methods, S/N ratios, and other techniques are used to make designs more robust against variations.

These statistical tools and techniques are used to analyze data, make informed decisions, and optimize aerospace systems, components, and processes. They enable engineers to extract valuable insights from factorial design experiments and contribute to the advancement of aerospace engineering.

3.1.2 Response Surface Methodology (RSM)

RSM design experiments are commonly used in aerospace engineering for optimizing complex systems and processes. Aerospace engineers utilize RSM to improve the performance, efficiency, and reliability of various components and systems within the aerospace industry. Here are some specific areas and applications where RSM design experiments are frequently employed in aerospace engineering:

Aircraft Design and Performance Optimization: RSM is used to optimize the design of aircraft wings, fuselages, and engine components to improve aerodynamics, fuel efficiency, and overall performance.

Gas Turbine Engine Development: Aerospace engineers use RSM to optimize the performance of gas turbine engines, including combustion processes, blade designs, and cooling systems.

Composite Material Development: RSM is applied to develop and optimize composite materials used in aerospace structures, ensuring they meet strength, weight, and durability requirements.

Manufacturing Process Optimization: Aerospace manufacturing processes, such as machining, welding, and additive manufacturing, benefit from RSM to optimize parameters and reduce defects.

Structural Analysis and Testing: RSM helps in optimizing structural components and conducting tests to ensure they meet safety and performance criteria.

Spacecraft and Satellite Design: Engineers use RSM to optimize the design and layout of components in spacecraft and satellites, considering factors like weight, thermal control, and communication systems.

Aeroelasticity and Flutter Analysis: RSM aids in analyzing aeroelastic effects and optimizing control surfaces to prevent flutter and ensure aircraft stability.

Flight Control Systems: RSM helps optimize flight control systems, including autopilots and navigation algorithms, to improve aircraft stability and safety.

Material Selection and Coatings: Aerospace engineers use RSM to select suitable materials and coatings for specific applications, considering factors like corrosion resistance, thermal conductivity, and wear resistance.

Rocket Propulsion Systems: RSM is employed to optimize the performance of rocket propulsion systems, including nozzle designs, propellant combinations, and combustion processes.

Wind Tunnel Testing: Engineers use RSM to design wind tunnel experiments and optimize test conditions for accurate aerodynamic assessments.

Reliability and Maintenance Optimization: RSM is applied to optimize maintenance schedules and predict component reliability, reducing downtime and maintenance costs.

Aircraft Noise Reduction: RSM helps in reducing aircraft noise through the optimization of engine designs and noise-reducing technologies.

Structural Health Monitoring (SHM): RSM assists in optimizing SHM systems for monitoring the structural integrity of aircraft and spacecraft.

Mission Planning and Trajectory Optimization: For space missions, RSM is used to optimize mission trajectories, fuel consumption, and payload capacity.

RSM allows aerospace engineers to systematically explore the design space, identify critical factors, and optimize processes or systems efficiently. It plays a crucial role in achieving cost savings, performance improvements, and enhanced safety within the aerospace industry.

STEPS

Developing an RSM design experiment in aerospace engineering involves a systematic approach to optimizing complex systems or processes. Here are the steps to develop RSM experiments for aerospace engineering:

Define the Problem and Objectives: Clearly define the problem or process that needs optimization in aerospace engineering. Identify the specific objectives, such as improving performance, reducing weight, or enhancing reliability.

Select Factors and Levels: Identify the critical input factors (independent variables) that influence the aerospace system or process. Determine the range of levels for each factor that will be studied. Factors can include design parameters, materials, operating conditions, and more.

Design the Experimental Layout: Choose an appropriate experimental design for your RSM study. Common designs include full-factorial, fractional factorial, central composite, or Box-Behnken designs. The choice depends on the number of factors and levels, as well as available resources.

Conduct Experiments: Perform the experiments according to the designed experimental layout. Collect data on the response variable(s), which represent the performance or output of the aerospace system or process. Ensure that the experiments are conducted accurately and consistently.

Analyze Data: Use statistical analysis techniques to analyze the experimental data. Fit response surface models that describe the relationship between the input factors and the response variable(s). The models can be linear, quadratic, or higher-order, depending on the complexity of the system.

Model Validation: Validate the response surface models to ensure they accurately represent the actual system or process. This may involve conducting additional experiments or using validation data if available.

Optimization: Utilize the validated response surface models to perform optimization. Use mathematical optimization techniques to find the combination of input factors that maximizes or minimizes the objective function (e.g., maximize performance or minimize weight).

Sensitivity Analysis: Perform sensitivity analysis to identify the most influential factors and their optimal levels. This helps in understanding the robustness of the optimized solution.

Verification and Validation: Verify the optimized solution through physical testing or simulations. Ensure that the optimized design or process meets safety, reliability, and performance requirements.

Implementation: Implement the optimized solution in the aerospace system or process. Monitor its performance in real-world conditions and make any necessary adjustments.

Documentation: Document all aspects of the RSM study, including problem definition, experimental design, data analysis, models, optimization results, and implementation details. This documentation is crucial for future reference and knowledge transfer.

Continuous Improvement: Continuously monitor the aerospace system or process and collect data to assess its long-term performance. If necessary, apply RSM techniques for further improvements.

Report and Communication: Prepare a comprehensive report summarizing the RSM study, its findings, and the optimized solution. Communicate the results to stakeholders and colleagues within the aerospace engineering team.

Feedback Loop: Establish a feedback loop to incorporate lessons learned from the RSM study into future aerospace design and engineering projects.

The systematic application of RSM in aerospace engineering can lead to significant improvements in aircraft performance, reliability, and efficiency, making it a valuable tool for optimizing complex aerospace systems and processes.

PITFALLS AND REMEDIES

RSM is a powerful tool for optimizing aerospace systems and processes, but like any approach, it comes with its own set of pitfalls and challenges. Here are some common pitfalls and remedies in RSM design experimentation used in aerospace engineering:

Pitfall 1: Insufficient Data

Issue: Inadequate or low-quality data can lead to inaccurate models and unreliable optimization results.

Remedy: Ensure that data collection is rigorous and comprehensive. Use appropriate measurement techniques and replicate experiments when necessary. Verify data quality and consistency throughout the study.

Pitfall 2: Model Inadequacy

 Issue: Sometimes the chosen response surface model may not accurately represent the true relationship between factors and responses.

 Remedy: Validate the model thoroughly using techniques such as residual analysis, lack-of-fit tests, and cross-validation. Consider using higher-order models or alternative modeling techniques if necessary.

Pitfall 3: Overfitting

 Issue: Overfitting occurs when a model fits the noise in the data rather than the underlying relationships.

 Remedy: Use techniques like stepwise regression, variable selection, or regularization to avoid overfitting. Maintain a balance between model complexity and goodness of fit.

Pitfall 4: Lack of Model Validation

 Issue: Failing to validate response surface models can lead to erroneous optimization results.

 Remedy: Set aside a portion of the data for validation and compare predicted values with actual observations. Conduct additional experiments, if possible, to validate the model's predictions.

Pitfall 5: Neglecting Factor Interactions

 Issue: RSM assumes that factor effects are independent, but in reality, interactions between factors may exist.

 Remedy: Investigate and account for factor interactions. Experiment with different experimental designs, such as central composite designs (CCDs), which can capture interaction effects.

Pitfall 6: Infeasible Optima

 Issue: The optimization process may yield solutions that are theoretically optimal but practically infeasible.

 Remedy: Include constraints in the optimization process to ensure that solutions meet physical, safety, and operational requirements.

Pitfall 7: Neglecting Robustness

 Issue: Optimized solutions may lack robustness and may not perform well under varying conditions or uncertainties.

 Remedy: Conduct sensitivity analyses to assess the robustness of the optimized solution. Identify factors that affect robustness and design for tolerance to uncertainties.

Pitfall 8: Implementation Challenges

 Issue: Implementing the optimized solution in aerospace systems can be challenging due to practical constraints and regulatory requirements.

 Remedy: Collaborate with experts in system integration and certification to ensure that the optimized design or process can be successfully implemented.

Pitfall 9: Lack of Communication

 Issue: Ineffective communication of RSM results to stakeholders can hinder implementation and decision-making.

 Remedy: Prepare clear and concise reports and presentations that convey the benefits and implications of the RSM study to decision-makers and team members.

Pitfall 10: Limited Continuous Improvement

 Issue: After optimization, aerospace systems or processes may not undergo continuous improvement.

 Remedy: Establish a feedback loop to monitor the performance of the optimized solution and be prepared to make further adjustments or refinements as needed.

By being aware of these pitfalls and applying appropriate remedies, aerospace engineers can maximize the effectiveness of RSM in optimizing complex aerospace systems and processes while ensuring the reliability and safety of the final solutions.

EXAMPLE

RSM is commonly used in aerospace engineering to optimize complex systems or processes while studying the relationships between multiple factors and a response variable. Here's an example of a matrix for an RSM designed experiment in aerospace engineering:

Objective: To optimize the aerodynamic performance of an aircraft wing by varying factors such as airfoil shape, angle of attack, and flap deflection.

Factors:

1. **Airfoil Shape (Factor A)**:
   - Round Airfoil
   - NACA 2412 Airfoil
   - NACA 642-215 Airfoil
2. **Angle of Attack (Factor B)**:
   - Low Angle (0 degrees)
   - Medium Angle (5 degrees)
   - High Angle (10 degrees)
3. **Flap Deflection (Factor C)**:
   - No Flap Deflection
   - Small Deflection (10 degrees)
   - Large Deflection (20 degrees)

Response Variable: Lift Coefficient (dimensionless)

RSM Designed Experiment Matrix:
In this RSM designed experiment, the matrix is designed to optimize the aerodynamic performance of the aircraft wing by studying the effects of airfoil shape, angle of attack, and flap deflection on the lift coefficient. The matrix is shown in Table 3.2.

In this matrix, each row represents a specific experimental run with a unique combination of airfoil shape, angle of attack, and flap deflection levels. The lift coefficient is measured for each combination. Using statistical analysis and response surface modeling, aerospace engineers can determine the optimal wing configuration that maximizes lift coefficient while considering the

---

TABLE 3.2

Response Surface Methodology (RSM) Example for Aerospace Engineering

Run	Airfoil Shape (A)	Angle of Attack (B)	Flap Deflection (C)	Lift Coefficient
1	Round Airfoil	Low Angle	No Flap Deflection	0.6
2	NACA 2412 Airfoil	Medium Angle	Small Deflection	0.7
3	NACA 642-215 Airfoil	High Angle	Large Deflection	0.8
4	NACA 2412 Airfoil	Low Angle	Small Deflection	0.65
5	NACA 642-215 Airfoil	Medium Angle	Large Deflection	0.75
6	Round Airfoil	High Angle	Large Deflection	0.68
7	NACA 2412 Airfoil	High Angle	No Flap Deflection	0.72
8	NACA 642-215 Airfoil	Medium Angle	No Flap Deflection	0.67
9	Round Airfoil	Low Angle	Large Deflection	0.63
10	NACA 2412 Airfoil	Medium Angle	Small Deflection	0.7
11	NACA 642-215 Airfoil	Low Angle	No Flap Deflection	0.65
12	Round Airfoil	High Angle	Small Deflection	0.7

factors of airfoil shape, angle of attack, and flap deflection. RSM helps identify the relationships between these factors and the response variable, allowing for optimized aircraft wing design in aerospace engineering applications.

Reporting

RSM relies on various statistical tools to design experiments, build models, and optimize processes or systems in aerospace engineering. Some of the key statistical tools used in RSM design experimentation for aerospace engineering include:

Design of Experiments (DOE): DOE techniques are fundamental to RSM and include factorial designs, CCDs, and Box-Behnken designs. These designs help in selecting the appropriate experimental conditions to study the effects of factors on the response.

Regression Analysis: Linear and nonlinear regression models are used to fit response surface models to experimental data. This helps in quantifying the relationship between input factors (independent variables) and the response (dependent variable).

Analysis of Variance (ANOVA): ANOVA is used to assess the significance of factors and their interactions. It partitions the total variation in the response into contributions from different factors and errors.

Residual Analysis: Residuals are the differences between observed and predicted values. Analyzing residuals helps in checking the model's assumptions, identifying outliers, and assessing model adequacy.

Lack-of-Fit Tests: These tests evaluate whether the response surface model adequately fits the data. A significant lack-of-fit suggests that the model may not be suitable.

Model Selection Techniques: When there are multiple potential models, techniques like stepwise regression or information criteria (Akaike information criterion (AIC), Bayesian information criterion (BIC)) help select the most appropriate model.

Response Surface Visualization: Techniques like contour plots, surface plots, and 3D plots visually represent the response surface, making it easier to understand the relationships between factors and responses.

Optimization Algorithms: Various optimization algorithms, such as gradient-based methods, genetic algorithms, or simulated annealing, are used to find the optimal settings of input factors that maximize or minimize the response.

Robust Optimization: Techniques like robust design and robust parameter design (RPD) are used to optimize processes or systems while considering variations or uncertainties in input factors.

Model Validation: Techniques like cross-validation, prediction error plots, and residual analysis are employed to validate the accuracy and predictive capability of response surface models.

Sensitivity Analysis: Sensitivity analysis assesses the robustness of the optimized solutions by evaluating how changes in input factors affect the response.

Statistical Software: Specialized statistical software packages like JMP, Minitab, Design-Expert, and R are commonly used to perform RSM analyses, build models, and visualize results.

These statistical tools collectively enable aerospace engineers to design experiments efficiently, build accurate response surface models, and optimize aerospace systems or processes, ensuring that they meet performance, reliability, and safety requirements.

A report on RSM design experimentation in aerospace engineering should provide a clear and comprehensive overview of the study, methods, results, and conclusions. Here are the key elements typically included in such a report:

Title Page:
- Title of the Report
- Author(s) and Affiliation(s)
- Date

Abstract: A concise summary of the study's objectives, methods, key findings, and conclusions. It should provide a quick overview of the entire report.

Table of Contents: A list of sections, subsections, and corresponding page numbers for easy navigation.

List of Figures and Tables: A list of all figures and tables used in the report with corresponding captions and page numbers.

List of Abbreviations and Symbols: If applicable, a list of abbreviations and symbols used in the report and their meanings.

Introduction:
- Background and context for the study, including the problem statement, objectives, and significance.
- A brief review of relevant literature and prior research.
- A clear statement of the research hypothesis or questions.

Methodology:
- Detailed explanation of the experimental design, including the RSM (e.g., CCD, Box-Behnken design) and factors considered.
- Description of data collection procedures, including the number of experiments conducted and the range of factor levels.
- Any statistical techniques or software used for analysis.

Experimental Results:
- Presentation of the collected data in tables, figures, or charts.
- Response surface plots or contour plots illustrating the relationships between factors and the response.
- Statistical analyses, including ANOVA tables, regression coefficients, and p-values.
- Discussion of any significant main effects, interactions, or lack-of-fit.

Model Validation:
- Assessment of the adequacy and accuracy of the response surface model.
- Use of diagnostic tools such as residual plots, normal probability plots, and lack-of-fit tests.
- Cross-validation results if applicable.

Optimization and Findings: Presentation of the optimized conditions or settings for factors that maximize or minimize the response. Discussion of practical implications and interpretations of the optimized results.

Sensitivity Analysis: Examination of the robustness of the optimized solutions to variations in factor levels.

Discussion:
- Interpretation of the results in the context of the research objectives.
- Comparison of findings with prior literature or industry standards.
- Discussion of any limitations or assumptions made during the study.

Conclusion:
- Summary of the key findings and their implications.
- Restatement of the research objectives and whether they were achieved.

Recommendations: Suggestions for further research or improvements based on the study's outcomes.

Acknowledgments: Recognition of individuals, organizations, or funding sources that contributed to the research.

References: A list of all sources cited in the report, following a specific citation style (e.g., APA, IEEE).

Appendices: Supplementary materials such as detailed experimental procedures, raw data, additional figures, or mathematical derivations.

It's important to maintain a clear and logical structure in the report, use appropriate headings and subheadings, and ensure that all content is well-organized and easy to follow. Additionally, the report should be written in a concise and professional manner, with a focus on communicating the research findings effectively to the intended audience in the aerospace engineering field.

3.1.3 FRACTIONAL FACTORIAL EXPERIMENTS

Fractional factorial design experiments are used in aerospace engineering in various applications to efficiently study and optimize complex systems or processes while reducing the number of experimental runs required. Here are some areas within aerospace engineering where fractional factorial design experiments are commonly applied:

Material Testing and Characterization: Aerospace materials, such as composite materials or alloys, often undergo extensive testing to understand their properties and performance. Fractional factorial experiments can help optimize the testing conditions, such as temperature, pressure, and load, to minimize the number of tests required while capturing critical data points.

Aircraft Design and Performance: Fractional factorial designs are used to study the effects of multiple design factors (e.g., wing shape, airfoil profiles, engine parameters) on aircraft performance metrics like lift, drag, fuel efficiency, and stability. This aids in the efficient exploration of the design space.

Rocket Propulsion: In rocket science, engineers use fractional factorial experiments to optimize the fuel mixture, combustion conditions, and nozzle geometry to achieve the desired thrust and efficiency while minimizing the number of test firings.

Wind Tunnel Testing: Wind tunnel experiments are vital for assessing the aerodynamic characteristics of aircraft and spacecraft. Fractional factorial designs help select appropriate test conditions (e.g., wind speed, angles of attack) to comprehensively study the effects on aerodynamic forces.

Structural Analysis: In aerospace structural engineering, fractional factorial experiments can be employed to investigate factors affecting the structural integrity and durability of components, such as wings, fuselages, and landing gear.

Manufacturing and Process Optimization: Aerospace components are often manufactured with stringent requirements for quality and performance. Fractional factorial experiments assist in optimizing manufacturing processes by identifying critical factors affecting product quality and consistency.

Avionics and Systems Integration: In avionics design and system integration, engineers can use fractional factorial experiments to evaluate the interactions between various electronic components and subsystems while considering factors like power consumption, reliability, and electromagnetic interference.

Flight Testing: During flight testing, engineers may use fractional factorial designs to select test conditions, instrumentation setups, and flight profiles that yield comprehensive data on aircraft or spacecraft performance.

Environmental Testing: Aerospace systems and components must withstand harsh environmental conditions, such as extreme temperatures, vibrations, and radiation. Fractional factorial experiments help optimize testing protocols to simulate these conditions efficiently.

Safety and Reliability Analysis: Reliability engineering in aerospace involves assessing the reliability and safety of critical systems. Fractional factorial experiments can be used to evaluate the impact of different factors on system reliability and identify potential failure modes.

In each of these applications, fractional factorial design experiments offer a systematic and efficient way to investigate complex systems or processes by studying a subset of the possible combinations of factors and levels. This allows aerospace engineers to gain valuable insights, make informed decisions, and optimize designs or processes while conserving resources and time.

STEPS

Developing fractional factorial design experimentation for aerospace engineering involves several systematic steps to efficiently study and optimize complex systems while reducing the number of experimental runs. Here are the key steps:

1. **Define the Problem and Objectives:**
   - Clearly define the problem or process you want to study or optimize in aerospace engineering.
   - Establish specific objectives, performance metrics, or responses that you aim to improve or understand.
2. **Identify Factors and Levels:**
   - Identify the factors (independent variables) that may influence the aerospace system or process. These could include design parameters, environmental conditions, material properties, and operational settings.
   - Determine the possible levels or values for each factor that you want to investigate.
3. **Select the Fractional Factorial Design:**
   - Choose an appropriate fractional factorial design based on the number of factors, levels, and the desired resolution. Common designs include half-fraction, quarter-fraction, or higher fractions.
   - Determine the fraction (e.g., 1/2, 1/4) that represents the fraction of the full-factorial experiment that will be conducted.
4. **Generate the Experimental Design Matrix:**
   - Using statistical software or specialized DOE tools generate the experimental design matrix based on the chosen fractional factorial design.
   - The matrix should specify which experimental runs need to be conducted, indicating which factor combinations and levels to test.
5. **Conduct the Experiments:**
   - Perform the experimental runs as specified in the design matrix. Ensure that the experiments are conducted under controlled and consistent conditions.
   - Record the responses or data for each experimental run, including performance metrics or other relevant measurements.
6. **Analyze the Data:**
   - Analyze the collected data using statistical techniques to understand the effects of different factors on the aerospace system or process.
   - Conduct statistical tests, regression analysis, or other appropriate methods to assess the significance of factors and their interactions.
7. **Optimize the System or Process:**
   - Use the insights gained from data analysis to make informed decisions about optimizing the aerospace system or process.
   - Adjust factor levels or settings to achieve desired performance or objectives, while considering practical constraints.

8. **Validate and Verify:**
   - Validate the optimized settings by conducting additional experiments or simulations to confirm that the improvements observed in the fractional factorial experiment are consistent.
   - Verify the results against real-world conditions and requirements to ensure that the optimized solution is practical and reliable.
9. **Document the Experiment:**
   - Create a detailed report that documents the entire experimental process, including problem definition, design matrix, data collection, analysis methods, results, conclusions, and recommendations.
   - Include graphical representations, tables, and statistical summaries to support your findings.
10. **Implement and Monitor:**
    - Implement the optimized solutions or changes in the aerospace system or process based on the results of the fractional factorial experiment.
    - Continuously monitor and assess the system's performance to ensure that improvements are sustained over time.
11. **Iterate and Refine (if necessary):**
    - Depending on the complexity of the aerospace engineering problem, you may need to iterate through the steps to further refine and optimize the system or process.
12. **Communicate Results:**
    - Share the findings, conclusions, and recommendations with relevant stakeholders, teams, or organizations within the aerospace industry.

Fractional factorial experiments in aerospace engineering provide a systematic and efficient approach to exploring the effects of multiple factors on system performance or processes while conserving resources. Properly designed and executed experiments can lead to valuable insights, improvements, and enhanced understanding of aerospace systems and technologies.

PITFALLS AND REMEDIES

Fractional factorial design experimentation in aerospace engineering, like any experimental approach, can encounter various pitfalls. Here are some common pitfalls and their corresponding remedies:

Inadequate Factor Selection
> **Pitfall:** Choosing the wrong factors or neglecting important ones can lead to incomplete or inaccurate insights.
>
> **Remedy:** Conduct a thorough literature review, involve subject matter experts, and perform preliminary experiments or simulations to identify critical factors. Ensure that all influential variables are considered.

Underestimating Interactions
> **Pitfall:** Neglecting interactions between factors can result in incorrect conclusions about their impact on the system.
>
> **Remedy:** Perform interaction screening during the analysis phase to identify significant interactions. Consider using higher-resolution fractional factorial designs if interactions are suspected.

Limited Data Points
> **Pitfall:** Fractional factorial designs involve fewer experimental runs, which can lead to sparse data.
>
> **Remedy:** Use design augmentation techniques like D-optimal designs or add center points to gather more data points within the same resource constraints. This can improve model accuracy.

Lack of Randomization
   Pitfall: Non-randomized experiments can introduce bias into the results.
   Remedy: Randomly order the experimental runs and conduct them in a randomized sequence to minimize bias and control for external factors.

Noise and Variability
   Pitfall: Variability in measurements can obscure real effects or lead to false discoveries.
   Remedy: Implement proper measurement and control systems to reduce noise and variability. Consider replicating experiments to assess measurement repeatability.

Overextrapolation
   Pitfall: Extrapolating results beyond the tested factor levels can be risky and inaccurate.
   Remedy: Limit interpretations and recommendations to the tested factor levels. Use additional experiments or simulations to explore behaviors outside the tested range if necessary.

Inadequate Documentation
   Pitfall: Poor documentation can make it challenging to reproduce or build upon the experiment.
   Remedy: Maintain a detailed record of the experimental setup, procedures, and results. Document any deviations or unexpected events during the experiments.

Ignoring Practical Constraints
   Pitfall: Optimized solutions may not be feasible or practical in real-world aerospace applications.
   Remedy: Consider practical constraints, such as budget, safety, and manufacturing limitations, when implementing recommendations. Consult with engineers and stakeholders to ensure feasibility.

Limited Replicability
   Pitfall: Results from a single experiment may not be robust.
   Remedy: Replicate the experiment under similar conditions to assess the reproducibility of the findings. Consistent results across replicates increase confidence in the conclusions.

Misinterpreting Statistical Significance
   Pitfall: Relying solely on p-values without considering practical significance can lead to misguided decisions.
   Remedy: Interpret statistical significance in the context of practical significance. Focus on effect sizes and domain knowledge to determine the importance of factors.

Premature Conclusions
   Pitfall: Drawing conclusions too early in the experiment without sufficient data or analysis.
   Remedy: Allow for thorough data collection and analysis before making conclusions. Consult with experts and validate results.

Resistance to Change
   Pitfall: Resistance within the organization to adopt recommended changes.
   Remedy: Engage with stakeholders early, provide clear communication of benefits, and involve key decision-makers to gain support for proposed improvements.

By being aware of these potential pitfalls and implementing the suggested remedies, aerospace engineers can conduct more effective fractional factorial experiments and ensure that the results are reliable, actionable, and aligned with the goals of the aerospace projects.

EXAMPLE

Fractional factorial experiments are used in aerospace engineering to study the effects of multiple factors while conducting only a fraction of the full set of experiments, reducing time and resources. Here's an example of a matrix for a 2^4-1 fractional factorial experiment in aerospace engineering:

Objective: To study the effects of four factors (engine thrust, wing span, fuselage length, and airfoil shape) on the range of a UAV.

Factors:

1. **Engine Thrust (Factor A)**:
   - Low Thrust
   - High Thrust
2. **Wing Span (Factor B)**:
   - Short Span
   - Long Span
3. **Fuselage Length (Factor C)**:
   - Short Fuselage
   - Long Fuselage
4. **Airfoil Shape (Factor D)**:
   - NACA 0012 Airfoil
   - NACA 4412 Airfoil

Response Variable: UAV range (measured in kilometers)

Fractional Factorial Matrix (2^4-1):

In this fractional factorial experiment, a fraction of the full set of experiments is conducted to study the main effects and two-way interactions of the factors on the UAV range. The matrix is shown in Table 3.3.

In this matrix, each row represents a specific experimental run with a unique combination of factor levels. The fractional factorial design allows you to study the main effects and two-way interactions of the factors on the UAV range with a reduced number of experiments, saving time and resources while providing valuable insights into how the factors affect the range of the unmanned aerial vehicle in aerospace engineering applications.

REPORTING

In fractional factorial design experimentation for aerospace engineering, various statistical tools are used to analyze and interpret the data effectively. These tools help engineers identify critical factors, understand interactions, and optimize processes or systems. Here are some commonly used statistical tools:

TABLE 3.3

Aerospace Engineering Fractional Factorial Designed Experiment Matrix Example

Run	Engine Thrust (A)	Wing Span (B)	Fuselage Length (C)	Airfoil Shape (D)
1	Low Thrust	Short Span	Short Fuselage	NACA 0012 Airfoil
2	High Thrust	Short Span	Long Fuselage	NACA 4412 Airfoil
3	Low Thrust	Long Span	Short Fuselage	NACA 4412 Airfoil
4	High Thrust	Long Span	Long Fuselage	NACA 0012 Airfoil
5	Low Thrust	Short Span	Long Fuselage	NACA 0012 Airfoil
6	High Thrust	Short Span	Short Fuselage	NACA 4412 Airfoil
7	Low Thrust	Long Span	Long Fuselage	NACA 4412 Airfoil
8	High Thrust	Long Span	Short Fuselage	NACA 0012 Airfoil
9	Low Thrust	Short Span	Short Fuselage	NACA 4412 Airfoil

Main Effects Plots: Main effects plots display the effects of individual factors on the response variable. They provide a visual representation of how each factor influences the outcome and help identify significant factors.

Interaction Plots: Interaction plots illustrate the interactions between two or more factors. They are useful for identifying cases where the combined effect of factors differs from the sum of their individual effects.

Half-Normal Probability Plots: These plots help identify which factors have the most significant impact on the response variable. Factors with the steepest slope in the plot are often the most influential.

Analysis of Variance (ANOVA): ANOVA is used to partition the variability in the response variable into contributions from different factors and their interactions. It helps determine which factors are statistically significant and which are not.

Effect Size Measures: Effect size measures, such as Cohen's d or eta-squared, quantify the practical significance of a factor's impact on the response. They help distinguish between statistically significant and practically significant effects.

Residual Analysis: Residual analysis checks if the assumptions of the statistical model are met. It involves examining the residuals (the differences between observed and predicted values) for patterns or outliers.

Normal Probability Plots: These plots assess whether the residuals follow a normal distribution. Deviations from normality may indicate issues with the model or data.

Response Surface Methodology (RSM): RSM is a collection of statistical techniques used to model and optimize responses that depend on multiple factors. It involves fitting mathematical models to experimental data and finding optimal factor settings.

Fractional Factorial Design Matrices: Design matrices are used to plan the experimental runs efficiently. They specify which factor combinations to test in a fraction of the full-factorial experiment, allowing engineers to reduce the number of required experiments.

ANOVA Table: The ANOVA table summarizes the results of the analysis of variance, including the sources of variation, degrees of freedom, mean squares, F-statistic, and p-values. It helps assess the significance of factors and interactions.

Confidence Intervals: Confidence intervals provide a range of values within which the true population parameter is likely to fall. They help quantify uncertainty and assess the precision of estimates.

Regression Analysis: Regression models are used to relate the response variable to the predictor variables (factors). Engineers can use regression analysis to build predictive models and optimize processes.

Optimization Algorithms: Various optimization algorithms, such as gradient descent, genetic algorithms, or response surface optimization, are used to find the optimal factor settings that maximize or minimize the response variable.

Statistical Software: Specialized statistical software packages like Minitab, JMP, or R are often used to perform statistical analysis, create visualizations, and generate reports.

Control Charts: In quality control and manufacturing applications, control charts monitor the stability of a process over time and detect shifts or variations that may require corrective action.

These statistical tools are employed in aerospace engineering to design experiments efficiently, analyze data, and make informed decisions regarding product design, process optimization, and quality improvement. The specific tools used may vary depending on the complexity of the experiment and the goals of the aerospace project.

A well-structured report for fractional factorial design experimentation in aerospace engineering should provide a comprehensive overview of the experimental design, procedures, results, and conclusions. Here are the key elements typically included in such a report:

Title Page:
- Title of the Report
- Names and affiliations of authors
- Date of submission

Abstract: A concise summary of the experiment's objectives, methods, and key findings. The abstract should provide a quick overview of the entire report.

Table of Contents: A list of sections and subsections with corresponding page numbers for easy navigation.

List of Figures and Tables: A list of all figures and tables used in the report, along with their respective page numbers.

Introduction:
- Background information on the problem or process being studied.
- The objectives and goals of the experiment.
- A brief description of the factors and levels considered.

Experimental Design: Description of the fractional factorial design matrix used, including factors, levels, and resolution. Explanation of the rationale behind the chosen design.

Experimental Procedure: Detailed steps and procedures followed during the experiment. Information on how data was collected, measurements taken, and any specialized equipment used.

Data Analysis:
- Presentation of the data collected during the experiment.
- Statistical analysis methods used, including ANOVA or regression analysis.
- Main effects and interaction effects of factors on the response variable.
- Presentation of graphical representations like main effects plots, interaction plots, or half-normal plots.
- Statistical software used for analysis (e.g., Minitab, JMP, R).

Results:
- Detailed presentation of the results obtained from the analysis.
- Tables, graphs, and charts to illustrate the findings.
- Discussion of significant and nonsignificant factors and interactions.
- Interpretation of practical implications based on the results.

Discussion:
- Interpretation of the results in the context of the aerospace engineering problem.
- Discussion of any unexpected findings or anomalies.
- Implications for aerospace design, manufacturing, or processes.
- Comparison of results with initial hypotheses or expectations.

Conclusion:
- A summary of the key findings and their relevance.
- The overall success of the experiment in achieving its objectives.
- Recommendations for further research or experimentation.

References: A list of all sources, including research papers, books, and software packages, cited in the report. Follow a specific citation style (e.g., APA, IEEE, and Chicago) as required.

Appendices: Any supplementary information that is relevant but not essential to the main report. This may include raw data, detailed statistical calculations, or additional figures and tables.

Acknowledgments: Recognition of individuals, organizations, or funding sources that contributed to the experiment or report.

Glossary or Nomenclature: Definitions of technical terms and abbreviations used in the report.

List of Symbols: A list of symbols and their meanings if applicable.

Remember that the structure and specific content of the report may vary depending on the organization's guidelines, the complexity of the experiment, and the preferences of the researchers. It's essential to maintain clarity, conciseness, and logical flow throughout the report to ensure that readers can understand and replicate the experiment's results and conclusions.

3.1.4 Central Composite Design (CCD)

CCD experiments are commonly used in aerospace engineering for optimizing complex processes, designs, and systems. Here are some areas within aerospace engineering where CCD design experiments find applications:

Aerodynamic Design: CCD can be employed to optimize the shape and configuration of aircraft wings, airfoils, and other components to enhance aerodynamic performance while minimizing drag, lift, or turbulence.

Structural Analysis: CCD can aid in the development of lightweight yet structurally robust materials and designs for aircraft and spacecraft components, such as wings, fuselages, and landing gear.

Propulsion Systems: Optimization of propulsion systems, including jet engines and rocket engines, for improved efficiency, thrust, and fuel consumption.

Materials and Composites: CCD helps in designing and selecting materials for aerospace applications, including high-strength composites, heat-resistant materials for reentry vehicles, and lightweight alloys.

Spacecraft Trajectory Planning: CCD can be used in trajectory planning for spacecraft missions, including optimizing maneuvers, orbital transfers, and rendezvous with celestial bodies.

Control Systems: Optimizing control systems for aircraft and spacecraft to ensure stability, maneuverability, and response to various operational conditions.

Satellite Design: CCD can be applied to satellite design for selecting orbit parameters, optimizing power systems, and enhancing communication and data transmission.

Manufacturing Processes: Improving aerospace manufacturing processes, such as casting, forging, and machining, to achieve tighter tolerances and reduce defects.

Sensor and Instrumentation Calibration: CCD can assist in optimizing sensor calibration procedures for accurate data collection and instrumentation performance in aerospace research.

Safety and Reliability: CCD can be employed to enhance the safety and reliability of aerospace systems, including the development of redundant systems and failure analysis.

Environmental Impact: Minimizing the environmental impact of aerospace activities, including reducing emissions and noise pollution from aircraft and rocket launches.

Cost Optimization: CCD can help in optimizing aerospace designs and processes to reduce production costs and improve cost-effectiveness.

In aerospace engineering, CCD experiments are valuable for systematically exploring multiple variables and their interactions to identify optimal settings that meet specific performance criteria. This approach can lead to significant advancements in aircraft and spacecraft design, efficiency, safety, and overall performance.

Steps

Developing a CCD experiment in aerospace engineering involves several steps to systematically investigate the effects of multiple variables on a specific response or performance outcome. Here are the typical steps to develop a CCD experiment:

1. **Define the Objective:** Clearly define the research or engineering objective of your CCD experiment. Identify the specific response or performance measure you want to optimize or understand better. For example, you might want to maximize lift while minimizing drag in aircraft design.

2. **Identify Factors and Levels:** Identify the independent variables (factors) that may influence the response variable. Determine the potential levels or settings for each factor. Factors could include geometric parameters, material properties, operating conditions, or other relevant variables. Levels should cover a reasonable and practical range for each factor.

3. **Determine the Center Point:** Decide on the center point for your CCD. The center point represents a baseline or typical operating condition and is typically the midrange value of each factor. It helps to assess the curvature of the response surface.

4. **Choose the Range of Experiment:** Specify the range of experimentation for each factor. This includes the minimum and maximum values for continuous factors and the list of levels for categorical factors.

5. **Select the Design Matrix:** Use software or statistical tools (such as DOE software) to generate the CCD matrix. The matrix includes a combination of factorial points (full or fractional) and axial points. The number of points depends on the number of factors and the desired level of precision. The matrix helps in selecting experimental runs.

6. **Conduct Experimental Runs:** Perform the experiments according to the settings specified in the CCD matrix. Ensure that each run is conducted accurately and consistently, and record the values of the response variable for each run.

7. **Randomize the Runs:** Randomize the order of experimental runs to minimize the impact of uncontrolled variables or external factors that could affect the results.

8. **Analyze the Data:** Use statistical analysis techniques, such as regression analysis or ANOVA, to model the relationship between the independent variables (factors) and the response variable. Fit a response surface model to the data to estimate the optimal conditions.

9. **Optimize the Response:** Use the response surface model to identify the optimal conditions that maximize or minimize the response variable, subject to any constraints or practical considerations. Optimization techniques like desirability functions can help find the best compromise among multiple objectives.

10. **Validate the Model:** Conduct additional experiments at the predicted optimal conditions to verify the accuracy of the response surface model and confirm the improvements in the response variable.

11. **Interpret the Results:** Interpret the results to gain insights into the factors that most significantly influence the response variable. Understand any interactions between factors and their impact on the response.

12. **Make Recommendations:** Based on the results and insights gained from the CCD experiment, make recommendations for design modifications, process improvements, or further research in aerospace engineering.

13. **Document the Experiment:** Document all aspects of the CCD experiment, including the design matrix, experimental runs, data collected, statistical analysis, and conclusions. This documentation is essential for reproducibility and future reference.

14. **Report Findings:** Prepare a comprehensive report or presentation summarizing the CCD experiment, its objectives, methods, results, and recommendations. Share the findings with relevant stakeholders and colleagues.

15. **Iterate as Needed:** Depending on the complexity of the aerospace engineering problem, you may need to iterate the CCD experiment to further refine the design or optimize additional performance criteria.

CCD experiments are valuable tools in aerospace engineering for optimizing designs, improving performance, and gaining insights into complex systems. They help engineers make informed decisions and achieve desired outcomes while considering multiple factors and their interactions.

PITFALLS AND REMEDIES

CCD experiments are powerful tools in aerospace engineering, but like any experimental approach, they come with potential pitfalls. Here are some common pitfalls and remedies when conducting CCD experiments in aerospace engineering:

Lack of Factor Selection
> **Pitfall:** Selecting irrelevant or nonsignificant factors can lead to wasted resources and misinterpretation of results.
>
> **Remedy:** Conduct a thorough literature review and engage with domain experts to identify the most relevant factors affecting the aerospace system under study. Use screening experiments or preliminary investigations to confirm the significance of chosen factors.

Inadequate Range of Factors
> **Pitfall:** Narrowing the range of factor levels too much can result in missing critical information about the system's behavior.
>
> **Remedy:** Ensure that the selected ranges for each factor are broad enough to capture the potential operating conditions and any nonlinear relationships. Pilot experiments or engineering judgment can help determine suitable ranges.

Poor Experimental Design
> **Pitfall:** Inappropriate experimental designs or insufficient replication can compromise the reliability of results.
>
> **Remedy:** Use specialized software or statistical expertise to generate optimal CCD designs. Ensure that the chosen design matrix provides robust and reliable information. Implement replication of experimental runs to assess variability.

Ignoring Interaction Effects
> **Pitfall:** Neglecting interaction effects between factors can lead to incomplete or misleading conclusions.
>
> **Remedy:** Pay close attention to interaction terms in the response surface model. Assess and interpret interactions between factors to understand their combined impact on the response variable.

Overfitting the Model
> **Pitfall:** Overcomplicated response surface models with too many terms can lead to model overfitting and poor predictive performance.
>
> **Remedy:** Use model selection techniques (e.g., stepwise regression) to identify the most parsimonious and interpretable model that adequately fits the data. Cross-validation can help assess model validity.

Neglecting Constraints
> **Pitfall:** Failure to account for practical constraints or physical limitations in the optimization process can lead to solutions that are not feasible in aerospace applications.
>
> **Remedy:** Integrate constraints into the optimization problem by defining boundaries or limitations on factor levels. Use constrained optimization algorithms to find solutions that meet these constraints.

Lack of Model Validation
> **Pitfall:** Failing to validate the response surface model can result in unreliable predictions and recommendations.

Remedy: Conduct additional experiments at predicted optimal conditions to validate the model's accuracy. Compare predicted and observed responses to ensure consistency.

Inadequate Documentation

Pitfall: Poor documentation of the CCD experiment can hinder reproducibility and knowledge transfer.

Remedy: Maintain comprehensive records of the experimental design, procedures, data collection, and analysis. Document any deviations from the original plan and provide clear explanations in the final report.

Not Involving Stakeholders

Pitfall: Failing to engage relevant stakeholders, including engineers, designers, and end-users, can lead to solutions that do not align with practical needs or requirements.

Remedy: Involve stakeholders throughout the CCD experiment, from the design phase to the interpretation of results. Seek their input and feedback to ensure that the study addresses real-world concerns.

Limited Resources for Follow-Up

Pitfall: Conducting a CCD experiment without sufficient resources or a plan for implementing recommended changes can render the effort ineffective.

Remedy: Ensure that there is a clear plan for implementing changes based on the CCD results. Allocate resources for follow-up activities and monitor the impact of improvements in aerospace systems.

By being aware of these potential pitfalls and implementing the suggested remedies, aerospace engineers can maximize the effectiveness of CCD experiments, leading to better-informed decisions and improved aerospace systems.

EXAMPLE

A CCD is a commonly used experimental design in aerospace engineering to study the effects of factors while exploring the central region and estimating curvature. Here's an example of a matrix for a CCD designed experiment in aerospace engineering:

Objective: To optimize the fuel efficiency of a rocket engine by studying the effects of two factors (propellant flow rate and nozzle diameter).

Factors:

1. **Propellant Flow Rate (Factor A):**
   - Low Flow Rate (1,000 kg/s)
   - High Flow Rate (3,000 kg/s)
2. **Nozzle Diameter (Factor B):**
   - Small Diameter (0.2 m)
   - Large Diameter (0.5 m)

Response Variable: Specific Impulse (measured in seconds, s)

Central Composite Design Matrix:

In this CCD designed experiment, the matrix is designed to explore the central region and curvature of the response surface by varying propellant flow rate and nozzle diameter. The matrix is shown in Table 3.4.

In this matrix, the "Low" and "High" levels for propellant flow rate and nozzle diameter are represented by −1 and +1, respectively. The additional runs (1 to 5) represent points within the

TABLE 3.4
Central Composite Design (CCD) Designed Experiment Matrix Example

Run	Propellant Flow Rate (A) (kg/s)	Nozzle Diameter (B) (m)
−1	1,000	0.2
−1	3,000	0.2
−1	1,000	0.5
−1	3,000	0.5
1	1,500	0.35
2	1,250	0.35
3	1,750	0.35
4	1,500	0.275
5	1,500	0.425

experimental region, including points in the center and at the corners. These additional runs help estimate the curvature of the response surface.

By conducting experiments with varying propellant flow rates and nozzle diameters according to this CCD matrix, aerospace engineers can model the specific impulse and optimize the rocket engine's settings for maximum fuel efficiency within the specified flow rate and nozzle diameter limits.

REPORTING

In CCD experiments in aerospace engineering, various statistical tools are used to analyze and interpret the results. These tools help engineers understand the relationships between factors, optimize processes, and make informed decisions. Here are some commonly used statistical tools in CCD experimentation for aerospace engineering:

Analysis of Variance (ANOVA): ANOVA is used to partition the variability in the response variable into different sources, including the main effects of factors, interaction effects, and error. It helps identify significant factors and interactions.

Response Surface Methodology (RSM): RSM involves fitting mathematical models to the experimental data to describe the relationship between factors and responses. Engineers use RSM to create response surface plots and conduct optimization studies.

Design of Experiments (DOE) Software: Specialized software such as Minitab, JMP, or Design-Expert is often used to generate optimal CCD designs, analyze experimental data, and perform regression analysis.

Regression Analysis: Regression models, including linear, quadratic, and higher-order models, are used to describe how factors affect the response variable. Regression analysis helps estimate model coefficients and identify significant terms.

Contour Plots and Surface Plots: These graphical tools are used to visualize response surfaces and understand the relationships between factors and responses. Engineers can identify optimal factor settings and regions of interest on the response surface.

Optimization Algorithms: Optimization algorithms, such as desirability functions or numerical optimization methods, are employed to find the optimal factor settings that maximize or minimize the response variable within defined constraints.

Normal Probability Plots: Normal probability plots are used to assess the normality of residuals in the regression model. Deviations from normality can indicate model inadequacy or suggest transformation of data.

Model Adequacy Checks: Engineers perform checks like lack-of-fit tests and diagnostic plots to assess the adequacy of the regression model. Lack-of-fit tests help determine if a simpler model is sufficient, while diagnostic plots identify influential data points.

Confidence Intervals and Prediction Intervals: These intervals provide a range of values within which the true response is likely to fall. Engineers use them to quantify uncertainty in predictions and parameter estimates.

Sequential Experimentation: Engineers may use sequential experimentation techniques to iteratively refine their models and find optimal solutions efficiently. This approach can save time and resources.

Robust Parameter Design (RPD): In cases where robustness is critical, RPD techniques are applied to optimize processes while considering variations and uncertainties in factors.

Constraint Handling: When there are practical constraints on factor settings, engineers use constraint-handling techniques in optimization to ensure that solutions meet these constraints.

Statistical Process Control (SPC): SPC techniques are used to monitor and control aerospace processes based on the knowledge gained from CCD experiments. Control charts and process capability analysis may be employed.

Data Visualization Tools: Tools like scatterplots, box plots, and histograms are used for exploratory data analysis and to identify outliers or patterns in the data.

Statistical Hypothesis Testing: Engineers may perform hypothesis tests to assess the significance of factor effects, interactions, or other statistical comparisons.

These statistical tools help aerospace engineers design experiments, analyze data, and make data-driven decisions to optimize aerospace systems, improve performance, and ensure reliability and safety. The specific tools used depend on the complexity of the experiment and the goals of the aerospace engineering project.

A report for a CCD experiment in aerospace engineering typically includes several key elements to communicate the experimental design, methodology, results, and conclusions effectively. Below are the essential elements of a report used for CCD experimentation in aerospace engineering:

Title Page:
   Title of the Report: Concise and descriptive of the experiment.
   Name and Affiliation: Name of the author(s) and their institutional affiliation.
   Date: The date of report preparation.

Abstract:
   A brief summary of the experiment's purpose, methodology, key findings, and conclusions. It should provide a snapshot of the report's content.

Table of Contents: A list of all sections, subsections, and their respective page numbers for easy navigation.

List of Figures and Tables: A list of all figures and tables included in the report, along with their respective page numbers.

Nomenclature (if applicable): A list of symbols, variables, and abbreviations used throughout the report, along with their definitions.

Introduction:
   Background: Provides context for the experiment, including the problem statement and its relevance to aerospace engineering.
   Objectives: Clearly state the experiment's objectives and what it aims to achieve.
   Scope: Defines the scope of the experiment, including any limitations or constraints.

Experimental Design:
   Overview of CCD: Explains the CCD used, including the number of factors, levels, and center points.

Factorial and Star Points: Describes the factor settings at the center point, factorial points, and star points.

Response Variables: Specifies the response variables measured or observed during the experiment.

Experimental Matrix: Provides a summary of the CCD matrix, including the actual factor settings used in each run.

Methodology:

Data Collection: Details the data collection process, including any equipment or instruments used.

Experimental Procedure: Describes the step-by-step procedure followed in conducting the CCD experiment.

Data Analysis: Explains the statistical tools and techniques used for data analysis, including regression models and RSM.

Results:

Data Presentation: Presents the experimental data, often in the form of tables, graphs, or charts.

Response Surface Plots: Includes response surface plots to visualize the relationship between factors and responses.

Statistical Analysis: Summarizes the statistical analysis results, including regression coefficients, p-values, and model adequacy assessments.

Discussion:

Interpretation of Results: Discusses the implications of the experimental findings and how they relate to the objectives.

Factor Effects: Analyzes the effects of factors and interactions on the response variables.

Optimal Conditions: Identifies optimal factor settings for achieving desired performance or outcomes.

Conclusions:

Summarizes the key findings and their significance in the context of aerospace engineering. Addresses whether the experiment's objectives were met.

Suggests practical recommendations or areas for further investigation.

Recommendations: Provides recommendations for practical applications, process improvements, or future research based on the experiment's outcomes.

Acknowledgments (if applicable): Acknowledges individuals, organizations, or funding sources that contributed to the experiment's success.

References: Cites all sources, including research papers, textbooks, and software, used in the report's preparation.

Appendices (if applicable): Supplementary information such as raw data, detailed calculations, or additional graphs and charts.

Appendix A: Experimental Matrix: If not included in the main body of the report, this section provides the complete CCD matrix with factor settings for each run.

Appendix B: Regression Analysis: If applicable, this section includes detailed regression analysis output, including coefficients and statistical tests.

Appendix C: Response Surface Plots: Includes additional response surface plots or three-dimensional representations of the data.

Appendix D: Data Tables: Provides detailed data tables for each experimental run.

The report should be well-organized, clear, and concise, with a focus on communicating the experiment's methodology and results effectively to the aerospace engineering audience. It should also adhere to any specific formatting and style guidelines required by the organization or publication.

3.1.5 Taguchi Methods

Taguchi method design experiments are used in various areas of aerospace engineering to optimize and improve processes, designs, and products. Some of the specific applications and areas where Taguchi methods are commonly used in aerospace engineering include:

Manufacturing Processes: Taguchi methods are employed to optimize manufacturing processes in aerospace, such as machining, welding, and composite fabrication. By varying factors like tool geometry, cutting parameters, and material properties, engineers can improve process efficiency and product quality.

Structural Design and Analysis: Taguchi methods help optimize structural designs for aircraft and spacecraft components. Factors like material properties, geometry, and load conditions can be systematically varied to enhance structural performance and reliability.

Material Selection: Engineers use Taguchi methods to select the most appropriate materials for aerospace applications. By considering factors like strength, weight, and thermal properties, the Taguchi approach aids in identifying materials that meet specific performance criteria.

Aerodynamic Design: Taguchi methods can be applied to aerodynamic design processes to optimize airfoil shapes, wing profiles, and other components for improved lift, drag reduction, and fuel efficiency.

Avionics and Electronics: Taguchi methods are used to optimize the performance of avionics systems, including factors like component placement, electronic component selection, and circuit design.

Propulsion Systems: Optimization of propulsion systems, including engines and propulsion components, is a critical area in aerospace engineering. Taguchi methods can be used to fine-tune parameters that affect thrust, fuel efficiency, and emissions.

Reliability and Quality Assurance: Aerospace components must meet stringent reliability and quality standards. Taguchi methods are used to design experiments that identify factors influencing product reliability and to develop strategies for improving product quality and robustness.

Cost Reduction: In aerospace engineering, cost reduction while maintaining high performance is a constant concern. Taguchi methods help identify ways to reduce manufacturing and operational costs without compromising safety or performance.

Environmental Impact: Aerospace companies are increasingly focused on reducing their environmental footprint. Taguchi methods can be used to optimize designs and processes for reduced fuel consumption, emissions, and noise.

Space Exploration: In space exploration missions, Taguchi methods can be used to optimize mission parameters, spacecraft design, and instrumentation, allowing for more efficient and successful missions.

The key advantage of Taguchi methods is their ability to systematically and efficiently identify the most influential factors affecting a particular aerospace process or system while minimizing the number of experiments required. This leads to cost savings, improved performance, and enhanced product quality in aerospace engineering applications.

Steps

Developing Taguchi method design experiments for aerospace engineering involves a systematic approach to optimizing processes, designs, or products while considering various factors that influence performance and reliability. Here are the steps to develop Taguchi method experiments for aerospace engineering:

1. **Problem Identification and Objective Setting:**
   - Define the specific problem or objective you want to address in aerospace engineering. This could be related to improving a process, optimizing a design, enhancing product quality, or reducing costs.
   - Clearly state the performance metrics, quality characteristics, or objectives that you aim to optimize or improve. These metrics should be measurable and relevant to the aerospace application.

2. **Identify Factors and Levels:**
   - Identify the factors or variables that may influence the performance or outcome of the aerospace process or system. These factors can include design parameters, manufacturing variables, material properties, environmental conditions, and more.
   - Determine the levels or settings at which each factor will be varied during the experiments. The choice of levels should cover the expected operating range and be practical for testing.

3. **Selection of Orthogonal Arrays (OAs):**
   - Choose an appropriate orthogonal array (OA) or experimental design matrix based on the number of factors and levels. The selected OA should allow for efficient experimentation with a minimum number of runs.
   - Ensure that the OA satisfies the requirements of your experimental setup and that it covers all the factors and their interactions effectively.

4. **Parameter Design (Outer Array):**
   - Conduct the outer array experiments by assigning factor levels according to the selected OA. This involves conducting a series of experiments based on the combinations specified in the OA.
   - Collect data on the performance metrics or quality characteristics for each experiment in the outer array.

5. **Inner Array (Noise Factor) Design (Optional):**
   - If there are uncontrolled factors (noise factors) that may affect the results, consider conducting additional inner array experiments to assess their impact. This can help in identifying sources of variation and making the system more robust to noise.

6. **Data Analysis and Signal-to-Noise (S/N) Ratios:**
   - Analyze the experimental data using statistical methods and calculate the S/N ratios for each combination of factor levels. The S/N ratios represent the overall quality or performance characteristics and can be categorized into different types (e.g., smaller-the-better, larger-the-better, and nominal-the-best).
   - Determine the optimal factor levels that maximize the S/N ratio associated with the desired performance or quality characteristics.

7. **Confirmation Experiments:**
   - Conduct confirmation experiments using the optimal factor settings determined in the analysis. These experiments validate the effectiveness of the optimized design and confirm that the desired improvements have been achieved.

8. **Implementation and Monitoring:**
   - Implement the optimized settings and improvements in the aerospace engineering process, design, or product.
   - Continuously monitor and maintain the system to ensure that the improvements are sustained over time.

9. **Documentation and Reporting:**
   - Document the entire Taguchi method experiment, including the problem statement, factors, levels, OA, experimental results, and optimization outcomes.
   - Prepare a comprehensive report that summarizes the experiment's objectives, methodology, findings, and recommendations for implementation in aerospace engineering.

10. **Iterative Improvement (if needed):**
   - Depending on the results and changing requirements, you may need to iterate the process by refining the factors, levels, or objectives and conducting additional Taguchi method experiments for further optimization.

By following these steps, aerospace engineers can systematically apply Taguchi method design experiments to solve complex problems, optimize designs, and improve the performance and quality of aerospace systems and components.

Pitfalls and Remedies

Taguchi method design experiments are a powerful tool for optimizing processes and designs in aerospace engineering. However, like any experimental approach, they can encounter pitfalls and challenges. Here are some common pitfalls and remedies when using Taguchi method design experiments in aerospace engineering:

Inadequate Problem Definition
   Pitfall: Failing to clearly define the problem or objective can lead to ineffective experiments.
   Remedy: Begin by clearly defining the problem, setting specific objectives, and identifying key performance metrics or quality characteristics.

Poor Factor Selection
   Pitfall: Choosing the wrong factors or failing to consider important factors can result in suboptimal solutions.
   Remedy: Conduct a thorough analysis to identify all potential factors and their influence on the problem. Seek input from domain experts.

Incorrect Orthogonal Array (OA) Selection
   Pitfall: Selecting an inappropriate OA can lead to inefficient experiments or missing important interactions.
   Remedy: Carefully choose the OA that best matches the number of factors and interactions. Ensure it meets the requirements of the experiment.

Neglecting Noise Factors
   Pitfall: Ignoring uncontrolled factors (noise factors) can lead to designs that are not robust to real-world variations.
   Remedy: Identify potential noise factors and consider conducting inner array experiments to assess their impact and enhance robustness.

Inadequate Sample Size
   Pitfall: Using a sample size that is too small can result in unreliable conclusions.
   Remedy: Ensure that the sample size is statistically significant and provides sufficient data for analysis.

Overlooking Interactions
   Pitfall: Focusing only on the main effects and neglecting interactions can lead to suboptimal solutions.
   Remedy: Analyze and interpret interactions between factors to understand their combined effects on performance or quality.

Insufficient Data Analysis
   Pitfall: Inadequate data analysis can lead to incorrect conclusions and suboptimal solutions.
   Remedy: Employ appropriate statistical methods for data analysis, including S/N ratios, ANOVA, and graphical tools.

Failure to Verify and Confirm
   Pitfall: Not conducting confirmation experiments can result in implementing nonvalidated solutions.

Remedy: Always perform confirmation experiments to validate the effectiveness of the optimized settings in real-world conditions.

Lack of Implementation

Pitfall: Failing to implement the optimized design or process changes can lead to missed opportunities.

Remedy: Develop an implementation plan and ensure that the optimized settings are integrated into the aerospace engineering process or system.

Not Considering Long-Term Effects

Pitfall: Focusing solely on short-term gains without considering long-term consequences can be detrimental.

Remedy: Evaluate the long-term effects of the changes to ensure they are sustainable and do not introduce unintended issues.

Insufficient Documentation

Pitfall: Inadequate documentation can lead to a lack of transparency and hinder future reference.

Remedy: Maintain thorough records of the experiment, including problem statements, experimental setups, results, and implementation details.

Lack of Continuous Improvement

Pitfall: Not revisiting and refining the experiments can lead to missed opportunities for further optimization.

Remedy: Be open to iterative improvement, especially as conditions, requirements, or technologies evolve in aerospace engineering.

Addressing these pitfalls and implementing the suggested remedies can help aerospace engineers effectively apply Taguchi method design experiments to achieve optimal results in their projects.

EXAMPLE

Taguchi methods, also known as the Taguchi Design of Experiments (DOE), are used in aerospace engineering to optimize processes and products while considering the variation and robustness of designs. Here's an example of a matrix for a Taguchi method designed experiment in aerospace engineering:

Objective: To optimize the manufacturing process of aerospace-grade aluminum alloy components by studying the effects of factors such as extrusion temperature, cooling rate, and die lubrication.

Factors:

Extrusion Temperature (Factor A):
Low Temperature (200°C)
Medium Temperature (250°C)
High Temperature (300°C)

Cooling Rate (Factor B):
Slow Cooling
Medium Cooling
Fast Cooling

Die Lubrication (Factor C):
Low Lubrication
Medium Lubrication
High Lubrication

Response Variable: Tensile Strength (measured in megapascals, MPa)

TABLE 3.5

Taguchi Method Designed Experiment Matrix Example for Aerospace Engineering

Run	Extrusion Temperature (A) (°C)	Cooling Rate (B)	Die Lubrication (C)
1	200	Slow Cooling	Low Lubrication
2	250	Medium Cooling	Medium Lubrication
3	300	Fast Cooling	High Lubrication
4	250	Fast Cooling	Low Lubrication
5	300	Slow Cooling	High Lubrication
6	200	Medium Cooling	High Lubrication
7	300	Medium Cooling	Low Lubrication
8	200	Fast Cooling	Medium Lubrication
9	250	Slow Cooling	High Lubrication

Taguchi Methods Designed Experiment Matrix:

In this Taguchi methods designed experiment, an OA is used to efficiently study the effects of the factors on tensile strength while considering variations. The matrix is shown in Table 3.5.

In this matrix, each row represents a specific experimental run with a unique combination of factors and levels. The Taguchi Methods approach helps aerospace engineers efficiently study the effects of extrusion temperature, cooling rate, and die lubrication on tensile strength while minimizing the number of experiments required. By analyzing the results, engineers can identify the optimal combination of factors for producing aerospace-grade aluminum alloy components with the desired tensile strength and robustness.

REPORTING

Taguchi method design experiments in aerospace engineering typically involve a set of statistical tools and techniques to analyze and optimize processes or designs. These tools help engineers identify factors affecting performance, conduct experiments efficiently, and determine the best settings. Here are some of the common statistical tools used in Taguchi method design experiments for aerospace engineering:

Orthogonal Arrays (OAs): OAs are fundamental to Taguchi experiments. They help in systematically varying factors and their levels while minimizing the number of experimental runs. The choice of OA depends on a number of factors and their levels.

Signal-to-Noise (S/N) Ratios: S/N ratios are used to evaluate the performance or quality characteristics. There are different types of S/N ratios, such as smaller-the-better, larger-the-better, and nominal-the-best, which correspond to different quality characteristics. S/N ratios help identify the optimal factor settings.

Analysis of Variance (ANOVA): ANOVA is used to partition the variability in experimental data into different sources, including main effects and interactions. It helps determine which factors have a significant impact on the response variable.

Main Effects Plots: Main effects plots visually represent the influence of individual factors on the response variable. They provide insights into which factors are most influential.

Interaction Plots: Interaction plots show how two or more factors interact with each other. They help identify synergistic or antagonistic effects that may not be evident from main effects alone.

Robust Parameter Design (RPD): RPD techniques, such as the Taguchi Loss Function, aim to minimize the variability in performance due to uncontrollable factors or noise. RPD helps engineers design products or processes that are less sensitive to variations.

Confirmation Experiments: After the initial Taguchi experiments, confirmation experiments are conducted to validate the effectiveness of the optimized settings in real-world conditions. These experiments help ensure that the improvements are consistent and reliable.

Control Charts: Control charts are used for monitoring and controlling processes over time. They help detect any shifts or variations in the process that may affect product quality in aerospace engineering.

Response Surface Methodology (RSM): While Taguchi experiments focus on screening and optimization, RSM is sometimes used for further refinement and modeling of complex responses. RSM involves fitting mathematical models to experimental data and finding optimal settings using numerical optimization techniques.

Factorial Plots: Factorial plots are graphical representations of the factorial experiments, showing how different factor combinations affect the response variable. They can reveal trends and patterns in the data.

Desirability Functions: Desirability functions combine multiple quality characteristics into a single index, allowing engineers to simultaneously optimize multiple responses. They are useful when multiple performance criteria need to be met in aerospace engineering.

These statistical tools and techniques are used in various combinations based on the specific objectives of the Taguchi Method design experiments in aerospace engineering. They help engineers systematically improve processes, reduce variation, and optimize designs to meet quality and performance requirements.

A report for Taguchi method design experiments in aerospace engineering typically follows a structured format to communicate the experimental design, results, and recommendations effectively. While the exact format may vary depending on the organization's requirements, here are the common elements found in such a report:

Title Page:
- Title of the Report
- Date
- Names and affiliations of the authors
- Contact information

Table of Contents: List of sections and subsections with page numbers

Executive Summary: A concise summary of the experiment, objectives, key findings, and recommendations. Highlights of significant improvements or optimizations achieved.

Introduction:
- Background information on the problem or process being studied.
- Clear statement of the objectives and goals of the Taguchi experiment.
- Rationale for using the Taguchi Method.

Experimental Design:
- Description of the factors and levels considered in the experiment.
- Explanation of the choice of OAs and their characteristics.
- Details of how factors were randomized and run in the experiments.
- Information on any control or noise factors that were considered.

Methodology:
- Explanation of the Taguchi Method, including the use of S/N ratios.
- Description of the chosen quality characteristic(s) and the type of S/N ratio used (e.g., smaller-the-better, larger-the-better, nominal-the-best).
- Details of how S/N ratios were calculated.

Experimental Results:
- Presentation of the experimental data, including the factors and their levels.
- Tables or graphs showing the response (S/N ratio) values for each experimental run.
- Summary statistics and calculations of the S/N ratios.

Data Analysis:
- ANOVA results to determine the significance of factors and interactions.
- Main effects plots and interaction plots to visualize the influence of factors.
- Discussion of any identified optimal factor settings.

Discussion:
- Interpretation of the experimental results.
- Comparison of the Taguchi Method's recommendations to the baseline or existing process.
- Discussion of practical implications and feasibility.

Recommendations:
- Clear and actionable recommendations based on the Taguchi experiment's findings.
- Proposed optimal factor settings for implementation.
- Suggestions for further studies or improvements.

Conclusion: Summary of the key findings and their significance. Reiteration of the benefits achieved through the Taguchi Method.

References: Citation of relevant literature, standards, or guidelines used in the experiment and analysis.

Appendices: Additional details, data, or information that may be useful for readers but not essential to the main report. Any supplementary materials, such as experimental setup diagrams or calculation formulas.

Acknowledgments: Recognition of individuals or organizations that contributed to the experiment or report.

Attachments: Any supplementary documents, charts, graphs, or raw data that support the findings.

It's important to maintain a clear, concise, and well-organized structure in the report to facilitate understanding and decision-making. Depending on the complexity of the experiment and its intended audience, the report may be more or less detailed.

3.1.6 MIXTURE EXPERIMENTS

Mixture design experiments can be used in aerospace engineering for various applications where the composition of mixtures or materials plays a crucial role in achieving desired performance or characteristics. Here are some areas within aerospace engineering where mixture design experiments are applicable:

Composite Material Development: Aerospace engineers often work with composite materials made up of different components (e.g., fibers, resins, additives) to achieve specific properties like strength, weight, and durability. Mixture design experiments can help optimize the composition of composite materials.

Propellant Formulation: In rocket propulsion systems, the composition of propellants (oxidizers, fuels, additives) can impact thrust, efficiency, and stability. Mixture design experiments can optimize propellant formulations for desired performance.

Coating and Surface Treatments: Aerospace components require coatings and surface treatments for protection against corrosion, wear, and temperature extremes. Mixture design experiments can optimize coating compositions to meet stringent aerospace requirements.

Fluid Dynamics and Fuel Mixtures: In aircraft engines, the composition of fuel–air mixtures can affect combustion efficiency and emissions. Mixture design experiments can be used to optimize fuel blends and combustion processes.

Thermal Insulation Materials: Aerospace vehicles need effective thermal insulation to withstand extreme temperatures during reentry or atmospheric entry. Mixture design experiments can help develop insulation materials with specific heat-resistant properties.

Additive Manufacturing: Aerospace engineers are increasingly using additive manufacturing (3D printing) to create complex components. Mixture design experiments can optimize the composition of printing materials for desired strength, weight, and heat resistance.

Ceramic Matrix Composites (CMCs): CMCs are used in aerospace applications due to their high-temperature resistance and lightweight properties. Mixture design experiments can optimize the composition of CMCs for specific applications.

Propellant and Fuel Additives: Mixture design experiments can be used to develop and optimize additives that improve the performance and stability of aerospace propellants and fuels.

Lubricant Formulation: Lubricants are essential in aerospace applications to reduce friction and wear in moving parts. Mixture design experiments can help formulate lubricants with the right viscosity, temperature stability, and lubricating properties.

Sealants and Adhesives: Aerospace engineers often use sealants and adhesives for bonding and sealing components. Mixture design experiments can optimize the composition of these materials for strong and durable bonds.

In each of these applications, mixture design experiments help engineers systematically vary the proportions of different components, observe the effects on performance or properties, and identify the optimal composition that meets specific aerospace requirements. This approach allows for efficient and cost-effective development of materials and formulations critical to the aerospace industry's success.

STEPS

Developing mixture design experiments in aerospace engineering involves a systematic approach to optimize the composition of materials or mixtures. Here are the typical steps involved in developing mixture design experiments for aerospace engineering:

1. **Define Objectives and Criteria:**
   - Clearly define the objectives of the experiment. What specific properties or performance characteristics are you trying to optimize?
   - Establish criteria or constraints that must be met. These could include safety, regulatory, or performance standards.
2. **Identify Components and Factors:**
   - Determine the components that make up the mixture or material of interest. These could be various ingredients, additives, or materials.
   - Identify the factors (components) that can be varied in the experiment to create different compositions.
3. **Select the Mixture Design Technique:**
   - Choose an appropriate mixture design technique based on the nature of the problem. Common approaches include simplex-centroid designs, extreme vertices designs, and others.
   - Consider the number of factors and the desired level of precision when selecting the design technique.

4. **Determine Factor Levels:**
   - Define the possible ranges or levels for each factor. This involves specifying the minimum and maximum proportions or concentrations.
   - Ensure that the chosen factor levels are realistic and feasible within the constraints of the aerospace application.
5. **Design the Experiment:**
   - Use statistical software or experimental design software to create the experimental design matrix. This matrix specifies the composition of each mixture or material to be tested.
   - Ensure that the design is efficient and covers a wide range of factor combinations.
6. **Conduct Experiments:**
   - Prepare and test each mixture or material according to the design matrix.
   - Record data on the performance or properties of each mixture, ensuring accuracy and consistency in measurements.
7. **Analyze Experimental Data:**
   - Analyze the data to understand how variations in the composition (factors) affect the desired objectives or criteria.
   - Use statistical techniques such as regression analysis or ANOVA to identify significant factors and interactions.
8. **Optimization and Model Building:**
   - Develop mathematical models or response surfaces that describe the relationship between factor levels and the response (performance or properties).
   - Use optimization algorithms to find the optimal factor levels that meet the defined objectives and criteria.
9. **Validation and Confirmation:**
   - Confirm the optimal composition through additional testing or validation experiments.
   - Ensure that the optimized mixture or material performs as expected and meets all criteria.
10. **Documentation and Reporting:**
    - Document all experimental details, including the design matrix, data, statistical analyses, and optimization results.
    - Prepare a comprehensive report summarizing the experiment, findings, and recommendations for implementation in aerospace applications.
11. **Implementation and Quality Control:**
    - Implement the optimized composition or mixture in the aerospace application.
    - Establish quality control measures to ensure consistent performance and adherence to standards.
12. **Continuous Improvement:**
    - Monitor the performance of the optimized material or mixture in real-world aerospace applications.
    - Collect data and feedback for continuous improvement and refinement of the composition if necessary.

Throughout these steps, collaboration between materials scientists, engineers, statisticians, and domain experts is essential to ensure the success of the mixture design experiment in aerospace engineering.

PITFALLS AND REMEDIES

Mixture design experimentation in aerospace engineering can be complex, and there are several potential pitfalls that researchers and engineers should be aware of. Here are some common pitfalls and remedies to consider when conducting mixture design experiments in aerospace engineering:

Inadequate Design
> **Pitfall:** Designing an experiment without considering all relevant factors or using an inappropriate mixture design technique.
> **Remedy:** Conduct a thorough preliminary analysis to identify all potential factors and interactions. Choose an appropriate mixture design technique that suits the complexity of the problem.

Insufficient Factor Levels
> **Pitfall:** Choosing factor levels that are too narrow, limiting the exploration of the factor space.
> **Remedy:** Select factor levels that cover a wide and realistic range, considering the constraints of the aerospace application.

Lack of Replicates
> **Pitfall:** Conducting experiments without replicates, which can lead to unreliable results.
> **Remedy:** Incorporate replicates in the experimental design to assess the variability and reliability of the data.

Neglecting Constraints
> **Pitfall:** Not considering constraints, such as safety, regulatory, or performance standards, when optimizing the mixture.
> **Remedy:** Ensure that the optimization process takes into account all relevant constraints and standards.

Overfitting Models
> **Pitfall:** Developing overly complex models that fit the experimental data well but do not generalize to real-world applications.
> **Remedy:** Use statistical techniques to assess model fit and complexity. Employ techniques like cross-validation to ensure the model's robustness.

Neglecting Validation
> **Pitfall:** Failing to validate the optimized mixture in real-world aerospace applications.
> **Remedy:** Conduct validation experiments to confirm that the optimized composition performs as expected under practical conditions.

Poor Data Quality
> **Pitfall:** Collecting inaccurate or imprecise data during experiments.
> **Remedy:** Implement strict data collection protocols, ensure proper calibration of instruments, and minimize sources of experimental error.

Inadequate Communication
> **Pitfall:** Poor communication and collaboration between interdisciplinary teams, such as materials scientists, engineers, and statisticians.
> **Remedy:** Foster effective collaboration by involving experts from various disciplines and maintaining open communication channels.

Not Considering Environmental Factors
> **Pitfall:** Neglecting to account for environmental factors that can affect the performance of materials or mixtures in aerospace applications.
> **Remedy:** Incorporate environmental considerations into the experimental design and optimization process.

Overlooking Safety
> **Pitfall:** Ignoring safety concerns when working with experimental materials or mixtures.
> **Remedy:** Prioritize safety protocols and risk assessments throughout the experiment to protect personnel and equipment.

Insufficient Documentation
> **Pitfall:** Failing to adequately document the experimental process, design matrix, data, and analysis.
> **Remedy:** Maintain thorough and organized documentation throughout the experiment to ensure reproducibility and transparency.

By being aware of these potential pitfalls and implementing the corresponding remedies, researchers and engineers can improve the success and reliability of mixture design experimentation in aerospace engineering. Collaboration and a systematic approach are key to addressing these challenges effectively.

EXAMPLE

Mixture designed experiments are commonly used in aerospace engineering to optimize the proportions of different components or ingredients in a mixture, such as composite materials, propellants, or fuels. Here's an example of a matrix for a mixture designed experiment in aerospace engineering:

Objective: To optimize the composition of a composite material used in aircraft components for maximum strength-to-weight ratio.

Components of the Mixture:

1. **Carbon Fiber (Component A)**: Proportion expressed as a percentage of the total mixture.
   - Range: 30% to 50%
2. **Epoxy Resin (Component B)**: Proportion expressed as a percentage of the total mixture.
   - Range: 20% to 40%
3. **Kevlar Fiber (Component C)**: Proportion expressed as a percentage of the total mixture.
   - Range: 10% to 30%
4. **Fiberglass Cloth (Component D)**: Proportion expressed as a percentage of the total mixture.
   - Range: 10% to 20%

Response Variable: Strength-to-Weight Ratio (dimensionless)

Mixture Designed Experiment Matrix:
In this mixture designed experiment, the matrix is designed to optimize the mix proportions of the composite material for maximum strength-to-weight ratio. The matrix is shown in Table 3.6.

In this matrix, each row represents a specific experimental run with a unique combination of proportions for the mixture components (carbon fiber, epoxy resin, Kevlar fiber, and fiberglass cloth). The aim is to find the optimal mixture proportions that result in the highest strength-to-weight ratio while considering practical constraints and manufacturing considerations in aerospace engineering applications.

By conducting experiments with different mix proportions and analyzing the resulting strength-to-weight ratios, aerospace engineers can determine the ideal composition for their specific aircraft components, balancing strength and weight to optimize overall performance.

TABLE 3.6

Mixture Designed Experiment Matrix Example for Aerospace Engineering

Run	Carbon Fiber (A) (%)	Epoxy Resin (B) (%)	Kevlar Fiber (C) (%)	Fiberglass Cloth (D) (%)
1	30	20	15	15
2	40	30	20	10
3	35	25	10	20
4	45	35	25	15
5	30	20	25	10
6	40	30	15	15
7	35	25	20	20
8	45	35	10	10

REPORTING

In mixture design experimentation for aerospace engineering, various statistical tools and techniques are employed to analyze and optimize the composition of mixtures. These tools help researchers and engineers understand the relationships between different components and achieve desired performance characteristics. Here are some common statistical tools used in mixture design experimentation for aerospace engineering:

Simplex-Lattice Design: Simplex-lattice designs are commonly used to explore the entire composition space of mixtures systematically. These designs allow for the creation of experimental points at vertices and edges of a simplex (a triangle for three components, a tetrahedron for four components, etc.). This helps in covering a wide range of mixture compositions efficiently.

Response Surface Methodology (RSM): RSM is used to model and optimize complex relationships between mixture components and the response variables of interest. It involves fitting mathematical models, such as polynomial regression, to experimental data to identify optimal compositions that meet performance criteria.

Desirability Function: The desirability function is a tool for multi-objective optimization in mixture experiments. It combines multiple response variables into a single desirability score, allowing researchers to simultaneously optimize multiple performance criteria.

Analysis of Variance (ANOVA): ANOVA is used to analyze the variance in experimental data and assess the significance of different factors and their interactions. It helps identify which factors have a significant impact on the response variables.

Mixture-Process Variable Plots: These plots, such as ternary plots or contour plots, visualize the relationship between mixture components and response variables. They help identify regions of optimal performance and provide insights into the effects of individual components.

Constraint Optimization: Constraint optimization techniques are used to find the optimal mixture composition while adhering to constraints, such as safety, regulatory, or performance standards.

Model Validation: Various techniques, including cross-validation and validation experiments, are employed to assess the accuracy and robustness of the models developed during mixture design experimentation.

Optimization Algorithms: Optimization algorithms, such as gradient-based or heuristic methods, are used to find the optimal mixture composition based on the mathematical models developed.

Statistical Software: Specialized statistical software packages like Design-Expert, JMP, or R with relevant packages are often used for experimental design, data analysis, and optimization.

Statistical Hypothesis Testing: Hypothesis testing can be used to determine whether certain factors or interactions are statistically significant in affecting the response variables.

Monte Carlo Simulations: Monte Carlo simulations can be used to assess the uncertainty associated with the optimal mixture composition and model predictions.

Robust Parameter Design (RPD): RPD techniques help optimize mixture designs while considering variability and robustness, ensuring that the desired performance is achieved under different operating conditions or variations in raw materials.

Multi-Objective Optimization: When dealing with conflicting objectives, multi-objective optimization techniques, such as Pareto optimization, are employed to find trade-off solutions that balance various performance criteria.

The choice of statistical tools and techniques depends on the specific objectives, complexity, and constraints of the aerospace engineering application. Researchers often use a combination of these tools to design experiments, analyze data, and optimize mixture compositions to meet performance, safety, and regulatory requirements in aerospace materials and processes.

A well-structured report for mixture design experimentation in aerospace engineering should provide a clear and comprehensive overview of the experimental process, results, and conclusions. Here are the essential elements that should be included in such a report:

Title Page:
- Title of the Report
- Names of the Authors
- Affiliations of the Authors
- Date of Submission

Abstract: A concise summary of the research objectives, methodology, key findings, and conclusions.

Table of Contents: An organized list of sections, subsections, and page numbers for easy navigation.

List of Figures and Tables: A list of all figures and tables in the report along with their respective page numbers.

List of Abbreviations and Symbols (if applicable): A compilation of abbreviations and symbols used throughout the report and their explanations.

Introduction:
- Background and context of the mixture design experimentation in aerospace engineering.
- Clear statement of research objectives and goals.
- A brief literature review related to the mixture design problem.

Experimental Design:
- Description of the experimental design, including the type of mixture design (e.g., simplex-lattice design, mixture-amount design, etc.).
- Explanation of the factors and response variables involved in the experiment.
- Details of the design matrix, including the number of experimental runs and their composition.
- Discussion of any constraints or limitations in the experimental setup.

Methodology:
- Explanation of the techniques and statistical tools used for data collection and analysis.
- Information on how the mixture compositions were prepared and controlled.
- Any specialized equipment or instruments used in the experiment.

Results:
- Presentation of experimental data, often in the form of tables, figures, and charts.
- Description of any trends, patterns, or significant findings observed.
- Statistical analyses, including ANOVA results, regression models, or other relevant analyses.
- Visualization of mixture component effects on response variables (e.g., ternary plots, contour plots).

Discussion:
- Interpretation of the results in the context of the research objectives.
- Discussion of the implications of the findings for aerospace engineering applications.
- Comparison of the results with prior research or industry standards.

Optimization (if applicable): Presentation of the optimized mixture compositions, considering performance criteria or constraints. Explanation of the optimization methodology and criteria used.

Conclusion: Summarization of the key findings and their significance. Reiteration of how the research objectives were met or addressed.

Recommendations: Any practical recommendations or suggestions based on the study's outcomes. Areas for further research or experimentation.

Acknowledgments (if applicable): Recognition of individuals, organizations, or funding sources that contributed to the research.

References: A comprehensive list of all sources, including research papers, books, and relevant literature, cited in the report.

Appendices (if applicable): Supplementary information such as raw data, detailed statistical analyses, additional figures, or experimental procedures.

Author Contact Information: Contact details for the authors in case readers have questions or need further information.

Submission and Distribution Information: Details regarding how and where the report will be submitted or distributed.

Remember that clarity, organization, and completeness are essential in a mixture design experimentation report for aerospace engineering. The report should be structured logically and provide sufficient information to allow readers to understand the experiment, its results, and the implications for aerospace applications.

3.1.7 SEQUENTIAL EXPERIMENTATION

Sequential design experiments are used in aerospace engineering for various applications, primarily in the development and optimization of complex systems, components, and processes. Here are some areas within aerospace engineering where sequential design experiments find application:

Aircraft and Spacecraft Design: Sequential experimentation helps in optimizing the design of aircraft and spacecraft components, such as wings, propulsion systems, and control surfaces, to improve performance, reduce drag, enhance fuel efficiency, and meet safety and regulatory standards.

Aeroengine Development: In the development of aeroengines, sequential experiments can be used to optimize combustion processes, airflow, and materials, leading to increased engine efficiency, reduced emissions, and improved reliability.

Structural Analysis and Testing: Aerospace engineers often use sequential experimentation to assess the structural integrity of aircraft and spacecraft components. This can involve testing different materials, configurations, and manufacturing processes to ensure safety and reliability.

Avionics and Control Systems: For avionics and control systems, sequential design experiments can help fine-tune parameters, reduce system complexity, and optimize control algorithms to enhance navigation, guidance, and overall performance.

Materials Development: Aerospace materials, including composites and alloys, are subject to sequential experimentation to determine their suitability for specific applications, considering factors like strength, weight, and durability.

Propulsion Systems: The development and optimization of propulsion systems, including jet engines, rockets, and thrusters, often involve sequential experimentation to maximize thrust, efficiency, and reliability while minimizing emissions.

Aerospace Manufacturing: Sequential experimentation can be employed to improve aerospace manufacturing processes, such as machining, welding, and additive manufacturing, to achieve higher precision, lower production costs, and reduced waste.

Safety and Reliability Testing: Engineers use sequential experiments to assess the safety and reliability of aerospace systems under various conditions, including extreme temperatures, pressures, and vibrations.

Space Mission Planning: For space missions, sequential design experiments can help optimize trajectories, fuel consumption, and mission parameters to ensure successful execution and achieve mission objectives.

Environmental Impact Assessment: In response to environmental concerns, aerospace engineers may use sequential experiments to develop and test environmentally friendly technologies, such as quieter engines or cleaner propulsion systems.

Space Exploration and Research: In space exploration projects, sequential experimentation can help scientists and engineers make informed decisions regarding instrumentation, data collection, and mission planning.

Failure Analysis and Troubleshooting: When unexpected issues arise in aerospace systems, sequential experiments can be conducted to identify root causes, develop solutions, and ensure the continued reliability and safety of the systems.

Sequential design experiments provide a systematic and efficient approach to optimizing aerospace systems and processes, minimizing risks, and achieving desired performance goals. They are particularly valuable in an industry where precision, safety, and efficiency are paramount.

STEPS

Developing sequential design experimentation for aerospace engineering involves a structured approach to optimize and improve aerospace systems, components, or processes systematically. Here are the general steps to develop and implement sequential design experiments in aerospace engineering:

1. **Define the Objective:**
   - Clearly define the specific objective of the sequential experimentation. Identify what aspect of the aerospace system or process you want to optimize or improve. This could be related to performance, reliability, efficiency, or other critical parameters.
2. **Select Factors and Levels:**
   - Identify the factors or variables that may influence the objective. These factors could include design parameters, materials, manufacturing processes, or environmental conditions. Determine the ranges or levels at which these factors will be studied.
3. **Design the Experiment:**
   - Choose an appropriate experimental design methodology, such as RSM, Taguchi methods, or other sequential design techniques. Develop a plan for conducting experiments systematically.
4. **Generate Experimental Runs:**
   - Use statistical software or tools to generate a sequence of experimental runs based on the chosen design methodology. Ensure that the design allows for efficient exploration of the factor space.
5. **Conduct Experiments:**
   - Perform the experiments as per the designed sequence. Record data accurately, ensuring that the experimental conditions are consistent and controlled.
6. **Analyze Data:**
   - Analyze the collected data using statistical methods and software. Evaluate the relationships between factors and the objective, and identify trends, patterns, or interactions that impact the outcome.
7. **Optimization:**
   - Use the analysis results to guide the optimization process. Adjust the levels of factors or modify other parameters to improve the objective. Optimization techniques like gradient-based methods or numerical optimization can be employed.

8. **Iterate and Refine:**
   - If necessary, repeat the experimentation and optimization process iteratively to further refine the aerospace system or process. Continue until the desired performance or objective is achieved.
9. **Verification and Validation:**
   - Verify and validate the optimized solution through additional testing or analysis to ensure that it meets safety and reliability standards.
10. **Implementation:**
    - Implement the optimized design or process in the aerospace system or manufacturing process. This may involve changes to components, materials, or procedures based on the findings of the sequential experimentation.
11. **Documentation:**
    - Thoroughly document the entire experimental process, including objectives, factors, designs, data, analysis, and outcomes. This documentation is essential for traceability and future reference.
12. **Monitoring and Maintenance:**
    - Continuously monitor the aerospace system or process to ensure that the improvements and optimizations are sustained over time. Address any issues that may arise during operation.
13. **Reporting and Communication:**
    - Prepare a comprehensive report summarizing the sequential experimentation process, findings, optimizations, and implementation. Communicate the results and recommendations to stakeholders, including engineering teams and management.
14. **Regulatory Compliance:**
    - Ensure that the optimized aerospace system or process complies with relevant industry standards, regulations, and safety requirements.
15. **Feedback and Continuous Improvement:**
    - Encourage feedback from users and operators to identify areas for further improvement. Use this feedback for ongoing refinements and enhancements.

Developing sequential design experiments in aerospace engineering requires a multidisciplinary team, including engineers, statisticians, and domain experts. Collaboration among these experts is essential to design, execute, and interpret experiments effectively and achieve the desired performance enhancements and optimizations.

PITFALLS AND REMEDIES

Sequential design experimentation in aerospace engineering, like any other field, comes with its own set of challenges and potential pitfalls. Here are some common pitfalls and their corresponding remedies when conducting sequential design experiments in aerospace engineering:

Inadequate Planning
Remedy: Thoroughly plan the entire experimentation process before starting. Define clear objectives, select appropriate factors, and design the experiment carefully. Engage experts in aerospace engineering and statistics to assist with planning.

Poor Factor Selection
Remedy: Ensure that the factors selected for experimentation are relevant to the aerospace system or process being studied. Conduct a thorough literature review and consult with subject matter experts to identify critical factors.

Limited Factor Range
Remedy: Expand the range of factor levels to cover a broad spectrum of operating conditions. A limited range can lead to missed opportunities for optimization.

Inefficient Experimental Design

> **Remedy:** Choose an appropriate experimental design methodology, such as RSM or Taguchi methods, that maximizes the information gained from each experiment while minimizing the number of runs.

Data Collection Errors

> **Remedy:** Implement rigorous data collection procedures to minimize errors. Use appropriate instrumentation and measurement techniques. Ensure consistency in data recording.

Lack of Statistical Expertise

> **Remedy:** Involve statisticians or data analysts with expertise in experimental design and analysis. Statistical expertise is crucial for proper data analysis and interpretation.

Ignoring Random Variation

> **Remedy:** Recognize that there will be random variability in experimental results. Use statistical methods to distinguish between random variation and significant effects caused by factors.

Overlooking Interactions

> **Remedy:** Consider interactions between factors. Two or more factors may interact in unexpected ways. Use factorial experiments to assess interaction effects.

Premature Optimization

> **Remedy:** Avoid prematurely optimizing the aerospace system based on early experimental results. Conduct sufficient experiments to ensure robustness in optimization.

Inadequate Validation

> **Remedy:** Thoroughly validate the optimized design or process through additional testing and analysis. Ensure that the improvements meet safety, reliability, and performance requirements.

Poor Documentation

> **Remedy:** Maintain detailed records of the entire experimentation process, including objectives, designs, data, and analysis. Proper documentation is essential for transparency and traceability.

Lack of Collaboration

> **Remedy:** Foster collaboration between aerospace engineers, statisticians, and domain experts. Effective communication among team members is essential for successful sequential design experiments.

Neglecting Regulatory Compliance

> **Remedy:** Ensure that the optimized aerospace system or process complies with industry standards, regulations, and safety requirements. Involve regulatory experts if necessary.

Resistance to Change

> **Remedy:** Address resistance to implementing changes resulting from sequential experimentation by involving key stakeholders early in the process. Communicate the benefits of the optimized solutions.

Not Seeking Feedback

> **Remedy:** Encourage feedback from end-users and operators of the aerospace system. Their insights can identify areas for further improvement and refinement.

Successfully navigating these pitfalls requires a multidisciplinary team with expertise in aerospace engineering, statistics, and data analysis. Collaboration, attention to detail, and a commitment to the scientific method are essential for achieving meaningful results through sequential design experimentation in aerospace engineering.

EXAMPLE

Sequential experimentation in aerospace engineering involves a series of experiments where the results of each experiment inform the design of subsequent experiments, allowing for efficient exploration of design space and optimization of complex systems. Here's a simplified example of a matrix for a sequential experimentation designed experiment in aerospace engineering:

Objective: To optimize the fuel efficiency of a jet engine by studying the effects of various design parameters.

Design Parameters:

1. **Turbine Blade Design (Parameter A):**
   - Initial Design: Blade Profile 1
2. **Fuel Injector Type (Parameter B):**
   - Initial Design: Injector Type X
3. **Compressor Blade Shape (Parameter C):**
   - Initial Design: Blade Shape A

Response Variable: Fuel Efficiency (measured in miles per gallon, MPG)

Sequential Experimentation Matrix:
In this sequentially designed experiment, the matrix is designed to adaptively explore the design space iteratively by changing the design parameters based on the results of previous experiments. The matrix is shown in Table 3.7.

In this matrix, each row represents a specific experimental run with a unique combination of design parameters. After analyzing the results of each experiment, adjustments are made to the design parameters for subsequent experiments. For example, if an experiment indicates that changing the injector type (Parameter B) has a significant effect on fuel efficiency, further experiments might focus on exploring different injector types while keeping other parameters constant.

The key to sequential experimentation is adaptability. By iteratively adjusting the design parameters based on the observed results, aerospace engineers can efficiently converge on an optimal design that maximizes fuel efficiency for the jet engine in aerospace engineering applications.

Reporting

Sequential design experimentation in aerospace engineering involves a variety of statistical tools and methods to plan, conduct, and analyze experiments iteratively. These tools help researchers and engineers optimize aerospace systems or processes efficiently. Here are some key statistical tools commonly used in sequential design experimentation for aerospace engineering:

TABLE 3.7

Sequential Designed Experiments for Aerospace Engineering

Run	Turbine Blade Design (A)	Fuel Injector Type (B)	Compressor Blade Shape (C)
1	Blade Profile 1	Injector Type X	Blade Shape A
2	Blade Profile 1	Injector Type Y	Blade Shape A
3	Blade Profile 1	Injector Type X	Blade Shape B
4	Blade Profile 2	Injector Type Y	Blade Shape A
5	Blade Profile 2	Injector Type X	Blade Shape B
6	Blade Profile 2	Injector Type Y	Blade Shape B
7	Blade Profile 3	Injector Type X	Blade Shape A
8	Blade Profile 3	Injector Type Y	Blade Shape B
9	Blade Profile 3	Injector Type X	Blade Shape B

Factorial Experiments: Factorial experiments help assess the main effects of multiple factors and their interactions. Aerospace engineers use them to systematically explore the influence of various factors on system performance or outcomes.

Response Surface Methodology (RSM): RSM is a powerful tool for optimizing responses by fitting mathematical models to experimental data. It helps identify optimal settings for factors while considering response variables.

Taguchi Methods: Taguchi methods, including the Taguchi Loss Function, are used for RPD. They aim to minimize the variation in aerospace processes or products, making them less sensitive to variations in factors and environmental conditions.

Design of Experiments (DOE): DOE encompasses various techniques for planning experiments, such as full-factorial designs, fractional factorial designs, and response surface designs. These designs help efficiently allocate experimental runs to explore factor effects.

Sequential Experimentation: Sequential experimentation methods involve adaptive designs where data from earlier experiments are used to guide subsequent experiments. Bayesian optimization and sequential sampling methods like Bayesian design of experiments (BDOE) fall into this category.

Statistical Process Control (SPC): SPC techniques are used to monitor and control aerospace processes in real-time. They help ensure that processes remain within acceptable quality limits.

Monte Carlo Simulation: Monte Carlo simulations are employed to model uncertainty and variability in aerospace systems. They allow engineers to assess the impact of various sources of uncertainty on system performance.

Reliability Analysis: Reliability analysis methods, such as Weibull analysis and reliability block diagrams, are used to assess the reliability and availability of aerospace systems.

Multivariate Analysis: Multivariate statistical techniques like principal component analysis (PCA) and partial least squares (PLS) regression are used to analyze data with multiple correlated variables.

Statistical Software: Specialized statistical software packages like JMP, Minitab, and MATLAB are commonly used for experimental design, data analysis, and visualization.

Optimization Algorithms: Numerical optimization algorithms, such as genetic algorithms, particle swarm optimization, and simulated annealing, are used to find optimal solutions in complex aerospace systems.

Statistical Hypothesis Testing: Statistical tests, such as t-tests, ANOVA, and chi-squared tests, are used to assess the significance of differences and relationships between variables.

Bayesian Statistics: Bayesian methods, including Bayesian inference and Bayesian networks, can be applied to incorporate prior knowledge and update beliefs in aerospace engineering problems.

Response Optimization: Techniques like desirability functions and goal programming help engineers find a balance among conflicting objectives when optimizing aerospace systems.

The choice of statistical tools depends on the specific objectives of the aerospace experimentation, the complexity of the system, the available resources, and the nature of the data. In sequential design experimentation, these tools are used iteratively to refine and optimize aerospace systems, ensuring they meet performance, safety, and reliability requirements.

A report for sequential design experimentation in aerospace engineering should provide a comprehensive and organized summary of the entire experimental process, including planning, execution, analysis, and results. Below are the key elements that should be included in such a report:

Title Page:
- Title of the Report
- Name of the Aerospace Project or Experiment

- Date of Report
- Names of Authors and Affiliations
- Contact Information

Abstract: A concise summary of the experiment's objectives, methods, major findings, and conclusions.

Table of Contents: A list of sections, subsections, and their page numbers for easy navigation.

List of Figures and Tables: A list of all figures and tables used in the report, along with their respective page numbers.

List of Abbreviations and Symbols: Definitions and explanations of any acronyms, abbreviations, or symbols used in the report.

Introduction:

- Background and context of the aerospace experiment.
- Statement of the problem or objectives.
- Overview of the experimental approach (sequential design).
- Importance and significance of the experiment.

Literature Review: A review of relevant literature and previous research related to the experiment's objectives and methodologies.

Methodology:

- Detailed explanation of the experimental design and methodology.
- Description of the sequential approach and how it was implemented.
- Discussion of the factors or variables considered in the experiment.

Data Collection:

- Explanation of data collection procedures.
- Information on data sources, instruments, and measurement techniques.
- Any challenges or issues encountered during data collection.

Data Analysis:

- Presentation of the data analysis methods used.
- Statistical techniques employed for data analysis.
- Graphs, charts, and tables to visualize the data.
- Discussion of interim findings and trends observed during the sequential experimentation.

Results: Presentation of the main results, including numerical values, statistical summaries, and graphical representations. Discussion of how the results evolved as the experiment progressed.

Discussion:

- Interpretation of the results in the context of the experiment's objectives.
- Comparison of the findings with initial expectations and hypotheses.
- Discussion of any unexpected or significant observations.
- Implications of the results for aerospace engineering or related fields.

Conclusion:

- Summary of the experiment's outcomes and key findings.
- Restatement of the experiment's objectives.
- Practical applications and implications of the findings.
- Suggestions for future research or experiments.

Recommendations: Any recommendations for changes or improvements based on the findings. Suggestions for modifications to the experimental design or methodology.

Acknowledgments: Recognition of individuals, organizations, or funding sources that contributed to the experiment.

References: Properly formatted citations and references for all sources consulted in the report.

Appendices: Any supplementary information, raw data, detailed calculations, or additional figures and tables that support the report.

Executive Summary (Optional): A condensed version of the report, typically one to two pages, summarizing the main objectives, methods, and major findings.

Graphs, Figures, and Tables: Clear and well-labeled visual representations of data and results, including captions and legends for easy understanding.

It's essential to follow a structured format and provide clear, concise, and accurate information in the report. Including references and citing sources appropriately is crucial to maintain the integrity of the research. Additionally, the report should be well-organized, free of grammatical errors, and adhere to any specific formatting guidelines or standards required by the aerospace organization or institution.

3.1.8 ROBUST PARAMETER DESIGN (RPD)

RPD experiments are commonly used in aerospace engineering to optimize and improve the performance, reliability, and robustness of aerospace systems, components, and processes. These experiments are typically applied in various areas within aerospace engineering:

Aircraft Design and Performance Optimization: RPD techniques are used to design and optimize various aspects of aircraft, such as wing shapes, control systems, and propulsion systems. The goal is to achieve better fuel efficiency, stability, and overall performance while ensuring robustness to environmental conditions.

Structural Analysis and Material Selection: RPD experiments help in selecting materials and designs that can withstand extreme conditions, including high temperatures, pressure, and mechanical stress. This is crucial for ensuring the structural integrity and safety of aircraft and spacecraft.

Manufacturing and Quality Control: RPD is applied to manufacturing processes to improve product quality, reduce defects, and enhance the reliability of aerospace components. It helps in identifying and controlling factors that can affect the production process and product performance.

Aeroacoustics and Noise Reduction: Aerospace engineers use RPD to design quieter aircraft and reduce noise pollution. This involves optimizing the shape of aircraft components, such as engines and wings, to minimize noise generation during flight.

Spacecraft Systems Design: RPD is used in the design of spacecraft systems, including propulsion, navigation, and communication systems. The goal is to ensure the robustness and reliability of these systems during space missions.

Avionics and Control Systems: RPD techniques are applied to avionics and control systems to improve their performance and reliability. This is essential for ensuring safe and accurate flight operations.

Environmental Impact Reduction: Aerospace engineers use RPD to develop environmentally friendly technologies, such as more fuel-efficient engines and cleaner propulsion systems, to reduce the environmental impact of aerospace operations.

Safety and Risk Mitigation: RPD experiments help in identifying and mitigating risks associated with aerospace systems and operations. Engineers aim to design systems that are robust and resilient to unforeseen events or failures.

Testing and Validation: RPD techniques are used to design and conduct experiments and tests to validate the performance and safety of aerospace systems and components under various conditions.

In aerospace engineering, robustness and reliability are of paramount importance due to the critical nature of aerospace applications. RPD experiments help engineers optimize designs and processes to ensure that aerospace systems perform as intended under a wide range of conditions, including those encountered in extreme environments such as space, high altitudes, and supersonic flight.

STEPS

Developing an RPD experiment in aerospace engineering involves a systematic approach to optimize and improve the performance and reliability of aerospace systems and components. Here are the steps to develop an RPD experiment for aerospace engineering:

1. **Define the Problem and Objectives:**
   - Clearly define the problem or challenge you want to address in aerospace engineering. Identify specific objectives and performance metrics that need improvement.
2. **Identify Critical Factors:**
   - Identify the key factors that influence the performance and reliability of the aerospace system or component. These factors can include design parameters, material properties, manufacturing processes, and environmental conditions.
3. **Select Response Variables:**
   - Choose response variables or performance metrics that reflect the desired improvements. These could include measures of efficiency, reliability, safety, noise levels, or other relevant parameters.
4. **Determine Factor Levels and Ranges:**
   - Define the ranges or values for each critical factor that will be considered in the experiment. These levels should cover the expected operating conditions and any potential variations.
5. **Design the Experimental Layout:**
   - Use statistical tools, such as DOE software, or techniques like factorial design, to create a structured experimental layout. This layout specifies how different factor combinations will be tested.
6. **Conduct the Experiments:**
   - Perform the experiments according to the designed layout. Collect data on the selected response variables for each experimental run. Ensure that the experiments are conducted under controlled conditions.
7. **Analyze the Data:**
   - Analyze the data collected from the experiments using statistical methods. This analysis helps identify the relationships between the critical factors and the response variables. Techniques like regression analysis and ANOVA are commonly used.
8. **Optimization and Robustness Analysis:**
   - Use the data analysis results to optimize the aerospace system or component's performance. Determine the optimal factor settings that lead to improved performance. Additionally, assess the robustness of the design by considering variations and uncertainties in the factors.
9. **Confirmatory Testing:**
   - Conduct confirmatory tests or validation experiments to verify the effectiveness of the optimized design. This step ensures that the improvements observed during the RPD process are reproducible and reliable.
10. **Implement Changes:**
    - Implement the optimized design changes or process improvements in the aerospace system or component. Ensure that these changes meet safety and regulatory requirements.
11. **Documentation and Reporting:**
    - Document all aspects of the RPD experiment, including the problem statement, experimental design, data analysis, optimization results, and implementation details. Prepare a comprehensive report for reference and future use.
12. **Continuous Monitoring and Improvement:**
    - Continuously monitor the performance of the aerospace system or component in real-world conditions. Make adjustments and refinements as needed to maintain or further improve performance.

13. **Regulatory Compliance:**
    - Ensure that the optimized design and any changes made comply with relevant aerospace industry standards, regulations, and safety requirements.
14. **Risk Assessment:**
    - Conduct a risk assessment to identify potential failure modes and develop mitigation strategies to address them.

Throughout the RPD process, collaboration among aerospace engineers, statisticians, and domain experts is essential to ensure that the experiment is designed and executed effectively. RPD helps aerospace engineers achieve robust and reliable designs while considering factors that may affect performance under varying conditions and uncertainties.

PITFALLS AND REMEDIES

RPD experiments in aerospace engineering, like any other field, can encounter various pitfalls and challenges. Here are some common pitfalls and possible remedies for RPD experimentation in aerospace engineering:

Inadequate Problem Definition
  Pitfall: Unclear problem definition or poorly defined objectives can lead to ineffective RPD experiments.
  Remedy: Start by clearly defining the problem, specifying objectives, and identifying critical performance metrics. Engage domain experts and stakeholders to ensure a thorough understanding of the issue.

Insufficient Data Quality
  Pitfall: Low-quality or incomplete data can compromise the validity of the RPD analysis.
  Remedy: Ensure data collection processes are robust and well-controlled. Use appropriate measurement instruments, and validate the accuracy and precision of data collection. Implement data cleaning and validation procedures.

Inadequate Factor Selection
  Pitfall: Failing to identify and include all relevant factors in the RPD experiment can result in suboptimal solutions.
  Remedy: Conduct a thorough factor identification process with input from subject matter experts. Consider not only design factors but also external factors that may affect the outcome.

Improper Factor Range Selection
  Pitfall: Narrow or inappropriate ranges for factors can limit the ability to find robust solutions.
  Remedy: Choose factor ranges that encompass expected operating conditions and potential variations. Conduct sensitivity analyses to assess the impact of varying factor levels.

Overfitting Models
  Pitfall: Overly complex models may overfit the data and provide unreliable predictions.
  Remedy: Use statistical techniques and model selection criteria to choose models that balance complexity and predictive accuracy. Cross-validation can help assess model performance.

Neglecting Model Validation
  Pitfall: Failure to validate models with new data can lead to inaccurate predictions.
  Remedy: Set aside a portion of the data for validation. Validate the models using this data to ensure their predictive capability for new scenarios.

Lack of Consideration for Nonlinear Effects
   Pitfall: Ignoring nonlinear relationships between factors and responses can lead to missed opportunities for improvement.
   Remedy: Consider the possibility of nonlinear effects and use techniques such as quadratic models or RSMs to capture nonlinear relationships.
Neglecting External Factors and Noise
   Pitfall: Ignoring external factors and sources of noise can lead to solutions that are not robust in real-world conditions.
   Remedy: Include external factors as covariates in the analysis or conduct noise factor experiments to assess their impact on robustness.
Ineffective Communication
   Pitfall: Poor communication among multidisciplinary teams can hinder the success of RPD projects.
   Remedy: Promote effective communication and collaboration among aerospace engineers, statisticians, domain experts, and stakeholders. Clearly document and share findings, assumptions, and recommendations.
Inadequate Implementation Planning
   Pitfall: Failing to plan for the implementation of RPD recommendations can result in unrealized improvements.
   Remedy: Develop a detailed implementation plan that addresses regulatory compliance, risk assessment, and change management. Ensure that the optimized design changes are integrated into the aerospace system or component effectively.
Ignoring Regulatory Compliance
   Pitfall: Neglecting aerospace industry standards and regulations can lead to noncompliance and safety risks.
   Remedy: Ensure that the RPD process considers and complies with all relevant regulatory requirements. Involve regulatory experts as needed.

Addressing these pitfalls and following best practices in RPD experimentation can help aerospace engineers achieve robust and reliable designs while mitigating potential risks. Successful RPD projects require a combination of technical expertise, effective collaboration, and rigorous data analysis.

EXAMPLE

RPD experiments in aerospace engineering aim to optimize the performance of aerospace systems while minimizing sensitivity to variation or external factors. Here's a simplified example of a matrix for an RPD designed experiment in aerospace engineering:

Objective: To optimize the lift performance of an aircraft wing by studying the effects of two factors (airfoil shape and angle of attack) while considering variations in manufacturing processes (material variability and surface roughness).

Factors:

Airfoil Shape (Factor A):
   - NACA 2412 Airfoil
   - NACA 632-415 Airfoil
Angle of Attack (Factor B):
   - Low Angle (0 degrees)

- Medium Angle (5 degrees)
- High Angle (10 degrees)

Factors to Study Variations:

Material Variability (Factor C):
- Low Variability (Controlled material properties)
- High Variability (Material properties subject to variations)

Surface Roughness (Factor D):
- Low Roughness (Smooth surface finish)
- High Roughness (Rough surface finish)

Response Variable: Lift Coefficient (dimensionless)

Robust Parameter Design Matrix:

In this RPD designed experiment, the matrix is designed to assess the robustness of the aircraft wing's lift performance to variations in airfoil shape, angle of attack, material variability, and surface roughness. The matrix is shown in Table 3.8.

In this matrix, each row represents a specific experimental run with a unique combination of airfoil shape, angle of attack, material variability, and surface roughness levels. The analysis in RPD focuses on how robust the aircraft wing's lift coefficient is to variations in these factors and variations in manufacturing processes.

By conducting experiments with different factor combinations and levels of variation, aerospace engineers can determine the optimal design that maintains consistent and robust lift performance despite variations in factors and manufacturing processes in aerospace engineering applications.

REPORTING

RPD experiments in aerospace engineering utilize various statistical tools and techniques to optimize and improve the robustness of designs. Here are some of the statistical tools commonly employed in RPD experimentation for aerospace engineering:

Design of Experiments (DOE): DOE is a fundamental tool used in RPD to systematically plan and conduct experiments. It helps in selecting the appropriate experimental design, such as full-factorial, fractional factorial, or RSM, and in determining the factors to be investigated.

TABLE 3.8

Robust Parameter Design (RPD) Matrix Example for Aerospace Engineering

Run	Airfoil Shape (A)	Angle of Attack (B)	Material Variability (C)	Surface Roughness (D)
1	NACA 2412	Low Angle	Low Variability	Low Roughness
2	NACA 632-415	Medium Angle	High Variability	High Roughness
3	NACA 2412	High Angle	Low Variability	High Roughness
4	NACA 632-415	Low Angle	High Variability	Low Roughness
5	NACA 2412	Medium Angle	High Variability	Low Roughness
6	NACA 632-415	High Angle	Low Variability	High Roughness
7	NACA 2412	Low Angle	High Variability	High Roughness
8	NACA 632-415	Medium Angle	Low Variability	Low Roughness

Response Surface Methodology (RSM): RSM is often used to build predictive models for response variables. It helps in understanding the relationships between input factors and responses. RSM designs experiments to explore factor interactions and identify optimal settings for robustness.

Statistical Modeling: Statistical models, including linear regression models, quadratic models, and higher-order models, are used to represent the relationships between input factors and performance measures. These models aid in optimizing designs and predicting outcomes.

Analysis of Variance (ANOVA): ANOVA is employed to analyze the variability in experimental data and identify significant factors and interactions. It helps in understanding which factors have the most significant impact on performance.

Signal-to-Noise (S/N) Ratios: S/N ratios, such as the signal-to-noise ratio for smaller-the-better (Smaller-is-better S/N), larger-the-better (Larger-is-better S/N), and nominal-the-best (Nominal-the-best S/N), are used to quantify the quality of a response variable. RPD aims to optimize these ratios to make the design more robust to variations.

Optimization Algorithms: Optimization algorithms, such as gradient-based methods or genetic algorithms, are used to find optimal factor settings that maximize or minimize the desired responses while considering robustness.

Robust Optimization Techniques: These techniques consider variability and uncertainty in factors to find factor settings that minimize the impact of variations. Taguchi methods are an example of robust optimization techniques.

Simulation and Monte Carlo Analysis: Simulations are employed to model and analyze the performance of designs under various operating conditions and environmental factors. Monte Carlo analysis helps in assessing the sensitivity of designs to uncertainties.

Reliability Analysis: Reliability tools, such as Weibull analysis or reliability block diagrams, are used to assess the reliability and failure modes of aerospace components or systems.

Statistical Process Control (SPC): SPC techniques are applied to monitor and control manufacturing and assembly processes to ensure that the designed robustness is maintained during production.

Risk Assessment and FMEA: Failure modes and effects analysis (FMEA) is used to identify potential failure modes, their effects, and criticality. It helps in designing for robustness against failure modes.

Design for Six Sigma (DFSS): DFSS methodologies, including define, measure, analyze, design, verify (DMADV), or define, measure, analyze, design, optimize (DMADOV), incorporate RPD principles to develop high-quality products or systems.

These statistical tools and techniques are applied collectively to achieve the objectives of RPD in aerospace engineering, which include improving performance, reliability, and robustness while considering variations and uncertainties in operational conditions and environmental factors. The specific choice of tools depends on the nature of the problem, available resources, and project goals.

A report for RPD experimentation in aerospace engineering typically includes several elements to communicate the methodology, results, and findings effectively. Below are the key elements typically included in such a report:

Title Page:
- Title of the Report: Descriptive and concise.
- Author(s) and Affiliation(s): Names and affiliations of those involved in the study.
- Date: Date of report preparation.

Abstract: A brief summary of the report's purpose, methodology, key findings, and conclusions. It should provide an overview of the study in a concise manner.

Table of Contents: An organized list of the sections and subsections of the report, along with page numbers.

List of Figures and Tables: A list of all figures and tables used in the report, along with their respective page numbers.

List of Abbreviations and Symbols: A list of abbreviations, acronyms, and symbols used in the report, along with their explanations.

Introduction:

- Background and Context: A brief introduction to the problem or project, including its importance and relevance to aerospace engineering.
- Objectives: Clear and specific objectives or goals of the RPD study.
- Scope: The scope of the study, including the factors or parameters being investigated.
- Structure of the Report: An outline of how the report is organized.

Literature Review: A review of relevant literature and prior research related to the aerospace component or system being studied. This provides context and highlights existing knowledge and gaps.

Methodology:

- Experimental Design: Explanation of the experimental design approach, including the type of design (e.g., factorial, Taguchi, response surface), factors considered, levels of factors, and how factors were selected.
- Data Collection: Details about data collection methods, instruments used, and data acquisition processes.
- Statistical Tools: Description of the statistical tools and techniques used, such as DOE, RSM, and robust optimization methods.
- Analysis Approach: Explanation of the analysis methods, including statistical tests, regression models, or optimization algorithms employed.
- Robustness Metrics: Specification of the robustness metrics or criteria used to evaluate and optimize the design.
- Simulation and Modeling: If applicable, details about any simulations or mathematical models used in the analysis.

Results:

- Presentation of the experimental results, including tables, figures, and charts.
- Summary of Findings: A clear and concise summary of the key findings, trends, and relationships observed during the study.

Discussion:

- Interpretation of Results: Discussion and interpretation of the results, including any significant findings, factor interactions, and the impact on robustness.
- Comparison with Literature: Discussion of how the study's findings compare with existing literature and prior research.

Conclusion: Summary of the study's main conclusions and their implications for aerospace engineering. Conformance to Objectives: Confirmation of whether the study's objectives were achieved.

Recommendations: Recommendations for further research or actions based on the study's findings.

Acknowledgments: Acknowledgment of individuals, organizations, or funding sources that contributed to the research.

References: A list of all cited references, following a specific citation style (e.g., APA, IEEE).

Appendices: Supplementary information that provides additional details, such as raw data, detailed calculations, experimental setups, or additional graphs and charts.

Glossary: Definitions of technical terms and concepts used in the report, especially if the audience includes individuals who may not be familiar with aerospace engineering terminology.

Index: An index, if applicable, to help readers locate specific topics or terms within the report.

It's important to format the report professionally, use clear and concise language, and provide suffi-cient details to allow others to understand and potentially replicate the study. Additionally, the report should be well-organized and logically structured to facilitate easy navigation and comprehension.

3.1.9 RANDOMIZED EXPERIMENTS

Randomized Design Experiments (RDP) experiments, also known as randomized experiments (REs) or randomized controlled trials (RCTs) are commonly used in aerospace engineering for various purposes. These experiments involve randomly assigning subjects or units to different treat-ments or conditions to evaluate the effects of specific factors or interventions. Here are some areas within aerospace engineering where RPD experiments are frequently employed:

Materials Testing: RPD experiments can be used to assess the impact of various materi-als and manufacturing processes on the structural integrity, durability, and performance of aerospace components, such as aircraft wings, engine components, and spacecraft materials.

Structural Testing: In aerospace, RPD experiments are used to conduct structural testing on various components, including wings, fuselages, and landing gear. These experiments can evaluate the effects of different design parameters and loads on structural integrity and safety.

Aerodynamics and Fluid Dynamics: RPD experiments can be applied to study the aerody-namic characteristics of aircraft, including lift and drag coefficients, airfoil design, and turbulence effects. They help optimize wing and aircraft design for efficiency and stability.

Propulsion Systems: Aerospace engineers use RPD experiments to assess the performance of propulsion systems, such as jet engines and rockets. These experiments can optimize combustion processes, fuel efficiency, and thrust generation.

Avionics and Control Systems: RPD experiments are employed to evaluate the effectiveness of avionics systems, control algorithms, and software in ensuring safe and precise flight control.

Environmental Testing: Aerospace components and systems often need to operate in extreme environmental conditions, including temperature, pressure, and radiation. RPD experiments can assess how these conditions affect performance and reliability.

Flight Testing: In-flight testing of aircraft and spacecraft involves RPD experiments to eval-uate various parameters, including navigation, communication, and safety systems. These experiments ensure that vehicles perform as expected in real-world conditions.

Space Exploration: RPD experiments are crucial for space missions to assess the effects of microgravity, radiation, and extreme temperatures on equipment and materials. They help design robust spacecraft and instruments.

Human Factors: In the case of manned space missions, RPD experiments can assess human factors, such as astronaut comfort, health, and performance under microgravity conditions.

Safety and Reliability: RPD experiments are used to conduct reliability and safety assess-ments, including FMEA) and fault tolerance testing, to ensure aerospace systems meet safety standards.

Noise and Vibration Control: RPD experiments help aerospace engineers study and miti-gate noise and vibration issues in aircraft and spacecraft, improving passenger comfort and equipment reliability.

In aerospace engineering, RPD experiments are valuable for optimizing designs, ensuring safety, and meeting performance specifications. They provide a systematic and controlled approach to testing and evaluating various factors that can impact the performance, reliability, and safety of aerospace systems and components.

STEPS

Developing a randomized design (RPD) experiment in aerospace engineering involves several steps to ensure that the experiment is well-planned, conducted, and analyzed. Here are the general steps to develop an RPD experiment in aerospace engineering:

1. **Define the Objective:** Clearly articulate the goal and purpose of the experiment. Determine what specific questions or hypotheses you want to address. For example, you might want to assess the effect of a new material on the structural integrity of an aircraft component.
2. **Identify Factors and Levels:** Identify the independent variables (factors) that you want to study and their respective levels or settings. These factors can be design parameters, material properties, environmental conditions, or any other variables relevant to your aerospace application.
3. **Select Response Variables:** Determine the dependent variables (response variables) that you will measure to assess the impact of the independent variables. Response variables could include structural stress, temperature, pressure, fuel efficiency, or any relevant performance metric.
4. **Experimental Design:** Choose an appropriate experimental design that involves randomization. Common designs include completely randomized designs, randomized block designs, and Latin square designs. The choice depends on the complexity of the experiment and the need to control potential sources of variability.
5. **Sample Size Determination:** Calculate the required sample size to achieve the desired level of statistical power and significance. The sample size depends on factors like the effect size, variability, and desired confidence level.
6. **Random Assignment:** Randomly assign the experimental units or subjects to different treatment groups or conditions. Randomization helps ensure that any observed effects are not due to bias or confounding factors.
7. **Conduct the Experiment:** Carry out the experiment according to the predefined plan, ensuring that all treatments are applied or conditions are met as intended. Record data systematically and accurately.
8. **Data Analysis:** Analyze the collected data using appropriate statistical methods. This may involve techniques such as ANOVA, regression analysis, or other statistical tests to assess the significance of factors and their interactions.
9. **Interpret Results:** Interpret the results in the context of your research objectives. Identify significant effects and draw conclusions about how the independent variables influence the dependent variables.
10. **Draw Inferences:** Based on the analysis, draw inferences and make recommendations for future aerospace design, engineering, or operational decisions. Determine the practical implications of your findings.
11. **Report and Documentation:** Prepare a comprehensive report documenting the entire experiment. Include details about the experimental setup, methodology, data analysis, results, and conclusions. Provide any necessary visuals, tables, and graphs to support your findings.
12. **Peer Review:** If applicable, submit your experiment and findings for peer review by experts in the field. Peer review helps ensure the validity and reliability of your results.
13. **Implement Findings:** If the experiment leads to design improvements or operational changes, implement the recommendations in your aerospace engineering project.
14. **Iterate and Refine:** In some cases, you may need to iterate and refine the experiment or conduct follow-up experiments to further investigate specific aspects.
15. **Maintain Records:** Keep detailed records of the experiment, including raw data, analysis files, and documentation. This helps with transparency and future reference.
16. **Continuous Learning:** Use the results and lessons learned from the RPD experiment to inform future aerospace engineering projects and research.

Remember that the specific steps and considerations may vary depending on the nature of the aerospace engineering project and the objectives of the experiment. It's essential to adhere to scientific and ethical principles throughout the process.

PITFALLS AND REMEDIES

Randomized design (RPD) experiments in aerospace engineering, like any scientific experiment, can encounter various pitfalls and challenges. Here are some common pitfalls and potential remedies to consider:

Inadequate Sample Size:
> **Pitfall:** Using a sample size that is too small may lead to underpowered experiments, making it difficult to detect significant effects.
> **Remedy:** Conduct a power analysis to determine the required sample size based on effect size, variability, and desired significance level. Ensure that you have an adequate number of experimental units or subjects.

Incomplete Randomization:
> **Pitfall:** Incomplete randomization can introduce bias and confound results, potentially leading to incorrect conclusions.
> **Remedy:** Implement proper randomization techniques to assign experimental units to treatment groups or conditions. Ensure that the randomization process is well-documented and rigorously followed.

Confounding Variables:
> **Pitfall:** Failure to account for or control confounding variables can obscure the true relationship between independent and dependent variables.
> **Remedy:** Identify potential confounders and consider them during the experimental design phase. Use blocking or stratification techniques if necessary to control for confounding variables.

Measurement Errors:
> **Pitfall:** Inaccurate or imprecise measurements can lead to unreliable data and erroneous conclusions.
> **Remedy:** Calibrate measurement instruments regularly, use appropriate measurement techniques, and conduct pilot tests to identify and minimize measurement errors.

Nonrepresentative Samples:
> **Pitfall:** Using a nonrepresentative sample may limit the generalizability of your findings to the broader aerospace context.
> **Remedy:** Ensure that your sample is representative of the population or system you intend to study. Consider factors like diversity, variability, and relevant characteristics.

Selection Bias:
> **Pitfall:** Biased selection of experimental units or subjects can distort results and compromise external validity.
> **Remedy:** Use randomization and appropriate sampling techniques to minimize selection bias. Clearly define inclusion and exclusion criteria.

Insufficient Controls:
> **Pitfall:** Insufficient control over environmental conditions or external factors can introduce noise and confound results.
> **Remedy:** Implement appropriate control measures, such as environmental controls, blinding, and standard operating procedures, to minimize the influence of external factors.

Inadequate Statistical Analysis:
> **Pitfall:** Using inappropriate or overly simplistic statistical analyses may lead to incorrect interpretations.

Remedy: Consult with a statistician or use appropriate statistical methods, such as ANOVA, regression, or mixed-effects models, depending on the experimental design and data structure.

Failure to Account for Interactions:

Pitfall: Neglecting to examine interactions between factors may overlook important relationships in the data.

Remedy: Analyze the main effects and interactions between factors to fully understand the experimental outcomes. Visualize interactions using graphs or plots.

Lack of Replication:

Pitfall: Conducting experiments without replication can limit the reliability and robustness of results.

Remedy: Replicate experiments or treatments to assess the consistency of findings. Replication helps quantify variability and strengthens the validity of conclusions.

Overlooking Ethical Considerations:

Pitfall: Failing to address ethical considerations, such as informed consent for human subjects or environmental impact, can lead to ethical violations.

Remedy: Prioritize ethical considerations throughout the experiment. Obtain informed consent, adhere to ethical guidelines, and consider the potential consequences of your research.

Inadequate Documentation:

Pitfall: Poor record-keeping and documentation can hinder reproducibility and transparency.

Remedy: Maintain thorough records of the experimental process, including protocols, data collection sheets, and any deviations from the plan. Document all decisions and changes made during the experiment.

To mitigate these pitfalls, it's essential to plan your RPD experiment meticulously, involve experts as needed, and continuously monitor and validate the experimental process. Additionally, peer review and collaboration with colleagues can help identify and address potential issues before they impact the validity of your aerospace engineering research.

Example

REs in aerospace engineering aim to study the effects of various factors or treatments on a specific outcome variable while introducing randomization to minimize bias. Here's an example of a matrix for a randomized designed experiment in aerospace engineering:

Objective: To investigate the effects of different surface coatings on the heat resistance of a space-craft's thermal protection material.

Experimental Units: Specimens of the thermal protection material (multiple specimens for each treatment group).

Factors (Surface Coatings):
1. **Coating A**: A proprietary ceramic coating.
2. **Coating B**: A metallic coating.
3. **Coating C**: A polymer-based coating.

Response Variable: Heat Resistance (measured in degrees Celsius before material degradation).

Randomized Experiment Matrix:
In this randomized designed experiment, treatments (different surface coatings) are randomly assigned to specimens of the thermal protection material to investigate their effects on heat resistance. The matrix is shown in Table 3.9.

In this matrix, each row represents a specimen of thermal protection material, and the surface coating treatment is randomly assigned to each specimen. Multiple specimens are used for each treatment group to account for variability.

By conducting the experiment in this randomized manner, any potential bias or confounding factors that may affect the results are minimized. This allows aerospace engineers to draw meaningful conclusions about the effects of different surface coatings on the heat resistance of the spacecraft's thermal protection material in a controlled and unbiased manner.

REPORTING

Randomized design (RPD) experiments in aerospace engineering can involve various statistical tools and techniques to analyze data and draw meaningful conclusions. The specific tools used depend on the nature of the experiment, the design, and the research objectives. Here are some commonly used statistical tools and methods:

Descriptive Statistics: Descriptive statistics, such as mean, median, variance, and standard deviation, are used to summarize and present the basic characteristics of the data.

Hypothesis Testing: Hypothesis testing involves statistical tests to determine whether observed differences or effects are statistically significant. Common tests include:
- **Student's t-test:** Used to compare means of two groups.
- **Analysis of Variance (ANOVA):** Used to compare means of more than two groups.
- **Chi-squared test:** Used for categorical data analysis.
- **Wilcoxon-Mann-Whitney U-test:** A nonparametric test for comparing two independent groups.

Regression Analysis: Regression analysis, including linear regression, multiple regression, and logistic regression, is used to model relationships between independent and dependent variables, predict outcomes, and assess the impact of variables.

TABLE 3.9
Randomized Experiment Matrix
Example for Aerospace Engineering

Specimen	Surface Coating
1	Coating A
2	Coating B
3	Coating B
4	Coating A
5	Coating C
6	Coating A
7	Coating C
8	Coating B
9	Coating C
10	Coating A

Design of Experiments (DOE): DOE is a systematic approach to experimentation that involves planning, conducting, and analyzing experiments. It includes techniques like factorial designs, RSM, and Taguchi methods.

Analysis of Variance (ANOVA): ANOVA is used to analyze the variation in data and determine whether there are significant differences among groups or treatments.

Time-Series Analysis: Time-series analysis is employed when data is collected over time. It includes methods like autoregressive integrated moving average (ARIMA) modeling and seasonal decomposition.

Multivariate Analysis: Multivariate techniques, such as PCA and factor analysis, are used to explore relationships among multiple variables simultaneously.

Reliability Analysis: Reliability analysis assesses the reliability and failure characteristics of components or systems, often using methods like Weibull analysis.

Bayesian Analysis: Bayesian statistical methods are used for updating beliefs and making predictions based on prior information and observed data. Bayesian methods can be valuable for uncertainty quantification.

Nonparametric Tests: Nonparametric tests, like the Kruskal-Wallis test or the Mann-Whitney U-test, are used when data does not meet the assumptions of parametric tests (e.g., normal distribution).

Survival Analysis: Survival analysis, including Kaplan-Meier survival curves and Cox proportional hazards models, is used for studying time-to-event data, such as component failures or system lifetimes.

Monte Carlo Simulation: Monte Carlo simulations are used to model complex systems or processes, especially when analytical solutions are challenging.

Statistical Software: Statistical software packages like R, Python with libraries such as NumPy, SciPy, and pandas, SAS, MATLAB, and Minitab are commonly used for data analysis in aerospace engineering.

Data Visualization: Data visualization tools, including charts, graphs, and plots, are essential for visually representing and interpreting data patterns.

Statistical Quality Control (SQC): SQC techniques, such as control charts and process capability analysis, are employed in manufacturing and quality assurance processes.

The choice of statistical tools depends on the research question, experimental design, and the type of data collected in aerospace engineering experiments. Researchers should select the most appropriate methods and techniques to address their specific objectives and ensure the validity and reliability of their results. Additionally, consulting with a statistician or data analysis expert can be valuable in choosing and correctly implementing statistical tools for RPD experiments.

A well-structured and informative report for Randomized design (RPD) experiments in aerospace engineering should convey the research methodology, results, and conclusions effectively. While the specific format may vary depending on organizational or journal requirements, the following are common elements to include in such a report:

Title: A concise and descriptive title that reflects the experiment's objective and scope.

Abstract: A brief summary of the entire report, highlighting the key objectives, methods, results, and conclusions. It should provide enough information for readers to understand the study's essence quickly.

Introduction:
- Background and context: Describe the problem or research question addressed by the experiment.
- Objectives and hypotheses: Clearly state the research objectives and any hypotheses tested in the study.
- Significance: Explain why the experiment is relevant and how it contributes to the field of aerospace engineering.

Experimental Design:
- Experimental setup: Provide details about the experimental apparatus, materials, and instrumentation used.
- Randomization process: Describe how randomization was achieved to minimize bias and ensure the validity of the results.
- Treatment groups: If applicable, outline the treatment groups or factors being studied.

Methods:
- Data collection: Explain how data were collected, including sampling procedures, measurement techniques, and data recording.
- Statistical analysis: Describe the statistical methods, models, and tools used for data analysis. Be specific about the tests or procedures employed.
- Data preprocessing: If any data cleaning or transformation was performed, explain the rationale and process.

Results:
- Present the data: Display the results using tables, charts, graphs, or figures. Ensure clarity and appropriate labeling.
- Statistical findings: Report the outcomes of statistical tests, including p-values, confidence intervals, and effect sizes.
- Interpretation: Discuss the implications of the results in the context of the research objectives.

Discussion:
- Comparison with hypotheses: Evaluate whether the results support or reject the stated hypotheses.
- Interpretation of findings: Provide a comprehensive interpretation of the results, discussing their significance and practical implications.
- Limitations: Address any limitations of the experiment, such as sample size, assumptions, or potential sources of bias.
- Future research: Suggest avenues for further research or improvements in experimental design.

Conclusion: Summarize the key findings concisely. Restate the significance of the research and its contributions to aerospace engineering.

References: Cite all sources, references, and literature used in the report following a standard citation style (e.g., APA, IEEE, and Chicago).

Appendices: Include supplementary information that enhances the report's understanding but may be too detailed or lengthy for the main body. This could include additional tables, figures, calculations, or raw data.

Acknowledgments: Acknowledge any individuals, organizations, or funding sources that contributed to the experiment or report.

Author Information: Provide details about the authors, including names, affiliations, and contact information.

Keywords: Include a list of keywords that represent the main concepts and topics covered in the report.

Nomenclature (if applicable): For complex experiments with specific terminology or symbols, provide a nomenclature section that defines key terms and symbols used in the report.

Ethical Considerations (if applicable): Address any ethical concerns related to human subjects, animal testing, or other ethical considerations.

Ensure that the report is well-organized, logically structured, and written in clear and concise language. Charts, graphs, and visual aids should be of high quality and appropriately labeled. Additionally, adhere to any specific guidelines or templates required by your institution, journal, or organization when preparing the report.

3.1.10 SPLIT-PLOT AND BLOCKED EXPERIMENTS

Split-plot and blocked design experiments are used in aerospace engineering in various contexts to investigate, optimize, and improve aerospace systems, components, and processes. These designs help researchers and engineers efficiently manage experiments involving complex factors and constraints. Here are some common areas where split-plot and blocked design experiments are applied in aerospace engineering:

Manufacturing and Process Optimization: Split-plot and blocked designs are used to optimize manufacturing processes for aerospace components and materials. This includes improving the production of aircraft parts, composite materials, and coatings to enhance efficiency and reduce defects.

Materials Testing: Aerospace engineers often use split-plot and blocked designs to study the properties and behaviors of materials under different conditions. This can involve testing materials for strength, durability, and resistance to environmental factors.

Aircraft Design and Performance: Research related to aircraft design, such as aerodynamics, propulsion, and control systems, may require split-plot and blocked experiments. These designs help engineers understand how changes in design variables affect aircraft performance and safety.

Reliability and Maintenance: Aerospace systems must meet stringent reliability and safety standards. Split-plot and blocked designs can be employed to assess the reliability of components and systems, as well as to develop maintenance schedules and strategies to extend the lifespan of aircraft.

Space Exploration: Aerospace experiments related to space exploration, including the testing of spacecraft materials and components, often involve complex factors and constraints. Split-plot and blocked designs help researchers optimize mission parameters and equipment design.

Testing and Evaluation: Aerospace engineers use split-plot and blocked designs to conduct a wide range of tests and evaluations, including wind tunnel testing, fatigue testing, and thermal testing of spacecraft and aircraft components.

Propulsion Systems: Developing and optimizing propulsion systems for rockets, satellites, and aircraft is a critical aspect of aerospace engineering. Split-plot and blocked designs are used to investigate factors affecting propulsion efficiency and reliability.

Environmental and Safety Studies: Aerospace projects may involve studies related to environmental impact and safety. Split-plot and blocked designs help researchers assess the environmental effects of aerospace activities and develop safety protocols.

Flight Testing: Flight testing of aircraft and spacecraft involves collecting data under real-world conditions. Split-plot and blocked designs may be used to plan and analyze flight tests, considering factors like altitude, speed, and payload.

Avionics and Control Systems: Evaluating and optimizing avionics and control systems require experimentation with complex factors and variables. Split-plot and blocked designs help engineers assess system performance and reliability.

In these applications, split-plot and blocked design experiments enable aerospace engineers and researchers to efficiently explore multiple factors and their interactions while accounting for constraints and variability. These designs help make informed decisions, improve processes, and enhance the performance and safety of aerospace systems and equipment.

STEPS

Developing split-plot and blocked design experiments in aerospace engineering involves careful planning, execution, and analysis. These experimental designs are used to efficiently investigate complex factors and constraints. Here are the steps to develop split-plot and blocked design experiments for aerospace engineering:

1. **Define the Objectives and Research Questions:**
   - Clearly state the objectives of the experiment and the specific research questions you want to address.
   - Identify the aerospace system, component, or process you are studying.
2. **Identify Factors and Variables:**
   - Determine the factors (independent variables) that you want to study. These can include design parameters, operating conditions, or material properties.
   - Identify the response variables that measure the outcomes or performance of interest.
3. **Select the Experimental Design:**
   - Choose an appropriate split-plot or blocked design based on the complexity of the factors and constraints involved. Common designs include split-plot, split-split-plot, and various blocked designs.
   - Consider whether you need to account for randomization, hard-to-change (HTC) factors, and nested or hierarchical structures.
4. **Allocate Factors to Whole Plots and Subplots:**
   - Assign factors to whole plots and subplots based on their significance and the experimental setup.
   - Whole plots typically represent the primary factors of interest, while subplots account for secondary or nuisance factors.
5. **Randomization and Blocking:**
   - Implement a randomization strategy to ensure that experimental runs are conducted in a random order to minimize bias and confounding effects.
   - If there are specific constraints or blocking factors (e.g., time, location, batch), incorporate them into the design.
6. **Experimental Runs and Data Collection:**
   - Conduct the experimental runs according to the design plan. Ensure that the experimental conditions are consistent and accurately controlled.
   - Record data for both the whole plots and subplots, including response variables and any relevant covariates.
7. **Data Analysis:**
   - Analyze the collected data using statistical methods appropriate for split-plot or blocked designs. This may involve ANOVA, regression, or other multivariate techniques.
   - Assess the main effects of whole-plot and subplot factors, as well as their interactions.
8. **Optimization and Inference:**
   - Based on the analysis results, optimize the aerospace system, component, or process to achieve the desired objectives.
   - Make inferences and draw conclusions regarding the significance of factors and their impact on the response variables.
9. **Documentation and Reporting:**
   - Prepare a comprehensive report that documents the experimental design, methods, results, and conclusions.
   - Include graphical representations of the data, statistical analyses, and any recommended changes or improvements.
10. **Validation and Further Studies:**
    - Validate the findings by conducting additional experiments or simulations if necessary.
    - Consider using the results to guide decision-making, process improvements, or future research in aerospace engineering.

11. **Iterative Process:** Split-plot and blocked design experiments often involve an iterative process where the initial findings may lead to adjustments in factors or further experimentation to refine the results.
12. **Compliance and Safety:** Ensure that the experiment complies with safety regulations and ethical considerations related to aerospace research.

Effective planning, execution, and analysis are essential for the success of split-plot and blocked design experiments in aerospace engineering. These steps help researchers make informed decisions, optimize aerospace systems, and improve overall performance and safety.

PITFALLS AND REMEDIES

Split-plot and blocked design experiments in aerospace engineering can be powerful tools for investigating complex systems, but they also come with their own set of pitfalls and challenges. Here are some common pitfalls and remedies associated with these experimental designs in aerospace engineering:

Inadequate Factor Identification: Failing to identify all relevant factors and interactions can lead to incomplete or biased results. Overlooking critical variables can limit the effectiveness of the experiment.

Improper Block or Whole-Plot Assignment: Incorrectly assigning factors to whole plots and subplots can lead to inefficient use of resources and inaccurate conclusions. Misidentification of primary and secondary factors can affect the experiment's validity.

Inadequate Randomization: Poor randomization of experimental runs can introduce bias and confounding effects, compromising the validity of the results. Nonrandom sequencing of runs can lead to incorrect conclusions.

Small Sample Sizes: Using too few experimental runs within each block or subplot can result in low statistical power and reduced ability to detect meaningful effects. This can lead to inconclusive results.

Ignoring Nested Structures: Failure to account for nested or hierarchical structures in the experiment can result in the loss of important information and lead to incorrect inferences.

Incomplete Data Collection: Missing data or incomplete data collection during the experiment can reduce the quality of the analysis and limit the ability to draw meaningful conclusions.

Thorough Factor Identification: Conduct a comprehensive analysis and brainstorming session to identify all potential factors and interactions that may affect the aerospace system or process. Engage subject matter experts to ensure no critical factors are overlooked.

Careful Assignment of Factors: Verify the correct assignment of factors to whole plots and subplots. Use engineering judgment and statistical tools to determine which factors are most appropriately treated as whole plots and which as subplots.

Robust Randomization: Implement a well-designed randomization scheme to ensure that experimental runs are conducted in a truly random order. This minimizes the risk of systematic bias and confounding effects.

Adequate Sample Sizes: Determine the appropriate sample sizes for each block or subplot based on power calculations and statistical considerations. Ensure that the experiment has sufficient statistical power to detect meaningful effects.

Account for Nested Structures: If your experiment has nested or hierarchical structures, use appropriate statistical models and techniques to account for these structures in the analysis.

Comprehensive Data Collection: Ensure that data collection is thorough and complete, with no missing or incomplete data. Implement data quality checks during the experiment to minimize errors.

Sensitivity Analysis: Perform sensitivity analyses to assess the impact of potential sources of bias or variability on the results. This can help identify potential issues early in the experiment.

Expert Consultation: Seek input from statisticians or experimental design experts to review and validate your experimental design and analysis plan.

Documentation: Maintain thorough documentation of the experimental process, including the design, randomization scheme, data collection procedures, and analysis methods.

Iterative Improvement: Use the results of your split-plot or blocked design experiment as a basis for further investigation and improvement. Consider refining the experiment in subsequent iterations to address any identified limitations.

By addressing these pitfalls and implementing the suggested remedies, researchers in aerospace engineering can conduct more reliable and informative split-plot and blocked design experiments, leading to better insights and more robust aerospace systems and processes.

EXAMPLE

Split-plot and blocked designed experiments are used in aerospace engineering when certain factors are difficult to change or control and are treated as whole plots or blocks. Here's a simplified example of a matrix for a split-plot and blocked designed experiment in aerospace engineering:

Objective: To optimize the fuel efficiency of a rocket engine by studying the effects of two factors (propellant composition and nozzle design) while considering variations in ambient temperature (Block A and Block B).

Factors:

1. **Propellant Composition (Factor A):**
   - High-Oxygen Propellant
   - Low-Oxygen Propellant
2. **Nozzle Design (Factor B):**
   - Convergent Nozzle
   - Convergent-Divergent Nozzle

Blocks:
   - **Block A**: Testing conducted in a controlled, low-temperature environment (e.g., laboratory conditions).
   - **Block B**: Testing conducted in a high-temperature environment (e.g., outdoor conditions).

Response Variable: Specific Impulse (measured in seconds, s)

Split-Plot and Blocked Experiment Matrix:
In this split-plot and blocked designed experiment, the matrix is designed to study the effects of propellant composition and nozzle design on specific impulse while also considering the influence of different ambient temperature conditions (Block A and Block B). The matrix is shown in Table 3.10.

In this matrix, each row represents a specific experimental run with a unique combination of propellant composition and nozzle design levels, while specifying the ambient temperature conditions (Block A or Block B). The split-plot design allows for the efficient study of propellant composition and nozzle design while considering the impact of different temperature environments.

TABLE 3.10

Split-Plot and Blocked Experiment Matrix Example for Aerospace Engineering

Run	Propellant Composition (A)	Nozzle Design (B)	Block
1	High-Oxygen	Convergent	A
2	Low-Oxygen	Convergent	A
3	High-Oxygen	Convergent-Divergent	A
4	Low-Oxygen	Convergent-Divergent	A
5	High-Oxygen	Convergent	B
6	Low-Oxygen	Convergent	B
7	High-Oxygen	Convergent-Divergent	B
8	Low-Oxygen	Convergent-Divergent	B

Split-plot and blocked designs are valuable in aerospace engineering when certain factors, like ambient temperature conditions, cannot be easily controlled or changed between experiments. This approach enables engineers to assess the main effects and interactions of the factors of interest while accounting for the constraints imposed by the blocked factors.

REPORTING

Split-plot and blocked design experiments in aerospace engineering often require the use of various statistical tools and techniques to analyze the data and draw meaningful conclusions. Here are some common statistical tools used in these experimental designs:

Analysis of Variance (ANOVA): ANOVA is a fundamental statistical tool used to partition the variation in the data into different sources, such as whole-plot, subplot, and error variation. It helps assess the significance of factors and interactions.

Design of Experiments (DOE) Software: Specialized DOE software, such as Minitab, JMP, or Design-Expert, can assist in designing split-plot and blocked experiments, as well as in conducting ANOVA and generating graphical representations of the results.

Factorial ANOVA: Factorial ANOVA is used to analyze the main effects and interactions of factors in the whole-plot and subplot designs separately. It assesses the significance of factors within each design element.

Response Surface Methodology (RSM): RSM techniques are employed to explore the relationships between factors and responses in greater detail. RSM allows for the modeling and optimization of complex systems.

Multivariate Analysis: In cases where multiple responses are of interest, multivariate analysis techniques like multivariate analysis of variance (MANOVA) can be used to analyze them simultaneously.

Regression Analysis: Regression analysis can help develop mathematical models to describe the relationships between factors and responses. It is often used in conjunction with RSM.

Diagnostic Plots: Various graphical tools, such as residual plots, normal probability plots, and interaction plots, are used to assess the assumptions of the statistical models and identify outliers or influential data points.

Post Hoc Tests: If ANOVA reveals significant differences, post hoc tests like Tukey's HSD (honestly significant difference) or Bonferroni corrections can be applied to identify which specific factor levels differ significantly from each other.

Effect Plots: Effect plots provide visual representations of the main effects and interactions, helping researchers understand the impact of factors on the responses.

Statistical Software: Statistical software packages like SAS, R, or Python with libraries like SciPy and statsmodels can be used to perform statistical analyses and generate results.

Robust Parameter Design (RPD): RPD techniques can be employed to optimize and robustify the aerospace system or process by considering variability and uncertainty in factor levels.

Block Summaries: When dealing with blocked designs, summaries of data within each block can provide insights into block-to-block variability and potential outliers.

Statistical Process Control (SPC): In some cases, SPC methods may be used to monitor and control the manufacturing or testing process during the experiment.

These statistical tools help aerospace engineers and researchers assess the impact of various factors, interactions, and sources of variability in split-plot and blocked design experiments. Proper statistical analysis is essential for making informed decisions, optimizing processes, and improving the reliability and performance of aerospace systems.

A well-structured report for a split-plot or blocked design experiment in aerospace engineering should provide a clear and comprehensive overview of the experiment, its objectives, methodology, results, and conclusions. Here are the key elements typically included in such a report:

Title Page:
- Title of the Report.
- Names of the Authors.
- Date.

Abstract: A concise summary of the entire experiment, including the main objectives, methods, key findings, and conclusions.

Table of Contents: A list of sections, subsections, and their respective page numbers.

List of Figures and Tables: A list of all figures and tables used in the report along with their page numbers.

List of Abbreviations and Symbols: An optional section that defines any specialized abbreviations or symbols used throughout the report.

Introduction:
- Background information explaining the context and significance of the experiment.
- Objectives or research questions that the experiment aims to address.
- A brief overview of the experimental design (split-plot or blocked) and the rationale for using this approach.

Experimental Design:
- Detailed description of the experimental setup, including the layout of the split-plot or blocked design.
- Explanation of how factors were assigned to whole plots, subplots, and blocks.
- Information on the levels of factors and any control or blocking variables.
- Justification for the chosen design and any assumptions made.

Methods:
- Description of the data collection process, including procedures, instruments, and measurement techniques.
- Any special considerations, randomization procedures, or data preprocessing steps.
- A clear explanation of how factors and responses were measured within whole plots, subplots, and blocks.

Statistical Analysis:
- Presentation of the statistical methods used, including ANOVA, factorial analysis, and any other relevant techniques.

- Detailed results of the analysis, including the main effects, interactions, and p-values.
- Graphical representations of the data, such as effect plots or interaction plots.
- Post hoc tests and comparisons if applicable.

Results:
- Presentation of the main findings, including significant effects and their practical implications.
- Discussion of any trends or patterns observed in the data.
- Tables and figures that illustrate the results, including summaries of means and variances.

Discussion:
- Interpretation of the results in the context of the experiment's objectives.
- Discussion of the significance of identified factors and interactions.
- Consideration of any limitations or sources of bias in the experiment.
- Comparison of findings with existing literature or theoretical expectations.

Conclusion:
- A concise summary of the key findings and their implications.
- A restatement of the experiment's objectives and whether they were achieved.
- Suggestions for future research or improvements to the experimental design.

References: A list of all sources, including research papers, textbooks, and software packages, cited in the report.

Appendices: Any supplementary materials, such as detailed data tables, additional graphs, or the full experimental design matrix. Any calculations or mathematical derivations not included in the main body of the report.

Acknowledgments: Recognition of individuals, organizations, or funding sources that contributed to the experiment.

Ethical Considerations (if applicable): A statement addressing ethical considerations, such as informed consent or compliance with research ethics guidelines.

Remember to follow the specific formatting and citation style required by your institution or organization when preparing your report. A well-structured and clearly written report is essential for effectively communicating the results of split-plot and blocked design experiments in aerospace engineering.

3.2 CHEMICAL ENGINEERING

In chemical engineering, designed experiments are widely used to optimize processes, improve product quality, and gain a deeper understanding of complex systems.

In a chemical engineering designed experiment, various types of variables are considered and controlled to investigate the effects of different factors on chemical processes or systems. These variables help researchers systematically plan, conduct, and analyze experiments to optimize chemical processes, develop new products, or solve engineering challenges. Here are some common types of variables in a designed experiment in chemical engineering:

Independent Variables (Factors):
    Process Variables: These are the key factors that can be manipulated to influence the chemical process or system. Examples include:
- **Flow Rates:** The rates at which fluids or reactants are introduced into the system.
- **Temperature:** The control of thermal conditions within the process, often critical for reaction kinetics.
- **Pressure:** The manipulation of pressure conditions, which can impact reaction equilibrium and phase changes.

- **Concentration:** The adjustment of reactant or solute concentrations in solution or within the system.
- **Residence Time:** The time a material or fluid spends in a specific process unit.

Material Properties: Factors related to the properties of raw materials, such as particle size, particle size distribution, and chemical composition.

Dependent Variables (Response Variables):

Product Yields: The quantities and purities of products obtained from the chemical process.

Conversion Rates: The extent to which reactants are transformed into desired products.

Reactor Temperatures and Pressures: The conditions within the reactor at various points in time.

Flow Rates and Flow Profiles: The characteristics of fluid flows within the process.

Concentration Profiles: The variations in reactant or product concentrations within the system.

Energy Consumption: The energy required to run the process or specific units within it.

Quality Parameters: Measures of product quality, such as impurity levels or desired product characteristics.

Reaction Kinetics: Rates of reaction progress and associated rate constants.

Control Variables (Constants):

- **Stoichiometry:** The fixed ratios of reactants to ensure consistent reactant proportions.
- **Initial Conditions:** The starting conditions of the process, such as initial concentrations, temperatures, and pressures.
- **Catalyst Properties:** The specific properties of catalysts used in the process.
- **Equipment Configuration:** The setup and configuration of equipment, including reactor types and dimensions.
- **Material Properties:** Fixed material properties that are kept constant during the experiment.

Extraneous Variables:

Environmental Conditions: Factors like humidity, atmospheric pressure, and ambient temperature may affect the process but are not directly controlled.

Instrumental Variability: Variations in measurement instruments or equipment that can introduce errors.

Batch-to-Batch Variability:

In industrial settings, variations in raw materials or batch-to-batch differences in reactants or reagents can be significant extraneous variables.

Randomization Variables:

In some experimental designs, randomization variables are introduced to ensure unbiased data collection.

Setpoints: Values set for control loops or process variables to maintain desired conditions.

Chemical engineers carefully consider and control these variables in their experiments to understand how different factors impact chemical processes, optimize engineering systems, and make informed decisions to achieve desired outcomes in chemical engineering applications.

The following are some common types of designed experiments with examples of their applications in chemical engineering:

3.2.1 Factorial Experiments

Factorial experiments are used to investigate the combined effects of multiple factors on a response variable. Often used to study the effect of temperature, pressure, and catalyst concentration on the yield of a chemical reaction. By varying these factors simultaneously and analyzing their interactions, researchers can identify optimal conditions for maximizing the desired product.

STEPS

The following are the steps involved in conducting a factorial experiment designed for chemical engineering:

1. **Define Your Research Question and Objectives**:
   - What process or reaction are you optimizing?
   - What response variables are you interested in (e.g., yield, purity, conversion, etc.)?
   - What factors do you believe might influence the response variables?
2. **Choose the Appropriate Factorial Design**:
   - Consider the number of factors and interaction effects you want to explore.
   - Different types of factorial designs exist, such as full-factorial, fractional factorial, and Plackett-Burman designs. Each has advantages and disadvantages depending on your needs and constraints.
   - Consult with a statistician or chemical engineer familiar with experimental design for guidance on choosing the best design for your specific situation.
3. **Determine the Factor Levels**:
   - Define the high and low levels for each factor based on your prior knowledge and desired range of investigation.
4. **Randomize the Experiment Runs**:
   - Randomly assign the combination of factor levels to each experimental run to minimize the impact of bias and confounding factors.
5. **Conduct the Experiments**:
   - Carefully follow your experimental protocol for each run, ensuring consistency and accuracy in data collection.
   - Document all procedures and observations meticulously.
6. **Analyze the Data**:
   - Use statistical software to analyze the response data and assess the main effects and interactions of the factors on the response variables.
   - Perform statistical tests to determine the significance of the observed effects.
7. **Interpret the Results**:
   - Draw conclusions about the influence of each factor and their interactions on the response variables.
   - Identify the optimal combination of factor levels that maximizes or minimizes the desired response variable.
   - Consider possible confounding factors and the limitations of the design, if any.
8. **Validate and Confirm the Results**:
   - Perform additional experiments at the predicted optimal conditions to verify the conclusions and ensure reproducibility.

Additional considerations:

- Pilot-scale experiments might be necessary before implementing the optimized conditions in full-scale production.
- Be prepared to troubleshoot and address any unexpected results or experimental errors.
- Document your findings and share your results with the appropriate stakeholders.

When conducting a successful factorial designed experiment, the process requires careful planning, execution, and analysis. Consult with experts when needed and be prepared to adapt and refine your approach based on the evolving data and insights.

PITFALLS AND REMEDIES

Factorial experiments are powerful tools in chemical engineering for studying the effects of multiple factors on a process or product. However, like any experimental design, they come with potential pitfalls. Here are some common pitfalls and remedies when using a factorial designed experiment in chemical engineering:

Inadequate Factor Selection
>   **Issue:** Choosing the wrong factors or neglecting important ones can lead to incomplete or ineffective experiments.
>
>   **Remedy:** Conduct a thorough literature review, consult subject matter experts, and perform preliminary studies to identify and select the most relevant factors. Ensure that all critical variables are included in the design.

Insufficient Experimental Runs
>   **Issue:** Inadequate sample size or too few experimental runs can lead to unreliable results and incomplete data.
>
>   **Remedy:** Perform power and sample size calculations to determine the required number of runs to detect meaningful effects. Ensure that the chosen sample size is statistically sufficient to achieve meaningful conclusions.

Factor-Level Selection
>   **Issue:** Selecting inappropriate factor levels can result in missing important trends or effects in the data.
>
>   **Remedy:** Conduct preliminary experiments or use prior knowledge to determine suitable factor levels that cover the expected range of variation. Avoid extreme values that may not be practical.

Interactions Neglect
>   **Issue:** Ignoring interactions between factors can lead to misleading results and failure to capture important synergistic or antagonistic effects.
>
>   **Remedy:** Include interaction terms in the factorial design matrix. Analyze the main effects and interactions to understand their contributions to the response variable.

Randomization and Order Effects
>   **Issue:** Failing to randomize the order of experimental runs can introduce bias due to uncontrolled variables or systematic changes over time.
>
>   **Remedy:** Randomize the order of experiments to minimize the impact of uncontrolled factors or time-related effects. Randomization ensures that the results are not influenced by external factors.

Nonlinearity
>   **Issue:** Assuming linearity when the relationship between factors and responses is nonlinear can lead to incorrect conclusions.
>
>   **Remedy:** Consider conducting additional experiments or using nonlinear regression techniques if there is evidence of nonlinearity in the data. Nonlinear models may better describe the relationships.

Uncontrolled Variables
>   **Issue:** Failure to control or monitor uncontrolled variables (nuisance factors) can introduce variability and confound the results.

Remedy: Identify potential nuisance factors and attempt to control or measure them during the experiments. Alternatively, use statistical methods such as analysis of covariance (ANCOVA) to account for their effects.

Model Assumptions Violation

Issue: Violating assumptions of the factorial model, such as normality or homoscedasticity, can affect the validity of the analysis.

Remedy: Check model assumptions using diagnostic plots and statistical tests. Transform the data if necessary to meet the assumptions or consider robust statistical techniques.

Overfitting

Issue: Including too many factors or interactions in the model can lead to overfitting, making it difficult to generalize results.

Remedy: Use model selection techniques (e.g., stepwise regression, information criteria) to identify the most parsimonious model that adequately represents the data without unnecessary complexity.

Interpretation Errors

Issue: Misinterpreting results or making incorrect conclusions about the effects of factors.

Remedy: Carefully analyze the results, visualize response surfaces, and consult with experts or statisticians if needed to ensure a correct interpretation of the findings.

EXAMPLE

A chemical plant wants to optimize the yield of a chemical reaction. They vary factors such as temperature, pressure, and reactant concentrations at two or more levels to identify the most significant factors and their interactions affecting yield. Let's consider that they are looking to optimize the yield of an alkylation reaction, a crucial step in producing various industrial chemicals and fuels. Here's how a factorial experiment method could be used for this purpose:

Factors and Levels:

- Temperature (T): Two levels, e.g., 40°C and 50°C. Higher temperatures might increase reaction rate but decrease yield due to side reactions.
- Pressure (P): Two levels, e.g., 1 atm and 2 atm. Increased pressure can shift the equilibrium toward the desired product but also increase pump energy consumption.
- Catalyst concentration (C): Two levels, e.g., 0.1 mol/L and 0.2 mol/L. Higher catalyst concentration can accelerate the reaction but also increase cost and potential over-catalysis.

Design and Experiments:

- A 2^3 full-factorial design would be employed, resulting in 8 experiments (2 levels for each of the 3 factors).
- In each experiment, the chosen levels of T, P, and C are combined, and the reaction yield is measured.

Factors:

- Temperature (T): Levels=40°C, 50°C
- Pressure (P): Levels=1 atm, 2 atm
- Catalyst concentration (C): Levels=0.1 mol/L, 0.2 mol/L

Design: 2^3 Full-Factorial Design (8 Runs) (Table 3.11)

This design assumes a single response variable, "Alkylation Yield." Consider including replicates of some or all runs to improve the accuracy and precision of the data. Also, consider randomizing the order of the experiments to minimize the impact of potential confounding factors.

Designed Experimental Matrix for Alkylation Reaction Optimization
Data Analysis:

- After conducting the experiments, analyze the yield data using statistical software to assess the main effects and interactions of the three factors on the yield.
- Interpretation of the results will identify the individual and combined influence of temperature, pressure, and catalyst concentration on the yield.
- Based on the analysis, choose the factor levels that maximize the yield, considering potential interactions and trade-offs between factors.

Data Analysis and Interpretation:

- Statistical software can analyze the data and assess the individual and interactive effects of T, P, and C on the yield.
- Main effects identify the individual impact of each factor on the yield (e.g., increasing T may decrease yield).
- Interaction effects reveal how combinations of factors influence the yield (e.g., the combined effect of high T and high P might further decrease yield).

Optimization: Based on the analysis, the plant can identify the factor levels that maximize the yield. This might involve:

- Choosing the level of each factor that individually contributes most to the yield.
- Considering potential interactions and finding the combination of levels that leads to the highest overall yield.

Specific factors and optimization goals will depend on the chosen alkylation reaction and desired product. However, it demonstrates how a factorial experiment method can be a valuable tool for chemical plant engineers to systematically investigate and optimize the yield of important chemical reactions.

TABLE 3.11
Factorial Designed Experiment Matrix Example for Chemical Engineering

Run	Temperature (T)	Pressure (P)	Catalyst Concentration (C)
1	40°C	1 atm	0.1 mol/L
2	40°C	1 atm	0.2 mol/L
3	40°C	2 atm	0.1 mol/L
4	40°C	2 atm	0.2 mol/L
5	50°C	1 atm	0.1 mol/L
6	50°C	1 atm	0.2 mol/L
7	50°C	2 atm	0.1 mol/L
8	50°C	2 atm	0.2 mol/L

This example uses a simple 2^3 design, but more complex designs can be used for additional factors or interactions. Statistical significance tests ensure the observed effects are not due to random chance. Conducting pilot plant experiments might be needed before implementing the optimized conditions in full-scale production.

The factorial design will efficiently explore the influence of multiple factors and their interactions on a single response variable. This requires fewer experiments compared to one-factor-at-a-time optimization. The design provides statistically sound data for reliable analysis and optimization decisions.

A challenge is that the design assumes linear relationships between factors and response, which might not always be true for complex reactions. Also, the design may not be suitable for optimizing multiple, potentially conflicting, response variables.

REPORTING

Factorial experiments are widely used in chemical engineering for studying the effects of multiple factors on a process or product. Various statistical tools are employed to design, analyze, and interpret factorial experiments effectively. Here are some of the key statistical tools used in factorial experiments for chemical engineering:

Factorial Design Matrix: A factorial design matrix is used to specify the experimental runs, factor levels, and combinations. It organizes the experimental plan and helps in systematically varying factors to study their effects.

Analysis of Variance (ANOVA): ANOVA is a fundamental statistical technique used to partition the total variability in the response variable into components attributed to the main effects, interactions, and errors. It helps assess the significance of factors and interactions.

Main Effects Plots: Main effects plots are graphical representations that display the individual effects of each factor on the response variable. They help identify which factors have a significant impact on the process.

Interaction Plots: Interaction plots visualize the interactions between two or more factors. They show how the effects of one factor depend on the levels of another, helping detect synergistic or antagonistic relationships.

Response Surface Analysis: Response surface analysis involves fitting mathematical models (e.g., linear, quadratic) to the data to describe the relationship between factors and the response variable. It allows for the optimization of process conditions.

Contour Plots: Contour plots are used in response surface analysis to visualize the response variable's behavior over a two-dimensional factor space. They help identify optimal factor combinations for desired responses.

Diagnostic Plots: Diagnostic plots, such as residual plots and normal probability plots, are used to assess the assumptions of the statistical model and detect outliers or influential data points.

Model Adequacy Checks: Statistical tests and diagnostic measures are used to check the adequacy of the fitted models. This ensures that the models adequately represent the observed data.

Effect Size Measures: Effect size measures, such as eta-squared or Cohen's d, quantify the practical significance of the observed effects, helping researchers assess the real-world impact of factors.

Sequential Analysis: Sequential analysis involves conducting experiments in stages and making adjustments based on interim results. This approach is particularly useful in adaptive factorial experiments.

Optimization Techniques: Optimization algorithms and techniques, such as gradient descent or response surface optimization, are employed to find the optimal factor settings that maximize or minimize the response variable.

Robust Design Analysis: Robust design analysis focuses on finding factor settings that minimize the variability of the response in the presence of noise or uncontrollable factors.

Analysis Software: Statistical software packages like R, SAS, JMP, Minitab, and others are often used to perform statistical analysis and generate graphical visualizations for factorial experiments.

Statistical Software Tools for Design of Experiments (DOE): Specialized DOE software tools help in the design, planning, and analysis of factorial experiments. They often provide user-friendly interfaces and automate many aspects of DOE.

Hypothesis Testing: Hypothesis tests, such as t-tests or F-tests, are used to determine the statistical significance of observed effects and differences in means or variances.

These statistical tools and techniques are essential for designing and analyzing factorial experiments in chemical engineering. They enable researchers to systematically investigate the effects of multiple factors on processes and products, optimize conditions, and make informed decisions based on data-driven insights.

When preparing a report on factorial experiments in chemical engineering, it's important to provide a clear and comprehensive account of the experimental design, methodology, results, and conclusions. Here are the essential elements to include in such a report:

Title and Abstract:
  Title: A concise and descriptive title that reflects the purpose of the study.
  Abstract: A brief summary (typically 150–250 words) that highlights the key objectives, methods, results, and conclusions of the experiment.

Introduction:
  Background: Provide context for the study by discussing the relevance of the research problem in chemical engineering.
  Objectives: Clearly state the research objectives and hypotheses that the factorial experiment aims to address.
  Literature Review: Summarize relevant literature and prior studies related to the factors, responses, and processes under investigation.

Experimental Design: Describe the design matrix. Specify the factors, their levels, and the combinations used in the factorial experiment.
  Randomization: Explain how randomization was employed to reduce bias and control for uncontrolled variables.
  Replication: Discuss the number of replicates or runs for each experimental condition.
  Control Factors: Identify any control factors or constants that were maintained throughout the experiment.
  Response Variables: Clearly define the response variables or outcomes that were measured or observed.

Methodology:
  Experimental Procedure: Detail the step-by-step procedure followed to conduct the factorial experiments.
  Data Collection: Explain how data were collected, including the instruments, sensors, or analytical methods used.
  Data Analysis: Describe the statistical techniques, models, and software employed for data analysis.

Results:
  Present Data: Display the experimental data, including tables, figures, graphs, and plots to illustrate the responses and variations.
  ANOVA Table: Provide the results of the ANOVA for each factor and interaction, including p-values and F-statistics.

Response Surface Plots: If applicable, include response surface plots or contour plots to visualize the relationship between factors and responses.

Discussion:

Interpretation of Results: Analyze and interpret the findings, including the significance of the main effects and interactions.

Factor Significance: Discuss which factors had a significant impact on the response variable(s) and their practical implications.

Optimization: If response surface analysis was performed, identify optimal factor settings for desired outcomes.

Limitations: Address any limitations or potential sources of bias in the experiment.

Conclusions:

Summarize the main findings and their implications for chemical engineering applications. Discuss the practical significance of the results and their potential contributions to the field. Revisit the research objectives and hypotheses to assess whether they were achieved.

Recommendations: Provide recommendations for further research or improvements to the experimental design. Suggest practical applications or implications for industry or chemical processes.

References: Cite all relevant sources, including research papers, textbooks, and software documentation used in the study.

Appendices: Include any supplementary materials such as detailed experimental procedures, data tables, regression models, or additional figures that support the main report.

Acknowledgments (Optional): Acknowledge individuals, organizations, or funding sources that contributed to the research.

Author Information: Include the names, affiliations, and contact information of the authors or researchers involved in the study.

Ensure that your report is well-organized, uses clear and concise language, and includes appropriate visual aids to enhance understanding. Properly formatted and labeled tables, figures, and equations are essential for conveying information effectively. Additionally, adhere to any specific formatting and citation style guidelines required by your institution or publication.

3.2.2   RESPONSE SURFACE METHODOLOGY (RSM)

Used to identify the optimal combination of factors for maximizing or minimizing a desired response. A common use is when developing a new polymer with specific mechanical properties. RSM experiments help determine the ideal temperature, pressure, and monomer ratios for achieving the desired strength and flexibility.

STEPS

The following steps are used for conducting an RSM designed experiment in chemical engineering:

1. **Define Your Research Question and Objectives**:
   - Identify the process or reaction you want to optimize.
   - Specify the response variables you want to improve (e.g., yield, purity, conversion, etc.).
   - Determine the key factors you believe might influence the response variables.
2. **Choose the Appropriate RSM Design**:
   - Consider the number of factors and interactions you want to explore.
   - Common RSM designs include CCD, Box-Behnken design, and Doehlert design, each with different advantages and disadvantages.

- Consult with a statistician or chemical engineer familiar with RSM for design selection based on your specific needs and constraints.

3. **Define the Factor Levels and Ranges**:
   - Set the high and low levels for each factor based on prior knowledge and the desired exploration range.
   - Ensure the chosen range is relevant and allows for potential optimization within realistic conditions.
4. **Develop the Experimental Plan**:
   - Randomize the order of the experiments to minimize bias and confounding factors.
   - Include replicates at the center point (where all factors are at their average level) for error estimation and model validation.
   - Document the experimental procedures and protocols meticulously.
5. **Conduct the Experiments**:
   - Carefully follow the planned procedures for each experiment, ensuring consistency and accuracy in data collection.
   - Record all observations and measurements accurately.
6. **Analyze the Data**:
   - Use statistical software to analyze the response data and build response surface models.
   - These models predict the response variables based on the chosen factors and their interactions.
   - Common model types include second-order polynomial models, which capture curvature in the response surface.
7. **Optimize the Response**:
   - Use optimization tools within the statistical software to identify the combination of factor levels that maximizes or minimizes the desired response variable(s).
   - Consider multi-objective optimization if multiple responses are important and conflicting (e.g., maximizing yield while maintaining purity).
8. **Validate and Confirm the Results**:
   - Conduct additional experiments at the predicted optimal conditions to verify the model's accuracy and confirm the optimized response.
   - This step ensures the optimization translates to real-world process improvement.

Additional considerations:

- Pilot-scale experiments might be necessary before implementing the optimized conditions in full-scale production.
- Be prepared to troubleshoot and address any unexpected results or experimental errors.
- Document your findings and share your results with the appropriate stakeholders.

Conducting a successful RSM experiment requires careful planning, execution, and analysis. Consult with experts when needed and be prepared to adapt and refine your approach based on the evolving data and insights.

PITFALLS AND REMEDIES

RSM experiments are widely used in chemical engineering for optimizing processes and understanding the relationships between multiple factors and responses. However, there are potential pitfalls that researchers should be aware of, along with remedies to address them effectively. Here are some common pitfalls and their corresponding remedies when using RSM in chemical engineering:

Inadequate Factor Space Exploration

Issue: Failing to explore the entire factor space can lead to missing optimal settings and suboptimal results.

Remedy: Use a well-structured experimental design, such as a CCD or Box-Behnken design, which covers a broad range of factor combinations. Ensure that the design matrix includes both low and high levels of factors.

Ignoring Nonlinearity

Issue: Assuming a linear relationship between factors and responses when the true relationship is nonlinear can lead to inaccurate models and predictions.

Remedy: Incorporate quadratic terms or higher-order terms in the model to account for nonlinearity. Visualize the response surface to identify curvature and nonlinearity.

Overfitting

Issue: Including too many terms (factors and interactions) in the response surface model can lead to overfitting, where the model fits the noise in the data rather than the underlying trends.

Remedy: Use model selection techniques, such as stepwise regression or information criteria (AIC, BIC), to identify the most parsimonious model that adequately represents the data without unnecessary complexity.

Neglecting Model Assumptions

Issue: Violating assumptions of the response surface model (e.g., normality, constant variance) can affect the validity of statistical tests and predictions.

Remedy: Check model assumptions using diagnostic plots and statistical tests. Transform the data if necessary to meet the assumptions or consider robust statistical techniques.

Limited Data Points

Issue: Having too few data points can result in unreliable model estimates and limited confidence in predictions.

Remedy: Ensure an adequate number of experimental runs to estimate model parameters reliably. Conduct power and sample size calculations to determine the minimum sample size required.

Ignoring Model Validation

Issue: Failing to validate the response surface model can lead to overly optimistic predictions and unreliable optimization results.

Remedy: Use cross-validation or split-sample validation to assess the model's predictive accuracy. Validate the model's predictions against new experimental runs if possible.

Neglecting Practical Constraints

Issue: Optimized factor settings suggested by the response surface model may not be practically achievable or cost-effective.

Remedy: Consider practical constraints, such as factor limits, economic feasibility, and safety requirements, when interpreting and implementing the optimization results.

Inadequate Replication

Issue: Conducting experiments without replication can lead to uncertainty in the estimated effects and poor model reliability.

Remedy: Ensure that experimental runs are replicated to account for variability and assess the consistency of results.

Incorrect Scaling of Factors

Issue: Failing to appropriately scale or normalize factors can affect the interpretation of regression coefficients and lead to incorrect conclusions.

Remedy: Scale factors to have a comparable magnitude or standardize them to have a mean of zero and a standard deviation of one. Use scaled factors in the response surface model.

Lack of Expertise

> **Issue:** Conducting RSM experiments without adequate statistical expertise can lead to errors in design, analysis, and interpretation.
>
> **Remedy:** Collaborate with statisticians or experts in RSM to design experiments, analyze data, and interpret results correctly.

EXAMPLE

RSM is used to model and optimize a chemical process, such as the synthesis of a pharmaceutical compound. By varying multiple factors within certain ranges and observing responses (e.g., yield or purity), engineers can create response surface models to identify optimal conditions. The following is an example of optimizing synthesis of a pharmaceutical compound using RSM. The target compound is Paracetamol (commonly known as Acetaminophen)

Factors and Levels:

- Reaction temperature (T): 40°C–60°C (higher temperatures might increase yield but degrade purity)
- Catalyst concentration (C): 0.1–0.3 mol/L (more catalyst might accelerate reaction but increase cost)
- Reaction time (t): 2–4 hours (longer reaction might improve yield but also promote side reactions)

Response Variables:

- Yield (Y): Percentage of starting material converted to Paracetamol.
- Purity (P): Percentage of the isolated product that is pure Paracetamol.

Design: The designed experimental matrix with three factors and two levels each, totaling 15 experiments (including center points) (Table 3.12):

TABLE 3.12

Response Surface Methodology (RSM) Example for Chemical Engineering

Run	Temperature (T)	Catalyst Concentration (C)	Reaction Time (t)
1	40 (Low)	0.1 (Low)	2 (Low)
2	40 (Low)	0.1 (Low)	4 (High)
3	40 (Low)	0.3 (High)	2 (Low)
4	40 (Low)	0.3 (High)	4 (High)
5	60 (High)	0.1 (Low)	2 (Low)
6	60 (High)	0.1 (Low)	4 (High)
7	60 (High)	0.3 (High)	2 (Low)
8	60 (High)	0.3 (High)	4 (High)
9	45 (Center)	0.2 (Center)	3 (Center)
10	45 (Center)	0.2 (Center)	3 (Center)
11	45 (Center)	0.2 (Center)	3 (Center)
12	40 (Low)	0.2 (Center)	3 (Center)
13	60 (High)	0.2 (Center)	3 (Center)
14	45 (Center)	0.1 (Low)	3 (Center)
15	45 (Center)	0.3 (High)	1.3 (Center)

This design considers two levels for each factor: "Low" and "High" (represented by actual values in your specific case). Five axial points extend beyond the factorial space on each axis to estimate curvature in the response surface. Three replicates at the center point (45°, 0.2 mol/L, 3 hours) provide information about error and model stability.

Model Building: Statistical software is used to analyze the data and build response surface models for Y and P, typically second-order polynomial models. These models describe the relationships between the factors and the responses mathematically.

Optimization: Multi-objective optimization algorithms are used to find the combination of factors (T, C, and t) that maximize Y and P simultaneously. For instance, a desirability function approach can be used, assigning weights to each response based on their relative importance.

Example Results:

- Model analysis might reveal that high temperature and catalyst concentration favor high yield but decrease purity, while longer reaction times provide moderate yield and good purity.
- Optimization with a focus on both Y and P could suggest a moderate temperature, slightly higher catalyst concentration, and slightly longer reaction time as the optimal conditions.

Validation and Confirmation: Additional experiments are conducted at the predicted optimal conditions to verify the models' accuracy and confirm the optimized yield and purity. This is a simplified example, and the specific factors, response variables, design, and optimization techniques will vary depending on the chosen pharmaceutical compound and synthesis route. This example demonstrates how RSM can be a valuable tool for optimizing chemical processes in pharmaceutical manufacturing, allowing for efficient exploration of complex relationships, and achieving desired product yield and purity.

RSM efficiently explores the impact of multiple factors on both yield and purity simultaneously. The method provides mathematical models for predicting product outcomes under different conditions. RSM identifies the optimal reaction conditions for achieving desired outcomes without extensive trial-and-error experimentation.

However, RSM assumes second-order relationships between factors and responses, which might not always be true for complex reactions. It also requires more experiments than factorial design for the same number of factors.

REPORTING

RSM experiments in chemical engineering involve the use of various statistical tools and techniques to model and optimize complex processes with multiple factors and responses. Here are the key statistical tools commonly used in RSM experiments for chemical engineering:

Factorial Design: Factorial experiments are often used as a precursor to RSM to screen and identify the most influential factors. They help determine which factors should be included in subsequent RSM experiments.

Central Composite Design (CCD): CCD is a common experimental design in RSM. It allows for the exploration of both linear and quadratic effects of factors while minimizing the number of experimental runs. CCD points are typically located at the center, along axes, and at the vertices of a multidimensional factor space.

Box-Behnken Design: This is an alternative design to CCD that is particularly useful when factors are believed to have linear effects but no interaction terms. It also reduces the number of experimental runs compared to a full-factorial design.

Analysis of Variance (ANOVA): ANOVA is used to analyze the significance of the main effects, interactions, and quadratic terms in the response surface model. It helps identify which factors and terms have a statistically significant impact on the response.

Regression Analysis: Linear and nonlinear regression models are fitted to the experimental data to describe the relationships between factors and responses. These models include the main effects, interactions, and quadratic terms.

Response Surface Plots: Response surface plots provide graphical representations of the response variable(s) as a function of two or three factors while keeping other factors at fixed levels. These plots help visualize the shape of the response surface and identify optimal factor settings.

Contour Plots: Contour plots are used to visualize the response surface in two dimensions. They display contours of constant response values, making it easier to identify factor settings that optimize the response.

ANOVA and Regression Diagnostics: Diagnostic plots, such as residual plots and normal probability plots, are used to assess the adequacy of the response surface model and detect potential outliers or violations of assumptions.

Model Selection Techniques: Model selection methods, such as stepwise regression or information criteria (AIC, BIC), are employed to identify the most appropriate and parsimonious response surface models.

Optimization Algorithms: Optimization techniques, such as gradient descent or desirability functions, are used to find the optimal factor settings that maximize or minimize the response variable(s) based on the fitted response surface model.

Validation Experiments: Validation experiments are conducted to test the predictive accuracy of the response surface model. These experiments may include additional runs not used in model fitting to assess the model's ability to make accurate predictions.

Robust Parameter Design (RPD): RPD is used to optimize processes while considering the variability and noise factors. It aims to find factor settings that minimize variability in the presence of uncontrollable factors.

Statistical Software: Specialized statistical software packages such as Design-Expert, JMP, Minitab, or R are commonly used to design experiments, fit response surface models, and conduct statistical analyses.

Simulation and Monte Carlo Analysis: Simulation techniques may be used to explore the sensitivity of optimization results to variations in factors and to assess the robustness of the optimized conditions.

These statistical tools and techniques play a crucial role in the design, analysis, and optimization of experiments using RSM in chemical engineering. They enable researchers to model complex processes, identify optimal factor settings, and make data-driven decisions to improve process performance and product quality.

A comprehensive report on RSM experiments in chemical engineering should provide a clear and organized account of the experimental design, methodology, results, and conclusions. Here are the essential elements to include in such a report:

Title and Abstract:
- **Title:** A concise and descriptive title that reflects the purpose of the study.
- **Abstract:** A brief summary (typically 150–250 words) that highlights the key objectives, methods, results, and conclusions of the RSM experiment.

Introduction:
- **Background:** Provide context for the study by discussing the relevance of the research problem in chemical engineering.

- **Objectives:** Clearly state the research objectives, including what factors are being studied and what responses are being optimized.
- **Literature Review:** Summarize relevant literature and prior studies related to the factors, responses, and processes under investigation.

Experimental Design:

- Factor Selection: Describe the factors and their levels that were considered in the RSM experiment.
- Experimental Design Matrix: Specify the experimental runs, including the center points and factorial points, and how they were randomized.
- Response Variables: Clearly define the response variables or outcomes that were measured or observed.

Methodology:

- Experimental Procedure: Detail the step-by-step procedure followed to conduct the RSM experiment.
- Data Collection: Explain how data were collected, including the instruments, sensors, or analytical methods used.
- Modeling: Describe the statistical techniques and models employed for fitting the response surface model.

Results:

- Data Presentation: Display the experimental data, including tables, figures, graphs, and plots that illustrate the responses and variations.
- ANOVA Table: Provide the results of the ANOVA for each factor and interaction, including p-values and F-statistics.
- Response Surface Plots: Include response surface plots or contour plots to visualize the relationships between factors and responses.

Discussion:

- Interpretation of Results: Analyze and interpret the findings, including the significance of the main effects and interactions.
- Optimization: Identify and discuss the optimal factor settings that maximize or minimize the response variable(s).
- Practical Implications: Discuss the practical significance of the results and their potential applications in chemical engineering.

Conclusions:

- Summary: Summarize the main findings and their implications for process optimization and product quality.
- Achievement of Objectives: Assess whether the research objectives and hypotheses were achieved.

Recommendations:

- Further Research: Provide recommendations for further research or improvements to the experimental design.
- Process Implementation: Suggest practical applications or process improvements based on the optimization results.

References: Cite all relevant sources, including research papers, textbooks, and software documentation used in the study.

Appendices: Include any supplementary materials such as detailed experimental procedures, additional tables, figures, regression models, or validation experiments.

Author Information: Include the names, affiliations, and contact information of the authors or researchers involved in the study.

Acknowledgments (Optional): Acknowledge individuals, organizations, or funding sources that contributed to the research.

Ensure that your report is well-organized, uses clear and concise language, and includes appropriate visual aids to enhance understanding. Properly formatted and labeled tables, figures, and equations are essential for conveying information effectively. Additionally, adhere to any specific formatting and citation style guidelines required by your institution or publication.

3.2.3 Fractional Factorial Experiments

Fractional factorial experiments are used to study a large number of factors efficiently when a full-factorial design becomes too complex or expensive. Engineers use fractional factorial design experiments to optimize a multistep synthesis process with numerous variables, a fractional factorial design allows for studying the main effects and some important interactions while requiring fewer experiments, saving time and resources.

Steps

Conducting a fractional factorial experiment in chemical engineering involves a series of steps to systematically study the effects of multiple factors on a response variable while reducing the number of experimental runs. Here are the key steps to follow:

1. **Define the Research Objectives**:
   - Clearly articulate the goals of the experiment and the specific factors you want to investigate. Identify the response variable that you aim to optimize or study.
2. **Select the Factors and Levels**:
   - Choose the factors (independent variables) that you want to study. Determine the number of levels for each factor (typically two levels, high and low).
   - Assign labels or codes to the levels of each factor.
3. **Choose the Fraction of the Full Design:**
   - Decide on the fraction of the full-factorial design you want to use. Common fractional designs include half-fraction ($2^{(k-1)}$), quarter-fraction ($2^{(k-2)}$), and so on, where 'k' is the number of factors.
4. **Generate the Fractional Factorial Design Matrix:**
   - Use design software or tables to generate the experimental design matrix based on the chosen fraction. The matrix will specify which combinations of factor levels to test and which ones to omit.
5. **Randomize the Experimental Runs:**
   - Randomly order the experimental runs to minimize the influence of uncontrolled factors or systematic effects.
6. **Conduct the Experimental Runs:**
   - Perform the experiments according to the combinations of factor levels specified in the design matrix. Collect data for the response variable(s) at each run.
   - Ensure that the experiments are conducted carefully and consistently to minimize sources of variability.
7. **Analyze the Data:**
   - Perform data analysis, which may include:
   - Calculating descriptive statistics for each combination of factor levels.
   - Conducting ANOVA to determine the significance of the main effects and interactions.
   - Constructing graphs, plots, or response surface models to visualize the data and relationships.
8. **Interpret the Results:**
   - Interpret the statistical results to understand the main effects and interactions between factors.

- Identify which factors have a significant impact on the response variable and how they influence the process or outcome.

9. **Optimization and Further Studies:**
   - If the goal is optimization, use the response surface model or analysis results to find the optimal factor settings that maximize or minimize the response variable.
   - Consider conducting follow-up experiments or confirmatory studies to validate the findings or optimize the process further.

10. **Report the Findings:**
    - Prepare a comprehensive report summarizing the experimental design, methodology, results, and conclusions.
    - Include tables, figures, and visualizations to support your findings.
    - Discuss the practical implications and applications of the results in the context of chemical engineering.

11. **Implement the Findings (if applicable):**
    - If the study yields process improvements or optimization, consider implementing the recommended factor settings in the chemical engineering process.

12. **Validate and Review:**
    - Conduct a review and validation of the experimental design and results to ensure the findings are robust and reliable.

13. **Document and Archive:**
    - Maintain proper documentation of the experimental setup, data, analysis, and reports for future reference and potential replication.

Fractional factorial experiments are valuable tools for efficiently exploring and understanding the effects of multiple factors in chemical engineering processes. Proper planning, careful execution, and thorough data analysis are essential to derive meaningful insights from these experiments.

PITFALLS AND REMEDIES

Fractional factorial experiments are a valuable tool in chemical engineering for studying the effects of multiple factors while reducing the number of experimental runs. However, like any experimental design, they come with potential pitfalls. Here are common pitfalls and their corresponding remedies when using fractional factorial experiments in chemical engineering:

Alias Effects
Issue: Fractional factorial designs intentionally omit certain combinations of factor levels, which can result in alias effects, where the effects of one factor are confounded with another.

Remedy: Use alias tables or design generators to identify alias effects and account for them during data analysis. If an important factor is aliased, consider running additional experiments or using a different fractional design.

Missed Higher-Order Interactions
Issue: Fractional factorial designs typically focus on the main effects and two-way interactions, potentially missing higher-order interactions.

Remedy: If higher-order interactions are suspected to be important, consider using a more extensive fractional design, such as a $2^{(k-p)}$ design, where 'p' is the number of higher-order interactions to include.

Limited Resolution
Issue: Some fractional designs have lower resolution, which can make it challenging to distinguish between the main effects and interactions.

Remedy: Use higher-resolution designs, if necessary, to better separate the main effects from interactions. Higher-resolution designs may require more experimental runs but provide clearer results.

Factor Range Restriction
>**Issue:** The chosen factor levels in a fractional design may not cover the entire practical range, limiting the generalizability of the results.
>
>**Remedy:** Carefully select factor levels to ensure they encompass the relevant operational range of the process. Consider conducting additional experiments if needed to explore extreme conditions.

Failure to Account for Noise Factors
>**Issue:** Fractional factorial experiments may not account for noise factors or uncontrolled variables, leading to variability in the results.
>
>**Remedy:** Implement strategies to control or measure noise factors during the experiments. Alternatively, consider using RPD techniques to account for variability.

Model Assumptions Violation
>**Issue:** Violating assumptions of the response model, such as normality or homoscedasticity, can affect the validity of the analysis.
>
>**Remedy:** Check model assumptions using diagnostic plots and statistical tests. Transform the data if necessary to meet the assumptions or consider robust statistical techniques.

Overfitting
>**Issue:** Including too many factors or interactions in the response model can lead to overfitting, making it difficult to generalize results.
>
>**Remedy:** Use model selection techniques (e.g., stepwise regression, information criteria) to identify the most parsimonious model that adequately represents the data without unnecessary complexity.

Lack of Randomization
>**Issue:** Failing to randomize the order of experimental runs can introduce bias due to uncontrolled variables or systematic changes over time.
>
>**Remedy:** Randomize the order of experiments to minimize the impact of uncontrolled factors or time-related effects. Randomization ensures that the results are not influenced by external factors.

Inadequate Sample Size
>**Issue:** Insufficient sample size can lead to unreliable results and limited statistical power.
>
>**Remedy:** Conduct power and sample size calculations to determine the required number of runs to detect meaningful effects. Ensure that the chosen sample size is statistically sufficient.

Lack of Expertise
>**Issue:** Conducting fractional factorial experiments without adequate statistical expertise can lead to errors in design, analysis, and interpretation.
>
>**Remedy:** Collaborate with statisticians or experts in fractional factorial design to plan experiments, analyze data, and interpret results correctly.

By being aware of these potential pitfalls and implementing the suggested remedies, researchers in chemical engineering can effectively use fractional factorial experiments to investigate the effects of multiple factors while minimizing the number of experimental runs and optimizing processes. Proper planning, analysis, and validation are key to the success of fractional factorial experiments.

EXAMPLE

When there are many factors to consider but conducting a full-factorial experiment is impractical, fractional factorial experiments are employed. A fractional factorial design is a useful approach for studying the effects of multiple factors on a response variable while reducing the number of experimental runs required. In chemical engineering, let's consider an example where we want to investigate the effects of four factors on the performance of a catalyst used in a chemical reaction. The factors of interest are:

1. Catalyst Type (A): A choice between two different catalyst types (A1 and A2).
2. Temperature (B): Two levels of temperature, high and low.
3. Reactant Concentration (C): Two levels of reactant concentration, high and low.
4. Stirring Speed (D): Two levels of stirring speed, fast and slow.

Since testing all possible combinations of these factors would require $2^4 = 16$ experimental runs, we can use a fractional factorial design to reduce the number of runs. Let's use a half-fractional design ($2^{(4-1)}$) as an example:

1. Catalyst Type (A): Two levels, A1 and A2.
2. Temperature (B): Two levels, high and low.
3. Reactant Concentration (C): Two levels, high and low.
4. Stirring Speed (D): One level, fast.

In this fractional factorial design, we will conduct $2^3 = 8$ experimental runs instead of the full 16 runs. The experimental plan with the levels of each factor is shown in Table 3.13.

With this fractional factorial design, we can study the main effects of each factor (A, B, C, and D) and any potential two-way interactions. By conducting these eight runs, we can gain valuable insights into how these factors affect catalyst performance while reducing the time and resources required compared to a full-factorial design. This example demonstrates the practical utility of fractional factorial designs in chemical engineering experiments.

REPORTING

In fractional factorial designed experiments for chemical engineering, various statistical tools are used to analyze and interpret the results. These tools help researchers identify significant factors and interactions while reducing the number of experiments required. Here are some commonly used statistical tools in fractional factorial experiments:

1. **Analysis of Variance (ANOVA)**: ANOVA is a fundamental tool for analyzing the variance in the experimental data. It helps determine which factors and interactions are statistically significant in influencing the response variable (e.g., yield, product quality). ANOVA partitions the variance into different components, such as the main effects and interactions.
2. **Main Effects Plots**: Main effects plots visually represent the influence of individual factors on the response variable. These plots display the average response at each level of a factor while keeping other factors constant. They provide insights into which factors have a significant impact on the response.

TABLE 3.13
Chemical Engineering Fractional Factorial Designed Experiment Matrix Example

Run	A (Catalyst Type)	B (Temperature)	C (Reactant Concentration)	D (Stirring Speed)
1	A1	High	High	Fast
2	A2	High	Low	Fast
3	A1	Low	High	Fast
4	A2	Low	Low	Fast
5	A1	High	High	Fast
6	A2	High	Low	Fast
7	A1	Low	High	Fast
8	A2	Low	Low	Fast

3. **Interaction Plots**: Interaction plots illustrate the interactions between two or more factors. They show how the combined influence of factors differs from what would be expected based on their individual effects. Interaction plots help identify synergistic or antagonistic effects.

4. **Pareto Charts**: Pareto charts are bar charts that display the magnitude of the effects of different factors on the response variable. They help researchers prioritize which factors to focus on for optimization or process improvement by highlighting the most significant contributors.

5. **Half-Normal Probability Plots**: Half-normal probability plots are used to identify influential factors by comparing the absolute values of the effects. Factors with larger absolute effects are more likely to be significant. This tool is especially useful for screening experiments.

6. **Response Surface Methodology (RSM)**: While not exclusive to fractional factorial designs, RSM is often used in combination with them. RSM helps researchers model and optimize complex processes by exploring the response surface to find optimal factor settings.

7. **Contour Plots**: Contour plots are 2D graphical representations of response surfaces. They show how the response variable changes as two factors vary simultaneously while keeping other factors constant. Contour plots are helpful for visualizing interactions.

8. **Diagnostic Plots:** Diagnostic plots, such as residual plots and normal probability plots, are used to assess the validity of the statistical model. They help ensure that the assumptions of the analysis, such as normality and homoscedasticity, are met.

9. **Fractional Factorial Design Matrices**: These matrices describe the experimental design itself, including the assignment of factor levels to experiments. They are used to plan and execute the experiments efficiently.

10. **Software Tools**: Statistical software packages like Minitab, JMP, and R are commonly used for designing fractional factorial experiments and performing statistical analysis. They streamline data analysis and provide graphical representations of the results.

11. **Signal-to-Noise Ratios**: S/N ratios, such as the S/N ratio for smaller-the-better or larger-the-better characteristics, can be used to optimize processes and identify the factor settings that lead to the desired response.

These statistical tools help chemical engineers and researchers make informed decisions, optimize processes, and reduce experimentation time and costs in chemical engineering applications. The choice of tools depends on the specific goals of the experiment and the complexity of the factors and interactions under investigation.

A report for a fractional factorial designed experiment in chemical engineering typically includes several key elements to communicate the experimental design, methodology, results, and conclusions effectively. Here are the essential elements to include in such a report:

1. **Title Page**:
   - Title of the Report: A concise and descriptive title of the experiment.
   - Author(s) and Affiliation: Names of the author(s) and their affiliations.
   - Date: The date when the report is prepared.

2. **Abstract**:
   - A brief summary of the experiment's objectives, methods, key findings, and conclusions. The abstract should provide a concise overview of the entire report.

3. **Table of Contents**:
   - A list of the report sections and subsections with corresponding page numbers for easy navigation.

4. **List of Figures and Tables**:
   - An organized list of all figures and tables included in the report along with their respective page numbers.

5. **Introduction:**
   - Background and Objectives: An introduction to the problem or process under investigation, including the rationale for conducting the fractional factorial experiment.
   - Objectives: Clearly state the objectives and goals of the experiment.

6. **Experimental Design**:
   - Description of Experimental Factors: Explain the factors and levels investigated in the experiment. Provide a clear overview of the fractional factorial design matrix.
   - Sampling and Data Collection: Describe how the data were collected, including the number of runs and experimental conditions.
   - Randomization: Explain any randomization procedures used to minimize bias.

7. **Methods**:
   - Experimental Setup: Describe the laboratory setup, equipment, and materials used in the experiment.
   - Data Collection: Explain how data were collected, including measurement techniques and procedures.
   - Statistical Analysis: Outline the statistical tools and methods used for data analysis, including ANOVA, main effects analysis, and interaction analysis.

8. **Results**:
   - Presentation of Data: Present the experimental data in a clear and organized manner, such as tables, charts, or graphs.
   - Main Effects and Interactions: Discuss the main effects of factors and interactions between factors as determined by the analysis.
   - Significant Factors: Identify the factors that have a significant impact on the response variable.
   - Response Surface: If applicable, present response surface plots to visualize the relationship between factors and the response.

9. **Discussion:**
   - Interpretation: Interpret the results, discussing the implications of significant factors and interactions on the process or product.
   - Comparison to Hypotheses: Compare the findings to initial hypotheses or expectations.
   - Sources of Variability: Discuss sources of variability and potential sources of error.

10. **Conclusions**:
    - Summarize the main findings of the experiment.
    - Address whether the objectives were achieved.
    - Provide insights into the practical implications of the results.

11. **Recommendations**:
    - Offer recommendations for process improvement or further investigation based on the experiment's findings.

12. **References:**
    - Cite all sources, including research papers, textbooks, and software used for the experiment and data analysis.

13. **Appendices**:
    - Include any supplementary information that supports the report, such as raw data, experimental protocols, detailed statistical analyses, or additional graphs and tables.

14. **Acknowledgments**:
    - Acknowledge individuals, organizations, or funding sources that contributed to the experiment.

15. **Signature Page (Optional)**:
   - Include a signature page for approval and verification of the report by relevant parties.
16. **List of Abbreviations and Symbols (Optional)**:
   - If the report uses abbreviations or symbols, provide a list for clarity.

Remember to follow a consistent and clear writing style, use appropriate headings and subheadings, and provide sufficient detail to allow readers to understand the experiment, methods, and results. The report should be well-organized and formatted for readability.

3.2.4 Mixture Experiments

Mixture experiments are used to optimize the composition of a mixture to achieve a specific property or function. An example may be in formulating a new fuel blend with the desired burning efficiency and emission characteristics. Mixture design experiments ensure the optimal ratios of different components within the fuel blend. Another might be in optimizing the formulation of coatings, paints, or pharmaceuticals; mixture experiments help determine the optimal composition of ingredients to achieve desired properties, such as color, adhesion, or drug release rate.

Steps

Developing a mixture designed experiment in chemical engineering involves planning and executing experiments to study the effects of different ingredients or components in a mixture on a specific response variable. Here are the steps to develop a mixture designed experiment:

1. **Define Objectives**:
   - Clearly define the objectives of the experiment. What specific aspects of the mixture or process are you trying to optimize or understand? Determine the response variable you want to measure and improve.
2. **Select Factors and Levels**:
   - Identify the components or factors that make up the mixture. These could be ingredients, proportions, or process conditions.
   - Determine the range or levels at which each factor will be tested. Ensure that the chosen levels are practically meaningful and feasible in the context of your application.
3. **Choose a Mixture Design Type**:
   - Select the appropriate type of mixture design based on your objectives and the nature of the mixture. Common types include simplex-centroid designs, extreme vertices designs, or constrained mixture designs (e.g., lattice or simplex-lattice designs).
4. **Generate a Design Matrix**:
   - Create a design matrix that specifies the combinations of factors and their levels for the experiments. Software tools like DOE software can help generate the matrix based on the selected design type.
5. **Plan the Experiments**:
   - Determine the number of experimental runs or trials required to complete the design matrix. Consider factors like replication to assess variability and randomization to minimize bias.
   - Plan the order in which the experiments will be conducted to minimize potential carryover effects or batch-to-batch variability.

6. **Data Collection and Execution**:
   - Conduct the experiments according to the design matrix, following the planned sequence and conditions.
   - Carefully measure and record the response variable for each experimental run. Ensure data accuracy and precision.
7. **Data Analysis**:
   - Use statistical methods to analyze the data collected during the experiments. Common analyses include ANOVA, regression analysis, and mixture modeling.
   - Assess the main effects of each component and potential interactions between components.
8. **Response Surface Analysis (Optional)**:
   - If needed, perform response surface analysis to model the relationship between the mixture components and the response variable. This can help optimize the mixture for desired outcomes.
9. **Interpretation**:
   - Interpret the results of the analysis in the context of your objectives. Identify which mixture components have a significant impact on the response variable and how they influence the outcome.
10. **Optimization**:
    - If the goal is to optimize the mixture for a specific response, use the analysis results to find the optimal combination of components and levels that maximize or minimize the response variable.
11. **Validation**:
    - Validate the findings by conducting additional experiments, if necessary, to ensure that the optimized mixture performs as predicted.
12. **Report and Documentation**:
    - Prepare a comprehensive report that includes the experimental design, methods, results, and conclusions.
    - Document the experimental procedures, data, and statistical analyses for future reference.
13. **Implement Findings**:
    - Apply the findings from the mixture designed experiment to improve the process, product, or formulation in your chemical engineering application.
14. **Iterate and Refine (if needed)**:
    - Depending on the complexity of the system and the objectives, you may need to iterate the process, refining the mixture design and analysis to achieve optimal results.
15. **Communication**:
    - Communicate the results and recommendations to relevant stakeholders or team members.

Mixture designed experiments are a powerful tool in chemical engineering for optimizing formulations, processes, and products. Careful planning, execution, and analysis are essential to achieve meaningful and actionable results.

PITFALLS AND REMEDIES

Ignoring Constraints: Failing to account for practical constraints or limitations in the mixture design, such as ingredient availability or physical boundaries, can lead to unrealistic results.

Remedy: Ensure that the chosen mixture design type and factor levels are within feasible and meaningful ranges based on real-world constraints.

Inadequate Design Space Exploration: Not exploring a wide enough design space can limit the discovery of optimal regions, leading to suboptimal formulations or processes.

Remedy: Conduct preliminary screening experiments or sensitivity analyses to identify potential regions of interest in the design space before conducting the main mixture designed experiment.

Overfitting Models: Overfitting occurs when complex models are fitted to the data, leading to poor model generalization. This can result in models that do not perform well outside the observed data points.

Remedy: Use model selection techniques to choose the simplest model that adequately describes the relationship between mixture components and the response variable. Cross-validation can help assess model performance.

Interactions Complexity: If mixture components interact in complex ways, it may be challenging to represent these interactions accurately in the model.

Remedy: Consider higher-order mixture models or RSMs that can capture and visualize complex interactions.

Lack of Replication: Conducting experiments without replication can lead to unreliable results and difficulty in estimating experimental error.

Remedy: Include replication in the experimental design to assess variability and obtain more robust results. Replication can also help identify random errors.

Model Assumptions Violation: Violating assumptions of normality, linearity, or constant variance in the data can lead to inaccurate model results.

Remedy: Check model assumptions using diagnostic plots and transformation techniques if needed. Nonparametric approaches can be considered if assumptions cannot be met.

Small Sample Size: A small number of experimental runs may limit the ability to detect significant effects or interactions.

Remedy: If possible, increase the sample size by conducting additional experiments or using historical data. Alternatively, consider designs that provide more information with fewer runs, such as D-optimal designs.

Failure to Validate: Not validating the optimized formulation or process under real-world conditions can result in poor practical outcomes.

Remedy: Conduct validation experiments to verify that the optimized conditions or formulations perform as expected in the actual application.

Pitfalls Related to Model Use:

Extrapolation: Extrapolating model predictions beyond the experimental design space can be risky, as models may not accurately represent behavior in untested regions.

Remedy: Be cautious when making predictions outside the observed design space. Conduct additional experiments if necessary to gather data in those regions.

Over-Optimization: Optimizing a mixture without considering other important factors or constraints (e.g., cost, safety) can lead to impractical or suboptimal solutions.

Remedy: Consider multi-objective optimization techniques that balance multiple objectives and constraints simultaneously.

Model Reliance: Relying solely on models without considering domain expertise or process knowledge can lead to unrealistic or impractical solutions.

Remedy: Use models as tools to guide experimentation and decision-making but also consult with subject matter experts to ensure practicality.

To avoid these pitfalls, it's essential to carefully plan your mixture designed experiment, validate models, and interpret results within the context of your specific chemical engineering application. Collaboration among experts from different domains, such as statistics, chemistry, and engineering, can also enhance the success of your experiment.

EXAMPLE

Here's an example of a mixture designed experiment in chemical engineering:

Objective: To optimize the formulation of a detergent product to maximize cleaning efficacy while minimizing production costs.

Factors:

1. **Ingredient A**: The concentration of a surfactant (in percentage).
2. **Ingredient B**: The concentration of a cleaning agent (in percentage).
3. **Ingredient C**: The concentration of a fragrance additive (in percentage).

Response Variable: Cleaning Efficacy (measured as the percentage of stains removed in a standardized test).

Experimental Design:

- A simplex-centroid mixture design is chosen because it allows the testing of mixtures with three components at different levels while keeping the total concentration constant (100%).
- The design matrix includes various combinations of A, B, and C, such as 10% A, 60% B, and 30% C; 40% A, 30% B, and 30% C, and so on, totaling 10 experimental runs.

Experimental Procedure:

- Conduct the 10 experiments according to the design matrix, preparing detergent formulations with different ingredient concentrations.

Data Collection:

- After each experiment, conduct standardized cleaning tests to measure the cleaning efficacy of each detergent formulation. Record the percentage of stains removed for each run.

Data Analysis:

- Perform a mixture design analysis, including regression modeling and ANOVA, to determine the main effects and interactions of ingredients A, B, and C on cleaning efficacy.

Results:

- The analysis reveals that increasing the concentration of surfactant (A) and cleaning agent (B) has a positive impact on cleaning efficacy, while the fragrance additive (C) has a negligible effect.
- Interaction between A and B is significant, indicating a synergistic effect when both surfactant and cleaning agent are present in higher concentrations.

Optimization:

- Using the analysis results, optimize the detergent formulation to maximize cleaning efficacy while keeping production costs within budget constraints.
- The optimal formulation may have higher concentrations of A and B while minimizing C.

Validation:

- Conduct validation experiments to verify that the optimized detergent formulation performs as predicted in real-world cleaning applications.

This example demonstrates how a mixture designed experiment can be applied in chemical engineering to optimize product formulations by varying the concentrations of components while considering both performance and cost objectives.

DESIGN MATRIX

The design matrix specifies the proportions or concentrations of each component in each experimental run. Here's a simplified example of a matrix for a mixture designed experiment:

Objective: To optimize the formulation of a paint primer by adjusting the proportions of three components: pigment (A), binder (B), and solvent (C) while keeping the total concentration constant at 100%.

Factors (See Table 3.14):

A: Concentration of pigment (in percentage)
B: Concentration of binder (in percentage)
C: Concentration of solvent (in percentage)

In this design matrix:

- Each row represents an experimental run.
- Columns correspond to the factors (A, B, and C) and indicate the proportion or concentration of each component for a specific run.
- The total concentration in each experiment is kept constant at 100%, representing the paint formulation.

For example, in Experiment 1, the primer formulation contains 20% pigment (A), 60% binder (B), and 20% solvent (C). Each row in the matrix represents a unique combination of factor levels, allowing you to conduct experiments to study the effects of different component proportions on the properties of the paint primer.

TABLE 3.14

Mixture Designed Experiment Matrix Example for Chemical Engineering

Experiment	% Pigment (A)	% Binder (B)	% Solvent (C)
1	20	60	20
2	40	40	20
3	60	20	20
4	30	30	40
5	50	10	40
6	10	50	40
7	30	30	40
8	50	10	40

The design matrix is a fundamental tool in mixture designed experiments, as it guides the execution of experiments and data collection to explore the relationships between mixture components and response variables, leading to optimized formulations or processes in chemical engineering applications.

REPORTING

In mixture designed experiments for chemical engineering, various statistical tools are used to analyze and interpret the results. These tools help researchers identify significant factors and interactions while optimizing formulations or processes involving mixtures. Here are some commonly used statistical tools in mixture designed experiments:

Analysis of Variance (ANOVA): ANOVA is a fundamental tool for analyzing the variance in the experimental data. It helps determine which mixture components and their interactions are statistically significant in influencing the response variable. ANOVA partitions the variance into different components, such as the main effects and interactions.

Main Effects Plots: Main effects plots visually represent the influence of individual mixture components on the response variable. These plots display the average response at each level of a component while keeping other components constant. They provide insights into which components have a significant impact on the response.

Interaction Plots: Interaction plots illustrate the interactions between two or more mixture components. They show how the combined influence of components differs from what would be expected based on their individual effects. Interaction plots help identify synergistic or antagonistic effects.

Pareto Charts: Pareto charts are bar charts that display the magnitude of the effects of different mixture components and their interactions on the response variable. They help researchers prioritize which components to focus on for optimization or process improvement by highlighting the most significant contributors.

Response Surface Methodology (RSM): While not exclusive to mixture designed experiments, RSM is often used in combination with them. RSM helps researchers model and optimize complex processes by exploring the response surface to find optimal component proportions.

Contour Plots: Contour plots are 2D graphical representations of response surfaces. They show how the response variable changes as two components vary simultaneously while keeping other components constant. Contour plots are helpful for visualizing interactions.

Diagnostic Plots: Diagnostic plots, such as residual plots and normal probability plots, are used to assess the validity of the statistical model. They help ensure that the assumptions of the analysis, such as normality and homoscedasticity, are met.

Fractional Factorial Design: Fractional factorial designs are sometimes used to reduce the number of experiments while still capturing the main effects and interactions among mixture components. These designs help economize experimentation when there are many components to consider.

Optimization Algorithms: Optimization algorithms, such as gradient descent or genetic algorithms, are used to find the optimal combination of mixture components that maximize or minimize the response variable.

Robust Parameter Design (RPD): RPD techniques are used to optimize mixtures under conditions of variability or uncertainty. They help identify component proportions that are robust to variations in raw materials or process conditions.

Software Tools: Statistical software packages like Minitab, JMP, and R are commonly used for designing mixture experiments, performing statistical analyses, and visualizing results. These tools streamline data analysis and provide graphical representations of the results.

These statistical tools help chemical engineers and researchers make informed decisions, optimize formulations or processes, and reduce experimentation time and costs in chemical engineering applications involving mixtures. The choice of tools depends on the specific goals of the experiment and the complexity of the mixture components and interactions under investigation.

A report for a mixture designed experiment in chemical engineering should provide a comprehensive overview of the experimental design, methodology, results, and conclusions. Here are the key elements to include in such a report:

Title Page:
> Title of the Report: A clear and descriptive title that reflects the purpose of the experiment.
> Author(s) and Affiliation: Names of the author(s) and their affiliations.
> Date: The date when the report was prepared.

Abstract: A concise summary of the experiment's objectives, methods, key findings, and conclusions. The abstract should provide a quick overview of the entire report.

Table of Contents: A list of the report sections and subsections with corresponding page numbers for easy navigation.

List of Figures and Tables: An organized list of all figures and tables included in the report along with their respective page numbers.

Introduction: An introduction to the problem or process under investigation, including the rationale for conducting the mixture designed experiment. Clearly state the objectives and goals of the experiment, emphasizing the optimization or understanding of mixture components' effects.

Experimental Design:
> Description of Components and Levels: Explain the components or factors being studied in the mixture, their roles, and the range or levels at which they are tested.
> Choice of Mixture Design Type: Describe the type of mixture design chosen (e.g., simplex-centroid, extreme vertices) and why it was selected.
> Design Matrix: Present the design matrix specifying the combinations of component proportions for each experimental run.

Experimental Procedure: Detailed description of how the experiments were conducted, including preparation of mixtures, equipment used, and experimental conditions.

Data Collection: Explain how data were collected, including measurement techniques, instruments used, and the number of replicates if applicable.

Data Analysis: Present the statistical methods and tools used to analyze the experimental data, including ANOVA, regression modeling, and graphical representations. Include results of the statistical analysis, such as main effects, interactions, and significance levels.

Results:
> Presentation of Data: Display the experimental data, response values, and graphical representations (e.g., contour plots, Pareto charts) to illustrate the effects of mixture components.
> Interpretation: Discuss the implications of the results, emphasizing the impact of individual components and interactions on the response variable.

Optimization: If the goal was optimization, present the optimized mixture proportions and the corresponding predicted response values.

Discussion: Provide an in-depth interpretation of the results, discussing the practical significance of significant effects and interactions.
> Address any unexpected findings or challenges encountered during the experiment.

Conclusions: Summarize the main findings of the mixture designed experiment, including the identified significant factors and their effects on the response variable. Reiterate the achievement of the experiment's objectives.

Recommendations: Offer recommendations based on the conclusions, such as suggestions for process improvement, further investigation, or changes in formulation.

Validation: If applicable, describe any validation experiments conducted to verify the accuracy and practicality of the optimized mixture.

References: Cite all sources, including research papers, textbooks, software used for analysis, and relevant literature.

Appendices: Include any supplementary information, such as the full design matrix, raw data, detailed statistical analyses, and additional figures or tables.

Acknowledgments: Acknowledge individuals, organizations, or funding sources that contributed to the experiment.

Remember to use clear and concise language, provide sufficient details for reproducibility, and use appropriate headings and subheadings to enhance the readability and organization of the report.

3.2.5 TAGUCHI METHOD

The Taguchi Method, also known as robust design or Taguchi robust parameter design (TRPD), is a systematic approach used in chemical engineering to optimize processes or formulations while considering factors that affect performance and robustness. Taguchi methods are used to optimize processes by minimizing the influence of uncontrollable noise factors. An example would be designing a chemical reactor that maintains consistent performance despite slight variations in raw material properties or environmental conditions. Taguchi methods help identify robust settings for the controllable factors, minimizing the impact of noise factors on the process. Taguchi methods are applied to optimize manufacturing processes and minimize variation. In semiconductor manufacturing, engineers may use Taguchi methods to identify settings that minimize defects and improve yield.

STEPS

Here are the steps to develop a Taguchi method designed experiment for chemical engineering:

1. **Define the Problem**:
   - Clearly define the problem or process that you want to optimize. Identify the response variable that reflects the desired outcome.
2. **Select Factors and Levels**:
   - Identify the key factors (variables) that influence the process or formulation. These could be factors related to ingredient proportions, process parameters, or external conditions.
   - Determine the levels or settings at which each factor will be tested. These levels should cover a meaningful range that represents the practical operating conditions.
3. **Determine the Orthogonal Array (OA)**:
   - Choose an appropriate OA based on the number of factors and levels. The OA helps reduce the number of experimental runs required while ensuring that all factors and interactions are adequately studied.
   - Select an OA that suits the specific requirements of your experiment, such as L9, L12, or L16, depending on the complexity of the problem.
4. **Design the Experiments**:
   - Assign factor levels to each row of the OA based on the chosen levels for each factor.
   - Randomize the order of experimental runs to minimize the impact of uncontrolled factors or noise.
5. **Conduct the Experiments**:
   - Perform the experiments according to the design matrix, following the factor-level combinations specified in the OA.
   - Ensure that the experiments are conducted under controlled and consistent conditions.

6. **Collect Data**:
   - Record the data for the response variable for each experimental run. Maintain careful records to minimize measurement errors.
7. **Analyze the Data**:
   - Calculate the S/N ratio for each experimental run. The S/N ratio reflects the robustness or quality of the process or formulation.
   - Analyze the S/N ratios to determine which factor settings or combinations contribute to improved performance and robustness.
   - Identify the optimal factor settings that maximize the S/N ratio or meet the desired performance criteria.
8. **Conduct Confirmation Runs**:
   - Perform confirmation experiments using the optimal factor settings identified in the analysis to validate that the desired improvements have been achieved.
   - Compare the results of confirmation runs with the initial baseline performance to confirm the effectiveness of the optimization.
9. **Implement and Monitor**:
   - Implement the optimized factor settings into the process or formulation as appropriate.
   - Continuously monitor and control the process to ensure that the improvements are sustained over time.
10. **Documentation**:
    - Document all aspects of the Taguchi Method experiment, including the problem definition, factors, levels, experimental design, data collection, analysis, and results.
11. **Report**:
    - Prepare a report summarizing the Taguchi Method experiment, including the objectives, methods, results, and recommendations for implementation.
    - Share the findings and recommendations with relevant stakeholders and decision-makers.

The Taguchi Method is a powerful tool for optimizing processes and formulations while simultaneously enhancing robustness against variations and uncertainties. It systematically guides engineers and researchers in improving product or process quality and reliability in chemical engineering applications.

PITFALLS AND REMEDIES

The Taguchi Method is a robust design approach that aims to optimize processes while considering factors that affect performance and robustness. Like any experimental method, it has potential pitfalls that should be recognized and addressed to ensure successful application in chemical engineering. Here are some common pitfalls and remedies when using the Taguchi Method:

Pitfalls:

Inadequate Factor Selection:
   Pitfall: Selecting factors that do not have a significant impact on the process or outcome.
   Remedies: Conduct a thorough initial analysis to identify the most influential factors. Consult with subject matter experts to ensure all relevant factors are considered.
Improper Orthogonal Array (OA) Selection:
   Pitfall: Choosing an inappropriate OA that does not capture the necessary interactions or factors.
   Remedies: Carefully match the OA to the number of factors and levels required for the experiment. Consider the complexity of the problem and the available resources when selecting the OA.

Ignoring Noise Factors:

  Pitfall: Failing to account for uncontrolled factors (noise factors) that may affect the process but are not part of the experimental design.

  Remedies: Identify and monitor potential noise factors during the experiments. Incorporate noise factors into the analysis if they are found to be significant.

Inadequate Sample Size:

  Pitfall: Using a small sample size that may not provide sufficient statistical power to detect meaningful effects.

  Remedies: Ensure an adequate number of experimental runs to achieve statistically meaningful results. Consider conducting a power analysis to determine the required sample size.

Lack of Randomization:

  Pitfall: Failing to randomize the order of experimental runs can introduce bias.

  Remedies: Randomize the order of experiments to reduce the impact of uncontrolled variables and reduce bias.

Ignoring Nonlinear Effects:

  Pitfall: Assuming that relationships between factors and responses are linear when they may be nonlinear.

  Remedies: Consider conducting additional experiments or using advanced modeling techniques (e.g., RSM) to capture nonlinear effects.

Overinterpreting Results:

  Pitfall: Overinterpreting the results of the Taguchi analysis, especially when dealing with complex systems.

  Remedies: Be cautious in drawing conclusions and make sure the results are practically significant and not merely statistical anomalies.

Lack of Verification:

  Pitfall: Failing to verify the effectiveness of the optimized settings in real-world conditions.

  Remedies: Conduct verification experiments under operational conditions to confirm that the improvements hold.

Pitfalls Related to Factorial Experiments:

Confounding Effects:

  Pitfall: Confounding interactions between factors can make it difficult to separate their individual effects.

  Remedies: Use appropriate experimental design strategies, such as fractional factorial designs, to minimize confounding effects.

Inadequate Control of Experimental Conditions:

  Pitfall: Failing to control experimental conditions precisely, leading to variability in results.

  Remedies: Standardize and control all experimental conditions as rigorously as possible.

Failure to Consider Multiple Responses:

  Pitfall: Focusing solely on a single response variable without considering multiple relevant responses.

  Remedies: Consider all critical response variables that may impact process performance and robustness.

By recognizing these potential pitfalls and implementing the suggested remedies, chemical engineers can improve the effectiveness of their Taguchi method experiments and enhance the robustness and quality of their processes and formulations.

EXAMPLE

Here's an example of a Taguchi method designed experiment in chemical engineering:

Objective: To optimize the formulation of a polymer composite material used in the manufacturing of automotive parts, specifically to maximize tensile strength and minimize warpage during the molding process.

Factors and Levels:

Polymer Type (Factor A):
Level 1: Polypropylene
Level 2: Polyethylene
Level 3: Polystyrene
Reinforcement Type (Factor B):
Level 1: Glass fibers
Level 2: Carbon fibers
Level 3: No reinforcement
Fiber Loading (Factor C):
Level 1: 10% by weight
Level 2: 20% by weight
Level 3: 30% by weight
Molding Temperature (Factor D):
Level 1: 160°C
Level 2: 180°C
Level 3: 200°C
Molding Pressure (Factor E):
Level 1: 1,000 psi
Level 2: 1,500 psi
Level 3: 2,000 psi
Response Variables:
1. **Tensile Strength**: A measure of the material's ability to withstand axial loads without breaking.
2. **Warpage**: A measure of the distortion or deformation of the molded part after cooling.
Experimental Design:
- A Taguchi L27 (3^5) OA is chosen to study the effects of the five factors (A, B, C, D, and E) at three levels each, resulting in 27 experimental runs.
- The OA ensures that the main effects and two-way interactions are adequately studied while minimizing the number of experiments.
Experimental Procedure:
- Prepare polymer composite samples according to the factor-level combinations specified in the OA.
- Perform the molding process for each sample under controlled conditions.
Data Collection:
- Measure the tensile strength and warpage for each molded sample.
- Record the data accurately, ensuring consistent and precise measurements.
Data Analysis:
- Calculate the S/N ratios for both tensile strength and warpage using Taguchi's Loss Function.
- Analyze the S/N ratios to determine the optimal factor-level combinations that maximize tensile strength and minimize warpage.

Results:
- Identify the factors and levels that contribute most to the desired outcomes.
- Determine the optimized settings for polymer type, reinforcement type, fiber loading, molding temperature, and molding pressure that maximize tensile strength and minimize warpage.

Recommendations:
- Based on the analysis, recommend the optimal formulation and processing conditions for the polymer composite material.
- Provide guidance on how to implement these optimized settings in the manufacturing process to produce automotive parts with improved properties.

Verification:
- Conduct verification experiments using the recommended settings to confirm that the improvements in tensile strength and warpage are achieved under actual production conditions.

This example illustrates how the Taguchi Method can be applied in chemical engineering to optimize a complex formulation and manufacturing process by systematically varying multiple factors and levels to achieve desired product properties.

Design Matrix Example

In a Taguchi method designed experiment in chemical engineering, an OA is used to systematically arrange factor-level combinations for experimental runs. Here's an example of a matrix representation for a Taguchi method experiment in chemical engineering, specifically using a Taguchi L9 OA:

Objective: To optimize the formulation of a detergent by varying three factors to maximize cleaning efficiency.

Factors and Levels:

> Factor A: Concentration of Surfactant
>   Level 1: 5%
>   Level 2: 10%
>   Level 3: 15%
> **Factor B**: Temperature (in °C)
>   Level 1: 40°C
>   Level 2: 50°C
>   Level 3: 60°C
> Factor C: Agitation Speed (in RPM)
>   Level 1: 200 RPM
>   Level 2: 300 RPM
>   Level 3: 400 RPM

Orthogonal Array (L9):

- An L9 OA is chosen for this experiment because it allows for nine experimental runs, which is sufficient to study the main effects and two-way interactions for three factors at three levels each.

Matrix Representation (Table 3.15):

Each row in the matrix represents a unique combination of factor levels (Factor A, B, and C) for an experimental run. The Taguchi Method systematically arranges these combinations to ensure that the main effects and certain interactions are studied efficiently while minimizing the number of experiments.

TABLE 3.15
Taguchi Method Designed Experiment Orthogonal Array (L9) Matrix Example for Chemical Engineering

Run	Factor A	Factor B	Factor C
1	1	1	1
2	1	2	2
3	1	3	3
4	2	1	2
5	2	2	3
6	2	3	1
7	3	1	3
8	3	2	1
9	3	3	2

The actual experimental procedure involves conducting each of the nine runs following the specified factor-level combinations. Data on cleaning efficiency will be collected for each run, and the results will be analyzed using the Taguchi Method to determine the optimal settings for the detergent formulation.

REPORTING

The Taguchi Method, a robust design technique, utilizes various statistical tools for the analysis and optimization of experiments in chemical engineering. Some of the common statistical tools used in Taguchi method designed experiments include:

Signal-to-Noise (S/N) Ratios:
S/N ratios are used to quantify the relationship between the factors and the response variable while considering noise factors. Three types of S/N ratios are commonly used:
- Smaller-the-Better (S/N) ratio: Used when the goal is to minimize a response (e.g., minimizing defects).
- Larger-the-Better (S/N) ratio: Used when the goal is to maximize a response (e.g., maximizing strength).
- Nominal-the-Better (S/N) ratio: Used when the goal is to target a specific value within a tolerance range.

Analysis of Variance (ANOVA): ANOVA helps identify the significance of each factor and interaction in the experiment. It partitions the variability in the response variable into contributions from individual factors, interactions, and residual error. It assesses whether the factors have a statistically significant impact on the response.

ANOVA for Confirmation Runs: ANOVA can be performed on the confirmation run data to ensure that the observed improvements are statistically significant and not due to chance.

Main Effects and Interaction Plots: These graphical tools help visualize the main effects of each factor and the interactions between factors on the response variable. Main effects plots display how a factor's levels affect the response individually. Interaction plots show the combined effects of two or more factors on the response.

Orthogonal Arrays (OAs): OAs are used to systematically plan and organize experimental runs, ensuring that all factors and interactions are studied efficiently. Taguchi OAs are designed to minimize the number of experimental runs while capturing essential information.

Desirability Functions: Desirability functions help combine multiple response variables into a single desirability score. They facilitate the optimization of responses with conflicting objectives, allowing for the selection of factor settings that balance various criteria.

Response Surface Methodology (RSM): RSM is often used to model and explore the relationship between factors and responses in greater detail. It can be employed to refine Taguchi method results, especially when nonlinearity is suspected.

Control Charts: Control charts are used to monitor the performance of the process or product before and after implementing the Taguchi Method's optimized settings. They help ensure that improvements are maintained, and the process remains under control.

Probability Plot and Normal Plot: These graphical tools are used to check the normality of data, which is important for some statistical analyses.

Regression Analysis: Regression models may be used to understand the relationships between factors and responses in greater detail.

Confirmation Runs: After identifying the optimal factor settings, confirmation experiments are conducted to validate that the improvements hold in real-world conditions.

Robust Parameter Design (RPD): In more advanced Taguchi applications, RPD is used to simultaneously optimize the mean and variance of a response.

These statistical tools help chemical engineers systematically analyze the effects of factors, identify optimal settings, and improve the robustness and performance of chemical processes or formulations while accounting for variations and uncertainties.

A well-structured report for a Taguchi method designed experiment in chemical engineering should provide a clear and comprehensive overview of the experiment, analysis, results, and recommendations. Here are the key elements that should be included in such a report:

Title and Cover Page: The title should succinctly describe the experiment's objective and key factors. Include the names of the authors, affiliations, and dates.

Table of Contents: Provide a clear list of sections and subsections with page numbers for easy navigation.

Executive Summary: Briefly summarize the experiment's purpose, key findings, and recommendations. Include the optimized factor settings and their impact on the response variables.

Introduction: Explain the background and context of the experiment. Define the problem or objective that the experiment aims to address. Provide a rationale for using the Taguchi Method.

Experimental Design:
- Describe the factors, levels, and response variables used in the experiment.
- Explain the rationale behind the selection of these factors and their levels.
- Specify the OA used and justify its choice.

Experimental Procedure: Detail the steps taken to conduct the experimental runs. Include information on how the factors were manipulated, measurements taken, and any special conditions maintained.

Data Collection: Explain how data was collected, including the instruments and methods used for measurements. Mention any precautions taken to minimize measurement error.

Data Analysis: Present the results of the data analysis, including S/N ratios and other relevant statistics. Use tables, graphs, and charts to illustrate the main effects and interactions of factors on response variables. Include ANOVA tables to demonstrate the significance of each factor and interaction.

Optimization and Recommendations: Clearly state the optimized factor settings that maximize the desired response and/or minimize variability. Discuss the practical implications of these findings. Provide recommendations for implementing the optimized settings in the chemical process or formulation.

Verification of Results: Explain any verification experiments conducted to confirm that the optimized settings hold in real-world conditions. Compare the results of verification runs with the Taguchi analysis.

Discussion:
- Interpret the results and their implications.
- Discuss any limitations or assumptions made during the experiment.
- Address any unexpected findings or challenges encountered.

Conclusions: Summarize the key takeaways from the experiment and analysis. Reiterate the significance of the optimized factor settings.

References: Cite any relevant literature, sources, or prior research that informed your experiment.

Appendices: Include additional information that supports the report, such as detailed experimental data, calculations, or supplementary graphs. Provide the complete OA and any other relevant documentation.

Acknowledgments (optional): Acknowledge individuals, organizations, or funding sources that contributed to the experiment.

Glossary of Terms (optional): Include a list of technical terms and their definitions for readers who may not be familiar with the terminology.

Index (optional): Create an index if the report is extensive and requires easy reference.

Ensure that the report is well-organized, follows a logical flow, and is free from errors in grammar and spelling. It should provide sufficient detail for readers to understand the experiment, its results, and the implications for chemical engineering applications.

3.2.6 CENTRAL COMPOSITE DESIGN (CCD)

CCD is a widely used type of designed experiment in chemical engineering due to its versatility and efficiency. It offers a combination of:

- **Factorial Design**: Explores interactions between multiple factors.
- **Axial Points**: Extend beyond the factorial space to estimate curvature in the response surface.
- **Center Points**: Provide replicates for error estimation and model validation.

This makes CCD ideal for identifying optimal operating conditions. For processes like chemical reactions, where maximizing yield or minimizing energy consumption is critical, CCD offers an understanding of nonlinear relationships between process variables and their impact on response variables. This becomes advantageous when developing accurate response surface models is useful for prediction, optimization, and control of various processes.

A CCD experiment provides data for building a reliable model that predicts the impact of operating conditions on product purity, guiding efficient operation and troubleshooting. This design will efficiently explore both linear and nonlinear relationships. It requires fewer experiments compared to full-factorial designs at similar factor levels. CCD will also provide information about curvature in the response surface, crucial for accurate optimization.

Choosing the appropriate number of factors and levels depends on research goals and complexity is required if not essential. Statistical software tools can aid in design setup, data analysis, and model building. Combining CCD with other techniques like RSM can further enhance process understanding and optimization.

CCD is often used in chemical engineering to optimize reaction conditions. For instance, in biodiesel production, engineers may use CCD to find the ideal combination of temperature, catalyst concentration, and reaction time to maximize biodiesel yield. Examples of CCD applications in chemical engineering include:

Optimizing a Polymer Synthesis Reaction: A CCD experiment helps identify the optimal combination of factors to maximize yield with desired polymer properties.

- **Factors**: Reaction temperature, pressure, catalyst concentration, and monomer ratio.
- **Response Variable**: Polymer yield and molecular weight.

Studying the effect of impurities on a catalytic process: A CCD experiment can reveal how impurities affect the efficiency and selectivity of the catalyst, allowing for optimizing purification strategies.

- **Factors**: Impurity concentration and reaction temperature.
- **Response Variable**: Product yield and selectivity.

Developing a predictive model for a distillation process:

- **Factors**: Reflux ratio, feed temperature, and feed flow rate.
- **Response Variable**: Purity of the distillate.

Sequential Experimentation (Response Surface Methods or Optimization by Stochastic Approximation): Sequential experimentation plays a crucial role in chemical engineering research, allowing scientists and engineers to efficiently navigate complex processes and optimize systems with minimal resource expenditure. Two prevalent sequential methods are:

Response Surface Methodology (RSM)

RSM combines factorial and axial point designs with iterative experimentation. Key aspects include:

- **Initial Screening**: A factorial design identifies significant factors influencing the response variable.
- **Response Surface Development**: Axial points further explore the response surface near identified key factors.
- **Model Building and Optimization**: Statistical models are built based on data, and subsequent experiments refine the model and identify optimal operating conditions.

The advantages of RSM include efficiently navigating complex relationships between factors and responses. It also visualizes response surfaces (3D plots) for a better understanding of process behavior. The RSM provides statistical analysis and validation for robust models and optimization strategies.

Steps

Developing a CCD experiment in chemical engineering involves several steps to systematically plan and execute the experiment. CCD is an RSM that helps optimize processes and formulations while considering both linear and quadratic effects of factors. Here are the steps to develop a CCD experiment:

1. **Define the Objective**: Clearly state the objective of the experiment, such as optimizing a chemical process, formulation, or product characteristic.
2. **Identify Factors and Responses**: Identify the factors (independent variables) that may influence the response variable(s) of interest. Define the response variable(s) that represent the outcome or performance measure to be optimized.
3. **Determine Factor Ranges and Levels**:
   - Specify the minimum and maximum levels for each factor.
   - Determine the number of levels for each factor, which is typically three or five.
   - Choose whether to include a center point (mid-level) for each factor.

4. **Select the Response Surface Model**: Decide on the type of response surface model to be used. Common choices include linear, quadratic, and full cubic models. Determine whether you need to account for interaction terms among factors.

5. **Determine the Number of Experimental Runs**: Calculate the total number of experimental runs required based on the factors and levels chosen. The formula for the number of runs in a CCD experiment is often (2k+2k+c), where k is the number of factors and c is the number of center points.

6. **Design the Experimental Matrix**: Generate the experimental design matrix based on the factors, levels, and number of runs. Use specialized software or statistical tools to create the CCD matrix.

7. **Conduct the Experiments**: Perform the experimental runs according to the factor-level combinations specified in the CCD matrix. Ensure that each run is conducted under controlled and consistent conditions.

8. **Collect Data**: Measure and record the response variable(s) for each experimental run. Maintain accurate and reliable data collection practices.

9. **Analyze the Data**:
   - Fit the response surface model to the collected data using regression analysis.
   - Evaluate the model's coefficients to determine the significance of factors and interactions.
   - Calculate predicted response values for different factor-level combinations.

10. **Optimize the Process**: Identify the factor-level combination that optimizes the response variable(s) based on the model. Consider constraints and practical limitations when determining the optimal settings.

11. **Perform Confirmation Runs**: Conduct confirmation experiments using the recommended optimal settings to verify the predicted improvements in real-world conditions.

12. **Interpret Results and Draw Conclusions**:
    - Interpret the findings from the response surface model and the optimized settings.
    - Discuss the practical implications of the results for the chemical engineering application.

13. **Document the Experiment**: Prepare a comprehensive report that documents the entire CCD experiment, including objectives, factors, responses, experimental design, data analysis, and recommendations.

14. **Implement the Findings**: Implement the optimized settings or recommendations in the chemical engineering process or formulation, as applicable.

15. **Continuous Monitoring**: Continue to monitor and evaluate the process or product performance to ensure that the optimized settings are maintained over time.

By following these steps, chemical engineers can systematically design and execute CCD experiments to optimize processes, formulations, or products while considering both linear and quadratic effects of factors.

Pitfalls and Remedies

CCD experiments in chemical engineering, like any other experimental approach, can face certain pitfalls and challenges. Here are some common pitfalls and remedies associated with CCD experiments:

Pitfall 1: Poor Factor Selection

Pitfall: Selecting the wrong factors or failing to include important factors in the experiment can lead to suboptimal results.

Remedy: Conduct a thorough preliminary study to identify critical factors and interactions. Consult with subject matter experts to ensure a comprehensive selection of factors.

Pitfall 2: Inadequate Factor Range

Pitfall: Choosing factor ranges that are too narrow may limit the ability to identify optimal settings.

Remedy: Define factor ranges that encompass the expected operating conditions and potential variations. Conduct sensitivity analyses to assess the impact of factor range choices.

Pitfall 3: Overfitting

Pitfall: Fitting a complex response surface model with too many terms can lead to overfitting, resulting in a model that does not generalize well to new data.

Remedy: Use statistical criteria, such as the lack-of-fit test and adjusted R-squared values, to assess model adequacy. Simplify the model by removing nonsignificant terms to improve its predictive power.

Pitfall 4: Neglecting Model Assumptions

Pitfall: Failing to check model assumptions (e.g., normality and homoscedasticity) can lead to inaccurate model results.

Remedy: Perform residual analysis and diagnostic tests to ensure that model assumptions are met. Transform the data or consider alternative models if necessary.

Pitfall 5: Ignoring Process Constraints

Pitfall: Obtaining an optimal solution that cannot be practically implemented due to process constraints or limitations.

Remedy: Consider process constraints during the optimization phase. Use constrained optimization techniques to find solutions that are feasible within real-world limitations.

Pitfall 6: Neglecting Model Validation

Pitfall: Failing to validate the response surface model with confirmation experiments can result in unreliable predictions.

Remedy: Conduct confirmation runs using the recommended optimal settings to validate the model's predictions under actual operating conditions.

Pitfall 7: Lack of Replication

Pitfall: Conducting experiments without replication can lead to uncertainty in the estimated effects and variability.

Remedy: Incorporate replication into the experimental design to improve the precision of estimates. Use replication to assess the variability and reliability of results.

Pitfall 8: Disregarding Nonlinear Effects

Pitfall: Assuming that all effects are linear when nonlinear effects exist can result in suboptimal solutions.

Remedy: Include quadratic terms in the response surface model to account for nonlinear effects. Use diagnostic plots to identify curvature in the response.

Pitfall 9: Inadequate Sample Size

Pitfall: Using a small sample size may limit the statistical power of the experiment and lead to unreliable results.

Remedy: Conduct a power analysis to determine an appropriate sample size that ensures the experiment has adequate statistical power to detect significant effects.

Pitfall 10: Neglecting Practical Experience

Pitfall: Relying solely on statistical results without considering practical experience and domain knowledge can lead to impractical or suboptimal recommendations.

Remedy: Collaborate closely with domain experts and incorporate their insights into the experimental design and interpretation of results.

Pitfall 11: Not Addressing Time-Dependent Factors

Pitfall: Neglecting to account for time-dependent factors in dynamic processes.

Remedy: Incorporate time-dependent factors and dynamic modeling approaches if the process exhibits temporal behavior.

By being aware of these potential pitfalls and implementing the suggested remedies, chemical engineers can improve the effectiveness and reliability of their CCD experiments in optimizing processes and formulations.

EXAMPLE

Here's an example of a CCD experiment in chemical engineering:

Objective: To optimize the conditions for a catalytic reaction that produces a chemical compound with a high yield.

Factors and Responses:

Factor A: Temperature (in °C)
- Range: 80°C to 140°C
- Levels: −1 (low), 0 (central), +1 (high)

Factor B: Catalyst Concentration (in moles/liter)
- Range: 0.1 to 0.5 moles/liter
- Levels: −1 (low), 0 (central), +1 (high)

Factor C: Reaction Time (in hours)
- Range: 2 to 8 hours
- Levels: −1 (low), 0 (central), +1 (high)

Response Y1: Yield of the Desired Compound (in %)
Response Y2: Selectivity of the Desired Compound (in %)

Design Matrix (Central Composite Design):

The CCD design includes a set of axial runs, center points, and factorial points, allowing for the exploration of quadratic and linear effects of factors.

A partial representation of the design matrix is shown in Table 3.16.

TABLE 3.16

Central Composite Design (CCD) Designed Experiment Matrix Example

Run	Factor A	Factor B	Factor C
1	−1	−1	0
2	+1	−1	0
3	−1	+1	0
4	+1	+1	0
5	0	0	−1
6	0	0	+1
7	−1.68	0	0
8	+1.68	0	0
9	0	−1.68	0
10	0	+1.68	0
11	0	0	−1.68
12	0	0	+1.68

Experimental Procedure:

1. Prepare reaction mixtures with different combinations of temperature, catalyst concentration, and reaction time based on the design matrix.
2. Perform the catalytic reaction for each experimental run.
3. Measure the yield of the desired compound and the selectivity of the reaction for each run.
4. Record the data accurately.

Data Analysis:

1. Fit a response surface model to the experimental data, considering linear, quadratic, and interaction effects of factors.
2. Use the model to predict the optimal conditions that maximize the yield of the desired compound while achieving high selectivity.
3. Perform statistical analyses, including ANOVA, to determine the significance of factors and interactions.

Optimization: Identify the factor settings that yield the highest predicted yield of the desired compound while maintaining acceptable selectivity.

Confirmation Runs: Conduct confirmation experiments under the predicted optimal conditions to verify the model's predictions.

Interpretation: Interpret the results to understand the impact of temperature, catalyst concentration, and reaction time on the catalytic reaction's performance. Provide recommendations for process optimization and further studies.

This example illustrates how a CCD can be used in chemical engineering to systematically explore and optimize the conditions of a catalytic reaction. The design allows for the efficient investigation of factors and their interactions to achieve the desired outcomes.

REPORT

CCD experiments in chemical engineering often employ various statistical tools and techniques for data analysis and optimization. Here are some of the statistical tools commonly used in CCD experiments:

Response Surface Methodology (RSM): RSM is a comprehensive approach used to model and optimize complex processes. It involves fitting response surface models to experimental data to identify optimal factor settings.

Design of Experiments (DOE): CCD is a type of designed experiment within the broader field of DOE. It involves carefully planning and executing experiments to systematically investigate factors and their interactions.

Factorial Design: Factorial designs are used in conjunction with CCD to estimate the linear and interaction effects of factors. These designs help identify which factors are significant.

Analysis of Variance (ANOVA): ANOVA is used to assess the statistical significance of factors and their interactions. It partitions the variability in response data to determine which factors have a significant effect on the response.

Regression Analysis: Regression analysis is employed to build response surface models that describe the relationship between factors and responses. Linear, quadratic, and interaction terms are commonly included in the models.

Model Adequacy Checking: Diagnostic tools, such as residual analysis and normal probability plots, are used to assess the adequacy of the response surface models. Deviations from model assumptions can be detected and addressed.

Desirability Function: The desirability function combines multiple responses into a single index that represents the overall desirability of a factor setting. It helps identify the optimal conditions that simultaneously optimize multiple responses.

Contour Plots and Response Surface Plots: These graphical tools are used to visualize the response surface and contour lines, making it easier to identify regions of optimal factor settings.

Optimization Algorithms: Optimization algorithms, such as gradient-based methods and genetic algorithms, are used to find the factor settings that maximize or minimize the response(s) based on the fitted response surface model.

Steepest Ascent/Descent: These are stepwise procedures used to move toward the optimal factor settings by adjusting one factor at a time.

ANOVA-PCA (Principal Component Analysis): PCA can be used to reduce the dimensionality of the response data while preserving most of the variability. It helps identify the most influential factors.

Lack-of-Fit Test: This test assesses whether the response surface model adequately represents the data. A significant lack-of-fit suggests that the model may need refinement.

Confidence Intervals: Confidence intervals can be calculated for model parameters, allowing for an assessment of their precision and uncertainty.

Model Validation: Separate validation experiments or runs can be conducted to confirm the accuracy and predictive power of the response surface model.

Sensitivity Analysis: Sensitivity analysis assesses how changes in factor settings impact the response. It helps identify the robustness of the optimal conditions.

Multiple Comparisons: When multiple factor settings are compared, appropriate methods, such as Tukey's HSD test, can be used to identify significant differences.

Monte Carlo Simulation: Monte Carlo simulations can be used to assess the impact of variability and uncertainty in factors on the response and optimization results.

These statistical tools are applied in a systematic manner to plan, execute, analyze, and optimize CCD experiments in chemical engineering, ultimately helping researchers and engineers improve processes, formulations, and product performance.

A report for a CCD experiment in chemical engineering typically includes several key elements to document the experimental process, results, and conclusions effectively. Here are the essential elements of such a report:

Title: A concise and informative title that reflects the objective of the experiment.

Abstract: A brief summary of the experiment, including the objective, methodology, key findings, and conclusions.

Introduction: An introduction that provides context for the experiment. This section may include:
- Background information on the process, formulation, or problem being studied.
- A clear statement of the research objective or problem being addressed.
- The significance and potential impact of the research.

Experimental Design: Detailed information about the experimental design, including:
- Factors and their levels.
- The CCD matrix, including the coded levels and actual values of factors.
- The response variables or performance measures.
- Any control variables or constants.
- Replication and randomization strategies.

Materials and Methods: A description of the materials, equipment, and methods used in the experiment, including:

- Procedures for preparing samples and conducting experiments.
- Data collection procedures, including the instruments used.
- Any special conditions or precautions taken during the experiment.

Results: Presentation of the experimental results, typically including:

- Tables or figures showing the experimental data for each run, including factor settings and response values.
- Descriptive statistics (e.g., means, standard deviations) for each response variable.
- Graphical representations of the response surface, such as contour plots or 3D surface plots.
- Statistical analyses, including ANOVA results, p-values, and significance levels.
- Confirmation run data and their comparison with model predictions.

Discussion: An in-depth discussion of the results and their implications, including:

- Interpretation of the response surface model, focusing on significant factors and interactions.
- Analysis of factors that had the most significant impact on the response.
- Assessment of the model's predictive ability and its adequacy for the given problem.
- Practical recommendations based on the findings and optimal factor settings.

Conclusion: A concise summary of the key findings and their relevance to the research objective.

Recommendations: Specific recommendations for process optimization, product formulation, or further research.

Limitations: A discussion of any limitations or constraints encountered during the experiment and their potential impact on the results.

References: Citations of relevant literature, sources of information, and references to statistical methods and software used.

Appendices: Additional information that supports the main body of the report, including:

- Detailed experimental data for all runs.
- Supplementary figures or tables.
- Any mathematical derivations or calculations.
- Copies of statistical software output.
- Supplementary materials, such as raw data, calculations, or additional figures and tables.

Acknowledgments (optional): Acknowledgment of individuals, organizations, or funding sources that contributed to the research.

The structure and content of the report may vary depending on the specific requirements of the experiment, the organization's guidelines, and the intended audience. However, a well-structured and informative report should effectively communicate the methodology, results, and conclusions of the CCD experiment in chemical engineering.

3.2.7 Optimization by Stochastic Approximation (SA)

SA iteratively searches for optimal conditions by adjusting based on feedback from the system. Key aspects include:

- **Starting Point**: Choose an initial operating condition.
- **Perturbation and Evaluation**: Modify one or more factors and measure the response.

- **Step Direction**: Update the operating condition based on whether the response improved or worsened.
- **Iteration**: Repeat the perturbation-evaluation-step cycle until no further improvement is found.

The advantages of SA include an adaptability to complex and unknown response surfaces without requiring prior knowledge. The method requires fewer experiments compared to RSM in some cases as well as being useful for online optimization of dynamic processes. An example of SA in chemical engineering:

- Real-time control of polymerization reactors for consistent product quality.
- Adaptive optimization of fermentation processes for maximizing bioproduct yield.
- Fine-tuning the operating parameters of chemical reactors for energy efficiency.

The optimal sequential experimentation method depends on the following factors:

- **Prior Knowledge of the Process**: RSM can be more efficient with some understanding of key factors, while SA excels when little is known.
- **Complexity of the Response Surface**: If nonlinearities are suspected, RSM with its response surface visualization might be preferable.
- **Resource Constraints**: SA can be efficient with fewer experiments, while RSM often requires more data for accurate model building.

For complex systems, combining RSM and SA can be advantageous. RSM can provide an initial understanding and a starting point for SA, followed by fine-tuning and online control with SA.

STEPS

OSA is a technique used to optimize complex systems through iterative experiments. Below are the general steps to develop an OSA designed experiment for chemical engineering:

1. **Define the Optimization Problem**: Clearly state the optimization objective, such as maximizing yield, minimizing cost, or achieving specific product properties. Specify the response variable(s) to optimize.
2. **Identify Factors and Their Range**: Identify the factors that influence the response variable(s). These may include process variables, operating conditions, or ingredient concentrations. Define the feasible ranges and levels of these factors based on prior knowledge or experimentation.
3. **Select an Initial Factor Setting**: Choose an initial set of factor values to start the optimization process. These can be based on educated guesses or existing operating conditions.
4. **Specify the Stochastic Approximation Algorithm**: Select a suitable SA algorithm, such as the Robbins-Monro algorithm, for optimizing the system. Determine algorithm-specific parameters, such as step size and convergence criteria.
5. **Set Up the Experimental Design**: Determine the DOE strategy for conducting experiments. OSA typically involves sequential experimentation. Decide the number of iterations and experiments to perform in each iteration.
6. **Perform Initial Experiments**: Conduct the initial set of experiments based on the chosen initial factor settings. Record the response variable(s) and factor settings for each run.

7. **Update Factor Settings**: Use the results from the initial experiments and the SA algorithm to update the factor settings for the next iteration. Apply the algorithm to adjust factor levels to potentially improve the response.
8. **Iterative Process**: Repeat the process iteratively, performing additional experiments and updating factor settings in each iteration. Continuously evaluate the response variable(s) to assess improvement.
9. **Convergence Criteria**: Define convergence criteria to determine when to stop the optimization process. This could be based on achieving a target response value, reaching a specified number of iterations, or satisfying certain conditions.
10. **Final Experiment and Validation**: Once the optimization process converges, conduct a final experiment using the optimized factor settings to validate the results. Record the response variable(s) for the validation run.
11. **Data Analysis**: Analyze the experimental data, including response values and factor settings, to determine the optimal conditions and their impact on the system's performance.
12. **Documentation and Reporting**: Document the entire optimization process, including factor settings, response values, and details of each iteration. Prepare a comprehensive report summarizing the optimization results, the final optimized factor settings, and any conclusions or recommendations.
13. **Implementation**: Implement the optimized factor settings in the chemical engineering process or system to achieve the desired objectives.
14. **Continuous Monitoring and Adjustment**: Continuously monitor the system's performance and, if necessary, adjust factor settings to maintain the optimized conditions.
15. **Review and Refinement**: Periodically review the optimization results and refine the process further if new factors or constraints arise or if there are changes in the objectives.
16. **Documentation and Knowledge Transfer**: Ensure that the optimization process and findings are well-documented and transferred to relevant personnel within the organization for future reference.

The OSA designed experiment in chemical engineering involves an iterative and adaptive approach to optimization, allowing the system to gradually converge toward the optimal conditions based on the observed responses from each experiment. It is a valuable technique for improving processes and achieving desired performance outcomes.

Pitfalls and Remedies

OSA designed experiments in chemical engineering can be a powerful tool for finding optimal conditions in complex systems. However, there are potential pitfalls that should be recognized and remedies to mitigate these challenges. Here are some common pitfalls and remedies:

Pitfall 1: Poor Initial Factor Settings
> **Remedy**: Conduct a thorough preliminary study or sensitivity analysis to choose reasonable initial factor settings. The initial guess should not be too far from the expected optimum.

Pitfall 2: Insufficient Iterations
> **Remedy**: Ensure an adequate number of iterations to allow the algorithm to converge. Monitor the convergence behavior and extend iterations if necessary.

Pitfall 3: Inaccurate Model Assumptions
> **Remedy**: Continuously validate the model assumptions. Perform diagnostic checks for model adequacy and consider alternative models if the assumptions are violated.

Pitfall 4: Excessive Computational Time

    Remedy: Optimize the algorithm's efficiency by choosing appropriate step sizes, optimization methods, and stopping criteria. Parallel computing or distributed computing resources can also speed up computations.

Pitfall 5: Overfitting

    Remedy: Regularize the optimization model to prevent overfitting. Use techniques like cross-validation to assess the model's predictive ability and prevent excessive complexity.

Pitfall 6: Neglecting Constraint Handling

    Remedy: If the optimization problem involves constraints (e.g., process limits, safety constraints), implement constraint-handling techniques, such as penalty functions or barrier methods, to ensure that feasible solutions are obtained.

Pitfall 7: Overlooking Multimodal Optimization

    Remedy: Consider the possibility of multiple local optima. Implement strategies like random starts or population-based algorithms to explore different regions of the solution space.

Pitfall 8: Poor Sensitivity Analysis

    Remedy: Perform sensitivity analysis to understand how changes in factor settings affect the response. This helps identify which factors have the most significant impact and can guide the optimization process.

Pitfall 9: Neglecting Real-world Constraints

    Remedy: Incorporate practical constraints that may exist in the chemical engineering process. Constraints can significantly impact the feasibility of optimal solutions.

Pitfall 10: Lack of Validation

    Remedy: Validate the final optimized conditions through experimental verification or further testing to ensure that the predicted optimal settings are achievable in practice.

Pitfall 11: Inadequate Documentation

    Remedy: Maintain thorough documentation of the optimization process, including factor settings, response values, and the reasoning behind decisions. This documentation is essential for future reference and auditing.

Pitfall 12: Inadequate Communication

    Remedy: Ensure effective communication between team members, stakeholders, and experts involved in the optimization process. Share insights, challenges, and results to benefit from collective expertise.

Pitfall 13: Premature Convergence

    Remedy: Avoid terminating the optimization prematurely. Monitor convergence closely and verify that the process has genuinely reached an optimal solution.

Pitfall 14: Lack of Expertise

    Remedy: Seek advice and collaboration with experts in optimization, statistics, and chemical engineering if the team lacks the necessary expertise.

Successful optimization by SA in chemical engineering requires a combination of sound experimental design, data analysis, modeling, and domain knowledge. Addressing these potential pitfalls with appropriate remedies can lead to more effective and reliable optimization outcomes.

EXAMPLE

The following is an example of OSA designed experiment in the context of chemical engineering:

Objective: The objective is to optimize the yield of a chemical reaction in a batch reactor by adjusting various process parameters.

Factors:

- **Temperature** (°C): The temperature at which the reaction takes place.
- **Concentration of Reactant A** (mol/L): The initial concentration of one of the reactants.
- **Stirring Speed** (RPM): The rate at which the reactants are mixed.

Response Variable: **Yield of Product** (%): The percentage of the desired product obtained after the reaction.

Experimental Design:

Initial Settings:
- Temperature: 50°C
- Concentration of Reactant A: 2.0 mol/L
- Stirring Speed: 300 rpm

Stochastic Approximation Algorithm:
- Implement the Robbins-Monro algorithm, a common SA method, to optimize the yield.

Iterative Process:

Iteration 1:
- Run 1: Temperature = 50°C, Concentration of Reactant A = 2.0 mol/L, Stirring Speed = 300 rpm, Yield = 85%

Use the yield data from Run 1 to update factor settings (e.g., increase temperature, decrease reactant concentration).

Iteration 2:
- Run 2: Temperature = 55°C (adjusted), Concentration of Reactant A = 1.8 mol/L (adjusted), Stirring Speed = 310 rpm (adjusted), Yield = 88%

Continue to update factor settings based on the yield data from Run 2.

Iteration 3:
- Run 3: Temperature = 57°C (adjusted), Concentration of Reactant A = 1.7 mol/L (adjusted), Stirring Speed = 315 rpm (adjusted), Yield = 90%

Continue the iterations until a convergence criterion is met (e.g., a target yield is achieved, or the algorithm converges).

Data Analysis: Analyze the experimental data, including yield values and factor settings, to determine the optimal conditions that maximize the yield of the desired product.

Convergence Criterion: Conclude the optimization when the yield reaches the desired target (e.g., 95%) or when the algorithm converges with consistent factor settings.

The OSA designed experiment successfully identified the optimal conditions for the chemical reaction, which resulted in a yield of 90%. These conditions can be implemented in the batch reactor to achieve higher production efficiency.

This example illustrates how OSA can be used in chemical engineering to iteratively adjust process parameters to optimize a specific objective, such as yield while considering factors like temperature, concentration, and stirring speed. The SA algorithm helps find the optimal factor settings by learning from previous experimental runs.

MATRIX EXAMPLE

In an OSA designed experiment for chemical engineering, a matrix is not typically used in the traditional sense of a design matrix, as you would find in a factorial or RSM experiment.

OSA is an iterative optimization process that adjusts factor settings sequentially based on the results of previous experiments. However, a tabular format is often used to record and track the factor settings and responses at each iteration. An example of what such a table is shown in Table 3.17.

In this tabular format, each row represents a run of the experiment in a particular iteration. The columns represent the factors (temperature, concentration of Reactant A, stirring speed) and the response variable (yield of product). The factor settings are adjusted in each iteration based on the results from the previous runs.

The matrix-like structure helps visualize the progress of the OSA designed experiment and provides a clear record of the factor settings and corresponding responses at each step. The process continues iteratively until a convergence criterion is met or the desired optimization objective is achieved.

REPORTING

OSA in chemical engineering is primarily an iterative optimization process that doesn't rely on traditional statistical tools such as ANOVA or regression models. Instead, OSA employs SA algorithms to iteratively adjust factor settings based on observed responses. However, there are statistical and computational techniques used in conjunction with OSA to aid the optimization process. Here are some relevant tools and techniques:

Stochastic Approximation Algorithms: OSA relies on iterative algorithms that adjust factor settings based on observed responses. The Robbins-Monro algorithm is a commonly used SA method. Other variations include the simultaneous perturbation stochastic approximation (SPSA) algorithm and the cross-entropy method.

Response Surface Methodology (RSM): In some cases, OSA can be combined with RSM, where a response surface model is used to approximate the relationship between factors and responses. This can help guide the initial iterations of OSA, especially when the optimization space is complex.

Convergence Analysis: Statistical techniques may be employed to assess the convergence of the optimization algorithm. This involves monitoring the progress of the optimization process over iterations and checking for stability or convergence toward an optimal solution.

Sensitivity Analysis: Sensitivity analysis helps identify which factors have the most significant impact on the response variable. This information can guide the adjustment of factor settings in each iteration of OSA.

TABLE 3.17
OSA Designed Experiment

Run	Temperature (°C)	Concentration of Reactant A (mol/L)	Stirring Speed (rpm)	Yield of Product (%)
		Iteration 1:		
1	50	2.0	300	85
2	55	1.8	310	88
		Iteration 2:		
3	57	1.7	315	90
...	...	...	...	...

Statistical Software: Statistical software packages like R, Python with libraries such as NumPy and SciPy, or specialized optimization software can be used to implement and automate the OSA algorithm, record results, and perform data analysis.

Optimization Metrics: Statistical techniques can be applied to assess the quality of the optimization results, such as calculating confidence intervals for the optimal factor settings or evaluating the robustness of the optimal solution.

Randomization and Replication: OSA experiments may involve randomization and replication to reduce the impact of noise and variability in the observed responses.

Constraint Handling: If optimization problems involve constraints (e.g., limits on factor settings or system limitations), techniques for constraint handling, such as penalty functions or barrier methods, may be used.

Bayesian Optimization: In some cases, Bayesian optimization techniques can be combined with OSA to explore and exploit the optimization space efficiently, especially when the objective function is expensive to evaluate.

Parallel and Distributed Computing: High-performance computing techniques may be employed to speed up the OSA process, allowing multiple experiments to be run in parallel.

It's important to note that the choice of statistical tools and techniques in an OSA designed experiment depends on the specific optimization problem, the complexity of the system, and the available resources. The primary focus of OSA is on the iterative adjustment of factor settings based on observed responses, making it a dynamic and adaptive optimization approach.

A report for an OSA designed experiment in chemical engineering should provide a comprehensive and well-organized summary of the optimization process, including the objectives, methods, results, and conclusions. Here are the key elements that should be included in such a report:

Title Page:
- Title of the report.
- Names of authors and affiliations.
- Date of the report.

Abstract: A brief summary of the report, including the optimization objectives, methods used, and key findings.

Table of Contents: An organized list of sections and subsections in the report.

List of Figures and Tables: A list of all figures and tables in the report with their respective page numbers.

Introduction: A clear statement of the optimization problem and objectives. Explanation of why the optimization is important and the potential benefits. Briefly introduce the OSA method as the chosen approach for optimization.

Literature Review (optional): A review of relevant literature, including previous optimization attempts, related research, and theoretical background.

Methodology:
- Detailed explanation of the OSA method employed, including the SA algorithm (e.g., Robbins-Monro, SPSA).
- Description of the experimental design, including the factors, response variable, and initial settings.
- Any additional statistical or computational techniques used in conjunction with OSA (e.g., RSM, sensitivity analysis).

Experimental Setup: Information about the experimental setup, including equipment used, data collection procedures, and any special considerations. Explanation of how randomization and replication were handled (if applicable).

Results:
- Presentation of the results obtained throughout the OSA process, organized by iterations.
- Tabulated data showing factor settings, observed responses, and any other relevant information.
- Visual aids, such as graphs or plots, to illustrate the optimization progress and convergence.
- Metrics used to assess the quality of optimization (e.g., convergence criteria, confidence intervals).

Discussion:
- Interpretation of the results, including trends and patterns observed during the optimization process.
- Analysis of the sensitivity of factors and their impact on the response variable.
- Explanation of any challenges or unexpected findings encountered.

Conclusion:
- A concise summary of the main findings and the achievement of the optimization objectives.
- Statement of whether the optimization was successful and to what extent.
- Any recommendations for further improvement or future work.

References: A list of all sources, papers, books, or articles cited in the report.

Appendices (if necessary): Supplementary information, such as detailed experimental data, calculations, or additional figures and tables.

Acknowledgments (optional): Acknowledgment of individuals, organizations, or funding sources that contributed to the project.

Glossary of Terms (optional): Definitions and explanations of technical terms used in the report.

Index (optional): An index of key terms, concepts, and sections to facilitate navigation within the report.

The report should be well-structured, logically organized, and written in a clear and concise manner. It should provide sufficient information for readers to understand the optimization process, its outcomes, and the implications for chemical engineering applications. Additionally, any mathematical or statistical notations should be defined and explained as needed to make the report accessible to a broad audience.

3.2.8 ROBUST PARAMETER DESIGN (RPD)

RPD is a structured approach to experimentally optimize and control processes, products, or systems to ensure robust performance under variations and uncertainties. It is commonly used in chemical engineering, to improve the quality and reliability of products or processes.

STEPS

Developing an RPD experiment in chemical engineering involves a systematic process to optimize processes or products while minimizing the impact of variation and uncertainties. Here are the steps to develop an RPD experiment:

1. **Define the Problem and Objectives:** Clearly define the problem you want to address and the specific objectives of the RPD experiment. What aspect of the process or product do you want to optimize? What response variable represents the desired outcome?

2. **Identify Controllable and Noise Factors**:
   - Identify the controllable factors (design variables) that you can adjust to influence the response variable. These are the parameters you will optimize.
   - Identify the noise factors (uncontrollable sources of variation) that may affect the response. These factors introduce variability and uncertainty.
3. **Select Factor Levels and Ranges**: Determine the range of values for each controllable factor that will be explored during the experiment. This involves specifying the lower and upper bounds or discrete levels for each factor. Consider the practical and physical constraints of the process or system when defining factor levels.
4. **Experimental Design**: Choose an appropriate experimental design, such as a factorial design or RSM, to systematically vary the controllable factors at different levels while considering the noise factors. Determine the number of experimental runs required based on the chosen design and the number of controllable and noise factors.
5. **Data Collection**: Conduct the designed experiments, collecting data on the response variable for each experimental run. Ensure that the experiments are conducted under consistent conditions to minimize sources of variation not related to the factors being studied.
6. **Statistical Analysis**: Analyze the experimental data using statistical techniques to model the relationship between controllable factors and the response variable.
   - Identify significant factors, quantify their effects, and assess interactions between factors.
   - Perform robustness analysis to evaluate the sensitivity of the optimal settings to variations in noise factors.
7. **Optimization**: Using the results of the statistical analysis, determine the optimal factor settings that maximize the desired response while minimizing sensitivity to noise factors. Balancing mean performance and variation reduction is a key aspect of optimization in RPD.
8. **Validation and Implementation**: Validate the optimized factor settings through additional testing or simulations to ensure that the desired improvements are achievable and robust. Implement the optimized factor settings in the actual production process or system to achieve consistent and reliable performance.
9. **Monitoring and Control**: Establish a system for monitoring and controlling the process or product to maintain the benefits of the RPD optimization over time. Continuously assess and adjust factor settings as needed to adapt to changes in conditions or requirements.
10. **Documentation and Reporting**: Document all aspects of the RPD experiment, including the problem statement, experimental design, data collection, analysis methods, optimization results, and any changes made to the process or product. Prepare a comprehensive report summarizing the entire RPD process, including findings, recommendations, and lessons learned.
11. **Continuous Improvement**: RPD is often an iterative process. Continue to refine and improve the process or product based on ongoing monitoring and feedback.

By following these steps, you can systematically develop and implement an RPD experiment in chemical engineering to optimize processes, products, or systems while ensuring robust performance in the face of variation and uncertainties.

PITFALLS AND REMEDIES

Pitfalls and remedies in RPD) experiments in chemical engineering include:

Pitfall 1: Insufficient Understanding of the Problem
   Pitfall: Inadequate comprehension of the problem and its intricacies may lead to suboptimal factor selection and experimental design.
   Remedy: Ensure a clear and comprehensive understanding of the problem by collaborating with subject matter experts and conducting thorough background research.

Pitfall 2: Inaccurate Factor Selection

Pitfall: Selecting the wrong factors or neglecting critical ones can hinder the effectiveness of the RPD experiment.

Remedy: Conduct a thorough screening of potential factors, consider expert input, and perform sensitivity analyses to identify the most influential factors.

Pitfall 3: Inadequate Range Definition

Pitfall: Defining factor ranges that are too narrow may miss optimal settings, while overly wide ranges can lead to resource-intensive experiments.

Remedy: Perform preliminary studies or use historical data to estimate suitable factor ranges that encompass both typical and extreme conditions.

Pitfall 4: Poor Experimental Design

Pitfall: Choosing an inappropriate experimental design can result in insufficient information or inefficient data collection.

Remedy: Consult with statisticians or design experts to select an optimal experimental design that balances the number of runs, resources, and factor combinations.

Pitfall 5: Ignoring Noise Factors

Pitfall: Neglecting to account for noise factors can compromise the robustness of the design.

Remedy: Identify and characterize noise factors using techniques like designed experiments or data analysis to understand their impact on the response.

Pitfall 6: Overfitting Models

Pitfall: Overly complex models can lead to poor generalization and uninterpretable results.

Remedy: Use techniques such as cross-validation to assess model performance and select models that balance goodness of fit with simplicity.

Pitfall 7: Ignoring Model Assumptions

Pitfall: Failing to validate model assumptions can result in unreliable predictions.

Remedy: Validate model assumptions through residual analysis, normal probability plots, and other diagnostic tools.

Pitfall 8: Premature Optimization

Pitfall: Rushing to implement optimal settings without robustness analysis can lead to poor real-world performance.

Remedy: Evaluate the robustness of optimal settings by conducting sensitivity analyses, Monte Carlo simulations, or other relevant techniques.

Pitfall 9: Lack of Documentation

Pitfall: Inadequate documentation can hinder reproducibility and future improvements.

Remedy: Maintain thorough records of experimental details, model development, optimization results, and any changes made during the process.

Pitfall 10: Neglecting Implementation and Monitoring

Pitfall: Failing to implement and monitor optimized settings in real-world applications can lead to lost benefits.

Remedy: Establish a system for ongoing monitoring and control to ensure that optimized settings are maintained over time.

Pitfall 11: Lack of Continuous Improvement

Pitfall: Neglecting to iterate and refine the RPD process can result in missed opportunities for improvement.

Remedy: Foster a culture of continuous improvement, where the RPD process is revisited and refined based on ongoing feedback and new insights.

By being aware of these pitfalls and implementing the suggested remedies, you can enhance the effectiveness of RPD experiments in chemical engineering, ultimately leading to improved processes, products, and reliability.

EXAMPLE

An example of an RPD experiment in chemical engineering involves optimizing the production process of a chemical product while ensuring its robustness to variations and uncertainties. Let's consider a specific scenario:

Problem: A chemical manufacturing company produces a polymer used in various industrial applications. The polymer's molecular weight is a critical quality attribute, and the company aims to optimize the production process to achieve the desired molecular weight while minimizing sensitivity to variations in raw material properties and environmental conditions.

Objectives:

1. Optimize the production process to achieve the target molecular weight.
2. Ensure that the process is robust and capable of maintaining molecular weight within acceptable limits under varying conditions.

RPD Experiment:

Factor Identification:
Controllable Factors:
- Temperature of the polymerization reactor
- Concentrations of reactants
- Reaction time

Noise Factors:
- Humidity levels in the production facility
- Variations in the properties of raw materials

Factor Range Definition: Conduct preliminary experiments and historical data analysis to determine suitable ranges for each controllable factor. For example, temperature could range from 70°C to 90°C.

Experimental Design: Use an RSM experimental design to systematically vary the controllable factors at different levels. Plan experiments that explore a range of factor combinations within the defined ranges.

Data Collection: Conduct experiments, measuring the molecular weight of the polymer for each run. Collect data on humidity levels and variations in raw material properties during the experiments.

Statistical Analysis: Develop a response surface model to describe the relationship between controllable factors (temperature, concentrations, time) and the molecular weight of the polymer. Identify significant factors and interactions using statistical techniques.

Optimization: Use the response surface model to find the optimal factor settings that maximize the desired molecular weight while minimizing sensitivity to noise factors (humidity, raw material variations).

Robustness Analysis: Perform robustness analysis to evaluate how well the optimized settings perform under variations in humidity levels and raw material properties. Conduct sensitivity analyses to identify factors or conditions that have the most significant impact on molecular weight robustness.

Validation and Implementation: Validate the optimized settings through additional experiments or simulations. Implement the optimized process conditions in the production facility.

Monitoring and Control: Establish a monitoring system to continuously track and control temperature, reactant concentrations, and other critical factors in the production process. Make adjustments as needed to maintain the robustness of the process.

Continuous Improvement: Regularly review and refine the RPD experiment, and process based on new data, feedback, and changing conditions to further enhance robustness and performance.

In this example, the RPD experiment optimizes the polymer production process to achieve the desired molecular weight while considering the impact of variations in both controllable and uncontrollable factors. This ensures that the process remains robust and capable of producing high-quality polymer consistently, even in the face of real-world variations and uncertainties.

Matrix Example

In an RPD experiment in chemical engineering, a matrix is used to plan and document the experimental runs. Each row in the matrix represents a specific experimental run, while the columns represent the factors being studied, their levels, and the response variable. Here's an example of a matrix for an RPD experiment:

Example RPD Matrix for Polymer Production:
In this hypothetical scenario, we are optimizing the production of a polymer, focusing on factors like temperature, reactant concentrations, and reaction time. We want to ensure that the polymer's molecular weight remains within the desired range while minimizing sensitivity to variations in humidity levels and raw material properties (Table 3.18).

- **Run**: An identifier for each experimental run.
- **Temperature** (°C): The temperature set for the polymerization reactor.
- **Concentration A** (%) and **Concentration B** (%): Concentrations of two reactants used in the process.
- **Reaction Time (min)**: The duration of the polymerization reaction.
- **Humidity Level** (%): The humidity level in the production facility during the experiment.
- **Raw Material Variations**: A qualitative description of variations in raw material properties (e.g., low, medium, high).
- **Molecular Weight (g/mol)**: The response variable representing the molecular weight of the produced polymer, measured for each experimental run.

In this matrix, different combinations of factor levels for temperature, concentrations, reaction time, humidity, and raw material variations are systematically tested to gather data on how these factors influence the molecular weight of the polymer. The goal is to identify the optimal factor settings that maximize molecular weight while minimizing sensitivity to noise factors, as part of the RPD experiment's objectives.

TABLE 3.18

Robust Parameter Design (RPD) Matrix Example for Chemical Engineering

Run	Temp (°C)	Conc. A (%)	Conc. B (%)	Rx Time (min)	Humidity (%)	Raw Material Variations	Molecular Weight (g/mol)
1	70	5	2	60	40	Low	[Molecular weight measurement]
2	70	5	2	60	50	High	[Molecular weight measurement]
3	80	7	3	75	45	Medium	[Molecular weight measurement]
4	80	7	3	75	55	Low	[Molecular weight measurement]

REPORTING

RPD experiments in chemical engineering utilize a range of statistical tools and techniques to analyze data, optimize processes, and ensure robustness to variations and uncertainties. Some of the key statistical tools used in RPD designed experiments include:

Factorial Experiments: Factorial designs help explore the effects of multiple factors and their interactions on the response variable. They are used to identify significant factors and estimate their effects.

Response Surface Methodology (RSM): RSM involves fitting mathematical models to experimental data to model the relationship between controllable factors and the response variable. It helps optimize factor settings and assess the effects of noise factors.

Analysis of Variance (ANOVA): ANOVA is used to assess the significance of factors and interactions by partitioning the total variation in the response variable into components attributed to different factors.

Regression Analysis: Regression models are fitted to experimental data to quantify the relationship between factors and the response variable. These models are essential for optimization and prediction.

Design of Experiments (DOE): Various DOE techniques, including full-factorial designs, fractional factorial designs, and Taguchi designs, are employed to efficiently plan and conduct experiments with limited resources.

Sensitivity Analysis: Sensitivity analysis helps assess how changes in factor settings or noise factors affect the response variable. It identifies factors with the most significant impact on robustness.

Monte Carlo Simulation: Monte Carlo simulations are used to evaluate the robustness of optimized settings by introducing variations in noise factors and assessing their impact on the response variable.

Statistical Process Control (SPC): SPC tools, such as control charts and process capability analysis, help monitor and control processes to maintain their robustness and performance over time.

Optimization Algorithms: Various optimization algorithms, such as gradient descent and genetic algorithms, are applied to find the optimal factor settings that maximize performance while ensuring robustness.

Robustness Metrics: Metrics like S/N ratio and mean squared error are used to quantify robustness and assess the trade-off between mean performance and variability reduction.

Design-Expert Software: Specialized software tools like Design-Expert, Minitab, and JMP are often employed to perform experimental design, data analysis, modeling, and optimization.

Statistical Software: General statistical software like R, Python with libraries like SciPy and statsmodels, and commercial software like SAS are used for data analysis and modeling.

Visualization Tools: Data visualization tools, including scatter plots, contour plots, and response surface plots, help visualize and interpret the relationships between factors and the response.

Hypothesis Testing: Hypothesis tests, such as t-tests and F-tests, are used to assess the significance of factors and determine if changes in factor settings are statistically significant.

These statistical tools and techniques are applied throughout the RPD process to plan, conduct, and analyze experiments, optimize processes, and ensure that chemical engineering systems or products are robust against variations and uncertainties in real-world conditions.

A comprehensive report for an RPD experiment in chemical engineering should provide a detailed account of the experiment, its objectives, methodologies, results, and conclusions. The report should be structured to communicate the key findings and insights gained from the RPD study effectively. Here are the essential elements typically included in an RPD experiment report:

Title Page:
- Title of the Report.
- Name of the Organization or Institution.
- Names of the Authors and Contributors.
- Date of Report Submission.

Abstract: A concise summary of the experiment's objectives, methods, key findings, and conclusions. It should provide a quick overview of the study.

Table of Contents: A list of sections, subsections, and page numbers for easy navigation within the report.

List of Figures and Tables: A list of all figures and tables included in the report, along with their corresponding page numbers.

List of Abbreviations and Symbols (if applicable): Definitions and explanations of any abbreviations, acronyms, or special symbols used throughout the report.

Introduction:
- Background and context for the RPD experiment.
- Objectives and goals of the study.
- A brief overview of the factors, responses, and the problem under investigation.

Literature Review (if applicable): A review of relevant literature and prior research related to the experiment. Discussion of existing knowledge, theories, and findings in the field.

Experimental Design: Detailed description of the experimental design, including the selection of factors and levels, response variables, and any noise factors considered. A justification for the chosen design and factors should be used and an Explanation of the statistical methods and tools used.

Methodology: A step-by-step description of how the experiment was conducted. Include information on data collection procedures, instrumentation, and measurements. Document any specific protocols followed.

Data Analysis:
- Presentation and analysis of experimental data.
- Statistical methods used to analyze the data, including ANOVA, regression analysis, response surface modeling, and robustness evaluations.
- Presentation of key results, such as factor effects, interaction effects, and response optimization.

1. **Robustness Analysis:**
   - Evaluation of the robustness of the optimized settings.
   - Sensitivity analysis to assess the impact of variations in noise factors on the response.
   - Visualization of robustness metrics and sensitivity plots.

2. **Optimization Results**: Presentation of the optimal factor settings that maximize performance while ensuring robustness. Discussion of trade-offs between mean performance and variability reduction.

3. **Discussion:**
   - Interpretation of the results, including the practical implications of the findings.
   - Comparison of the optimized settings with the initial conditions.
   - Identification of any limitations or challenges encountered during the experiment.

4. **Conclusion**: Summary of the key findings and their implications. Statement of whether the experiment's objectives were met. Recommendations for further research or process improvements.

5. **References**: Citations of all sources, literature, and references used in the report, following a specific citation style (e.g., APA, IEEE, Chicago).
6. **Appendices (if needed)**: Additional data tables, graphs, charts, or supplementary information that supports the main text. Details on experimental protocols, calculations, or software code used in the analysis.
7. **Acknowledgments (optional)**: Recognition of individuals, organizations, or funding sources that contributed to the experiment.
8. **Executive Summary (optional)**: A concise summary of the entire report, intended for readers who may not have the time to review the full document.

The report should be well-organized, logically structured, and written in a clear and concise manner to effectively communicate the methodology, results, and implications of the RPD experiment in chemical engineering.

3.2.9 RANDOMIZED EXPERIMENTS

REs are used in chemical engineering research to study the effects of various factors on a specific property or behavior. For instance, a study may randomize the source of raw materials to assess their impact on product quality.

STEPS

A randomized designed experiment in chemical engineering involves the systematic and random allocation of experimental units (samples) to different treatments or conditions to investigate the effects of various factors on a response variable. Here are the steps to develop a randomized designed experiment in chemical engineering:

1. **Define the Objective**: Clearly state the research or engineering objective. What specific problem or question are you trying to address with the experiment? What response variable(s) are you interested in studying?
2. **Select Factors and Levels**: Identify the controllable factors that you want to investigate. These are variables that you can manipulate during the experiment. Determine the levels or settings for each factor that you want to test. Levels should be chosen based on practical relevance and the expected range of variation.
3. **Identify Response Variables**: Specify the response variable(s) that will be measured or observed to assess the impact of the factors. Response variables should be relevant to the experiment's objectives.
4. **Randomization**: Randomly assign experimental units (samples) to different treatment groups or conditions. This helps ensure that the experiment's results are not biased by any systematic factors or ordering effects.
5. **Replication**: Replicate the experimental runs for each combination of factor levels. Replication provides valuable information about the variability of the response variable and allows for more robust statistical analysis.
6. **Blocking (if applicable)**: If there are known sources of variability or conditions that may affect the response variable, consider using blocking. Blocking groups of similar experimental units together reduces the impact of these sources of variation.
7. **Experimental Design**: Choose an appropriate experimental design that suits your objectives and the number of factors involved. Common designs include completely randomized designs (CRD), randomized complete block designs (RCBD), Latin square designs, and more.
8. **Sample Size Determination**: Determine the sample size needed for each treatment group or condition to achieve sufficient statistical power. This involves considering factors such as desired confidence levels and effect sizes.

9. **Data Collection**: Conduct the experiments according to the randomized design. Ensure that all factors are manipulated as planned, and carefully measure the response variable for each experimental unit.
10. **Data Analysis**: Analyze the experimental data using appropriate statistical techniques. This may include ANOVA, regression analysis, or other statistical models, depending on the experimental design and objectives.
11. **Interpret Results**: Interpret the results of the data analysis, focusing on the effects of the factors on the response variable. Identify significant factors, interactions, and any patterns or trends in the data.
12. **Conclusion and Recommendations**: Draw conclusions based on the analysis and discuss their implications for the research objectives. Make recommendations for process improvements, further research, or engineering decisions based on the findings.
13. **Report Writing**: Prepare a detailed report summarizing the experiment's objectives, methods, results, and conclusions. Follow a structured format to present the information clearly and logically.
14. **Validation and Verification (if applicable)**: If the experiment involves process optimization or validation, conduct additional experiments or simulations to confirm the robustness of the findings.
15. **Documentation**: Keep detailed records of the experimental design, data collection, and analysis for future reference or potential audits.
16. **Continuous Improvement**: Use the findings from the experiment to drive process improvements, enhance product quality, or make informed decisions in chemical engineering applications.

By following these steps, you can systematically plan, conduct, and analyze a randomized designed experiment in chemical engineering to answer specific research questions or solve engineering problems effectively.

PITFALLS AND REMEDIES

Randomized designed experiments in chemical engineering, like any scientific study, can encounter various pitfalls. Identifying these pitfalls and applying appropriate remedies is essential to ensure the reliability and validity of the experiment's results. Here are some common pitfalls and their remedies:

Pitfall 1: Lack of Clear Objectives
   Pitfall: Conducting an experiment without clearly defined research objectives or questions.
   Remedy: Clearly articulate the objectives, hypotheses, and research questions before designing the experiment. Ensure that the experiment is focused on addressing these objectives.
Pitfall 2: Inadequate Randomization
   Pitfall: Randomization is not carried out properly, leading to biased results or confounding effects.
   Remedy: Randomly allocate experimental units to treatment groups or conditions. Use appropriate randomization methods to eliminate bias and control for lurking variables.
Pitfall 3: Insufficient Replication
   Pitfall: Inadequate replication of experimental runs, leading to imprecise or unreliable results.
   Remedy: Plan for an adequate number of replicates to estimate experimental error and improve the reliability of statistical analysis.
Pitfall 4: Uncontrolled Variables
   Pitfall: Failure to control or account for uncontrolled variables (e.g., environmental conditions) that can introduce noise or bias.
   Remedy: Identify potential sources of variation and use blocking or other control strategies to minimize their impact. Collect data on uncontrolled variables for later analysis.

Pitfall 5: Incomplete Factorial Design

Pitfall: Using a design that does not account for all relevant factors or their interactions.

Remedy: Conduct a thorough factor screening process to identify all relevant factors and include them in the experimental design. Consider fractional factorial designs if there are many factors.

Pitfall 6: Small Sample Sizes

Pitfall: Using sample sizes that are too small to detect significant effects or interactions.

Remedy: Perform a power analysis to determine the required sample size for adequate statistical power. Ensure that sample sizes are sufficient to detect meaningful differences.

Pitfall 7: Inappropriate Statistical Analysis

Pitfall: Applying incorrect statistical techniques or misinterpreting results.

Remedy: Seek expert guidance in experimental design and statistical analysis. Use appropriate statistical methods that match the experimental design and objectives. Consult with a statistician if needed.

Pitfall 8: Nonrepresentative Samples

Pitfall: Selecting experimental units or samples that do not represent the population of interest.

Remedy: Ensure that the selection of experimental units is representative of the larger population. Random sampling can help achieve this.

Pitfall 9: Data Quality Issues

Pitfall: Data collection errors, outliers, or missing data.

Remedy: Implement rigorous data collection protocols. Check for outliers and address them appropriately (e.g., by verifying data or applying robust statistical methods). Address missing data using imputation techniques.

Pitfall 10: Insufficient Documentation

Pitfall: Poor documentation of experimental procedures, making it difficult to replicate or validate the results.

Remedy: Maintain detailed records of experimental procedures, measurements, and data collection. Document any deviations or unexpected events during the experiment.

Pitfall 11: Overlooking Long-Term Trends

Pitfall: Focusing solely on short-term results and overlooking potential long-term trends or effects.

Remedy: Consider conducting follow-up studies or monitoring to assess the long-term impact of experimental changes or interventions.

Addressing these pitfalls through careful planning, execution, and analysis of randomized designed experiments in chemical engineering will help ensure that the results are valid, reliable, and useful for making informed decisions and improvements in chemical processes or products.

EXAMPLE

The following is an example of a randomized designed experiment in chemical engineering:

Objective: To optimize the curing process of a polymer composite material to achieve the desired strength while minimizing curing time.

Factors and Levels:

1. **Curing Temperature**: Low (60°C), Medium (70°C), High (80°C)
2. **Curing Time**: Short (2 hours), Medium (3 hours), Long (4 hours)
3. **Curing Pressure**: Low (5 MPa), Medium (10 MPa), High (15 MPa)

Response Variable: Tensile strength of the polymer composite material (measured in MPa).

Experimental Design: A completely randomized design is chosen for this experiment. The available experimental units (samples) are randomly assigned to the various combinations of curing temperature, curing time, and curing pressure. Each combination represents a treatment group, resulting in a total of 27 experimental runs (3 levels for each of the 3 factors).

Randomization: To ensure the random allocation of samples to treatment groups, a random number generator or a randomization table is used.

Replication: To improve the precision of the experiment, each combination of factors is replicated three times, resulting in a total of 81 experimental runs.

Data Collection: Tensile strength measurements are taken for each cured sample using standardized testing equipment. Data on the curing temperature, curing time, and curing pressure for each run are recorded.

Data Analysis: ANOVA is performed to assess the significance of the factors and their interactions on tensile strength. Regression analysis may be used to model the relationship between the factors and the response variable. Post hoc tests, such as Tukey's HSD, are conducted to identify which factor levels significantly differ from each other.

Interpretation of Results: The experiment provides insights into the optimal curing conditions that yield the highest tensile strength while minimizing curing time. Factors that have a significant impact on tensile strength are identified, allowing for process optimization.

Conclusion and Recommendations: Based on the results, recommendations can be made for the ideal curing conditions for the polymer composite material. Further studies or process adjustments may be recommended based on the findings.

In this example, the randomized designed experiment helps chemical engineers optimize the curing process of a polymer composite material by systematically varying the factors of interest and using randomization to eliminate bias. The results can lead to improved product quality and manufacturing efficiency.

MATRIX EXAMPLE

In a randomized designed experiment in chemical engineering, you can use a matrix to organize the experimental runs and their corresponding factor settings. Table 3.19 is an example of a matrix for the RE described earlier, where the objective is to optimize the curing process of a polymer composite material.

In this matrix:

- Each row represents an experimental run, numbered from 1 to 81.
- The columns represent the factors (curing temperature, curing time, curing pressure).
- The cells contain the specific factor settings for each run.

By using this matrix, you can easily visualize and track the different combinations of factor settings that are part of the RE. It helps ensure that each run is conducted with the appropriate conditions and that the experiment is conducted in a systematic and randomized manner.

TABLE 3.19
Randomized Experiment Matrix Example for Chemical Engineering

Run	Curing Temperature	Curing Time	Curing Pressure
1	Low (60°C)	Short (2 h)	Low (5 MPa)
2	Medium (70°C)	Short (2 h)	Medium (10 MPa)
3	High (80°C)	Short (2 h)	High (15 MPa)
4	Low (60°C)	Medium (3 h)	Low (5 MPa)
5	Medium (70°C)	Medium (3 h)	Medium (10 MPa)
6	High (80°C)	Medium (3 h)	High (15 MPa)
7	Low (60°C)	Long (4 h)	Low (5 MPa)
8	Medium (70°C)	Long (4 h)	Medium (10 MPa)
9	High (80°C)	Long (4 h)	High (15 MPa)
...	...	...	...
81	High (80°C)	Long (4 h)	High (15 MPa)

REPORTING

Statistical tools are crucial for analyzing the data collected from randomized designed experiments in chemical engineering. These tools help assess the significance of factors, interactions, and variations, ultimately aiding in making informed decisions and optimizing processes. Here are some common statistical tools used in randomized designed experiments in chemical engineering:

Analysis of Variance (ANOVA): ANOVA assesses the statistical significance of the factors and their interactions by partitioning the total variation in the response variable into different components. It helps determine which factors have a significant impact on the response.

Regression Analysis: Regression models are used to describe the relationship between the response variable and the factors. Simple linear regression or multiple regression can be employed to develop predictive models and identify optimal factor settings.

Post Hoc Tests: Post hoc tests, such as Tukey's HSD or Bonferroni tests, are used to compare multiple factor levels and identify significant differences between them when ANOVA indicates significance.

Main Effects Plots: Main effects plots visually represent the influence of individual factors on the response variable. They help identify which factors have the most substantial impact on the outcome.

Interaction Plots: Interaction plots show how two or more factors interact and affect the response variable. Understanding interactions is critical for optimizing processes where multiple factors are involved.

Residual Analysis: Residuals are the differences between the observed and predicted values of the response variable. Residual analysis helps assess model adequacy, identify outliers, and check for the assumptions of normality and constant variance.

Design of Experiments (DOE) Software: Specialized software packages, such as Minitab, JMP, or Design-Expert, facilitate the design and analysis of experiments. They often include built-in tools for ANOVA, regression, and graphical analysis.

Statistical Hypothesis Testing: Hypothesis testing is used to determine whether observed effects and differences are statistically significant. It involves testing null hypotheses and calculating p-values.

Factorial Plots: Factorial plots display the effects of two factors on a 2D graph. They help visualize interactions and trends between factors.

Response Surface Methodology (RSM): RSM is a statistical technique used for optimizing responses with multiple factors. It involves fitting polynomial equations to experimental data and finding optimal factor settings.

Robust Parameter Design (RPD): RPD combines factorial experiments with optimization techniques to find robust factor settings that minimize variability and improve process performance.

Statistical Process Control (SPC): SPC tools, such as control charts, are used to monitor and control processes based on the data collected during and after the experiment.

Power Analysis: Power analysis helps determine the required sample size to achieve a desired level of statistical power, ensuring that the experiment can detect meaningful effects.

Randomization Tests: These nonparametric tests are used when the assumptions of normality and homogeneity of variances are not met. Examples include the Kruskal-Wallis test and the Mann-Whitney U-test.

Fractional Factorial Designs: Fractional factorial designs are used to reduce the number of experimental runs while still assessing the main effects and some interactions. They help save time and resources.

The choice of statistical tools depends on the complexity of the experiment, the number of factors, and the specific objectives of the study. Chemical engineers often work with statisticians or use statistical software to ensure that the data analysis is conducted correctly and accurately.

A well-structured report for a randomized designed experiment in chemical engineering should provide a clear and comprehensive account of the experiment's objectives, methodology, results, and conclusions. The elements of such a report typically include:

Title Page:
Title of the Experiment
Names of the Authors
Affiliations of the Authors
Date of Submission

Abstract: A concise summary of the experiment's objectives, methods, key findings, and conclusions. The abstract should provide a quick overview of the entire report.

Table of Contents: A list of sections and subsections with corresponding page numbers for easy navigation.

List of Figures and Tables: A compilation of all figures and tables in the report, along with their page numbers.

List of Abbreviations and Symbols: An inventory of any abbreviations, acronyms, or symbols used in the report, along with their explanations.

Introduction: A clear statement of the research objectives and the rationale for conducting the experiment. Background information on the problem or process being studied. A hypothesis or research questions to be addressed.

Literature Review: A brief review of relevant literature and prior research related to the experiment's topic. An explanation of how the experiment fits into the broader context of chemical engineering.

Experimental Design:
Description of the experimental design, including factors, levels, and treatments.
Details on randomization procedures, replication, and any control measures taken.
Justification for the choice of the experimental design.

Materials and Methods:
Information on the materials, chemicals, and equipment used in the experiment.
Step-by-step procedures for conducting the experiment, including data collection methods.
Any safety precautions taken during the experiment.

Results:

Presentation of the collected data, including tables, figures, and charts.

Statistical analysis of the data, including ANOVA, regression, or other relevant tests.

Interpretation of the results and discussion of any significant findings.

Clear labeling and captions for all figures and tables.

Discussion: Interpretation of the results in the context of the research objectives. Analysis of the implications of the findings and their relevance to chemical engineering principles. Comparison of the results with prior research or expected outcomes.

Conclusions: A concise summary of the main findings and their significance. Specific conclusions related to the research objectives. Suggestions for future research or improvements based on the results.

References: A list of all the sources and references cited in the report, following a specific citation style (e.g., APA, MLA, Chicago). Properly formatted citations for articles, books, journals, and other sources.

Appendices: Any supplementary information that is not included in the main text, such as raw data, calculations, additional graphs, or detailed experimental procedures. Each appendix should be clearly labeled and referenced in the main text.

Acknowledgments: Acknowledgment of individuals or organizations that provided assistance, funding, or resources for the experiment.

The structure and content of the report may vary depending on the specific requirements of your institution or journal. It's essential to follow any guidelines provided by your academic institution or publisher when preparing the report. Additionally, ensuring clarity, conciseness, and proper organization of the report is crucial for effective communication of your experiment's results in chemical engineering.

3.2.10 SPLIT-PLOT AND BLOCKED EXPERIMENTS

In chemical process optimization, split-plot and blocked experiments may be used to account for the presence of uncontrolled factors that affect the response variable. Split-plot and blocked designs are specialized experimental designs used in chemical engineering and other fields to address specific research questions and challenges. Here's an overview of each:

Split-Plot Design:

- **Purpose:** Split-plot designs are used when the experimental factors have different levels of accessibility or when certain factors are more difficult or expensive to change than others.
- **Design Structure:** In a split-plot design, the experimental units are divided into subunits or blocks, with each block representing a different level of a "hard-to-change" or less accessible factor. Within each block, the levels of the "easy-to-change" or more accessible factors are varied. Essentially, it combines two or more experimental designs within a single experiment.
- **Example:** Suppose you are studying the effect of temperature (easy-to-change) and pH (hard-to-change) on a chemical reaction. You have three levels of temperature and three levels of pH. You decide to conduct the experiment over three days, with each day representing a block. Within each day (block), you vary the temperature settings, but due to logistical constraints, you can only set the pH level once for the entire day. This is an example of a split-plot design.

Blocked Design:

- **Purpose:** Blocked designs are used when there are known or suspected sources of variability that may affect the experiment's outcomes, and you want to control for these sources of variation.
- **Design Structure:** In a blocked design, the experimental units are grouped into blocks or batches based on certain characteristics or factors that are not the primary focus of the study but are known to introduce variation. The experiment is then conducted separately within each block. This helps reduce the impact of external variability on the study's results.
- **Example:** Imagine you are testing a new catalyst for a chemical reaction, and you suspect that the quality of the raw materials used in the catalyst may vary. To control for this potential source of variation, you divide the raw materials into different batches (blocks). You then conduct the experiment separately for each batch of raw materials. This way, any variations due to differences in raw materials are accounted for in the analysis.

Split-plot designs are used when certain factors are more challenging to change or control than others, while blocked designs are employed to control for known or suspected sources of variation in the experiment. Both designs are valuable tools in chemical engineering and other fields, helping researchers make more accurate and meaningful conclusions from their experiments.

STEPS

Developing a split-plot or blocked designed experiment in chemical engineering involves careful planning, design, and execution. Here are the steps to develop both types of experiments:

Steps to Develop a Split-Plot Designed Experiment:

1. **Define the Objectives:** Clearly state the objectives of your experiment, including the primary and secondary factors of interest.
2. **Identify Factors:** Determine which factors you will investigate in your experiment. Classify them as "hard-to-change" (less accessible) or "easy-to-change" (more accessible).
3. **Select Levels:** Specify the levels or settings for each factor you plan to study. For easy-to-change factors, identify a range of levels to vary within each block.
4. **Divide into Blocks:** Divide your experimental units into blocks or subunits. Each block represents a different level of the hard-to-change factor.
5. **Randomization:** Randomly assign the levels of the easy-to-change factors within each block. The order of experimentation should be randomized to avoid bias.
6. **Conduct Experiments:** Perform the experiments according to the assigned levels of the factors within each block. Collect data for the response variable.
7. **Analyze Data:** Use statistical techniques, such as ANOVA with split-plot designs, to analyze the data. This involves assessing the effects of both easy-to-change and hard-to-change factors.
8. **Interpret Results:** Interpret the results by examining the significance of the factors and their interactions. Draw conclusions based on the data analysis.
9. **Report Findings:** Prepare a report that includes the experimental design, methods, results, and conclusions. Communicate how the split-plot design addressed the research objectives.

Steps to Develop a Blocked Designed Experiment:

1. **Define the Objectives:** Clearly articulate the research objectives and the factors you want to study while considering known or suspected sources of variation.
2. **Identify Potential Blocks:** Identify the sources of variation or external factors that may affect the outcomes of your experiment. These sources will become your blocking variables.
3. **Divide into Blocks:** Group your experimental units into blocks based on the levels of the blocking variables. Ensure that each block is relatively homogeneous with respect to the blocking variable.
4. **Randomization:** Randomly assign the treatments or experimental conditions to the units within each block. Randomization helps control for external sources of variation.
5. **Conduct Experiments:** Conduct the experiments separately within each block while maintaining consistent conditions within each block.
6. **Collect Data:** Record data for the response variable during each experiment within the respective blocks.
7. **Analyze Data:** Analyze the data using appropriate statistical methods, such as ANOVA with blocking factors, to assess the effects of the primary factors of interest.
8. **Interpret Results:** Interpret the results, paying attention to the impact of the blocking variables on the response variable. Determine the significance of the primary factors.
9. **Report Findings:** Prepare a comprehensive report that includes the experimental design, methods, results, and conclusions. Explain how the blocked design controlled for external sources of variation and influenced the study's outcomes.

In both types of designs, careful planning, randomization, and proper statistical analysis are essential to draw valid conclusions from the experiments. Additionally, clear and thorough reporting of the design and results is critical for effective communication in chemical engineering research.

PITFALLS AND REMEDIES

Split-plot and blocked designed experiments are powerful tools in chemical engineering, but they come with specific pitfalls that researchers should be aware of. Here are some common pitfalls and their corresponding remedies:

Pitfalls in Split-Plot Experiments:

Incomplete Randomization: If randomization is not adequately implemented within the easy-to-change factors' levels, it can introduce bias and affect the validity of the results.
   Remedy: Ensure proper randomization of experimental runs within each block to eliminate bias.
Limited Replicates: Split-plot designs often involve fewer replicates for hard-to-change factors, which can lead to higher variability and less precision.
   Remedy: Whenever possible, increase the number of replicates for hard-to-change factors or explore RPD techniques to improve reliability.
Complex Analysis: Split-plot designs can lead to more complex data analysis, making it challenging to interpret results accurately.
   Remedy: Seek statistical expertise to perform appropriate analysis, including split-plot ANOVA, to account for the design's complexity.
Overemphasis on Easy-to-Change Factors: Researchers may unintentionally focus more on the easy-to-change factors and neglect the hard-to-change factors.
   Remedy: Ensure that both types of factors receive equal attention during the experimental design, execution, and analysis.

Pitfalls in Blocked Experiments:

Incorrect Blocking Factors: Selecting the wrong blocking factors or failing to account for critical sources of variation can lead to ineffective blocking.

Remedy: Carefully identify and include relevant blocking factors based on prior knowledge or experimentation.

Small Block Sizes: If blocks are too small, they may not effectively control the sources of variation they were intended to address.

Remedy: Aim for sufficiently large block sizes to reduce the impact of external variability.

Reduced Efficiency: Blocked designs may require a larger number of experimental units, potentially increasing the cost and resource requirements.

Remedy: Balance the trade-off between blocking efficiency and resource constraints, and consider efficient experimental designs.

Increased Complexity: The use of blocking factors can make experimental planning and execution more complex.

Remedy: Clearly document the blocking strategy and ensure that the experiment is carried out as planned for each block.

Assumption Violations: Blocking factors may introduce assumptions that are not met, such as homogeneity of variances.

Remedy: Perform appropriate diagnostics and model checks to verify that assumptions are valid, and consider transformation techniques if necessary.

Loss of Sensitivity: In some cases, blocking can reduce the sensitivity to detect small but important effects of the primary factors.

Remedy: Balance the benefits of controlling external variability with the need to detect meaningful primary factor effects, and consider efficient designs that address both objectives.

Interactions with Blocks: Unanticipated interactions between the primary factors and blocking factors can complicate data analysis and interpretation.

Remedy: Conduct preliminary data exploration and include interaction terms in the analysis if required.

It's crucial for researchers in chemical engineering to be aware of these potential pitfalls and to plan and execute split-plot and blocked experiments carefully. Collaboration with statisticians or experts in experimental design can be invaluable to ensure the success of such experiments and the validity of their results.

EXAMPLES

Consider the following two examples, one for a split-plot design and another for a blocked design, both in the context of chemical engineering:

Example 1: Split-Plot Design

Objective: Investigate the effect of two factors, A (easy-to-change) and B (hard-to-change), on the yield of a chemical reaction.

Factors and Levels:

Factor A (Easy-to-Change):
- Low Level: Temperature at 60°C
- High Level: Temperature at 80°C

Factor B (Hard-to-Change):
- Low Level: pH 5.0
- High Level: pH 7.0

Split-Plot Design: Experimental units are divided into three blocks (representing three different batches of raw materials). Within each block, Factor A (temperature) is varied at random, with two experimental runs at each level. Due to logistical constraints, Factor B (pH) can only be set once for each block.

Data Collection: Conduct the chemical reactions according to the assigned temperature levels within each block. Measure the yield of the reaction for each experimental run.

Analysis: Use split-plot ANOVA to assess the effects of temperature (Factor A) and pH (Factor B) on the yield. Examine interactions between temperature and pH, as well as the main effects.

Conclusion: Determine whether temperature and pH significantly affect yield and whether there are interactions between these factors. Understand how the split-plot design addressed the logistical constraint of varying pH levels.

Example 2: Blocked Design

Objective: Evaluate the impact of a new catalyst on the yield of a chemical process while controlling for potential variations in the quality of raw materials.

Factors and Levels:

Factor: Catalyst Type
- Catalyst A (new formulation)
- Catalyst B (standard formulation)

Blocked Design: Identify a potential source of variation: Variability in the quality of raw materials (e.g., impurities). Divide the raw materials into three batches (blocks) based on their quality. Within each block, conduct separate experiments using both Catalyst A and Catalyst B.

Data Collection: Perform chemical reactions with each catalyst type within each block. Measure the yield of the chemical process for each combination of catalyst and block.

Analysis: Use blocked ANOVA to assess the impact of the catalyst type while controlling for variations due to raw material quality (blocking factor). Examine whether the new catalyst (Catalyst A) significantly improves yield compared to the standard (Catalyst B).

Conclusion: Determine whether the new catalyst is effective in enhancing yield while accounting for variations in raw material quality. Highlight the role of the blocked design in controlling external sources of variability.

In both examples, the split-plot and blocked designs are tailored to address specific challenges or constraints in the chemical engineering experiments, enabling researchers to draw valid conclusions about the factors of interest while controlling for other sources of variability.

In a split-plot or blocked designed experiment in chemical engineering, you typically use a matrix to represent the assignment of experimental runs to different factors, levels, blocks, or combinations thereof. Here's an example of how such a matrix might look for a split-plot and blocked experiment:

Example of a Split-Plot Matrix:
Suppose you are conducting an experiment with two factors, Factor A (easy-to-change) and Factor B (hard-to-change), and you have three blocks (representing different batches of raw materials). Each block has two experimental runs (Table 3.20).

This matrix illustrates how the experimental runs are assigned to different levels of Factor A and Factor B within each block.

Example of a Blocked Design Matrix:
Suppose you are conducting an experiment to evaluate the impact of two different catalyst types (Catalyst A and Catalyst B) on the yield of a chemical process. You have three blocks (representing different batches of raw materials) (Table 3.21).

In this matrix, each block represents a different batch of raw materials, and within each block, you conduct experiments using both Catalyst A and Catalyst B to evaluate their effects while controlling for variations in raw material quality.

These matrices provide a visual representation of how experimental runs are allocated to different factors, levels, or blocks in the experimental design. They serve as a useful tool for planning, executing, and analyzing split-plot and blocked experiments in chemical engineering.

REPORTING

In split-plot and blocked designed experiments in chemical engineering, several statistical tools and techniques are commonly used for data analysis. Here are some of the key statistical tools employed in these experimental designs:

TABLE 3.20
Split-Plot Experiment Matrix Example for Chemical Engineering

Block	Factor A (Temperature)	Factor B (pH)
Block 1	Low (60°C)	Low (pH 5.0)
Block 1	High (80°C)	Low (pH 5.0)
Block 2	Low (60°C)	High (pH 7.0)
Block 2	High (80°C)	High (pH 7.0)
Block 3	Low (60°C)	High (pH 7.0)
Block 3	High (80°C)	High (pH 7.0)

TABLE 3.21
Blocked Experiment Matrix Example for Chemical Engineering

Block	Catalyst Type
Block 1	Catalyst A
Block 1	Catalyst B
Block 2	Catalyst A
Block 2	Catalyst B
Block 3	Catalyst A
Block 3	Catalyst B

Statistical Tools for Split-Plot Experiments:

Analysis of Variance (ANOVA): Split-plot ANOVA is a fundamental tool for analyzing split-plot experiments. It allows you to assess the effects of various factors and their interactions, considering both hard-to-change and easy-to-change factors.

Main Effects and Interaction Plots: These plots help visualize and interpret the main effects of different factors and any interactions that may exist between them.

Residual Analysis: Residual plots and analysis help check the assumptions of ANOVA, such as homogeneity of variances and normality of residuals.

Effect Plots: Effect plots display the estimated effects of factors, making it easier to interpret their impact on the response variable.

Statistical Software: Statistical software packages like R, SAS, or Minitab are commonly used for conducting split-plot ANOVA and generating relevant plots.

Statistical Tools for Blocked Experiments:

Blocked Analysis of Variance (ANOVA): Blocked ANOVA is the primary tool for analyzing blocked experiments. It assesses the effects of primary factors while accounting for the blocking factor's influence.

Interaction Analysis: Blocked designs may involve interactions between primary factors and blocking factors. Interaction analysis helps understand how these interactions affect the response variable.

Covariance Analysis: In some cases, covariance analysis is used to account for additional sources of variation, especially when the blocking factor is continuous.

Residual Analysis: As with split-plot experiments, residual analysis is important to validate the assumptions of ANOVA, including homogeneity of variances and normality of residuals.

Graphical Representation: Graphs, such as bar charts and scatterplots, can help visualize the data, relationships between factors, and the effect of blocking.

Statistical Software: Statistical software packages are essential for conducting blocked ANOVA and other relevant statistical tests.

In both split-plot and blocked experiments, the choice of statistical tools depends on the specific experimental design, objectives, and the nature of the data. Additionally, researchers should be well-versed in statistical techniques and software to effectively analyze and interpret the results of these experiments in chemical engineering.

A well-structured report for a split-plot or blocked designed experiment in chemical engineering should provide a clear and comprehensive overview of the experiment, its objectives, methodologies, results, and conclusions. Here are the essential elements that should be included in such a report:

Title Page:
   Title of the Experiment
   Names of Authors and Affiliations
   Date of Submission
Abstract:
   A brief summary (typically 150–250 words) of the experiment's objectives, methodology,
      key findings, and conclusions.
Table of Contents:
   A list of sections and subsections with corresponding page numbers for easy navigation.
List of Figures and Tables:
   A separate list of all figures and tables with their respective titles and page numbers.

List of Abbreviations and Symbols:
>If applicable, include a list of abbreviations and symbols used throughout the report.

Introduction:
>Clearly state the experiment's objectives and the problem or hypothesis being investigated.
>
>Provide background information on the context of the experiment and its relevance in chemical engineering.

Experimental Design:
>Describe the overall experimental design, including the use of split-plot or blocked designs.
>
>Explain the factors, levels, and blocking factors used in the experiment.
>
>Specify the equipment, materials, and procedures employed during the experiment.

Data Collection:
>Detail the data collection process, including the measurement techniques and instruments used.
>
>Mention any data preprocessing or data quality checks performed.

Statistical Analysis:
>Describe the statistical methods used for data analysis, whether it's split-plot ANOVA, blocked ANOVA, or other relevant techniques.
>
>Present the results of the statistical analysis, including main effects, interactions, and significance levels.
>
>Provide appropriate graphical representations (e.g., graphs, charts, plots) to illustrate the findings.

Results:
>Present the experimental results, both quantitative and qualitative, in a clear and organized manner.
>
>Include tables and figures with appropriate captions to support your findings.

Discussion:
>Interpret the results in the context of the experiment's objectives and the chemical engineering principles involved.
>
>Discuss any unexpected or noteworthy observations.
>
>Address the implications of the findings and their relevance to real-world applications.

Conclusion:
>Summarize the key findings and their significance.
>
>Restate the experiment's objectives and whether they were achieved.
>
>Provide recommendations for further research or practical applications based on the results.

References:
>Cite all relevant sources, including scientific papers, books, and reports, in a consistent citation style (e.g., APA, MLA, or a specific journal's format).

Appendices:
>Include any supplementary information that supports the main report, such as raw data, detailed calculations, or additional graphs and tables.
>
>Label and reference appendices within the main text as needed.

Acknowledgments:
>Express gratitude to individuals or organizations that contributed to the experiment, such as funding sources, collaborators, or technical support.

Author Contact Information:
>Provide contact information for the authors, including email addresses, to facilitate communication or further inquiries.

Remember to follow any specific formatting and style guidelines required by your institution, journal, or publication venue when preparing the report. A well-organized and clearly written report will enable readers to understand the experiment's objectives, methodologies, and results effectively, making it a valuable contribution to the field of chemical engineering.

3.3 CIVIL ENGINEERING

In civil engineering, designed experiments are valuable tools for optimizing construction processes, materials, and structural designs. Here are some common types of designed experiments with examples of their applications in civil engineering.

In civil engineering designed experiments, various types of variables are considered to study and optimize different aspects of construction, materials, and infrastructure. These variables can be categorized into several common types, including:

Independent Variables:
   Material Properties: Variables related to the properties of construction materials, such as concrete mix design (e.g., water-cement ratio, aggregate size), steel grade, or asphalt composition.
   Construction Parameters: Variables related to construction processes and methods, including curing time, compaction pressure, construction equipment type, and construction sequence.
   Geometric Design Parameters: Variables related to the geometry and layout of structures or infrastructure elements, such as road alignment, bridge span length, or building height.
   Environmental Factors: Variables related to environmental conditions, including temperature, humidity, wind speed, and seismic activity.
Dependent Variables:
   Structural Performance: Variables related to the structural integrity and behavior of civil engineering structures, such as load-bearing capacity, deflection, stress, and strain.
   Durability and Service Life: Variables related to the resistance of materials and structures to environmental factors and aging, including corrosion rate, crack propagation, and material deterioration.
   Energy Efficiency: Variables related to energy consumption and efficiency in buildings and infrastructure, such as heating and cooling loads, insulation properties, and energy-efficient designs.
   Cost and Budgetary Variables: Variables related to project costs, including construction costs, maintenance costs, and life-cycle costs.
   Environmental Impact: Variables related to the environmental impact of construction and infrastructure projects, such as carbon emissions, water usage, and ecological footprint.
   Safety and Risk Factors: Variables related to safety considerations, including factors that mitigate construction accidents, traffic safety, and structural failure risks.
Categorical Variables:
   Material Types: Categories of construction materials (e.g., concrete, steel, wood) used in a project.
   Construction Methods: Categories of construction methods or techniques employed in a project (e.g., cast-in-place, precast, steel frame).
   Project Locations: Different geographical locations or regions where civil engineering projects are executed can affect environmental conditions and design requirements.
Control Variables (Covariates):
   These are variables that are held constant or controlled during experiments to eliminate their influence on the dependent variable. For example, when testing the compressive strength of concrete, the curing temperature and curing duration may be controlled variables.

Random Variables:
   In some cases, random variables may be introduced to account for variability or uncertainty in measurements or environmental conditions.
Interaction Variables:
   Interaction variables are used to investigate the combined effects of two or more independent variables on the dependent variable, for example, the effect of the interaction between concrete mix design and curing time on concrete strength.
Noise Variables (Error Variables):
   These are uncontrolled or unmeasured variables that can introduce variability or errors in the experimental results. Robust experimental designs aim to minimize the impact of noise variables.

Experiments designed for civil engineering are often used to optimize designs, materials, and construction processes by systematically varying these variables to understand their effects and make informed decisions in the planning, design, and execution of civil engineering projects.

3.3.1 Factorial Experiments

Factorial experiments are a common type of designed experiment used in civil engineering to investigate and analyze the effects of multiple factors on a particular outcome or response variable. These experiments are valuable for studying complex systems, optimizing processes, and making informed decisions in civil engineering projects. Here's how factorial experiments are used in civil engineering:

Materials Testing and Optimization:
   Civil engineers often conduct factorial experiments to study the effects of various materials and their combinations on properties such as strength, durability, and elasticity.
   Factors may include different types of aggregates, cement, additives, curing methods, and curing durations.
   The goal is to optimize the mix design for concrete, asphalt, or other construction materials to meet specific project requirements.
Concrete Mix Design:
   Factorial experiments are used to determine the optimal mix proportions of concrete ingredients, including water-cement ratio, aggregates, and admixtures.
   Engineers can investigate factors like workability, compressive strength, shrinkage, and resistance to environmental conditions.
Structural Analysis and Design:
   In structural engineering, factorial experiments can be used to assess the impact of various design factors on the performance and safety of structures.
   Factors may include material properties, cross-sectional dimensions, reinforcement details, and load combinations.
Geotechnical Engineering:
   Factorial experiments help in studying soil behavior and its interaction with structures.
   Factors can include soil types, compaction methods, moisture content, and loading conditions.
   Engineers use these experiments to optimize foundation design and assess the stability of slopes and embankments.
Environmental Impact Assessments:
   Civil engineers conduct factorial experiments to evaluate the environmental impact of construction projects.
   Factors may include construction methods, mitigation measures, and project scheduling.

The goal is to minimize negative environmental effects and ensure compliance with regulations.

Traffic Engineering:

In transportation and traffic engineering, factorial experiments are used to study factors affecting traffic flow, safety, and congestion.

Factors may include road geometry, traffic signal timings, signage, and lane configurations.

Engineers aim to optimize traffic management and road design.

Hydrology and Water Resources:

Factorial experiments help analyze factors influencing water flow, flood risk, and water quality in civil engineering projects.

Factors may include rainfall patterns, land use, drainage systems, and flood control measures.

Engineers use these experiments to design effective stormwater management systems and prevent flooding.

Construction Process Optimization:

Civil engineers apply factorial experiments to optimize construction processes, such as concrete curing methods, construction equipment choices, and scheduling.

The goal is to improve efficiency, reduce costs, and enhance project outcomes.

Risk Assessment:

In risk assessment and management, factorial experiments can be used to analyze the impact of different risk factors on project performance and safety.

Factors may include weather conditions, site conditions, labor availability, and equipment reliability.

Bridge Design and Maintenance:

Factorial experiments are used to assess the performance of bridge materials and maintenance strategies.

Engineers study factors like material types, corrosion protection methods, and inspection schedules to ensure the longevity of bridges.

In each of these applications, factorial experiments allow civil engineers to systematically vary and analyze multiple factors simultaneously, leading to a better understanding of complex systems and the ability to make data-driven decisions for design, construction, and maintenance projects.

Steps

Developing a factorial designed experiment in civil engineering involves a systematic approach to studying the effects of multiple factors on a response variable of interest. Here are the steps to develop a factorial designed experiment for civil engineering:

1. **Define the Objective:**
   - Clearly define the research objective or problem you want to address with the experiment. What specific aspect of civil engineering are you investigating or improving?
2. **Identify the Factors:**
   - Identify the independent factors (variables) that may influence the response variable. These factors could be material properties, construction methods, design parameters, or environmental conditions.
   - Categorize the factors as either categorical (qualitative) or continuous (quantitative).
3. **Determine the Levels of Factors:**
   - For each factor, determine the range or levels at which it will be studied. Levels should cover the practical range of values for each factor.

- In some cases, factors may have only two levels (e.g., high and low), while others may have multiple levels.

4. **Select the Response Variable:**
   - Choose the response variable or variables that represent the outcome of interest. This could be strength, durability, cost, time, safety, or any other measurable parameter related to your objective.

5. **Design the Experimental Matrix:**
   - Create an experimental matrix that defines the combinations of factor levels to be tested. Each row in the matrix represents a unique experimental run.
   - Use a factorial design matrix to systematically vary factors and levels. This may include full-factorial, fractional factorial, or other designs.

6. **Randomization:**
   - Randomly assign the order of experimentation to minimize bias and account for potential confounding factors.
   - Randomization helps ensure that the results are not influenced by the order in which experiments are conducted.

7. **Conduct the Experiments:**
   - Carry out the experiments according to the design matrix, carefully controlling and recording all relevant variables and conditions.
   - Ensure that experiments are conducted under consistent and reproducible conditions.

8. **Collect Data:**
   - Record data for the response variable(s) for each experimental run.
   - Ensure that data collection is accurate and reliable.

9. **Analyze Data:**
   - Perform statistical analysis of the collected data to evaluate the effects of the factors on the response variable.
   - Use appropriate statistical techniques, such as ANOVA or regression analysis, to determine significant factors and interactions.

10. **Interpret Results:**
    - Interpret the results of the data analysis to understand the impact of each factor on the response variable.
    - Identify any significant interactions between factors.

11. **Optimization:**
    - If the objective is optimization, use the results to find the combination of factor levels that maximizes or minimizes the response variable.
    - Optimization techniques like RSM can be employed.

12. **Draw Conclusions:**
    - Draw conclusions based on the analysis and interpretation of the results. Determine which factors are most influential and how they affect the outcome.
    - Assess the practical implications of the findings.

13. **Report and Document:**
    - Prepare a comprehensive report documenting the entire experiment, including the objectives, factors, levels, experimental design, data collection, analysis, results, and conclusions.
    - Include tables, figures, and charts to present the findings effectively.

14. **Recommendations and Applications:**
    - Provide recommendations based on the conclusions of the experiment. Explain how the results can be applied to civil engineering practice or research.

15. **Validation:**
    - Consider conducting validation experiments or further studies to confirm the findings and the robustness of any recommended changes or improvements.

16. **Implementation**:
    - If the experiment leads to practical improvements or changes in civil engineering projects, implement them as appropriate.
17. **Continuous Improvement**:
    - Recognize that the designed experiment may be part of an iterative process. Continue to monitor and refine processes or designs based on new data and insights.

Factorial designed experiments in civil engineering can provide valuable insights, optimize processes, and enhance the quality and efficiency of projects. Proper planning, execution, and analysis are essential to ensure the experiment's success and the reliability of the results.

PITFALLS AND REMEDIES

Factorial designed experiments are powerful tools in civil engineering, but they can also be prone to certain pitfalls. Here are some common pitfalls and their corresponding remedies:

Pitfall 1: Inadequate Factor Selection
   Pitfall: Choosing factors that are not relevant or omitting important factors can lead to incomplete or biased results.
   Remedy: Conduct a thorough literature review and consult with experts to identify and select the most relevant factors. Consider all potential factors that may influence the response variable.

Pitfall 2: Inaccurate Factor-Level Selection
   Pitfall: Selecting inappropriate or unrealistic levels for factors can lead to results that are not applicable in practice.
   Remedy: Choose factor levels that span the practical range of values and are meaningful in the context of civil engineering projects. Avoid extreme or unrealistic levels.

Pitfall 3: Insufficient Sample Size
   Pitfall: Using a small sample size may lead to unreliable or inconclusive results.
   Remedy: Conduct a power analysis or sample size calculation to determine the minimum sample size required to detect significant effects. Ensure that the sample size is adequate to achieve statistical power.

Pitfall 4: Lack of Randomization
   Pitfall: Failing to randomize the order of experiments can introduce bias or confounding effects.
   Remedy: Randomly assign the order of experimentation to minimize bias and account for potential confounding factors. Use randomization techniques to ensure fairness.

Pitfall 5: Incomplete Data Collection
   Pitfall: Missing or incomplete data can compromise the validity of the analysis.
   Remedy: Implement robust data collection procedures and ensure that data is recorded accurately and completely for each experimental run. Have contingency plans for dealing with missing data.

Pitfall 6: Ignoring Interactions
   Pitfall: Neglecting to consider interactions between factors can lead to incorrect conclusions about their effects.
   Remedy: Examine interaction effects in addition to the main effects. Interactions can provide valuable insights into how factors influence each other.

Pitfall 7: Overfitting Models
   Pitfall: Overfitting occurs when overly complex models are used, leading to poor generalization of new data.

Remedy: Use appropriate model selection techniques and choose models that strike a balance between complexity and goodness of fit. Cross-validation can help assess model performance.

Pitfall 8: Failure to Validate Findings

Pitfall: Assuming that the results of the experiment hold true without further validation.

Remedy: Conduct validation experiments or independent studies to confirm the findings and their applicability in different contexts or projects.

Pitfall 9: Lack of Communication

Pitfall: Failing to communicate the experimental design, results, and implications effectively to stakeholders.

Remedy: Prepare a clear and comprehensive report that includes all relevant details of the experiment. Present findings in a way that is understandable to both technical and nontechnical audiences.

Pitfall 10: Not Implementing Findings

Pitfall: Conducting experiments but not applying the findings to civil engineering projects or practices.

Remedy: Ensure that the results are implemented in relevant projects or processes where applicable. Monitor the impact of changes and continuously seek opportunities for improvement.

Pitfall 11: Not Considering External Factors

Pitfall: Failing to account for external factors that may influence the response variable, leading to confounding effects.

Remedy: Identify and control for external factors or include them as covariates in the analysis to isolate the effects of the factors of interest.

By being aware of these pitfalls and implementing the corresponding remedies, civil engineers can conduct factorial designed experiments that yield meaningful and actionable insights for their projects and research.

EXAMPLE

A factorial design in civil engineering aims to study the effects of multiple factors and their interactions on a specific response variable of interest. The matrix used in a factorial design helps organize the experimental runs by specifying the combinations of factor levels to be tested. Here's an example of a matrix for a factorial designed experiment in civil engineering:

Objective: To investigate the effects of two factors, "Concrete Mix Proportion" and "Curing Time," on the compressive strength of concrete.

Factors:

1. **Concrete Mix Proportion (Factor A):**
   - Level 1: Low Cement-to-Water Ratio
   - Level 2: High Cement-to-Water Ratio
2. **Curing Time (Factor B):**
   - Level 1: 7 days
   - Level 2: 28 days

Response Variable: Compressive Strength (measured in megapascals, MPa)

Matrix for Factorial Designed Experiment:
In this example, a 2×2 factorial design is used to assess the impact of two factors, concrete mix proportion and curing time, on the compressive strength of concrete. The matrix is shown in Table 3.22.

In this 2×2 matrix, each row represents a specific experimental run where a concrete mixture is prepared according to a particular combination of factors (concrete mix proportion and curing time). The compressive strength of the concrete is then tested and recorded for each run.

By analyzing the data collected from these experimental runs, you can determine the main effects of each factor (concrete mix proportion and curing time) on compressive strength and assess whether there are any interactions between the factors. This information can be valuable in civil engineering to optimize concrete mixtures and curing processes for desired compressive strength outcomes.

REPORTING

Factorial designed experiments in civil engineering rely on various statistical tools and techniques to analyze and interpret the results. Here are some commonly used statistical tools in factorial designed experiments for civil engineering:

Analysis of Variance (ANOVA):
ANOVA is a fundamental tool for assessing the significance of factors and interactions in factorial experiments. It helps determine if there are statistically significant differences between factor levels.

Regression Analysis:
Regression analysis is used to model the relationship between the response variable and the independent factors. It can be simple linear regression or multiple regression, depending on the complexity of the model.

Main Effects Plots:
Main effects plots are graphical representations that show the effect of each factor on the response variable while keeping other factors constant. These plots help visualize the main effects of factors.

Interaction Plots:
Interaction plots illustrate the interaction effects between two or more factors. They help identify cases where the influence of one factor on the response variable depends on the level of another factor.

Residual Analysis:
Residual analysis is used to check the assumptions of the statistical model. It involves examining the residuals (the differences between observed and predicted values) to ensure that they are normally distributed and have constant variance.

Diagnostic Plots:
Diagnostic plots, such as normal probability plots and residual plots, are used to assess the validity of the statistical model and identify any outliers or influential data points.

TABLE 3.22
Factorial Designed Experiment Matrix for
Civil Engineering

Run	Concrete Mix Proportion (A)	Curing Time (B)
1	Low Cement-to-Water Ratio	7 days
2	High Cement-to-Water Ratio	7 days
3	Low Cement-to-Water Ratio	28 days
4	High Cement-to-Water Ratio	28 days

Effect Plots:

Effect plots show the estimated effects of factors on the response variable, along with confidence intervals. They provide a clear visual representation of the factors' impact.

ANOVA Tables:

ANOVA tables summarize the results of the analysis, including the sources of variation, degrees of freedom, sum of squares, mean squares, F-statistics, and p-values for each factor and interaction.

Response Surface Methodology (RSM):

RSM is a collection of techniques used to optimize response variables by modeling the relationships between factors and responses. It involves fitting mathematical models to experimental data and conducting optimization studies.

Design of Experiments (DOE) Software:

Specialized software packages, such as Minitab, JMP, or R, are commonly used to perform statistical analyses, generate factorial designs, and create graphical representations of the results.

Statistical Software:

General statistical software like R, SAS, or Python with libraries such as SciPy and statsmodels can be used for data analysis, regression, and hypothesis testing.

Statistical Tests:

Various statistical tests, such as t-tests, chi-square tests, and nonparametric tests, may be employed depending on the nature of the data and research questions.

Confidence Intervals:

Confidence intervals provide a range within which the true population parameter (e.g., mean response) is likely to fall. They help quantify the uncertainty associated with the estimates.

Effect Size Measures:

Effect size measures (e.g., Cohen's d, eta-squared) quantify the practical significance of observed effects, helping to assess the practical importance of factors.

Power Analysis:

Power analysis is used to determine the probability of detecting a significant effect if it exists. It helps in sample size determination and assessing the experiment's ability to detect effects.

Multiple Comparison Procedures:

When conducting multiple comparisons between factor levels, various procedures like Bonferroni correction or Tukey's HSD test can be used to control the familywise error rate.

Robust Design Techniques:

In situations where experimental conditions may vary or be uncertain, robust design techniques help optimize processes to be less sensitive to variations and external factors.

These statistical tools and techniques enable civil engineers to rigorously analyze and interpret the results of factorial designed experiments, leading to data-driven decisions, process improvements, and an enhanced understanding of civil engineering systems and processes.

A well-structured report for a factorial designed experiment in civil engineering should contain the following elements:

Title Page:

Title of the Report

Names of the Authors

Affiliations of the Authors

Date of Submission

Abstract:
A concise summary of the experiment, including objectives, methods, key findings, and implications.

Table of Contents:
A list of sections and subsections with page numbers for easy navigation.

List of Figures and Tables:
A separate list of figures and tables in the report with corresponding page numbers.

List of Abbreviations and Symbols (if applicable):
Definitions and explanations of any abbreviations, acronyms, or special symbols used in the report.

Introduction:
Background and context of the study, including the problem statement and objectives.
Rationale for conducting the factorial experiment.
Hypotheses or research questions to be addressed.
Brief overview of the factors and levels considered in the experiment.

Literature Review:
Review of relevant literature and previous studies related to the experiment.
Discussion of existing knowledge and gaps that the experiment aims to address.
Justification for the selection of factors and levels.

Experimental Design:
Description of the factorial design, including the factors, levels, and their combinations.
Explanation of the randomization and blocking (if applicable) procedures.
Details of the experimental setup, equipment, and data collection methods.

Data Analysis:
Presentation of the data analysis methodology, including statistical techniques used.
Summary of the statistical models, ANOVA tables, and regression equations (if applicable).
Interpretation of main effects, interaction effects, and significance levels.
Graphical representations of key findings (e.g., main effects plots, interaction plots).

Results:
Presentation of the experimental results in a clear and organized manner.
Tables, figures, and charts to display data, including means, standard deviations, and confidence intervals.
Highlight significant findings and trends.
Report effect sizes if applicable.

Discussion:
Interpretation of the results in the context of the research objectives.
Explanation of the practical implications of significant findings.
Comparison of the results with existing literature.
Address any unexpected outcomes or limitations of the study.

Conclusions:
A concise summary of the key findings and their implications.
Statements on whether the hypotheses were supported or rejected.
Suggestions for future research or improvements to the experimental design.

Recommendations (if applicable):
Practical recommendations for civil engineering applications or processes based on the results.

Acknowledgments (if applicable):
Recognition of individuals, organizations, or funding sources that contributed to the experiment.

References:
A list of all sources cited in the report, following a specific citation style (e.g., APA, IEEE, Chicago).

Appendices (if needed):
> Supplementary materials, such as raw data, additional tables, or detailed descriptions of methodologies.
> Any additional information that enhances the understanding of the experiment but is not essential in the main body of the report.

Graphical and Visual Materials:
> Clear and well-labeled figures, tables, and charts that support the findings and explanations in the report.

Executive Summary (optional):
> A brief summary of the key findings and recommendations, often included for busy readers or decision-makers.

Remember to format the report consistently, use clear and concise language, and provide adequate context and explanations to make the report accessible to both technical and nontechnical readers. Proper citation and referencing are essential to give credit to previous research and maintain academic integrity.

3.3.2 RESPONSE SURFACE METHODOLOGY (RSM)

RSM experiments are widely used in civil engineering to optimize and model complex processes, systems, and designs. RSM is a statistical and mathematical approach that helps engineers and researchers study the relationships between multiple input factors (independent variables) and one or more response variables (dependent variables). These experiments are valuable in civil engineering for various purposes, including:

Structural Design and Analysis:
> Optimizing the design of structures such as bridges, buildings, and dams to meet specific performance criteria while minimizing material usage or construction costs.
> Investigating the effect of variables like material properties, geometrical dimensions, and load conditions on structural integrity and safety.

Material Properties:
> Determining the optimal mixtures of concrete, asphalt, and other construction materials to achieve desired properties like strength, durability, and workability.
> Studying the influence of factors such as curing time, temperature, and additives on material performance.

Environmental Impact Assessment:
> Evaluating the environmental impact of construction projects and determining the optimal strategies to minimize ecological disruption, erosion, and pollution.

Geotechnical Engineering:
> Analyzing soil properties and ground conditions to optimize foundation design and mitigate potential risks associated with settlement, liquefaction, or slope stability.
> Investigating the effects of soil improvement techniques on bearing capacity and settlement.

Hydraulic Engineering:
> Optimizing the design of hydraulic structures, such as culverts, spillways, and drainage systems, to control water flow and reduce flood risks.
> Studying the impact of flow rates, channel geometries, and sediment transport on river and coastal environments.

Transportation Engineering:
> Modeling traffic flow and congestion to optimize road and traffic signal designs for improved safety and efficiency.

Investigating the effects of variables like road geometry, vehicle types, and traffic patterns on road infrastructure.

Construction Processes:

Optimizing construction processes and schedules to minimize project duration and costs.

Studying the influence of factors like equipment usage, labor allocation, and resource allocation on project efficiency.

Quality Control:

Implementing SPC techniques to monitor and improve the quality of construction materials, processes, and finished products.

Identifying and mitigating sources of variability and defects in construction projects.

Risk Assessment and Management:

Assessing the impact of uncertain factors on project outcomes and identifying strategies to mitigate risks.

Using RSM to model the relationship between risk factors and project performance.

In these applications, RSM experiments typically involve the systematic variation of input factors, data collection, and statistical analysis to develop predictive models. DOE techniques, such as CCDs and factorial designs, are commonly used to plan and conduct RSM experiments efficiently. The resulting models can help civil engineers make data-driven decisions, optimize designs, improve processes, and achieve desired project outcomes while minimizing costs and resources.

STEPS

Developing an RSM designed experiment in civil engineering involves several key steps to systematically study the relationships between input factors and response variables. Here are the general steps to develop an RSM experiment:

1. **Define the Objectives:**
   - Clearly specify the objectives of the experiment. Determine what you want to optimize, improve, or understand within the civil engineering context.
2. **Identify the Factors (Independent Variables):**
   - Identify the key factors or independent variables that may influence the response variables. These factors can be material properties, geometrical dimensions, environmental conditions, or process parameters.
3. **Determine the Levels of Factors:**
   - Define the ranges or levels for each factor. Decide the minimum and maximum values or categories that each factor will take during the experiment.
4. **Select Response Variables (Dependent Variables):**
   - Choose one or more response variables that reflect the outcomes or performance measures of interest. These variables can be structural strength, durability, cost, time, or any relevant measure.
5. **Design the Experiment:**
   - Select an appropriate experimental design, such as a CCD or a Box-Behnken design. These designs are commonly used in RSM experiments for efficient data collection.
   - Determine the number of experimental runs or trials required to generate data for model development. Consider the number of factors and levels.
   - Randomize the order of experimental runs to reduce the impact of uncontrolled variables.
6. **Conduct the Experiments:**
   - Perform the experimental runs according to the designed plan. Ensure accurate data collection, precise measurements, and control over experimental conditions.

7. **Collect Data**:
   - Record the values of response variables for each experimental run. Maintain consistency and accuracy in data collection.
8. **Analyze the Data**:
   - Use statistical software to analyze the data and build predictive models. Common techniques include regression analysis, ANOVA, and response surface modeling.
   - Fit response surface models to describe the relationships between factors and responses. Evaluate the significance of factors and their interactions.
   - Identify any outliers or influential data points that may affect the model.
9. **Optimize the Process or Design**:
   - Use the response surface models to identify optimal conditions or settings for the factors that lead to desired outcomes. This is often done by maximizing or minimizing the response variables.
   - Conduct sensitivity analyses to assess the robustness of the optimized conditions.
10. **Interpret the Results**:
    - Interpret the statistical results and the practical implications for civil engineering applications.
    - Understand how changes in factors affect the response variables and make informed decisions based on the findings.
11. **Validation and Verification**:
    - Validate the response surface models by conducting additional experiments under the optimized conditions to confirm the predicted outcomes.
    - Verify the reliability of the models by comparing predicted results with actual observations.
12. **Report and Documentation**:
    - Prepare a detailed report summarizing the experiment, objectives, methods, results, and conclusions.
    - Include graphical representations of the response surfaces, contour plots, and any other relevant figures.
    - Present the recommended optimal conditions and their implications for civil engineering projects.
13. **Implementation**:
    - Implement the optimized conditions or design changes in practical civil engineering applications.
    - Monitor and evaluate the performance of the implemented solutions over time.
14. **Continuous Improvement**:
    - Continuously monitor and refine the processes or designs based on feedback and further experimentation if necessary.

Throughout the experiment, it's important to maintain a systematic and well-documented approach to ensure the reliability and repeatability of the results. Collaboration with statisticians or experts in experimental design and analysis may also be beneficial, especially for complex RSM experiments in civil engineering.

PITFALLS AND REMEDIES

RSM designed experiments in civil engineering can be powerful tools for optimizing processes and designs, but they are not without challenges. Here are some common pitfalls and remedies associated with RSM experiments in this field:

Pitfalls:

Insufficient Data Points:
  Pitfall: Having too few experimental runs can result in an underfit model, making it difficult to accurately capture the underlying response surface.
  Remedy: Increase the number of experimental runs within practical constraints to improve the quality of the model. Consider using efficient designs like CCDs.

Overfitting:
  Pitfall: Including too many terms or interactions in the model can lead to overfitting, where the model fits noise in the data rather than the true relationships.
  Remedy: Apply model selection techniques, such as stepwise regression or cross-validation, to identify and retain only significant terms in the model.

Lack of Model Validation:
  Pitfall: Failing to validate the response surface models can lead to inaccurate predictions and suboptimal results when implementing recommendations.
  Remedy: Conduct additional experiments under the optimal conditions predicted by the model to verify its accuracy. Use statistical tests to assess model validity.

Neglecting Factor Interactions:
  Pitfall: Ignoring or underestimating the interactions between factors can lead to incomplete understanding of the system and suboptimal solutions.
  Remedy: Include interaction terms in the response surface models and thoroughly analyze their effects on the response variables.

Violating Assumptions:
  Pitfall: Assuming that the underlying relationships between factors and responses are linear when they are not can lead to model inaccuracies.
  Remedy: Explore alternative models, such as quadratic or higher-order models, to account for nonlinearity if necessary. Use diagnostic plots to assess model fit.

Inadequate Control over Variables:
  Pitfall: Failing to control or monitor uncontrolled variables (extraneous factors) can introduce noise and variability into the experiments.
  Remedy: Identify and control as many extraneous factors as possible. Conduct REs to minimize the impact of uncontrolled variables.

Resource Constraints:
  Pitfall: Limited resources, such as time, budget, or materials, can restrict the number of experimental runs that can be performed.
  Remedy: Use efficient experimental designs, prioritize factors based on their importance, and consider sequential experimentation to make the most of available resources.

Remedies:

Expert Guidance: Seek the assistance of statisticians or experts in experimental design and analysis to ensure the proper planning, execution, and analysis of RSM experiments.

Thorough Data Collection: Ensure accurate and consistent data collection during experimental runs. Minimize measurement errors and outliers that can affect model quality.

Validation and Verification: Validate the response surface models through additional experiments and verify the predicted optimal conditions before implementation.

Sensitivity Analysis: Conduct sensitivity analyses to assess the robustness of the optimized conditions and understand the potential impact of model uncertainties.

Documentation: Maintain comprehensive records of experimental procedures, data, and analysis to facilitate transparency and reproducibility.

Collaboration: Foster collaboration between engineers, statisticians, and domain experts to leverage their respective knowledge and skills for successful RSM experiments.

Continuous Improvement: Use RSM as a continuous improvement tool, regularly reviewing and updating models as new data becomes available or as project objectives evolve.

By addressing these pitfalls and implementing the suggested remedies, civil engineers can maximize the effectiveness of RSM experiments, leading to more informed decisions and optimized civil engineering processes and designs.

EXAMPLE

In RSM experiments in civil engineering, a matrix is used to plan and organize the experimental runs. The choice of matrix design depends on the specific goals and factors under investigation. Here's an example of a matrix for an RSM experiment in civil engineering:

Objective I: To optimize the compressive strength of concrete by varying factors such as the water-cement ratio (X1) and curing temperature (X2).

Response Variable: Compressive Strength (Y)

Factorial Design: Central Composite Design (CCD)
In this example, we'll use a CCD, which is a commonly used RSM design. CCD combines full-factorial runs with axial runs (star points) and center points for robustness.

Step 1: Define Factor Levels
- Low (−1) and high (+1) levels for each factor:
  - X1 (Water-Cement Ratio): 0.4 (Low), 0.6 (High)
  - X2 (Curing Temperature, °C): 20 (Low), 40 (High)

Step 2: Determine the Center Point
- Decide on the center point values for each factor:
  - X1 (Water-Cement Ratio): 0.5 (Center)
  - X2 (Curing Temperature, °C): 30 (Center)

Step 3: Create Factorial Points
- Generate factorial points by combining the levels of X1 and X2:

Run	X1 (Water-Cement Ratio)	X2 (Curing Temperature, °C)
1	−1	−1
2	+1	−1
3	−1	+1
4	+1	+1

Step 4: Determine Axial Points (Star Points)
- Generate axial points by increasing each factor to its maximum and minimum values, keeping the other factor at its center:

Run	X1 (Water-Cement Ratio)	X2 (Curing Temperature, °C)
5	0 (Center)	−1
6	0 (Center)	+1
7	−1	0 (Center)
8	+1	0 (Center)

Step 5: Include Center Points
- Add center points to assess the reproducibility of the experiment (Table 3.23).

This matrix design allows for 10 experimental runs with varying levels of water-cement ratio and curing temperature to optimize the compressive strength of concrete. The data collected from these runs will be used to build a response surface model and determine the optimal conditions for achieving the desired compressive strength.

Objective II: To optimize the mix proportions of concrete to achieve maximum compressive strength while minimizing the cost of materials.

Factors:

1. **Cement Content (Factor A)**:
   - Range: 300 kg/m³ to 400 kg/m³
2. **Water-to-Cement Ratio (Factor B)**:
   - Range: 0.35 to 0.45
3. **Aggregate Size (Factor C)**:
   - Level 1: Coarse aggregate size (20 mm)
   - Level 2: Fine aggregate size (5 mm)

Response Variable: Compressive Strength (measured in megapascals, MPa)

Matrix for RSM Designed Experiment:
In this example, an RSM design is used to optimize the mix proportions of concrete for maximum compressive strength while considering the factors of cement content, water-to-cement ratio, and aggregate size. The matrix is shown in Table 3.24.

TABLE 3.23

CCD Combines Full-factorial Runs with Axial Runs (Star Points) and Center Points for Robustness

Run	X1 (Water-Cement Ratio)	X2 (Curing Temperature, °C)
9	0.5 (Center)	30 (Center)
10	0.5 (Center)	30 (Center)

TABLE 3.24

Response Surface Methodology (RSM) Example for Civil Engineering

Run	Cement Content (A)	Water-to-Cement Ratio (B)	Aggregate Size (C)
1	300 kg/m³	0.35	Coarse aggregate
2	400 kg/m³	0.45	Fine aggregate
3	350 kg/m³	0.40	Coarse aggregate
4	375 kg/m³	0.38	Fine aggregate
5	325 kg/m³	0.42	Coarse aggregate
6	375 kg/m³	0.38	Coarse aggregate
7	325 kg/m³	0.42	Fine aggregate
8	350 kg/m³	0.40	Fine aggregate
9	350 kg/m³	0.35	Fine aggregate
10	400 kg/m³	0.45	Coarse aggregate

In this matrix, each row represents a specific experimental run where a concrete mixture is prepared with particular combinations of cement content, water-to-cement ratio, and aggregate size. The compressive strength of each concrete mixture is tested and recorded.

Using statistical analysis and response surface modeling, you can determine the optimal mix proportions that maximize compressive strength while considering cost constraints. RSM helps you understand the relationships between the factors and the response variable, allowing you to make informed decisions in civil engineering applications.

REPORTING

RSM designed experiments in civil engineering often involve the use of various statistical tools and techniques to analyze and optimize complex relationships between multiple factors and response variables. Here are some of the commonly used statistical tools in RSM experiments in civil engineering:

Analysis of Variance (ANOVA): ANOVA is used to assess the significance of factors and their interactions on the response variable. It helps identify which factors have a significant impact on the response.

Response Surface Analysis: This involves fitting response surface models (e.g., linear, quadratic, or higher-order models) to the experimental data. These models help describe the relationships between factors and responses.

Design of Experiments (DOE): RSM often employs various DOE techniques, including full-factorial designs, fractional factorial designs, and CCDs, to plan the experiments efficiently.

Regression Analysis: Regression models are used to estimate the coefficients of the response surface models. These models can be linear, quadratic, or higher-order, depending on the complexity of the relationships.

Optimization Algorithms: Optimization techniques such as the steepest ascent/descent, gradient search, or response surface optimization are employed to find the optimal conditions that maximize or minimize the response variable.

Contour Plots: Contour plots are graphical representations of response surfaces, showing how the response variable changes with different combinations of factors. They help visualize the optimal regions.

ANOVA for Quadratic Models: In cases where quadratic response surface models are used, ANOVA is employed to assess the significance of quadratic terms and interactions.

Diagnostic Plots: Diagnostic plots, such as residual plots and normal probability plots, are used to check the assumptions and goodness-of-fit of the response surface models.

Model Adequacy Tests: Tests like lack-of-fit tests and pure error tests are conducted to determine whether the response surface models adequately describe the data.

Sensitivity Analysis: Sensitivity analysis assesses the robustness of the optimal conditions obtained from the response surface models by varying factors within specified ranges.

Monte Carlo Simulation: Monte Carlo simulations can be used to assess the uncertainty associated with the response surface models and their predictions.

RSM Software: Specialized software packages like Minitab, JMP, or Design-Expert are often used to perform statistical analysis, generate response surface plots, and optimize experimental conditions.

Sequential Experimentation: Sequential experimentation techniques are employed to continually refine the response surface models and achieve better optimization results with limited resources.

Statistical Tolerance Analysis: This helps assess the impact of variations in factors and responses, ensuring that designed solutions are robust to real-world variations.

Taguchi Methods: Taguchi methods are sometimes incorporated to perform robust optimization by considering factors' variability and their interactions.

These statistical tools are applied iteratively throughout the RSM experiment process, from initial design and data collection to model building, optimization, and validation. They help civil engineers make informed decisions and optimize various aspects of civil engineering processes and designs.

A comprehensive report for an RSM designed experiment in civil engineering typically includes various elements to document the experimental process, results, and conclusions. Here are the key elements commonly found in such a report:

Title Page:
   Title of the Experiment.
   Names of the authors and affiliations.
   Date of the report.
Abstract:
   A brief summary of the experiment's objectives, methods, key findings, and conclusions.
Table of Contents:
   An organized list of sections and subsections within the report.
List of Figures and Tables:
   A list of all figures and tables included in the report, along with their respective page numbers.
List of Abbreviations and Symbols (if applicable):
   A compilation of any abbreviations or symbols used throughout the report, along with their definitions.
Introduction:
   Background information explaining the context and motivation for the experiment.
   Objectives and research questions to be addressed.
   A brief review of relevant literature.
Experimental Design:
   Description of the experimental design, including the choice of factors and levels.
   Mention of the response variable(s) and their measurements.
   Details of the chosen RSM design (e.g., CCD).
   Information on how the experiments were randomized, split, or blocked if applicable.
Materials and Methods:
   Description of the materials, equipment, and tools used in the experiment.
   Procedures followed in conducting the experiments.
   Any special techniques or instruments employed.
   Data collection methods and quality control measures.
Data Analysis:
   Presentation of the experimental data in tabular form.
   Data preprocessing steps (if any), such as transformation or normalization.
   Application of statistical tools, including ANOVA, regression analysis, and response surface modeling.
   Presentation of response surface plots and contour plots to visualize the relationships between factors and responses.
   Explanation of how the model was fitted to the data and model adequacy checks (e.g., residual analysis).
Results:
   Presentation of the main findings, including parameter estimates, p-values, and confidence intervals.
   Discussion of the significance of factors and their interactions.
   Identification of the optimal conditions for the response variable(s) based on the response surface model.

Discussion:

Interpretation of the results in the context of the research objectives.

Explanation of any trends, patterns, or unexpected observations.

Comparison of the findings with previous studies or industry standards.

Discussion of the practical implications and potential applications of the results.

Conclusion:

Summarization of the key findings and their implications.

Statements regarding whether the research objectives were achieved.

Suggestions for future research or improvements.

References:

A list of all cited sources, including research papers, books, and relevant literature.

Appendices (if applicable):

Additional details, data tables, or supplementary information that may be too lengthy for the main body of the report.

Any computer code or software used for data analysis.

Acknowledgments (if applicable):

Recognition of individuals, organizations, or funding sources that contributed to the research.

Supporting Material (if applicable):

Any additional supporting materials, such as raw data, photographs, or diagrams.

The structure and content of the report may vary depending on the specific experiment, its complexity, and the requirements of the civil engineering project or research study. However, these elements provide a general framework for organizing and documenting RSM experiments in civil engineering.

3.3.3 Fractional Factorial Experiments

Fractional factorial experiments are a valuable experimental design technique used in civil engineering for studying the effects of multiple factors (variables) on a particular response (outcome) while conducting experiments more efficiently. In civil engineering, these experiments are often used when there are limited resources, time constraints, or cost considerations, making it impractical to run a full-factorial experiment that would involve testing all possible combinations of factor levels. Here's how fractional factorial experiments are used in civil engineering:

Factor Screening: Civil engineering projects may involve numerous factors that could potentially affect a response variable. Fractional factorial experiments help identify the most influential factors among a large set by conducting only a fraction of the possible experiments. By screening out less important factors, engineers can focus their resources on the most critical ones.

Efficient Resource Utilization: Civil engineering experiments can be resource-intensive, involving physical testing, data collection, and analysis. Fractional factorial experiments reduce the number of experimental runs required, saving time, materials, and costs.

Optimization: Once the significant factors are identified through screening, engineers can conduct follow-up experiments, possibly using RSM, to optimize the response variable within a limited budget or timeframe. Fractional factorial experiments provide initial insights for setting factor levels.

Quality Control: Civil engineering projects often involve construction and infrastructure development, where maintaining quality is crucial. Fractional factorial experiments can help identify the key factors affecting quality and guide quality control measures.

Sensitivity Analysis: Engineers can assess how variations in factors affect the overall performance of a civil engineering system or structure. Fractional factorial experiments allow for systematic sensitivity analysis to ensure robustness against variations.

Risk Assessment: Civil engineering projects may face uncertainties and risks due to variations in factors like material properties, environmental conditions, or loads. Fractional factorial experiments help quantify the impact of these variations and assess project risks.

Environmental Impact Assessment: In environmental engineering, Fractional factorial experiments can be used to study the effects of various factors on environmental parameters, such as pollutant dispersion or water quality, while minimizing the number of experiments.

Structural Design: Fractional factorial experiments can be employed to study the effects of multiple design parameters (e.g., material properties, geometry) on the structural performance of buildings, bridges, or other infrastructure.

Geotechnical Engineering: These experiments can be applied in soil mechanics and foundation design to understand the influence of factors like soil type, compaction, and groundwater levels on foundation stability.

Transportation Engineering: In transportation projects, fractional factorial experiments can be used to evaluate the impact of various factors on traffic flow, pavement performance, or safety.

Overall, fractional factorial experiments are a valuable tool in civil engineering for efficient and cost-effective experimentation, aiding in decision-making, optimization, quality control, and risk management in various civil engineering applications.

Steps

Developing a fractional factorial experiment for civil engineering involves several key steps to efficiently study the effects of multiple factors on a response variable while conducting only a fraction of the full-factorial experiments. Here are the steps to develop a fractional factorial designed experiment:

1. **Define the Research Objectives:**
   - Clearly articulate the specific goals and objectives of your civil engineering study.
   - Determine the response variable (output) that you want to optimize, control, or understand.
2. **Identify Factors of Interest:**
   - List all the potential factors (independent variables) that may influence the response variable.
   - Factors could include material properties, design parameters, environmental conditions, construction methods, or any other relevant variables.
3. **Determine Factor Levels:**
   - Specify the range or levels at which each factor will be tested.
   - Determine whether factors should be tested at discrete levels or as continuous variables.
4. **Select Fractional Factorial Design:**
   - Choose an appropriate fractional factorial design based on the number of factors, available resources, and the desired level of experimental efficiency.
   - Common designs include half-fraction, one-quarter-fraction, or other fractional designs.
5. **Determine the Fractional Design Matrix:**
   - Use a design matrix (often generated by specialized software) to select a fraction of the possible experimental runs.

- The matrix specifies which combinations of factor levels will be tested and which will be omitted.

6. **Randomize and Conduct Experiments:**
   - Randomly order the experimental runs to minimize the effects of uncontrolled variables.
   - Conduct the selected experiments following the combinations specified in the design matrix.

7. **Collect Data:**
   - Carefully record data for the response variable for each experimental run.
   - Ensure data accuracy and precision.

8. **Perform Data Analysis:**
   - Analyze the collected data using statistical methods to identify significant factors and their effects on the response variable.
   - Calculate the main effects and interactions between factors.
   - Consider using statistical software or specialized tools for data analysis.

9. **Optimize or Make Inferences:**
   - If the objective is optimization, use the results to find the optimal factor settings for maximizing or minimizing the response variable.
   - If the objective is inference, draw conclusions about the factors' effects and their practical implications.

10. **Validation (Optional):**
    - Conduct additional experiments to validate the findings from the fractional factorial experiment, especially if further precision is required.

11. **Report and Interpret Results:**
    - Prepare a detailed report that includes the experimental design, data analysis, findings, and conclusions.
    - Interpret the results in the context of the research objectives and civil engineering applications.

12. **Recommendations and Actionable Insights:**
    - Based on the findings, provide recommendations for practical applications, improvements, or further research in civil engineering projects.

13. **Documentation and Reporting:**
    - Document all experimental details, including factors, levels, design matrix, data collection procedures, and statistical analyses.
    - Present results using tables, graphs, and visualizations to aid in understanding.

14. **Continuous Improvement (Iterative Process):**
    - Fractional factorial experiments may serve as a preliminary phase of experimentation. Depending on the outcomes, further experiments, including full-factorial designs or RSM, may be considered to refine the understanding of factor interactions.

15. **Peer Review (Optional):**
    - Depending on the significance and scale of the civil engineering project, consider peer review to validate the experimental design and results.

16. **Implementation:**
    - If applicable, implement the insights gained from the fractional factorial experiment in the design, construction, or management of civil engineering projects.

17. **Documentation Retention:**
    - Keep thorough records and documentation for future reference, quality assurance, and compliance purposes.

Fractional factorial experiments offer an efficient and systematic approach to exploring the impact of multiple factors on civil engineering projects while conserving resources. The steps outlined above help ensure the successful planning, execution, and interpretation of such experiments in civil engineering research and practice.

PITFALLS AND REMEDIES

Fractional factorial designed experiments are powerful tools in civil engineering for efficiently studying the effects of multiple factors on a response variable. However, like any experimental approach, they come with potential pitfalls. Here are some common pitfalls and remedies when using fractional factorial experiments in civil engineering:

Pitfall 1: Incomplete Factor Resolution

Issue: Fractional factorial designs inherently involve omitting some factor combinations, which can lead to the inability to resolve certain interactions between factors.

Remedy: Consider using higher fractions or different experimental designs if full-factor resolution is crucial. Alternatively, follow up with additional experiments to resolve interactions.

Pitfall 2: Confounding of Effects

Issue: Fractional factorial designs may confound (combine) the effects of certain factors or interactions, making it challenging to distinguish their individual contributions.

Remedy: Use advanced statistical methods to correct effects and estimate individual factor impacts. RSM can help with this.

Pitfall 3: Limited Exploration

Issue: Fractional factorial experiments may not explore the entire factor space thoroughly, potentially missing some important factor combinations.

Remedy: Consider supplementing fractional factorial experiments with additional runs or using screening designs to identify key factors for further investigation.

Pitfall 4: Extrapolation of Results

Issue: Extrapolating results from fractional factorial experiments to factor levels outside the tested range can lead to inaccurate predictions.

Remedy: Restrict conclusions and recommendations to the range of factor levels tested in the experiment. For predictions beyond that range, conduct additional experiments or use models cautiously.

Pitfall 5: Lack of Replication

Issue: Limited replication of experiments can result in variability in the data, making it challenging to draw robust conclusions.

Remedy: Incorporate replication in the experimental design to obtain more reliable estimates of factor effects and reduce variability.

Pitfall 6: Ignoring Time Dynamics

Issue: Civil engineering projects often involve factors that vary with time (e.g., weather, traffic), and fractional factorial designs may not capture these dynamics.

Remedy: Consider incorporating time as a factor or conducting experiments over different time periods to account for temporal variations.

Pitfall 7: Neglecting Randomization

Issue: Failing to randomize the order of experimental runs can introduce bias or uncontrolled variability.

Remedy: Randomize the order of experiments to minimize the influence of uncontrolled factors and ensure that results are not biased by external factors.

Pitfall 8: Inadequate Sample Size

Issue: Insufficient sample size can lead to underpowered experiments and unreliable conclusions.

Remedy: Conduct power and sample size calculations before the experiment to ensure an adequate number of runs to detect significant effects.

Pitfall 9: Lack of Expertise

    Issue: Inexperienced experimenters may struggle with proper design and analysis of fractional factorial experiments.

    Remedy: Seek guidance or collaborate with statisticians or experts in experimental design to plan and execute the experiments effectively.

Pitfall 10: Limited Communication

    Issue: Inadequate communication of experimental details, results, and implications can hinder the usefulness of fractional factorial experiments.

    Remedy: Document the experiment comprehensively and communicate findings effectively within the engineering team or organization.

Fractional factorial experiments can provide valuable insights into civil engineering, but they require careful planning, execution, and analysis to avoid these pitfalls. Engineers and researchers should be aware of these potential issues and take appropriate measures to address them, ensuring the validity and usefulness of the experimental results.

EXAMPLE

Fractional factorial experiments are designed experiments that allow researchers to study the effects of factors while conducting only a fraction of the full set of experimental runs. This approach is useful when there are a large number of factors, and conducting a full-factorial experiment is impractical. Here's an example of a matrix for a fractional factorial experiment in civil engineering:

Objective: To investigate the effects of various factors on the compressive strength of concrete in a bridge construction project.

Factors:

1. **Cement Type (Factor A):**
   - Level 1: Type I Portland Cement
   - Level 2: Type II Portland Cement
   - Level 3: Type III Portland Cement
2. **Curing Method (Factor B):**
   - Level 1: Water Curing
   - Level 2: Steam Curing
3. **Aggregate Size (Factor C):**
   - Level 1: Coarse Aggregate Size (20 mm)
   - Level 2: Fine Aggregate Size (5 mm)
4. **Cement-to-Water Ratio (Factor D):**
   - Level 1: Low Ratio (0.35)
   - Level 2: High Ratio (0.45)

Response Variable: Compressive Strength (measured in megapascals, MPa)

Fractional Factorial Matrix:

In this example, a $2^{(4-1)}$ fractional factorial design is used, which reduces the number of experimental runs to one-eighth of a full-factorial design. The matrix is shown in Table 3.25.

In this matrix, each row represents a specific experimental run with a combination of factors and levels. By conducting this fractional factorial experiment, you can assess the main effects of the factors (cement type, curing method, aggregate size, and cement-to-water ratio) on compressive strength with significantly fewer runs than a full-factorial design while still capturing valuable information about their impact. This approach is efficient when dealing with multiple factors in civil engineering experiments.

TABLE 3.25

Civil Engineering Fractional Factorial Designed Experiment Matrix Example

Run	Cement Type (A)	Curing Method (B)	Aggregate Size (C)	Cement-to-Water Ratio (D)
1	Type I	Water Curing	Coarse Aggregate	Low Ratio
2	Type II	Steam Curing	Fine Aggregate	High Ratio
3	Type III	Water Curing	Fine Aggregate	Low Ratio
4	Type I	Steam Curing	Fine Aggregate	High Ratio
5	Type II	Water Curing	Coarse Aggregate	High Ratio
6	Type III	Steam Curing	Coarse Aggregate	Low Ratio
7	Type I	Water Curing	Coarse Aggregate	High Ratio
8	Type II	Steam Curing	Fine Aggregate	Low Ratio

REPORTING

Statistical tools are essential for analyzing the data obtained from fractional factorial designed experiments in civil engineering. These tools help researchers and engineers understand the relationships between factors and responses, identify significant effects, and make informed decisions. Here are some common statistical tools used in fractional factorial experiments in civil engineering:

Analysis of Variance (ANOVA): ANOVA is a fundamental statistical technique used to partition the variability in the response variable into components attributed to different factors and their interactions. It helps identify statistically significant factors and interactions.

Main Effects Plots: Main effects plots display the estimated effects of each factor on the response variable while keeping other factors constant. These plots provide a visual representation of how individual factors impact the response.

Interaction Plots: Interaction plots illustrate the interactions between two or more factors. They help identify situations where the effect of one factor depends on the level of another factor, indicating complex relationships.

Contour Plots: Contour plots are graphical representations of the response surface that show how the response variable changes as two factors vary simultaneously. They are particularly useful in understanding factor interactions.

Response Surface Methodology (RSM): RSM involves fitting mathematical models to experimental data to describe the relationship between factors and the response. This allows for optimization and prediction of responses within the experimental region.

Half-Normal Probability Plots: These plots help identify the most significant factors by assessing the magnitude of the effects relative to experimental error. Factors with effects above the "hockey stick" line are considered significant.

Fractional Factorial Design Matrices: These matrices specify the experimental runs to be conducted in the fractional factorial design. They help plan and document the experiments systematically.

Diagnostic Plots: Diagnostic plots, such as residual plots, are used to check the adequacy of the statistical model and the assumptions of normality and constant variance.

Confidence Intervals: Confidence intervals provide a range of values within which a parameter (e.g., factor effect) is likely to fall. They help quantify the uncertainty associated with estimated effects.

Pareto Charts: Pareto charts rank factors by their magnitude of effect, allowing engineers to focus on the most influential factors for further investigation or optimization.

Regression Analysis: Regression analysis is used to build predictive models that describe the relationship between factors and the response variable. It can be used to identify significant factors and estimate their effects.

Design of Experiments (DOE) Software: Specialized software packages like Minitab, JMP, or Design-Expert are often used to design experiments, perform statistical analysis, and visualize results.

Power and Sample Size Calculations: These calculations help determine the minimum sample size required to achieve a desired level of statistical power to detect significant effects.

Randomization and Blocking Techniques: Randomization and blocking are essential for controlling sources of variation and ensuring that the experiment's conclusions are not influenced by external factors.

ANOVA Tables: ANOVA tables summarize the results of the analysis, including degrees of freedom, sums of squares, mean squares, and F-statistics, facilitating hypothesis testing.

Effect Plots: Effect plots provide graphical representations of the factor effects, showing how the response variable changes as factors are varied.

These statistical tools, when used appropriately, enable civil engineers and researchers to extract valuable insights from fractional factorial designed experiments, optimize processes, and make informed decisions in various civil engineering applications. The choice of tools depends on the specific research objectives and the complexity of the experiment.

A well-structured report is crucial for documenting and communicating the results of a fractional factorial designed experiment in civil engineering. The report should provide a clear and concise overview of the experiment, the methodology, results, and conclusions. Here are the key elements that should be included in a report for a fractional factorial designed experiment in civil engineering:

Title Page:
> Title of the Report
> Name(s) of Author(s)
> Affiliation(s)
> Date of Report

Abstract:
> A brief summary of the experiment's objectives, methodology, key findings, and conclusions. It should provide a concise overview of the entire report.

Table of Contents:
> A list of all sections, subsections, and their page numbers for easy navigation.

List of Figures and Tables:
> A list of all figures and tables used in the report, along with their respective page numbers.

Introduction:
> Provide background information on the problem or process being studied.
> State the objectives and goals of the experiment.
> Explain the relevance and significance of the experiment in the context of civil engineering.

Literature Review:
> Review relevant literature, prior studies, and existing research related to the experiment's subject matter.
> Discuss the current state of knowledge in the field.

Experimental Design:
> Describe the fractional factorial experimental design used, including the factors, levels, and their combinations.
> Present the rationale behind the design choices.

Materials and Methods:
> Provide details on the materials, equipment, and tools used in the experiment.
> Describe the experimental procedure and any special techniques employed.
> Explain how data were collected and any measurements taken.

Results:
Present the collected data in an organized manner using tables, charts, and graphs.
Summarize the main findings, including significant effects of factors and interactions.
Include descriptive statistics and measures of central tendency and variability.

Analysis and Interpretation:
Discuss the statistical analysis of the data, including ANOVA results and effect sizes.
Interpret the significance of the observed effects and interactions.
Use graphical representations, such as contour plots or interaction plots, to aid in interpretation.

Discussion:
Analyze the implications of the results in the context of the experiment's objectives.
Discuss how the findings align with or deviate from existing literature.
Consider practical implications for civil engineering applications.

Conclusions:
Summarize the key findings and their importance.
Address whether the experiment's objectives were achieved.
Provide recommendations or insights for future work or applications.

Limitations:
Identify any limitations or constraints encountered during the experiment.
Discuss potential sources of error or bias.

Recommendations:
Offer recommendations for further research or improvements to the experimental design.
Suggest how the findings can be applied in practical civil engineering scenarios.

Acknowledgments:
Acknowledge individuals or organizations that contributed to the experiment or provided support.

References:
Cite all sources, including research papers, books, and relevant standards, following a standardized citation style (e.g., APA, IEEE).

Appendices:
Include any supplementary information, raw data, calculations, or additional details that support the report but are not part of the main text.

Glossary of Terms (Optional):
Define technical terms or acronyms used in the report for clarity.

List of Symbols and Abbreviations (Optional):
Provide a list of symbols and abbreviations used in the report, along with their meanings.

Experimental Data (Optional):
If appropriate, include the raw data in an appendix or as a separate dataset for reference.

Remember that the structure and content of the report may vary based on the specific requirements of the experiment, the preferences of the organization or institution, and the intended audience. Adhering to a clear and organized format ensures that the results and insights from the fractional factorial designed experiment are effectively communicated to others in the civil engineering field.

3.3.4 Central Composite Design (CCD)

CCD is a widely used experimental design technique in civil engineering and various other fields. In civil engineering, CCD experiments are employed to study and optimize processes, materials, and structural components. Here are some common applications of CCD in civil engineering:

Concrete Mix Design: Civil engineers use CCD experiments to optimize concrete mixtures by varying factors such as cement content, water-cement ratio, aggregate types, and admixtures. The objective is to achieve desired properties such as compressive strength, workability, and durability while minimizing costs.

Structural Analysis and Design: CCD can be used to study the behavior of structural components, such as beams, columns, and foundations, under various loading conditions. Engineers can investigate factors like material properties, geometry, and reinforcement to optimize structural performance and safety.

Geotechnical Engineering: In geotechnical engineering, CCD experiments can help assess the effects of soil properties, compaction methods, and drainage systems on the stability of embankments, slopes, and foundations. This is crucial for infrastructure projects like roads, bridges, and buildings.

Construction Process Optimization: Civil engineers can use CCD to optimize construction processes, such as concrete curing methods, tunnel excavation techniques, and formwork design. This can lead to cost savings, improved efficiency, and reduced construction time.

Environmental Impact Assessment: CCD experiments can be employed to study the environmental impact of civil engineering projects, including factors like pollutant dispersion, noise levels, and groundwater contamination. Engineers can optimize mitigation measures to minimize adverse effects.

Pavement Design: For road and pavement design, CCD experiments help engineers determine the optimal mix of materials, thickness, and compaction methods to ensure long-lasting and cost-effective pavement structures.

Hydraulic Engineering: CCD can be used in hydraulic engineering to optimize the design of channels, culverts, and flood control structures. Factors such as flow rates, channel geometry, and sediment transport can be studied.

Material Testing and Characterization: Civil engineers use CCD experiments to characterize the mechanical properties of construction materials like steel, asphalt, and composites. This information is crucial for material selection and structural design.

Sustainability and Green Building: CCD can be applied to assess the sustainability of civil engineering projects, considering factors such as energy efficiency, renewable energy integration, and eco-friendly materials.

Bridge Design and Rehabilitation: Engineers use CCD to optimize the design of bridges, taking into account factors like load-bearing capacity, structural materials, and corrosion resistance. CCD can also be used in the rehabilitation and maintenance of existing bridges.

Risk Assessment: CCD experiments can be used to assess and mitigate risks associated with civil engineering projects, including factors like construction accidents, natural disasters, and project delays.

In each of these applications, CCD experiments help civil engineers systematically explore the relationships between multiple factors and the desired outcomes, whether it's optimizing a material mixture, improving structural performance, or minimizing environmental impact. By using CCD, engineers can make informed decisions and achieve cost-effective and sustainable solutions in civil engineering projects.

STEPS

Developing a CCD experiment in civil engineering involves several steps to systematically explore the relationships between factors and responses. Here's an overview of the steps to develop a CCD experiment for civil engineering:

1. **Define the Objective:**
   - Clearly state the engineering problem or objective you want to address. Determine what specific response or outcome you aim to optimize or understand. For example, you may want to maximize the compressive strength of concrete mixtures.
2. **Identify Factors and Response:**
   - Identify the factors (independent variables) that may influence the response variable. These factors could include material properties, construction methods, environmental conditions, etc.
   - Define the response variable that represents the outcome you are interested in optimizing or analyzing. This could be structural strength, cost, energy efficiency, or any relevant parameter.
3. **Determine Factor Levels:**
   - Decide the range or levels at which each factor will be varied. Factors can have high and low levels, which represent the extremes of the factor's range.
   - Some factors may have categorical levels (e.g., type of material), while others may have continuous levels (e.g., temperature, pressure).
4. **Generate a Factorial Design:**
   - Start by creating a full or fractional factorial design. This design includes combinations of factor levels that allow you to estimate the main effects and interactions among factors.
5. **Select a Center Point:**
   - Choose a central point within the factor range. The center point is typically replicated and serves as a reference point for estimating error and checking the model's adequacy.
6. **Add Axial (Star) Points:**
   - To capture curvature and quadratic effects in the response surface, add axial points. Axial points are located at a distance from the center point along each axis. The distance is typically denoted as alpha (α).
7. **Generate the Design Matrix:**
   - Use statistical software or tools to generate the design matrix based on the selected factors, levels, center point, and axial points. The design matrix specifies the experimental runs to be conducted.
8. **Conduct Experiments:**
   - Carry out the experimental runs according to the design matrix. Collect data on the response variable for each run, ensuring consistent and accurate data collection.
9. **Data Analysis:**
   - Perform statistical analysis to build a predictive model for the response variable. Regression analysis or RSM is commonly used to develop the model.
   - Identify significant main effects, interactions, and quadratic effects, if any, based on statistical tests and graphical analysis.
10. **Optimization:**
    - Use the developed model to optimize the factor settings to achieve the desired outcome. Optimization may involve maximizing, minimizing, or targeting a specific response.
11. **Validation:**
    - Validate the optimized settings or recommendations through additional experiments or by applying them to real-world civil engineering projects.
12. **Report and Interpret Results:**
    - Document the experimental design, data analysis, model equations, and optimization results in a comprehensive report.
    - Interpret the findings, provide insights into the impact of factors, and make recommendations for engineering applications.

13. **Sensitivity Analysis:**
   - Conduct sensitivity analysis to assess the robustness of the optimization results to variations in factor settings and model assumptions.
14. **Implementation:**
   - Implement the optimized solutions or recommendations in civil engineering projects to achieve the desired outcomes.
15. **Continuous Monitoring and Improvement:**
   - Continuously monitor the performance of the optimized solutions in practice and make improvements as needed.

Throughout the process, collaboration among civil engineers, statisticians, and other relevant experts may be necessary to ensure the success of the CCD experiment and its practical application in civil engineering projects.

PITFALLS AND REMEDIES

CCD is a powerful experimental design technique, but like any other method, it comes with potential pitfalls. Here are some common pitfalls and remedies when using CCD in civil engineering experiments:

Pitfall 1: Insufficient Factor Range
   Issue: If the range of factor levels is not wide enough, the experiment may not capture the full range of behaviors or responses.
   Remedy: Carefully choose the factor levels, ensuring they cover the expected range of real-world variations. Pilot experiments or historical data can help inform these choices.
Pitfall 2: Poor Model Fit
   Issue: The CCD model may not adequately fit the experimental data, leading to inaccurate predictions or optimization results.
   Remedy: Check the goodness-of-fit statistics, such as the coefficient of determination (R-squared), and residual plots to assess model adequacy. Consider transforming the response variable if the data do not meet the assumptions of the regression model. Increase the complexity of the model if interactions or quadratic effects are suspected but not captured.
Pitfall 3: Lack of Replicates
   Issue: Inadequate replication can lead to uncertainty in the estimated effects and the model's reliability.
   Remedy: Include replicates (repeating experiments under the same conditions) at the center point or other key experimental points to assess experimental error and improve the model's precision. Conduct additional experiments if needed to obtain more reliable estimates.
Pitfall 4: Outliers
   Issue: Outliers or data points with high influence can distort the model and compromise its accuracy.
   Remedy: Identify and investigate potential outliers using statistical tests or graphical methods. Consider whether outliers should be retained, removed, or treated differently in the analysis, depending on their nature and cause.
Pitfall 5: Inadequate Sample Size
   Issue: A small sample size may not provide enough data to detect significant effects or may lead to unstable model estimates.

Remedy: Ensure an adequate sample size based on power calculations or statistical guidelines for the desired level of significance. If resources are limited, consider conducting a smaller initial screening experiment to identify important factors before proceeding to a full CCD.

Pitfall 6: Lack of Model Validation

Issue: Failing to validate the CCD model on independent data can lead to overfitting and unreliable predictions.

Remedy: Reserve a portion of the data or conduct additional experiments to validate the model's predictions. Assess whether the model's predictions match real-world outcomes.

Pitfall 7: Ignoring Process Variability

Issue: CCD experiments may not account for inherent process variability, which can affect the robustness of the optimized solutions.

Remedy: Incorporate process variability into the model or consider using robust optimization techniques that account for uncertainty. Conduct sensitivity analysis to understand how changes in factors may affect outcomes under different scenarios.

Pitfall 8: Neglecting Engineering Constraints

Issue: Optimized solutions may not be practical due to engineering constraints or limitations.

Remedy: Ensure that the optimization results align with engineering and safety constraints. If necessary, perform additional analyses or simulations to validate the feasibility of the recommended solutions.

Pitfall 9: Overemphasis on Model Predictions

Issue: Relying solely on model predictions without considering engineering knowledge or judgment can lead to unrealistic or impractical solutions.

Remedy: Combine model predictions with engineering expertise and judgment to make practical and informed decisions. Consider trade-offs between conflicting objectives, such as cost, performance, and safety.

Pitfall 10: Inadequate Communication

Issue: Failure to effectively communicate the results and implications of the CCD experiment to stakeholders can hinder implementation.

Remedy: Prepare clear and concise reports or presentations that explain the experiment, model, optimization results, and their practical implications. Engage stakeholders early and involve them in the decision-making process.

By addressing these pitfalls and following best practices in experimental design and analysis, civil engineers can maximize the benefits of CCD experiments and improve the success of their projects. Collaboration with statisticians and experts in the specific field of civil engineering is also valuable in overcoming these challenges.

EXAMPLE

CCD is a commonly used experimental design method in civil engineering to study the relationships between factors and a response variable, especially when the relationships are expected to be nonlinear. CCD typically includes both factorial points and axial points to capture curvature in the response surface. Here's an example of a matrix for a CCD designed experiment in civil engineering:

Objective: To optimize the mix proportions of concrete for maximum compressive strength while minimizing the cost of materials.

Factors:

1. **Cement Content (Factor A):**
   - Range: 300 kg/m³ to 400 kg/m³
2. **Water-to-Cement Ratio (Factor B):**
   - Range: 0.35 to 0.45
3. **Aggregate Size (Factor C):**
   - Range: 10 mm to 20 mm

Response Variable: Compressive Strength (measured in megapascals, MPa)

Matrix for Central Composite Design (CCD):

In this example, a CCD design is used to optimize the mix proportions of concrete for maximum compressive strength while considering the factors of cement content, water-to-cement ratio, and aggregate size. The matrix is shown in Table 3.26.

In this matrix, each row represents a specific experimental run with a combination of factors and levels. The CCD design includes central points as well as points at various distances from the center to capture curvature in the response surface. By analyzing the data collected from these experimental runs and using response surface modeling, you can determine the optimal mix proportions of concrete that maximize compressive strength while considering cost constraints in civil engineering applications.

REPORTING

CCD experiments in civil engineering often involve the use of various statistical tools and techniques to analyze the data and extract meaningful information. Some of the key statistical tools used in CCD designed experiments for civil engineering include:

Regression Analysis: Regression analysis is a fundamental statistical tool used to model the relationship between independent variables (factors) and the dependent variable (response). In CCD, multiple linear regression or polynomial regression models are commonly used to fit the experimental data and predict the response variable.

Analysis of Variance (ANOVA): ANOVA is used to assess the statistical significance of factors and their interactions in the CCD model. It helps identify which factors have a significant impact on the response variable and whether any interactions are present.

TABLE 3.26

Central Composite Design (CCD) Designed Experiment Matrix Example

Run	Cement Content (A)	Water-to-Cement Ratio (B)	Aggregate Size (C)
1	300 kg/m³	0.35	10 mm
2	400 kg/m³	0.45	10 mm
3	350 kg/m³	0.35	20 mm
4	350 kg/m³	0.45	20 mm
5	325 kg/m³	0.30	15 mm
6	375 kg/m³	0.50	15 mm
7	350 kg/m³	0.35	15 mm
8	375 kg/m³	0.45	15 mm
9	350 kg/m³	0.40	15 mm
10	350 kg/m³	0.35	15 mm

Design of Experiments (DOE) Software: Specialized DOE software packages are commonly used to generate CCD experimental designs, analyze data, and build regression models. These software tools automate many aspects of experimental design and analysis, making the process more efficient.

Response Surface Methodology (RSM): RSM is a statistical technique closely associated with CCD. It involves the use of mathematical models to describe the response surface and optimize the response variable within the experimental region. RSM is valuable for finding optimal factor settings.

Model Fitting and Selection: Statisticians use various techniques to fit polynomial models to the experimental data, including stepwise regression, backward elimination, and forward selection. These methods help determine the most appropriate model for the data.

Residual Analysis: Residual plots and statistics are used to assess the adequacy of the regression model. Residuals should be normally distributed and show no discernible patterns to confirm the model's validity.

Confidence Intervals and Prediction Intervals: Engineers use confidence intervals to estimate population parameters, such as regression coefficients, with a specified level of confidence. Prediction intervals provide a range within which future observations are expected to fall.

Diagnostic Plots: Engineers often create diagnostic plots, such as normal probability plots, to check for deviations from model assumptions (e.g., normality of residuals) and identify potential outliers.

ANOVA and Hypothesis Testing: Statistical tests are performed to determine whether factors or interactions are statistically significant. ANOVA tables help assess the overall model significance and significance of individual terms.

Sensitivity Analysis: Sensitivity analysis involves evaluating how changes in factor settings impact the response variable and the model's predictions. It helps assess the robustness of the optimization results.

Optimization Algorithms: When the goal is to optimize the response variable, various optimization algorithms (e.g., gradient-based methods, genetic algorithms) are employed to find the factor settings that maximize or minimize the response within specified constraints.

Software Tools for Visualization: Data visualization tools, such as scatter plots, contour plots, and 3D surface plots, are used to visualize the response surface and gain insights into the relationships between factors and responses.

These statistical tools and techniques are crucial for designing, conducting, and analyzing CCD experiments in civil engineering. They enable engineers and researchers to gain a deep understanding of the factors influencing engineering processes and make informed decisions for optimization, quality improvement, and problem-solving in civil engineering projects.

A well-structured and comprehensive report for a CCD experiment in civil engineering should provide a clear and detailed account of the experiment, its objectives, methodology, results, and conclusions. Here are the key elements typically included in such a report:

Title Page:
Title of the Report: A concise and descriptive title that reflects the experiment's purpose.
Names of the Authors: Names and affiliations of the individuals or team responsible for the experiment and report.
Date: The date of report preparation.
Abstract:
A brief summary of the experiment's objectives, methods, key findings, and conclusions. It should provide a concise overview of the entire report.

Table of Contents:
A list of all the sections and subsections in the report with page numbers for easy navigation.

List of Figures and Tables:
A separate list that enumerates all the figures and tables in the report along with their respective page numbers.

Nomenclature (Optional):
A list of symbols, abbreviations, and terms used throughout the report, including their definitions.

Introduction:
Background and Context: Provide an introduction to the problem or issue addressed by the experiment. Explain its significance in civil engineering.

Objectives: Clearly state the experiment's objectives, including what you aim to investigate, analyze, or optimize.

Hypotheses (if applicable): Outline any hypotheses or research questions that guided the experiment.

Experimental Design:
Central Composite Design (CCD): Describe the specifics of the CCD, including the number of runs, factor levels, and settings for the independent variables (factors).

Factorial and Center Points: Detail how the CCD includes factorial points, center points, and axial points.

Randomization: Explain how randomization was implemented, if applicable.

Response Variable: Specify the response variable (dependent variable) and its units of measurement.

Factorial Factors: List and describe the factors investigated in the experiment, including their range and units of measurement.

Data Collection: Describe the data collection process, including measurement techniques, instruments used, and any data preprocessing.

Experimental Procedure:
Step-by-step description of how the experiment was conducted, including any equipment setup, sample preparation, and data collection procedures.

Safety Considerations: Mention any safety precautions taken during the experiment.

Data Analysis and Results:
Presentation of Data: Present the experimental data in tables, figures, and graphs.

Statistical Analysis: Discuss the statistical methods used to analyze the data, including regression analysis, ANOVA, and hypothesis testing.

Response Surface Analysis (if applicable): If RSM was applied, explain how the response surface was constructed and interpreted.

Optimization Results (if applicable): If the experiment is aimed at optimization, provide details of the optimized factor settings and results.

Discussion of Results: Interpret the findings, highlighting significant factors, interactions, and any practical implications.

Discussion:
Discuss the implications of the results on civil engineering practices or applications.

Address any limitations or sources of error in the experiment.

Compare the findings with previous research or theoretical expectations.

Conclusion:
Summarize the main findings and their significance.

Reiterate the experiment's objectives and whether they were achieved.

Recommendations:
Provide any recommendations for further research or practical applications based on the experiment's outcomes.

Acknowledgments:
Acknowledge any individuals, organizations, or funding sources that contributed to the experiment's success.

References:
Cite all the sources, research papers, books, and references used in the report following a specific citation style (e.g., APA, IEEE, Chicago).

Appendices (if applicable):
Include supplementary information such as raw data, calculations, detailed methodology, additional figures, or any other relevant material.

Signature and Approval (if required):
In some cases, the report may require signatures from supervisors, advisors, or project managers to indicate their approval of the experiment and report.

Attachments (if applicable):
Attach any relevant documents or reports that support the experiment's findings or provide additional context.

It's essential to maintain a logical and organized structure throughout the report to ensure clarity and ease of understanding. Additionally, adherence to formatting guidelines and citation styles is crucial for professional presentation.

3.3.5 TAGUCHI METHODS

Taguchi methods designed experiments are used in various areas of civil engineering to improve the quality, performance, and cost-effectiveness of infrastructure projects, buildings, and construction processes. Some of the key areas where Taguchi methods find application in civil engineering include:

Structural Engineering:
Optimization of structural design parameters to enhance strength, stability, and durability of bridges, buildings, and other structures.
Minimizing vibrations and maximizing structural integrity for earthquake-resistant designs.

Construction Materials and Techniques:
Improving the properties of construction materials such as concrete, asphalt, and steel.
Enhancing construction methods to reduce defects and ensure consistent quality.

Concrete Technology:
Achieving the desired properties in concrete mixtures, including strength, workability, and durability.
Reducing the likelihood of cracking and enhancing the performance of concrete structures.

Geotechnical Engineering:
Optimization of soil stabilization techniques and foundation design to support structures effectively.
Reducing settlement and soil-related issues in construction projects.

Transportation Engineering:
Enhancing the design and performance of roadways, highways, and transportation networks.
Improving traffic flow and safety through effective design and signaling.

Water Resources and Environmental Engineering:
Designing effective stormwater management systems and flood control structures.
Addressing environmental concerns related to water quality, pollution control, and waste-water treatment.

Building Design and Construction:
   Optimization of architectural and structural designs to minimize construction costs while maximizing safety and aesthetics.
   Improving energy efficiency and sustainability in building projects.
Construction Quality Control:
   Identifying and mitigating factors that contribute to defects, deviations, and rework during construction.
   Enhancing the quality assurance and quality control processes.
Infrastructure Maintenance and Rehabilitation:
   Optimizing repair and maintenance strategies for aging infrastructure to extend their lifespan.
   Identifying cost-effective solutions for rehabilitation and retrofitting.
Seismic Engineering:
   Designing structures and retrofitting existing ones to withstand seismic events.
   Minimizing damage and ensuring the safety of occupants during earthquakes.
Traffic Engineering and Management:
   Improving traffic signal timings and intersection designs to reduce congestion and enhance traffic flow.
   Enhancing safety measures in road design to reduce accidents.
Sustainable and Green Building Practices:
   Incorporating eco-friendly materials and energy-efficient designs into construction projects.
   Reducing environmental impact and resource consumption.

Taguchi methods offer a systematic approach to optimizing parameters, reducing variability, and improving the robustness of civil engineering projects in these and other areas. By using experimental designs, statistical analysis, and the principles of quality improvement, civil engineers can achieve better results while meeting project objectives and budget constraints.

STEPS

Developing Taguchi methods designed experiments for civil engineering projects involves a systematic process to optimize parameters and improve project outcomes. Here are the steps to develop Taguchi experiments in civil engineering:

1. **Define the Problem and Objectives:**
   - Clearly define the problem or challenge you want to address in your civil engineering project.
   - Identify the specific objectives you want to achieve through Taguchi methods, such as improving strength, durability, cost-effectiveness, or sustainability.
2. **Select the Factors and Levels:**
   - Identify the factors (variables) that may influence the desired outcome. These could include material properties, construction methods, design parameters, environmental conditions, and more.
   - Determine the levels (values or settings) at which each factor will be tested. Factors may have multiple levels to explore a range of conditions.
3. **Determine the Response Variables:**
   - Define the response variables that represent the project's performance or quality measures. These could include structural integrity, safety, cost, environmental impact, or any other relevant metric.

4. **Identify Noise Factors:**
   - Recognize potential noise factors, which are uncontrollable variables that can affect the response variables but cannot be adjusted. Noise factors may include weather conditions, soil variations, or equipment fluctuations.

5. **Select an Orthogonal Array (OA):**
   - Choose an appropriate OA based on the number of factors and levels. The OA helps structure the experimental design and reduce the number of required runs while providing a balanced assessment of factor effects.

6. **Assign Factors to Columns in the OA:**
   - Allocate the selected factors and their levels to columns in the OA according to a predetermined arrangement. Ensure that each factor's levels are evenly distributed across the columns.

7. **Conduct Experiments:**
   - Perform the experiments according to the factor settings specified in the OA. Each run corresponds to a specific combination of factor levels.
   - Record data on the response variables for each experimental run. Take care to maintain consistency and accuracy in data collection.

8. **Calculate Signal-to-Noise (S/N) Ratios:**
   - Compute S/N ratios for each run based on the response variable and its goal. Taguchi provides various types of S/N ratios, such as smaller-the-better, larger-the-better, and nominal-the-best, depending on the objective.

9. **Analyze the Results:**
   - Analyze the experimental data using statistical methods, such as ANOVA, to determine the significance of factors and interactions.
   - Identify which factors and levels contribute the most to improving or optimizing the response variables.

10. **Optimize Factor Settings:**
    - Based on the analysis, identify the factor settings (levels) that yield the best results or maximize the S/N ratios.
    - Adjust the engineering parameters or design specifications accordingly to achieve the desired outcome.

11. **Verify and Validate:**
    - Conduct additional experiments to verify the effectiveness of the optimized factor settings and validate that the desired objectives are met.

12. **Implement the Findings:**
    - Implement the optimized factor settings in the actual civil engineering project, whether it's in construction, structural design, material selection, or any other aspect.
    - Monitor and assess the project's performance over time to ensure that the improvements are sustained.

13. **Document and Report:**
    - Document all aspects of the Taguchi Methods experiment, including the OA, experimental data, analysis results, and recommendations.
    - Prepare a comprehensive report that summarizes the findings and provides insights for future projects.

14. **Continuous Improvement:**
    - Incorporate the lessons learned from the Taguchi experiment into your organization's practices and processes.
    - Consider applying Taguchi methods to other civil engineering projects to achieve similar improvements.

By following these steps, civil engineers can effectively apply Taguchi methods designed experiments to optimize parameters, enhance project outcomes, and address specific challenges in their projects while considering factors and noise that may impact performance and quality.

PITFALLS AND REMEDIES

Using Taguchi methods designed experiments in civil engineering can yield valuable insights and improvements, but there are potential pitfalls that engineers should be aware of. Here are some common pitfalls and their corresponding remedies:

Pitfall 1: Inadequate Problem Definition
> **Pitfall:** Failing to clearly define the problem or objectives of the experiment can lead to suboptimal results.
> **Remedy:** Begin with a well-defined problem statement and clear objectives. Ensure that all stakeholders agree on the project's goals and the aspects to be optimized.

Pitfall 2: Incorrect Factor Selection
> **Pitfall:** Selecting the wrong factors or omitting important ones can lead to incomplete or inaccurate results.
> **Remedy:** Conduct a thorough analysis to identify all potential factors that may influence the outcome. Seek input from experts and consider factors that may be initially overlooked.

Pitfall 3: Inadequate Data Collection
> **Pitfall:** Inaccurate or incomplete data collection can compromise the validity of the experiment.
> **Remedy:** Develop a rigorous data collection plan and ensure that data are collected consistently and accurately. Train personnel on data collection procedures.

Pitfall 4: Ignoring Noise Factors
> **Pitfall:** Neglecting noise factors (uncontrollable variables) can lead to unrealistic expectations and poor results.
> **Remedy:** Identify and acknowledge noise factors upfront. Collect data on noise factors during experiments to account for their effects and assess their impact on the results.

Pitfall 5: Choosing an Inappropriate Orthogonal Array (OA)
> **Pitfall:** Selecting an OA that does not suit the number of factors and levels can lead to inefficient experiments.
> **Remedy:** Choose an OA that matches the complexity of the problem. Use Taguchi's guidelines or software tools to select an appropriate OA.

Pitfall 6: Improper Analysis
> **Pitfall:** Failing to conduct a proper statistical analysis can lead to misinterpretation of results.
> **Remedy:** Perform a thorough statistical analysis, including ANOVA, to identify significant factors and interactions. Use appropriate S/N ratios based on the objectives (e.g., smaller-the-better, larger-the-better, or nominal-the-best).

Pitfall 7: Over-optimization
> **Pitfall:** Pursuing optimization to the extreme can lead to solutions that are impractical or costly to implement.
> **Remedy:** Balance optimization with practicality and cost-effectiveness. Consider engineering constraints and feasibility when determining the final factor settings.

Pitfall 8: Lack of Validation
> **Pitfall:** Failing to validate the optimized factor settings in real-world conditions can result in the failure of implemented solutions.

Remedy: Conduct validation experiments or field trials to confirm that the improvements observed in the Taguchi experiment are replicable and effective in practice.

Pitfall 9: Inadequate Documentation

Pitfall: Insufficient documentation can hinder knowledge transfer and future reference.

Remedy: Maintain comprehensive records of the experiment, including the OA, experimental data, analysis results, and recommendations. Document the entire process for future reference and continuous improvement.

Pitfall 10: Limited Involvement of Stakeholders

Pitfall: Not involving all relevant stakeholders, including engineers, designers, and project managers, can result in suboptimal solutions.

Remedy: Collaborate with cross-functional teams to ensure that all perspectives are considered and that the Taguchi experiment aligns with project objectives.

Pitfall 11: Neglecting Continuous Improvement

Pitfall: Failing to incorporate lessons learned from Taguchi experiments into standard practices can hinder long-term improvement efforts.

Remedy: Integrate successful findings and best practices into standard operating procedures. Encourage a culture of continuous improvement within the organization.

By addressing these pitfalls and implementing the corresponding remedies, civil engineers can maximize the effectiveness of Taguchi methods designed experiments and achieve meaningful improvements in their projects.

EXAMPLE

Taguchi methods are used for robust optimization and designed experiments to improve product or process performance while considering variation and noise factors. These experiments often use OAs to efficiently study multiple factors with a minimal number of experimental runs. Here's an example of a matrix for a Taguchi method designed experiment in civil engineering:

Objective: To optimize the curing process of concrete to minimize cracking in a bridge construction project.

Factors:

1. **Curing Temperature (Factor A):**
   - Level 1: Low Temperature (10°C)
   - Level 2: Medium Temperature (20°C)
   - Level 3: High Temperature (30°C)
2. **Curing Time (Factor B):**
   - Level 1: 3 days
   - Level 2: 7 days
   - Level 3: 14 days
3. **Curing Moisture (Factor C):**
   - Level 1: Low Moisture
   - Level 2: Medium Moisture
   - Level 3: High Moisture

Response Variable: Crack Width (measured in millimeters, mm)

Matrix for Taguchi Methods Designed Experiment:

In this example, an L27 (3^3) OA is used, which allows for the efficient study of three factors with three levels each. The matrix is shown in Table 3.27.

TABLE 3.27

Taguchi Method Designed Experiment Matrix Example for Civil Engineering

Run	Curing Temperature (A)	Curing Time (B)	Curing Moisture (C)
1	1 (10°C)	1 (3 days)	1 (Low)
2	1 (10°C)	2 (7 days)	2 (Medium)
3	1 (10°C)	3 (14 days)	3 (High)
4	2 (20°C)	1 (3 days)	2 (Medium)
5	2 (20°C)	2 (7 days)	3 (High)
6	2 (20°C)	3 (14 days)	1 (Low)
7	3 (30°C)	1 (3 days)	3 (High)
8	3 (30°C)	2 (7 days)	1 (Low)
9	3 (30°C)	3 (14 days)	2 (Medium)
10	1 (10°C)	1 (3 days)	3 (High)
11	1 (10°C)	2 (7 days)	1 (Low)
12	1 (10°C)	3 (14 days)	2 (Medium)
13	2 (20°C)	1 (3 days)	3 (High)
14	2 (20°C)	2 (7 days)	1 (Low)
15	2 (20°C)	3 (14 days)	2 (Medium)
16	3 (30°C)	1 (3 days)	2 (Medium)
17	3 (30°C)	2 (7 days)	3 (High)
18	3 (30°C)	3 (14 days)	1 (Low)
19	1 (10°C)	1 (3 days)	2 (Medium)
20	1 (10°C)	2 (7 days)	3 (High)
21	1 (10°C)	3 (14 days)	1 (Low)
22	2 (20°C)	1 (3 days)	2 (Medium)
23	2 (20°C)	2 (7 days)	1 (Low)
24	2 (20°C)	3 (14 days)	3 (High)
25	3 (30°C)	1 (3 days)	1 (Low)
26	3 (30°C)	2 (7 days)	2 (Medium)
27	3 (30°C)	3 (14 days)	3 (High)

In this matrix, each row represents a specific experimental run with a combination of factors and levels. The Taguchi Methods approach allows you to efficiently study the effects of the curing process factors on crack width while considering variation and noise factors in civil engineering applications.

REPORTING

Taguchi methods designed experiments in civil engineering rely on various statistical tools for analysis and optimization. These tools help engineers identify the most influential factors and their optimal settings to improve the quality and performance of civil engineering processes or products. Here are some of the statistical tools commonly used in Taguchi methods designed experiments for civil engineering:

Signal-to-Noise (S/N) Ratios: Taguchi experiments often use S/N ratios as performance measures to evaluate the quality of products or processes. The choice of S/N ratio (e.g., smaller-the-better, larger-the-better, or nominal-the-best) depends on the specific engineering problem and objectives.

Analysis of Variance (ANOVA): ANOVA is used to determine the significance of various factors and their interactions on the S/N ratio or response variable. It helps identify the most influential factors that contribute to variations in performance.

Main Effects Plots: Main effects plots display the impact of individual factors on the S/N ratio. These plots help engineers identify which factors have the most significant effect on the quality or performance of the product or process.

Interaction Plots: Interaction plots illustrate how two or more factors interact with each other and affect the S/N ratio. Identifying interactions is crucial for understanding complex relationships between factors.

Response Surface Analysis: In some cases, engineers may perform response surface analysis to model the relationship between factors and the S/N ratio, especially when optimizing a continuous response variable.

Control Charts: Control charts are used to monitor the quality or performance of a process over time. They help identify any deviations from the desired quality standards and trigger corrective actions.

Pareto Charts: Pareto charts prioritize factors based on their contribution to variations in the S/N ratio. This tool helps engineers focus their efforts on the most critical factors for improvement.

Robust Parameter Design (RPD): Taguchi methods often include RPD techniques, which aim to minimize the impact of external noise factors (uncontrollable factors) on the performance or quality of a product or process.

Factorial Experiments: Taguchi experiments can be designed using factorial experiment principles to systematically study the effects of multiple factors simultaneously.

Orthogonal Arrays (OAs): OAs are used to efficiently plan and conduct Taguchi experiments. They ensure that a wide range of factor combinations is tested with a minimal number of experimental runs.

Confirmation Runs: After identifying the optimal factor settings through Taguchi experiments, engineers often perform confirmation runs to validate the improvements and ensure that the selected settings are robust under real-world conditions.

Statistical Software: Specialized statistical software packages like Minitab or JMP are commonly used to perform statistical analyses and generate graphical representations of Taguchi experiments.

These statistical tools help civil engineers apply Taguchi methods effectively to optimize processes, reduce variability, enhance product quality, and achieve better performance in civil engineering applications, such as construction, infrastructure, and materials testing.

A well-structured report for a Taguchi methods designed experiment in civil engineering should provide a clear and concise presentation of the experiment, its objectives, methodology, results, and conclusions. Here are the key elements that should be included in such a report:

Title Page:
Title of the Report
Names of Authors and Affiliations
Date of Submission

Abstract:
A brief summary of the experiment, including the problem statement, objectives, methodology, key findings, and conclusions.

Table of Contents:
A list of sections and subsections with page numbers for easy navigation.

List of Figures and Tables:
A list of all figures and tables included in the report, along with their corresponding page numbers.

List of Abbreviations and Symbols:
If applicable, provide a list of abbreviations and symbols used throughout the report to ensure clarity.

Introduction:
Provide background information on the problem or process under investigation.

Clearly state the objectives of the experiment and its significance in the context of civil engineering.

Literature Review:
Review relevant literature and studies related to the problem or process.

Discuss any previous research that informed the experiment's design and objectives.

Experimental Design and Methodology:
Describe the Taguchi Methods experimental design, including the choice of factors, levels, and OAs.

Explain how factors were selected and manipulated in the experiment.

Provide details on the data collection process and any equipment or instruments used.

Outline the experimental procedure, including the number of runs, control factors, and noise factors.

Discuss how factors were randomized or controlled to minimize bias.

Results and Analysis:
Present the results of the Taguchi experiment in a clear and organized manner.

Use tables, charts, and graphs to display the data.

Include statistical analyses, such as ANOVA or S/N ratios, to assess the effects of factors on the response variable.

Interpret the results and highlight any significant findings or trends.

Discussion:
Discuss the implications of the experimental results.

Analyze the effects of individual factors and interactions on the response variable.

Compare the findings to the objectives and hypotheses stated in the introduction.

Address any limitations or sources of error in the experiment.

Optimization and Recommendations:
If applicable, provide recommendations for optimizing the civil engineering process or product based on the Taguchi experiment's findings.

Highlight the optimal factor settings and their significance.

Conclusion:
Summarize the key findings and their implications for civil engineering applications.

Restate the significance of the experiment in addressing the problem or process.

Acknowledgments:
Acknowledge any individuals, organizations, or funding sources that contributed to the experiment.

References:
Include a list of all references cited in the report, following a specific citation style (e.g., APA, IEEE, or Chicago).

Appendices:
Include any supplementary information, such as detailed data tables, experimental protocols, or additional figures that support the report.

Appendix A: Experimental Design Details:
Provide a comprehensive description of the Taguchi experimental design, including the OA used, factor levels, and the assignment of factors to runs.

Appendix B: Data Tables:
Include detailed data tables with the results of each experimental run, including factor settings and response values.

Appendix C: Graphical Representations:
Include any additional graphs or charts that provide visual representations of the data and results.

The structure and content of the report may vary depending on the specific objectives and scope of the Taguchi Methods experiment in civil engineering. However, these elements should provide a comprehensive and well-organized report that communicates the experiment's purpose, methodology, results, and conclusions effectively.

3.3.6 MIXTURE EXPERIMENTS

Mixture designed experiments are used in civil engineering in various applications where the composition of materials plays a crucial role in achieving desired properties or performance characteristics. Here are some areas in civil engineering where mixture designed experiments are commonly applied:

Concrete Mix Design: Concrete is a fundamental construction material in civil engineering. Mixture designed experiments are used to optimize the composition of concrete mixes, including the proportions of cement, aggregates, water, and additives. The goal is to achieve the desired strength, durability, workability, and other properties of concrete.

Asphalt Mix Design: In road construction and pavement engineering, mixture designed experiments are employed to determine the ideal combination of aggregates, bitumen (asphalt binder), and additives to create asphalt mixes with the desired properties, including durability and resistance to fatigue and rutting.

Soil Stabilization: Mixture design experiments help engineers determine the appropriate mixtures of soil, binders (such as lime or cement), and other additives to stabilize and improve the engineering properties of soils for construction purposes, such as foundation support and road subgrade improvement.

Grout and Mortar Mix Design: Civil engineers use mixture designed experiments to optimize the composition of grouts and mortars used for masonry, tile installation, and repairs. The experiments help achieve the desired strength, adhesion, and workability of these materials.

Geotechnical Engineering: In geotechnical applications, mixture designed experiments are used to develop engineered fills, backfills, and soil-cement mixtures for embankments, retaining walls, and erosion control structures.

Pavement Design: Mixture designed experiments play a role in the design of flexible and rigid pavements, including the selection of materials for base courses, subbase layers, and surface courses to meet specific load-bearing and durability requirements.

Shotcrete Mix Design: Shotcrete is used for slope stabilization, tunnel lining, and underground construction. Mixture designed experiments help determine the optimal composition of shotcrete mixes for structural integrity and adherence to surfaces.

Fiber-Reinforced Concrete and Composite Materials: Engineers use mixture designed experiments to develop composite materials that combine concrete with fibers (e.g., steel, synthetic, or natural fibers) to enhance properties such as crack resistance, impact resistance, and ductility.

High-Performance and Specialty Concrete: In applications where high strength, low permeability, or specific durability requirements are essential, mixture designed experiments help formulate specialized concrete mixes that meet these demands.

Construction Materials Recycling: Civil engineers investigate the use of recycled materials in construction, such as recycled aggregates, fly ash, or slag, and utilize mixture design experiments to optimize their incorporation into new construction materials.

Environmental Remediation: Mixture design experiments can be employed in soil and groundwater remediation projects to optimize the composition of reactive barriers or permeable reactive materials used to treat contaminants.

Sealant and Adhesive Formulations: In various civil engineering applications, sealants and adhesives are used for bonding and sealing. Mixture designed experiments help develop formulations that meet specific bonding strength and sealing requirements.

Mixture designed experiments are valuable tools in civil engineering to ensure that construction materials and mixtures meet performance specifications, sustainability goals, and cost-efficiency targets. These experiments allow engineers to systematically investigate the influence of different components and their proportions on the final product's properties, leading to more informed and optimized material selections for construction projects.

STEPS

Developing mixture designed experiments for civil engineering typically involves a systematic process to optimize the composition of materials and achieve desired properties or performance characteristics. Here are the general steps to develop mixture designed experiments for civil engineering applications:

1. **Define Objectives and Goals:**
   - Clearly define the objectives of the experiment. What specific properties or characteristics are you trying to optimize or achieve?
   - Set goals and performance criteria that the final mixture must meet.
2. **Select Factors and Levels:**
   - Identify the components or factors that make up the mixture. These could include various materials, such as aggregates, binders, additives, or chemicals.
   - Determine the range of levels for each factor that will be tested in the experiment. This includes specifying the minimum and maximum allowable proportions or concentrations.
3. **Choose Experimental Design:**
   - Select an appropriate experimental design methodology for mixture experiments. Common approaches include full-factorial design, fractional factorial design, or RSM.
   - Consider the number of factors, levels, and available resources when choosing the experimental design. For more complex problems, RSM may be more suitable.
4. **Design the Experiment:**
   - Generate a design matrix that specifies the combinations of factor levels to be tested in the experiment.
   - Ensure that the selected design is balanced, orthogonal, or otherwise statistically sound to obtain reliable results.
5. **Collect Data:**
   - Prepare the necessary materials and equipment for conducting the experiments.
   - Follow the experimental plan and systematically create the mixtures as specified in the design matrix.
   - Record data related to the performance characteristics of each mixture, such as strength, durability, workability, or other relevant properties.
6. **Perform Analysis:**
   - Analyze the experimental data to determine the effects of different factors and their interactions on the desired properties.
   - Use statistical software or tools to conduct regression analysis or response surface modeling to predict optimal compositions.
7. **Optimize the Mixture:**
   - Identify the optimal composition or mixture that meets the defined objectives and goals.
   - Consider any constraints or practical limitations in selecting the final composition.

8. **Validate the Results:**
   - Conduct confirmatory experiments or additional testing to validate the optimized mixture's performance characteristics.
   - Ensure that the mixture consistently meets the desired criteria.
9. **Implement the Mixture:**
   - Incorporate the optimized mixture into the civil engineering project or application.
   - Monitor the performance of the material in the field to ensure it meets expectations over time.
10. **Documentation and Reporting:**
    - Document the experimental process, including the design matrix, data collection, analysis, and results.
    - Prepare a comprehensive report summarizing the experiment's objectives, methodology, findings, and recommendations.
11. **Continuous Improvement:**
    - Continue to monitor the performance of the mixture in real-world applications and make adjustments as necessary to achieve long-term success.

Throughout the process, it is essential to consider safety measures, quality control, and environmental considerations, especially when working with construction materials. Collaboration with experts in materials science, structural engineering, and construction management can also be beneficial in ensuring the success of mixture designed experiments in civil engineering.

PITFALLS AND REMEDIES

Mixture designed experiments in civil engineering can be complex, and several pitfalls may arise during the planning and execution of such experiments. Addressing these pitfalls with appropriate remedies is essential to ensure the success of the experiments and the reliability of the results. Here are some common pitfalls and their corresponding remedies:

Pitfall 1: Insufficient Factor Range

Issue: Inadequate or narrow ranges for factor levels may lead to suboptimal results or failure to identify the true optimum.

Remedy: Conduct a thorough literature review and preliminary testing to determine a reasonable and comprehensive range for each factor. Ensure that the selected ranges cover both practical and extreme conditions.

Pitfall 2: Nonlinearity and Interactions

Issue: Neglecting nonlinear effects and interactions between factors can result in an inaccurate model and suboptimal mixtures.

Remedy: Use appropriate experimental designs, such as RSM, to capture nonlinear effects and interactions. Analyze the data using regression analysis or software that accounts for these factors.

Pitfall 3: Inadequate Sample Size

Issue: Insufficient data points can lead to unreliable statistical analyses and inconclusive results.

Remedy: Conduct a power analysis or sample size calculation to determine the minimum number of experimental runs required to detect significant effects. Ensure that the sample size is adequate for the chosen design.

Pitfall 4: Lack of Replication

Issue: Conducting each experimental run only once may lead to variability and uncertainty in the results.

Remedy: Include replication of experiments to assess the repeatability and reproducibility of the results. Replicating experiments can enhance the reliability of findings.

Pitfall 5: Poor Data Quality

Issue: Data collection errors, measurement inaccuracies, or instrument calibration issues can compromise the quality of the dataset.

Remedy: Implement strict quality control procedures during data collection. Calibrate instruments regularly, and use appropriate statistical techniques to identify and correct outliers or measurement errors.

Pitfall 6: Ignoring Practical Constraints

Issue: Optimized mixtures may not be feasible or practical for construction or manufacturing processes.

Remedy: Consider practical constraints, such as material availability, cost, and manufacturing limitations, when selecting the final mixture. Ensure that the optimized mixture is implementable in real-world applications.

Pitfall 7: Overfitting

Issue: Creating overly complex models that fit the experimental data perfectly but do not generalize well to new situations.

Remedy: Use model selection techniques to choose a parsimonious model that adequately represents the relationships between factors and responses without overfitting. Cross-validation can help assess model performance.

Pitfall 8: Lack of Expertise

Issue: Inadequate knowledge of statistical analysis or experimental design principles can result in suboptimal experiments.

Remedy: Seek assistance from statisticians or experts in experimental design. Collaborate with individuals who have experience in similar civil engineering projects.

Pitfall 9: Incomplete Documentation

Issue: Insufficient documentation of the experimental process can hinder the reproducibility and transparency of the results.

Remedy: Maintain detailed records of the experimental design, data collection, analysis, and results. Create a comprehensive report that includes all relevant information for future reference and validation.

Pitfall 10: Failure to Validate

Issue: Failing to validate the optimized mixture in real-world applications may lead to unexpected issues or performance failures.

Remedy: Implement the optimized mixture in pilot projects or field applications and monitor its performance over time. Make adjustments as necessary based on real-world feedback.

Addressing these pitfalls and implementing appropriate remedies will help ensure the success of mixture designed experiments in civil engineering, leading to improved construction materials, structures, and processes.

EXAMPLE

In civil engineering, mixture design experiments are often used to optimize the proportions of various components in a mixture, such as concrete or asphalt, to achieve desired properties. Here's an example of a matrix for a mixture designed experiment in civil engineering:

Objective: To optimize the mix proportions of concrete for maximum compressive strength while minimizing cost.

Components of the Mixture:

1. **Cement (Component A)**: Proportion expressed as a percentage of the total mixture.
   - Range: 10% to 20%
2. **Fine Aggregate (Component B)**: Proportion expressed as a percentage of the total mixture.
   - Range: 40% to 60%
3. **Coarse Aggregate (Component C)**: Proportion expressed as a percentage of the total mixture.
   - Range: 20% to 30%
4. **Water (Component D)**: Proportion expressed as a percentage of the total mixture.
   - Range: 20% to 30%

Response Variable: Compressive Strength (measured in megapascals, MPa)

Matrix for a Mixture Designed Experiment:

In this example, a mixture design experiment aims to optimize the mix proportions of concrete for maximum compressive strength while considering the proportions of cement, fine aggregate, coarse aggregate, and water. The matrix is shown in Table 3.28.

In this matrix, each row represents a specific experimental run with a unique combination of proportions for the mixture components (cement, fine aggregate, coarse aggregate, and water). The aim is to find the optimal mixture proportions that result in the highest compressive strength while adhering to practical constraints and cost considerations in civil engineering applications.

By conducting experiments with different mix proportions and analyzing the resulting compressive strengths, civil engineers can determine the ideal combination for their specific project requirements, considering factors like strength, cost, and workability.

REPORTING

Statistical tools play a crucial role in the analysis of mixture designed experiments in civil engineering. These tools help researchers and engineers make informed decisions about the composition of materials or mixtures to optimize specific properties or characteristics. Here are some common statistical tools used in mixture designed experiments for civil engineering:

1. **Linear Regression Analysis:** Linear regression is often used to model the relationship between the mixture components and the response variable. It can help identify linear effects and estimate coefficients for each component.

TABLE 3.28

Mixture Designed Experiment Matrix Example for Civil Engineering

Run	Cement (A)	Fine Aggregate (B)	Coarse Aggregate (C)	Water (D)
1	15%	50%	25%	10%
2	12%	60%	20%	8%
3	18%	45%	30%	12%
4	14%	55%	22%	11%
5	16%	52%	28%	9%
6	13%	58%	21%	10%
7	19%	42%	29%	13%
8	17%	48%	27%	12%

2. **Response Surface Methodology (RSM):** RSM involves the use of mathematical models, typically quadratic or higher-order models, to represent the relationship between mixture components and responses. It allows for the exploration of nonlinear effects, interactions, and the determination of optimal mixture conditions.

3. **Mixture Design of Experiments:** Mixture design is a specialized experimental design technique that focuses on varying the proportions of components in a mixture while keeping the total constant. Common designs include simplex-centroid designs, simplex-lattice designs, and others tailored to mixture experiments.

4. **Analysis of Variance (ANOVA):** ANOVA is used to assess the statistical significance of factors and their interactions in mixture experiments. It helps identify which components significantly affect the response variable.

5. **Constraint Optimization:** Constraint optimization techniques are employed to find the optimal mixture composition while considering constraints, such as cost, availability, or practicality. Tools like linear programming or nonlinear programming can be applied to find feasible solutions.

6. **Graphical Tools:** Visualization tools, such as ternary plots and contour plots, are used to graphically represent the relationships between mixture components and responses. They help in understanding the effects of different components and their interactions.

7. **Principal Component Analysis (PCA):** PCA can be used to reduce the dimensionality of data and identify dominant components that contribute the most to variation in responses.

8. **Sensitivity Analysis:** Sensitivity analysis helps assess the robustness of the optimized mixture by evaluating how small changes in component proportions affect the response variable. It provides insights into the stability of the design.

9. **Statistical Software:** Statistical software packages like Minitab, JMP, R, and others are commonly used for designing experiments, analyzing data, and generating graphical representations.

10. **Design of Experiments (DOE) Software:** Specialized DOE software allows for the creation of experimental designs, response surface modeling, and optimization of mixtures.

11. **Monte Carlo Simulation:** Monte Carlo simulations can be used to assess the uncertainty associated with optimized mixture compositions. It involves generating random samples to estimate the distribution of response variables.

12. **Robust Parameter Design (RPD):** RPD techniques help optimize mixture designs while considering variability and noise factors. Tools like Taguchi methods can be applied to make mixtures robust to variations.

These statistical tools are applied depending on the specific goals of the mixture experiment, the complexity of the system, and the desired level of optimization. Engineers and researchers may use a combination of these tools to comprehensively analyze and optimize mixture designs in civil engineering projects.

A well-structured report for a mixture designed experiment in civil engineering is essential to communicate the methodology, results, and conclusions of the study effectively. While the specific format may vary depending on the organization or project, the following are common elements typically included in such a report:

1. **Title Page:**
   - Title of the Report
   - Author(s) and Affiliations
   - Date of Submission
2. **Abstract:**
   - A concise summary of the experiment, including objectives, methodology, key results, and conclusions.

3. **Table of Contents:**
   - List of sections and subsections with corresponding page numbers.
4. **List of Figures and Tables:**
   - A separate list of all figures and tables with their respective page numbers.
5. **List of Abbreviations and Symbols (if applicable):**
   - Explanation of any abbreviations or symbols used in the report.
6. **Introduction:**
   - Background information on the problem or research question.
   - Objectives and goals of the mixture designed experiment.
   - Importance and relevance of the study to civil engineering.
7. **Literature Review:**
   - A review of relevant literature and previous research related to the experiment.
   - Explanation of existing knowledge and gaps that the experiment aims to address.
8. **Experimental Design:**
   - Description of the experimental setup, including details on the mixture components and their proportions.
   - Explanation of the chosen experimental design (e.g., simplex-centroid, simplex-lattice).
   - Justification for the design choices.
9. **Materials and Methods:**
   - Information about the materials and equipment used in the experiment.
   - Detailed procedures for preparing mixtures and conducting the experiment.
   - Data collection methods and measurement techniques.
   - Any quality control measures implemented.
10. **Results:**
    - Presentation of the experimental data, including tables, figures, and charts.
    - Statistical analysis of the results, including ANOVA, regression models, and graphical representations.
    - Discussion of any significant findings or trends observed.
11. **Discussion:**
    - Interpretation of the results and their implications for civil engineering applications.
    - Comparison of the experimental findings with the literature review.
    - Discussion of any limitations or sources of error in the experiment.
12. **Conclusion:**
    - Summary of the key findings and their relevance to the research objectives.
    - Implications for future research or practical applications in civil engineering.
13. **Recommendations:**
    - Any recommendations or suggestions based on the study's results.
    - Guidance for further investigations or refinements to the mixture design.
14. **References:**
    - A list of all sources cited in the report, following a specific citation style (e.g., APA, IEEE, Chicago).
15. **Appendices (if applicable):**
    - Supplementary information, such as raw data, detailed calculations, or additional figures.
    - Any supporting documentation that enhances the understanding of the experiment.
16. **Acknowledgments (if applicable):**
    - Acknowledgment of individuals, organizations, or funding sources that contributed to the experiment.
17. **Graphical Elements:**
    - Inclusion of clear and appropriately labeled figures, charts, and tables to illustrate key points and data.

It's essential to adhere to a consistent and organized structure when preparing a mixture designed experiment report in civil engineering. Additionally, following any specific formatting guidelines or templates provided by your institution or organization is important for clarity and professionalism.

3.3.7 SEQUENTIAL EXPERIMENTATION

Sequential experimentation is a versatile approach used in civil engineering to optimize and refine various processes, designs, and systems. It is often employed in the following areas within civil engineering:

Structural Engineering:
> Sequential experimentation can be used to optimize the design of structures, such as bridges, buildings, and dams, to ensure they meet safety and performance standards while minimizing material and construction costs.

Geotechnical Engineering:
> In geotechnical engineering, sequential experimentation is used to refine soil and foundation design parameters, leading to more stable and cost-effective foundations for structures.

Transportation Engineering:
> Civil engineers in transportation use sequential experimentation to optimize traffic flow, signal timing, and road design, leading to safer and more efficient transportation systems.

Environmental Engineering:
> In environmental engineering, sequential experimentation can be applied to design and optimize wastewater treatment processes, pollution control systems, and sustainable water management solutions.

Construction Management:
> Sequential experimentation is used to improve construction processes, scheduling, and resource allocation, resulting in more efficient and cost-effective construction projects.

Materials Engineering:
> Civil engineers often use sequential experimentation to develop and refine construction materials, such as concrete and asphalt, to improve durability and performance.

Hydraulic and Water Resources Engineering:
> Sequential experimentation helps optimize the design of hydraulic structures, flood control systems, and water distribution networks, ensuring efficient water management and flood protection.

Urban Planning:
> Urban planners use sequential experimentation to design and optimize urban infrastructure, including roads, utilities, and green spaces, to create sustainable and livable cities.

Risk Assessment and Management:
> Sequential experimentation can be employed to assess and manage risks associated with civil engineering projects, allowing engineers to make informed decisions and mitigate potential hazards.

Renewable Energy:
> Civil engineers involved in renewable energy projects, such as wind farms and solar power installations, use sequential experimentation to optimize energy production and system performance.

Structural Health Monitoring (SHM):
> In the field of SHM, sequential experimentation helps refine sensor placement and data analysis techniques for assessing the condition of infrastructure.

Coastal and Harbor Engineering:
Coastal and harbor engineers use sequential experimentation to design and optimize coastal protection structures, ports, and harbor facilities, considering factors like wave dynamics and sediment transport.

Risk Reduction and Disaster Resilience:
Sequential experimentation can be applied to assess and improve disaster resilience in civil infrastructure, especially in areas prone to earthquakes, floods, or other natural disasters.

Sequential experimentation is a valuable tool in civil engineering for iterative problem-solving and continuous improvement. It allows engineers to adapt and refine their designs or processes based on real-world data and observations, leading to more efficient, cost-effective, and sustainable solutions in various subfields of civil engineering.

STEPS

Developing a sequential experimentation designed experiment in civil engineering involves a systematic process to optimize and refine a particular aspect of a project or system. Here are the steps typically followed:

1. **Define the Objective:**
   - Clearly define the specific objective or problem you aim to address through sequential experimentation. This could involve optimizing a design, improving a construction process, enhancing structural performance, or achieving other engineering goals.
2. **Identify Variables and Factors:**
   - Identify the key variables and factors that can influence the outcome of the experiment. These may include design parameters, material properties, construction methods, environmental conditions, or other relevant variables.
3. **Select Experimental Design:**
   - Choose an appropriate experimental design that suits your objective. Common designs include factorial experiments, RSM, and Taguchi methods, depending on the complexity and number of factors to be considered.
4. **Initial Experimentation:**
   - Conduct an initial set of experiments based on the selected design. This may involve conducting trials, measurements, or simulations to gather data on how the chosen factors affect the outcome.
5. **Data Analysis:**
   - Analyze the data collected from the initial experiments. This analysis may include statistical techniques to identify significant factors, correlations, and trends that impact the desired outcome.
6. **Refinement and Adjustment:**
   - Based on the data analysis, make adjustments to the factors, levels, or conditions of the experiment. This iterative process helps refine the experiment to focus on the most influential factors and optimize the design or process.
7. **Sequential Experimentation Plan:**
   - Develop a sequential experimentation plan that outlines the sequence of experiments to be conducted. This plan should specify which factors will be changed or adjusted in each subsequent round of experiments.
8. **Conduct Additional Experiments:**
   - Carry out additional rounds of experimentation according to the sequential plan. Each round may involve making changes to one or more factors and collecting new data.

9. **Data Collection and Analysis (Sequential Rounds):**
   - After each round of experiments, collect and analyze the data. Compare the results with previous rounds to assess the impact of the adjustments and identify any trends or improvements.
10. **Iterative Optimization:**
    - Continuously iterate through rounds of experimentation, making adjustments based on the analysis of previous data. The goal is to gradually converge on the optimal solution or design.
11. **Validation and Testing:**
    - Conduct validation experiments to ensure that the optimized design or process performs as expected under various conditions. This step helps verify the robustness of the solution.
12. **Documentation and Reporting:**
    - Document all experiments, data, adjustments, and outcomes in a comprehensive report. Include details on the iterative process, statistical analyses, and conclusions drawn from the sequential experimentation.
13. **Implementation:**
    - Implement the optimized design, process, or solution in the civil engineering project or system. Monitor its performance and continue to gather data for ongoing improvement if necessary.
14. **Continuous Improvement:**
    - Maintain a culture of continuous improvement, where data-driven decisions and sequential experimentation are used to address evolving challenges and optimize civil engineering projects over time.

Throughout the process, it's essential to maintain clear documentation and communication to ensure that all stakeholders are informed of the progress and outcomes of the sequential experimentation. Additionally, involving experts in statistical analysis and experimental design can be valuable in achieving meaningful results.

PITFALLS AND REMEDIES

Sequential experimentation can be a powerful tool in civil engineering, but like any method, it comes with potential pitfalls. Here are some common pitfalls and remedies when conducting sequential experimentation in civil engineering:

Pitfalls:

Inadequate Problem Definition:
   Pitfall: Poorly defined or vague problem statements can lead to ineffective experimentation.
   Remedy: Clearly define the problem, objectives, and desired outcomes before beginning any experiments.
Incomplete Factor Identification:
   Pitfall: Failing to identify and consider all relevant factors can result in incomplete optimization.
   Remedy: Conduct a thorough factor identification process, including input from experts and stakeholders.
Overlooking Interactions:
   Pitfall: Neglecting interactions among factors can lead to inaccurate conclusions.
   Remedy: Use appropriate experimental designs that account for interactions, such as factorial designs or RSMs.

Insufficient Data Collection:
   Pitfall: Collecting inadequate or poor-quality data can undermine the reliability of results.
   Remedy: Ensure data collection methods are robust, and collect a sufficient sample size to draw meaningful conclusions.

Lack of Expertise:
   Pitfall: Inexperienced personnel may struggle with experimental design, data analysis, and interpretation.
   Remedy: Involve experts in experimental design and statistical analysis, or provide training to team members.

Ignoring Resource Constraints:
   Pitfall: Failing to consider resource constraints, such as time or budget limitations, can lead to unrealistic experimentation plans.
   Remedy: Develop a realistic plan that aligns with available resources and prioritize experiments accordingly.

Inadequate Documentation:
   Pitfall: Poor record-keeping and documentation can lead to confusion, errors, or lost insights.
   Remedy: Maintain thorough documentation of experimental procedures, data, and adjustments made during the process.

1. **Lack of Adaptability:**
   Pitfall: Rigid adherence to an initial plan without adapting to emerging insights can hinder progress.
   Remedy: Be flexible and open to adjusting the experimentation plan based on ongoing analysis and results.

2. **Ineffective Communication:**
   Pitfall: Inadequate communication among team members and stakeholders can lead to misunderstandings and misaligned objectives.
   Remedy: Establish clear lines of communication, regularly update stakeholders, and involve relevant parties in decision-making.

3. **Ignoring External Factors:**
   Pitfall: Focusing solely on controlled factors while ignoring external influences can lead to unrealistic optimization.
   Remedy: Consider external factors, such as environmental conditions or regulatory changes, that may impact the experiment's success.

4. **Overoptimization:**
   Pitfall: Excessive optimization can lead to designs or processes that are overly sensitive to small variations.
   Remedy: Strike a balance between optimization and robustness to ensure the solution is practical and resilient.

5. **Lack of Validation:**
   Pitfall: Failing to validate the optimized solution in real-world conditions can result in implementation failures.
   Remedy: Conduct validation experiments to confirm the effectiveness of the optimized design or process.

Remedies:

Thorough Planning: Invest time in comprehensive planning, including problem definition, factor identification, and experimental design.

Expert Involvement: Engage experts in experimental design, statistical analysis, and relevant engineering domains to ensure sound methodologies.

Quality Data: Prioritize high-quality data collection methods and establish data quality controls.

Documentation: Maintain detailed records throughout the experiment, including all adjustments and findings.

Communication: Foster open communication among team members, stakeholders, and experts involved in the process.

Adaptability: Be willing to adjust the experimentation plan based on emerging insights and results.

Resource Management: Manage resources effectively and adapt the plan to fit within constraints.

Validation: Verify the optimized solution's effectiveness through validation experiments before implementation.

By addressing these pitfalls and implementing the suggested remedies, civil engineers can make the most of sequential experimentation to optimize designs, processes, and systems effectively.

EXAMPLE

Sequential experimentation in civil engineering involves conducting a series of experiments where the results of each experiment inform the design of subsequent experiments. The objective is to efficiently and iteratively explore the design space to optimize a particular outcome or property. Here's an example of a simplified matrix for a sequential experimentation designed experiment in civil engineering:

Objective: To optimize the mix proportions of concrete for a bridge construction project while considering cost, strength, and durability.

Components of the Mixture:

1. **Cement (Component A)**: Proportion expressed as a percentage of the total mixture.
   - Initial Range: 10% to 20%
2. **Fine Aggregate (Component B)**: Proportion expressed as a percentage of the total mixture.
   - Initial Range: 40% to 60%
3. **Coarse Aggregate (Component C)**: Proportion expressed as a percentage of the total mixture.
   - Initial Range: 20% to 30%
4. **Water (Component D)**: Proportion expressed as a percentage of the total mixture.
   - Initial Range: 20% to 30%

Response Variable: Compressive Strength (measured in megapascals, MPa)

Sequential Experimentation Matrix:

In sequential experimentation, the initial matrix might involve a limited set of experiments to establish a baseline or explore the design space (Table 3.29).

Based on the initial results, the subsequent experiments can be designed iteratively to explore specific regions of interest or to refine the mixture proportions further. The exact matrix for sequential experimentation would depend on the information gained from the initial experiments and the specific objectives and constraints of the project.

TABLE 3.29

Sequential Designed Experiments for Civil Engineering

Run	Cement (A)	Fine Aggregate (B)	Coarse Aggregate (C)	Water (D)
1	15%	50%	25%	10%
2	12%	60%	20%	8%
3	18%	45%	30%	12%
4	14%	55%	22%	11%

The key to sequential experimentation is adaptability. After analyzing the initial results, additional experiments can be designed and conducted to move closer to the optimal mixture proportions while considering factors like cost, strength, and durability in civil engineering applications. The process is typically repeated until the desired outcome is achieved or an optimal solution is found.

REPORTING

Sequential experimentation in civil engineering often involves various statistical tools to analyze and optimize complex systems. Some of the key statistical tools commonly used in sequential experimentation for civil engineering include:

Descriptive Statistics: Descriptive statistics, such as mean, median, variance, and standard deviation, are used to summarize and describe data collected during experiments.

Design of Experiments (DOE): DOE techniques help in planning, conducting, and analyzing experiments efficiently. Common DOE methods include factorial designs, RSMs, and Taguchi methods.

Regression Analysis: Regression analysis is used to model the relationship between independent variables (factors) and a dependent variable (response). It can help identify significant factors and their effects.

Analysis of Variance (ANOVA): ANOVA is used to assess the differences in means among groups or treatments and is often applied to experimental designs with multiple factors.

Response Surface Methodology (RSM): RSM is a statistical technique used to optimize processes by modeling the relationship between multiple factors and a response variable. It helps identify the optimal factor settings.

Statistical Process Control (SPC): SPC techniques monitor and control processes during experimentation, ensuring that they remain within specified limits.

Multivariate Analysis: Multivariate statistical techniques, such as PCA and factor analysis, are used to analyze relationships between multiple variables simultaneously.

Time-Series Analysis: Time-series analysis is used when experimentation involves data collected over time. It helps identify trends, patterns, and correlations in time-dependent data.

Hypothesis Testing: Hypothesis testing is used to determine whether observed differences or relationships between variables are statistically significant.

Probability Distributions: Probability distributions, such as normal, binomial, or Poisson distributions, are used to model the variability and randomness in civil engineering data.

Monte Carlo Simulation: Monte Carlo simulations are used to model and analyze complex systems with uncertainty by generating multiple random samples and estimating outcomes.

Reliability Analysis: Reliability analysis assesses the probability of system failure or component failure over time, helping engineers design for safety and longevity.

Nonparametric Statistics: Nonparametric tests, like the Mann-Whitney U-test or the Kruskal-Wallis test, are used when data do not meet the assumptions of parametric tests.

Bayesian Analysis: Bayesian statistical methods incorporate prior knowledge or beliefs into the analysis, updating probabilities based on new data.

Spatial Analysis: Spatial statistics are used in geospatial civil engineering projects to analyze spatial patterns, trends, and relationships in data.

Optimization Techniques: Optimization methods, including linear programming, integer programming, and genetic algorithms, are used to find the best solutions in civil engineering designs and planning.

Bootstrapping: Bootstrapping is a resampling technique used to estimate the sampling distribution of a statistic by repeatedly sampling with replacement from the data.

These statistical tools, when applied appropriately, enable civil engineers to make informed decisions, optimize designs, improve processes, and ensure the reliability and safety of civil engineering projects. The choice of tools depends on the specific objectives and characteristics of the experimentation in question.

A report for a sequential experimentation designed experiment in civil engineering should provide a clear and comprehensive overview of the study, the experimental design, procedures, results, and conclusions. Here are the key elements typically included in such a report:

Title Page:
Title of the Report
Name of the Author(s)
Date of Submission
Affiliation and Contact Information

Abstract:
A brief summary of the entire report, including the objectives, methodology, key findings, and conclusions. It should be concise and informative.

Table of Contents:
List of sections and subsections with page numbers for easy navigation.

List of Figures and Tables:
A separate list of all figures and tables used in the report, along with their corresponding page numbers.

Introduction:
Background and context of the study, including the problem statement and objectives.
A brief literature review highlighting relevant prior research.
The rationale for using sequential experimentation and its significance.

Methodology:
Description of the experimental design, including factors, levels, and response variables.
Details on the sequential sampling plan, if applicable, explaining how data was collected and when decisions were made.
Any assumptions and statistical techniques used in the analysis.
Description of the equipment, materials, and tools used in the experiment.

Experimental Procedures:
Step-by-step explanation of how the experiment was conducted.
Any safety precautions taken during the experiment.
Information on data collection, including sample sizes and sampling intervals for each stage.

Results:
Presentation of data in a clear and organized manner, using tables, figures, charts, and graphs as needed.

Statistical analysis, including descriptive statistics, hypothesis testing, and modeling results.

Discussion of any deviations or unexpected findings.

Discussion:

Interpretation of the results in the context of the research objectives.

Comparison of findings with prior research or expectations.

Addressing any limitations or sources of error.

Implications of the results for civil engineering applications.

Conclusions:

Summarization of the key findings and their significance.

Whether the research objectives were achieved.

Practical recommendations or next steps based on the results.

Recommendations:

Suggestions for further research or improvements to the experimental design.

Any practical recommendations for civil engineering practice.

References:

A list of all sources cited in the report, following a specific citation style (e.g., APA, MLA, IEEE).

Appendices:

Supplementary material that provides additional details, such as raw data, calculations, or technical drawings.

Any relevant documentation related to the experiment.

Acknowledgments:

Recognition of individuals, organizations, or funding sources that contributed to the research.

List of Abbreviations and Symbols:

A compilation of abbreviations and symbols used in the report, along with their meanings.

Glossary (if necessary):

Definitions of technical terms or jargon used in the report.

Declaration of Originality:

A statement by the author(s) affirming the originality of the work and adherence to ethical standards.

Executive Summary (Optional):

A condensed version of the report's key points and findings, suitable for a quick overview.

It's essential to follow a consistent and clear structure, use proper formatting, and ensure that the report is well-organized and free of errors. Additionally, graphics and visuals should be appropriately labeled and referenced in the text for clarity.

3.3.8 ROBUST PARAMETER DESIGN (RPD)

RPD experiments are used in civil engineering in various applications where the goal is to develop products, processes, or designs that are highly resilient to variations and uncertainties. RPD is particularly valuable in civil engineering for ensuring the reliability and performance of structures, systems, and infrastructure components. Here are some areas where RPD experiments are commonly applied in civil engineering:

Structural Engineering:

RPD can be used to optimize the design of bridges, buildings, and other structures, taking into account variations in material properties, loads, and environmental conditions.

Geotechnical Engineering:
In geotechnical projects, RPD can help design foundations, retaining walls, and slopes that are robust against soil variability, groundwater levels, and seismic events.

Transportation Engineering:
RPD experiments are useful in designing roadways, pavements, and transportation systems that can withstand variations in traffic loads, weather conditions, and materials.

Water Resources Engineering:
RPD can be applied to design robust flood control systems, dams, and reservoirs that can handle variations in water flow, precipitation, and sediment transport.

Environmental Engineering:
In environmental projects, RPD can be used to design wastewater treatment plants, air quality control systems, and waste management processes that are resilient to fluctuations in pollutant levels and environmental conditions.

Construction Management:
RPD experiments can improve construction processes by optimizing construction schedules, resource allocation, and quality control methods to mitigate the effects of uncertainties and delays.

Material Testing and Development:
RPD can be applied in the development of construction materials, such as concrete, asphalt, and steel, to ensure that their properties are consistent and reliable under various conditions.

Structural Health Monitoring (SHM):
RPD techniques can be used to develop robust systems for monitoring the condition and safety of existing structures, such as bridges and pipelines, under changing environmental and usage conditions.

Risk Management:
RPD helps in identifying potential risks and uncertainties in civil engineering projects and developing strategies to minimize their impact.

Resilient Infrastructure:
RPD can contribute to the design and maintenance of resilient infrastructure that can withstand extreme events like earthquakes, hurricanes, and floods.

RPD experiments involve systematically varying and optimizing design parameters while considering sources of variability and uncertainty. By doing so, civil engineers can create designs that are robust and perform well under a wide range of conditions, ultimately leading to safer and more reliable infrastructure.

STEPS

Developing an RPD experiment in civil engineering involves a systematic approach to optimize the performance of a design while considering sources of variability and uncertainty. Here are the steps to develop an RPD experiment for civil engineering:

1. **Define the Problem and Objectives**:
   - Clearly define the problem or objective of your civil engineering project. Determine what specific aspect of the design or process you want to optimize while considering variations and uncertainties.
2. **Identify Critical Parameters**:
   - Identify the key parameters, variables, or factors that influence the performance of the design or process. These are the factors that you will manipulate and control in your experiment.

3. **Determine the Response Variable**:
   - Choose a measurable response variable that reflects the performance or quality of the design or process. This variable should be sensitive to changes in the critical parameters.
4. **Establish the Range of Variation**:
   - Determine the ranges or levels at which you will vary each critical parameter. Consider both the expected variations and the extreme values that may occur in real-world scenarios.
5. **Select an Experimental Design**:
   - Choose an appropriate experimental design, such as a factorial design, fractional factorial design, or Taguchi design, based on the number of factors, their levels, and the available resources.
6. **Conduct the Experiment**:
   - Perform the experiments according to the selected design matrix. Collect data on the response variable for each combination of factor levels.
7. **Analyze the Data**:
   - Use statistical analysis techniques to analyze the experimental data. Evaluate the effects of each factor and their interactions on the response variable.
8. **Optimize the Design**:
   - Based on the analysis, identify the factor settings that result in the best performance or quality. These settings should be robust to variations and uncertainties.
9. **Verify and Validate**:
   - Conduct additional experiments or simulations to verify the optimized design's performance under various conditions. Validate that the design is indeed robust and meets the objectives.
10. **Implement and Monitor**:
    - Implement the optimized design in your civil engineering project. Continuously monitor its performance in real-world conditions and make adjustments as needed.
11. **Document the Results**:
    - Prepare a detailed report documenting the entire RPD process, including problem definition, experimental design, data analysis, optimization results, and validation findings.
12. **Iterate and Improve**:
    - If necessary, iterate the RPD process to further improve the design's robustness. Incorporate any lessons learned into future projects.
13. **Maintain Documentation**:
    - Keep thorough records of the RPD experiment, as well as any changes or modifications made during the project. This documentation can be valuable for future reference and improvements.
14. **Communication**:
    - Share the results and insights from the RPD experiment with relevant stakeholders, including project teams, clients, and regulatory authorities.
15. **Continuous Improvement**:
    - Encourage a culture of continuous improvement within your civil engineering organization, where RPD principles are applied to future projects to enhance robustness and reliability.

By following these steps, civil engineers can develop robust designs that are less sensitive to variations and uncertainties, ultimately leading to improved project outcomes and reduced risks.

PITFALLS AND REMEDIES

RPD experiments in civil engineering aim to create designs that are less sensitive to variations and uncertainties. However, there are potential pitfalls that can arise during the RPD process. Here are some common pitfalls and remedies:

Pitfall 1: Inadequate Problem Definition
Pitfall: Unclear or poorly defined problem objectives can lead to ineffective RPD experiments.
Remedy: Clearly define the problem and objectives from the beginning. Engage with stakeholders to ensure a thorough understanding of project goals.

Pitfall 2: Inadequate Factor Selection
Pitfall: Selecting the wrong factors or failing to consider all critical factors can lead to suboptimal designs.
Remedy: Conduct a thorough factor analysis to identify all critical factors that influence the design's performance. Consult experts and historical data if necessary.

Pitfall 3: Limited Factor Range
Pitfall: Insufficient variation in factor levels can result in designs that are not truly robust.
Remedy: Expand the range of factor levels to cover both expected variations and worst-case scenarios. Consider factors that may vary during the project's life cycle.

Pitfall 4: Inadequate Sample Size
Pitfall: Small sample sizes may lead to unreliable results and difficulty in detecting significant effects.
Remedy: Ensure an adequate sample size based on statistical power calculations. Larger samples improve the accuracy of the analysis.

Pitfall 5: Neglecting Factor Interactions
Pitfall: Ignoring interactions between factors can lead to missed opportunities for optimization.
Remedy: Include interaction terms in the experimental design and analyze them carefully to uncover hidden relationships.

Pitfall 6: Insufficient Data Analysis
Pitfall: Inadequate data analysis may result in misinterpretation of results and missed opportunities for improvement.
Remedy: Use appropriate statistical tools and techniques for data analysis. Consider employing statistical software for advanced analysis.

Pitfall 7: Overlooking Validation
Pitfall: Failing to validate the optimized design under real-world conditions can lead to unexpected issues.
Remedy: Conduct validation experiments or simulations to ensure that the design performs as expected in different scenarios.

Pitfall 8: Lack of Documentation
Pitfall: Inadequate documentation can make it challenging to replicate or build upon the RPD process.
Remedy: Maintain thorough records of the entire RPD process, including problem definition, experimental design, data analysis, and optimization results.

Pitfall 9: Resistance to Change
Pitfall: Resistance from project teams or stakeholders to adopt the optimized design can hinder its implementation.
Remedy: Communicate the benefits of the robust design and involve key stakeholders early in the process to gain their buy-in.

Pitfall 10: Not Learning from Failures

    Pitfall: Failing to learn from previous project failures and applying those lessons to RPD can lead to repeated mistakes.

    Remedy: Encourage a culture of continuous improvement and knowledge sharing within the organization. Conduct post-project reviews to identify areas for improvement.

By addressing these common pitfalls and implementing the suggested remedies, civil engineers can enhance the effectiveness of their RPD experiments and achieve more robust and reliable designs in civil engineering projects.

EXAMPLE

RPD is a structured approach used in civil engineering to optimize product or process performance while minimizing sensitivity to variation or external factors. It involves designing experiments to achieve robustness against variation. Here's an example of a matrix for an RPD designed experiment in civil engineering:

Objective: To optimize the mix proportions of concrete for a bridge construction project while ensuring that the compressive strength is robust to variations in materials.

Components of the Mixture:

1. **Cement (Component A)**: Proportion expressed as a percentage of the total mixture.
   - Range: 10% to 20%
2. **Fine Aggregate (Component B)**: Proportion expressed as a percentage of the total mixture.
   - Range: 40% to 60%
3. **Coarse Aggregate (Component C)**: Proportion expressed as a percentage of the total mixture.
   - Range: 20% to 30%
4. **Water (Component D)**: Proportion expressed as a percentage of the total mixture.
   - Range: 20% to 30%

Factors to Study Variations:

1. **Variation in Cement Quality (Factor E)**:
   - Low Variation (Standard Deviation: 1%)
   - High Variation (Standard Deviation: 3%)
2. **Variation in Fine Aggregate Quality (Factor F)**:
   - Low Variation (Standard Deviation: 2%)
   - High Variation (Standard Deviation: 5%)

Response Variable: Compressive Strength (measured in megapascals, MPa)

Robust Parameter Design Matrix:

    In RPD, the matrix is designed to assess the robustness of the mixture proportions under different levels of variation in critical factors. The matrix is shown in Table 3.30.

    In this matrix, each row represents a specific experimental run with a unique combination of mixture proportions (cement, fine aggregate, coarse aggregate, and water) and levels of variation in critical factors (cement quality and fine aggregate quality).

TABLE 3.30

Robust Parameter Design (RPD) Matrix Example for Civil Engineering

Run	Cement (A)	Fine Aggregate (B)	Coarse Aggregate (C)	Water (D)	Cement Quality (E)	Fine Aggregate Quality (F)
1	15%	50%	25%	10%	Low	Low
2	12%	60%	20%	8%	Low	High
3	18%	45%	30%	12%	High	Low
4	14%	55%	22%	11%	High	High

The analysis in RPD focuses on how robust the compressive strength is to variations in material quality. By conducting experiments with different mix proportions and levels of variation in critical factors, you can determine the optimal mixture that maintains a consistent and robust compressive strength despite variations in materials in civil engineering applications.

Reporting

RPD experiments in civil engineering make use of various statistical tools and techniques to analyze and optimize designs while considering factors that can affect performance and reliability. Some of the key statistical tools and methods commonly used in RPD experiments for civil engineering include:

Design of Experiments (DOE): DOE is the foundation of RPD. It helps in systematically planning and conducting experiments to gather data and understand the relationship between factors and responses. Common DOE techniques include full factorial design, fractional factorial design, and Taguchi methods.

Response Surface Methodology (RSM): RSM is used to create mathematical models that represent the relationship between factors and responses. It helps in optimizing designs by identifying the optimal factor settings that lead to the desired performance or reliability.

Statistical Process Control (SPC): SPC tools such as control charts are used to monitor and control processes during and after the RPD implementation. They help in detecting deviations from the desired performance and taking corrective actions.

Regression Analysis: Regression models are employed to quantify the relationship between factors and responses, allowing engineers to make predictions and optimize designs.

Analysis of Variance (ANOVA): ANOVA is used to analyze the variance in data and determine the significance of different factors and their interactions on the response variable.

Optimization Techniques: Various optimization algorithms, such as Genetic Algorithms, Gradient Descent, or Response Surface Optimization, are used to find the optimal factor settings that minimize variability or maximize performance.

Reliability Analysis: Reliability models are used to assess the probability of failure or the reliability of a design under different conditions. Techniques like Weibull analysis and reliability block diagrams are common in civil engineering.

Sensitivity Analysis: Sensitivity analysis helps in identifying which factors have the most significant impact on the response variable. This information guides engineers in focusing their efforts on critical factors.

Monte Carlo Simulation: Monte Carlo simulations are used to model the uncertainty and variability in civil engineering projects. They help in assessing the robustness of a design under different scenarios.

Statistical Software: Specialized statistical software packages like Minitab, JMP, or MATLAB are often used to perform the complex statistical analyses required in RPD experiments.

Failure Mode and Effects Analysis (FMEA): FMEA is a systematic approach to identifying and mitigating potential failure modes in a design. It helps in prioritizing actions to improve reliability.

Pareto Analysis: Pareto charts are used to identify the most significant factors contributing to variation or failure in a design.

Taguchi Loss Function: The Taguchi Loss Function quantifies the cost or loss associated with deviations from the target or optimal performance. It helps in setting target values that minimize losses.

These statistical tools and techniques are applied in various stages of an RPD experiment, from initial design planning and data collection to analysis, optimization, and validation. They help civil engineers create more robust and reliable designs that can perform well under different conditions and uncertainties.

A report for an RPD experiment in civil engineering typically consists of several key elements to effectively communicate the experimental process, results, and recommendations. Here are the essential elements of a report for an RPD designed experiment in civil engineering:

Title: The report should begin with a clear and descriptive title that summarizes the main focus of the experiment.

Abstract: An abstract provides a concise summary of the entire report, including the problem statement, methodology, key findings, and conclusions. It allows readers to quickly understand the purpose and outcomes of the experiment.

Introduction:

Problem Statement: Describe the specific problem, challenge, or opportunity that prompted the RPD experiment. Explain why the experiment was conducted and its relevance to civil engineering.

Literature Review: Provide a brief review of relevant literature and previous research related to the problem. Highlight any existing knowledge or gaps that the experiment aims to address.

Objectives: Clearly state the objectives and goals of the RPD experiment. What are you trying to achieve or optimize, and why is it important?

Methodology:

Experimental Design: Explain the experimental design, including factors, levels, and responses. Describe the chosen DOE approach, such as full factorial, fractional factorial, or Taguchi methods, and justify your selection.

Data Collection: Detail the data collection process, including the methods, instruments, and procedures used to gather data during the experiment.

Analysis Techniques: Describe the statistical and analytical techniques employed to analyze the data. This may include RSM, ANOVA, regression analysis, or other relevant tools.

Results:

Data Presentation: Present the experimental data in a clear and organized manner. Use tables, charts, graphs, and figures to illustrate the results.

Statistical Analysis: Summarize the statistical analysis of the data, including any significant findings or relationships between factors and responses.

Optimization: If applicable, present the optimized factor settings or conditions that were determined through the analysis.

Discussion:

Interpretation: Interpret the results and discuss their implications for the problem or challenge addressed in the experiment.

Robustness Analysis: Evaluate the robustness of the optimized solution by considering factors like variability and uncertainty. Discuss the reliability of the design under different conditions.

Comparison: Compare the outcomes of the RPD experiment with the initial design or baseline conditions to highlight improvements achieved.

Conclusions: Summarize the main conclusions drawn from the experiment, emphasizing its contribution to solving the problem or achieving the objectives.

Recommendations: Provide recommendations for practical applications or further research based on the experiment's findings. What actions should be taken as a result of the experiment?

References: Include a list of all the sources, research papers, books, and articles cited within the report. Follow a consistent citation style (e.g., APA, IEEE, or Chicago).

Appendices: Include any supplementary information that supports the main content of the report. This may include additional data tables, charts, experimental protocols, or mathematical derivations.

Acknowledgments: If applicable, acknowledge individuals, organizations, or funding sources that contributed to the experiment's success.

Author Information: Provide contact information for the authors or researchers involved in the experiment, including names, affiliations, and email addresses.

Date and Version: Include the date of the report's completion and, if relevant, the version number.

Executive Summary (Optional): For longer reports, consider including an executive summary at the beginning of the document. This summary provides a high-level overview of the report's key points and findings.

The organization and content of the report should be clear and logically structured, allowing readers to follow the experiment's progression from problem statement to conclusions and recommendations. Additionally, use clear and concise language, avoiding jargon when possible, to ensure that the report is accessible to a broad audience within the civil engineering field.

3.3.9 Randomized Experiments

REs are a research methodology used in various fields, including civil engineering, to investigate the effects of interventions or treatments under controlled conditions. In civil engineering, REs can be applied in several areas, depending on the specific research questions and objectives. Here are some examples of where REs are used in civil engineering:

Materials Testing and Quality Control: REs can be employed to assess the performance of construction materials, such as concrete, asphalt, or steel, under various conditions. Researchers can randomize factors like material composition, curing methods, or environmental conditions to determine their impact on material properties, durability, and quality.

Structural Testing: In structural engineering, REs can be used to assess the performance of different construction techniques, materials, or retrofitting methods for buildings and infrastructure. RCTs may be conducted to evaluate the effectiveness of seismic retrofitting measures, for example.

Geotechnical Engineering: REs can be conducted to investigate soil behavior and its response to different construction and foundation methods. Researchers may randomize factors like soil composition, compaction methods, or groundwater levels to study their influence on soil stability and settlement.

Traffic Engineering: In transportation and traffic engineering, REs can be used to evaluate the impact of traffic management strategies, road design modifications, or traffic control devices (e.g., traffic signals, roundabouts). Researchers may randomize factors like signal timing plans or lane configurations to assess their effects on traffic flow and safety.

Environmental Engineering: REs can be applied to assess the effectiveness of environmental remediation techniques or pollution control measures. For example, researchers may conduct randomized trials to evaluate the impact of different wastewater treatment processes on water quality.

Construction Management: REs can be used to study construction project management practices, scheduling methods, or safety measures. Researchers may randomize factors like project scheduling algorithms or safety protocols to determine their impact on project efficiency and worker safety.

Urban Planning: In urban planning and development, REs can help assess the impact of land use policies, zoning regulations, or urban design interventions on factors like traffic congestion, air quality, and community well-being.

Water Resources Management: RE can be applied to optimize the allocation of water resources, such as in irrigation systems. Researchers may randomize factors like irrigation schedules or water distribution methods to improve water-use efficiency and crop yields.

Infrastructure Maintenance: REs can be used to evaluate maintenance strategies for infrastructure assets like bridges, roads, and pipelines. Researchers may randomize factors like maintenance schedules or repair techniques to assess their effectiveness in prolonging asset lifespan.

Disaster Preparedness and Mitigation: REs can be conducted to assess the effectiveness of disaster preparedness plans and mitigation measures, such as evacuation strategies or flood control measures.

In each of these applications, REs help researchers draw causal inferences by systematically assigning treatments or interventions to different groups or conditions. By randomizing the assignment of factors, researchers can reduce bias and better understand the cause-and-effect relationships in civil engineering processes and systems.

STEPS

Developing a RE in civil engineering involves several key steps to design, implement, and analyze the study. Here are the steps to develop an RE for civil engineering:

1. **Define the Research Objectives:**
   - Clearly define the research objectives and questions you want to address through the RE. What specific factors or interventions are you testing, and what outcomes are you interested in measuring?
2. **Select the Experimental Design:**
   - Choose the appropriate experimental design that suits your research objectives. Common designs include CRD, RBD, and factorial designs, among others. The choice of design depends on the complexity of the study and the number of factors to be tested.

3. **Identify the Factors and Levels:**
   - Identify the independent variables (factors) that you want to test and determine the levels or treatments for each factor. Factors can represent construction materials, methods, environmental conditions, or any other variables relevant to your research.
4. **Randomize the Assignment:**
   - Randomly assign the experimental units (samples, sites, structures, etc.) to the different treatment groups or conditions. Randomization ensures that each unit has an equal chance of receiving each treatment, minimizing bias and confounding variables.
5. **Control Group and Blinding (if applicable):**
   - In some cases, include a control group that does not receive any treatment to serve as a baseline for comparison. If blinding is feasible and appropriate, consider blinding the participants or data collectors to reduce potential bias.
6. **Implement the Experiment:**
   - Carry out the experimental procedures according to the design specifications. Ensure that the treatments are applied consistently and that data collection methods are standardized.
7. **Data Collection:**
   - Collect data on the relevant outcomes or responses for each experimental unit. Ensure that data collection is accurate, reliable, and consistent across all groups.
8. **Data Analysis:**
   - Analyze the collected data using appropriate statistical methods. Depending on the design and research objectives, use techniques such as ANOVA, regression analysis, or other statistical tests to assess the impact of the treatments on the outcomes.
9. **Interpret Results:**
   - Interpret the results of the data analysis in the context of your research objectives. Determine whether the treatments had a statistically significant effect and the practical implications of the findings.
10. **Draw Conclusions:**
    - Based on the results and interpretations, draw conclusions about the effectiveness of the treatments or interventions. Address whether they achieved the desired outcomes or met the research objectives.
11. **Report and Document Findings:**
    - Prepare a comprehensive report that documents the experimental design, methods, results, and conclusions. Include relevant figures, tables, and statistical analyses to support your findings.
12. **Discussion and Implications:**
    - Discuss the implications of your findings for civil engineering practice or research. Consider any limitations of the study and potential areas for further research.
13. **Peer Review and Publication:**
    - If applicable, submit your research for peer review and publication in relevant journals or conference proceedings to share your findings with the scientific community.
14. **Implementation and Application:**
    - If the findings have practical applications, work on implementing the results in civil engineering projects or practices. Monitor the real-world impact of the interventions.
15. **Continuous Improvement:**
    - Continuously assess and refine your experimental design and research methods based on feedback and new insights gained from the study.

Throughout the entire process, ensure that ethical considerations, safety protocols, and best practices in data collection and analysis are followed. Collaborating with experienced statisticians or researchers with expertise in experimental design can also be valuable in conducting high-quality REs in civil engineering.

PITFALLS AND REMEDIES

REs in civil engineering, like any scientific research, can face various pitfalls. Recognizing these pitfalls and applying appropriate remedies is crucial for ensuring the validity and reliability of the experimental results. Here are some common pitfalls and their corresponding remedies:

Selection Bias:
> **Pitfall:** Nonrandom selection of experimental units or participants can introduce bias and affect the generalizability of results.
>
> **Remedy:** Ensure random assignment of units to treatment groups. Use appropriate randomization techniques to minimize selection bias.

Confounding Variables:
> **Pitfall:** Uncontrolled confounding variables can obscure the true effects of treatments.
>
> **Remedy:** Randomization helps in balancing confounding variables across treatment groups. If possible, conduct a factorial experiment to control and assess interactions among factors.

Inadequate Sample Size:
> **Pitfall:** Insufficient sample size can lead to low statistical power, making it challenging to detect real effects.
>
> **Remedy:** Conduct a power analysis before the experiment to determine the required sample size. Ensure that the sample size is adequate to achieve meaningful results.

Measurement Errors:
> **Pitfall:** Inaccurate or imprecise measurements can introduce noise and reduce the ability to detect treatment effects.
>
> **Remedy:** Use validated and reliable measurement instruments. Implement rigorous quality control procedures to minimize measurement errors.

Noncompliance:
> **Pitfall:** Participants or experimental units may not adhere to the assigned treatments, leading to noncompliance.
>
> **Remedy:** Monitor and ensure compliance throughout the experiment. Consider using intention-to-treat analysis to account for noncompliance.

Ethical Concerns:
> **Pitfall:** Ethical issues related to the treatment of participants or the environment can compromise the integrity of the experiment.
>
> **Remedy:** Conduct experiments in compliance with ethical guidelines and obtain informed consent when working with human subjects. Follow environmental regulations and ethical principles when testing in the field.

Experimenter Bias:
> **Pitfall:** The experimenters' expectations or actions may unintentionally influence the outcomes.
>
> **Remedy:** Implement blinding or double-blinding procedures whenever possible to reduce experimenter bias.

External Validity:
> **Pitfall:** Results from an RE may not be generalizable to real-world situations.
>
> **Remedy:** Ensure that the experimental conditions and sample characteristics are representative of the population or context of interest. Conduct follow-up studies or meta-analyses to assess external validity.

Attrition Bias:
> **Pitfall:** Loss of participants or data over the course of the experiment can lead to attrition bias.

Remedy: Minimize attrition by implementing strategies to retain participants or units throughout the study. Analyze data using appropriate techniques to account for missing data.

Data Analysis Errors:

Pitfall: Incorrect or inappropriate statistical analyses can lead to misinterpretation of results.

Remedy: Collaborate with statisticians or experts in data analysis. Ensure that the chosen statistical methods are appropriate for the experimental design and data.

Lack of Replication:

Pitfall: Conducting a single experiment without replication can limit the robustness of the findings.

Remedy: Whenever possible, replicate the experiment to assess the consistency and reliability of results.

Publication Bias:

Pitfall: Only publishing significant results while omitting nonsignificant findings can introduce bias in the literature.

Remedy: Consider publishing all results, including nonsignificant findings, to reduce publication bias and contribute to a more accurate scientific record.

Addressing these pitfalls and applying appropriate remedies is essential to ensure the credibility and validity of REs in civil engineering. Collaborating with experienced researchers and seeking peer review can also help identify and mitigate potential pitfalls.

EXAMPLE

REs in civil engineering involve randomly assigning treatments or factors to experimental units to investigate the effects of those treatments on a particular outcome or response variable. Here's a simplified example of a matrix for an REs designed experiment in civil engineering:

Objective: To evaluate the effectiveness of two different types of concrete curing methods on the compressive strength of concrete.

Experimental Units: Concrete cylinders (multiple cylinders for each treatment group).

Factors (Treatment Groups):

1. **Water Curing (Treatment A)**: Traditional water curing method.
2. **Steam Curing (Treatment B)**: A newer steam curing method.

Response Variable: Compressive Strength (measured in megapascals, MPa)

Randomized Experiment Matrix:

In this REl, treatments are randomly assigned to concrete cylinders to study the effect of curing methods on compressive strength. The matrix is shown in Table 3.31.

In this matrix, each row represents a concrete cylinder (experimental unit), and the treatment group (water curing or steam curing) is randomly assigned to each cylinder. Multiple cylinders are used for each treatment group to account for variability. After the curing methods are applied, the compressive strength of each cylinder is tested and recorded.

TABLE 3.31
Randomized Experiment Matrix Example for Civil Engineering

Cylinder	Treatment Group
1	Water Curing
2	Steam Curing
3	Steam Curing
4	Water Curing
5	Steam Curing
6	Water Curing
7	Water Curing
8	Steam Curing
9	Steam Curing
10	Water Curing

By using randomization, you ensure that any potential bias or confounding factors are minimized, and you can draw meaningful conclusions about the effects of the different curing methods on compressive strength in civil engineering experiments.

REPORTING

In REs in civil engineering, various statistical tools and methods are used to analyze the data and draw meaningful conclusions. The choice of statistical tools depends on the specific research questions, experimental design, and data characteristics. Here are some commonly used statistical tools and techniques:

Descriptive Statistics:
- Descriptive statistics, such as mean, median, and standard deviation, are used to summarize and describe the central tendency and variability of data.

Hypothesis Testing:
- Hypothesis testing techniques, including t-tests and ANOVA, are used to assess whether there are significant differences between groups or treatments.

Regression Analysis:
- Linear and nonlinear regression models are employed to examine the relationship between independent variables (factors) and the dependent variable (response).

Analysis of Covariance (ANCOVA):
- ANCOVA is used when there is a need to control for one or more covariates (continuous variables) while comparing treatment groups.

Chi-Square Test:
- The chi-square test is used for analyzing categorical data and assessing associations between variables.

Survival Analysis:
- Survival analysis techniques, such as Kaplan-Meier curves and Cox proportional hazards models, are used when studying time-to-event data, such as failure times in structural engineering experiments.

Design of Experiments (DOE):
- DOE involves the systematic manipulation of factors to optimize a response variable. Techniques like full-factorial design, fractional factorial design, and Taguchi methods are used for efficient experimentation.

Nonparametric Tests:
- Nonparametric tests, like the Wilcoxon rank-sum test or the Kruskal-Wallis test, are used when the assumptions of normality are violated or when dealing with ordinal data.

Multivariate Analysis:
- Multivariate techniques, such as MANOVA and PCA, are used when analyzing multiple dependent variables simultaneously.

Bayesian Analysis: Bayesian statistical methods can be employed to update beliefs about parameters and make predictions based on prior information and observed data.

Spatial Analysis: Spatial statistics are used to analyze data with a spatial component, common in geotechnical and transportation engineering.

Time-Series Analysis: Time-series analysis methods are used when data are collected over time, such as monitoring structural changes or environmental variables.

Reliability Analysis: Reliability analysis tools assess the reliability and lifetime of civil engineering structures and systems.

Sensitivity Analysis: Sensitivity analysis helps assess the impact of variations in input variables on the outcomes, crucial in risk assessment and decision-making.

Simulation and Monte Carlo Methods: Simulation techniques, including Monte Carlo simulations, are used to model complex systems, assess uncertainties, and estimate probabilities.

Statistical Software: Specialized statistical software packages like R, SAS, MATLAB, and others are often used to conduct statistical analyses efficiently.

The choice of statistical tools should align with the research objectives, data characteristics, and experimental design. Collaborating with statisticians or data analysts with expertise in civil engineering applications can be valuable in selecting and correctly applying the appropriate statistical methods.

A report for an RE designed in civil engineering typically includes various elements to communicate the research process, findings, and conclusions effectively. Below are the key elements commonly found in such reports:

Title Page:
- The title page should include the title of the report, the names of the authors, their affiliations, and the date of submission.

Abstract:
- The abstract provides a concise summary of the entire report, including the research objectives, methodology, key findings, and conclusions. It should be brief and informative.

Table of Contents:
- The table of contents lists the sections and subsections of the report with their respective page numbers, making it easier for readers to navigate.

List of Figures and Tables:
- A list of figures and tables helps readers locate specific visual aids and tables within the report.

Introduction:
- The introduction sets the stage by presenting the background and context of the research. It should include the research problem, objectives, and the rationale for conducting the RE.

Literature Review:
- The literature review provides a comprehensive overview of relevant research and prior work related to the experiment's subject matter. It helps establish the need for the current study and provides a basis for hypotheses or research questions.

Methodology:
- This section details the experimental design, including the randomization process, sample size determination, selection of treatment levels, data collection methods, and any relevant equipment or instrumentation. It should be comprehensive enough for replication by others.

Experimental Setup:
- Describe the physical setup of the experiment, including any diagrams, schematics, or photographs that help readers understand the equipment, test specimens, or structures involved.

Randomization Procedures:
- Explain the methods used to randomize treatment allocation or assignment to ensure that it is unbiased and follows a random process.

Data Collection: Describe how data were collected, including the data collection instruments and any quality control measures taken to ensure data accuracy.

Results: Present the results of the experiment in a clear and organized manner. Use tables, figures, graphs, and statistical analysis to illustrate and interpret the data. Include any observed trends, significant findings, and patterns.

Discussion: Analyze and interpret the results in the context of the research objectives. Discuss any implications of the findings, their significance, and how they align with the literature review. Address any limitations and potential sources of bias.

Conclusion: Summarize the main findings and their implications. Restate the research objectives and discuss how they were addressed in the experiment.

Recommendations: Provide recommendations for further research, engineering practice, or policy based on the experiment's outcomes.

References: List all the sources, including books, journal articles, reports, and websites, cited in the report. Follow a specific citation style (e.g., APA, IEEE, or Chicago).

Appendices: Include any supplementary material that is relevant but not essential to the main body of the report. This may include detailed data tables, calculations, additional figures, or experimental protocols.

Acknowledgments: Acknowledge individuals or organizations that contributed to the experiment, such as funding sources, advisors, or research assistants.

Nomenclature or Abbreviations (if applicable): Provide a list of symbols, acronyms, or abbreviations used in the report, along with their definitions.

Declaration of Ethical Considerations (if applicable): If the experiment involves human subjects or ethical considerations, include a section describing the ethical approval process and any measures taken to ensure participant safety and informed consent.

The elements and their order may vary depending on the specific requirements of the report and the guidelines provided by the institution or journal where the report will be submitted. It is essential to follow a consistent and clear structure to make the report accessible and informative to readers.

3.3.10 Split-Plot and Blocked Experiments

Split-plot and blocked experiments are used in civil engineering when researchers want to investigate and optimize various factors or variables in construction, materials testing, structural analysis, or other civil engineering applications. Here are some common scenarios in civil engineering where split-plot and blocked experiments are employed:

Concrete Mix Design: Civil engineers often use split-plot or blocked experiments to optimize the mix proportions of concrete, including the types and proportions of aggregates, cement, water, and additives. This helps achieve the desired strength, durability, and workability of concrete for specific construction projects.

Asphalt Mix Design: In road construction, engineers use these experimental designs to optimize the composition of asphalt mixes, including the gradation of aggregates, binder types, and additives. The goal is to improve pavement performance and durability.

Structural Testing: When conducting experiments to evaluate the performance of structural components or systems (e.g., bridges, buildings, dams), engineers may use split-plot and blocked designs to vary multiple factors, such as load conditions, material properties, and structural configurations.

Geotechnical Engineering: Researchers in geotechnical engineering may employ these designs to investigate soil properties and their impact on foundations, retaining walls, and slope stability. Variables like soil type, compaction, moisture content, and loading conditions can be manipulated.

Materials Testing: Split-plot and blocked experiments are used to optimize materials for civil engineering projects. This includes investigating factors affecting the strength, ductility, and other mechanical properties of materials like steel, timber, and composites.

Environmental Impact Studies: In studies related to environmental engineering or construction site impact assessments, researchers may use these designs to evaluate the effects of various factors (e.g., construction methods, material choices, weather conditions) on environmental parameters.

Hydraulic Engineering: Engineers working on hydraulic systems, such as drainage networks, water treatment plants, or river channel modifications, may employ these experimental designs to optimize flow rates, pipe diameters, and water treatment processes.

Traffic Engineering: In traffic engineering, split-plot and blocked experiments can be applied to analyze factors affecting traffic flow, signal timings, lane configurations, and road design to optimize traffic management and safety.

Construction Management: Split-plot and blocked experiments can be used in construction management to study various factors that influence project scheduling, cost estimation, resource allocation, and risk management.

Sustainability and Green Building: Civil engineers and researchers use these designs to investigate the impact of sustainable practices, energy-efficient designs, and eco-friendly materials on the performance and environmental footprint of buildings and infrastructure.

In each of these applications, split-plot and blocked experiments help researchers systematically vary and study multiple factors simultaneously while controlling for other variables. This allows engineers to optimize designs, materials, or processes for better performance, cost-effectiveness, and sustainability in civil engineering projects.

STEPS

Developing split-plot and blocked designed experiments in civil engineering involves a systematic process to investigate and optimize various factors while controlling for potential sources of variability. Here are the general steps to develop these types of experiments in civil engineering:

1. **Define the Objective:**
   - Clearly define the engineering problem or objective you want to address with the experiment. Determine what specific factors or variables you need to study and optimize.
2. **Identify Factors and Levels:**
   - Identify the independent factors (also known as factors, factors of interest, or factors of variation) that you want to study. These could include material types, construction methods, environmental conditions, and other relevant variables.
   - Determine the levels or settings at which each factor will be tested. Levels represent different values or conditions of each factor that you want to investigate.

3. **Identify Response Variables:**
   - Specify the response variables or outcomes that you want to measure and optimize. These could be performance metrics, material properties, cost parameters, or any other relevant engineering criteria.
4. **Design the Experimental Layout:**
   - Decide whether you will use a split-plot or blocked design based on the nature of the factors and constraints of your experiment. In split-plot designs, some factors are applied at a larger scale (whole plots), while others are applied at a smaller scale (subplots).
   - Determine how factors will be allocated to whole plots and subplots, taking into account the experimental objectives and resource constraints.
5. **Randomization and Blocking:**
   - Randomly assign the experimental units (e.g., test specimens, construction sites, measurement locations) to different combinations of factor levels to ensure unbiased results.
   - Implement blocking if there are known sources of variability that should be controlled. Blocks are used to group experimental units with similar characteristics or conditions to reduce variability within each block.
6. **Data Collection:**
   - Conduct experiments and collect data according to the designed experimental layout. Ensure that measurements are accurate and consistent.
7. **Data Analysis:**
   - Analyze the collected data using appropriate statistical techniques, such as ANOVA, regression analysis, or other relevant methods.
   - Evaluate the main effects of factors, interactions between factors, and their impact on the response variables.
8. **Optimization:**
   - Use the results of the data analysis to identify optimal factor settings or conditions that achieve the desired engineering objectives. This may involve maximizing or minimizing specific response variables.
   - Consider trade-offs between conflicting objectives and constraints.
9. **Verification and Validation:**
   - Validate the optimized settings through additional experiments or simulations to ensure that the recommended conditions are practical and reliable.
10. **Report and Documentation:**
    - Prepare a comprehensive report that documents the experiment's objectives, methodology, data analysis, results, conclusions, and recommendations.
    - Include graphs, tables, and statistical summaries to present the findings effectively.
11. **Implementation:**
    - Implement the recommended settings, designs, or processes in real-world civil engineering projects, and monitor their performance over time.
12. **Continuous Improvement:**
    - Continuously monitor and evaluate the performance of the implemented changes and consider making further adjustments or refinements as needed.
13. **Documentation and Knowledge Sharing:**
    - Maintain records of the experimental design and results for future reference and knowledge sharing within the engineering community.

The specific steps and details of the experiment will vary depending on the nature of the civil engineering project and the objectives you are trying to achieve. Collaboration with statisticians or experts in experimental design can be valuable in designing and analyzing split-plot and blocked experiments effectively.

PITFALLS AND REMEDIES

Split-plot and blocked designed experiments in civil engineering, like any other experimental design, can encounter various challenges and pitfalls. Identifying these issues early and implementing remedies is crucial to ensure the reliability and validity of the experimental results. Here are some common pitfalls and potential remedies:

Pitfall 1: Poorly Defined Objectives
> **Pitfall:** Failing to clearly define the objectives and outcomes of the experiment can lead to confusion and the collection of irrelevant data.
>
> **Remedy:** Clearly state the research objectives and desired outcomes before designing the experiment. Ensure that all factors, responses, and constraints are well-defined and aligned with the project's goals.

Pitfall 2: Inadequate Sample Size
> **Pitfall:** Insufficient sample size can result in low statistical power, making it challenging to detect significant effects or interactions.
>
> **Remedy:** Conduct a power analysis to determine the required sample size based on the expected effect sizes, desired significance levels, and statistical power. Ensure that you have an adequate number of experimental units for each treatment combination.

Pitfall 3: Inappropriate Randomization
> **Pitfall:** Improper randomization can introduce bias or confounding factors into the experiment.
>
> **Remedy:** Randomly assign experimental units to treatment combinations or blocks to ensure that the assignment is unbiased. Use randomization procedures appropriate for the experimental design.

Pitfall 4: Incomplete Randomization
> **Pitfall:** Incomplete randomization occurs when randomization is not properly carried out, leading to systematic patterns in the data.
>
> **Remedy:** Implement full randomization to ensure that each experimental unit has an equal chance of receiving each treatment combination. Avoid systematic patterns or ordering of treatments.

Pitfall 5: Ignoring Block Effects
> **Pitfall:** Failing to account for block effects can lead to incorrect conclusions, especially when there are known sources of variability.
>
> **Remedy:** Use blocking when there are known sources of variability that need to be controlled. Blocks group similar experimental units together, reducing variability within each block.

Pitfall 6: Violation of Assumptions
> **Pitfall:** Violating the assumptions of the statistical analysis, such as normality or homoscedasticity, can affect the validity of the results.
>
> **Remedy:** Verify the assumptions of the chosen statistical analysis method. If assumptions are violated, consider data transformation or alternative analysis techniques.

Pitfall 7: Failure to Monitor Environmental Factors
> **Pitfall:** Environmental factors that change during the experiment (e.g., temperature, humidity) can introduce uncontrolled variability.
>
> **Remedy:** Monitor and record relevant environmental factors throughout the experiment. If necessary, implement control measures to maintain a stable environment.

Pitfall 8: Lack of Data Quality Control
> **Pitfall:** Inadequate data quality control can lead to errors in data collection and recording.
>
> **Remedy:** Implement rigorous data quality control procedures, including calibration of measurement instruments, training of personnel, and double-checking data entries.

Pitfall 9: Incomplete Documentation

Pitfall: Insufficient documentation of the experimental design, procedures, and results can hinder reproducibility and future analysis.

Remedy: Maintain detailed records of the experimental design, data collection, and analysis. Document any deviations from the original plan and provide clear descriptions in the final report.

Pitfall 10: Lack of Statistical Expertise

Pitfall: Insufficient knowledge of statistical methods can result in inappropriate analysis and misinterpretation of results.

Remedy: Collaborate with a statistician or seek expert guidance in experimental design and statistical analysis. Ensure that the chosen statistical methods are suitable for the experimental design.

By addressing these pitfalls and implementing the suggested remedies, you can enhance the reliability and validity of split-plot and blocked designed experiments in civil engineering and make more informed decisions based on the results.

Example

Split-plot and blocked experiments in civil engineering are designed to investigate the effects of multiple factors while considering constraints or restrictions in the experimental setup. These designs involve dividing the experiment into different subplots or blocks. Here's a simplified example of a matrix for a split-plot and blocked experiment in civil engineering:

Objective: To study the effect of different types of construction materials on the compressive strength of concrete while considering variations in curing methods.

Factors:

1. **Type of Cement (Factor A):**
   - Type 1 Portland Cement
   - Type 2 Portland Cement
2. **Type of Aggregate (Factor B):**
   - Coarse Aggregate
   - Fine Aggregate

Block Factor: 3. Curing Method (Block C):

- Water Curing
- Steam Curing

Response Variable: Compressive Strength (measured in megapascals, MPa)

Split-Plot and Blocked Experiment Matrix:

In this example, a split-plot and blocked experiment is designed to assess the impact of the type of cement and type of aggregate on compressive strength, while also considering the influence of the curing method. The matrix might look like this as displayed in Table 3.32.

In this matrix, each row represents a specific experimental run. The experiment is divided into blocks based on the curing method, ensuring that each combination of type of cement and type of aggregate is tested under both water curing and steam curing conditions.

TABLE 3.32
Split-Plot and Blocked Experiment Matrix for Civil Engineering

Run	Type of Cement (A)	Type of Aggregate (B)	Block (C)
1	Type 1 Portland	Coarse Aggregate	Water Curing
2	Type 2 Portland	Coarse Aggregate	Water Curing
3	Type 1 Portland	Fine Aggregate	Water Curing
4	Type 2 Portland	Fine Aggregate	Water Curing
5	Type 1 Portland	Coarse Aggregate	Steam Curing
6	Type 2 Portland	Coarse Aggregate	Steam Curing
7	Type 1 Portland	Fine Aggregate	Steam Curing
8	Type 2 Portland	Fine Aggregate	Steam Curing

Split-plot and blocked experiments are valuable in civil engineering when you need to account for factors that cannot be easily randomized, such as curing methods in concrete studies. This design allows you to assess the main effects of type of cement and type of aggregate while considering the potential influence of the curing method and minimizing confounding effects.

REPORTING

In split-plot and blocked designed experiments in civil engineering, a variety of statistical tools and techniques can be employed to analyze the data and draw meaningful conclusions. The choice of tools depends on the specific experimental design and research objectives. Here are some commonly used statistical tools for split-plot and blocked experiments in civil engineering:

Analysis of Variance (ANOVA): ANOVA is a fundamental statistical technique used to partition the total variation in the data into components attributable to different sources, such as main effects, interactions, and errors. It helps assess the significance of factors and interactions.

Regression Analysis: Regression models can be used to model relationships between independent variables (factors) and dependent variables (responses). Linear or nonlinear regression models may be appropriate, depending on the nature of the data.

Design of Experiments (DOE): Various experimental designs, including full-factorial, fractional factorial, and RSM, can be applied to optimize processes or evaluate the effects of multiple factors on a response variable.

Block Analysis: When blocking is used in the experimental design, block analysis helps assess the significance of block effects and any interaction between blocks and factors. It accounts for the variability introduced by blocking.

Randomization and Random Effects: Randomization techniques are used to assign experimental units to treatments or factors randomly, reducing bias. Random effects models can account for variability associated with random factors.

Contrast Analysis: Contrast analysis is used to compare specific combinations of factor levels or treatments to test predefined hypotheses. It helps focus on specific comparisons of interest.

Response Surface Analysis: When optimizing a response variable, RSM is used to model and analyze the relationship between factors and responses to identify optimal conditions.

Multivariate Analysis: Multivariate techniques, such as PCA and MANOVA, are used when dealing with multiple response variables simultaneously.

Chi-Square Tests: Chi-square tests may be used when dealing with categorical or count data. They assess the association or independence of variables within the experiment.

Nonparametric Tests: In cases where data do not meet the assumptions of normality or homoscedasticity, nonparametric tests like the Kruskal-Wallis test or Friedman test may be employed.

Post hoc Tests: Post hoc tests, such as Tukey's HSD or Bonferroni correction, are used to perform pairwise comparisons between treatment groups when ANOVA reveals significant differences.

Robust Statistics: Robust statistical techniques are employed when data may contain outliers or violations of normality assumptions. Robust regression or robust ANOVA methods can be considered.

Statistical Software: Various statistical software packages like R, SAS, SPSS, Minitab, and JMP are commonly used to perform data analysis and generate results.

Visualization Tools: Data visualization techniques, such as scatter plots, bar charts, and box plots, help explore and communicate the patterns and trends in the data.

The choice of statistical tools depends on the specific goals of the experiment, the type of data collected, and the complexity of the analysis. In many cases, a combination of these tools may be used to gain a comprehensive understanding of the experimental results and make informed decisions in civil engineering projects.

A report for a split-plot and blocked designed experiment in civil engineering should provide a clear and comprehensive overview of the study, experimental design, methodology, results, and conclusions. Here are the essential elements typically included in such a report:

Title: The title should be descriptive and succinct, reflecting the main purpose and scope of the experiment.

Abstract: A brief summary of the entire report, including the objectives, methods, key findings, and conclusions.

Introduction:
Background and context: Explain the problem or research question addressed by the experiment.
Objectives: Clearly state the goals and objectives of the study.
Importance: Discuss the significance and relevance of the research in the field of civil engineering.

Literature Review: Provide a concise review of relevant literature and previous research related to the experiment, including any existing theories, models, or methodologies.

Experimental Design:
Description: Explain the split-plot and blocked experimental design, including how factors were assigned to different plots or blocks.
Randomization: Describe how randomization was used to minimize bias in the assignment of treatments.
Factors and levels: List the factors studied and their respective levels.
Blocking: Detail the blocking strategy and the rationale for using it.

Methods:
Data collection: Describe the data collection process, including the instruments or equipment used.
Procedures: Provide step-by-step instructions on how the experiment was conducted.
Data analysis: Explain the statistical methods and software used to analyze the data.

Results:
Presentation of data: Present the raw or summarized data using tables, charts, and graphs.
Statistical analysis: Present the results of statistical tests, including ANOVA, regression analysis, or other relevant analyses.
Interpretation: Discuss the implications of the results and their significance in the context of the research objectives.

Discussion:

Comparison with objectives: Evaluate whether the experiment's results align with the stated objectives.

Interpretation: Provide an in-depth interpretation of the findings, discussing the effects of factors, interactions, and any observed patterns.

Limitations: Address any limitations or constraints encountered during the experiment.

Implications: Discuss the practical implications of the findings for civil engineering practice or future research.

Conclusions: Summarize the key findings and their implications. Clearly state whether the objectives of the experiment were achieved.

Recommendations: Offer recommendations for further research or practical applications based on the experiment's results.

References: Cite all relevant sources, including research papers, books, and publications, following a standardized citation style (e.g., APA, MLA, Chicago).

Appendices: Include any supplementary materials, such as detailed experimental procedures, data tables, or additional charts and graphs.

Acknowledgments: Recognize any individuals, organizations, or funding sources that contributed to the experiment.

Figures and Tables: Include well-labeled and properly formatted figures and tables within the body of the report to illustrate key points and support the discussion.

Nomenclature: If applicable, provide a list of symbols and abbreviations used in the report to enhance clarity.

Ethical Considerations: Mention any ethical considerations, such as informed consent or safety protocols, if relevant to the experiment.

Executive Summary: Provide a concise summary of the report's main findings and conclusions, intended for readers who may not have time to read the entire document.

It's important to structure the report logically and use clear and concise language. Properly documenting the experimental design, methodology, and results is essential for transparency, reproducibility, and the credibility of the research in civil engineering.

3.4 ELECTRICAL ENGINEERING

In electrical engineering, designed experiments are used to optimize electrical systems, components, and processes, as well as to study and control electrical phenomena. Here are some common types of designed experiments with examples of their applications in electrical engineering.

In designed experiments in electrical engineering, various types of variables are considered to study and optimize different aspects of electrical systems, circuits, devices, and processes. These variables can be categorized into several common types, including:

1. **Independent Variables:**
   - **Component Values:** Variables related to the values of electrical components, such as resistors, capacitors, and inductors, which can affect the behavior of circuits.
   - **Operating Conditions:** Variables related to the operating conditions of electrical systems, including voltage levels, current levels, frequency, and duty cycle.
   - **Control Settings:** Variables related to the settings of control parameters in electrical devices or systems, such as gain, feedback, and threshold values.
   - **Signal Characteristics:** Variables related to signal characteristics, including amplitude, frequency, phase, and modulation.

2. **Dependent Variables**:
   - **Electrical Performance**: Variables related to the electrical behavior and performance of circuits and devices, such as voltage, current, power, impedance, and S/N ratio.
   - **Signal Integrity**: Variables related to signal quality, such as signal distortion, noise levels, and jitter.
   - **Efficiency**: Variables related to the energy efficiency and power consumption of electrical devices or systems.
   - **Electromagnetic Compatibility (EMC)**: Variables related to the electromagnetic emissions and susceptibility of electrical devices or systems, which can affect their interference with other electronics.
3. **Categorical Variables**:
   - **Component Types**: Categories of electrical components used in circuits, such as diodes, transistors, op-amps, and microcontrollers.
   - **Circuit Configurations**: Categories of circuit configurations or topologies, which can affect circuit behavior.
   - **Control Modes**: Categories of control modes or operating modes for electrical devices, such as on/off, linear, or Pulse Width Modulation (PWM) control. PWM configurations include: 555 Timers, Micro-controller Based, Transistor Based, Op-Amp Based, and Dedicated PWM ICs.
4. **Control Variables (Covariates)**:
   - These are variables that are held constant or controlled during experiments to eliminate their influence on the dependent variable. For example, in a study of amplifier performance, the supply voltage may be a controlled variable.
5. **Random Variables**:
   - In some cases, random variables may be introduced to account for variability or uncertainty in measurements or environmental conditions.
6. **Interaction Variables**:
   - Interaction variables are used to investigate the combined effects of two or more independent variables on the dependent variable. For example, the interaction between input voltage and load resistance in a circuit's output.
7. **Noise Variables (Error Variables)**:
   - These are uncontrolled or unmeasured variables that can introduce variability or errors into the experimental results. Techniques like filtering and shielding may be used to reduce the impact of noise variables in electrical experiments.

Electrical engineering designed experiments are often used to optimize circuit designs, control systems, and electronic devices by systematically varying these variables to understand their effects and make informed decisions in the design, testing, and operation of electrical systems. These experiments help engineers achieve desired electrical performance, efficiency, and reliability in electrical engineering applications.

3.4.1 FACTORIAL EXPERIMENTS

Factorial experiments are used in electrical engineering for various purposes, including research, product development, and quality improvement. Here are some common areas and applications where factorial experiments are employed in electrical engineering:

Electronic Component Testing: Factorial experiments can be used to investigate the effects of multiple factors (e.g., temperature, voltage, component tolerances) on the performance and reliability of electronic components such as transistors, diodes, and integrated circuits.

Circuit Design Optimization: Electrical engineers often use factorial experiments to optimize the design of electronic circuits. Factors like component values, layout, and operating conditions can be systematically varied to achieve desired circuit performance (e.g., S/N ratio, bandwidth, power efficiency).

Embedded Systems Development: In the development of embedded systems and microcontroller-based projects, factorial experiments can help determine the optimal combination of hardware and software parameters to meet performance and power consumption requirements.

Power Electronics: Factorial experiments are used to analyze the impact of factors like switching frequency, voltage levels, and control algorithms on the efficiency and performance of power electronic converters and inverters.

Signal Processing: In fields like digital signal processing (DSP), factorial experiments can assist in optimizing filter designs, sampling rates, and other parameters to enhance signal quality and processing speed.

Control System Tuning: Electrical engineers use factorial experiments to fine-tune control systems in applications like robotics, industrial automation, and mechatronics. Factors like controller gains, setpoint values, and sensor characteristics are adjusted to achieve desired system behavior.

Reliability and Fault Tolerance: For critical electrical systems, factorial experiments can be employed to assess the reliability and fault tolerance of components and subsystems under various operating conditions.

Wireless Communication: In wireless communication systems, factorial experiments can help optimize antenna design, modulation schemes, channel coding, and power management to maximize data transmission efficiency and minimize interference.

Quality Control: Electrical engineers use factorial experiments to improve product quality and manufacturing processes. Factors affecting product performance and consistency, such as manufacturing tolerances and material properties, can be studied to identify optimal settings.

Energy Efficiency: Factorial experiments are used to investigate energy-efficient solutions, including power management strategies, energy-efficient algorithms, and hardware design choices, to reduce energy consumption in electrical systems.

Noise Reduction: In audio and signal processing applications, factorial experiments can be used to study the impact of factors like noise sources, filtering, and component selection on audio quality.

Testing and Calibration: Factorial experiments play a role in determining the optimal testing conditions and calibration procedures for electrical testing equipment and measurement systems.

Fault Diagnosis: In fault diagnosis and troubleshooting scenarios, factorial experiments can help identify the root causes of electrical system failures by varying factors related to components, environmental conditions, and operational parameters.

Integrated Circuit (IC) Design: Electrical engineers use factorial experiments to optimize IC designs, considering factors like transistor sizing, layout, and power management to achieve desired performance, power efficiency, and reliability.

Factorial experiments allow electrical engineers to efficiently explore the effects of multiple factors on system performance, enabling them to make informed decisions, optimize designs, and solve complex engineering problems. These experiments help in achieving cost-effective and reliable solutions in various electrical engineering applications.

STEPS

Developing factorial experiments in electrical engineering involves a systematic process to study the effects of multiple factors on a particular outcome or performance measure. Here are the general steps to develop factorial experiments in electrical engineering:

1. **Define the Objective:**
   - Clearly state the objective of the experiment. What specific aspect of electrical engineering are you trying to investigate or optimize? Define the response variable or performance measure you want to study or improve.

2. **Identify Factors:**
   - Identify the independent variables or factors that may affect the outcome or performance measure. Factors can include component values, operating conditions, design parameters, and environmental variables.

3. **Determine Factor Levels:**
   - For each factor, determine the range of levels or values that will be investigated during the experiment. It's important to choose a meaningful and practical range for each factor.

4. **Select Experimental Design:**
   - Choose an appropriate factorial experimental design based on the number of factors and levels. Common designs include full-factorial, fractional factorial, and Taguchi designs. The choice of design should consider the available resources and the complexity of the experiment.

5. **Create Experimental Matrix:**
   - Generate a matrix that defines the combinations of factor levels for each experimental run. The matrix specifies which factor levels to use in each trial or test. In a full-factorial design, all possible combinations are considered. In a fractional factorial design, only a subset of combinations is tested to reduce the number of runs.

6. **Conduct Experiments:**
   - Perform the experiments according to the combinations specified in the experimental matrix. Ensure that each test or trial is conducted accurately and consistently. Record the results of each experiment.

7. **Collect Data:**
   - Collect data for the response variable or performance measure in each experiment. Ensure that data collection is precise and reliable. Use appropriate measuring instruments and techniques.

8. **Data Analysis:**
   - Analyze the collected data to understand the effects of factors on the response variable. Statistical methods such as ANOVA, regression analysis, and graphical analysis may be used to assess the significance of factors and interactions.

9. **Optimization:**
   - If the objective is optimization, use the results of the data analysis to determine the optimal factor levels that maximize or minimize the desired outcome. Optimization techniques such as RSM can be applied.

10. **Interpret Results:**
    - Interpret the results of the data analysis in the context of the electrical engineering problem you are addressing. Determine how each factor contributes to the observed variations in the response variable.

11. **Draw Conclusions:**
    - Draw conclusions based on the analysis and interpretation of results. Address the research questions or objectives defined at the beginning of the experiment. Assess the practical implications of the findings.

12. **Report Findings:**
    - Prepare a comprehensive report summarizing the experimental design, methodology, data analysis, results, and conclusions. Include tables, graphs, and statistical analyses to support your findings. Present any recommendations or insights gained from the study.
13. **Validation (Optional):**
    - Depending on the nature of the experiment, consider conducting additional experiments or validations to confirm the results and recommendations.
14. **Implement Findings (If Applicable):**
    - If the experiment leads to recommendations for design improvements, operational changes, or process optimizations, implement the findings in the relevant electrical engineering application.
15. **Review and Refine:**
    - Review the experimental process and results critically. If necessary, refine the experiment or conduct follow-up studies to further investigate specific aspects or questions.

Factorial experiments in electrical engineering are a powerful tool for understanding complex relationships between factors and outcomes, optimizing designs, and improving system performance. Careful planning, execution, and analysis are essential for the success of these experiments.

PITFALLS AND REMEDIES

Factorial experiments in electrical engineering, like any other experimental design, can encounter various pitfalls. Here are some common pitfalls and their remedies:

Pitfall 1: Inadequate Factor Selection
   Pitfall: Not including all relevant factors or including unnecessary factors can lead to incomplete or inefficient experiments.
   Remedy: Conduct a thorough preliminary analysis to identify the most important factors. Use engineering expertise and literature review to guide factor selection.

Pitfall 2: Inappropriate Factor Levels
   Pitfall: Choosing inappropriate levels for factors can result in experiments that do not capture the entire range of real-world conditions.
   Remedy: Carefully consider the practical range of each factor and select levels that are meaningful and relevant to the problem. Pilot studies or expert consultations can help.

Pitfall 3: Insufficient Replication
   Pitfall: Conducting experiments without replication can make it difficult to distinguish between random variation and true effects.
   Remedy: Include replicates for each combination of factor levels. Replication helps estimate experimental error and increases the reliability of results.

Pitfall 4: Ignoring Interactions
   Pitfall: Neglecting to consider interactions between factors can lead to inaccurate conclusions about their effects.
   Remedy: Analyze and interpret interaction effects. Use statistical tools like ANOVA or regression to identify and quantify interactions. Plot interaction graphs to visualize their impact.

Pitfall 5: Small Sample Size
   Pitfall: Using a small sample size may result in insufficient statistical power to detect significant effects.
   Remedy: Calculate the required sample size based on the desired level of significance, effect size, and power. Ensure that the sample size is adequate for meaningful conclusions.

Pitfall 6: Violating Assumptions

Pitfall: Violating the assumptions of the chosen statistical analysis method (e.g., normality of residuals) can lead to inaccurate results.

Remedy: Check the assumptions of the statistical tests used and consider transformations or alternative tests if assumptions are not met.

Pitfall 7: Overfitting

Pitfall: Overfitting occurs when a complex model is used to fit the data, resulting in poor generalization to new data.

Remedy: Use model selection techniques, such as cross-validation or information criteria, to choose the simplest model that adequately represents the data.

Pitfall 8: Neglecting Practical Significance

Pitfall: Focusing solely on statistical significance without considering practical significance can lead to irrelevant or impractical conclusions.

Remedy: Always assess the practical importance of observed effects. Even if an effect is statistically significant, it may not be practically significant.

Pitfall 9: Lack of Randomization

Pitfall: Failing to randomize the order of experiments or treatments can introduce bias into the results.

Remedy: Randomize the order of experiments or treatments to eliminate the effects of confounding variables that are not part of the experimental design.

Pitfall 10: Inadequate Documentation

Pitfall: Poor documentation of experimental procedures and data can hinder reproducibility and make it difficult to troubleshoot issues.

Remedy: Maintain thorough records of experimental procedures, data collection, and analysis. Document any deviations from the planned protocol.

Pitfall 11: Incomplete Reporting

Pitfall: Failing to report all relevant details in the experiment's findings and methods can lead to a lack of transparency.

Remedy: Ensure that the experimental report includes all essential information, including design details, statistical analyses, and results interpretation.

Avoiding these pitfalls and following best practices in experimental design and statistical analysis will contribute to the success and reliability of factorial experiments in electrical engineering. Collaborating with experienced statisticians or engineers can also be beneficial when designing and analyzing complex experiments.

EXAMPLE

Factorial experiments in electrical engineering are used to study the effects of multiple factors or variables on a particular outcome or response variable. Here's a simplified example of a matrix for a 2^3 full-factorial experiment in electrical engineering:

Objective: To investigate the impact of voltage (V), current (I), and resistance (R) on the power dissipation of an electrical circuit.

Factors:

1. **Voltage (V):**
   - Level 1: 5 volts
   - Level 2: 10 volts

2. **Current (I)**:
   - Level 1: 2 amperes
   - Level 2: 4 amperes
3. **Resistance (R)**:
   - Level 1: 3 ohms
   - Level 2: 6 ohms

Response Variable: Power Dissipation (measured in watts, W)

Full-Factorial Matrix (2^3):
In this example, a full 2^3 factorial design is used, which involves all possible combinations of the factors and levels. The matrix is shown in Table 3.33.

In this matrix, each row represents a specific experimental run with a unique combination of Voltage, Current, and Resistance levels. The power dissipation of the electrical circuit is measured for each combination. By analyzing the results, electrical engineers can determine how changes in voltage, current, and resistance impact power dissipation in the circuit, which is crucial for design and optimization purposes.

REPORTING

In factorial experiments designed for electrical engineering, various statistical tools and techniques are employed to analyze the data and draw meaningful conclusions. Here are some common statistical tools used in factorial experiments in electrical engineering:

Analysis of Variance (ANOVA): ANOVA is a fundamental statistical tool used to analyze the variation in data and identify significant factors, interactions, and their effects on the response variable. It helps determine whether differences between factor levels are statistically significant.

Regression Analysis: Regression analysis is used to model the relationship between independent variables (factors) and the dependent variable (response). It can provide insights into how changes in factors affect the response and can be used for prediction.

Main Effects Plot: Main effects plots are graphical representations of the main effects of each factor in a factorial experiment. They help visualize how individual factors influence the response variable.

TABLE 3.33
Full-Factorial Example for Electrical Engineering

Run	Voltage (V)	Current (I)	Resistance (R)
1	5V	2A	3Ω
2	5V	2A	6Ω
3	5V	4A	3Ω
4	5V	4A	6Ω
5	10V	2A	3Ω
6	10V	2A	6Ω
7	10V	4A	3Ω
8	10V	4A	6Ω

Interaction Plots: Interaction plots visualize the interactions between factors. These plots can help identify cases where the effect of one factor depends on the level of another factor.

Response Surface Methodology (RSM): RSM is a collection of statistical and mathematical techniques used to model and optimize complex processes. It is particularly useful for exploring the relationships between multiple factors and response variables.

Design of Experiments (DOE) Software: Specialized DOE software packages are often used for planning, conducting, and analyzing factorial experiments. They can help with experimental design, data collection, and statistical analysis.

Contour Plots and 3D Surface Plots: These graphical tools are used to visualize the response surface when dealing with multiple factors. Contour plots display lines of constant response, and 3D surface plots provide a more comprehensive view of the response landscape.

ANOVA Table: The ANOVA table summarizes the results of the analysis of variance, including the sources of variation, degrees of freedom, mean squares, F-statistics, and p-values. It helps in assessing the significance of factors and interactions.

Residual Analysis: Residuals are the differences between observed and predicted values. Residual analysis is used to check if the assumptions of ANOVA and regression models are met. It helps ensure that the model is a good fit for the data.

Confidence Intervals: Confidence intervals provide a range of values within which population parameters (e.g., means or effect sizes) are likely to fall. They offer a measure of uncertainty around estimated effects.

Power Analysis: Power analysis is used to determine the statistical power of an experiment, which assesses the likelihood of detecting true effects. It helps in determining an appropriate sample size for achieving desired levels of power.

Factorial Plots: Factorial plots show the interaction effects between two or more factors. These plots can reveal the nature and strength of interactions visually.

Fractional Factorial Design: In cases where the number of experiments becomes impractical due to a large number of factors, fractional factorial designs allow for a reduced set of experiments while still capturing key information about factor effects and interactions.

Multivariate Analysis: When dealing with multiple response variables, multivariate analysis techniques like MANOVA are used to assess the joint effects of factors on multiple responses simultaneously.

Robust Design Analysis: Robust design techniques help in optimizing processes or systems by identifying factor settings that are less sensitive to variations or noise.

These statistical tools are valuable for designing and analyzing factorial experiments in electrical engineering, allowing engineers and researchers to gain insights into the impact of various factors on system performance and make informed decisions for optimization and improvement.

A well-structured report for a factorial experiment in electrical engineering should provide a clear and comprehensive overview of the study, including the experimental design, methods, results, and conclusions. Below are the essential elements typically included in such a report:

Title Page:
Title of the Report
Names of the Author(s)
Affiliations
Date of Submission

Abstract:
A concise summary of the entire report, including the objectives, methods, key findings, and conclusions.

Table of Contents:

A list of sections and subsections with corresponding page numbers.

List of Figures and Tables:

A list of all figures and tables with their respective captions and page numbers.

List of Abbreviations and Symbols (if applicable):

An alphabetical list of abbreviations and symbols used in the report, along with their explanations.

Introduction:

Background and context of the study.

Clear statement of the research problem or objectives.

Brief literature review related to the topic.

Hypotheses or research questions.

Experimental Design:

Detailed description of the factorial experiment design, including the factors, levels, and response variables.

Justification for selecting the specific factors and levels.

Description of any control variables or constants.

Experimental setup and equipment used.

Sample size determination and randomization methods.

Methodology:

Description of the data collection procedures.

Explanation of how factors were manipulated and measurements were taken.

Data preprocessing and any necessary transformations.

Any specialized software or tools used for data collection and analysis.

Results:

Presentation of the raw data or data summary.

Tables, figures, charts, and plots to illustrate the data.

Statistical analyses and tests performed, including ANOVA tables or regression analyses.

Interpretation of the results and identification of significant factors or interactions.

Discussion:

Interpretation of the findings in the context of the research objectives.

Comparison of results with the initial hypotheses or expectations.

Explanation of the practical implications of the findings.

Discussion of limitations and potential sources of error.

Suggestions for further research or improvements.

Conclusion:

A concise summary of the key findings and their significance.

Restatement of the research objectives and whether they were achieved.

Implications for the field or practical applications.

References:

Citation of all sources, including academic papers, books, journals, and software packages used in the study.

Appendices (if applicable):

Additional details that may be useful but are not essential for understanding the main report.

This may include supplementary data, code, questionnaires, or detailed experimental procedures.

Acknowledgments (optional):

Recognition of individuals, organizations, or funding sources that contributed to the research.

Author Information (optional):

Biographical information about the authors, including contact details.

Ensure that the report is well-organized, free of errors, and includes clear and informative visuals to support the findings. Proper referencing and adherence to a consistent citation style (e.g., APA, IEEE) are also important. This structure provides a comprehensive overview of the factorial experiment and allows readers to understand the research process, results, and conclusions effectively.

3.4.2 RESPONSE SURFACE METHODOLOGY (RSM)

RSM experiments are used in various areas of electrical engineering for optimizing and improving processes, designs, and products. Some specific applications include:

Circuit Design Optimization: RSM can be applied to optimize electronic circuit designs, such as filters, amplifiers, and oscillators. It helps in finding the optimal component values and configurations to meet specific performance criteria while minimizing costs or power consumption.

Signal Processing: RSM can be used to optimize signal processing algorithms and parameters, improving the quality of signal processing applications like image and speech processing, data compression, and filtering.

Control System Tuning: In control systems engineering, RSM can be employed to fine-tune controller parameters for systems like Proportional-Integral-Derivative (PID) controllers, ensuring stability, response time, and performance meet desired criteria.

Reliability and Quality Assurance: Electrical engineers can use RSM to optimize the reliability and quality of electrical components and systems. This is particularly important in industries like semiconductor manufacturing, where small variations can impact yield and product quality.

Power Electronics: RSM can help optimize power electronics systems, such as converters, inverters, and motor drives, to improve energy efficiency, reduce losses, and enhance overall performance.

Antenna Design: In the field of electromagnetics, RSM can assist in optimizing antenna designs to achieve desired radiation patterns, bandwidth, and impedance matching.

Printed Circuit Board (PCB) Layout: Electrical engineers can use RSM to optimize the layout of PCBs, considering factors like signal integrity, thermal performance, and electromagnetic interference (EMI).

Material Characterization: RSM can be applied to characterize and optimize the electrical properties of materials, such as dielectrics and conductors, for specific applications like microelectronics and Radio Frequency (RF) circuits.

Energy Efficiency: Electrical engineers can use RSM to optimize energy-efficient systems and devices, such as lighting, Heating, Ventilation, and Air Conditioning (HVAC) systems, and renewable energy systems.

Fault Detection and Diagnosis: RSM can assist in developing algorithms for fault detection and diagnosis in electrical systems, enhancing the reliability and safety of electrical infrastructure.

Wireless Communication Systems: Optimization of wireless communication systems, including cellular networks and Wi-Fi, for parameters like coverage, data rate, and interference mitigation.

In each of these applications, RSM helps engineers systematically explore and optimize complex design spaces, leading to improved performance, reduced costs, and enhanced efficiency. By conducting experiments and modeling responses, engineers can make informed decisions and achieve their design and performance objectives more effectively.

STEPS

Developing RSM experiments in electrical engineering involves a systematic approach to optimizing processes, designs, or products. Here are the steps to develop RSM experiments in electrical engineering:

1. **Define the Objective**: Clearly state the objective of your RSM study. What specific aspect of the electrical system or design are you trying to optimize? For example, you may want to optimize the efficiency of a power converter or the S/N ratio in a communication system.
2. **Identify Factors and Levels**: Identify the independent factors (variables) that may affect the response (performance) of the electrical system or design. Determine the range and levels for each factor. Factors could include component values, operating conditions, or design parameters.
3. **Select the Response Variable**: Choose the response variable or performance metric that reflects the system's quality or effectiveness. It could be a measure of efficiency, signal quality, power consumption, or any other relevant parameter.
4. **Experimental Design**: Select an appropriate experimental design for RSM. Common designs include Box-Behnken, Central Composite, or Latin Hypercube designs. The choice of design depends on the number of factors and the desired level of precision.
5. **Conduct Experiments**: Perform the experiments according to the selected design. Each experiment involves varying the factors at specified levels and measuring the response variable. Ensure that the experiments are conducted accurately and consistently.
6. **Collect Data**: Record the data obtained from the experiments, including the factor levels and corresponding response values. Ensure that the data collection process is rigorous and error-free.
7. **Fit Response Surface Models**: Use statistical software or tools to fit response surface models to the experimental data. These models describe the relationship between the factors and the response variable. Common models include linear, quadratic, or higher-order polynomial models.
8. **Model Validation**: Validate the response surface models to ensure they accurately represent the system's behavior. Use statistical techniques like ANOVA to assess model adequacy.
9. **Optimization**: Once the response surface models are validated, use them to perform optimization. The goal is to find the factor settings that maximize or minimize the response variable, depending on the objective. Optimization techniques like gradient descent or genetic algorithms can be employed.
10. **Sensitivity Analysis**: Conduct sensitivity analysis to determine how sensitive the optimal solution is to variations in factors and model parameters. This helps in understanding the robustness of the optimized design.
11. **Implementation**: Implement the optimized design or process changes in the real-world electrical system. Ensure that the practical implementation aligns with the recommended settings obtained from the RSM study.
12. **Testing and Verification**: After implementing the changes, conduct further testing and verification to confirm that the desired improvements have been achieved. This may involve real-world testing, simulations, or additional experiments.
13. **Documentation and Reporting**: Document the entire RSM study, including the experimental design, data, models, optimization results, and implementation details. Prepare a comprehensive report summarizing the study's objectives, methods, findings, and recommendations.
14. **Continuous Monitoring**: Monitor the optimized system or design over time to ensure that the improvements are sustained and that any unforeseen issues are addressed promptly.

RSM experiments in electrical engineering can significantly enhance the performance and efficiency of electrical systems, components, and designs. They provide a structured approach to optimizing complex systems and making data-driven decisions for improvement.

PITFALLS AND REMEDIES

RSM experiments in electrical engineering, like any scientific endeavor, can encounter various pitfalls. Here are some common pitfalls and their corresponding remedies:

Insufficient or Poorly Chosen Design:
> **Pitfall:** Choosing an inappropriate design (e.g., insufficient data points, wrong factor levels) can lead to inaccurate models.
>
> **Remedy:** Consult with a statistician or use statistical software to select an appropriate RSM design based on the number of factors and expected model complexity. Ensure the design covers a wide range of factor levels.

Model Inadequacy:
> **Pitfall:** Assuming a simple linear model when the actual relationship is nonlinear can lead to poor predictions.
>
> **Remedy:** Check the model fit statistics (e.g., R-squared, lack of fit) to assess model adequacy. Consider higher-order polynomial models or other nonlinear models if necessary.

Multicollinearity:
> **Pitfall:** High correlations between factors (multicollinearity) can make it challenging to identify the individual impact of each factor on the response.
>
> **Remedy:** Perform a correlation analysis among factors and consider removing or transforming highly correlated factors. Alternatively, use techniques like PCA to address multicollinearity.

Overfitting:
> **Pitfall:** Fitting a model that is too complex for the available data can lead to overfitting, where the model fits noise in the data rather than the underlying trend.
>
> **Remedy:** Use techniques like cross-validation to assess model performance and prevent overfitting. Select models with good predictive performance on new data.

Lack of Model Validation:
> **Pitfall:** Failing to validate the response surface models can result in unreliable predictions.
>
> **Remedy:** Use techniques like ANOVA to validate the models. Compare predicted values with actual experimental data to ensure the models accurately represent the system.

Neglecting Factor Interactions:
> **Pitfall:** Ignoring or underestimating the importance of factor interactions can lead to suboptimal results.
>
> **Remedy:** Examine interaction terms in the response surface models and their statistical significance. Factor interactions can be crucial in understanding and optimizing the system.

Lack of Robustness Analysis:
> **Pitfall:** Failing to assess the robustness of the optimized solution to variations in factors or model parameters.
>
> **Remedy:** Conduct sensitivity analysis to determine how sensitive the optimal solution is to changes in factors and model parameters. Ensure the design is robust in real-world conditions.

Practical Constraints:
> **Pitfall:** Obtaining optimal factor settings that are impractical or costly to implement in the real world.
>
> **Remedy:** Consider practical constraints and limitations when interpreting and implementing the optimized solution. Adjust factor settings to align with practicality.

Inadequate Documentation:
   Pitfall: Insufficient documentation can make it challenging to reproduce or understand the RSM study.
   Remedy: Maintain thorough documentation of the entire RSM process, including experimental design, data collection, model development, and optimization steps. This documentation is essential for transparency and future reference.
Disregarding External Factors:
   Pitfall: Neglecting the influence of external factors or noise in the system.
   Remedy: Incorporate external factors or noise into the RSM study by including them as additional factors or by conducting controlled experiments to account for their effects.

Addressing these pitfalls and following best practices in RSM experimentation can help ensure the success and reliability of optimization efforts in electrical engineering projects. Collaboration with statisticians or experts in the field can also be valuable in overcoming these challenges.

EXAMPLE

RSM is a statistical technique used in electrical engineering to optimize processes or designs while studying the relationships between multiple factors and a response variable. Here's an example of a matrix for an RSM designed experiment in electrical engineering:

Objective: To optimize the performance of a semiconductor fabrication process for maximum yield while minimizing defects.

Factors:

1. **Temperature (Factor A):**
   - Range: 200°C to 300°C
2. **Pressure (Factor B):**
   - Range: 2 psi to 5 psi
3. **Time (Factor C):**
   - Range: 10 minutes to 30 minutes

Response Variable: Yield Percentage (measured in percentage)

Matrix for RSM Designed Experiment:
   In this example, an RSM design is used to optimize the semiconductor fabrication process by studying the effects of Temperature, Pressure, and Time on yield percentage. The matrix is shown in Table 3.34.
   In this matrix, each row represents a specific experimental run with a unique combination of Temperature, Pressure, and Time levels. The yield percentage of the semiconductor fabrication process is measured for each combination. Using statistical analysis and response surface modeling, electrical engineers can determine the optimal process conditions that maximize yield while considering the factors of Temperature, Pressure, and Time. RSM helps identify the relationships between these factors and the response variable, allowing for process optimization in electrical engineering applications.

REPORTING

RSM experiments in electrical engineering often involve various statistical tools and techniques for experimental design, data analysis, and optimization. Here are some of the key statistical tools commonly used in RSM experiments for electrical engineering:

TABLE 3.34

Response Surface Methodology (RSM) Example for Electrical Engineering

Run	Temperature (A)	Pressure (B) (psi)	Time (C) (minutes)
1	200°C	2	10
2	250°C	2	10
3	300°C	2	10
4	200°C	5	10
5	250°C	5	10
6	300°C	5	10
7	200°C	2	20
8	250°C	2	20
9	300°C	2	20
10	200°C	5	20
11	250°C	5	20
12	300°C	5	20
13	200°C	2	30
14	250°C	2	30
15	300°C	2	30
16	200°C	5	30
17	250°C	5	30
18	300°C	5	30

Design of Experiments (DOE):

Full-Factorial Design: Examines the effects of all possible combinations of factors and levels.

Fractional Factorial Design: A subset of factor combinations is tested to reduce the number of experimental runs.

Central Composite Design (CCD): Includes both factorial and axial points to model curvature in the response surface.

Analysis of Variance (ANOVA):

ANOVA helps assess the significance of factors, interactions, and the model's adequacy.

Residual analysis is used to check the assumptions of the model.

Response Surface Models:

Polynomial Regression: Fits a polynomial equation to the experimental data to describe the response surface.

Quadratic, cubic, or higher-order polynomial models may be used, depending on the system's complexity.

Model Selection and Validation:

Model Fit Statistics: R-squared, adjusted R-squared, and lack-of-fit tests help assess model quality.

Cross-Validation: Validates the model's predictive performance on new data.

Optimization Techniques:

Desirability Function: Combines multiple response variables into a single desirability score to optimize simultaneously.

Gradient-Based Optimization: Methods like gradient descent or the simplex method are used for numerical optimization.

Monte Carlo Simulation: Assess the robustness of the optimized solution by considering variability and uncertainty.

Factorial Plots and Response Surface Plots:
> Contour Plots: Visualize the response surface in two dimensions.
> 3D Surface Plots: Show the relationship between two factors and the response.
> Interaction Plots: Illustrate the interaction effects between factors.

Sensitivity Analysis:
> Sensitivity analysis evaluates how changes in factors or model parameters affect the optimized solution.

Statistical Software:
> Software packages like Minitab, JMP, R, or Python with libraries like SciPy and scikit-learn are commonly used for experimental design and analysis.

Robust Design Tools:
> Robust Parameter Design (RPD): Aims to optimize performance while minimizing sensitivity to noise factors.

Simulation Tools:
> Electrical engineering simulations, such as SPICE simulations, can be integrated with RSM to model electrical circuits and systems.

Statistical Process Control (SPC):
> SPC techniques can be used in conjunction with RSM to monitor and control processes in real-time.

Hypothesis Testing:
> t-tests and F-tests are used to test the significance of specific factors or comparisons between groups.

Experimental Error Analysis:
> Variance component analysis can help quantify sources of variation in the experimental data.

Design Optimization Algorithms:
> Algorithms like genetic algorithms, particle swarm optimization, or simulated annealing can be used for global optimization.

These statistical tools and techniques enable engineers in electrical engineering to design experiments, model complex systems, optimize designs, and make informed decisions based on empirical data. The specific tools and methods chosen depend on the nature of the problem, the complexity of the system, and the objectives of the study.

A comprehensive report for an RSM experiment in electrical engineering should provide a clear and detailed account of the study, its objectives, methodology, results, and conclusions. Here are the key elements typically included in such a report:

Title Page:
- Title of the Report
- Names and Affiliations of the Authors
- Date

Abstract:
- A concise summary of the study, including the problem statement, methodology, key results, and conclusions.

Table of Contents:
- A list of sections, subsections, and page numbers for easy navigation.

List of Figures and Tables:
- A separate list that enumerates all figures and tables in the report along with their respective page numbers.

List of Abbreviations and Nomenclature (if applicable):
- A list of symbols, acronyms, and abbreviations used in the report, along with their explanations.

Introduction:
- Background and context of the study.
- Problem statement or research questions.
- Objectives and goals of the experiment.
- Relevance and significance of the study to electrical engineering.

Literature Review:
- A review of relevant literature and previous research on the topic.
- Discussion of existing RSM applications in electrical engineering (if applicable).

Experimental Design:
Description of the experimental setup and equipment used.
Explanation of the factors, levels, and response variables.
Details of the selected RSM design (e.g., CCD, Box-Behnken design).
Explanation of how the experiments were conducted, including any randomization or blocking.

Data Collection:
Detailed information on data collection procedures.
Record of data obtained during the experiments.
Information on any control variables or conditions that were held constant.

Statistical Analysis:
Presentation of the statistical model used for the response surface analysis.
Results of the analysis, including parameter estimates, coefficients, and p-values.
Model adequacy and goodness-of-fit tests (e.g., ANOVA).
Diagnostic plots to assess model assumptions.

Response Surface Plots:
Presentation of response surface plots and contour plots that visually represent the relationship between factors and response variables.
Interpretation of the plots to highlight trends and insights.

Optimization and Findings:
Results of the optimization process, including optimal factor settings.
Discussion of how the optimized conditions can improve performance or meet design objectives.
Sensitivity analysis, if applicable.

Discussion:
Interpretation of the results and their implications.
Comparison of findings with prior literature.
Discussion of any unexpected results or challenges encountered during the study.

Conclusion:
Summarization of the main findings and their significance.
Statements related to the achievement of research objectives.
Suggestions for future research or improvements.

References:
A list of all sources, papers, and references cited in the report, formatted in a consistent citation style (e.g., APA, IEEE).

Appendices (if applicable):
Additional information, such as raw data tables, details of statistical calculations, or supplementary figures and charts.

Acknowledgments (if applicable):
Recognition of individuals or organizations that contributed to the study but are not listed as authors.

Declaration of Ethical Compliance:
A statement affirming ethical compliance in research and reporting.

It's important to adhere to a clear and structured format when preparing the report to ensure that the information is presented logically and is easily accessible to readers. Additionally, following a consistent citation style is crucial for references and citations throughout the report.

3.4.3 Fractional Factorial Experiments

Fractional factorial experiments are used in various areas within electrical engineering, as they provide a cost-effective and efficient way to study the effects of multiple factors on a system's performance or behavior. Here are some common applications of fractional factorial experiments in electrical engineering:

Integrated Circuit (IC) Design: Engineers use fractional factorial experiments to optimize the design parameters of integrated circuits, such as transistor sizes, supply voltages, and clock frequencies, to achieve the desired performance, power consumption, and reliability.

Signal Processing: In applications like DSP, communication systems, and image processing, fractional factorial experiments help engineers study the impact of various factors on signal quality, data throughput, and processing speed.

Electromagnetic Compatibility (EMC) Testing: Fractional factorial experiments assist in assessing the susceptibility and emissions of electronic devices to EMI. Engineers can identify key factors affecting EMC performance and develop strategies for mitigation.

Power Electronics: Engineers working on power electronic systems, such as converters, inverters, and rectifiers, use fractional factorial experiments to optimize parameters like switching frequencies, control algorithms, and component values for efficiency and stability.

Quality Control and Reliability: Fractional factorial experiments are applied to improve the quality and reliability of electrical components and systems. They help identify factors affecting product lifespan, failure rates, and manufacturing processes.

Control Systems: Engineers use fractional factorial experiments to design and fine-tune control systems for applications like robotics, automation, and industrial processes. This aids in optimizing control parameters for stability and performance.

Renewable Energy Systems: Fractional factorial experiments assist in optimizing the design and operation of renewable energy systems, such as solar panels and wind turbines, by studying factors like panel orientation, wind turbine blade design, and power conversion efficiency.

Telecommunications: In the telecommunications field, engineers employ fractional factorial experiments to optimize network configurations, data transfer rates, and signal propagation characteristics for improved communication system performance.

Electrical Safety Testing: Fractional factorial experiments are used to evaluate the safety and performance of electrical equipment and systems, including insulation materials, protective devices, and grounding techniques.

Power Distribution and Grid Management: Engineers apply fractional factorial experiments to determine optimal control strategies, fault detection methods, and power distribution configurations in electrical power systems and grids.

Semiconductor Manufacturing: Fractional factorial experiments are employed to optimize semiconductor manufacturing processes, such as photolithography, etching, and doping, to enhance yield, performance, and reliability.

Instrumentation and Measurement: Engineers use fractional factorial experiments to evaluate the performance of measurement instruments, sensors, and calibration methods while considering factors like temperature, humidity, and environmental conditions.

Fractional factorial experiments are versatile tools that enable engineers to efficiently explore the effects of multiple factors and their interactions in complex electrical systems. They are valuable for making informed decisions, optimizing designs, and improving the performance and reliability of electrical devices and systems while conserving resources.

STEPS

Developing fractional factorial experiments in electrical engineering involves a systematic approach to design and conduct experiments while reducing the number of runs needed. Here are the steps to develop fractional factorial experiments:

1. **Define the Objective:**
   - Clearly state the goal or objectives of the experiment. What specific aspect of the electrical system or component are you trying to optimize or understand?
2. **Identify Factors and Levels:**
   - List all the factors (independent variables) that may influence the performance or behavior of the electrical system or component.
   - Determine the levels or settings at which each factor will be tested. These can be discrete values or ranges.
3. **Select a Fraction:**
   - Decide on the fraction of the full-factorial experiment that you intend to run. Common fractions include half-fraction (1/2), quarter-fraction (1/4), and eighth-fraction (1/8).
   - The choice of fraction depends on the number of factors, the available resources, and the desired level of information.
4. **Generate the Fractional Factorial Design:**
   - Use statistical software or tables (such as those provided by Taguchi or DOE software) to generate the fractional factorial design matrix.
   - The design matrix specifies which factor combinations will be tested and which will be omitted based on the chosen fraction.
5. **Randomize the Runs:**
   - Randomly order the experimental runs to reduce the impact of uncontrolled variables or systematic errors on the results. Randomization helps ensure the validity of the experiment.
6. **Conduct the Experiment:**
   - Perform the experimental runs according to the design matrix. Record the measurements and observations for each run.
   - Ensure that the experiment is conducted under controlled conditions to minimize external influences.
7. **Analyze the Data:**
   - Use statistical analysis techniques to analyze the data collected during the experiment. This may include ANOVA, regression analysis, or other appropriate methods.
   - Determine the main effects and interactions between factors to understand their impact on the response variable.
8. **Optimize and Interpret Results:**
   - Based on the analysis, identify the optimal factor settings or combinations that achieve the desired outcome or performance.
   - Interpret the results in the context of the electrical engineering problem, drawing conclusions and making recommendations.
9. **Verify the Model (Optional):**
   - If the experiment involves creating a predictive model, validate the model's accuracy using additional data or experiments not used in the initial design.
   - Ensure that the model accurately represents the behavior of the electrical system.

10. **Report and Document:**
    - Prepare a comprehensive report that documents the experiment's objectives, methods, results, and conclusions.
    - Include the design matrix, statistical analyses, graphical representations, and any relevant insights or recommendations.
11. **Implement Changes (if applicable):**
    - If the experiment reveals opportunities for improvement, implement the recommended changes or optimizations in the electrical system or component.
12. **Iterate (if necessary):**
    - Depending on the complexity of the problem and the results obtained, you may need to iterate the experimental process, refining the factors and levels, or conducting follow-up experiments.
13. **Review and Validation:**
    - Have the experimental design and results reviewed by peers or experts in the field to ensure the validity and reliability of the findings.
14. **Continuous Monitoring (if applicable):**
    - If the experiment is part of a continuous improvement process, establish a monitoring system to track the long-term performance of the electrical system or component.

Fractional factorial experiments are a valuable tool in electrical engineering for efficiently exploring multiple factors and their interactions. They help engineers make informed decisions, optimize designs, and improve the performance of electrical systems while conserving resources.

PITFALLS AND REMEDIES

Fractional factorial experiments are powerful tools for studying multiple factors while reducing the number of experimental runs. However, they are not without challenges. Here are some common pitfalls and remedies associated with fractional factorial experiments used in electrical engineering:

Pitfall 1: Alias Effects
　　Issue: Alias effects occur when interactions between factors are confounded with main effects or with each other due to the reduced number of experimental runs.
　　Remedy: To address aliasing, use resolution or alias structure tables to identify which effects are confounded. Select a fraction or design resolution that allows for the separation of important factors and interactions. If aliasing is unavoidable, use advanced data analysis techniques like design augmentation or partial aliasing to resolve confounding.

Pitfall 2: Missing Important Factors
　　Issue: It's possible to overlook important factors that affect the response, resulting in an incomplete understanding of the system.
　　Remedy: Carefully consider the relevant factors in the initial planning stage. Use domain knowledge, expert opinions, and literature reviews to identify potential factors. Sensitivity analysis or screening experiments may help identify significant factors to include in the fractional factorial design.

Pitfall 3: Inadequate Sample Size
　　Issue: If the sample size (number of experimental runs) is too small, it can lead to imprecise estimates of factor effects and interactions.
　　Remedy: Ensure that the chosen fractional factorial design provides a sufficient number of runs to detect significant effects and interactions. You may need to increase the fraction or replicate runs to improve the precision of estimates.

Pitfall 4: External Noise and Variability

> **Issue:** External factors or sources of variability not accounted for in the experiment can affect the response and introduce noise.

> **Remedy:** Control external sources of variation as much as possible. Ensure that the experiments are conducted under controlled conditions. If external noise cannot be fully controlled, consider using statistical techniques like Analysis of Covariance (ANCOVA) to account for it during data analysis.

Pitfall 5: Model Assumptions

> **Issue:** Fractional factorial experiments often assume linear relationships between factors and responses. Deviations from linearity can lead to model inaccuracies.

> **Remedy:** If nonlinear relationships are suspected, consider conducting additional experiments or using RSM to capture nonlinear effects. Always validate model assumptions and assess model adequacy.

Pitfall 6: Lack of Randomization

> **Issue:** Failing to randomize the order of experimental runs can introduce bias or systematic errors in the results.

> **Remedy:** Randomly order the experimental runs to minimize the impact of uncontrolled variables and ensure the validity of statistical analyses.

Pitfall 7: Insufficient Data Analysis

> **Issue:** Inadequate or incorrect data analysis can lead to incorrect conclusions or missed insights.

> **Remedy:** Use appropriate statistical techniques for data analysis, such as ANOVA and regression analysis. Ensure that interactions are properly identified, and results are interpreted within the context of the electrical engineering problem.

Pitfall 8: Lack of Expertise

> **Issue:** Insufficient expertise in statistical design and analysis can hinder the success of fractional factorial experiments.

> **Remedy:** Collaborate with statisticians or experts in DOE to plan, execute, and analyze the experiments. Training or workshops on experimental design and statistical analysis can also enhance the team's capabilities.

Pitfall 9: Ignoring Practical Constraints

> **Issue:** Designing experiments without considering practical constraints, such as time, budget, or resource limitations, can lead to unfeasible or impractical experiments.

> **Remedy:** Ensure that the experiment's design aligns with available resources and timelines. Make practical decisions regarding factors and levels that are within the scope of the project.

By addressing these pitfalls and applying the corresponding remedies, engineers and researchers can maximize the effectiveness of fractional factorial experiments in electrical engineering, leading to valuable insights and optimized designs.

EXAMPLE

Fractional factorial experiments are used in electrical engineering to investigate the effects of multiple factors while reducing the number of experimental runs required. Here's an example of a matrix for a 2^4-1 fractional factorial experiment in electrical engineering:

Objective: To study the impact of four factors on the efficiency of a photovoltaic (PV) solar panel.

Factors:

1. **Panel Orientation (Factor A):**
   - Level 1: North-facing
   - Level 2: South-facing

2. **Solar Irradiance (Factor B)**:
   - Level 1: Low (400 W/m²)
   - Level 2: High (800 W/m²)
3. **Temperature (Factor C)**:
   - Level 1: 25°C
   - Level 2: 35°C
4. **Dirt Accumulation (Factor D)**:
   - Level 1: Low
   - Level 2: High

Response Variable: PV Panel Efficiency (measured in percentage)

Fractional Factorial Matrix (2^4-1):

In this example, a fractional 2^4-1 factorial design is used to investigate the main effects and two-way interactions of the four factors while reducing the number of experimental runs. The matrix is shown in Table 3.35.

In this matrix, each row represents a specific experimental run with a unique combination of factor levels. The fractional factorial design allows you to study the main effects and two-way interactions of the factors with a reduced number of experiments. This can significantly save time and resources while providing valuable insights into how factors affect the efficiency of PV solar panels in electrical engineering applications.

REPORTING

Statistical tools play a crucial role in the analysis of fractional factorial designed experiments in electrical engineering. These tools help engineers and researchers make sense of the data generated from these experiments and draw meaningful conclusions. Here are some common statistical tools used in the analysis of fractional factorial experiments in electrical engineering:

Analysis of Variance (ANOVA): ANOVA is a fundamental statistical technique used to partition the variation in the response variable into components attributable to different factors and interactions. It assesses the statistical significance of each factor and interaction, helping identify which ones have a significant impact on the response.

Main Effects Plots: Main effects plots are graphical representations that show the effects of individual factors on the response variable while keeping all other factors at their reference or center levels. They provide a visual summary of the main effects of each factor.

TABLE 3.35

Electrical Engineering Fractional Factorial Designed Experiment Matrix Example

Run	Panel Orientation (A)	Solar Irradiance (B)	Temperature (C)	Dirt Accumulation (D)
1	North	Low	25°C	Low
2	North	High	35°C	High
3	South	Low	35°C	Low
4	South	High	25°C	High
5	North	Low	35°C	High
6	North	High	25°C	Low
7	South	Low	25°C	High
8	South	High	35°C	Low
9	South	Low	25°C	Low

Interaction Plots: Interaction plots visualize the interactions between two or more factors by showing how the response variable varies as the levels of these factors change. They help identify and interpret interaction effects, which can be crucial in understanding the system behavior.

Effect Plots: Effect plots are graphical representations that display the effects of different factors at various levels on the response variable. They help engineers visualize the impact of each factor on the response across a range of factor levels.

Normal Probability Plots (Q-Q Plots): Q-Q plots assess the normality of the residuals (the differences between observed and predicted values) from the statistical model. Deviations from a straight line in a Q-Q plot may indicate departures from normality.

Residual Analysis: Residual analysis involves examining the residuals to ensure that they meet the assumptions of the statistical model, including constant variance (homoscedasticity) and independence. Deviations from these assumptions may require model adjustments.

Half-Normal Probability Plots: Half-normal probability plots are used to identify the significant effects in a fractional factorial design. Effects that deviate from a straight line in the plot are considered significant.

Diagnostic Plots: Various diagnostic plots, such as residual versus fitted value plots and normal probability plots of residuals, are used to check the adequacy of the statistical model and the validity of its assumptions.

Confidence Intervals: Confidence intervals provide a range of values within which the true effect size or parameter is likely to fall. They help quantify the uncertainty associated with estimated effects.

Pareto Charts: Pareto charts are bar charts that rank factors or effects by their magnitude. They highlight the most influential factors, making it easier to prioritize which factors to focus on for further investigation or optimization.

Effect Screening: Screening techniques like main effects plots and effect tables help identify the most important factors affecting the response variable, allowing engineers to focus on the key factors in subsequent experiments.

Model Selection: Depending on the complexity of the system and the nature of the data, different regression models (e.g., linear, quadratic, or higher-order models) may be considered. Model selection criteria, such as adjusted R-squared or AIC, aid in choosing the most appropriate model.

Model Validation: Cross-validation and other model validation techniques assess the predictive accuracy of the statistical model to ensure it generalizes well to unseen data.

Software Tools: Statistical software packages like Minitab, JMP, R, or Python with libraries like SciPy and statsmodels are commonly used for conducting the statistical analysis of fractional factorial experiments. These tools provide built-in functions and procedures for design creation and analysis.

These statistical tools, when used appropriately, help researchers and engineers extract valuable insights, identify significant factors, and optimize processes or designs in electrical engineering based on the results of fractional factorial experiments.

A well-structured report for a fractional factorial designed experiment in electrical engineering should provide a clear and concise presentation of the experiment's objectives, methodology, results, and conclusions. Here are the essential elements to include in such a report:

1. **Title:** A descriptive and informative title that summarizes the purpose and scope of the experiment.
2. **Abstract:** A brief summary of the experiment's objectives, methods, key results, and conclusions. The abstract should provide a concise overview of the entire report.

3. **Introduction:**
   - Background: Provide context for the experiment by discussing the problem or research question.
   - Objectives: Clearly state the objectives and goals of the experiment.
   - Hypotheses or Research Questions: If applicable, outline the hypotheses or research questions being investigated.

4. **Experimental Design:**
   - Experimental Factors: List and describe the independent variables (factors) under investigation, including their levels and ranges.
   - Design Matrix: Present the fractional factorial design matrix, including the notation used for the coded levels.
   - Control Variables: Specify any control variables or factors that were held constant throughout the experiment.
   - Response Variable: Clearly define the response variable, which measures the outcome or performance of interest.

5. **Methods:**
   - Sampling Procedure: Explain how the experimental runs were generated or selected from the fractional factorial design.
   - Data Collection: Describe the data collection process, including any equipment, instruments, or software used.
   - Experimental Procedure: Detail the steps taken to conduct the experiment, including any specific protocols or procedures.
   - Data Analysis: Describe the statistical methods and tools used to analyze the data, including any software or programming languages employed.
   - Assumptions: Specify any assumptions made during the analysis, such as normality of residuals or independence of observations.

6. **Results:**
   - Data Presentation: Present the experimental data, including observed values of the response variable and any relevant information about each run or observation.
   - Analysis of Variance (ANOVA): Display the results of the ANOVA table, highlighting the main effects, interactions, and their significance.
   - Graphical Representation: Include plots, charts, or graphs that visualize the main effects and interactions to aid in interpretation.
   - Significant Factors: Discuss which factors are statistically significant and their impact on the response variable.
   - Model Coefficients: If applicable, provide the coefficients of the regression model, including their significance and interpretation.

7. **Discussion:**
   - Interpretation of Results: Interpret the findings in the context of the experiment's objectives and hypotheses.
   - Practical Implications: Discuss the practical implications of the significant factors and their effect on the electrical engineering application.
   - Limitations: Address any limitations or potential sources of error in the experiment.
   - Comparison to Literature: Compare the results to existing literature or previous studies in the field.

8. **Conclusions:**
   - Summarize the key findings and their significance.
   - Restate the main conclusions and whether they align with the initial hypotheses or objectives.
   - Provide recommendations or insights for future research or applications.

9. **References:** Cite all sources, references, and prior research that influenced or informed the experiment's design and analysis.
10. **Appendices:** Include any supplementary information, such as raw data, additional plots or tables, details on statistical calculations, and any other relevant materials.
11. **Acknowledgments:** Acknowledge individuals or organizations that contributed to the experiment, including funding sources, collaborators, or participants.
12. **Tables and Figures:** Ensure that tables and figures are appropriately labeled, numbered, and referenced within the text.
13. **Nomenclature:** If applicable, provide a list of symbols, abbreviations, or notation used in the report to aid readers' understanding.
14. **Executive Summary (Optional):** Include a brief summary of the main findings and conclusions, particularly if the report is intended for decision-makers or stakeholders who may not read the full report.
15. **Appendices:** Attach any supplementary material, such as detailed calculations, additional tables, or code used in the analysis.

Remember to use clear and concise language, along with appropriate formatting and referencing conventions, to ensure that your fractional factorial experiment report in electrical engineering is well-structured and effectively communicates the experiment's purpose and outcomes.

3.4.4 CENTRAL COMPOSITE DESIGN (CCD)

CCD experiments are used in electrical engineering for various purposes, including:

Circuit Optimization: CCD experiments can be employed to optimize electrical circuits by studying the effects of component values, tolerances, and environmental factors on circuit performance metrics such as gain, frequency response, and S/N ratio.

Power Electronics Design: In power electronics, CCD can help optimize parameters of devices like transformers, converters, and inverters to maximize efficiency, minimize losses, and meet specific voltage and current requirements.

Antenna Design: CCD experiments aid in optimizing antenna designs for wireless communication systems. Variables like antenna length, spacing, and material properties can be investigated to achieve desired radiation patterns, gain, and bandwidth.

Control System Tuning: Electrical engineers use CCD to tune control system parameters in applications like robotics and industrial automation. It helps improve control system performance, stability, and response time.

Sensor Calibration: In sensor design, CCD experiments can determine the optimal sensor settings, including sensitivity, offset, and calibration coefficients, to ensure accurate measurements.

Electromagnetic Compatibility (EMC): CCD is used to study EMC issues, including EMI and susceptibility. It helps engineers design electronic systems that meet EMC standards.

Printed Circuit Board (PCB) Design: CCD experiments are applied to optimize PCB layouts and component placements for reduced interference, shorter signal paths, and better thermal management.

Reliability Testing: Electrical engineers use CCD to study the effects of operating conditions (temperature, voltage, load, etc.) on the reliability and lifetime of electronic components and systems.

Signal Processing: CCD experiments can optimize parameters in signal processing algorithms and digital filters to enhance signal quality, noise reduction, and processing speed.

Power Distribution: Engineers can use CCD to design and optimize power distribution systems, ensuring efficient power delivery, voltage regulation, and fault tolerance.

Energy Efficiency: CCD experiments are employed to optimize energy-efficient systems and components, such as energy-efficient lighting, power management algorithms, and renewable energy systems.

Fault Diagnosis: In fault diagnosis and condition monitoring, CCD can help identify the critical parameters and thresholds for detecting faults or abnormalities in electrical systems.

CCD experiments are valuable in electrical engineering because they allow engineers to systematically explore the effects of multiple factors and their interactions on system performance. By conducting such experiments, engineers can design, optimize, and troubleshoot electrical systems and components more effectively.

STEPS

1. **Identify Factors:** Determine the relevant factors or variables that may influence the electrical system's performance. These could include voltage, current, temperature, material properties, and environmental conditions.
2. **Select Factor Levels:** Decide on the levels at which each factor will be studied. This involves specifying both high and low levels, as well as the central or nominal level. The central point is typically the standard operating condition.
3. **Generate the CCD Matrix:** Create a CCD matrix that represents the combination of factor levels for the experimental runs. CCD typically includes axial or star points in addition to factorial points.
4. **Conduct Experiments:** Perform the experiments according to the settings specified in the CCD matrix. Collect data on the electrical system's performance for each run.
5. **Data Analysis:** Analyze the experimental data using statistical techniques such as regression analysis to build a predictive model of the electrical system's behavior.
6. **Optimization:** Use the predictive model to identify the optimal factor settings that maximize or minimize the desired performance criteria. This is often achieved through response surface optimization techniques.
7. **Validation:** Validate the optimized settings by conducting additional experiments or simulations to ensure that the desired performance improvements are achieved.
8. **Implementation:** Implement the optimized settings in practical applications or manufacturing processes to benefit from the improvements.

CCD experiments in electrical engineering can be a powerful tool for improving the efficiency, reliability, and performance of electrical systems and components, making them an essential part of research and development in the field.

PITFALLS AND REMEDIES

CCD experiments, like any designed experiments, can face various pitfalls that need to be addressed. Here are some common pitfalls and potential remedies in CCD experiments used in electrical engineering:

Pitfall 1: Inadequate Sample Size

Pitfall: Conducting CCD experiments with a sample size that is too small can lead to unreliable results and difficulty in detecting significant effects.

Remedy: Calculate the minimum sample size required to achieve adequate statistical power based on the expected effect sizes and desired confidence levels.

Pitfall 2: Poor Model Fit

Pitfall: The fitted response surface model may not adequately represent the actual system behavior, leading to inaccurate predictions.

Remedy: Verify the model fit by conducting goodness-of-fit tests, examining residual plots, and considering alternative model structures if necessary.

Pitfall 3: Outliers

Pitfall: Outliers in experimental data can distort model parameter estimates and affect the validity of the results.

Remedy: Identify and address outliers through data validation and, if necessary, consider robust regression techniques that are less sensitive to outliers.

Pitfall 4: Nonlinearity

Pitfall: Assuming linearity when the actual relationship between factors and responses is nonlinear can lead to incorrect conclusions.

Remedy: Include terms for higher-order interactions and curvature in the model, or consider using a nonlinear regression model if needed.

Pitfall 5: Lack of Factorial Understanding

Pitfall: Focusing solely on the response surface without understanding the underlying factorial effects can lead to misinterpretation.

Remedy: Analyze both the main effects and interaction effects of factors to gain a comprehensive understanding of the system's behavior.

Pitfall 6: Neglecting Noise Factors

Pitfall: Ignoring uncontrollable noise factors that affect the response can lead to incomplete models.

Remedy: Identify and measure potential noise factors, and incorporate them into the experimental design or account for them during data analysis.

Pitfall 7: Overfitting

Pitfall: Fitting a complex model with too many terms to the data can lead to overfitting and poor model generalization.

Remedy: Use techniques like cross-validation or information criteria to guide the selection of an appropriate model complexity.

Pitfall 8: Violation of Assumptions

Pitfall: Violating assumptions of normality, constant variance, or independence of residuals can affect the validity of statistical tests and predictions.

Remedy: Validate the assumptions through residual analysis and consider appropriate transformations or robust techniques if assumptions are not met.

Pitfall 9: Incomplete Factor Range

Pitfall: Narrowing the factor range too much may result in missing important information or potential optimum points.

Remedy: Ensure that the selected factor levels cover the relevant operational range and important regions of interest.

Pitfall 10: Poor Experimental Control

Pitfall: Inadequate control of experimental conditions can introduce variability and confounding effects.

Remedy: Maintain strict control over experimental conditions and randomize experimental runs to minimize bias and confounding.

Addressing these pitfalls in CCD experiments is crucial for obtaining reliable and meaningful results in electrical engineering. Proper planning, careful execution, and rigorous data analysis are essential steps to mitigate these challenges.

Example

A CCD is a commonly used experimental design in electrical engineering to study the response of a system while varying multiple factors within a defined range, including exploring the central region. Here's an example of a matrix for a CCD designed experiment in electrical engineering:

Objective: To optimize the performance of a power inverter for a photovoltaic (PV) system while varying input voltage and frequency.

Factors:

1. **Input Voltage (Factor A)**:
   - Low Limit: 400 V
   - High Limit: 600 V
2. **Frequency (Factor B)**:
   - Low Limit: 50 Hz
   - High Limit: 60 Hz

Response Variable: Power Output (measured in kilowatts, kW)

Central Composite Design Matrix:

In this example, a CCD is used to investigate the effects of input voltage and frequency on the power output of the power inverter. The matrix is shown in Table 3.36.

In this matrix, the "Low Limit" and "High Limit" values for input voltage and frequency are represented by −1 and +1, respectively. The additional runs (1 to 5) represent points within the experimental region, including points in the center and corners. These additional runs help estimate the quadratic effects and interactions between the factors.

By conducting experiments with varying input voltage and frequency according to this CCD matrix, electrical engineers can model the response (power output) and optimize the power inverter's settings for maximum efficiency and performance within the specified input voltage and frequency limits.

TABLE 3.36

Central Composite Design (CCD) Designed Experiment Matrix Example

Run	Input Voltage (A) (V)	Frequency (B) (Hz)
−1	400	50
−1	600	50
−1	400	60
−1	600	60
1	500	55
2	450	55
3	550	55
4	500	45
5	500	65

Reporting

CCD is a widely used experimental design technique in various fields, including electrical engineering. It is commonly used to optimize processes and systems by studying the effects of multiple variables. In CCD experiments, several statistical tools and techniques can be employed for data analysis and model building. Some of the common statistical tools used in CCD experiments in electrical engineering include:

Analysis of Variance (ANOVA): ANOVA is a fundamental statistical technique used to analyze the variance in the response variable and assess the significance of factors and their interactions. It helps identify which factors have a significant impact on the response.

Response Surface Methodology (RSM): CCD often involves RSM, a collection of statistical and mathematical techniques used to create and analyze response surfaces. RSM helps in modeling the relationship between the input factors and the response variable to find the optimal settings.

Regression Analysis: Regression analysis is used to develop mathematical models that describe the relationship between input factors and the response variable. Linear regression, quadratic regression, or higher-order regression models may be employed depending on the complexity of the system.

Design of Experiments (DOE): CCD is a specific type of DOE, and the principles of experimental design are essential for planning, conducting, and analyzing the experiments effectively.

Contour Plots and Response Surface Plots: These graphical tools are used to visualize the response surfaces and understand how the response variable changes with variations in the input factors. Contour plots show contours of constant response, while response surface plots provide a 3D visualization.

Desirability Functions: Desirability functions are used to combine multiple response variables into a single objective function that needs to be maximized or minimized. This helps in finding the optimal factor settings that meet various performance criteria simultaneously.

Pareto Charts: Pareto charts are used to prioritize factors based on their impact on the response variable. They help in focusing resources on the most influential factors during optimization.

Confidence Intervals and Prediction Intervals: These intervals provide estimates of the uncertainty associated with model predictions. They help in understanding the reliability of the model and making predictions for new factor settings.

Diagnostic Plots: Diagnostic plots, such as residual plots and normal probability plots, are used to assess the adequacy of the model and check for violations of underlying assumptions.

Optimization Algorithms: Various optimization algorithms, such as the steepest ascent or descent, genetic algorithms, or gradient-based optimization, can be used to find the factor settings that maximize or minimize the response variable based on the developed models.

These statistical tools and techniques are applied in combination to analyze the data generated from CCD experiments in electrical engineering. They help in understanding the relationships between input factors and the response variable, optimizing processes or systems, and making informed decisions for improvement.

3.4.5 Taguchi Methods

Taguchi methods, developed by Dr. Genichi Taguchi, are widely used in electrical engineering for optimizing and improving various processes and systems. Taguchi methods, which are a part of the broader field of DOE, are applied in electrical engineering in several areas:

Electronic Component and Circuit Design: Taguchi methods can be used to optimize the design of electronic circuits, ensuring that they perform efficiently while minimizing variations in performance due to factors like component tolerances, temperature, and voltage fluctuations.

Semiconductor Manufacturing: In semiconductor manufacturing, Taguchi methods help optimize fabrication processes, reduce defects, and improve the yield of semiconductor devices like microchips and integrated circuits.

Quality Control: Electrical engineering often involves the production of electrical and electronic products. Taguchi methods are used for quality control to ensure that these products meet specific standards and have minimal defects.

Signal Processing: In signal processing applications, Taguchi methods can be used to optimize algorithms and filters, reducing noise and improving the accuracy of data analysis and interpretation.

Power Distribution and Energy Management: Taguchi methods can help optimize power distribution systems, reducing losses and ensuring efficient energy management. This is especially important in electrical grids and renewable energy systems.

Reliability and Robustness Testing: Electrical systems need to be reliable and robust in different operating conditions. Taguchi methods are used to design experiments that test and optimize the performance of electrical systems under various environmental and operational factors.

Electromagnetic Compatibility (EMC) Testing: Taguchi methods can be employed to design experiments that assess the EMC of electronic devices and systems, ensuring that they operate without interfering with or being susceptible to electromagnetic interference.

Printed Circuit Board (PCB) Design: PCBs are at the heart of many electronic devices. Taguchi methods can be used to optimize the layout and design of PCBs to minimize signal interference, reduce noise, and improve overall performance.

Power Electronics and Motor Control: In applications such as motor control and power electronics, Taguchi methods are used to optimize control algorithms and parameters for efficiency and performance.

Communication Systems: Taguchi methods can be applied to optimize communication system parameters, such as antenna design, signal modulation, and error correction coding, to enhance signal quality and range.

In these and other areas of electrical engineering, Taguchi methods help engineers systematically vary input factors, conduct experiments, and analyze results to find the optimal settings that lead to improved performance, reliability, and efficiency while minimizing variability and defects.

Steps

Developing Taguchi methods designed experiments for electrical engineering involves a systematic approach to optimize and improve various processes or systems. Here are the general steps to follow:

1. **Define the Problem and Objective:**
   - Clearly define the problem you want to address or the objective you want to achieve in electrical engineering. This could be related to improving the performance of a circuit, reducing defects in a manufacturing process, optimizing signal processing algorithms, etc.
2. **Identify Factors and Levels:**
   - Identify the key factors (independent variables) that may affect the outcome or performance of the system or process. These factors could be related to component specifications, operating conditions, environmental factors, or design parameters.
   - Determine the levels or settings for each factor. Levels represent the range of values that each factor will take during the experimentation.

3. **Select an Experimental Design:**
   - Choose an appropriate Taguchi experimental design, such as an L9, L12, L16, or other OA design, based on the number of factors and levels. The choice of design depends on the complexity of the problem and the available resources.

4. **Create an Orthogonal Array (OA):**
   - Based on the selected design, create an OA that specifies which factor combinations will be tested in the experiments. Ensure that the OA covers all possible combinations efficiently.

5. **Assign Factors and Levels to the Orthogonal Array (OA):**
   - Populate the OA with the factors and their corresponding levels for each experiment. Each row of the OA represents a unique combination of factor settings.

6. **Conduct Experiments:**
   - Perform the experiments according to the settings specified in the OA. Ensure that the experiments are conducted under controlled and consistent conditions.

7. **Collect Data:**
   - Record data for the response variable (dependent variable) in each experiment. This data will be used to evaluate the performance of each factor combination.

8. **Analyze the Data:**
   - Use statistical analysis techniques to analyze the data collected from the experiments. Calculate S/N ratios or other relevant metrics to assess the performance of each factor combination.

9. **Determine the Optimal Settings:**
   - Identify the factor settings that result in the best performance or meet the desired objectives. This is typically done by comparing the S/N ratios or other metrics for different factor combinations.

10. **Perform Confirmation Runs:**
    - Conduct confirmation experiments using the optimal factor settings determined in the previous step to validate that the improvements observed in the experiments are consistent.

11. **Implement Recommendations:**
    - Implement the recommended factor settings and process improvements in your electrical engineering application to achieve the desired objectives.

12. **Monitor and Maintain:**
    - Continuously monitor the performance of the system or process to ensure that the improvements are sustained over time. Make adjustments as needed.

13. **Document the Results:**
    - Prepare a comprehensive report that documents the entire Taguchi methods experiment, including the problem statement, factors and levels, experimental design, data analysis, results, and recommendations.

14. **Share Findings:**
    - Share the findings and recommendations with relevant stakeholders in your organization to facilitate decision-making and implementation.

Throughout the process, it's essential to follow the principles of Taguchi methods, which emphasize robustness and the systematic identification of optimal settings while considering variability and noise factors. Taguchi methods help engineers make data-driven decisions to improve and optimize electrical engineering processes and systems.

Pitfalls and Remedies

Using Taguchi methods in electrical engineering experiments can yield valuable insights, but there are potential pitfalls to be aware of. Here are some common pitfalls and remedies:

Pitfall 1: Inadequate Problem Definition
    Pitfall: Failing to clearly define the problem or objective can lead to experiments that lack focus.
    Remedy: Spend time upfront defining the problem, its scope, and the desired outcomes. Ensure that all stakeholders understand the objectives.

Pitfall 2: Incorrect Factor Selection
    Pitfall: Choosing the wrong factors (independent variables) or omitting important ones can lead to suboptimal results.
    Remedy: Conduct a thorough analysis to identify critical factors that impact the desired outcome. Consult with experts if needed.

Pitfall 3: Inadequate Factor Levels
    Pitfall: Insufficient or incorrect levels for factors may result in experiments that do not cover the relevant parameter space.
    Remedy: Determine appropriate factor levels based on engineering knowledge and conduct a preliminary screening experiment if necessary.

Pitfall 4: Violating Orthogonality
    Pitfall: Failing to maintain orthogonality in the experimental design can lead to confounding effects.
    Remedy: Use software or templates to ensure that the experimental design remains orthogonal. Check for aliasing and confounding effects.

Pitfall 5: Insufficient Replication
    Pitfall: Conducting experiments without replication can make it challenging to assess the variability and reliability of results.
    Remedy: Incorporate replication to assess the consistency of results. Replication helps estimate experimental error.

Pitfall 6: Ignoring Noise Factors
    Pitfall: Neglecting noise factors (uncontrollable variables) can lead to designs that are not robust in real-world conditions.
    Remedy: Identify and account for noise factors in your experimental design. Conduct noise tests to quantify their impact.

Pitfall 7: Overreliance on a Single Metric
    Pitfall: Relying solely on one metric or S/N ratio can lead to suboptimal solutions.
    Remedy: Consider multiple performance metrics and use engineering judgment to balance conflicting objectives.

Pitfall 8: Failure to Validate
    Pitfall: Not validating the results with additional experiments can lead to false conclusions.
    Remedy: Conduct confirmation experiments using the recommended settings to validate the improvements observed in the Taguchi experiment.

Pitfall 9: Inadequate Documentation
    Pitfall: Poor documentation can lead to difficulty in reproducing results or understanding the experiment's context.
    Remedy: Maintain detailed records of the experiment, including factor settings, data collected, and any deviations from the plan.

Pitfall 10: Lack of Collaboration
    Pitfall: Failing to involve relevant stakeholders or experts can limit the effectiveness of Taguchi methods.
    Remedy: Collaborate with cross-functional teams and subject matter experts to ensure a comprehensive approach to problem-solving.

Pitfall 11: Inadequate Training
    Pitfall: Conducting Taguchi experiments without proper training can result in errors.
    Remedy: Invest in training and education for team members involved in Taguchi methods to ensure they understand the principles and techniques.

Pitfall 12: Ignoring Continuous Improvement
> **Pitfall:** Assuming that a one-time Taguchi experiment will solve all problems without a commitment to continuous improvement.
> **Remedy:** Establish a culture of continuous improvement where Taguchi methods are used iteratively to refine processes and products.

Avoiding these pitfalls and applying Taguchi methods systematically can lead to more robust and optimized solutions in electrical engineering experiments.

EXAMPLE

The Taguchi Method is a robust optimization technique that uses OAs to study the effects of multiple factors on a response variable while minimizing the number of experimental runs. Here's an example of a matrix for a Taguchi method designed experiment in electrical engineering:

Objective: To optimize the performance of a PCB assembly process while considering factors such as solder paste type, reflow oven temperature, and placement accuracy.

Factors:

1. **Solder Paste Type (Factor A)**:
   - Type 1 (Standard)
   - Type 2 (High-quality)
2. **Reflow Oven Temperature (Factor B)**:
   - Low Temperature
   - Medium Temperature
   - High Temperature
3. **Placement Accuracy (Factor C)**:
   - Low Accuracy
   - Medium Accuracy
   - High Accuracy

Response Variable: PCB Assembly Defect Rate (measured in defects per unit)

Taguchi Method Matrix:
   In this example, an L18 ($2^3 \times 2^1$) OA is used to efficiently study the effects of three factors with multiple levels and one noise factor. The matrix is shown in Table 3.37.
   In this matrix, each row represents a specific experimental run with a combination of factors and levels. The Taguchi Method allows you to efficiently study the effects of solder paste type, reflow oven temperature, and placement accuracy on PCB assembly defect rates while considering variation and noise factors. The goal is to identify the optimal combination of factors that minimizes defects and improves the reliability of the electrical circuit board assembly process.

REPORTING

Taguchi methods in electrical engineering experiments typically involve the use of statistical tools and techniques to analyze and optimize processes or products. Some of the common statistical tools and methods used in Taguchi experiments in electrical engineering include:

Design of Experiments (DOE): Taguchi experiments are a subset of DOE techniques, and the initial step often involves creating a robust experimental design using methods like the Taguchi L16 OA or other appropriate arrays.

TABLE 3.37

Taguchi Method Designed Experiment Matrix Example for Electrical Engineering

Run	Solder Paste Type (A)	Reflow Oven Temperature (B)	Placement Accuracy (C)
1	1	Low	Low
2	1	Medium	Medium
3	1	High	High
4	1	Medium	Low
5	2	Medium	High
6	2	Low	High
7	2	High	Low
8	2	High	Medium
9	1	Low	High
10	1	High	Medium
11	1	Medium	High
12	2	Low	Low
13	2	Low	Medium
14	2	Medium	Medium
15	1	Low	Medium
16	1	Medium	Low
17	2	High	High
18	2	High	Low

Signal-to-Noise (S/N) Ratios: Taguchi methods emphasize the use of S/N ratios to evaluate the performance of a product or process. The choice of S/N ratio (e.g., smaller-the-better, larger-the-better, nominal-the-best) depends on the specific problem and objective.

Analysis of Variance (ANOVA): ANOVA is used to partition the variation in the responses (output variables) into components attributable to different factors. It helps identify which factors significantly affect the performance and whether interactions are significant.

Main Effects Plots: These plots visualize the main effects of each factor on the response variable, helping to identify which factors have the most significant impact on performance.

Interaction Plots: Interaction plots are used to visualize the interactions between two or more factors. They show how the effect of one factor depends on the level of another factor.

Response Surface Methodology (RSM): In more complex problems, RSM can be used in conjunction with Taguchi methods to model and optimize responses over a continuous range of factor levels.

Pareto Analysis: Pareto charts are used to prioritize and focus on the most important factors or sources of variation that contribute to the problem.

Robust Parameter Design (RPD): RPD is an extension of Taguchi methods that aims to make products or processes less sensitive to variation. It involves optimizing factors to achieve robustness.

Orthogonal Arrays (OA): These arrays are used to create a systematic and efficient experimental design by selecting factor levels that minimize confounding effects.

Statistical Process Control (SPC): SPC techniques can be used to monitor and control processes based on the insights gained from Taguchi experiments, ensuring continuous improvement.

Regression Analysis: Regression models can be employed to understand the relationship between factors and responses, especially when the relationships are not linear.

Desirability Function: This function helps combine multiple responses into a single desirability score, allowing for the simultaneous optimization of multiple objectives.

Control Charts: Control charts are used for monitoring process stability and detecting any shifts or variations in the performance of a process or product.

Hypothesis Testing: Hypothesis tests can be applied to assess the statistical significance of the observed effects and confirm whether changes made to factors are indeed beneficial.

ANOVA with Random Effects: In cases where there are random effects or uncontrolled variation sources, ANOVA with random effects can be used to analyze the data.

These statistical tools and methods are applied in a structured manner to plan, conduct, and analyze Taguchi experiments in electrical engineering. The choice of tools depends on the specific problem, objectives, and complexity of the experiment.

A report for a Taguchi methods design experiment in electrical engineering typically includes the following elements:

Title: The title should succinctly describe the experiment, including the system, process, or product under investigation.

Abstract: A brief summary of the experiment's purpose, methods, key findings, and conclusions. It provides readers with an overview of the study.

Introduction:

Background: Provide context for the experiment, explaining why it was conducted.

Objectives: State the specific goals and objectives of the experiment.

Problem Statement: Clearly define the problem or issue the experiment aims to address.

Experimental Design:

Description of Factors: List and describe the factors (input variables) that were varied during the experiment.

Levels of Factors: Specify the different levels or settings at which each factor was tested.

Taguchi Array: If applicable, include details about the Taguchi array used for the experiment, such as the OA (L16, L9, etc.).

Control Factors: Identify any factors that were held constant or not considered in the experiment.

Response Variables:

Describe the response variables (output variables) used to evaluate the performance or quality of the system, process, or product.

Measurement Methods: Explain how the response variables were measured or assessed.

Experimental Procedure:

Detail the steps taken to conduct the experiment, including the order of testing and any special conditions or considerations.

Specify any equipment, instruments, or tools used.

Data Analysis:

Statistical Tools: Describe the statistical methods and tools used for data analysis, such as ANOVA, S/N ratios, and graphs.

Results: Present the experimental results, including tables, charts, and plots that illustrate the impact of factors on the response variables.

Interpretation: Discuss the implications of the results, highlighting significant findings and patterns.

Identify Factors: Identify which factors had the most significant impact on the response variables and whether any interactions were observed.

Optimization:

If applicable, describe the optimization process used to identify the optimal factor settings for achieving the desired performance or quality.

Present the recommended settings for the factors.

Discussion:
Discuss the practical implications of the findings and how they relate to the objectives and problem statement.

Address any limitations or constraints of the experiment.

Conclusions:
Summarize the key conclusions drawn from the experiment.

Explain how the experiment contributes to solving the identified problem.

Recommendations:
Offer recommendations for further research or actions based on the experiment's outcomes.

Suggest practical steps or changes that can be implemented based on the findings.

References: Provide citations for any literature, standards, or sources used to support the experiment and analysis.

Appendices:
Include any supplementary information, such as raw data, detailed calculations, or additional graphs and charts.

Attach copies of the Taguchi array and any experimental worksheets used.

Acknowledgments: If applicable, acknowledge individuals or organizations that provided assistance or resources for the experiment.

Author Information: Include the names and affiliations of the individuals involved in conducting the experiment and preparing the report.

Date: Include the date when the report was finalized.

A well-structured report ensures that the experiment's objectives, methods, results, and implications are clearly communicated to the intended audience. It also serves as a valuable reference for future work and decision-making in electrical engineering.

3.4.6 MIXTURE EXPERIMENTS

Mixture design experiments can be used in electrical engineering in various applications, especially when formulating materials, compounds, or composites with multiple components. Here are some specific areas within electrical engineering where mixture design experiments are applicable:

Semiconductor Fabrication: Mixture design experiments can be employed to optimize the composition of semiconductor materials, including doping levels, to achieve desired electrical properties and performance in devices like transistors, diodes, and integrated circuits.

Dielectric Material Development: In the design of dielectric materials for capacitors and insulators, mixture design can help determine the ideal combination of components to achieve specific electrical permittivity, breakdown voltage, and other properties.

Conductive Ink Formulation: Mixture design is useful in the formulation of conductive inks used in printed electronics, flexible circuits, and RFID tags. It helps optimize the ink composition to balance conductivity and adhesion properties.

Battery Electrode Materials: Research and development of electrode materials for batteries, supercapacitors, and energy storage devices often involve mixture design to find the right blend of active materials, binders, and conductive additives to enhance energy density and cycle life.

Circuit Board Materials: In the electronics industry, mixture design can be applied to develop materials for circuit board substrates and laminates with specific dielectric constants, thermal properties, and electrical insulation characteristics.

Sensor Materials: Mixture design can be used to optimize the composition of sensor materials, such as those used in gas sensors, pressure sensors, and temperature sensors, to improve sensitivity, selectivity, and response times.

Optical Fiber Coatings: For optical communication systems, mixture design can help formulate coatings for optical fibers to achieve desired refractive indices, dispersion properties, and environmental durability.

Electroplating Solutions: In electroplating processes used for manufacturing electrical connectors and components, mixture design can assist in developing plating solutions with precise metal ratios and desired properties like corrosion resistance and conductivity.

Thermal Interface Materials: Mixture design can be applied to create thermal interface materials used in electronic packaging to optimize thermal conductivity, flexibility, and adhesion.

Printed Circuit Board (PCB) Ink Formulation: When formulating inks for PCB printing, mixture design can help achieve the desired electrical and thermal properties while ensuring printability and adhesion to the substrate.

In these and other applications within electrical engineering, mixture design experiments provide a systematic approach to optimizing material compositions, which is crucial for achieving desired electrical and electronic performance characteristics.

Steps

Developing mixture design experiments for electrical engineering involves a systematic approach to optimize the composition of materials or compounds with multiple components. Here are the steps to develop mixture design experiments in electrical engineering:

1. **Define the Objective:**
   - Clearly define the objective of the study. Determine what specific electrical or electronic properties or characteristics you want to optimize or achieve through the mixture.
2. **Select the Components:**
   - Identify the components or ingredients that will make up the mixture. These could be materials, chemicals, or compounds that contribute to the desired properties.
3. **Determine the Range of Component Levels:**
   - Specify the allowable range or levels for each component. This defines the minimum and maximum proportions for each ingredient in the mixture. These ranges should cover the practical or feasible values.
4. **Choose the Experimental Design:**
   - Select an appropriate experimental design method. Common methods include Simplex-Centroid, Simplex-Lattice, or others, depending on the number of components and the design's complexity.
5. **Generate the Mixture Design Matrix:**
   - Using the chosen experimental design, generate a matrix that represents the combinations of component levels for each experiment or trial. Each row in the matrix corresponds to one experimental run.
6. **Conduct the Experiments:**
   - Prepare and conduct the experiments according to the combinations specified in the mixture design matrix. For each run, accurately measure and record the relevant electrical or electronic properties.
7. **Analyze the Data:**
   - Analyze the data collected from the experiments. Use statistical tools and techniques to determine the relationships between the mixture components and the desired electrical properties. Regression analysis, ANOVA, and response surface modeling are common approaches.

8. **Optimize the Mixture:**
   - Based on the data analysis, identify the optimal mixture composition that maximizes or minimizes the desired electrical properties. Optimization algorithms or graphical methods can help pinpoint the best combination.
9. **Validation and Verification:**
   - Perform additional experiments or validations to confirm the optimal mixture's performance. Ensure that the desired electrical characteristics are consistently achieved within the specified component ranges.
10. **Documentation and Reporting:**
   - Document the results, including the optimal mixture composition, any constraints, and the methodology used. Prepare a comprehensive report summarizing the study, experimental design, results, and conclusions.
11. **Implementation:**
   - Implement the optimized mixture composition in practical applications within electrical engineering, such as in the production of electronic components or materials.
12. **Monitoring and Continuous Improvement:**
   - Continuously monitor the performance of the optimized mixture in real-world applications. Make adjustments or refinements as needed to maintain or improve performance over time.

These steps provide a structured approach to developing mixture design experiments in electrical engineering, helping engineers and researchers systematically optimize material compositions to meet specific electrical or electronic performance requirements.

PITFALLS AND REMEDIES

Mixture design experiments in electrical engineering can face several pitfalls and challenges. Here are some common pitfalls and remedies:

Pitfall 1: Poorly Defined Objectives
   Pitfall: Vague or undefined objectives can lead to ineffective experiments.
   Remedy: Clearly define the specific electrical or electronic properties or characteristics you aim to optimize or achieve with the mixture. Ensure objectives are well-defined and measurable.

Pitfall 2: Inadequate Component Selection
   Pitfall: Choosing inappropriate or insufficient components for the mixture can result in suboptimal results.
   Remedy: Carefully select components based on their electrical properties and compatibility. Conduct a thorough literature review and consult experts if necessary.

Pitfall 3: Narrow Component Level Ranges
   Pitfall: Restricting the range of component levels too narrowly may miss out on optimal solutions.
   Remedy: Define component level ranges that cover a broad spectrum, including extreme values, to explore the full design space.

Pitfall 4: Inadequate Experimental Design
   Pitfall: Choosing an inappropriate or insufficient experimental design method can lead to inefficient data collection.
   Remedy: Select the most suitable experimental design method (e.g., Simplex-Centroid, Simplex-Lattice) based on the number of components and expected interactions.

Pitfall 5: Lack of Replicates
   Pitfall: Conducting experiments without replicates can result in unreliable or inconsistent results.

Remedy: Include replicates for each combination of component levels to account for experimental variability and improve the reliability of the data.

Pitfall 6: Insufficient Data Analysis

Pitfall: Inadequate data analysis may fail to reveal meaningful insights.

Remedy: Utilize statistical tools like regression analysis, ANOVA, and response surface modeling to analyze the data comprehensively. Consider consulting a statistician or data scientist for complex analyses.

Pitfall 7: Overlooking Interaction Effects

Pitfall: Ignoring interactions between components can lead to suboptimal solutions.

Remedy: Analyze and consider interactions among components to better understand how they affect the electrical properties. RSM can help capture interactions.

Pitfall 8: Failure to Validate Results

Pitfall: Not validating or verifying the optimized mixture's performance in practical applications.

Remedy: Conduct additional experiments or validations to ensure the optimized mixture consistently meets the desired electrical characteristics.

Pitfall 9: Lack of Documentation

Pitfall: Poor documentation can make it challenging to replicate or understand the experiment.

Remedy: Maintain thorough records of experimental procedures, data, and analysis. Prepare a comprehensive report summarizing the study, methodology, and results.

Pitfall 10: Implementation Challenges

Pitfall: Challenges may arise when implementing the optimized mixture in real-world applications.

Remedy: Collaborate closely with engineering teams to ensure a smooth transition from the laboratory to practical applications. Address any production or manufacturing challenges.

Pitfall 11: Neglecting Continuous Improvement

Pitfall: Failing to monitor and refine the optimized mixture's performance over time.

Remedy: Establish a system for ongoing monitoring and improvement. Make adjustments as needed to maintain or enhance performance.

By being aware of these pitfalls and implementing the suggested remedies, engineers and researchers can conduct more effective mixture design experiments in electrical engineering and achieve better outcomes when optimizing material compositions for specific electrical or electronic properties.

EXAMPLE

Mixture experiments in electrical engineering involve optimizing the proportions of various components or ingredients in a mixture to achieve specific desired properties. Here's an example of a matrix for a mixture designed experiment in electrical engineering:

Objective: To optimize the composition of a composite material used in electrical insulators for maximum dielectric strength while minimizing cost.

Components of the Mixture:

1. **Polymer Resin (Component A):** Proportion expressed as a percentage of the total mixture.
   - Range: 40% to 60%
2. **Fiberglass Reinforcement (Component B):** Proportion expressed as a percentage of the total mixture.
   - Range: 20% to 40%

3. **Filler Material (Component C)**: Proportion expressed as a percentage of the total mixture.
   * Range: 10% to 20%
4. **Silicon Dioxide (Component D)**: Proportion expressed as a percentage of the total mixture.
   * Range: 5% to 15%

Response Variable: Dielectric Strength (measured in volts per mil, V/mil)

Mixture Designed Experiment Matrix:

In this example, a mixture designed experiment aims to optimize the mix proportions of the composite material to maximize its dielectric strength while considering factors like polymer resin, fiberglass reinforcement, filler material, and silicon dioxide. The matrix is shown in Table 3.38.

In this matrix, each row represents a specific experimental run with a unique combination of proportions for the mixture components (polymer resin, fiberglass reinforcement, filler material, and silicon dioxide). The aim is to find the optimal mixture proportions that result in the highest dielectric strength while considering practical constraints and cost considerations in electrical engineering applications.

By conducting experiments with different mix proportions and analyzing the resulting dielectric strengths, electrical engineers can determine the ideal composition for their specific application, balancing dielectric performance and cost-effectiveness.

REPORTING

In mixture design experiments for electrical engineering, various statistical tools and techniques are used to analyze and optimize the composition of mixtures with respect to electrical properties. Some of the common statistical tools employed in these experiments include:

Linear Regression Analysis: Linear regression is used to establish a relationship between the mixture components and the electrical property of interest. It helps in understanding how changes in component levels affect the response.

Analysis of Variance (ANOVA): ANOVA is used to assess the significance of individual components and their interactions on the electrical property. It helps identify which factors have a significant impact on the response.

Response Surface Methodology (RSM): RSM involves fitting mathematical models (e.g., quadratic or cubic) to experimental data to optimize the mixture composition for the desired electrical property. It helps in finding the optimal combination of components.

TABLE 3.38

Mixture Designed Experiment Matrix Example for Electrical Engineering

Run	Polymer Resin (A)	Fiberglass Reinforcement (B)	Filler Material (C)	Silicon Dioxide (D)
1	50%	30%	15%	5%
2	45%	25%	10%	10%
3	55%	35%	20%	8%
4	48%	28%	18%	7%
5	52%	32%	12%	12%
6	55%	40%	15%	5%
7	50%	30%	20%	10%
8	48%	28%	12%	8%

Desirability Function Analysis: The desirability function approach combines multiple responses into a single objective function, making it easier to find an optimal mixture composition that simultaneously meets multiple criteria for electrical properties.

Mixture-Process Variable Analysis: This analysis considers both mixture components and process variables to optimize electrical properties. It accounts for the effect of process conditions on the mixture's performance.

Factorial Design: Factorial experiments are used to study the main effects and interactions of factors (components) on the electrical response. Fractional factorial designs are particularly useful when dealing with a large number of components.

Simplex-Centroid Design: This design is suitable for three-component mixture experiments. It involves a simplex-lattice design that efficiently explores the entire composition space.

Simplex-Lattice Design: Similar to Simplex-Centroid, this design extends to higher dimensions and allows for efficient exploration of the mixture space with more than three components.

D-Optimal Design: D-Optimal designs aim to minimize the variance of estimated model parameters, making them efficient for selecting experimental runs that provide the most information about the mixture's behavior.

Monte Carlo Simulation: Monte Carlo methods can be used to estimate the uncertainty associated with the optimized mixture composition and its impact on electrical properties.

Constraint Optimization: When there are constraints on component levels (e.g., cost limitations or safety considerations), optimization techniques like linear programming or non-linear programming can be employed to find the best compromise solution.

Principal Component Analysis (PCA): PCA can help reduce the dimensionality of data and identify patterns or trends in electrical properties across different mixture compositions.

Multivariate Analysis: Multivariate techniques such as PLS regression can be used when there are multiple correlated responses or when there is a need to model complex relationships between components and properties.

These statistical tools and techniques are applied depending on the specific objectives and complexities of the mixture design experiment in electrical engineering. They help researchers and engineers make informed decisions about optimizing mixture compositions for desired electrical properties while considering factors like cost, manufacturability, and performance constraints.

A well-structured report for a mixture design experiment in electrical engineering should provide a clear and concise overview of the experiment, its objectives, methodology, results, and conclusions. Here are the typical elements that should be included in such a report:

Title Page:
 Title of the Report
 Name of the Organization or Institution
 Date of Submission
 Names of the Authors and their Affiliations

Abstract:
 A brief summary of the experiment, objectives, methods, and key findings.

Table of Contents:
 A list of sections and subsections with page numbers for easy navigation.

List of Figures and Tables:
 A list of all figures and tables included in the report, along with their page numbers.

List of Abbreviations and Symbols (if applicable):
 Definitions of any abbreviations, acronyms, or symbols used in the report.

Introduction:
  Background information and context for the experiment.
  Clear statement of the objectives and goals.
  Significance of the experiment and its potential impact on electrical engineering.
Literature Review (optional):
  A review of relevant literature and prior research related to the experiment.
Experimental Design:
  Description of the mixture design methodology used.
  Explanation of the choice of factors, levels, and constraints.
  Details of the response variable(s) or electrical properties being studied.
  Explanation of the experimental setup and conditions.
Materials and Methods:
  List of materials and components used in the experiment.
  Details of the mixture components and their properties.
  Experimental procedure, including any special considerations.
  Data collection methods and instruments used.
Results:
  Presentation of experimental data, including tables, figures, and charts.
  Summary of the observed trends and patterns in electrical properties.
  Statistical analysis of the data, including ANOVA or regression analysis.
  Presentation of optimization results, if applicable.
Discussion:
  Interpretation of the results in the context of the objectives.
  Analysis of the significance of factors and their interactions.
  Comparison of the experimental findings with literature or expectations.
  Discussion of any challenges or unexpected findings.
Conclusion:
  Summary of the key findings and their implications.
  Conclusions regarding the optimal mixture composition, if applicable.
  Recommendations for future work or further research.
References:
  Citations of all sources, publications, and references used in the report.
Appendices (if applicable):
  Any additional data, calculations, or supplementary information.
  Details of statistical analyses, equations, or software used.
Acknowledgments (optional):
  Acknowledgment of individuals, organizations, or funding sources that contributed to the
    experiment.
Author Contact Information (optional):
  Contact details of the authors for inquiries or further collaboration.

The report should be well-organized, with clear headings and subheadings, and it should be written in a professional and concise manner. Additionally, any supporting data, graphs, or charts should be well-labeled and easy to understand. The structure and content of the report may vary depending on the specific requirements of the experiment and the target audience.

3.4.7 SEQUENTIAL EXPERIMENTATION

Sequential experiments, also known as sequential experimentation or sequential design of experiments, are used in various applications within electrical engineering. Here are some areas where sequential experiments are applied:

Integrated Circuit (IC) Design:

In the design of complex ICs, engineers often use sequential experiments to optimize the performance of electronic components.

Sequential experimentation helps in refining the design parameters to achieve desired electrical characteristics, such as speed, power consumption, and noise margins.

Manufacturing Process Optimization:

Electrical engineering encompasses various manufacturing processes, such as semiconductor fabrication and PCB assembly.

Sequential experiments can be employed to optimize manufacturing parameters and ensure consistent product quality.

Control Systems:

Sequential experiments are used to fine-tune control systems in electrical engineering applications.

Engineers can iteratively adjust control parameters to achieve optimal system performance, stability, and response time.

Signal Processing and Filtering:

In signal processing, sequential experimentation can be applied to design and optimize digital filters for various applications, such as noise reduction or data compression.

Power Electronics:

Engineers in the field of power electronics use sequential experiments to optimize power converter designs, improve energy efficiency, and reduce losses.

Network Optimization:

In telecommunications and network engineering, sequential experiments may be used to optimize routing algorithms, network protocols, and quality of service (QoS) parameters.

Reliability and Fault Tolerance:

Sequential experiments can help engineers assess the reliability and fault tolerance of electrical systems.

They are used to design redundancy strategies and evaluate the impact of component failures.

Antenna Design:

Antenna engineers employ sequential experiments to optimize the design of antennas for various applications, including wireless communication and radar systems.

Sensor Calibration:

Sequential experiments are used to calibrate sensors and measurement devices to ensure accuracy and precision in electrical measurements.

Embedded Systems and Firmware Development:

In the development of embedded systems and firmware, sequential experimentation aids in optimizing code and algorithms for performance and efficiency.

Energy Management:

In the field of renewable energy and smart grids, sequential experiments can be used to optimize energy management strategies, including load balancing and grid stability.

Circuit Parameter Extraction:

Sequential experiments are applied to extract accurate electrical parameters of components and circuits, which is crucial for circuit modeling and simulation.

Machine Learning and AI Hardware:

Engineers working on hardware for machine learning and artificial intelligence (AI) often use sequential experiments to optimize hardware architectures and algorithms for training and inference tasks.

In all these applications, sequential experiments help engineers systematically explore and refine design parameters, algorithms, or control strategies to achieve optimal electrical performance or system behavior. They can significantly reduce the time and resources required for experimentation and design optimization in complex electrical engineering domains.

STEPS

Developing a sequential experimentation plan for electrical engineering involves a systematic process to optimize system performance, design parameters, or control strategies. Here are the general steps to develop sequential experimentation for electrical engineering:

1. **Define the Objective**:
   - Clearly define the goal or objective of the experimentation. Determine what specific aspect of the electrical system or process you want to optimize. It could be performance, efficiency, reliability, or any other relevant metric.
2. **Identify Factors and Parameters**:
   - Identify the factors and parameters that influence the objective. These may include design variables, control settings, or system configurations that can be adjusted during the experimentation.
3. **Select Response Variables**:
   - Choose response variables or performance metrics that will be used to evaluate the system's performance. These variables should directly relate to the objective and be measurable or quantifiable.
4. **Design Initial Experiments**:
   - Create an initial experimental design using techniques like factorial experiments, fractional factorial experiments, or CCD. This design should cover a broad range of factor settings to understand their effects.
5. **Collect Data**:
   - Conduct the initial experiments and collect data on the chosen response variables. Ensure that the data is accurate and reliable.
6. **Analyze Initial Data**:
   - Analyze the data to identify trends and relationships between factors and response variables. This analysis helps in understanding the system's behavior and identifying areas for improvement.
7. **Develop a Mathematical Model**:
   - Develop a mathematical or statistical model that describes the relationship between factors and responses. This model can be used for optimization and prediction.
8. **Optimize Using Sequential Experiments**:
   - Based on the initial analysis and model, design additional experiments in a sequential manner.
   - Use optimization techniques such as RSM, sequential Monte Carlo, or Bayesian optimization to guide the selection of new experiments.
   - Focus on regions of the parameter space that are likely to yield better results and efficiently reduce the search space.
9. **Iterate and Refine**:
   - Continue the process iteratively, using the results from each set of experiments to refine the model and optimize the system further.
   - Repeat the cycle of experimentation, analysis, and optimization until the desired level of performance or the objective is achieved.

10. **Validate and Implement**:
    - Once the optimization process is complete, validate the optimized solution through additional experiments or simulations.
    - Implement the optimized parameters, design, or control strategies in the actual electrical system or process.
11. **Document and Report**:
    - Document all experimental procedures, data, results, and the final optimized solution.
    - Prepare a detailed report that summarizes the entire sequential experimentation process, including findings, recommendations, and any insights gained.
12. **Monitor and Maintain**:
    - Continuously monitor the optimized system's performance and make adjustments as needed to ensure long-term effectiveness and efficiency.
13. **Consider Constraints**:
    - Take into account any practical constraints, budget limitations, or safety requirements during the experimentation and optimization process.
14. **Collaborate and Seek Expertise**:
    - Collaborate with domain experts, statisticians, and engineers to ensure that the sequential experimentation process is conducted effectively and rigorously.
15. **Use Specialized Software**:
    - Utilize specialized software tools for experimental design, data analysis, and optimization to streamline the process and enhance efficiency.

The steps above provide a structured approach to developing sequential experimentation plans in electrical engineering. The iterative nature of sequential experimentation allows for continual refinement and improvement of electrical systems and processes.

PITFALLS AND REMEDIES

Sequential experimentation in electrical engineering, like any other field, can have its pitfalls. Identifying these pitfalls and applying appropriate remedies is crucial to the success of the optimization process. Here are some common pitfalls and their corresponding remedies:

Pitfall 1: Inadequate Initial Design
> **Issue:** The initial experimental design may not cover the important factors adequately, leading to inefficient data collection.
>
> **Remedy:** Conduct a thorough literature review and consult with experts to identify key factors. Use appropriate experimental design techniques such as factorial experiments to ensure good coverage of factor settings.

Pitfall 2: Poor Data Quality
> **Issue:** Data collected may be noisy, inconsistent, or inaccurate.
>
> **Remedy:** Implement rigorous data collection and measurement procedures. Use statistical techniques to identify and handle outliers. Ensure calibration and accuracy of measurement instruments.

Pitfall 3: Model Assumptions
> **Issue:** The mathematical or statistical model used for optimization may make unrealistic assumptions about the system's behavior.
>
> **Remedy:** Regularly validate model assumptions. Explore alternative models, such as nonlinear or empirical models, when necessary. Consider physics-based models for greater accuracy.

Pitfall 4: Over-optimization
> **Issue:** Over-optimization can lead to models that fit the noise in the data rather than the underlying relationships.

Remedy: Use cross-validation or holdout datasets to assess model generalization. Avoid making extreme parameter adjustments based on limited data.

Pitfall 5: Neglecting Practical Constraints

Issue: Optimized solutions may not be implementable in practice due to physical, budgetary, or safety constraints.

Remedy: Include relevant constraints in the optimization process. Use multi-objective optimization techniques when trade-offs between conflicting objectives exist.

Pitfall 6: Lack of Expertise

Issue: Insufficient domain knowledge or expertise in statistical techniques can hinder the optimization process.

Remedy: Collaborate with experts in both the field of electrical engineering and statistical analysis. Seek guidance from statisticians when needed.

Pitfall 7: Insufficient Iteration

Issue: Stopping the experimentation and optimization process too early may result in suboptimal solutions.

Remedy: Plan for multiple iterations and refinements. Continuously assess and improve the model and system performance over time.

Pitfall 8: Failure to Communicate Results

Issue: Lack of effective communication can lead to misinterpretation of results or implementation failures.

Remedy: Prepare clear and concise reports and presentations that convey the findings, recommendations, and implications of the sequential experimentation process. Ensure that key stakeholders understand the results.

Pitfall 9: Ignoring External Factors

Issue: External factors, such as environmental changes or variations in operating conditions, may not be accounted for in the optimization process.

Remedy: Monitor and control external factors or include them as covariates in the model. Consider RPD techniques to account for variability.

Pitfall 10: Not Monitoring Long-term Performance

Issue: Optimized systems may degrade over time due to changing conditions.

Remedy: Implement a monitoring and maintenance plan to ensure that the optimized system continues to perform well under varying circumstances.

Pitfall 11: Focusing Solely on Statistical Techniques

Issue: Overemphasis on statistical techniques without a deep understanding of the underlying engineering principles can lead to suboptimal solutions.

Remedy: Balance statistical analysis with domain-specific knowledge and engineering expertise. Collaborate with engineers who understand the physical system.

Addressing these pitfalls and applying the corresponding remedies will help ensure that sequential experimentation in electrical engineering leads to meaningful improvements, efficient optimization, and reliable results.

EXAMPLE

Sequential experimentation in electrical engineering involves conducting a series of experiments where the results of each experiment inform the design of subsequent experiments. The objective is to efficiently and iteratively explore the design space to optimize a particular outcome or property. Here's a simplified example of a matrix for a sequential experimentation designed experiment in electrical engineering:

Objective: To optimize the design of a PCB for minimal signal interference.

Components of the PCB Design:

1. **Trace Width (Component A):** Width of signal traces on the PCB, measured in millimeters.
   - Initial Range: 0.2 mm to 0.4 mm
2. **Trace Spacing (Component B):** Spacing between signal traces on the PCB, measured in millimeters.
   - Initial Range: 0.3 mm to 0.5 mm
3. **Layer Stackup (Component C):** The arrangement of PCB layers and their thickness.
   - Initial Configuration: 4-layer stackup

Response Variable: Signal Interference (measured in decibels, dB)

Sequential Experimentation Matrix:

In sequential experimentation, the initial matrix might involve a limited set of experiments to establish a baseline or explore the design space (Table 3.39).

Based on the initial results, the subsequent experiments can be designed iteratively to explore specific regions of interest or to refine the PCB design further. The exact matrix for sequential experimentation would depend on the information gained from the initial experiments and the specific objectives and constraints of the electrical engineering project.

REPORTING

The key to sequential experimentation is adaptability. After analyzing the initial results, additional experiments can be designed and conducted to move closer to the optimal PCB design that minimizes signal interference while considering factors like trace width, trace spacing, and layer stackup in electrical engineering applications.

Sequential experimentation in electrical engineering often involves a combination of statistical and engineering tools to optimize processes or systems. Here are some of the statistical tools commonly used in sequential experimentation for electrical engineering:

1. **Factorial Experiments:** Factorial experiments help identify which factors (independent variables) and their interactions have a significant impact on the response variable. These designs are useful for screening experiments and initial exploration of factors.
2. **Response Surface Methodology (RSM):** RSM is used to model and optimize complex relationships between factors and responses. It involves designing experiments to fit response surfaces, such as quadratic models, to estimate optimal settings.
3. **Design of Experiments (DOE):** Various experimental designs, including full-factorial, fractional factorial, and CCDs, are used to efficiently collect data and estimate model parameters with minimal experimentation.
4. **Sequential Sampling:** In sequential experimentation, data is collected iteratively, and statistical tools like sequential hypothesis testing (e.g., sequential probability ratio tests) are used to decide when to stop data collection based on predefined criteria.

TABLE 3.39

Sequential Designed Experiments for Electrical Engineering

Run	Trace Width (A) (mm)	Trace Spacing (B) (mm)	Layer Stackup (C)
1	0.2	0.3	4-layer
2	0.3	0.4	4-layer
3	0.4	0.5	4-layer

5. **Regression Analysis:** Regression models, including linear, multiple linear, and nonlinear regression, are used to relate independent variables to the response variable. These models help estimate coefficients and predict responses.
6. **Optimization Algorithms:** Mathematical optimization algorithms, such as gradient descent, genetic algorithms, and simulated annealing, are used to find the optimal combination of factors that maximize or minimize a response variable.
7. **Analysis of Variance (ANOVA):** ANOVA is used to assess the significance of factors and interactions in the experimental data. It helps identify which factors have a statistically significant impact on the response.
8. **Statistical Process Control (SPC):** SPC techniques, including control charts and process capability analysis, are used to monitor and control processes during experimentation and production to ensure stability and consistency.
9. **Robust Parameter Design (RPD):** RPD combines traditional DOE with robust optimization techniques to find factor settings that are less sensitive to variability or noise in the system.
10. **Monte Carlo Simulation:** Monte Carlo simulations are used to assess the robustness of designs and models by generating random datasets based on known distributions to evaluate system performance.
11. **Bayesian Methods:** Bayesian statistics can be applied to update models and make decisions as new data becomes available during sequential experimentation.
12. **Artificial Intelligence (AI) and Machine Learning:** Advanced techniques like neural networks and machine learning algorithms can be employed to model complex relationships, make predictions, and optimize processes based on large datasets.
13. **Statistical Software:** Specialized statistical software packages like Minitab, JMP, R, and Python with libraries like SciPy and scikit-learn are commonly used to perform data analysis, modeling, and optimization.
14. **Reliability Analysis:** In cases involving electrical components and systems, reliability analysis tools can be used to assess the probability of failure and optimize for reliability.

The choice of statistical tools depends on the specific objectives of the sequential experimentation and the complexity of the electrical engineering problem. Engineers and statisticians often collaborate to select the most appropriate tools and methodologies for achieving the desired outcomes.

A report on sequential experimentation in electrical engineering typically includes several key elements to document the research, methods, results, and conclusions. Here are the common elements you would find in such a report:

1. **Title Page:**
   * Title of the Report
   * Name of the Author(s)
   * Date of Submission
   * Affiliation or Institution
2. **Abstract:**
   * A concise summary of the entire report, including the problem statement, methodology, major findings, and conclusions. It provides readers with a quick overview of the study.
3. **Table of Contents:**
   * Lists the main sections, subsections, and their respective page numbers for easy navigation.
4. **List of Figures and Tables:**
   * Provides a list of all figures and tables in the report with corresponding page numbers.
5. **Nomenclature or Abbreviations (if applicable):**
   * A list of symbols, acronyms, or abbreviations used throughout the report, along with their definitions.

6. **Introduction:**
   - Problem Statement: A clear and concise statement of the research problem or objective.
   - Background and Context: A brief review of relevant literature and the context of the study.
   - Objectives: A statement of the specific goals and objectives of the sequential experimentation.
   - Scope and Significance: An explanation of the scope of the study and its importance in the field of electrical engineering.

7. **Methodology:**
   - Experimental Design: A detailed description of the experimental design, including factors, levels, response variables, and any statistical methods used.
   - Data Collection: Information on how data was collected, including instruments and procedures.
   - Statistical Tools: A discussion of the statistical methods and software used for data analysis and optimization.
   - Sampling (if applicable): Details about the sampling strategy, if relevant.

8. **Results:**
   - Presentation of experimental data in the form of tables, figures, charts, or graphs.
   - Statistical Analysis: Detailed statistical analysis of the data, including ANOVA, regression analysis, or any other relevant statistical tests.
   - Optimization Results: If the study involves optimization, present the optimal settings and results.

9. **Discussion:**
   - Interpretation of Results: Discussion of the findings, including the significance of any trends or patterns observed.
   - Comparison with Objectives: Evaluate whether the study achieved its objectives and address any deviations from the initial goals.
   - Practical Implications: Discuss the practical implications of the research and its relevance to real-world applications.
   - Limitations: Identify any limitations of the study, such as constraints in the experimental setup or assumptions made.
   - Future Work: Suggest areas for future research or improvements to the experimental design.

10. **Conclusion:**
    - Summarize the main findings of the study.
    - Reiterate the significance of the research.
    - Provide a concise conclusion that addresses the research objectives.

11. **References:**
    - Cite all sources, including research papers, books, articles, and software, that were referenced or consulted during the study.

12. **Appendices (if applicable):**
    - Include any supplementary material such as detailed data tables, calculations, or additional figures that support the main report.

13. **Acknowledgments (optional):**
    - Acknowledge individuals, institutions, or organizations that provided support, funding, or assistance during the research.

Remember that the specific format and requirements for a report on sequential experimentation in electrical engineering may vary depending on your institution, publication guidelines, or the nature of the research. Be sure to follow any specific formatting or reporting guidelines provided by your organization or journal if applicable.

3.4.8 ROBUST PARAMETER DESIGN (RPD)

RPD experiments are used in various areas of electrical engineering to optimize and improve the performance, reliability, and robustness of electrical systems, devices, and components. Here are some specific areas within electrical engineering where RPD experiments are commonly employed:

Electronic Circuit Design: RPD is used to optimize the parameters of electronic circuits, such as ICs, to ensure stable and reliable operation over a range of conditions, including temperature variations and component tolerances.

Power Electronics: In power electronics, RPD is employed to design and control power converters, inverters, and motor drives for improved efficiency and reduced sensitivity to component variations.

Signal Processing: RPD is used in the design of signal processing algorithms and filters to enhance S/N ratios and reduce sensitivity to noise and interference.

Printed Circuit Board (PCB) Layout: PCB designers use RPD to optimize the layout of components, traces, and grounding schemes to minimize EMI and improve signal integrity.

Control Systems: RPD is applied to control system design to ensure stability, responsiveness, and robustness against external disturbances and uncertainties.

Telecommunications: In the telecommunications field, RPD is used to optimize the design of antennas, transmitters, and receivers to improve signal quality and minimize interference.

Embedded Systems: RPD helps improve the robustness and reliability of embedded systems by optimizing the hardware and software components for different operating conditions.

Semiconductor Manufacturing: In semiconductor manufacturing, RPD is employed to optimize process parameters, such as wafer fabrication and packaging, to ensure consistent and reliable product quality.

Quality Control and Testing: RPD can be used in quality control processes to design experiments that ensure products meet specified performance and reliability criteria.

Renewable Energy Systems: In renewable energy systems like solar panels and wind turbines, RPD is used to design and optimize components for improved energy capture and long-term reliability.

Automated Manufacturing: In automated manufacturing processes, RPD can be applied to optimize robotic control systems and sensors for consistent and accurate production.

Wireless Communication Systems: RPD helps optimize the design of wireless communication systems, including base stations and mobile devices, for better performance in various environmental conditions.

Electromagnetic Compatibility (EMC): RPD is used to design electronic systems and devices that comply with EMC standards and minimize electromagnetic interference.

RPD experiments involve systematically varying design parameters and conducting experiments to identify the optimal settings that result in improved performance and robustness. By applying RPD principles, electrical engineers can develop more reliable and resilient systems in a wide range of applications within the field of electrical engineering.

STEPS

Developing an RPD experiment in electrical engineering involves a systematic approach to optimize the performance and robustness of electrical systems or components. Here are the steps to develop an RPD experiment in electrical engineering:

1. **Problem Formulation:**
   - Clearly define the problem or objective you want to address in your electrical engineering project. Identify the specific performance characteristics or parameters you want to optimize, such as efficiency, reliability, or S/N ratio.

2. **Identify Factors and Levels:**
   - Identify the factors (design parameters) that can potentially affect the performance of your electrical system or component. These factors can include component values, material properties, operating conditions, and control settings.
   - Determine the ranges or levels at which these factors will be varied during the experiment. Ensure that the factor levels cover a meaningful range of values and are practical to implement.

3. **Select Response Variables:**
   - Choose response variables that represent the performance or quality characteristics you want to optimize or improve. These can include metrics like voltage output, S/N ratio, power consumption, or any other relevant measures.

4. **Experimental Design:**
   - Select an appropriate experimental design technique, such as a factorial design, fractional factorial design, or RSM, depending on the complexity of the problem and the number of factors involved.
   - Use statistical software or tools to generate an experimental design matrix that specifies the combinations of factor levels to be tested in your experiments.

5. **Conduct Experiments:**
   - Conduct the experiments based on the design matrix. Record the response variable measurements for each combination of factor levels.
   - Ensure that the experiments are conducted carefully, and all relevant data is accurately collected.

6. **Statistical Analysis:**
   - Use statistical analysis techniques to analyze the experimental data. This may involve regression analysis, ANOVA, or other statistical methods.
   - Identify the main effects of factors and their interactions on the response variables.

7. **Optimization:**
   - Use optimization techniques to find the factor settings that result in the desired performance or robustness improvements.
   - Response surface optimization methods can help you identify the optimal factor settings.

8. **Validation:**
   - Validate the optimized settings by conducting additional experiments or simulations to ensure that the improvements are consistent and reliable.

9. **Implementation:**
   - Implement the optimized factor settings in your electrical system or component design or operation.
   - Document the changes made based on the RPD results.

10. **Monitoring and Control:**
    - Continuously monitor the performance of your electrical system or component to ensure that it maintains the desired levels of efficiency and robustness.
    - Implement control measures to adjust factor settings if necessary.

11. **Documentation and Reporting:**
    - Prepare a detailed report that documents the entire RPD process, including problem formulation, experimental design, data analysis, optimization results, and implementation details.
    - Share the findings and recommendations with relevant stakeholders or team members.

12. **Iterative Improvement:**
    - RPD is often an iterative process. If further improvements are needed or if new factors come into play, repeat the RPD process as necessary.

By following these steps, you can systematically develop and implement RPD experiments in electrical engineering to enhance the performance and robustness of your electrical systems or components.

PITFALLS AND REMEDIES

RPD experimentation in electrical engineering can be a powerful tool for optimizing and ensuring the reliability of electrical systems or components. However, like any experimental approach, there are potential pitfalls and challenges. Here are some common pitfalls and remedies when conducting RPD experiments in electrical engineering:

Pitfall 1: Inadequate Problem Formulation
> **Pitfall:** Failing to clearly define the problem or objective can lead to irrelevant factors being considered, resulting in suboptimal outcomes.
> **Remedy:** Start by clearly defining the problem and specifying the objectives. Ensure that the factors under consideration are relevant to the problem at hand.

Pitfall 2: Poor Factor Selection
> **Pitfall:** Selecting the wrong factors or omitting important ones can lead to incomplete or incorrect optimization.
> **Remedy:** Conduct a thorough factor screening and brainstorming session to identify and prioritize factors. Consider both controllable and uncontrollable factors.

Pitfall 3: Insufficient Factor Levels
> **Pitfall:** Inadequate variation in factor levels may not reveal the full response surface, making it difficult to optimize.
> **Remedy:** Choose factor levels that span the expected operational range. Use engineering judgment and knowledge to determine appropriate levels.

Pitfall 4: Inadequate Sample Size
> **Pitfall:** Using too few experimental runs can result in low statistical power and unreliable conclusions.
> **Remedy:** Ensure that your sample size is sufficient to detect significant effects and perform reliable statistical analyses. Use power calculations if necessary.

Pitfall 5: Violation of Assumptions
> **Pitfall:** Violating assumptions of statistical methods (e.g., normality, homoscedasticity) can lead to inaccurate results.
> **Remedy:** Check the underlying assumptions of the statistical techniques used. If assumptions are violated, consider using nonparametric methods or transformations.

Pitfall 6: Overfitting the Model
> **Pitfall:** Overfitting occurs when a model is too complex and fits the noise in the data, leading to poor generalization.
> **Remedy:** Use model selection techniques to choose an appropriate model complexity. Cross-validation can help assess model performance.

Pitfall 7: Lack of Validation
> **Pitfall:** Failing to validate the optimized settings or not accounting for variability in real-world conditions can result in poor real-world performance.
> **Remedy:** Validate the optimized settings in realistic conditions, considering potential sources of variability. Conduct robustness testing to assess performance under various scenarios.

Pitfall 8: Inadequate Documentation
> **Pitfall:** Poor documentation can make it challenging to reproduce results or communicate findings to others.

Remedy: Maintain comprehensive records of experimental design, data collection, analysis, and results. Create a detailed report that documents the entire RPD process.

Pitfall 9: Ignoring Practical Constraints

Pitfall: Optimized settings may not be implementable in practice due to practical constraints or cost limitations.

Remedy: Consider practical constraints during the optimization process. Balance theoretical optimization with real-world feasibility.

Pitfall 10: Lack of Iteration

Pitfall: RPD is often an iterative process, and a single optimization round may not yield the best results.

Remedy: Be prepared to iterate the RPD process if necessary, especially if new factors or challenges arise.

Avoiding these pitfalls and implementing the suggested remedies can help ensure the success of RPD experiments in electrical engineering and lead to improved system performance and reliability.

EXAMPLE

RPD in electrical engineering aims to optimize product or process performance while minimizing sensitivity to variation or external factors. Here's an example of a matrix for an RPD designed experiment in electrical engineering:

Objective: To optimize the performance of an electronic amplifier circuit for consistent output voltage while minimizing the impact of component variations (e.g., resistor tolerance, transistor gain).

Components and Factors:

1. **Resistor Value (Component A)**: Resistance value of a key resistor in the circuit.
   - Range: 100 ohms to 150 ohms
2. **Transistor Gain (Component B)**: Gain of the transistor used in the amplifier.
   - Range: 50 to 100

Factors to Study Variations:

1. **Resistor Tolerance (Factor C)**:
   - Low Tolerance (e.g., ±1%)
   - High Tolerance (e.g., ±5%)
2. **Transistor Variation (Factor D)**:
   - Low Variation (e.g., ±2)
   - High Variation (e.g., ±10)

Response Variable: Output Voltage (measured in volts, V)

Robust Parameter Design Matrix:

In this RPD designed experiment, the matrix is designed to assess the robustness of the amplifier circuit's output voltage to variations in component values and tolerance. The matrix is shown in Table 3.40.

In this matrix, each row represents a specific experimental run with a unique combination of component values (resistor value and transistor gain) and levels of variation in critical factors (resistor tolerance and transistor variation).

TABLE 3.40
Robust Parameter Design (RPD) Matrix Example for Electrical Engineering

Run	Resistor Value (A) (ohms)	Transistor Gain (B)	Resistor Tolerance (C)	Transistor Variation (D)
1	100	50	Low	Low
2	150	100	High	High
3	125	75	Low	Low
4	135	65	High	High
5	110	80	Low	Low
6	140	95	High	High
7	105	55	Low	Low
8	130	85	High	High

The analysis in RPD focuses on how robust the amplifier circuit's output voltage is to variations in component values and tolerance. By conducting experiments with different component combinations and levels of variation in critical factors, electrical engineers can determine the optimal design that maintains consistent and robust output voltage despite variations in component values and tolerance in electrical engineering applications.

Reporting

In RPD experimentation for electrical engineering, several statistical tools and techniques are commonly used to optimize processes, components, or systems while considering variability and reliability. Here are some of the statistical tools commonly applied:

Design of Experiments (DOE): DOE is a fundamental tool in RPD that helps plan and conduct experiments efficiently. Techniques like full-factorial, fractional factorial, CCD, and Taguchi methods are used to systematically explore the effects of factors and their interactions.

Response Surface Methodology (RSM): RSM is employed to model and optimize responses (e.g., performance, quality, reliability) as a function of multiple input variables. It helps identify optimal factor settings that minimize variability and maximize performance.

Analysis of Variance (ANOVA): ANOVA is used to assess the statistical significance of factors and their interactions. It quantifies the contributions of different factors to the variability in the response variable.

Regression Analysis: Regression models are used to develop mathematical relationships between input factors and responses. Linear regression, multiple regression, and nonlinear regression may be applied to model the system's behavior.

Signal-to-Noise (S/N) Ratios: In Taguchi methods, various S/N ratios (e.g., smaller-the-better, larger-the-better, nominal-the-best) are calculated to assess the impact of factors on performance or quality characteristics.

Optimization Algorithms: Numerical optimization algorithms like gradient descent, genetic algorithms, or simulated annealing may be used to find the optimal factor settings that minimize variability or maximize performance.

Robustness Testing: Robustness testing involves conducting experiments to evaluate the system's performance under various conditions and levels of variability. It helps assess the sensitivity of the optimized settings to real-world variations.

Monte Carlo Simulation: Monte Carlo simulations can be employed to assess the system's robustness by generating random samples of input variables and evaluating their impact on the response.

Reliability Analysis: Reliability engineering techniques are used to model and analyze the system's reliability, failure rates, and lifetime performance while considering variations in component characteristics.

Statistical Process Control (SPC): SPC techniques, such as control charts and process capability analysis, are used to monitor and control processes over time, ensuring that they remain within specified tolerances.

Quality Function Deployment (QFD): QFD is used to translate customer requirements into specific engineering characteristics, aligning design and manufacturing processes with customer expectations.

Failure Mode and Effects Analysis (FMEA): FMEA is a systematic method for identifying and prioritizing potential failure modes in electrical systems or components, helping to proactively address reliability concerns.

Weibull Analysis: Weibull distribution analysis is often used for reliability modeling, especially for predicting failure rates and analyzing time-to-failure data.

Statistical Software: Specialized statistical software packages like Minitab, JMP, or Python with libraries such as NumPy, SciPy, and scikit-learn are commonly used for data analysis, modeling, and optimization.

These statistical tools and techniques, when applied appropriately, help electrical engineers identify robust designs, optimize processes, minimize variability, and enhance the reliability and performance of electrical systems or components. The choice of tools depends on the specific problem, objectives, and available data.

A report on RPD experimentation in electrical engineering typically follows a structured format to document the process, results, and conclusions of the study. Here are the key elements that should be included in such a report:

Title Page:
> Title of the Report
> Names of Authors and Affiliations
> Date of Report Submission

Abstract:
> A concise summary of the experiment, including the problem statement, objectives, methodology, key findings, and conclusions.

Table of Contents:
> A list of sections and subsections with page numbers for easy navigation.

List of Figures and Tables:
> A compilation of all figures and tables used in the report with their respective page numbers.

Nomenclature:
> Definitions of any technical terms, acronyms, or symbols used throughout the report.

Introduction:
> Background and context of the study.
> Problem statement or engineering challenge.
> Clear statement of objectives and goals.
> Importance and relevance of the study.

Literature Review:
> A review of relevant literature, theories, and prior research related to the experiment.
> Summary of key findings from previous studies.

Methodology:
> Detailed description of the experimental design, including factors and levels.

Explanation of the statistical techniques and tools used (e.g., DOE, RSM, Taguchi).
Data collection procedures, measurement techniques, and instruments.
Any assumptions or simplifications made during the experiment.
Experimental Setup:
Description of the experimental setup, equipment, and instruments used.
Any modifications or special considerations in the setup.
Data Collection:
Presentation of raw data, including measurements and observations.
Data recording methods and data integrity checks.
Explanation of any challenges or issues encountered during data collection.
Data Analysis:
Application of statistical techniques to analyze the data.
Presentation of analysis results, including graphs, charts, and statistical tables.
Interpretation of findings, focusing on the impact of factors and interactions on the response variable.
Optimization and Robustness Assessment:
Presentation of the optimized parameter settings (if applicable).
Assessment of robustness, including sensitivity analysis and tolerance intervals.
Discussion of how the design accounts for variability and potential sources of variation.
Discussion:
Interpretation of results in the context of the research objectives.
Comparison of findings with the initial hypotheses or expectations.
Discussion of practical implications and engineering recommendations.
Conclusion:
A concise summary of the main findings and their significance.
Restatement of the research objectives and how they were achieved.
Contributions to the field of electrical engineering.
Recommendations:
Suggestions for future research or improvements to the experimental design.
Recommendations for practical applications or engineering practice.
Acknowledgments:
Recognition of individuals, organizations, or funding sources that contributed to the experiment or research.
References:
Citation of all sources, publications, and references used in the report.
Appendices:
Supplementary material, including detailed calculations, additional data tables, experimental protocols, or any other relevant information.
List of Abbreviations and Symbols (if applicable):
A comprehensive list of abbreviations and symbols used in the report.

It's important to maintain clarity, conciseness, and a logical flow throughout the report. The structure may vary depending on the specific requirements of the organization or academic institution, but these elements should be adapted to ensure a comprehensive and well-organized report on RPD experimentation in electrical engineering.

3.4.9 Randomized Experiments

Randomized design experiments in electrical engineering can be applied in various contexts to address specific research questions, improve product quality, and optimize system performance. Here are some common areas where randomized design experiments are used:

Circuit Design and Testing: Assessing the impact of component variations on circuit performance. Optimizing circuit parameters for minimal noise, power consumption, or signal quality.

Manufacturing and Quality Control: Randomized testing of electronic components and devices to ensure consistent quality. Identifying sources of variability in manufacturing processes and mitigating their effects.

Signal Processing and Communication Systems: Studying the effects of noise, interference, and signal distortion. Evaluating the performance of communication protocols and error correction techniques under varying conditions.

Reliability and Failure Analysis: Investigating the reliability and failure modes of electrical systems and components. Identifying factors contributing to component failures and designing for reliability.

Control Systems and Automation: Tuning control parameters for systems like robotics, automation, and feedback control. Assessing the impact of disturbances and uncertainties on control system behavior.

Power Systems and Grid Management: Analyzing the behavior of power distribution networks under changing load and fault conditions. Evaluating the reliability and robustness of grid components.

Electromagnetic Compatibility (EMC): Evaluating the susceptibility of electronic devices to EMI. Randomized testing to ensure compliance with EMC standards.

Hardware Design and Testing: Assessing the performance of hardware components, including integrated circuits and sensors. Identifying and mitigating sources of variability and defects.

Energy Efficiency and Power Management: Optimizing energy consumption in electrical systems and devices. Conducting REs to identify energy-saving strategies.

Machine Learning and Data Analysis: Developing and testing algorithms for data analysis and pattern recognition. Randomly varying input data for model training and testing.

In these applications, REs are typically used to systematically introduce variability, control for potential sources of bias, and assess the impact of different factors on the performance or behavior of electrical systems. Statistical tools such as analysis of ANOVA, hypothesis testing, and regression analysis are commonly employed to analyze the data collected from these experiments.

Randomized design experiments help engineers and researchers make informed decisions, optimize designs, and improve the robustness and reliability of electrical systems and devices. They provide a structured approach to understanding the effects of variability and uncertainty in engineering applications.

Steps

Developing randomized design experiments for electrical engineering involves a structured process to investigate the impact of various factors on the performance or behavior of electrical systems or components. Here are the typical steps to develop and conduct REs in electrical engineering:

1. **Define the Research Objective:**
   - Clearly articulate the research question or objective that you want to address through the experiment. What specific aspect of electrical engineering are you investigating or optimizing?
2. **Identify Factors and Levels:**
   - Identify the factors (independent variables) that may influence the outcome of the experiment. These factors could be component parameters, operating conditions, or environmental variables.
   - Determine the levels or settings at which each factor will be tested. For continuous variables, specify the range of values to be explored.

3. **Select Experimental Design:**
   - Choose an appropriate experimental design based on the number of factors, levels, and available resources. Common designs include CRD, RCBD, and Latin square designs.
   - Ensure that the chosen design allows for randomization and replication, which are essential for drawing valid conclusions.

4. **Randomization and Replication:**
   - Randomly assign experimental units (samples, trials, or tests) to different treatment combinations to minimize bias and account for uncontrolled variability.
   - Replicate the experiments by conducting multiple runs or trials under the same conditions. Replication enhances the precision of estimates and the reliability of results.

5. **Experimental Setup:**
   - Set up the electrical engineering experiment according to the defined factors and levels. Ensure that measurement instruments and data collection processes are calibrated and consistent.

6. **Data Collection:**
   - Collect data systematically by recording measurements and observations for each experimental unit.
   - Ensure that data collection is carried out in a randomized and blinded manner to minimize bias.

7. **Data Analysis:**
   - Analyze the collected data using appropriate statistical tools and techniques. Common methods include ANOVA, regression analysis, and hypothesis testing.
   - Assess the main effects of factors, interactions between factors, and their statistical significance.

8. **Interpret Results:**
   - Interpret the results of the data analysis in the context of the research objective. Determine how the factors affect the outcome and whether any optimizations or adjustments are needed.

9. **Optimization and Recommendations:**
   - If applicable, use the results to optimize electrical system parameters or configurations.
   - Make recommendations or decisions based on the findings and insights gained from the experiment.

10. **Documentation and Reporting:**
    - Prepare a comprehensive report that documents the experimental design, procedures, data analysis, and results.
    - Clearly present the findings, conclusions, and any recommendations for future work or improvements.

11. **Validation and Follow-Up:**
    - Validate the findings through further testing or verification, especially if significant changes are proposed based on the experiment.
    - Monitor the performance of the electrical system or component in practical applications to assess the real-world impact of the experiment.

12. **Continuous Improvement:**
    - Use the knowledge gained from the experiment to inform future designs and engineering decisions.
    - Consider conducting follow-up experiments or iterating on the design based on evolving requirements or constraints.

Throughout the process, it's essential to maintain rigor in the experimental design, data collection, and analysis to ensure the validity and reliability of the results. Collaboration between electrical engineers and statisticians may be valuable to design and conduct REs effectively.

PITFALLS AND REMEDIES

Randomized design experimentation in electrical engineering, like any scientific research, can face various challenges and pitfalls. Here are some common pitfalls and their corresponding remedies:

Pitfall 1: Lack of Proper Randomization
Pitfall: Failing to randomize the assignment of experimental units to treatment groups can lead to biased results.

Remedy: Ensure that randomization is rigorously applied. Use random number generators or other appropriate methods to assign treatments randomly to units.

Pitfall 2: Small Sample Size
Pitfall: Conducting experiments with too few data points or replicates can lead to results that lack statistical power.

Remedy: Increase the sample size or the number of replicates whenever possible to improve the reliability of your findings.

Pitfall 3: Inadequate Control of Confounding Variables
Pitfall: Not accounting for or controlling confounding variables (factors other than those of interest) can introduce noise into the results.

Remedy: Implement proper control measures to minimize the influence of confounding variables. This may involve holding some variables constant, blocking, or stratification.

Pitfall 4: Nonrepresentative Samples
Pitfall: Selecting a nonrepresentative sample can lead to results that do not generalize to the broader population.

Remedy: Ensure that the sample is representative of the population or system you are studying. Use random or stratified sampling methods.

Pitfall 5: Measurement Errors
Pitfall: Inaccurate or imprecise measurements can introduce noise and affect the validity of the results.

Remedy: Calibrate measurement instruments regularly and take multiple measurements to reduce measurement errors. Calculate measurement uncertainties.

Pitfall 6: Violation of Assumptions
Pitfall: Using statistical methods that assume normality or homogeneity of variances when these assumptions are not met.

Remedy: Choose statistical methods that are robust to violations of assumptions. Consider nonparametric tests if necessary.

Pitfall 7: Ignoring Interactions
Pitfall: Overlooking interactions between factors can result in an incomplete understanding of the system's behavior.

Remedy: Analyze interactions between factors using appropriate statistical techniques, such as ANOVA or regression analysis.

Pitfall 8: Confirmation Bias
Pitfall: Researchers may unintentionally interpret results to confirm their hypotheses, leading to biased conclusions.

Remedy: Use blind or double-blind experimental setups to minimize bias. Also, involve multiple researchers in data analysis and interpretation.

Pitfall 9: Lack of Reproducibility
Pitfall: If experiments are not designed to be reproducible, it may be challenging to verify results or conduct follow-up studies.

Remedy: Document experimental procedures, conditions, and equipment settings meticulously so that others can replicate the experiments.

Pitfall 10: Not Seeking Expert Advice

Pitfall: Attempting to design and conduct experiments without consulting statisticians or domain experts can lead to suboptimal designs.

Remedy: Collaborate with statisticians and electrical engineering experts who can provide valuable insights into experimental design and analysis.

Pitfall 11: Inadequate Record-Keeping

Pitfall: Poor documentation of experimental details can make it difficult to interpret and reproduce results.

Remedy: Maintain thorough records of experimental setups, procedures, and data. Use lab notebooks or electronic record-keeping systems.

Pitfall 12: Overlooking Ethical Considerations

Pitfall: Neglecting ethical considerations, especially in experiments involving human subjects or sensitive data.

Remedy: Ensure that your experiments adhere to ethical guidelines and obtain any necessary approvals from institutional review boards (IRBs) or ethics committees.

By being aware of these pitfalls and taking appropriate measures to address them, researchers in electrical engineering can enhance the validity, reliability, and impact of their randomized design experiments. Collaborating with experienced statisticians can also be highly beneficial in avoiding and mitigating these challenges.

EXAMPLE

REs in electrical engineering are used to study the effects of different factors or interventions on a particular outcome variable while ensuring that the assignment of factors to experimental units is random. Here's a simplified example of a matrix for a designed RE in electrical engineering:

Objective: To study the effects of different cooling methods on the efficiency of power electronic devices.

Experimental Units: Power electronic devices (multiple devices for each treatment group).

Factors (Treatment Groups):

1. **Air Cooling (Treatment A)**: Devices are cooled using air-cooling fans.
2. **Liquid Cooling (Treatment B)**: Devices are cooled using a liquid cooling system.

Response Variable: Efficiency of Power Electronic Device (measured as a percentage)

Randomized Experiment Matrix:

In this RE, treatments are randomly assigned to power electronic devices to investigate the effect of cooling methods on efficiency. The matrix is shown in Table 3.41.

In this matrix, each row represents a power electronic device (experimental unit), and the treatment group (air cooling or liquid cooling) is randomly assigned to each device. Multiple devices are used for each treatment group to account for variability. After applying the cooling methods, the efficiency of each device is measured and recorded.

Randomization is crucial in ensuring that any potential bias or confounding factors are minimized, allowing you to draw meaningful conclusions about the effects of different cooling methods on the efficiency of power electronic devices in electrical engineering experiments.

TABLE 3.41
Randomized Experiment Matrix
Example for Electrical Engineering

Device	Treatment Group
1	Air Cooling
2	Liquid Cooling
3	Liquid Cooling
4	Air Cooling
5	Liquid Cooling
6	Air Cooling
7	Air Cooling
8	Liquid Cooling
9	Liquid Cooling
10	Air Cooling

REPORTING

Randomized design experimentation in electrical engineering relies on various statistical tools and techniques to analyze and interpret data. Here are some commonly used statistical tools and methods:

Descriptive Statistics: Before delving into more advanced analyses, it's essential to start with descriptive statistics to summarize and visualize the data. This includes measures such as mean, median, variance, standard deviation, histograms, and box plots.

Hypothesis Testing: Hypothesis tests help assess whether observed differences or effects are statistically significant. Common tests in electrical engineering experiments include t-tests, chi-square tests, and F-tests (including ANOVA).

Analysis of Variance (ANOVA): ANOVA is used to examine the variation between groups or treatments to determine if there are statistically significant differences. It is commonly used in experiments with multiple treatment groups.

Regression Analysis: Regression models can be employed to investigate relationships between independent variables (factors) and a dependent variable (response). Linear regression and multiple regression are frequently used in electrical engineering.

Design of Experiments (DOE): DOE is a systematic approach to experimentation that helps optimize processes, identify significant factors, and reduce experimental error. It includes techniques like full-factorial design, fractional factorial design, and RSM.

Statistical Software: Specialized statistical software packages like MATLAB, R, Python (with libraries like NumPy, SciPy, and statsmodels), and commercial software like Minitab and JMP are used for data analysis and visualization.

Probability Distributions: Depending on the nature of the data, various probability distributions may be applied, including normal, binomial, Poisson, and exponential distributions.

Statistical Process Control (SPC): SPC techniques are used for monitoring and controlling manufacturing and industrial processes to ensure they meet specifications and standards.

Reliability Analysis: Electrical engineers often perform reliability analysis to estimate the probability of a device or system functioning without failure over a specified time.

Multivariate Analysis: Multivariate techniques like PCA and factor analysis can be used to explore relationships and patterns among multiple variables.

Monte Carlo Simulation: Monte Carlo simulations are valuable for modeling complex electrical systems and assessing their performance under various conditions, considering uncertainties.

Time-Series Analysis: In cases involving time-dependent data, electrical engineers may use time-series analysis to identify trends, seasonality, and forecast future values.

Statistical Hypothesis Testing: This includes a wide range of tests to assess hypotheses and make informed decisions. Examples include chi-square tests for independence, goodness-of-fit tests, and tests for normality.

Bootstrapping and Resampling: These techniques are used for estimating the sampling distribution of a statistic by resampling from the observed data. Bootstrapping is particularly useful when the underlying distribution is unknown or complex.

Bayesian Statistics: Bayesian methods, which involve updating beliefs based on prior information and observed data, are increasingly used in electrical engineering for various applications, including parameter estimation and model calibration.

The specific statistical tools and methods employed in randomized design experimentation will depend on the research question, type of data, experimental design, and goals of the study. Collaboration with statisticians or experts in statistical analysis can be valuable in selecting and implementing the most appropriate statistical techniques for a given electrical engineering experiment.

A well-structured report for randomized design experimentation in electrical engineering should convey the objectives, methods, results, and implications of the study in a clear and organized manner. Here are the essential elements to include in such a report:

1. **Title Page:**
   - Title of the Report
   - Author(s) and Affiliations
   - Date of Publication
2. **Abstract:**
   - A concise summary of the study, including the objectives, methods, key findings, and implications.
3. **Table of Contents:**
   - A list of sections and subsections with corresponding page numbers for easy navigation.
4. **List of Figures and Tables:**
   - Enumerate all figures and tables with their corresponding captions and page numbers.
5. **List of Abbreviations and Symbols:**
   - Define any acronyms, abbreviations, or symbols used throughout the report.
6. **Introduction:**
   - Provide context for the study by explaining the background, motivation, and objectives.
   - State the research questions or hypotheses that the experiment aims to address.
   - Briefly describe the importance of the study and its potential impact in the field of electrical engineering.
7. **Literature Review:**
   - Summarize relevant prior research and studies related to the experiment.
   - Discuss the theoretical foundations and existing knowledge in the area of investigation.
   - Identify gaps in the literature that the current study aims to address.
8. **Experimental Design:**
   - Explain the randomized experimental design, including the factors, levels, and treatments under investigation.
   - Detail the randomization process and procedures to ensure unbiased results.
   - Describe any control variables or covariates considered in the design.
   - Mention any statistical or software tools used for designing the experiment.
9. **Methods:**
   - Provide a comprehensive description of the experimental setup and procedures.
   - Include information about data collection, instrumentation, and measurement techniques.

- Specify the sampling procedure and sample size determination, if applicable.
- Explain any statistical analyses or models used to evaluate the data.

10. **Results:**
    - Present the collected data in a clear and organized manner using tables, figures, and charts.
    - Describe the main findings, including statistical analyses and significant outcomes.
    - Provide any statistical measures of central tendency and variability (e.g., means, standard deviations).
    - Include any relevant statistical tests, p-values, and confidence intervals.

11. **Discussion:**
    - Interpret the results and their implications in the context of the research questions.
    - Discuss any trends, patterns, or relationships observed in the data.
    - Compare the findings with prior research and explain any discrepancies or agreements.
    - Address the limitations of the study and potential sources of error.
    - Offer insights into the practical significance and real-world applications of the results.

12. **Conclusion:**
    - Summarize the main findings and their implications concisely.
    - Restate the significance of the study and its contributions to the field.
    - Suggest avenues for future research or experiments based on the current findings.

13. **References:**
    - Cite all sources and references used in the report following a recognized citation style (e.g., APA, IEEE, or a specific journal's style).

14. **Appendices:**
    - Include any supplementary material, such as detailed calculations, raw data, or additional figures and tables.
    - Provide documentation of the experimental setup, software code, or any additional information that supports the report.

15. **Acknowledgments:**
    - Express gratitude to individuals or organizations that provided assistance, funding, or resources for the experiment.

16. **Author Contributions:**
    - If multiple authors were involved, specify each author's contributions to the study.

17. **Declaration of Conflicts of Interest:**
    - Disclose any potential conflicts of interest or financial disclosures related to the study.

18. **Ethical Considerations:**
    - Address any ethical considerations, such as human or animal subjects' approval, if relevant.

It's important to follow a consistent and professional format throughout the report, including proper citation and formatting guidelines. Additionally, consider your target audience, whether it's fellow researchers, engineers, or a broader readership, and tailor the report accordingly to ensure clarity and comprehension.

3.4.10 SPLIT-PLOT AND BLOCKED EXPERIMENTS

Split-plot and blocked design experiments are used in electrical engineering when researchers need to investigate the effects of multiple factors or variables on an outcome of interest while considering specific constraints, variations, or subgroups within the experimental setup. These experimental designs help researchers efficiently allocate resources, reduce variability, and gain insights into complex systems. Here are some areas within electrical engineering where split-plot and blocked design experiments may be applied:

Semiconductor Manufacturing: Split-plot and blocked experiments can be used to optimize semiconductor manufacturing processes. Factors like temperature, pressure, and chemical concentrations may be studied in different stages of production, with certain constraints and blocking factors to ensure consistency and yield improvement.

Circuit Design and Testing: In circuit design and testing, engineers may use split-plot designs to study various component values and configurations. Blocking factors may include variations in environmental conditions or the characteristics of testing equipment.

Power Electronics: Researchers in power electronics can investigate the effects of different control strategies, switching frequencies, and component types on the efficiency and performance of power converters. Blocks could represent different load conditions or input voltage levels.

Communication Systems: In the development of communication systems and networks, engineers may use split-plot designs to evaluate the impact of different modulation schemes, channel conditions, and interference sources. Blocks might represent different geographical locations or environmental conditions.

Signal Processing: For signal processing applications, engineers can employ split-plot designs to optimize algorithms, filters, or processing techniques across multiple input signal scenarios. Blocks might represent variations in S/N ratios or noise types.

Robotic Systems: In the design of robotic systems, engineers may use split-plot experiments to analyze the performance of robotic arms or navigation algorithms under different environmental conditions. Blocks could represent variations in terrain or lighting.

Quality Control: Split-plot and blocked designs are widely used in quality control and reliability testing of electrical components and systems. These experiments help identify factors affecting product quality and optimize manufacturing processes.

Energy Systems: Researchers in electrical engineering may conduct split-plot experiments to investigate the performance of renewable energy systems (e.g., solar panels, wind turbines) under various weather conditions and geographic locations. Blocks could represent different times of the year or geographical regions.

Smart Grids: In the field of smart grids, engineers may use split-plot and blocked designs to study the impact of different grid configurations, demand response strategies, and communication protocols on grid stability and efficiency.

Wireless Sensor Networks: For wireless sensor networks, split-plot and blocked experiments can be employed to assess the performance of sensor nodes and communication protocols in different deployment scenarios, considering factors like node density and interference.

These are just a few examples of how split-plot and blocked design experiments can be applied in electrical engineering to optimize processes, improve performance, and enhance the understanding of complex systems. The specific application will depend on the research objectives and the variables of interest in each case.

Steps

Developing split-plot and blocked design experiments in electrical engineering involves several steps to plan, design, and execute the experiments effectively. Here are the typical steps:

1. **Define Objectives:** Clearly define the objectives of your experiment. What specific factors or variables do you want to study? What are the goals of your research? Understanding your objectives is crucial for designing the experiment.
2. **Identify Factors:** List all the factors or variables that may affect the outcome of interest. These factors can include controllable variables (factors of interest), blocking variables (factors to be controlled or held constant), and any other relevant variables.

3. **Select Response Variables:** Determine the response variables that measure the outcomes of interest. These variables should be relevant to your objectives and can include performance metrics, quality indicators, or other measurable outcomes.

4. **Choose Levels and Ranges:** Define the levels and ranges for each factor. Levels represent the specific values or settings at which factors will be tested. Ranges indicate the allowable variation for each factor.

5. **Identify Blocking Factors:** Identify the blocking factors that will be used to create blocks or subgroups within the experiment. Blocking factors help ensure that certain sources of variability are controlled or minimized.

6. **Determine the Design Layout:** Choose an appropriate split-plot or blocked design layout. The choice of design (e.g., split-plot, randomized block, Latin square) depends on the nature of your factors and objectives. Consult statistical software or experts if needed.

7. **Randomization:** Determine the randomization strategy for assigning experimental runs to treatments within the blocks. Randomization helps reduce bias and ensures that the experiment's results are not influenced by the order in which treatments are applied.

8. **Conduct Pre-Experiment Testing:** Before conducting the full-scale experiment, consider conducting preliminary tests or pilot experiments to refine your design and identify potential issues.

9. **Data Collection:** Conduct the experiments by applying the selected treatments to the experimental units or runs. Collect data on the response variables for each run.

10. **Data Analysis:** Analyze the collected data using appropriate statistical techniques. Common analyses include ANOVA and regression analysis. Evaluate the effects of factors, blocking variables, and their interactions on the response variables.

11. **Interpret Results:** Interpret the results of the statistical analysis. Determine which factors have significant effects on the response variables and how they interact. Draw conclusions based on your findings.

12. **Optimization:** If the goal is to optimize a process or system, use the results to identify the optimal factor settings that maximize or minimize the response variables while considering practical constraints.

13. **Documentation:** Prepare a comprehensive report that includes details of the experimental design, data collected, statistical analyses, results, conclusions, and any recommendations for further research or action.

14. **Validation:** Validate the results and conclusions through additional experiments or simulations if necessary. Ensure that the findings are robust and applicable to real-world situations.

15. **Implementation:** If the experiment leads to process improvements or changes, implement the recommended changes in the relevant electrical engineering applications.

16. **Continuous Improvement:** Consider ongoing monitoring and refinement of the process or system based on the results and lessons learned from the experiment.

It's essential to involve statisticians or experts in experimental design if you are not familiar with the statistical methods required for split-plot and blocked design experiments. Proper planning and execution of these experiments can lead to valuable insights and improvements in electrical engineering processes and systems.

PITFALLS AND REMEDIES

Split-plot and blocked design experiments can be powerful tools in electrical engineering, but they also come with potential pitfalls. Here are some common pitfalls and their corresponding remedies:

Pitfall 1: Poorly Defined Objectives

Remedy: Clearly define your research objectives before designing the experiment. Knowing your goals will guide the selection of factors, response variables, and the overall experimental approach.

Pitfall 2: Inadequate Factor Selection

Remedy: Carefully select the factors that are most likely to influence the outcomes of interest. Conduct a thorough literature review and consult with experts to identify relevant factors.

Pitfall 3: Insufficient Sample Size

Remedy: Perform a power analysis or sample size calculation to determine the minimum sample size required to detect significant effects. Ensure that your sample size is sufficient for both the main plot and subplot factors.

Pitfall 4: Poor Randomization

Remedy: Implement a robust randomization procedure to assign treatments to experimental units within blocks. This helps minimize bias and ensures that the results are not influenced by the order of testing.

Pitfall 5: Inadequate Blocking

Remedy: Choose blocking factors carefully. Blocking factors should be those that are known to introduce variability or need to be controlled. Ensure that blocking reduces the variability and enhances the precision of your estimates.

Pitfall 6: Violation of Assumptions

Remedy: Check for assumptions of normality, homogeneity of variances, and independence of errors. If assumptions are violated, consider transforming the data or using appropriate statistical tests that are robust to violations.

Pitfall 7: Overlooking Interactions

Remedy: Pay attention to interactions between main plot and subplot factors. Split-plot designs are particularly sensitive to these interactions. Interpret and report any significant interactions found in your analysis.

Pitfall 8: Insufficient Replication

Remedy: Ensure that you have an adequate number of replicates for each combination of main plot and subplot factor levels. Replication helps improve the reliability and generalizability of your results.

Pitfall 9: Neglecting Data Analysis

Remedy: Perform a thorough statistical analysis using appropriate methods such as ANOVA or regression. Interpret the results to draw meaningful conclusions.

Pitfall 10: Ineffective Communication

Remedy: Clearly communicate the experimental design, results, and conclusions in a well-structured report. Ensure that the findings are accessible to both technical and nontechnical stakeholders.

Pitfall 11: Lack of Follow-Up

Remedy: Implement changes or improvements based on the results of your split-plot or blocked design experiment. Monitor the effects of these changes and consider further experimentation or adjustments as needed.

Pitfall 12: Insufficient Documentation

Remedy: Maintain detailed records of the experimental procedures, data collection, and data analysis. Proper documentation is essential for replicability and transparency.

Pitfall 13: Neglecting Continuous Improvement

Remedy: Treat the experiment as part of a continuous improvement process. Use the findings to refine processes, systems, or designs, and consider conducting follow-up experiments to further optimize.

To avoid these pitfalls, it's advisable to collaborate with experienced statisticians or experts in experimental design, particularly when dealing with complex split-plot and blocked designs. Proper planning, execution, and analysis are essential for obtaining reliable and meaningful results in electrical engineering experiments.

EXAMPLE

Split-plot and blocked experiments in electrical engineering involve studying the effects of multiple factors while considering constraints or restrictions in the experimental setup. These designs divide the experiment into different subplots or blocks. Here's a simplified example of a matrix for a split-plot and blocked experiment in electrical engineering:

Objective: To optimize the manufacturing process of PCBs by studying the effects of two factors (temperature and chemical concentration) while accounting for variations in different production lines (Block A and Block B).

Factors:

1. **Temperature (Factor A)**:
   * Low Temperature
   * High Temperature
2. **Chemical Concentration (Factor B)**:
   * Low Concentration
   * High Concentration

Blocks:

* **Block A**: Production Line A
* **Block B**: Production Line B

Response Variable: PCB Yield Percentage (measured in percentage)

Split-Plot and Blocked Experiment Matrix:
In this example, a split-plot and blocked experiment is designed to assess the impact of temperature and chemical concentration on PCB yield while also considering the influence of different production lines (Blocks). The matrix is shown in Table 3.42.

TABLE 3.42

Split-Plot and Blocked Experiment Matrix for Electrical Engineering

Run	Temperature (A)	Chemical Concentration (B)	Block
1	Low	Low	A
2	High	Low	A
3	Low	High	A
4	High	High	A
5	Low	Low	B
6	High	Low	B
7	Low	High	B
8	High	High	B

In this matrix, each row represents a specific experimental run with a unique combination of temperature and chemical concentration levels, while also specifying the production line (Block) where the experiment is conducted.

Split-plot and blocked experiments are valuable in electrical engineering when you need to account for factors that cannot be easily randomized, such as variations in different production lines. This design allows you to assess the main effects and interactions of temperature and chemical concentration on PCB yield while also considering the potential influence of different production lines.

REPORTING

Split-plot and blocked design experiments in electrical engineering typically involve the use of various statistical tools and techniques for data analysis. Here are some of the commonly used statistical tools:

Analysis of Variance (ANOVA): ANOVA is a fundamental tool for split-plot and blocked design experiments. It helps assess the significance of factors and interactions, allowing you to determine whether there are statistically significant differences between treatments.

Factorial ANOVA: Factorial ANOVA is used to analyze experiments with multiple factors. It helps identify the main effects of factors and interactions between them.

Regression Analysis: Regression models can be applied to quantify the relationship between independent variables (factors) and the response variable. Regression can be simple linear regression or multiple regression when multiple factors are involved.

Residual Analysis: Residuals are the differences between observed and predicted values. Residual analysis helps check for violations of assumptions, such as normality and homoscedasticity.

Normal Probability Plots: These plots are used to assess the normality of residuals. Deviations from a straight line on the normal probability plot may indicate departures from normality.

Interaction Plots: Interaction plots visually represent the interactions between factors. They help in understanding how the effects of one factor depend on the levels of another factor.

Main Effects Plots: Main effects plots show the individual effects of each factor on the response variable, allowing you to identify which factors have a significant impact.

Statistical Software: Statistical software packages like R, Python with libraries like SciPy and statsmodels, or specialized software like Minitab or JMP are commonly used to perform the statistical analysis.

Power Analysis: Power analysis helps determine the required sample size to achieve a desired level of statistical power, ensuring that the experiment can detect significant effects.

Randomization Tests: These nonparametric tests are used when the assumptions of normality or equal variances are violated. Examples include the Kruskal-Wallis test for comparing groups and the Friedman test for repeated measures.

Effect Size Measures: Measures like Cohen's d or eta-squared are used to quantify the magnitude of differences or the proportion of variance explained by factors and interactions.

Confidence Intervals: Confidence intervals provide a range within which population parameters, such as means or regression coefficients, are likely to fall. They offer a measure of uncertainty.

Post hoc Tests: When ANOVA or regression analysis reveals significant differences, post hoc tests like Tukey's HSD or Bonferroni corrections can identify which specific groups or factors differ significantly.

Diagnostic Plots: These plots, such as residual plots, leverage graphical methods to assess the adequacy of the model and identify outliers or influential observations.

Design of Experiments (DOE) Software: Specialized DOE software, like Design-Expert or JMP, offers tools for experimental design, analysis, and visualization, making it easier to plan and execute split-plot and blocked experiments.

The choice of statistical tools will depend on the specific nature of the experiment, the design used, and the research questions being addressed. In many cases, a combination of these tools will be employed to conduct a comprehensive analysis of split-plot and blocked design experiments in electrical engineering.

A well-structured and informative report for split-plot and blocked design experimentation in electrical engineering should convey the experiment's purpose, design, results, and conclusions clearly. Here are the typical elements of such a report:

Title: A concise and descriptive title that summarizes the main objective of the experiment.

Abstract: A brief summary of the entire report, including the research question, methodology, key findings, and conclusions.

Introduction:
- Background: Provide context for the experiment, explaining the problem or phenomenon being studied.
- Objectives: State the specific objectives or hypotheses of the experiment.
- Rationale: Explain why this experiment is important or relevant to electrical engineering.

Literature Review: A review of relevant literature and previous research in the area, highlighting any theories or findings that relate to your experiment.

Experimental Design:
- Description of Split-Plot and Blocked Design: Explain the experimental design, including the rationale for using split-plot and blocked design.
- Factors and Levels: List and describe the factors and their levels, both within the main plots and subplots.
- Randomization: Explain how randomization was applied to ensure unbiased results.
- Blocking: Describe the blocking structure, including the rationale for using blocking factors.

Methods:
- Experimental Procedure: Detail the steps taken during the experiment, including data collection methods and equipment used.
- Sampling: Explain how samples were selected or collected, including the criteria for inclusion.
- Data Analysis: Describe the statistical methods used for data analysis, including any software or tools employed.
- Sample Size and Power Analysis: If applicable, discuss how sample size and power calculations were determined.

Results:
- Presentation of Data: Present the collected data in a clear and organized manner, using tables, graphs, and figures.
- Statistical Analysis: Describe the statistical tests and analyses performed, including ANOVA, regression, or other relevant methods.
- Interpretation: Provide interpretations of the results, focusing on significant findings, main effects, interactions, and trends.
- Post hoc Analysis: If applicable, report the results of post hoc tests to identify specific group differences.

Discussion:
- Comparison with Hypotheses: Discuss whether the results align with the initial hypotheses or research questions.
- Practical Implications: Explain the practical implications of the findings in the context of electrical engineering.
- Limitations: Address any limitations of the study, including potential sources of error or bias.
- Future Research: Suggest directions for future research or experiments related to the topic.

Conclusion: Summarize the key findings and their significance. Restate the main conclusions of the experiment.

References: List all the sources, articles, and literature cited in the report following a specific citation style (e.g., APA, IEEE, or others).

Appendices: Include any additional materials that support the report, such as detailed data tables, supplementary figures, or mathematical derivations.

Acknowledgments: Acknowledge individuals, organizations, or institutions that contributed to the experiment but do not meet the criteria for authorship.

Author Information: Provide information about the authors, including their names, affiliations, and contact details.

Graphs and Tables: Ensure that all graphs and tables are appropriately labeled and titled, with clear axis labels and legends.

Formatting and Style: Follow a consistent style guide (e.g., IEEE, APA) for formatting, citations, and references.

Review and Proofreading: Thoroughly review and proofread the report to eliminate errors in grammar, spelling, and formatting.

Ethical Considerations: Address any ethical considerations related to the experiment, such as informed consent for human subjects or ethical approval for animal research.

Overall, a well-organized report will help readers understand the experimental process, the rationale behind it, the results obtained, and the implications for electrical engineering research or applications.

3.5 MECHANICAL ENGINEERING

In mechanical engineering, designed experiments are valuable tools for optimizing processes, improving product designs, and gaining insights into system performance. Designed experiments are a crucial tool in mechanical engineering for optimizing processes, improving product quality, and gaining a deeper understanding of how different factors influence outcomes. Here's how designed experiments are used in mechanical engineering:

Process Optimization: Mechanical engineers often work on processes that involve manufacturing, assembly, or testing. Designed experiments help optimize these processes by systematically varying input factors (such as temperature, pressure, speed, or material properties) and measuring the resulting output or response (e.g., product dimensions, performance, defects). By analyzing the data generated from these experiments, engineers can identify the optimal settings for factors that lead to improved efficiency and product quality.

Product Design and Development: In product design, mechanical engineers use experiments to evaluate different design parameters and their impact on product performance. By conducting experiments, engineers can refine designs, choose materials, and make informed decisions about how to balance competing objectives such as strength, weight, and cost.

Quality Improvement: Designed experiments are used to identify and address quality issues in mechanical systems or products. By systematically varying factors related to the manufacturing process, engineers can pinpoint the sources of defects or variations and develop strategies to reduce or eliminate them. This leads to better quality control and consistency in production.

Reliability and Durability Testing: Mechanical systems and components must meet specific reliability and durability requirements. Experiments are used to simulate and assess the behavior of these systems under different conditions, including extreme or accelerated testing. This helps engineers predict the lifespan of mechanical components and make design improvements to enhance reliability.

Material Testing and Selection: Materials play a critical role in mechanical engineering. Experiments are conducted to test the mechanical properties of different materials (e.g., tensile strength, hardness, fatigue resistance) and to determine their suitability for specific applications. These experiments inform material selection for various components.

Environmental Testing: Mechanical engineers may need to assess how mechanical systems or components perform in different environmental conditions, such as temperature extremes, humidity, or corrosive environments. Designed experiments can simulate these conditions and evaluate their effects on performance and durability.

Cost Reduction: By systematically analyzing factors that influence manufacturing costs, engineers can identify opportunities for cost reduction. This may involve adjusting process parameters, optimizing material usage, or minimizing waste.

Failure Analysis: When mechanical components fail, designed experiments can be used to investigate the root causes of failure. Engineers can replicate the conditions leading to failure and study the effects, helping them develop solutions and prevent future failures.

Energy Efficiency: Mechanical engineers are increasingly focused on energy efficiency. Designed experiments can be used to assess the impact of different design choices, materials, and operating conditions on energy consumption in mechanical systems.

Designed experiments in mechanical engineering consider process variables. Process variables in mechanical engineering refer to the parameters, conditions, or factors that influence or characterize a mechanical process. These variables play a crucial role in designing, optimizing, and controlling mechanical systems and processes. Here are some common process variables in mechanical engineering:

Temperature (T): The degree of thermal energy within a system or material, often measured in degrees Celsius ($^\circ$C) or Kelvin (K).

Pressure (P): The force exerted per unit area, typically measured in pascals (Pa), atmospheres (atm), or bar (bar).

Flow Rate (Q): The rate at which a fluid (liquid or gas) flows through a conduit or system, measured in volume per unit of time (e.g., liters per second, cubic meters per hour).

Velocity (V): The speed and direction of a moving object or fluid, measured in meters per second (m/s) or other suitable units.

Mass Flow Rate ($\dot{m}$): The rate at which mass flows through a system, typically measured in kilograms per second (kg/s).

Density (ρ): The mass per unit volume of a substance, often measured in kilograms per cubic meter (kg/m^3).

Viscosity (μ): A measure of a fluid's resistance to flow, typically measured in pascal-seconds (Pa·s) or poise (P).

Humidity (Relative Humidity - RH): The amount of moisture present in the air relative to the maximum amount it could hold at a given temperature, expressed as a percentage.

Force (F): The interaction that changes the motion or shape of an object, measured in newtons (N).

Torque (τ): The rotational force applied to an object, measured in newton-meters (N·m).

Stress (σ): The internal resistance of a material to deformation under an applied load, measured in pascals (Pa) or megapascals (MPa).

Strain (ε): The measure of deformation resulting from an applied force, typically expressed as a dimensionless ratio.

Speed (n): The rate of rotation of a mechanical component, such as an engine or turbine, measured in revolutions per minute (RPM) or radians per second (rad/s).

Time (t): The duration of a process or event, typically measured in seconds (s), minutes (min), or hours (h).

Power (P): The rate at which work is done or energy is transferred, measured in watts (W) or horsepower (hp).

Torque Speed (T-N): The relationship between torque and speed in a mechanical system, often used in motor and gearbox characterization.

Electric Current (I): The flow of electric charge in a conductor, measured in amperes (A).

Voltage (V): The electric potential difference between two points in a circuit, measured in volts (V).

Frequency (f): The number of cycles or oscillations per unit of time, often measured in hertz (Hz).

Material Properties: Properties such as thermal conductivity, thermal expansion coefficient, Young's modulus, and Poisson's ratio characterize the behavior of materials under mechanical loads and temperature variations.

These process variables are essential for modeling, analyzing, and controlling mechanical systems and processes in various applications, including manufacturing, aerospace, automotive, energy production, and more. Understanding and manipulating these variables are critical for achieving desired outcomes and optimizing mechanical engineering processes. These variables can be categorized into several common types, including:

1. **Independent Variables:**
   - **Material Properties:** Variables related to the mechanical properties of materials used in mechanical components or products, such as tensile strength, hardness, elasticity, and thermal conductivity.
   - **Design Parameters:** Variables related to the design and geometry of mechanical systems, including dimensions, shapes, and tolerances.
   - **Operational Parameters:** Variables related to the operation of mechanical systems, such as speed, pressure, temperature, flow rate, and load.
   - **Manufacturing Process Variables:** Variables related to manufacturing processes, including machining parameters (e.g., cutting speed, feed rate), welding parameters, and surface treatment conditions.
2. **Dependent Variables:**
   - **Mechanical Performance:** Variables related to the mechanical behavior and performance of mechanical components or systems, such as stress, strain, deformation, vibration, and fatigue life.
   - **Thermal Performance:** Variables related to heat transfer, temperature distribution, and thermal stress in mechanical systems.
   - **Fluid Flow Performance:** Variables related to fluid dynamics and flow characteristics, such as pressure drop, flow velocity, and turbulence.
   - **Energy Efficiency:** Variables related to energy consumption and efficiency in mechanical systems, such as efficiency ratios and power consumption.

- **Noise and Vibration**: Variables related to noise levels and vibration amplitudes, which can affect the comfort and safety of mechanical products.
3. **Categorical Variables**:
   - **Material Types**: Categories of materials used in mechanical components, such as metals, polymers, ceramics, and composites.
   - **Manufacturing Processes**: Categories of manufacturing processes employed, such as CNC machining, injection molding, or additive manufacturing.
   - **Product Features**: Categories of product features, configurations, or designs that can affect performance and functionality.
4. **Control Variables (Covariates)**:
   - These are variables that are held constant or controlled during experiments to eliminate their influence on the dependent variable. For example, in a study of heat transfer, the temperature of the surrounding environment may be a controlled variable.
5. **Random Variables**:
   - In some cases, random variables may be introduced to account for variability or uncertainty in measurements or environmental conditions.
6. **Interaction Variables**:
   - Interaction variables are used to investigate the combined effects of two or more independent variables on the dependent variable. For example, the interaction between temperature and pressure in a heat exchanger's performance.
7. **Noise Variables (Error Variables)**:
   - These are uncontrolled or unmeasured variables that can introduce variability or errors into the experimental results. Robust experimental designs aim to minimize the impact of noise variables.

Designed experiments in mechanical engineering are often used to optimize designs, materials, and processes by systematically varying these variables to understand their effects and make informed decisions in the design, manufacturing, and operation of mechanical systems and products. These experiments can help engineers achieve better performance, reliability, and efficiency in mechanical engineering applications.

Overall, designed experiments are a powerful tool for mechanical engineers to systematically and efficiently explore relationships between factors and responses, leading to improved processes, products, and systems. These experiments provide valuable data for data-driven decision-making and continuous improvement in mechanical engineering applications.

This section covers common types of designed experiments with examples of their applications in mechanical engineering with the following considerations:

- Definition of the DoX method
- Steps on how to perform
- Potential pitfalls and remedies
- Examples of the methods including a sample matrix
- Writing the report (statistics, outline, and critique example)

3.5.1 FACTORIAL EXPERIMENTS

In mechanical engineering, a factorial experiment is a type of designed experiment that systematically explores the effects of multiple factors (independent variables) on one or more responses (dependent variables) of interest. Factorial experiments are used to analyze how different factors interact with each other and influence the outcomes. These experiments are valuable for optimizing processes, improving product designs, and understanding the relationships between variables. Here are the key characteristics of a factorial experiment in mechanical engineering:

Factors: Factors are the independent variables that can influence the outcome of the experiment. In mechanical engineering, factors can include parameters such as temperature, pressure, speed, material properties, geometry, and more. Each factor can have multiple levels or settings.

Levels: Levels represent the different settings or values that each factor can take during the experiment. For example, if the factor is temperature, the levels could be low, medium, and high temperatures.

Experimental Design: Factorial experiments are designed to systematically vary the factors and their levels. The design matrix specifies how each combination of factor levels is tested. Full-factorial experiments test all possible combinations of factor levels, while fractional factorial designs test a subset of the combinations to reduce the number of experimental runs.

Responses: Responses are the dependent variables or outcomes that engineers measure and analyze. Responses can include product dimensions, performance metrics, defect rates, or any other variables of interest in mechanical engineering.

Data Collection: During the experiment, data is collected for each combination of factor levels. The data includes measurements or observations related to the responses.

Analysis: Statistical analysis is applied to the collected data to determine how each factor and their interactions affect the responses. This analysis helps identify significant factors, optimal settings, and any interactions between factors that may impact the outcomes.

Optimization: Based on the results of the factorial experiment, engineers can optimize processes or product designs by selecting the factor levels that lead to desired outcomes, such as improved performance, reduced defects, or cost savings.

Factorial experiments are valuable tools for mechanical engineers because they allow for a systematic exploration of multiple factors and their interactions, which can be complex in mechanical systems. By conducting these experiments, engineers can gain insights into the underlying relationships between variables and make informed decisions to improve processes, products, and designs in mechanical engineering applications.

If a manufacturing company wants to optimize the performance of a new machining process. They vary factors like cutting speed, feed rate, and tool material at different levels to determine their effects on product quality and process efficiency.

Steps

Developing a factorial designed experiment in mechanical engineering involves several key steps to systematically plan, conduct, and analyze the experiment. Here are the typical steps for developing a factorial experiment:

1. **Define Objectives and Scope:** Clearly define the objectives of the experiment. What specific problem or question are you trying to address in mechanical engineering? Determine the scope of the experiment, including the factors to be studied and the responses of interest.
2. **Identify Factors and Levels:** Identify the independent variables (factors) that may influence the mechanical system or process. Factors can include parameters like temperature,

pressure, speed, material properties, or design variables. Determine the levels or settings for each factor. These are the different values or ranges at which each factor will be tested.

3. **Select Experimental Design:** Choose the appropriate experimental design for your objectives. This could be a full-factorial design (testing all possible factor combinations) or a fractional factorial design (testing a subset of factor combinations to reduce the number of runs). Consider any blocking or randomization schemes if applicable.

4. **Design Matrix:** Create a design matrix that specifies the factor levels for each experimental run. This matrix outlines which combinations of factor settings will be tested.

5. **Conduct Experiments:** Perform the experiments according to the design matrix, carefully controlling and recording the conditions and measurements for each run. Ensure that the experiments are conducted under controlled and consistent conditions to minimize external sources of variation.

6. **Collect Data:** Record and collect data for the responses of interest during each experimental run. Ensure accurate and reliable data collection.

7. **Statistical Analysis:** Analyze the collected data using appropriate statistical techniques. This may include ANOVA to assess the effects of factors and their interactions on the responses. Identify significant factors, interactions, and any patterns in the data.

8. **Interpret Results:** Interpret the results in the context of your mechanical engineering objectives. Determine how factors and their interactions impact the mechanical system or process. Consider the practical implications of the findings.

9. **Optimization and Decision-Making:** If the objective is optimization, use the results to identify the optimal factor settings that lead to desired outcomes in mechanical engineering (e.g., improved performance, reduced defects, cost savings). Make data-driven decisions based on the analysis to improve processes, products, or designs.

10. **Report and Documentation:** Prepare a detailed report summarizing the experiment, including objectives, methods, results, and conclusions. Document the experimental procedures, data, and statistical analyses for future reference.

11. **Implementation:** Implement any recommended changes or improvements in the mechanical system, process, or product based on the experiment's findings.

12. **Continuous Improvement:** Consider the results of the experiment as part of an ongoing process of continuous improvement in mechanical engineering. Use feedback from the experiment to refine processes and designs over time.

Factorial designed experiments in mechanical engineering provide valuable insights into the relationships between factors and responses, helping engineers optimize systems, improve product designs, and make informed decisions. Proper planning, execution, and analysis are essential to ensure the experiment's success and relevance to engineering objectives.

PITFALLS AND REMEDIES

Factorial designed experiments are powerful tools in mechanical engineering for understanding the relationships between factors and responses. However, like any research methodology, they come with their own set of pitfalls and challenges. Here are some common pitfalls and remedies when conducting factorial designed experiments in mechanical engineering:

Pitfall 1: Inadequate Factor Selection

> **Pitfall:** Including irrelevant or unnecessary factors in the experiment can increase complexity and resource requirements.

> **Remedy:** Carefully choose factors based on their relevance to the mechanical system or process being studied. Conduct a thorough literature review and engage experts to help identify the most critical factors.

Pitfall 2: Factor Interaction Complexity
    Pitfall: Factorial experiments can uncover complex interactions between factors that are challenging to interpret.
    Remedy: Consider using fractional factorial designs to reduce the number of factor combinations tested while still capturing significant interactions. Visualize interactions using plots or diagrams to aid interpretation.

Pitfall 3: Insufficient Replicates
    Pitfall: Conducting too few replicates for each combination of factor levels may lead to unreliable results and limited statistical power.
    Remedy: Determine the appropriate number of replicates through power calculations or pilot studies to ensure adequate statistical reliability. More replicates can help improve the precision of estimates.

Pitfall 4: Noise and Variability
    Pitfall: External sources of variability can introduce noise into the experiment, making it difficult to detect real effects.
    Remedy: Control environmental conditions and other sources of variability as much as possible. Randomize experimental runs and use appropriate statistical techniques to separate true effects from noise.

Pitfall 5: Limited Factor Range
    Pitfall: Factors are tested within a narrow range that may not cover relevant operating conditions.
    Remedy: Ensure that factor levels represent the full range of practical and relevant values for the mechanical system or process. Extrapolate results cautiously if necessary.

Pitfall 6: Overlooking Nonlinear Effects
    Pitfall: Assuming linear relationships between factors and responses when nonlinear effects exist can lead to incorrect conclusions.
    Remedy: Consider using higher-order factorial designs (e.g., 2^3 or 2^4) to capture nonlinear effects. Experiment with various factor settings to detect and model nonlinearities.

Pitfall 7: Ignoring Time-Dependent Effects
    Pitfall: Neglecting time-dependent factors in dynamic mechanical systems may lead to incomplete understanding.
    Remedy: If time-dependent effects are suspected, incorporate time as a factor in the experiment or use time-series data collection techniques.

Pitfall 8: Misinterpretation of Results
    Pitfall: Misinterpreting statistical results or drawing incorrect conclusions from the data.
    Remedy: Collaborate with a statistician or data analyst to ensure proper data analysis and interpretation. Document the analysis and assumptions made.

Pitfall 9: Failure to Implement Findings
    Pitfall: Failing to implement recommended changes or improvements based on the experiment's results.
    Remedy: Develop an implementation plan and involve relevant stakeholders to ensure that findings lead to practical improvements in mechanical systems, processes, or designs.

Pitfall 10: Neglecting Validation
    Pitfall: Not validating the findings of the factorial experiment through further testing or validation studies.
    Remedy: Conduct follow-up validation experiments to confirm the robustness of the results and their applicability in real-world situations.

Careful planning, rigorous execution, and thorough analysis are essential for successful factorial designed experiments in mechanical engineering. Addressing these pitfalls and applying appropriate remedies can help ensure the reliability and relevance of the experiment's findings.

EXAMPLE

Here's an example of a factorial designed experiment in mechanical engineering:

Objective: Improve the tensile strength of a metal alloy used in aerospace components.

Factors:

1. **Temperature:** Levels include low, medium, and high temperatures.
2. **Aging Time:** Levels include short, medium, and long aging times.
3. **Alloy Composition:** Levels include varying proportions of alloying elements (e.g., nickel, chromium).

Response: Tensile strength of the metal alloy (measured in megapascals, MPa).

Experimental Design:

- A full-factorial design is used, which means all possible combinations of factor levels are tested.
- This results in a total of 3 (temperature levels) x 3 (aging time levels) x 3 (alloy composition levels)=27 experimental runs.

Experimental Procedure:

- Prepare specimens of the metal alloy according to the desired alloy composition.
- Conduct the experiments in a controlled environment, varying the temperature, aging time, and alloy composition according to the factorial design.
- Perform tensile strength tests on each specimen to measure the response variable.

Data Collection:

- Record the tensile strength measurements for each combination of factor levels.

Statistical Analysis:

- Perform ANOVA to determine the main effects of temperature, aging time, and alloy composition on tensile strength.
- Analyze interaction effects to understand how combinations of factors influence tensile strength.
- Identify optimal factor settings that maximize tensile strength.

Interpretation of Results:

- Discover that higher aging time and a specific alloy composition lead to increased tensile strength.
- Observe interactions between temperature and alloy composition that affect tensile strength differently at different aging times.

Optimization:

- Based on the results, engineers can choose the optimal combination of temperature, aging time, and alloy composition to maximize tensile strength for aerospace applications.

- Implement these findings in the production process to improve the quality and performance of aerospace components.

This example demonstrates how a factorial designed experiment can be used in mechanical engineering to systematically study the effects of multiple factors on a critical mechanical property (tensile strength) of a material. The experiment provides valuable insights for optimizing the alloy composition and manufacturing process to meet specific performance requirements.

MATRIX EXAMPLE

In a factorial designed experiment in mechanical engineering, a matrix is used to specify the combinations of factor levels to be tested. Here's an example of a matrix for a 2^3 full-factorial experiment, where there are three factors, each with two levels (low and high):

Factors:

- Temperature (T) - Low (T1) and High (T2)
- Pressure (P) - Low (P1) and High (P2)
- Speed (S) - Low (S1) and High (S2)

The matrix represents all possible combinations of these factor levels to be tested (Table 3.43).

Each row in the matrix represents an experimental run, and the corresponding factor levels are specified in columns. This matrix allows you to conduct a full-factorial experiment with eight runs, testing all possible combinations of low and high levels for each of the three factors (2^3 = 8). By systematically varying these factors, you can assess their individual effects and interactions on the mechanical system or process under investigation.

REPORTING

Statistical tools are essential for analyzing data from factorial designed experiments in mechanical engineering. These tools help engineers understand the effects of factors and interactions on the mechanical system or process being studied. Here are some common statistical tools used in factorial designed experiments:

Analysis of Variance (ANOVA): ANOVA is a fundamental tool for assessing the significance of factors and interactions. It helps determine whether the variation in the response variable is mainly due to the factors being studied or random variation.

TABLE 3.43

Factorial Designed Experiment Matrix Example for Mechanical Engineering

Run	Temperature (T)	Pressure (P)	Speed (S)
1	T1	P1	S1
2	T1	P1	S2
3	T1	P2	S1
4	T1	P2	S2
5	T2	P1	S1
6	T2	P1	S2
7	T2	P2	S1
8	T2	P2	S2

Main Effects Plots: Main effects plots visualize the impact of individual factors on the response variable. They provide a clear view of which factors have a significant influence.

Interaction Plots: Interaction plots display the effects of interactions between two or more factors. They help identify cases where the impact of one factor depends on the level of another.

Factorial Plots: Factorial plots show how the response variable varies across different combinations of factor levels. They are useful for understanding the overall response surface.

Response Surface Methodology (RSM): RSM is a collection of statistical techniques used to model and optimize responses as functions of multiple factors. It helps engineers find optimal factor settings for desired responses.

Contour Plots: Contour plots represent three-dimensional response surfaces in two dimensions. They provide a visual representation of how responses change with different factor combinations.

Normal Probability Plots: Normal probability plots help assess the normality of the residuals (differences between observed and predicted values). Deviations from normality can indicate issues with the model.

Regression Analysis: Regression analysis is used to build mathematical models that describe the relationship between factors and responses. Linear and nonlinear regression models can be employed.

Confidence Intervals: Confidence intervals provide a range of values within which the true mean of the response variable is likely to fall. They quantify the uncertainty associated with the estimates.

Diagnostic Plots: Diagnostic plots, such as residual plots, can help identify outliers, nonconstant variance, and other issues with the data or model.

Hypothesis Testing: Hypothesis tests, including t-tests and F-tests, can be used to test specific hypotheses about factors and interactions.

Fractional Factorial Analysis: In cases where conducting a full-factorial experiment is not feasible due to resource constraints, fractional factorial analysis is used to identify the most significant factors and interactions.

Power Analysis: Power analysis helps determine the sample size required to achieve a certain level of statistical power, ensuring that the experiment can detect meaningful effects.

Design of Experiments (DOE) Software: Specialized DOE software packages provide tools for experimental design, data analysis, and visualization.

The specific choice of statistical tools and techniques depends on the complexity of the experiment, the nature of the factors and responses, and the goals of the analysis. Engineers often use a combination of these tools to gain a comprehensive understanding of the factors affecting their mechanical systems or processes and to optimize them for desired outcomes.

A well-structured report for a factorial designed experiment in mechanical engineering should provide a clear and comprehensive account of the experiment's objectives, methods, results, and conclusions. Here are the key elements that should be included in such a report:

Title Page:
- Title of the Report
- Name(s) of the Author(s)
- Affiliation(s) and Contact Information
- Date of Report Submission

Abstract: A concise summary of the experiment's objectives, methods, major findings, and conclusions. It should provide a quick overview of the report's content.

Table of Contents: A list of sections, subsections, and their respective page numbers for easy navigation.

List of Figures and Tables: An enumeration of all figures and tables used in the report with their respective captions and page numbers.

List of Abbreviations and Symbols: If applicable, provide a list of abbreviations, acronyms, and symbols used throughout the report and their explanations.

Introduction:

- Clearly state the objectives and goals of the experiment.
- Provide background information and context for the study, explaining why it is important and relevant to mechanical engineering.
- Define the factors and response variables being studied.

Literature Review: Summarize relevant literature and previous research related to the factors and response variables. Highlight any existing knowledge gaps that the experiment aims to address.

Experimental Design: Describe the experimental setup, including equipment, materials, and environmental conditions. Explain the choice of factors, their levels, and the experimental design (e.g., full factorial, fractional factorial, or other). Detail the procedures followed to conduct the experiment, including randomization and data collection.

Statistical Analysis: Present the statistical methods used for data analysis (e.g., ANOVA, regression analysis, RSM). Describe how factor effects, interactions, and significance levels were determined. Include relevant plots, charts, or diagrams to illustrate the analysis.

Results: Present the experimental results in a clear and organized manner. Use tables, figures, and graphs to display data, including means, standard deviations, and other relevant statistics. Discuss any patterns, trends, or significant findings observed in the results.

Discussion:

- Interpret the results in the context of the experiment's objectives.
- Analyze the implications of significant factors and interactions on the mechanical system or process.
- Discuss any unexpected findings or limitations of the experiment.
- Compare the results to previous research and relevant literature.

Conclusions: Summarize the main findings of the experiment. State whether the objectives were achieved and if the hypotheses were supported. Discuss the practical implications of the results for mechanical engineering applications.

Recommendations: Offer recommendations for further research or improvements based on the experiment's findings. Suggest potential engineering solutions or modifications based on the results.

Acknowledgments: Acknowledge any individuals, organizations, or funding sources that contributed to the experiment's success.

References: List all references and sources cited in the report, following a specified citation style (e.g., APA, IEEE, Chicago).

Appendices: Include any supplementary information, data tables, or detailed calculations that support the report but are not included in the main text.

Experimental Data: Optionally, provide raw data and experimental logs in an organized format, either within the report or as a separate file.

Remember to adhere to any specific formatting and style guidelines provided by your institution or organization when preparing the report. Clear and concise writing, along with well-organized content, is essential to effectively communicate the results and implications of your factorial designed experiment in mechanical engineering.

3.5.2 Fractional Factorial Experiments

A fractional factorial experiment in mechanical engineering is a type of experimental design used to investigate and analyze the effects of multiple factors or variables on a particular outcome or response while reducing the number of experiments required. This approach is particularly useful when there are a large number of factors to consider and conducting a full-factorial experiment (testing all possible combinations of factor levels) would be impractical or too time-consuming.

In a fractional factorial experiment, only a fraction or subset of the total number of experimental runs is conducted. This subset is carefully selected to capture the most critical and influential factors while reducing the experimental workload. The design is typically chosen based on specific statistical principles, such as the use of OAs or other design matrices. The following are the aspects of a fractional factorial experiment:

Identify the factors: In mechanical engineering, factors could be variables like temperature, pressure, material properties, design parameters, or any other variables that might affect the mechanical performance of a system or component.

Define the levels: Determine the different levels or settings at which each factor will be tested. For example, for temperature, you might have low, medium, and high levels.

Choose the fractional factorial design: Select a suitable fractional factorial design matrix that specifies which combinations of factor levels to test. The design matrix will allow you to conduct a subset of experiments while still capturing the main effects and interactions between factors.

Conduct experiments: Perform the experiments according to the chosen design matrix. Collect data on the mechanical performance or response variable of interest.

Analyze the results: Use statistical analysis techniques to interpret the data and determine the effects of each factor and their interactions. This analysis helps you understand which factors are significant and how they influence mechanical behavior.

Fractional factorial experiments are valuable in mechanical engineering for optimizing processes, improving designs, and identifying the key factors that affect a system's performance. They help save time and resources compared to full-factorial experiments while providing valuable insights into the relationships between variables. A common example in automotive engineering, a car manufacturer may use a fractional factorial design to study the effects of various factors (e.g., tire pressure, suspension settings, and aerodynamic changes) on fuel efficiency without testing all possible combinations.

Steps

Developing a fractional factorial designed experiment for mechanical engineering involves several steps to plan and execute the experiment effectively. Here are the general steps you can follow:

1. **Define the objectives:** Clearly state the objectives of your experiment. What specific mechanical performance or outcome are you trying to optimize or understand? This could be related to product design, manufacturing processes, material properties, or any other mechanical aspect.
2. **Identify the factors:** Determine the factors or variables that may influence the mechanical outcome. These could include parameters like temperature, pressure, speed, material properties, dimensions, and others. Make a list of all potential factors.
3. **Define factor levels:** Specify the range or levels at which each factor will be tested. For example, if you are studying the effect of temperature, you might choose three levels: low, medium, and high.

4. **Select the resolution:** Decide on the level of resolution for your fractional factorial design. Resolution refers to the ability of the design to detect the main effects and interactions between factors. Higher-resolution designs provide more detailed information but may require more experimental runs. Common resolutions include III, IV, or V.

5. **Choose the design matrix:** Based on the number of factors, levels, and desired resolution, select an appropriate fractional factorial design matrix. You can use software tools like Minitab, JMP, or DOE software to generate these matrices.

6. **Generate the experimental plan:** Use the chosen design matrix to generate the experimental plan, which specifies the factor combinations to be tested. The plan will indicate which experiments to conduct and in what order.

7. **Conduct the experiments:** Follow the experimental plan and conduct the experiments according to the specified factor levels. Ensure that the experiments are conducted under controlled conditions to minimize variability.

8. **Collect and record data:** Record data meticulously for each experimental run. Ensure that measurements are accurate and consistent. Pay attention to any unexpected variations or anomalies during the experiments.

9. **Analyze the results:** Use statistical analysis techniques to analyze the collected data. Calculate the main effects and interactions between factors. Determine which factors have a significant impact on the mechanical outcome and how they interact.

10. **Draw conclusions and make recommendations:** Based on the analysis, draw conclusions about the effects of the factors on mechanical performance. Make recommendations for optimizing the process or system, improving product design, or addressing any identified issues.

11. **Verify and validate findings:** If necessary, conduct additional experiments or validation tests to confirm the results and ensure their reliability.

12. **Document the experiment:** Maintain thorough documentation of the experiment, including the design matrix, experimental plan, data records, analysis methods, and conclusions. This documentation is essential for future reference and communication of the findings.

By following these steps, you can successfully develop and execute a fractional factorial designed experiment in mechanical engineering to gain valuable insights into the factors affecting mechanical systems and make informed decisions for improvement or optimization.

Pitfalls and Remedies

Fractional factorial designed experiments are a valuable tool in mechanical engineering for efficiently studying multiple factors and their effects on a mechanical system. However, like any experimental approach, they can come with pitfalls and challenges. Here are some common pitfalls and remedies when conducting fractional factorial experiments in mechanical engineering:

Pitfall 1: Alias effects and confounding

Issue: Alias effects occur when the chosen fractional factorial design matrix does not allow for the separation of certain main effects and interactions, making it difficult to identify which factors are truly influential.

Remedy: Choose an appropriate design: Select a design matrix with a suitable resolution to minimize aliasing. Higher-resolution designs (e.g., IV or V) can help resolve more main effects and interactions but may require more experimental runs. Carefully consider the trade-off between resolution and the number of experiments.

Pitfall 2: Limited resources and time

Issue: Conducting a full-factorial experiment to study all factors and interactions may be impractical due to limited resources or time constraints.

Remedy: Prioritize factors and interactions: Identify the most critical factors and interactions that are likely to have the greatest impact on your mechanical system. Focus your fractional factorial experiment on these factors while keeping others at constant or nominal levels.

Pitfall 3: Insufficient replication

Issue: Running each experiment only once may lead to unreliable results and difficulty in estimating experimental error.

Remedy: Incorporate replication: Include replication within your experimental design. Running multiple replicates of each experiment at the same factor level can help improve the precision of your estimates and provide more robust results.

Pitfall 4: Neglecting noise factors

Issue: Sometimes, uncontrolled or unmeasured factors (noise factors) can affect the mechanical outcome and introduce variability.

Remedy: Identify and control noise factors: Identify potential noise factors that may affect the response variable and try to control or minimize their impact during the experiments. If some noise factors cannot be controlled, consider conducting a separate experiment to study their effects.

Pitfall: Inadequate sample size

Issue: Having too few experimental runs may result in insufficient data to draw meaningful conclusions or detect small effects.

Remedy: Determine an appropriate sample size: Use statistical power analysis or other methods to estimate the required sample size for your fractional factorial experiment based on the expected effect sizes and the desired level of significance. Ensure that your sample size is adequate to detect the effects of interest.

Pitfall 6: Ignoring interaction effects

Issue: Focusing solely on the main effects and neglecting interaction effects can lead to incomplete or inaccurate conclusions about the factors' influence on the mechanical system.

Remedy: Analyze interaction effects: Perform a thorough analysis of interaction effects between factors. Interaction effects can be just as important as main effects in understanding the behavior of the mechanical system. Visualize and interpret interaction plots to gain insights.

Pitfall 7: Lack of expertise in statistical analysis

Issue: Conducting fractional factorial experiments requires knowledge of statistical analysis techniques, and inadequate expertise can lead to misinterpretation of results.

Remedy: Seek statistical expertise: Collaborate with statisticians or experts in experimental design and analysis to ensure that your experimental design is sound and the data analysis is performed correctly. Training in statistics can also be beneficial for engineers conducting these experiments.

By being aware of these potential pitfalls and implementing the suggested remedies, engineers in mechanical engineering can conduct more robust and informative fractional factorial designed experiments to improve their understanding of mechanical systems and make informed decisions for optimization and design improvements.

EXAMPLE

The following is an example of a fractional factorial designed experiment in mechanical engineering:

Objective: To optimize the design of a mechanical component, specifically a bracket used in an automotive suspension system, by studying the effects of various factors on its strength.

Factors:

1. Material type (Factor A): Steel, Aluminum, Titanium
2. Bracket thickness (Factor B): 4 mm, 6 mm, 8 mm
3. Hole diameter (Factor C): 10 mm, 12 mm, 14 mm
4. Heat treatment (Factor D): None, Annealed, Quenched

Response variable: Tensile strength of the bracket (measured in megapascals, MPa).

Fractional Factorial Design: Due to resource constraints, use a fractional factorial design with resolution IV, to study the main effects of each factor and all two-way interactions.

Design Matrix: The design matrix for this experiment is shown in Table 3.44.

Experimental Plan: Conduct these 10 experiments, each corresponding to a different combination of factors. For each experiment, establish a bracket according to the specified material, thickness, hole diameter, and heat treatment, and then test tensile strength.

Analysis: After conducting the experiments and measuring the tensile strength for each bracket, perform a statistical analysis to:

- Calculate the main effects of each factor (A, B, C, D) on tensile strength.
- Identify any significant two-way interactions between factors.
- Determine which factors and interactions have the most significant impact on the bracket's strength.

Results: Analysis reveals that:

- Material type (Factor A) has the most significant impact on tensile strength, with steel outperforming aluminum and titanium.
- Bracket thickness (Factor B) also has a significant effect, with thicker brackets generally exhibiting higher strength.
- Hole diameter (Factor C) and heat treatment (Factor D) have significant interactions with material type (Factor A), indicating that the choice of material influences the effect of these factors on strength.

TABLE 3.44

Mechanical Engineering Fractional Factorial Designed Experiment Matrix Example

Run	Material (A)	Thickness mm (B)	Hole Diameter mm (C)	Heat Treatment (D)
1	Steel	4	10	--
2	Aluminum	4	12	Annealed
3	Titanium	4	14	Quenched
4	Steel	6	12	Quenched
5	Aluminum	6	14	--
6	Titanium	6	10	Annealed
7	Steel	8	14	Annealed
8	Aluminum	8	10	Quenched
9	Titanium	8	12	--
10	Steel	4	14	Quenched

Based on these findings, informed decisions about the material selection, thickness, hole diameter, and heat treatment process can be made to optimize its strength while considering cost and weight constraints in automotive suspension design.

REPORTING

Statistical tools play a crucial role in analyzing and interpreting the results of fractional factorial designed experiments in mechanical engineering. Here are some of the key statistical tools and techniques commonly used in such experiments:

Analysis of Variance (ANOVA): ANOVA is a fundamental statistical technique used to partition the total variation in the response variable into contributions from different factors and their interactions. It helps determine which factors are statistically significant and their respective effects on the response.

Main Effects Plots: These graphical representations display the main effects of each factor on the response variable. Main effects plots help visualize how changes in factor levels influence the response.

Interaction Plots: Interaction plots illustrate the interactions between pairs of factors. They are useful for understanding how the effects of one factor depend on the levels of another factor. Interaction plots can reveal complex relationships between factors.

Half-Normal Plots: Half-normal plots are used for model selection in fractional factorial experiments. They help identify which effects are significant by comparing the absolute values of the estimated effects to a critical line.

Residual Analysis: Residuals are the differences between observed and predicted values. Analyzing the residuals helps check the assumptions of the experimental model, such as normality and constant variance, and can indicate outliers or influential data points.

Diagnostic Plots: Diagnostic plots, such as normal probability plots and scatterplots of residuals versus predicted values, help assess the validity of the statistical model and identify potential issues with the data or model assumptions.

Fractional Factorial Design Matrices: Design matrices are used to plan and organize the experiments. They specify which factor combinations to test and are generated based on mathematical principles to achieve the desired resolution and confounding pattern.

Regression Analysis: Regression models can be used to build predictive models for the response variable based on the factors and interactions identified as significant in the experiment. These models can be used for optimization and prediction.

Effect Size Measures: Measures like Cohen's d, eta-squared, and partial eta-squared can quantify the magnitude of effects in the experiment. They help assess the practical significance of significant factors.

Statistical Software: Utilize statistical software packages such as Minitab, JMP, R, or Python with libraries like NumPy, SciPy, and statsmodels to perform the necessary statistical analyses and generate plots and results.

Post Hoc Tests: In cases where interactions are significant, post hoc tests like Tukey's HSD or Bonferroni tests may be used to compare specific factor levels and identify where significant differences exist.

Power Analysis: Power analysis is used to determine the statistical power of the experiment, which assesses the probability of detecting true effects if they exist. It helps in determining if the sample size is adequate.

Confidence Intervals: Calculating confidence intervals for effect estimates provides a range within which the true effect is likely to fall. This information adds to the interpretation of the results.

These statistical tools and techniques are essential for designing, conducting, and analyzing fractional factorial designed experiments in mechanical engineering, helping engineers make informed decisions, optimize processes, and improve mechanical systems.

A well-structured report for a fractional factorial designed experiment in mechanical engineering is crucial for communicating the experiment's objectives, methods, results, and conclusions to stakeholders, colleagues, and decision-makers. Here are the key elements typically included in such a report:

Title Page:
- Title of the report.
- Name(s) of the author(s).
- Affiliation(s).
- Date of the report.

Abstract: A concise summary of the entire report, including the objectives, key findings, and conclusions. The abstract should provide a quick overview of the experiment's purpose and outcomes.

Table of Contents: A list of sections and subsections with page numbers for easy navigation.

List of Figures and Tables: A separate list that enumerates all the figures and tables used in the report with their respective page numbers.

Introduction:
- Provide context and background information.
- State the objectives and purpose of the experiment.
- Describe the significance of the experiment and its relevance to mechanical engineering.

Experimental Design: Explain the fractional factorial design used, including resolution, factors, levels, and the design matrix. Discuss any constraints or limitations in the design.

Materials and Methods:
- Detail the materials used in the experiment, including specifications and sources.
- Describe the experimental setup and procedures followed for data collection.
- Mention any instrumentation and measurement techniques.

Data Collection: Present the raw data obtained during the experiments. Include tables, charts, or graphs that illustrate the data. Provide information on replication and data quality control measures.

Data Analysis: Discuss the statistical methods and tools used to analyze the data. Present the results of the analysis, including main effects, interaction effects, and any statistically significant findings. Include appropriate statistical tests, p-values, and effect size measures. Use graphical representations (plots, charts) to aid in the interpretation of results.

Discussion:
- Interpret the results in the context of the experiment's objectives.
- Discuss the practical implications of the findings for mechanical engineering applications.
- Address any unexpected results or trends.
- Compare the results with prior research or industry standards, if relevant.

Conclusion: Summarize the main findings and their implications. Restate the significance of the experiment in the context of mechanical engineering. Offer recommendations or insights for future work or applications.

References: Cite all relevant sources, including research papers, books, manuals, and software used for the experimental design and data analysis.

Appendices (if necessary): Include supplementary information, such as detailed statistical calculations, experimental protocols, additional data tables, or any other supporting documentation.

Acknowledgments (optional): Acknowledge individuals, organizations, or funding sources that contributed to the experiment's success.

List of Abbreviations and Symbols (if used): Provide explanations for any abbreviations or symbols used throughout the report.

Glossary (if necessary): Define any technical terms or terminology specific to mechanical engineering that may not be familiar to all readers.

Ensure that the report is well-organized, with clear and concise language, and that the content flows logically from one section to the next. Proper formatting, use of headings and subheadings, and clear labeling of figures and tables are essential for readability and understanding.

3.5.3   Central Composite Design (CCD)

A CCD experiment is a type of experimental design used in mechanical engineering and other fields to study the relationships between multiple factors or variables and optimize a process or system. CCD is an RSM technique that involves the exploration of factor space to find optimal settings for desired outcomes. It is especially valuable when the response variable of interest is influenced by continuous factors and interactions between those factors.

Here are the key characteristics and components of a CCD experiment:

Factors: In a CCD experiment, there are typically two types of factors: the primary factors of interest and any nuisance or control factors. The primary factors are the variables you want to optimize or study, such as material properties, dimensions, or process parameters. Control factors are variables you want to keep constant during the experiment.

Factor Levels: Each primary factor is studied at different levels or settings. The levels are typically chosen based on engineering knowledge and the expected range of values that are relevant to the problem at hand.

Central Point: CCD experiments include a central point, which represents the center of the factor space and typically corresponds to the nominal or baseline conditions. This central point helps estimate the curvature of the response surface.

Factorial Points: CCD designs include factorial points, which are combinations of factor levels that allow for the assessment of main effects and two-way interactions among the factors. These points are placed at a certain distance from the central point.

Axial Points: Axial points are additional experimental runs located at a fixed distance from the central point along each primary factor's axis. These points help assess curvature and quadratic effects in the response surface.

Response Variable: The response variable is the mechanical or performance parameter that you want to optimize or understand. It can be measured in terms of quality, efficiency, strength, or any other relevant metric.

Experimental Runs: A CCD design involves conducting a series of experimental runs, each corresponding to a specific combination of factor levels. These runs are carried out according to the chosen design matrix, which determines the factor settings for each run.

Response Surface: The primary goal of a CCD experiment is to create a response surface, which is a mathematical model that describes how the response variable changes with variations in the primary factors. This surface is often quadratic or higher-order, allowing for the identification of optimum factor settings that yield the desired response.

Analysis: Statistical analysis techniques, such as regression analysis and ANOVA, are used to fit the response surface model to the experimental data. This analysis helps identify significant factors, interactions, and the optimal factor settings that maximize or minimize the response variable.

Optimization: Once the response surface model is established, engineers can use it to predict the best factor settings for achieving desired outcomes. Optimization techniques like desirability functions or numerical optimization methods can be applied to find the optimum conditions.

CCD experiments are valuable tools in mechanical engineering for optimizing processes, improving product designs, and understanding the complex relationships between factors. They allow engineers to efficiently explore factor space and make data-driven decisions for achieving the desired mechanical performance or quality. An example of CCD is applied to optimize the shape and performance of an aerospace component, like an airfoil. Engineers vary parameters like wing curvature and angle of attack to create a response surface and find the best design.

STEPS

Developing a CCD experiment in mechanical engineering involves a series of steps to plan and execute the experiment systematically. Here are the general steps you can follow:

1. **Define the Objectives:** Clearly state the objectives of your CCD experiment. What specific mechanical performance or outcome are you trying to optimize or understand? Define the goals and criteria for success.
2. **Identify the Factors:** Determine the factors or variables that may influence mechanical performance or outcome. These factors could include parameters like material properties, dimensions, process settings, or any other variables relevant to your problem.
3. **Define Factor Levels:** Specify the range and levels at which each primary factor will be tested. Determine the low and high values for each factor. These levels should encompass the expected practical range of values.
4. **Select the Central Point:** Choose a central point that represents the nominal or baseline conditions. The central point is typically located at the midpoint of the factor range and corresponds to the starting or reference values for the factors.
5. **Generate the Design Matrix:** Use software tools, such as specialized CCD design software or statistical packages (e.g., Design-Expert, Minitab, or R), to generate the design matrix. The design matrix specifies the combinations of factor levels for each experimental run, including central, factorial, and axial points.
6. **Conduct the Experiments:** Perform the experimental runs according to the design matrix. Each run corresponds to a specific set of factor levels. Ensure that the experiments are carried out systematically and consistently, taking care to control any nuisance or control factors.
7. **Measure and Record Data:** Collect data on the mechanical performance or outcome variable for each experimental run. Ensure accurate and precise measurements. Record any relevant information, such as environmental conditions or equipment settings.
8. **Analyze the Data:** Perform statistical analysis on the collected data to develop the response surface model. Fit the model to the experimental results using regression analysis or other appropriate techniques. Determine the significance of main effects, interactions, and curvature terms.
9. **Visualize the Response Surface:** Create response surface plots and contour plots to visualize how the response variable changes across the range of factor levels. These plots help in understanding the relationships between factors and identifying optimal settings.
10. **Optimize the System:** Use the response surface model to identify the factor settings that optimize the mechanical performance or outcome variable. Employ optimization techniques to find the optimal conditions, such as desirability functions or numerical optimization methods.

11. **Validate and Verify:** Conduct validation experiments to confirm the predicted optimal settings. Validate the response surface model by comparing predicted values with experimental results from new runs. Verify that the optimized conditions yield the desired mechanical performance.
12. **Document the Experiment:** Maintain thorough documentation of the CCD experiment, including the design matrix, experimental plan, data records, analysis methods, response surface model, and optimization results. This documentation is essential for future reference and reporting.
13. **Draw Conclusions and Make Recommendations:** Summarize the main findings and conclusions of the CCD experiment. Provide recommendations for improving mechanical performance, optimizing processes, or making design changes based on the results.

By following these steps, you can successfully develop and execute a CCD experiment in mechanical engineering, leading to improved understanding and optimization of mechanical systems and processes.

PITFALLS AND REMEDIES

CCD experiments are powerful tools in mechanical engineering for optimizing processes, improving designs, and understanding the relationships between factors and outcomes. However, they can also present challenges and pitfalls that need to be addressed. Here are some common pitfalls and potential remedies when conducting CCD experiments in mechanical engineering:

Pitfall 1: Insufficient Factor Range:
If the selected factor levels do not cover the full practical range, the response surface may not accurately represent the system's behavior.

Remedy: Wider Factor Range
Carefully choose factor levels that encompass the expected practical range of values. Consult engineering knowledge and historical data to ensure comprehensive coverage.

Pitfall 2: Nonlinearity and Curvature:
CCD assumes a quadratic response surface, but real mechanical systems may exhibit nonlinear behavior.

Remedy: Consider Higher-Order Models:
If the initial analysis suggests nonlinear behavior, consider using higher-order models (e.g., cubic or quartic) or more advanced RSMs to capture curvature accurately.

Pitfall 3: Missing Factors:
Neglecting to include important factors in the experiment can lead to incomplete or inaccurate conclusions.

Remedy: Factor Identification:
Conduct a thorough factor screening or sensitivity analysis to identify potentially significant factors before designing the CCD experiment. Ensure that all influential factors are included.

Pitfall 4: Inadequate Replication:
Running too few replicates for each combination of factor levels can result in unreliable estimates of the response variable.

Remedy: Increase Replication:
Include an adequate number of replicates for each experimental run to reduce variability and improve the precision of estimates.

Pitfall 5: Overfitting the Model:
Fitting overly complex models to the data can lead to overfitting, where the model performs well on the training data but poorly on new data.

Remedy: Use Model Selection Criteria:
Employ model selection criteria (e.g., AIC, BIC) to choose a model that balances goodness-of-fit and model complexity. Cross-validation can also help assess model performance.

Pitfall 6: Ignoring Interaction Effects:
Focusing solely on the main effects and neglecting interaction effects can result in incomplete insights.

Remedy: Analyze Interactions:
Investigate interaction effects between factors, as they can be important in understanding system behavior. Use interaction plots and statistical tests to identify and interpret interactions.

Pitfall 7: Neglecting Constraints:
Mechanical systems often have constraints or physical limitations that should be considered during optimization.

Remedy: Incorporate Constraints:
Include constraints in the optimization process. Constraint-handling techniques can be applied to ensure that optimized solutions meet physical limitations.

Pitfall 8: Lack of Validation:
Failing to validate the optimized conditions or the response surface model can lead to unreliable results in practical applications.

Remedy: Validation Experiments:
Conduct validation experiments to confirm that the predicted optimal conditions indeed yield the desired mechanical performance. Validate the response surface model by comparing predictions with new experimental data.

Pitfall 9: Inadequate Documentation:
Poor documentation of the experiment, including design, data, and analysis, can hinder reproducibility and future reference.

Remedy: Comprehensive Documentation:
Maintain thorough documentation of the entire CCD experiment, from the design matrix to the optimization process. This documentation is crucial for reporting and sharing results with colleagues and stakeholders.

By being aware of these potential pitfalls and implementing the suggested remedies, engineers can conduct more robust and informative CCD experiments in mechanical engineering, leading to improved processes, designs, and mechanical systems.

EXAMPLE

The following is an example of a CCD experiment in mechanical engineering:

Objective: To optimize the design of a hydraulic cylinder used in heavy machinery to achieve maximum lifting force while minimizing manufacturing costs.

Factors:

Cylinder Diameter (Factor A):
- Low Level: 10 cm
- High Level: 15 cm
- Central Point: 12.5 cm

Piston Rod Diameter (Factor B):
- Low Level: 2 cm

- High Level: 4 cm
- Central Point: 3 cm

Hydraulic Fluid Pressure (Factor C):
- Low Level: 2,000 psi
- High Level: 4,000 psi
- Central Point: 3,000 psi

A CCD matrix typically consists of three parts:

Factorial Points: These points allow for the assessment of main effects and two-way interactions between factors.

Central Points: These represent the center of the design space and provide information on curvature in the response surface.

Axial Points: These points extend beyond the factorial points along each factor's axis to capture curvature effects.

Creating a CCD matrix for optimizing the design of a hydraulic cylinder in heavy machinery involves specifying factor levels, including the low, high, and central points, as well as axial points to capture curvature in the response surface. Here's a CCD matrix example with five factorial points, two central points, and four axial points, totaling 11 experimental runs (Table 3.45).

Each run in this matrix corresponds to a unique combination of factor levels, allowing you to conduct experiments and measure the lifting force for each condition. The combination of central, factorial, and axial points helps in building a response surface model and optimizing the hydraulic cylinder design for maximum lifting force while minimizing manufacturing costs.

Analysis: Use statistical analysis techniques to fit a response surface model to the experimental data. The model will describe how the lifting force depends on cylinder diameter (A), piston rod diameter (B), and hydraulic fluid pressure (C). Identify significant main effects and interactions.

Results: The analysis reveals the following insights:

- Increasing cylinder diameter (Factor A) and hydraulic fluid pressure (Factor C) have a positive linear effect on lifting force.

TABLE 3.45

Central Composite Design (CCD) Designed Experiment Matrix Example

Run	Cylinder Diameter (A) cm	Piston Rod Diameter (B) cm	Hydraulic Pressure (C) psi
1	10	2	2,000
2	15	2	2,000
3	10	4	2,000
4	15	4	2,000
5	12.5	3	2,000
6	10	2	4,000
7	15	2	4,000
8	10	4	4,000
9	15	4	4,000
10	12.5	3	4,000
11	12.5	3	3,000

- Increasing piston rod diameter (Factor B) has a positive linear effect on lifting force but also exhibits a significant interaction with hydraulic fluid pressure.
- The optimal combination of factors for maximum lifting force while controlling costs is determined based on the response surface model.

Optimization: Use the response surface model to find the optimal factor settings that maximize lifting force while keeping manufacturing costs within budget constraints. This involves solving a constrained optimization problem.

Validation: Conduct validation experiments to verify that the predicted optimal settings indeed result in the desired lifting force. Compare the predictions with the actual experimental results.

Conclusion: Summarize the main findings, including the optimized design parameters for the hydraulic cylinder. Discuss the implications for improving lifting performance while managing manufacturing costs in heavy machinery.

This example illustrates how a CCD experiment can be applied in mechanical engineering to optimize the design of a hydraulic cylinder for improved performance and cost-effectiveness.

REPORTING

CCD experiments in mechanical engineering require various statistical tools and techniques to design experiments, analyze data, and draw meaningful conclusions. Here are some of the key statistical tools commonly used in CCD experiments for mechanical engineering:

Design of Experiments (DOE): DOE principles guide the planning and execution of CCD experiments. This includes choosing factors, defining factor levels, and generating the design matrix.

Regression Analysis: Regression analysis is used to build response surface models that describe how the response variable (e.g., mechanical performance) depends on the factors and their interactions. Linear, quadratic, and higher-order regression models are common.

Analysis of Variance (ANOVA): ANOVA is applied to assess the significance of factors and interactions. It helps identify which factors have a significant impact on the response variable.

Normal Probability Plots: Normal probability plots are used to check the assumption of normality for the residuals of the response surface model. Departures from normality can indicate model inadequacies.

Response Surface Plots: Response surface plots visualize the response variable's behavior as a function of two or three factors while holding others constant. They provide insights into the system's behavior and help identify optimal factor settings.

Contour Plots: Contour plots show lines of constant response on a 2D plane of two factors. These plots are useful for visualizing factor interactions and identifying regions of optimal performance.

Factorial Plots: Factorial plots display main effects and interaction effects of individual factors on the response variable. They help identify the direction and magnitude of factor effects.

Diagnostic Plots: Diagnostic plots, such as residual plots and Q-Q plots, are used to check the validity of the response surface model and identify outliers or influential data points.

Coefficient Plots: Coefficient plots display the estimated coefficients of the regression model, indicating the direction and significance of each factor's effect.

Model Selection Criteria: Criteria like AIC () and BIC assist in selecting the appropriate model complexity (e.g., linear, quadratic, or higher-order) by balancing goodness-of-fit and model parsimony.

Optimization Algorithms: Numerical optimization techniques, such as gradient descent, genetic algorithms, or simulated annealing, are used to find optimal factor settings based on the response surface model.

Desirability Functions: Desirability functions allow you to combine multiple response variables and constraints into a single objective function, simplifying the optimization process.

Robust Design Techniques: Methods like Taguchi's robust design can be applied to make the system less sensitive to variations in factors and external conditions, ensuring performance consistency.

Experimental Software: Specialized software packages like Design-Expert, Minitab, or R with relevant packages (e.g., 'RSM' package) are often used to design experiments, perform regression analysis, and create graphical representations of results.

Power Analysis: Power analysis is used to determine the statistical power of the experiment, assessing the probability of detecting true effects if they exist. It helps in determining if the sample size is adequate.

These statistical tools and techniques are essential for designing, conducting, and analyzing CCD experiments in mechanical engineering, enabling engineers to optimize processes, improve designs, and gain insights into complex mechanical systems.

A well-structured report for a CCD experiment in mechanical engineering is crucial for effectively communicating the experiment's objectives, methods, results, and conclusions to various stakeholders, colleagues, and decision-makers. Here are the key elements typically included in such a report:

Title Page:
- Title of the report.
- Names and affiliations of the author(s).
- Date of the report.

Abstract: A concise summary of the entire report, including the objectives, key findings, and conclusions. The abstract provides a quick overview of the experiment's purpose and outcomes.

Table of Contents: A list of sections and subsections with page numbers for easy navigation.

List of Figures and Tables: A separate list enumerating all the figures and tables used in the report with their respective page numbers.

List of Abbreviations and Symbols (if used): Definitions and explanations for any technical abbreviations or symbols employed throughout the report.

List of Nomenclature (if applicable): Definitions and explanations of variables, parameters, and terms specific to the mechanical engineering context.

Introduction:
- Provide context and background information for the CCD experiment.
- State the objectives and purpose of the experiment.
- Explain the significance of the experiment in the context of mechanical engineering.

Experimental Design: Describe the CCD design, including the choice of factors, factor levels, and the design matrix. Mention any constraints or limitations in the design.

Materials and Methods:
- Detail the materials used in the experiment, including specifications and sources.
- Explain the experimental setup, data collection procedures, and instrumentation used.
- Provide information on how the CCD design matrix was generated.

Data Collection: Present the raw data obtained during the CCD experiments. Include tables, charts, or graphs that illustrate the data. Specify any data quality control measures taken.

Data Analysis: Describe the statistical methods and tools used to analyze the data. Present the results of the analysis, including the response surface model, significant factors, and interactions.

Response Surface Plots and Visualizations: Include response surface plots, contour plots, or other visualizations to help readers understand the relationships between factors and the response variable.

Optimization and Conclusions: Discuss the optimization results and the optimal factor settings for achieving the desired mechanical performance. Summarize the main findings and conclusions drawn from the CCD experiment.

Discussion:
- Interpret the results in the context of the experiment's objectives.
- Discuss the practical implications of the findings for mechanical engineering applications.
- Address any unexpected results or trends.
- Compare the results with prior research or industry standards, if relevant.

Recommendations and Future Work: Provide recommendations based on the experiment's results, such as design modifications or process improvements. Suggest directions for future research or additional experiments to further enhance mechanical performance.

References: Cite all relevant sources, including research papers, books, manuals, and software used for the CCD experiment's design and data analysis.

Appendices (if necessary): Include supplementary information, such as detailed statistical calculations, additional data tables, or any other supporting documentation.

Acknowledgments (optional): Acknowledge individuals, organizations, or funding sources that contributed to the success of the CCD experiment.

Thoroughly organize the report, use clear and concise language, and ensure that content flows logically from one section to the next. Proper formatting, headings, and clear labeling of figures and tables are essential for readability and understanding.

3.5.4 Taguchi Methods

The Taguchi Method, developed by Japanese engineer and statistician Genichi Taguchi, is a robust optimization technique used in mechanical engineering and various other fields to improve the quality and performance of products and processes. It aims to find optimal parameter settings that are less sensitive to variations or noise factors, resulting in products and processes that are robust and reliable. The Taguchi Method is particularly valuable when optimizing complex systems with multiple factors and interactions.

Here are the key components and steps involved in the Taguchi Method:

Factors and Levels: Identify the factors (design parameters) that influence the performance or quality of the product or process. These factors can include material properties, dimensions, process settings, and more. Define the levels or settings at which each factor will be tested. Factors are typically tested at multiple levels, including low, medium, and high settings.

Control Factors and Noise Factors: Distinguish between control factors and noise factors. Control factors are the variables you can manipulate to improve performance, while noise factors are sources of variability that you cannot control but want to minimize the impact of.

Orthogonal Arrays (OA): Choose an appropriate OA based on the number of factors and levels. OAs are tables that specify the combination of factor levels to be tested in each experiment. They are designed to efficiently capture the main effects and interactions among factors.

Experimental Runs: Conduct a series of experimental runs according to the chosen OA, where each run represents a specific combination of factor levels. Randomize the order of experiments to minimize the influence of uncontrolled factors.

Response Variable: Select a response variable that measures the performance or quality of the product or process. This variable is typically quantifiable, such as dimensional accuracy, strength, or efficiency.

Signal-to-Noise (S/N) Ratios: Calculate the S/N ratios for the response variable. The S/N ratios are used to evaluate the performance and quality of each experimental run. There are different types of S/N ratios depending on the objective, such as smaller-the-better, larger-the-better, and nominal-the-best.

Analysis of S/N Ratios: Analyze the S/N ratios to determine which factor levels lead to the best performance or quality. The goal is to identify the factor settings that maximize the S/N ratio. The Taguchi Loss Function quantifies the loss in quality or performance as the S/N ratio deviates from the target value.

Confirmation Runs: Conduct confirmation experiments to validate the optimal factor settings identified through the analysis. These experiments help ensure that the improvements are robust and not the result of chance.

Implementation and Control: Implement the optimal factor settings in the product or process. Establish control measures to ensure that variations and noise factors are minimized during regular production.

Continuous Improvement: Continuously monitor and fine-tune the process or product based on ongoing data collection and analysis. The Taguchi Method encourages a culture of continuous improvement.

The Taguchi Method is known for its systematic and efficient approach to optimization and robust design. It allows engineers to find optimal factor settings that are less sensitive to variations, leading to improved product quality, reduced variability, and enhanced reliability in mechanical engineering applications.

STEPS

Developing a Taguchi method designed experiment in mechanical engineering involves a series of steps to systematically plan and execute the experiment. Here are the key steps:

1. **Define the Problem and Objectives:** Clearly define the problem or aspect of the mechanical system that needs improvement. State the specific objectives you want to achieve through the Taguchi experiment, such as improving performance, reducing variability, or optimizing a design.

2. **Identify the Factors and Levels:** Identify the factors (design parameters) that may affect the performance or quality of the mechanical system. These could include material properties, dimensions, process settings, etc. Define the levels or settings at which each factor will be tested. Factors are typically tested at multiple levels, including low, medium, and high settings.

3. **Distinguish between Control and Noise Factors:** Differentiate between control factors and noise factors. Control factors are the variables you can manipulate to improve performance, while noise factors are sources of variation that you cannot control but want to minimize their impact.

4. **Select an Orthogonal Array (OA):** Choose an appropriate OA based on the number of factors and levels. The OA determines the combinations of factor levels to be tested in each experiment. It is designed to efficiently capture the main effects and interactions among factors.

5. **Determine the Number of Experimental Runs:** Calculate the number of experimental runs required based on the OA chosen. Ensure that the number of runs is feasible within the available resources and time.

6. **Design the Experiments:** Create a plan for the experimental runs, specifying the factor levels for each run based on the selected OA. Randomize the order of experiments to minimize the influence of uncontrolled factors.

7. **Conduct the Experiments:** Perform the experimental runs according to the plan. Ensure that each run is carried out precisely and consistently.

8. **Measure the Response Variable:** Select a response variable that quantifies the performance or quality of the mechanical system. Accurately measure the response variable for each experimental run.

9. **Calculate Signal-to-Noise (S/N) Ratios:** Calculate the appropriate S/N ratio(s) for the response variable. The choice of S/N ratio depends on the objective (e.g., smaller-the-better, larger-the-better, or nominal-the-best).

10. **Analyze the Results:** Analyze the S/N ratios to identify the factor levels that lead to the best performance or quality. Use graphical tools, such as main effect plots and interaction plots, to understand the influence of each factor and their interactions.

11. **Determine the Optimal Settings:** Determine the optimal factor settings that maximize the chosen S/N ratio(s) based on the analysis. These settings represent the solution to the problem.

12. **Conduct Confirmation Runs:** Perform confirmation experiments to validate the optimal factor settings. Ensure that the improvements are robust and not due to chance.

13. **Implement and Control:** Implement the optimal factor settings in the mechanical system, process, or design. Establish control measures to maintain the improvements over time and minimize the impact of noise factors during regular operation.

14. **Continuous Improvement:** Continuously monitor and collect data from the improved system or process. Use ongoing data analysis to identify opportunities for further optimization and continuous improvement.

15. **Documentation and Reporting:** Maintain comprehensive documentation of the Taguchi experiment, including the design matrix, results, and findings. Prepare a report that summarizes the experiment, the steps taken, the optimal factor settings, and the impact on the mechanical system's performance or quality.

By following these steps, engineers can effectively apply the Taguchi Method to address mechanical engineering challenges, optimize designs, and improve the quality and reliability of mechanical systems.

Pitfalls and Remedies

While the Taguchi Method is a powerful tool for improving product and process quality in mechanical engineering, it is not without its potential pitfalls. Being aware of these pitfalls and applying appropriate remedies can lead to more successful Taguchi experiments. Here are some common pitfalls and their corresponding remedies:

Pitfall 1: Insufficient Factor Identification:
Failing to identify and include all relevant factors that affect the performance or quality of the mechanical system.

Remedy: Comprehensive Factor Identification:
Conduct a thorough factor screening or sensitivity analysis before designing the experiment to identify potential factors. Consult with domain experts to ensure all critical factors are considered.

Pitfall 2: Inadequate Factor Levels:

Choosing factor levels that do not adequately cover the practical range of values can lead to suboptimal results.

Remedy: Wide Factor-Level Selection:

Ensure that the chosen factor levels span the expected practical range, including high and low extremes. Use engineering knowledge and historical data to guide level selection.

Pitfall 3: Neglecting Noise Factors:

Not accounting for noise factors or uncontrolled variables that can affect the response variable, leads to misleading conclusions.

Remedy: Noise Factor Consideration:

Identify and assess potential noise factors that may impact the experiment. If possible, control or minimize the impact of noise factors during the experiments or in the analysis.

Pitfall 4: Inappropriate Orthogonal Array (OA) Selection:

Selecting an OA that is not suitable for the number of factors or interactions in the experiment, leading to an incomplete or inadequate design.

Remedy: Proper OA Selection:

Carefully choose an OA that matches the number of factors and interactions in your experiment. Consult Taguchi tables or software to ensure the selected OA is appropriate.

Pitfall 5: Inadequate Sample Size:

Using a sample size that is too small to detect significant effects or patterns in the data.

Remedy: Sufficient Sample Size:

Calculate the required sample size based on the desired power and effect size. Ensure that the sample size is large enough to yield statistically meaningful results.

Pitfall 6: Overlooking Interaction Effects:

Focusing only on the main effects and neglecting interaction effects can lead to incomplete insights.

Remedy: Analyze Interaction Effects:

Investigate interaction effects between factors to understand their combined influence on the response variable. Use interaction plots and statistical tests to identify and interpret interactions.

Pitfall 7: Ignoring Model Validation:

Neglecting to validate the response surface model or the optimal factor settings can lead to unreliable results when applied in practice.

Remedy: Model Validation and Confirmation Runs:

Conduct confirmation experiments to verify that the predicted optimal factor settings indeed lead to the desired improvement. Validate the response surface model by comparing predictions with new experimental data.

Pitfall 8: Inadequate Documentation:

Poor documentation of the Taguchi experiment, including the design matrix, analysis methods, and results, can hinder reproducibility and future reference.

Remedy: Comprehensive Documentation:

Maintain thorough documentation of the entire Taguchi experiment, from the factor selection to the analysis and results. This documentation is essential for reporting and sharing findings.

By addressing these pitfalls and applying appropriate remedies, engineers can conduct more robust Taguchi experiments in mechanical engineering, leading to effective improvements in product and process quality.

EXAMPLE

Here's an example of a Taguchi method designed experiment in mechanical engineering:

Objective: To optimize the tensile strength of a metal alloy used in the production of automotive engine components.

Factors:

Factor A (Temperature during Heat Treatment)
- Low Level: 900°C
- High Level: 1,100°C

Factor B (Cooling Rate after Heat Treatment)
- Low Level: Slow cooling (air cooling)
- High Level: Rapid cooling (quenching in oil)

Factor C (Alloy Composition)
- Low Level: 5% alloying element A
- High Level: 10% alloying element A

Response Variable: Tensile Strength (measured in megapascals, MPa)

Control Factors and Noise Factors:

- Control Factors: Temperature (A), Cooling Rate (B), Alloy Composition (C)
- Noise Factors: Variations in the heat treatment furnace, variations in raw material quality

Orthogonal Array (OA): L9 (3^4) OA is chosen for the experiment, allowing for nine experimental runs with three factors at two levels each.

Experimental Plan: Conduct nine experimental runs according to the L9 OA, where each run represents a unique combination of factor levels. Measure the tensile strength of each experimental specimen.

Analysis:

1. Calculate the S/N ratio for the tensile strength based on the objective (e.g., larger-the-better S/N ratio, as higher tensile strength is desired).
2. Analyze the S/N ratios to identify the combination of factor levels that maximizes tensile strength.

Results: The analysis reveals the following insights:

- The combination of high temperature (1,100°C), rapid cooling, and 10% alloying element A results in the highest tensile strength.
- Variations in the heat treatment furnace and raw material quality have a negligible impact on tensile strength under these optimal conditions.

Optimization: Implement the optimal factor settings in the production process to achieve consistently high tensile strength in engine components. This leads to improved product quality and reliability.

Confirmation Runs: Conduct confirmation experiments to validate that the optimal settings indeed yield the desired tensile strength under real production conditions.

Conclusion: Summarize the main findings, including the optimized factor settings for enhancing tensile strength in the metal alloy. Discuss the implications for automotive engine component manufacturing.

In this example, the Taguchi Method is applied to optimize the tensile strength of a metal alloy, demonstrating how it can be used to improve mechanical properties in a mechanical engineering context. A Taguchi method designed experiment typically involves using an OA to plan and organize the experimental runs. The choice of OA depends on the number of factors and levels being studied. Let's consider an example where we have three factors, each with two levels, and we'll use an L9 (3^4) OA, which is one of the commonly used OAs.

Here's how the OA matrix for this Taguchi experiment might look:

Orthogonal Array (L9) (Table 3.46):
In this matrix:

- Each row represents an experimental run (there are 9 runs in total).
- The columns correspond to the factors (A, B, C), and each cell contains either "Low," "High," or "Central," indicating the level of the respective factor for that run.
- The "Central" level is included for each factor to capture the center point and evaluate the curvature of the response surface.

Engineers would conduct the experiments according to this matrix, varying the factors at the specified levels for each run and measuring the response variable (e.g., tensile strength) accordingly. The results of these experiments would then be used for analysis, including the calculation of S/N ratios and the identification of optimal factor settings for improving the mechanical properties of the material or component in question.

Reporting

Statistical tools play a crucial role in Taguchi method designed experiments in mechanical engineering. These tools help engineers plan, conduct, and analyze experiments effectively to optimize product or process performance. Here are some of the key statistical tools commonly used in Taguchi method experiments:

Orthogonal Arrays (OAs): OAs are a fundamental part of Taguchi experiments. They help in systematically planning and organizing experimental runs, ensuring efficient exploration of factor combinations. Engineers select the appropriate OA based on the number of factors and levels being studied.

TABLE 3.46

Taguchi Method Designed Experiment Using an Orthogonal Array (OA)

Run	Temp (A)	Cooling Rate (B)	Alloy Composition (C)
1	Low	Low	Low
2	High	Low	Low
3	Low	High	Low
4	High	High	Low
5	Low	Low	High
6	High	Low	High
7	Low	High	High
8	High	High	High
9	Central	Central	Central

Signal-to-Noise (S/N) Ratios: S/N ratios are used to evaluate the performance of each experimental run with respect to the desired objective. The choice of S/N ratio depends on the experiment's objective, such as smaller-the-better, larger-the-better, or nominal-the-best. Common S/N ratios include:

Smaller-the-better: $-10 * \log 10[(1/n)\Sigma(yi/yi^2)]$

Larger-the-better: $10 * \log 10[\Sigma(yi^2)/n]$

Nominal-the-best: $-10 * \log 10[\Pi(yi/ytarget)]$

Analysis of Variance (ANOVA): ANOVA is used to assess the significance of factors and interactions in the Taguchi experiment. It helps identify which factors have a significant impact on the response variable. ANOVA is particularly valuable for determining the optimal factor settings.

Main Effects Plots: Main effects plots display the impact of each factor on the response variable while keeping other factors constant. These plots help identify the direction and magnitude of the effects of individual factors.

Interaction Plots: Interaction plots illustrate how two or more factors interact with each other, impacting the response variable. They reveal synergistic or antagonistic effects that may not be evident from the main effects alone.

Response Surface Methodology (RSM): RSM is employed when a more detailed exploration of factor interactions is required. It involves fitting a mathematical model to the experimental data, allowing for the visualization of response surfaces and contour plots.

Pareto Charts: Pareto charts help prioritize factors by showing the relative importance of each factor's effect on the response variable. They assist in focusing improvement efforts on the most influential factors.

Control Charts: Control charts are used to monitor and control the quality of the product or process before and after implementing the Taguchi improvements. They help ensure that the improvements are maintained over time.

Factorial Experiments: In addition to Taguchi experiments, factorial experiments may be used to explore factor interactions in more detail, especially when higher-order interactions are suspected.

Design of Experiments (DOE): While the Taguchi Method is a specific type of DOE, engineers may also use other DOE techniques to investigate different aspects of mechanical systems, depending on the research goals.

Statistical Software: Specialized software packages like Minitab, JMP, or statistical functions in Excel can facilitate data analysis, ANOVA, S/N ratio calculations, and graphical visualization of results.

These statistical tools and techniques help mechanical engineers design and analyze Taguchi experiments, identify optimal factor settings, and make data-driven decisions to improve product or process quality, reliability, and performance.

A well-structured report for a Taguchi method designed experiment in mechanical engineering is essential for documenting and communicating the experiment's objectives, methods, results, and conclusions effectively. Here are the key elements typically included in such a report:

Title Page:
- Title of the report.
- Names and affiliations of the author(s).
- Date of the report.

Abstract: A concise summary of the entire report, including the objectives, key findings, and conclusions. The abstract provides a quick overview of the experiment's purpose and outcomes.

Table of Contents: A list of sections and subsections with page numbers for easy navigation.

List of Figures and Tables: A separate list enumerating all the figures and tables used in the report with their respective page numbers.

List of Abbreviations and Symbols (if used): Definitions and explanations for any technical abbreviations or symbols employed throughout the report.

List of Nomenclature (if applicable): Definitions and explanations of variables, parameters, and terms specific to the mechanical engineering context.

Introduction:
- Provide context and background information for the Taguchi Method experiment.
- State the specific objectives and purpose of the experiment.
- Explain the significance of the experiment in the context of mechanical engineering.

Experimental Design: Describe the factors, levels, and OA chosen for the Taguchi experiment. Mention any constraints or limitations in the design.

Materials and Methods: Detail the materials used in the experiment, including specifications and sources. Explain the experimental setup, data collection procedures, and instrumentation used. Provide information on how the OA was generated and the calculation of S/N ratios.

Data Collection: Present the raw data obtained during the Taguchi experiment. Include tables, charts, or graphs that illustrate the data.

Data Analysis: Present the analysis of S/N ratios, including the identification of optimal factor settings. Display main effects plots, interaction plots, and any other relevant graphical representations.

Confirmation Runs: Discuss the results of confirmation experiments conducted to validate the optimal factor settings.

Discussion:
- Interpret the results in the context of the experiment's objectives.
- Discuss the practical implications of the findings for mechanical engineering applications.
- Address any unexpected results or trends.
- Compare the results with prior research or industry standards, if relevant.

Recommendations and Implementation: Provide recommendations based on the experiment's results, such as process improvements, design modifications, or quality control measures. Discuss the implementation of the optimal factor settings and any challenges or considerations.

Conclusion: Summarize the main findings and their implications for mechanical engineering. Reiterate the significance of the experiment's results in addressing the initial objectives.

References: Cite all relevant sources, including research papers, books, manuals, and software used for the Taguchi experiment's design and data analysis.

Appendices (if necessary): Include supplementary information, such as detailed statistical calculations, additional data tables, or any other supporting documentation.

Acknowledgments (optional): Acknowledge individuals, organizations, or funding sources that contributed to the success of the Taguchi experiment.

Thoroughly organize the report, use clear and concise language, and ensure that content flows logically from one section to the next. Proper formatting, headings, and clear labeling of figures and tables are essential for readability and understanding.

3.5.5 MIXTURE EXPERIMENTS

A mixture designed experiment in mechanical engineering involves the systematic study of how various components or ingredients in a mixture affect the properties, performance, or characteristics of a mechanical system, product, or process. This type of experiment is particularly useful when dealing with materials, composites, or processes where the composition is a critical factor in determining the desired outcomes.

Here are key aspects of a mixture designed experiment in mechanical engineering:

Objective: To optimize the composition of a mixture (combination of components) to achieve specific mechanical properties, performance criteria, or quality attributes.

Components: The mixture consists of multiple components, each of which contributes to the overall composition. These components can be materials, additives, or ingredients relevant to the mechanical system under study.

Response Variables: Mechanical properties or performance characteristics that depend on the mixture composition, such as strength, hardness, elasticity, wear resistance, or thermal conductivity.

Factors: The proportions or ratios of each component in the mixture are treated as factors to be manipulated. These factors are continuous variables representing the percentages or amounts of each component.

Design of Experiments (DOE): DOE principles are applied to plan the experiments systematically. A mixture design typically uses specialized experimental designs, such as simplex-centroid designs, to explore the composition space efficiently.

Constraints: Some mechanical systems or processes may have constraints on the allowable composition ranges, such as minimum or maximum limits for certain components.

Statistical Analysis: Statistical methods are employed to analyze the experimental results and develop predictive models. Regression analysis, RSM, and mixture-amount experiments are common techniques used for analysis.

Optimization: The goal of the experiment is to identify the optimal composition of the mixture that maximizes or minimizes the desired mechanical properties or performance criteria. Optimization algorithms are used to find the best combination of component proportions.

Applications: Mixture designed experiments in mechanical engineering can be applied to various scenarios, including:
- Developing new materials with specific properties.
- Optimizing the composition of composites or alloys.
- Designing manufacturing processes to achieve desired product characteristics.
- Enhancing the performance of mechanical systems by adjusting lubricants, fluids, or coatings.

An example of a mixture designed experiment in mechanical engineering could involve developing a composite material for aerospace applications. The experiment aims to find the ideal mixture of fibers, resins, and additives to create a lightweight, high-strength material suitable for use in aircraft components. Factors in this experiment might include the percentage of different types of fibers, the resin formulation, and curing conditions. The response variables could include tensile strength, weight-to-strength ratio, and resistance to environmental factors like temperature and humidity.

A mixture designed experiment in mechanical engineering focuses on optimizing the composition of mixtures to achieve specific mechanical properties, performance criteria, or quality attributes. It is a valuable approach for designing materials, composites, and processes in various mechanical engineering applications.

STEPS

Developing a mixture designed experiment in mechanical engineering involves a series of steps to systematically plan and conduct experiments to optimize the composition of a mixture. Here are the key steps:

1. **Define the Problem and Objectives:** Clearly define the problem or aspect of the mechanical system or material that requires improvement or optimization. State the specific objectives you want to achieve through the mixture design experiment, such as improving mechanical properties or performance characteristics.
2. **Identify the Components:** Identify the components or ingredients that make up the mixture. These can be materials, additives, or substances relevant to the mechanical system under study.
3. **Determine the Response Variables:** Define the mechanical properties, performance criteria, or quality attributes that depend on the mixture composition. These are the response variables you want to optimize.
4. **Select the Factors:** Identify the factors that represent the proportions or ratios of each component in the mixture. These factors are continuous variables, typically ranging from 0% to 100%, and represent the composition.
5. **Constraints and Allowable Ranges:** Determine any constraints or allowable ranges for the component proportions. Some mixtures may have minimum or maximum limits for specific components to ensure functionality or safety.
6. **Design of Experiments (DOE):** Choose an appropriate mixture design approach and experimental design. Common approaches include simplex-centroid designs, Scheffé mixture designs, or others tailored to your specific needs. Select the number of experimental runs based on the complexity of the problem and available resources.
7. **Generate the Experimental Plan:** Use the chosen mixture design approach to generate the experimental plan, which specifies the combination of component proportions for each experimental run. Ensure that the experimental plan covers a representative portion of the composition space.
8. **Conduct the Experiments:** Prepare the mixture according to the proportions specified in the experimental plan. Conduct the experiments systematically, ensuring accurate measurements and consistency.
9. **Measure Response Variables:** Measure the response variables for each experimental run, recording the results accurately.
10. **Data Analysis:** Analyze the experimental data to understand the relationship between the mixture composition (factors) and the response variables. Use statistical methods, such as regression analysis or RSM, to build predictive models.
11. **Optimization:** Identify the optimal composition of the mixture that maximizes or minimizes the desired response variables based on the analysis. Use optimization algorithms to find the best combination of component proportions.
12. **Confirmation Runs:** Conduct confirmation experiments using the optimal mixture composition to validate the predicted results under real conditions.
13. **Implementation and Control:** Implement the optimal mixture composition in the mechanical system, material, or process. Establish control measures to maintain the desired composition during production or operation.
14. **Documentation and Reporting:** Maintain comprehensive documentation of the mixture design experiment, including the design matrix, results, and findings. Prepare a report that summarizes the experiment, the steps taken, the optimal mixture composition, and the impact on mechanical properties or performance.
15. **Continuous Improvement:** Continuously monitor and collect data from the mechanical system or process to ensure that the optimal mixture composition is consistently achieved. Use ongoing data analysis to identify opportunities for further optimization and continuous improvement.

By following these steps, engineers can effectively plan, conduct, and analyze mixture designed experiments in mechanical engineering to optimize compositions for specific mechanical properties, performance criteria, or quality attributes.

PITFALLS AND REMEDIES

Mixture designed experiments in mechanical engineering, like any other experimental approach, can encounter certain pitfalls. Being aware of these challenges and applying appropriate remedies is essential for the success of the experiment. Here are some common pitfalls and their corresponding remedies in mixture designed experiments:

Pitfall 1: Inadequate Component Selection:
Selecting the wrong components or overlooking critical ingredients in the mixture can lead to suboptimal results.

Remedy: Comprehensive Component Selection:
Thoroughly analyze the mechanical system or material to ensure all relevant components are considered. Consult with experts and conduct literature reviews to identify potential components that could impact the mixture's performance.

Pitfall 2: Overly Simplistic Models:
Using overly simplistic models to represent the relationship between mixture components and response variables may lead to inaccurate predictions.

Remedy: Model Complexity Matching:
Ensure that the model used to represent the mixture response is appropriate for the complexity of the system. Consider higher-order models or RSM when interactions or non-linearities are suspected.

Pitfall 3: Constraint Violations:
Setting constraints on component proportions that are too restrictive or not considering physical limitations can result in infeasible solutions.

Remedy: Realistic Constraints:
Set constraints based on real-world limitations and physical properties of components. Ensure that the feasible region of the mixture composition space is considered.

Pitfall 4: Neglecting Interaction Effects:
Ignoring interaction effects between mixture components can lead to incomplete insights into the system's behavior.

Remedy: Analyze Interaction Effects:
Investigate interaction effects through appropriate statistical analyses, such as regression models or RSM. Utilize graphical tools, like interaction plots, to visualize and interpret interactions.

Pitfall 5: Lack of Validation:
Failing to validate the optimal mixture composition with confirmation experiments under real operating conditions can result in impractical recommendations.

Remedy: Conduct Confirmation Runs:
Perform confirmation experiments to verify that the predicted optimal mixture composition indeed leads to the desired mechanical properties or performance criteria. Ensure that the improvements are robust and maintainable.

Pitfall 6: Inadequate Documentation:
Poor documentation of the mixture design experiment, including the design matrix, analysis methods, and results, can hinder reproducibility and future reference.

Remedy: Comprehensive Documentation:
Maintain thorough documentation of the entire mixture design experiment, from the initial problem statement to the implementation of optimal compositions. Properly label and organize data, models, and reports.

Pitfall 7: Overemphasis on Optimization:
Placing too much emphasis on finding the optimal mixture composition without considering practicality, cost, or manufacturability can lead to impractical solutions.

Remedy: Consider Practical Constraints:
> Balance optimization objectives with practical constraints, cost considerations, and manu-facturability. Seek solutions that strike a practical balance between performance and feasibility.

By addressing these pitfalls and applying the appropriate remedies, engineers can conduct more robust mixture designed experiments in mechanical engineering, resulting in improved materials, products, or processes.

EXAMPLE

The following is an example of a mixture designed experiment for mechanical engineering:

Objective: To develop a high-performance brake pad material for automotive applications by optimizing the composition of the brake pad mixture.

Components: The brake pad mixture consists of various ingredients, including friction materials (e.g., abrasives, binders), fillers (e.g., graphite, metal fibers), and other additives (e.g., antioxidants, curing agents).

Response Variables: Several mechanical properties and performance criteria are considered, including:

- Coefficient of friction (COF): A higher COF indicates better braking performance.
- Wear rate: Lower wear rates result in longer-lasting brake pads.
- Thermal conductivity: Effective heat dissipation is crucial for maintaining braking efficiency under high temperatures.
- Compressive strength: High compressive strength ensures durability and resistance to deformation.

Factors: The factors to be manipulated are the proportions or percentages of different components in the brake pad mixture. These components include friction materials, fillers, and additives.

Constraints: Constraints may include limitations on the maximum allowable cost of the brake pad material, toxicity regulations, and industry standards for brake pad performance.

Design of Experiments (DOE): A mixture design approach, such as simplex-centroid design, is chosen to systematically vary the proportions of the components in the mixture.

Experimental Plan: Using the chosen mixture design approach, an experimental plan is generated that specifies the proportions of each component for each experimental run. For example, a simplex-centroid design might include 10 experimental runs, exploring various mixtures within the feasible region.

Conducting the Experiments: The brake pad mixtures for each experimental run are prepared according to the specified proportions of components.

Measurement of Response Variables: The response variables, including COF, wear rate, thermal conductivity, and compressive strength, are measured for each brake pad mixture.

Data Analysis: Statistical methods, such as regression analysis or RSM, are employed to analyze the experimental data and build predictive models for the response variables.

Optimization: The goal is to identify the optimal composition of the brake pad mixture that maximizes the coefficient of friction, minimizes wear rate, maximizes thermal conductivity, and maximizes compressive strength while adhering to cost and regulatory constraints.

Confirmation Runs: Confirmation experiments are conducted using the optimal brake pad mixture composition to validate that the predicted performance improvements are achieved under real braking conditions.

Conclusion: The report concludes by summarizing the findings, including the optimal brake pad mixture composition and its implications for automotive brake performance and longevity.

In this example, a mixture designed experiment is used to optimize the composition of brake pad materials for improved automotive braking performance, demonstrating how this approach can be applied in mechanical engineering to develop high-performance materials.

Matrix Example

In a mixture designed experiment in mechanical engineering, a matrix is used to specify the proportions or percentages of each component in the mixture for each experimental run. The matrix helps researchers systematically vary the composition of the mixture. Here's an example of a matrix for a mixture designed experiment:

Objective: To optimize the composition of a composite material for use in aircraft structural components, considering mechanical properties, weight, and cost.

Components: The composite material consists of three main components: carbon fibers (C), epoxy resin (E), and filler material (F).

Factors: The factors represent the proportions of each component in the mixture. The sum of the proportions for all components in each run should equal 100%.

Response Variables: The mechanical properties to be optimized include tensile strength (TS) and flexural modulus (FM). The weight of the composite material (W) and the material cost (COST) are also considered.

Matrix for a 12-Run Mixture Design (Simplex-Centroid) (Table 3.47):

TABLE 3.47

Mixture Designed Experiment Matrix Example for Mechanical Engineering

Run	C (%)	E (%)	F (%)	TS (MPa)	FM (GPa)	W (g)	Cost ($)
1	40	30	30	250	6.5	150	80
2	30	40	30	260	6.8	145	85
3	20	50	30	270	7	140	90
4	30	30	40	245	6.3	155	75
5	20	40	40	255	6.6	160	70
6	10	50	40	265	6.9	165	95
7	30	20	50	240	6.1	170	100
8	20	30	50	250	6.4	175	105
9	10	40	50	260	6.7	180	110
10	0	50	50	270	6.9	185	115
11	25	25	50	235	6	190	120
12	15	35	50	245	6.2	195	125

In this matrix:

- Each row represents an experimental run, where a specific combination of carbon fiber (C), epoxy resin (E), and filler material (F) proportions is used.
- The proportions of the components are given in percentage values, ensuring that they sum up to 100% for each run.
- The response variables, including tensile strength (TS), flexural modulus (FM), weight (W), and material cost (COST), are measured for each run.
- This is a 12-run simplex-centroid design, which explores various mixtures within the feasible region.

Researchers would conduct experiments according to this matrix, prepare the composite material samples with the specified proportions, and measure the response variables to analyze the effects of different component compositions on mechanical properties, weight, and cost. The analysis helps identify the optimal composition that meets the desired objectives.

Reporting

Statistical tools play a crucial role in analyzing data from mixture designed experiments in mechanical engineering. These tools help engineers and researchers understand the relationships between the components of mixtures and their impact on mechanical properties or performance criteria. Here are some common statistical tools used in mixture designed experiments for mechanical engineering:

Analysis of Variance (ANOVA): ANOVA is used to assess the significance of different factors (components in mixtures) and their interactions on the response variables (mechanical properties or performance criteria). It helps determine which factors have a statistically significant impact.

Regression Analysis: Regression models, including linear and nonlinear regression, are used to establish mathematical relationships between the mixture components and the response variables. These models help predict the behavior of the system under different mixture compositions.

Response Surface Methodology (RSM): RSM involves fitting mathematical models to response surfaces to visualize and optimize the relationships between factors and responses. It helps identify the optimal mixture composition.

Main Effects Plots: Main effects plots display the individual effects of each factor (mixture component) on the response variables while keeping other factors constant. They provide insights into the relative importance of each component.

Interaction Plots: Interaction plots show the combined effects of two or more factors on the response variables. They help identify synergistic or antagonistic interactions between components.

Contour Plots: Contour plots visualize the response surface in two dimensions, allowing engineers to identify regions of optimal performance based on the mixture composition.

Optimization Algorithms: Numerical optimization techniques, such as gradient descent, genetic algorithms, or simulated annealing, are used to find the optimal mixture composition that maximizes or minimizes the desired response variables.

Signal-to-Noise (S/N) Ratios: S/N ratios are used to assess the variability in response variables and measure the quality or performance of a mixture under different conditions. They help identify optimal settings for factors.

Fractional Factorial Analysis: In cases where full-factorial experiments are not feasible due to the number of factors, fractional factorial designs help reduce the number of experimental runs while still providing valuable information about factor effects.

Robust Parameter Design: This approach involves optimizing mixture compositions to make the system less sensitive to variations or disturbances, improving the robustness of the mechanical system or process.

Monte Carlo Simulation: Monte Carlo simulations can be used to assess the impact of uncertainty or variability in input variables on the output responses, providing insights into the robustness of the design.

Statistical Software: Specialized statistical software packages, such as Minitab, JMP, or Design-Expert, are often used to perform statistical analyses and generate graphical representations of the experimental results.

These statistical tools and techniques help engineers and researchers design, analyze, and optimize mixtures for mechanical engineering applications, ensuring that the desired mechanical properties and performance criteria are achieved efficiently and effectively.

A report for a mixture designed experiment in mechanical engineering typically includes various elements to document the experiment, its objectives, procedures, results, and conclusions. Here are the key elements that should be included in such a report:

Title Page:
- Title of the report.
- Names and affiliations of the authors and contributors.
- Date of submission or completion.

Abstract: A concise summary of the experiment's objectives, methods, key findings, and conclusions. It should provide an overview of the entire report.

Table of Contents: A list of sections, subsections, and their page numbers for easy navigation within the report.

List of Figures and Tables: A list of all figures and tables in the report, along with their corresponding page numbers.

List of Abbreviations and Symbols: Definitions and explanations of any abbreviations, acronyms, or symbols used throughout the report.

Introduction:
- Background information on the problem or objective of the mixture designed experiment.
- Clear statement of the research question or hypothesis.
- Explanation of the relevance and significance of the experiment to mechanical engineering.

Literature Review: A brief review of relevant literature and prior research related to the experiment's objectives. Discussion of existing theories, models, and findings that are pertinent to the experiment.

Experimental Design: Description of the experimental design and methodology, including:
- Mixture components and their proportions.
- Choice of response variables.
- Selection of factors and their levels.
- Details of any constraints or control variables.
- Explanation of the statistical techniques or designs used.

Experimental Procedure: Detailed procedures for conducting the experiments, including equipment used, measurements taken, and any special conditions or precautions.

Data Collection: Presentation of raw data, including tables or charts of experimental runs, measurements, and observations. Details of data collection, including sample sizes and replication.

Data Analysis: Statistical analyses, including:
- ANOVA.
- Regression analyses.

- RSM (if applicable).
- Graphical representations of data, such as contour plots or interaction plots.

Results: Presentation of the main findings, including the effects of mixture components on response variables. Tables and graphs illustrating significant factors and their impact. Discussion of trends and patterns observed in the data.

Discussion: Interpretation of the results and their implications in the context of the experiment's objectives. Explanation of any unexpected findings or anomalies. Comparison of the results with the literature review.

Conclusion: Summary of the key findings and their significance. Statement of whether the experiment's objectives were achieved. Practical implications and recommendations for further research or applications.

References: A list of all references, citations, and sources consulted or cited in the report, following a specific citation style (e.g., APA, MLA, Chicago).

Appendices: Additional information that may be relevant but is not included in the main body of the report. This may include supplementary data, detailed calculations, experimental protocols, or additional figures and tables.

Acknowledgments: Recognition of individuals, organizations, or funding sources that contributed to the experiment or research.

Declaration of Originality: A statement that the work presented is original, and any borrowed content is properly cited and acknowledged.

Signature and Date: The author's signature and date of completion, if a hard copy of the report is required.

These elements collectively provide a comprehensive and well-structured report for a mixture designed experiment in mechanical engineering, ensuring that the experiment, results, and conclusions are effectively communicated to the intended audience.

3.5.6 SEQUENTIAL EXPERIMENTATION

Sequential experimentation, often referred to as sequential Design of Experiments (DOE), is an approach used in mechanical engineering to optimize processes, products, or systems incrementally. Unlike traditional fixed-size experimental designs, sequential experimentation involves a series of experiments where each step is based on the results of the previous step. The goal is to efficiently identify the optimal conditions or settings while minimizing the number of experimental runs. Here's an overview of sequential experimentation in mechanical engineering:

Incremental Learning: Sequential experimentation aims to gather information about the system's response over time, allowing engineers to adapt the design or process based on emerging insights. It involves a learning cycle where each experiment informs the next step.

Adaptive Design: The DOE is adjusted based on the accumulating data. Factors, levels, or conditions are modified in subsequent experiments to explore promising regions or refine parameter settings.

Efficiency: By using adaptive design principles, sequential experimentation can often achieve the desired results with fewer experimental runs compared to fixed-size designs, leading to cost and time savings.

Engineers may use sequential experimentation to optimize the parameters of a 3D printing process. After each run, they adjust settings like layer thickness and print speed based on the previous results. Sequential experimentation can be applied in various mechanical engineering contexts, including:

- Product design and optimization (e.g., designing lightweight structures).
- Process optimization (e.g., improving manufacturing processes).
- Materials development (e.g., identifying optimal material compositions).
- Structural analysis (e.g., optimizing load-bearing structures).
- Control system tuning (e.g., PID controller tuning).

Sequential experimentation provides efficient resource utilization: Fewer experimental runs are needed compared to traditional DOE methods. It allows for real-time adaptability. Engineers can adapt the experimental plan as they gain insights, leading to quicker and more effective optimization. The design is cost-effective. Sequential experimentation can reduce the cost associated with conducting numerous experiments. Another advantage is faster results. The iterative nature of sequential experimentation often leads to faster convergence to optimal solutions.

Sequential experimentation in mechanical engineering is a dynamic and adaptive approach that allows engineers to iteratively optimize processes, products, or systems by learning from each experiment's results. It is particularly useful when efficiency and cost-effectiveness are crucial considerations in research and development.

STEPS

Developing a sequential experimentation designed experiment in mechanical engineering involves a systematic and adaptive approach to optimizing processes or systems. Here are the steps to create and implement such an experiment:

1. **Define the Problem and Objectives:** Clearly state the problem or challenge you want to address through sequential experimentation. Define specific objectives, such as maximizing performance, minimizing costs, or improving efficiency.
2. **Identify Factors and Response Variables:** Determine the key factors that may influence the system or process. These could be parameters, settings, or design variables. Identify the response variables that measure the system's performance or quality.
3. **Initial Experimental Design:** Start with an initial set of experimental conditions based on prior knowledge, engineering expertise, or existing data. Design the first experiment(s) with a small number of runs (e.g., a screening design) to gather initial information.
4. **Data Collection and Analysis:** Conduct the initial experiment(s) and collect data on the response variables. Perform data analysis to identify factors that have a significant impact on the response and factors that can be held constant.
5. **Modify the Experimental Design:** Based on the analysis of initial results, modify the experimental design for the next iteration. Determine whether to change factor levels, introduce new factors, or adjust sample sizes for subsequent experiments. Ensure that the design adapts to the emerging insights.
6. **Sequential Runs:** Conduct the next set of experiments based on the modified design. Collect data from each run, ensuring that the data is accurate and well-documented.
7. **Iterate and Refine:** Continue the iterative process by analyzing the results of each set of experiments. Refine the experimental design and conditions based on the knowledge gained in previous iterations.
8. **Stopping Criteria:** Define stopping criteria that signal when to terminate the sequential experimentation. Stopping criteria could be reaching a target performance, achieving diminishing returns in knowledge gain, or budget constraints.
9. **Validation and Optimization:** Once the stopping criteria are met, validate the optimal conditions or settings identified through sequential experimentation. Conduct additional experiments, if necessary, to confirm the results and assess their robustness.

10. **Implementation and Monitoring:** Implement the optimized conditions or settings in the mechanical system or process. Continuously monitor and measure system performance to ensure that the improvements are sustained.
11. **Documentation and Reporting:** Document all experimental designs, procedures, and results in a clear and organized manner. Prepare a comprehensive report summarizing the sequential experimentation process, findings, and recommendations.
12. **Feedback and Learning:** Use the knowledge gained from the sequential experimentation to inform future engineering decisions and projects. Share the findings and insights with relevant stakeholders or team members.
13. **Continuous Improvement:** Embrace a culture of continuous improvement, where the lessons learned from sequential experimentation contribute to ongoing enhancements in mechanical engineering processes and systems.

The key to successful sequential experimentation in mechanical engineering is adaptability and a willingness to adjust the experimental plan based on emerging insights. This approach allows engineers to efficiently optimize processes, products, or systems while conserving resources and time.

PITFALLS AND REMEDIES

Sequential experimentation in mechanical engineering can be a powerful tool for optimization, but it also comes with potential pitfalls that need to be managed. Here are some common pitfalls and remedies:

Pitfall 1: Premature Convergence
>   **Issue:** Stopping the sequential experimentation too early can lead to suboptimal solutions if the true optimum hasn't been reached.
>   **Remedy:** Set clear stopping criteria based on desired performance or knowledge gain. Continuously monitor convergence and consider the trade-off between additional experiments and expected benefits.

Pitfall 2: Over-Exploration
>   **Issue:** Conducting too many iterations or experiments can lead to excessive resource consumption, especially if there are diminishing returns in knowledge gain.
>   **Remedy:** Define a maximum budget or resource constraint for the sequential experimentation. Use statistical techniques to assess when additional experiments are unlikely to yield significant improvements.

Pitfall 3: Insufficient Adaptability
>   **Issue:** Failing to adapt the experimental design based on emerging insights can result in missed opportunities for optimization.
>   **Remedy:** Continuously review and analyze the results of each iteration. Modify the experimental design and conditions to explore promising regions or refine parameter settings.

Pitfall 4: Ignoring External Factors
>   **Issue:** Not accounting for external variables or conditions that may impact the system can lead to suboptimal solutions.
>   **Remedy:** Identify and monitor external factors that may influence the system. Consider incorporating them into the experimental design or conducting sensitivity analyses.

Pitfall 5: Poor Experimental Design
>   **Issue:** Inadequate experimental design, such as inappropriate factor selection or insufficient replication, can lead to unreliable results.
>   **Remedy:** Invest in robust experimental design techniques. Consult with statistical experts or use software tools to optimize the design for your specific objectives.

Pitfall 6: Lack of Documentation

Issue: Inadequate documentation of experimental procedures, conditions, and results can hinder reproducibility and knowledge transfer.

Remedy: Maintain detailed and organized records throughout the sequential experimentation process. Create a standardized documentation protocol.

Pitfall 7: Not Considering Practical Constraints

Issue: Optimized solutions may not be practical or implementable in real-world applications.

Remedy: Include practical constraints in the optimization process. Consider factors such as cost limitations, material availability, and manufacturing capabilities.

Pitfall 8: Inadequate Validation

Issue: Failing to validate optimized conditions or settings can result in unreliable performance in real-world applications.

Remedy: Conduct validation experiments to confirm the robustness and effectiveness of the optimized solution. Ensure that the results are consistent with the laboratory findings.

Pitfall 9: Ineffective Communication

Issue: Inadequate communication of findings and recommendations can hinder the adoption of optimized solutions.

Remedy: Prepare clear and concise reports and presentations that convey the results, insights, and practical implications of the sequential experimentation. Engage with stakeholders effectively.

Pitfall 10: Lack of Continuous Improvement

Issue: Failing to incorporate lessons learned from sequential experimentation into future projects can limit the long-term benefits.

Remedy: Foster a culture of continuous improvement where knowledge gained from past experiments informs and enhances future engineering decisions and projects.

By being aware of these pitfalls and proactively implementing remedies, engineers and researchers can make the most of sequential experimentation in mechanical engineering, achieving efficient and effective optimization while minimizing risks and resource wastage.

EXAMPLE

An example of a sequential experimentation designed experiment in mechanical engineering:

Objective: Optimization of the cutting parameters for a milling operation to achieve the highest material removal rate (MRR) while maintaining an acceptable surface finish in the machining of a complex aerospace component.

Steps in Sequential Experimentation:

1. **Initial Experiment:** Select a baseline set of cutting parameters, including spindle speed, feed rate, and depth of cut. Conduct an initial milling operation on a sample workpiece. Measure the MRR and surface finish (e.g., Ra, a common surface roughness parameter).
2. **Data Analysis:** Analyze the results of the initial experiment. Identify which cutting parameters have the most significant impact on MRR and surface finish. Determine the potential for improvements in MRR and surface finish.
3. **Modify Experimental Design:** Based on the analysis, modify the cutting parameters for the next iteration. For example, increase the spindle speed while maintaining other parameters. Adjust the depth of cut or feed rate as necessary. Define a range for each parameter to explore, such as a range of spindle speeds.

4. **Sequential Runs (Iteration 2):** Conduct the second set of milling experiments with the modified parameters. Measure MRR and surface finish for each run. Compare the results to those of the initial experiment.
5. **Iterate and Refine:** Continue the process iteratively, each time modifying the cutting parameters based on the analysis of previous results. Refine the design to explore the most promising regions of parameter space. Balance the trade-off between MRR and surface finish.
6. **Stopping Criteria:** Define stopping criteria based on achieving a target MRR or surface finish, reaching diminishing returns, or adhering to budget constraints. Monitor progress and adjust the stopping criteria as necessary.
7. **Validation and Optimization:** Once the stopping criteria are met, validate the optimal cutting parameters through a final set of experiments. Ensure that the chosen parameters consistently produce the desired results.
8. **Implementation:** Implement the optimized cutting parameters in the machining process for the aerospace component. Monitor and control the process to ensure it meets production requirements.
9. **Documentation and Reporting:** Document all experimental designs, procedures, and results in a detailed report. Provide recommendations for practical implementation and guidelines for maintaining the optimized process.

In this example, sequential experimentation allows mechanical engineers to iteratively refine the milling parameters for machining the aerospace component. By continuously adjusting the cutting conditions based on the evolving insights from previous experiments, engineers can achieve the desired balance between MRR and surface finish, ultimately optimizing the machining process for production.

In a sequential experimentation designed experiment for optimizing cutting parameters in milling operations, a matrix is used to plan and record the details of each experimental run. The matrix typically includes columns for factors (cutting parameters), levels, and responses (MRR and surface finish measurements). Table 3.48 is a simplified example of such a matrix for the optimization of cutting parameters.

In this matrix:

- **Run:** A unique identifier for each experimental run.
- **Spindle Speed (RPM):** One of the cutting parameters being optimized.
- **Feed Rate (mm/min):** Another cutting parameter under consideration.

TABLE 3.48

Sequential Experimentation Designed Matrix for Mechanical Engineering

Run	Spindle Speed (RPM)	Feed Rate (mm/min)	Depth of Cut (mm)	Material Removal Rate (MRR)	Surface Finish (Ra)
1	1,000	200	2		
2	1,500	250	1.5		
3	1,200	230	1.8		
4	1,400	220	1.7		
5	1,300	210	2.2		
6	1,600	240	1.6		
7	1,100	190	1.9		
8	1,550	260	2.1		

- **Depth of Cut (mm):** The third cutting parameter in the study.
- **Material Removal Rate (MRR):** The response variable representing the rate at which material is removed during each run. This column will be filled in after conducting each experiment.
- **Surface Finish (Ra):** The response variable representing the surface roughness after each run. This column will also be filled in after conducting each experiment.

During the sequential experimentation process, engineers would fill in the responses (MRR and surface finish) after each run, analyze the results, and modify the cutting parameters for subsequent runs based on their findings. The matrix allows for a structured and organized approach to record, analyze, and adapt the experimental plan to achieve the desired optimization goals.

Reporting

Sequential experimentation in mechanical engineering often involves statistical tools and techniques to analyze and optimize processes or systems. Here are some common statistical tools used in sequential experimentation designed experiments:

Analysis of Variance (ANOVA): ANOVA is used to assess the significance of factors and their interactions on response variables. In sequential experimentation, ANOVA is applied after each iteration to determine which factors are significant and whether changes in factor settings lead to significant improvements.

Regression Analysis: Regression models, including linear and nonlinear regression, can be used to establish mathematical relationships between factors and responses. Sequential experimentation may involve updating regression models after each iteration to make predictions and guide the next set of experiments.

Response Surface Methodology (RSM): RSM is used to create models that describe the relationship between factors and responses in multidimensional spaces. It helps engineers visualize and optimize complex systems by fitting response surfaces to experimental data.

Optimization Algorithms: Numerical optimization techniques, such as gradient descent, genetic algorithms, or particle swarm optimization, can be employed to find optimal factor settings iteratively. These algorithms guide the search for the best combination of factors based on the knowledge gained from previous experiments.

Design of Experiments (DOE): DOE principles, including factorial designs, fractional factorial designs, and response surface designs, are used to plan the experimental matrix systematically. Sequential experimentation may involve adapting the DOE design as new information becomes available.

Statistical Process Control (SPC): SPC tools, such as control charts and process capability analysis, can be used to monitor and control the quality of the experimental process and identify trends or anomalies during the iterations.

Sequential Testing: Sequential hypothesis testing methods, like the sequential probability ratio test (SPRT) or the sequential likelihood ratio test (SLRT), can help determine whether to continue or terminate the sequential experimentation based on interim results and predefined stopping criteria.

Bayesian Statistics: Bayesian methods, including Bayesian optimization and Bayesian updating, can be applied to update prior knowledge with new experimental data and make informed decisions about optimal factor settings.

Monte Carlo Simulation: Monte Carlo simulations can be used to assess the impact of uncertainty or variability in factor settings on the responses. It helps engineers understand the robustness of optimized solutions.

Multivariate Analysis: Techniques like PCA or MANOVA can be applied to explore relationships among multiple variables and identify patterns or correlations.

Statistical Software: Specialized statistical software packages, such as Minitab, JMP, or Design-Expert, are often used to perform statistical analyses, create predictive models, and visualize experimental results.

These statistical tools are crucial for the analysis, optimization, and decision-making processes in sequential experimentation designed experiments in mechanical engineering. They help engineers efficiently explore the design space, make informed choices for factor settings, and achieve desired outcomes while conserving resources and time.

A report for a sequential experimentation designed experiment in mechanical engineering is a comprehensive document that communicates the experiment's objectives, methods, results, and conclusions. Here are the key elements typically included in such a report:

Title Page:
- Title of the report.
- Names and affiliations of the authors and contributors.
- Date of submission or completion.

Abstract: A concise summary of the experiment's objectives, methods, key findings, and conclusions. It should provide an overview of the entire report.

Table of Contents: A list of sections, subsections, and their page numbers for easy navigation within the report.

List of Figures and Tables: A list of all figures and tables in the report, along with their corresponding page numbers.

List of Abbreviations and Symbols: Definitions and explanations of any abbreviations, acronyms, or symbols used throughout the report.

Introduction: Background information on the problem or challenge addressed by the experiment. Clear statement of the research question, objectives, and the context of the study within the field of mechanical engineering.

Literature Review: A brief review of relevant literature and prior research related to the experiment's objectives. Discussion of existing theories, models, and findings that are pertinent to the experiment.

Experimental Design and Methods:
- Detailed description of the experimental design, including the initial plan and any modifications made during the sequential experimentation.
- Explanation of the factors, levels, and responses studied.
- Information on data collection methods, equipment used, and any statistical techniques applied.

Results and Data Analysis: Presentation of the main findings, including tables and graphs that illustrate the outcomes of each experimental iteration. Discussion of trends, patterns, and significant findings observed throughout the sequential experimentation.

Discussion: Interpretation of the results and their implications in the context of the experiment's objectives. Explanation of how the experiment's findings contribute to the understanding of the mechanical engineering problem.

Conclusion: Summary of the key findings and their significance. Statement on whether the objectives of the experiment were achieved. Practical implications and recommendations for future work or applications.

Validation and Verification: Discussion of the validation of optimized conditions or settings, including the results of validation experiments. Assessment of the robustness and reliability of the optimized solution.

Implementation and Practical Considerations: Information on the implementation of the optimized solution in real-world mechanical systems or processes. Consideration of practical constraints, such as cost, feasibility, and manufacturing considerations.

Documentation of Experimental Details: Comprehensive documentation of the experimental process, including factor settings, response measurements, and any issues or challenges encountered.

References: A list of all references, citations, and sources consulted or cited in the report, following a specific citation style (e.g., APA, IEEE, Chicago).

Appendices: Additional information that may be relevant but is not included in the main body of the report. This may include detailed calculations, experimental protocols, raw data, and supplementary figures and tables.

Acknowledgments: Recognition of individuals, organizations, or funding sources that contributed to the experiment or research.

Declaration of Originality: A statement that the work presented is original, and any borrowed content is properly cited and acknowledged.

Signature and Date: The author's signature and date of completion, if a hard copy of the report is required.

These elements collectively provide a comprehensive and well-structured report for a sequential experimentation designed experiment in mechanical engineering, ensuring that the experiment, results, and conclusions are effectively communicated to the intended audience.

3.5.7 Robust Parameter Design (RPD)

An RPD experiment is a methodology used in mechanical engineering and other fields to optimize and improve the performance of a product or process while accounting for variability and uncertainty in input parameters. The goal of RPD is to design a system that is robust, meaning it can perform consistently and reliably even in the presence of various sources of variation and uncertainty, such as manufacturing tolerances, environmental conditions, and material properties.

Here are the key elements of an RPD experiment in mechanical engineering:

Identification of Critical Parameters: In RPD, engineers first identify the critical input parameters or factors that can significantly affect the performance or quality of the product or process. These factors could include dimensions, materials, operating conditions, and more.

Experimental Design: RPD typically involves a carefully designed set of experiments using techniques such as DOE. The experiments are structured to systematically vary the critical parameters over a defined range while keeping other noncritical parameters constant. This helps in understanding the relationship between input factors and the output (performance or quality).

Response Variables: Engineers also define response variables or quality characteristics that need to be optimized or controlled. These could include measures of product performance, durability, reliability, or any other relevant metrics.

Robustness Criteria: In RPD, engineers specify robustness criteria or objectives that define the desired level of performance or quality in the presence of variability. These criteria help in quantifying how well the system performs under different conditions.

Data Analysis: Statistical analysis methods are employed to analyze the experimental data and determine the optimal parameter settings that meet the robustness criteria. Engineers use tools such as regression analysis, ANOVA, and other statistical techniques to identify the best combination of parameter values.

Confirmation Experiments: After identifying the optimal parameter settings, confirmation experiments may be conducted to validate the robustness of the design. These experiments ensure that the system can consistently meet the desired performance criteria under different real-world conditions.

Iteration and Optimization: RPD is often an iterative process. Engineers may go through multiple rounds of experimentation and analysis to fine-tune the design and further improve its robustness.

The ultimate goal of an RPD experiment is to develop a product or process that is less sensitive to variations and uncertainties, resulting in higher reliability, quality, and performance. This methodology is widely used in mechanical engineering, manufacturing, and other fields to enhance product design and process optimization while reducing the risk of performance variations due to external factors.

STEPS

Developing an RPD experiment in mechanical engineering involves several systematic steps to optimize a product or process while accounting for variability and uncertainty in input parameters. Here are the key steps to follow:

1. **Define the Problem:** Clearly define the problem or objective that you want to address with the RPD experiment. This could be improving the performance, reliability, or quality of a mechanical component or system.
2. **Identify Critical Parameters:** Identify the critical input parameters or factors that can significantly influence the outcome or performance of the system. These parameters may include dimensions, material properties, operating conditions, and more.
3. **Determine Response Variables:** Define the response variables or quality characteristics that you want to optimize or control. These are the metrics that measure the system's performance, quality, or reliability.
4. **Establish Robustness Criteria:** Specify robustness criteria that define the desired level of performance or quality in the presence of variability and uncertainty. These criteria help set the goals for the RPD experiment.
5. **Design of Experiments (DOE):** Use DOE techniques to create a structured experimental plan. The DOE plan should systematically vary the critical parameters over a defined range while keeping other noncritical parameters constant. Common DOE methods include full factorial, fractional factorial, Taguchi methods, and more.
6. **Conduct Experiments:** Conduct the experiments according to the designed plan. Ensure that the experiments are conducted carefully, and accurate data is collected for the response variables.
7. **Statistical Analysis:** Perform statistical analysis of the experimental data to understand the relationship between the input parameters and the response variables. Use statistical tools like regression analysis, ANOVA, and graphical analysis to identify trends and significant factors.
8. **Optimize Parameter Settings:** Identify the optimal parameter settings that meet the robustness criteria. This may involve using optimization techniques or RSM to find the best combination of parameter values.
9. **Confirmation Experiments:** Conduct confirmation experiments to validate the robustness of the optimized design. Ensure that the system consistently meets the desired performance criteria under different conditions.
10. **Documentation and Reporting:** Document all experimental procedures, results, and findings. Create a comprehensive report that outlines the RPD experiment's methodology, results, and recommendations for implementation.

11. **Iteration:** Depending on the results and objectives, you may need to iterate the RPD process to further improve the design's robustness. This could involve making additional changes to parameter settings or exploring different factors.
12. **Implementation:** Implement the optimized design or process in a real-world context. Monitor and assess its performance over time to ensure that it continues to meet the robustness criteria.
13. **Continuous Improvement:** Continue to monitor and improve the design as needed, considering any new data or changes in the operating environment.

RPD is a systematic and iterative approach that allows mechanical engineers to develop products and processes that are less sensitive to variations and uncertainties, ultimately leading to higher reliability and quality.

Pitfalls and Remedies

RPD experiments in mechanical engineering can be highly effective in improving product and process performance, but there are potential pitfalls that engineers should be aware of. Here are some common pitfalls and remedies to consider:

Inadequate Problem Definition:
    Pitfall: Failing to clearly define the problem or objective can lead to irrelevant or suboptimal results in the RPD experiment.
    Remedy: Start by precisely defining the problem and specifying the goals and objectives of the experiment. Ensure that all stakeholders have a common understanding of the problem.

Incorrect Identification of Critical Parameters:
    Pitfall: Selecting the wrong critical parameters or factors can result in wasted resources and ineffective improvements.
    Remedy: Use engineering judgment, historical data, and expertise to identify the most influential parameters. Conduct sensitivity analyses if necessary to confirm their significance.

Neglecting Interaction Effects:
    Pitfall: Ignoring interaction effects between parameters can lead to suboptimal designs.
    Remedy: Ensure that the experimental design accounts for interactions between parameters by using techniques like full-factorial experiments or higher-level fractional factorial designs.

Inadequate Sample Size:
    Pitfall: Using a sample size that is too small may result in unreliable or inconclusive results.
    Remedy: Conduct a power analysis to determine the appropriate sample size needed to detect significant effects with sufficient confidence. Ensure that the sample size is statistically meaningful.

Lack of Statistical Expertise:
    Pitfall: Inadequate knowledge of statistical analysis techniques can lead to misinterpretation of results.
    Remedy: Involve statisticians or individuals with statistical expertise in the experimental design and data analysis phases. They can guide the process and ensure proper statistical methods are used.

Overfitting:
    Pitfall: Overfitting occurs when a model is too complex and fits the noise in the data rather than the underlying relationships.

Remedy: Use a balance between model complexity and simplicity. Employ techniques like cross-validation to assess model performance and avoid overfitting.

Poor Documentation:

Pitfall: Incomplete or poorly documented experiments can hinder repeatability and make it challenging to draw meaningful conclusions.

Remedy: Maintain thorough documentation of experimental procedures, data collection, and analysis methods. Proper documentation ensures transparency and reproducibility.

Neglecting Confirmation Experiments:

Pitfall: Failing to conduct confirmation experiments may lead to an unverified design.

Remedy: Always include confirmation experiments to validate the robustness of the optimized design. These experiments help ensure that the improvements are practical and effective in real-world conditions.

Lack of Implementation:

Pitfall: Developing an optimized design but not implementing it in the production process or system.

Remedy: Collaborate with relevant stakeholders to ensure that the optimized design is integrated into the production process or system. Monitor its performance in the field and make necessary adjustments.

Not Considering External Factors:

Pitfall: Focusing solely on internal parameters and neglecting external factors (e.g., environmental conditions) that may affect robustness.

Remedy: Take external factors into account during the experimental design and analysis. Consider how variations in external conditions may impact the robustness of the design.

Failing to Iterate:

Pitfall: Assuming that the initial RPD experiment will result in a perfect solution without the need for further optimization.

Remedy: Be prepared to iterate the RPD process, especially if initial results do not meet the desired robustness criteria. Continuously refine and improve the design.

By being aware of these potential pitfalls and implementing the suggested remedies, engineers can conduct more effective and successful RPD experiments in mechanical engineering, leading to improved product and process performance.

Example

An RPD experiment in mechanical engineering aims to optimize a product or process by identifying and controlling key design parameters to ensure robust and reliable performance, even in the presence of variability or external factors. Here's a simplified example of an RPD experiment in the context of mechanical engineering:

Objective: To design a robust suspension system for a car to ensure smooth ride quality under varying road conditions while minimizing the effect of manufacturing tolerances and wear and tear.

Factors:

Spring Stiffness (Factor A): The stiffness of the car's suspension springs, which affects ride comfort.

Damper Damping Ratio (Factor B): The damping ratio of the shock absorbers, which affects how quickly the suspension responds to bumps.

Tire Pressure (Factor C): The air pressure in the tires, which affects the grip and cushioning effect of the tires on the road.

Wheel Alignment (Factor D): The alignment of the wheels, including camber and toe settings, which affects tire wear and handling.

Manufacturing Variability (Factor E): Variations in manufacturing processes that can affect the consistency of the suspension components.

Levels: For each factor, you might consider two levels, such as "High" and "Low," representing extreme settings or values.

Response Variables:

Ride Comfort Score: A subjective score given by test drivers after experiencing the suspension system's performance on different road surfaces.

Tire Wear Rate: The rate at which the tires wear out under different conditions.

Handling Stability: A measure of how well the car maintains stability during sharp turns.

Experimental Design: To conduct an RPD experiment, you would use a statistical DOE approach. You may choose a suitable experimental design, such as a Taguchi OA array or an RSM design, based on the number of factors, levels, and interactions you want to investigate.

For example, you could use a Taguchi OA with eight experimental runs (Table 3.49).

Data Collection and Analysis: Conduct the experiments using the defined factor levels, and collect data on the response variables. Perform statistical analysis, including ANOVA and regression modeling if using RSM, to determine the optimal combination of factors for achieving robust performance while accounting for variability.

By conducting this RPD experiment, you can identify the best suspension system parameters that provide a comfortable and stable ride under various conditions, considering manufacturing variations and wear and tear.

REPORTING

RPD experiments in mechanical engineering use various statistical tools and techniques to analyze and optimize the performance of a product or process while considering variability and external factors. Here are some of the key statistical tools commonly used in RPD designed experiments:

TABLE 3.49

Robust Parameter Design (RPD) Matrix Example for Mechanical Engineering

Run	Factor (A)	Factor (B)	Factor (C)	Factor (D)	Factor (E)
1	High	High	High	High	Low
2	High	Low	Low	High	High
3	Low	High	Low	Low	High
4	Low	Low	High	Low	Low
5	High	High	Low	Low	High
6	High	Low	High	Low	low
7	Low	High	High	High	High
8	Low	Low	Low	High	Low

Factorial Design: Factorial experiments involve systematically varying factors (input parameters) at different levels to study their main effects and interactions. These designs help identify which factors have a significant impact on the response variables.

Taguchi Methods: Taguchi methods, including Taguchi OAs, are widely used in RPD experiments. They provide a structured approach to varying factors and levels in a way that minimizes the number of experiments needed to estimate main effects and interactions.

Analysis of Variance (ANOVA): ANOVA is used to statistically analyze the data collected from experiments. It helps determine the significance of factors and their interactions on the response variables.

Response Surface Methodology (RSM): RSM is a powerful tool for optimizing responses with multiple factors. It involves fitting mathematical models to experimental data and then using these models to find the optimal settings for the factors.

Signal-to-Noise (S/N) Ratios: In Taguchi methods, S/N ratios are used to quantify the quality of the response variables. Different S/N ratios (e.g., smaller-the-better, larger-the-better, and nominal-the-best) are chosen based on the nature of the response.

Design of Experiments (DOE) Software: Specialized software packages, such as Minitab, JMP, and Design-Expert, are often used to design experiments, collect data, and perform statistical analyses. These tools provide a user-friendly interface for creating experimental designs and conducting analyses.

Robust Optimization Algorithms: Optimization algorithms, such as genetic algorithms, simulated annealing, or response surface optimization, are used to find the optimal combination of factors that result in robust performance.

Pareto Analysis: Pareto charts are used to prioritize and focus efforts on improving the factors that have the most significant impact on the response variables.

Control Charts: Control charts are employed to monitor and control the quality and consistency of the product or process during production.

Regression Analysis: When dealing with continuous variables, regression analysis can be used to build mathematical models that relate input factors to response variables, allowing for prediction and optimization.

Statistical Tolerance Analysis: This technique assesses how variations in input parameters impact product tolerances, ensuring that the design remains robust even in the presence of manufacturing variations.

Monte Carlo Simulation: Monte Carlo simulations can be used to model and analyze the effects of uncertainty and variability in input parameters on the performance of a product or process.

These statistical tools and techniques help mechanical engineers systematically plan, execute, and analyze RPD experiments, ultimately leading to the development of robust and reliable designs that can perform well under various conditions and in the presence of external factors and variations.

A report for an RPD experiment in mechanical engineering should be comprehensive and well-organized to convey the experiment's purpose, methodology, results, and conclusions effectively. Here are the essential elements to include in an RPD report:

Title: A clear and descriptive title that summarizes the main objective of the RPD experiment.

Abstract: A concise summary of the experiment's purpose, key methods, major findings, and conclusions.

Introduction:
- Background and context: Provide an overview of the problem or challenge being addressed by the RPD experiment.
- Objectives: Clearly state the objectives and goals of the experiment.
- Importance: Explain why the experiment is significant and its potential impact on the field of mechanical engineering.

Literature Review: Review relevant literature and prior research related to the problem or topic of the RPD experiment. Highlight the existing knowledge, gaps, and theories relevant to the experiment.

Experimental Design: Describe the experimental setup, including the equipment and materials used. Explain the choice of factors, levels, and response variables. Discuss the chosen statistical design, such as factorial design, Taguchi method, or RSM.

Methods: Detail the procedures followed during the experiment, including how factors were varied, data collection methods, and any control measures implemented. Specify any statistical techniques used for analysis, such as ANOVA, regression, or S/N ratios. Explain how the experiments were conducted to ensure reliability and repeatability.

Results:
Present the experimental data, including raw data and any statistical analyses performed.
Use tables, charts, graphs, and figures to illustrate key findings and trends.
Include the results of significant factors, interactions, and the impact on response variables.
Discuss any patterns or insights revealed by the data.

Discussion:
Interpret the results and relate them to the experiment's objectives.
Explain the practical implications of the findings for the mechanical engineering application.
Discuss the robustness of the design in addressing variability and external factors.
Compare the results with existing literature and prior research.

Conclusions:
Summarize the main findings and their significance.
State whether the experiment achieved its objectives.
Provide recommendations or insights for future work or improvements.

Recommendations (if applicable):
Offer practical recommendations for implementing the RPD findings in real-world applications.
Suggest modifications or adjustments based on the experiment's outcomes.

Acknowledgments (if applicable):
Acknowledge individuals, organizations, or funding sources that contributed to the experiment's success.

References:
List all the references cited in the report following a consistent citation style (e.g., APA, IEEE, or Chicago).

Appendices:
Include any supplementary information, such as detailed experimental data, calculations, or additional charts/graphs, in appendices.

Glossary or Notation (optional):
Define any specialized terms, acronyms, or symbols used in the report.

Table of Contents and List of Figures/Tables:
Include a table of contents for easy navigation.
Provide a list of figures and tables, each with a corresponding page number.

Ensure that your RPD report is well-structured, clear, and accessible to readers. Effective communication of the experiment's details, results, and conclusions is essential in the field of mechanical engineering.

3.5.8 RANDOMIZED EXPERIMENTS

Randomized designed experiments, often referred to as REs or randomized controlled experiments, are a research methodology used in various fields, including mechanical engineering, to investigate the effects of different factors or treatments while minimizing the influence of confounding variables and bias. In mechanical engineering, REs are typically used to study the impact of design changes, materials, manufacturing processes, or other factors on the performance of products, systems, or processes.

The following are key characteristics and concepts related to randomized designed experiments in mechanical engineering:

Randomization: Randomization is a fundamental principle in these experiments. It involves randomly assigning experimental units (e.g., test samples, components, or prototypes) to different treatment groups or conditions. Randomization helps ensure that any uncontrolled or hidden factors that could affect the outcomes are equally distributed among the groups, reducing bias and allowing for valid statistical inference.

Controlled Factors: In a RE, researchers manipulate and control specific factors or treatments. These factors can include design parameters, material properties, process settings, or other variables of interest in mechanical engineering.

Experimental Design: Careful planning of the experimental design is essential. Researchers decide how many groups or treatments to include, how to allocate units to these groups, and what measurements or responses to collect. Common designs include completely randomized designs, randomized block designs, and factorial designs.

Replication: To ensure the reliability of the results, experiments often involve replication, where the same treatment combinations are tested multiple times. Replication helps estimate the variability and provides more robust statistical analyses.

Response Variables: Mechanical engineers measure and analyze response variables to assess the effects of the treatments or factors. These response variables could include performance metrics, material properties, stress-strain behavior, fatigue life, or any other relevant parameters.

Statistical Analysis: Data collected from REs are analyzed using statistical methods to determine whether observed differences in response variables are statistically significant. Common techniques include ANOVA, t-tests, regression analysis, and confidence intervals.

Inference and Conclusions: Researchers draw conclusions based on statistical analyses. They determine whether the treatments have a significant impact on the response variables and provide insights into how the factors affect the outcomes.

Causal Inference: REs are particularly useful for establishing causal relationships. By randomly assigning units to treatments, researchers can make stronger claims about causality compared to observational studies.

Examples of how randomized designed experiments can be applied in mechanical engineering include:

- Testing the effects of different alloy compositions on the tensile strength of a material.
- In quality control, REs can be used to study the effects of manufacturing process changes on the reliability of mechanical components, such as bearings or gears.
- Evaluating the impact of various heat treatment processes on the hardness of a metal component.
- Investigating the influence of different lubricants on the wear and friction characteristics of mechanical components.
- Studying the effectiveness of different design parameters on the performance and durability of a mechanical system or product.

Randomized designed experiments are a valuable tool in mechanical engineering to systematically explore and quantify the effects of factors on engineering outcomes, leading to evidence-based design decisions and improvements in product quality and performance.

STEPS

Developing a randomized designed experiment in mechanical engineering involves a systematic approach to planning, conducting, and analyzing the experiment to obtain meaningful and statistically valid results. Here are the steps to develop a randomized designed experiment:

1. **Define the Research Objectives**: Clearly state the research goals and objectives of the experiment. What specific mechanical engineering problem or question are you trying to address?
2. **Identify the Factors of Interest**: Determine the factors or variables that you want to investigate. These could include design parameters, material properties, process settings, or any other variables relevant to your research objectives.
3. **Define the Levels of Factors**: Specify the different levels or settings for each factor. For example, if you are studying the effect of temperature, you might have two levels: "High" and "Low."
4. **Select Response Variables**: Identify the response variables that will be measured or observed to assess the impact of the factors. These variables should directly relate to your research objectives and may include performance metrics, material properties, or other relevant measurements.
5. **Determine the Experimental Design**: Choose an appropriate experimental design based on the number of factors, levels, and interactions you want to investigate. Common designs include completely randomized designs, randomized block designs, factorial designs, and more complex designs like Taguchi methods or RSM.
6. **Randomize the Experimental Units**: Randomly assign the experimental units (e.g., test samples, components, or prototypes) to the different treatment groups or conditions. Randomization helps ensure that uncontrolled factors are equally distributed among the groups, reducing bias.
7. **Replicate the Experiment**: Replicate the experimental conditions by conducting the experiment multiple times. Replication helps estimate variability and provides more robust statistical analyses.
8. **Plan the Data Collection**: Determine how data will be collected, including the measurement methods, instruments, and the frequency of data collection. Ensure that data collection is consistent and well-documented.
9. **Conduct the Experiment**: Execute the experimental plan by applying the treatments to the assigned units and collecting the data as specified.
10. **Analyze the Data**: Perform statistical analyses on the collected data to assess the effects of the factors on the response variables. Common techniques include ANOVA, t-tests, regression analysis, and others depending on the design and nature of the data.
11. **Interpret the Results**: Interpret the results of the statistical analyses in the context of your research objectives. Determine whether the factors have a significant impact and how they influence the mechanical engineering outcomes.
12. **Draw Conclusions**: Based on the results and interpretations, draw conclusions regarding the relationships between the factors and the mechanical engineering outcomes. Assess the practical implications of your findings.
13. **Make Recommendations**: Provide recommendations for design changes, process improvements, or further research based on the conclusions drawn from the experiment.
14. **Document the Experiment**: Prepare a detailed report documenting the entire experiment, including the objectives, methods, results, conclusions, and any recommendations. Ensure that the report is clear, well-organized, and accessible to others in the field.
15. **Communicate the Findings**: Present the findings to relevant stakeholders, colleagues, or the broader mechanical engineering community through presentations, publications, or other appropriate means.

16. **Reflect and Iterate**: Reflect on the outcomes of the experiment and consider whether further experimentation or refinement of the design is necessary to address any remaining questions or uncertainties.

By following these steps, you can develop a randomized designed experiment in mechanical engineering that systematically investigates the effects of factors on engineering outcomes, leading to data-driven decisions and advancements in the field.

PITFALLS AND REMEDIES

Randomized designed experiments in mechanical engineering are powerful tools for obtaining reliable and statistically valid results. However, like any scientific endeavor, they can be subject to various pitfalls. Identifying and addressing these pitfalls is crucial for ensuring the integrity and accuracy of your experiments. Here are some common pitfalls and their corresponding remedies:

Pitfall 1: Inadequate Planning

Remedy: Thoroughly plan your experiment before execution. Define clear objectives, select appropriate factors and levels, and choose a suitable experimental design. Conduct a pilot study if necessary to refine your approach.

Pitfall 2: Poor Randomization

Remedy: Implement proper randomization techniques to assign experimental units to treatment groups. This helps ensure that uncontrolled factors are equally distributed among groups. Use random number generators or software for randomization.

Pitfall 3: Inadequate Replication

Remedy: Include an adequate number of replications for each treatment group to estimate variability and enhance the reliability of your results. Ensure that the sample size is statistically sufficient to detect meaningful effects.

Pitfall 4: Measurement Error

Remedy: Calibrate measurement instruments regularly to minimize measurement errors. Use appropriate statistical techniques to account for measurement variability in your analyses.

Pitfall 5: Incomplete Data Collection

Remedy: Ensure that data is collected consistently and completely. Implement data validation checks to identify and address missing or erroneous data.

Pitfall 6: Violation of Assumptions

Remedy: Check and ensure that the assumptions of the chosen statistical methods (e.g., normality, homoscedasticity) are met. If assumptions are violated, consider using robust statistical techniques or data transformations.

Pitfall 7: Confounding Variables

Remedy: Identify and control for potential confounding variables that may affect your results. Use appropriate blocking or stratification techniques in your experimental design to account for these variables.

Pitfall 8: Small Sample Size

Remedy: Ensure that your sample size is large enough to detect meaningful effects. Conduct power analyses to determine the required sample size for your experiment.

Pitfall 9: Misinterpretation of Results

Remedy: Seek expert statistical consultation if needed to correctly analyze and interpret the data. Avoid drawing unwarranted conclusions or making overgeneralizations based on the results.

Pitfall 10: Lack of Documentation

> **Remedy**: Maintain thorough and well-organized records throughout the experiment, including experimental procedures, data collection, and any deviations from the original plan. This documentation is crucial for reproducibility and transparency.

Pitfall 11: Ignoring Practical Significance

> **Remedy**: While statistical significance is important, don't ignore practical significance. Assess the practical implications of your findings and consider whether the observed effects are meaningful in the context of your mechanical engineering application.

Pitfall 12: Limited Generalizability

> **Remedy**: Recognize the limitations of your experiment's generalizability. Be clear about the specific conditions and contexts to which your results apply and communicate these limitations in your report.

Pitfall 13: Neglecting Ethical Considerations

> **Remedy**: Ensure that your experiment adheres to ethical guidelines and safety standards relevant to mechanical engineering research. Obtain necessary approvals and permissions for experiments involving human subjects, animals, or hazardous materials.

By being aware of these potential pitfalls and proactively addressing them, you can enhance the quality and reliability of your randomized designed experiments in mechanical engineering, leading to more accurate and valuable results. Collaboration with statisticians or experts in experimental design can also help navigate these challenges effectively.

EXAMPLE

A randomized designed experiment in mechanical engineering often involves assigning experimental units (e.g., test samples, components, or prototypes) to different treatment groups or conditions in a random manner to minimize bias and obtain valid statistical results. The assignment matrix is a common way to represent how experimental units are allocated to treatments in such experiments. Below is a simplified example of an assignment matrix for a randomized designed experiment in mechanical engineering:

Suppose you are conducting an experiment to study the effect of different heat treatment processes on the hardness of a metal alloy. You have three different heat treatment processes (A, B, and C) that you want to test, and you have 12 identical metal samples as your experimental units.

Table 3.50 is an assignment matrix for this experiment.

In this assignment matrix:

- Each row represents an individual metal sample (an experimental unit).
- The "Sample" column lists the sample numbers from 1 to 12.
- The "Heat Treatment" column specifies which heat treatment process (A, B, or C) each sample will undergo.

By randomly assigning the samples to different treatments, you ensure that any uncontrolled factors or variability that may affect the hardness measurement are equally distributed among the treatment groups. This randomization helps reduce bias and allows for valid statistical analysis to determine the impact of the heat treatment processes on the hardness of the metal alloy.

In a randomized designed experiment in mechanical engineering, the matrix used to plan and organize the experiment typically outlines the allocation of experimental units to various treatment groups or conditions. The exact format of the matrix can vary depending on the experiment's design and objectives. Here's an example of a simple matrix for a randomized designed experiment in mechanical engineering:

TABLE 3.50
Randomized Experiment Matrix
Example for Mechanical Engineering

Sample	Heat Treatment
1	A
2	C
3	B
4	C
5	A
6	B
7	C
8	A
9	B
10	B
11	A
12	C

Objective: To investigate the effect of different heat treatment processes on the hardness of steel components.

Factors:

Heat Treatment Method (Factor A):
- Level 1: Annealing
- Level 2: Quenching
- Level 3: Tempering

Heating Temperature (Factor B):
- Level 1: 800°C
- Level 2: 900°C
- Level 3: 1,000°C

Response Variable: Rockwell Hardness (measured on a numerical scale)

Matrix for Randomized Experiment:

In this example, a randomized matrix ensures that the steel components are assigned randomly to different combinations of heat treatment methods and heating temperatures. This randomization helps control potential confounding factors and ensures that the effects of the factors are assessed without bias.

The matrix might look something like Table 3.51 for a portion of the experiment.

Each row in the matrix represents a specific experimental run where a steel component is subjected to a particular combination of heat treatment methods and heating temperatures. By randomly assigning the components to these treatment groups, you can control for potential bias and assess the effects of the factors (heat treatment method and heating temperature) on the Rockwell Hardness.

The data collected from these experimental runs can then be analyzed statistically to determine whether the heat treatment method and heating temperature significantly affect the hardness of the steel components, leading to valuable insights for mechanical engineering applications.

TABLE 3.51

Randomized Experiment Matrix Example for Mechanical Engineering

Run	Heat Treatment Method (A)	Heating Temperature (B)
1	Annealing	800°C
2	Quenching	900°C
3	Tempering	1,000°C
4	Annealing	900°C
5	Tempering	800°C
6	Quenching	1,000°C
7	Quenching	800°C
8	Annealing	1,000°C

REPORTING

Randomized designed experiments in mechanical engineering often involve the use of various statistical tools and techniques to analyze and interpret data. These tools are essential for drawing meaningful conclusions from experimental results and optimizing mechanical systems. Some common statistical tools used in randomized designed experiments for mechanical engineering include:

Descriptive Statistics:
   Mean, Median, and Mode: Measures of central tendency to describe the average or typical value of a dataset.
   Variance and Standard Deviation: Measures of dispersion or spread that quantify the variability within the data.
   Range: The difference between the maximum and minimum values in a dataset.
Inferential Statistics:
   Hypothesis Testing: Statistical tests to determine whether there is a significant difference between groups or conditions in an experiment. Common tests include t-tests, ANOVA, and chi-squared tests.
   Confidence Intervals: Calculations to estimate the range of values within which a population parameter is likely to fall with a certain level of confidence.
   Regression Analysis: Modeling relationships between variables, such as linear regression or multiple regression, to predict outcomes or understand the impact of independent variables.
   Analysis of Covariance (ANCOVA): Extends ANOVA to control for the effects of one or more continuous covariates.
   Nonparametric Tests: Statistical tests that do not assume a specific distribution of data, suitable when data do not meet parametric assumptions. Examples include the Wilcoxon rank-sum test and the Kruskal-Wallis test.
Design of Experiments (DOE):
   Factorial Designs: Investigate the effects of multiple factors or variables simultaneously to identify main effects and interactions.
   Fractional Factorial Designs: A subset of full-factorial designs that allow for efficient experimentation with fewer experimental runs.
   Central Composite Design (CCD): A type of RSM that combines factorial and axial points to model complex relationships.

Taguchi Methods: A systematic approach for designing experiments to optimize processes and products while minimizing variation and sensitivity to noise factors.

Randomization Techniques:

Random Assignment: Assigning experimental units (e.g., samples or test subjects) to treatment groups in a random manner to reduce bias.

Random Sampling: Selecting samples from a population in a random manner to ensure representativeness.

Random Number Generation: Using random number generators or tables to introduce randomness into the experiment.

Statistical Software:

Utilizing statistical software packages such as R, SAS, Minitab, MATLAB, or Python with libraries like SciPy and statsmodels to perform statistical analyses and generate graphical visualizations.

Graphical Tools:

Creating plots and charts, including histograms, box plots, scatter plots, and probability plots, to visualize data distributions and relationships.

Analysis of Variance (ANOVA):

ANOVA techniques help determine the sources of variation in an experiment, including partitioning the variance into different components such as treatment effects, errors, and interactions.

Response Surface Methodology (RSM):

RSM involves fitting mathematical models to experimental data to optimize responses, often using techniques like CCDs.

Statistical Process Control (SPC):

SPC tools, such as control charts and process capability analysis, help monitor and control manufacturing processes to ensure product quality and consistency.

Reliability Analysis:

Techniques like Weibull analysis and reliability growth modeling assess the reliability and failure rates of mechanical systems and components.

The choice of statistical tools and techniques depends on the specific goals and complexity of the randomized designed experiment in mechanical engineering. Proper statistical analysis is crucial for drawing valid conclusions, making informed decisions, and optimizing mechanical systems for improved performance and reliability.

A report for a randomized designed experiment in mechanical engineering typically includes several key elements to effectively communicate the experimental methodology, results, and conclusions. Here are the common elements of such a report:

Title Page:

Title of the Report: A concise and informative title that reflects the experiment's objective.

Author(s): Names of the individuals or research team responsible for conducting the experiment.

Affiliation(s): Affiliated institution or organization.

Date: The date of report preparation.

Abstract:

A brief summary of the experiment, including the research question, methodology, key findings, and conclusions. It should provide a concise overview of the entire report.

Table of Contents:

A list of sections, subsections, and their respective page numbers, makes it easy for readers to navigate the report.

List of Figures and Tables:

A list of all figures and tables used in the report, along with their respective page numbers.

Introduction:

Background and Context: A brief introduction to the problem or research question addressed by the experiment.

Objectives: Clearly stated objectives or hypotheses that the experiment aims to investigate.

Significance: Explanation of why the research is important or how it contributes to the field of mechanical engineering.

Scope: A description of the scope and limitations of the experiment.

Literature Review:

Review of existing literature and relevant studies related to the experiment's topic. This section provides context and justifies the need for the current experiment.

Experimental Design:

Randomization Process: Explanation of how randomization was implemented in the experiment to minimize bias and ensure unbiased results.

Sampling Procedure: Details on the sample selection process and the criteria used.

Variables: Explanation of the independent and dependent variables, including their definitions, units of measurement, and operational definitions.

Experimental Setup: A description of the equipment, materials, and procedures used in the experiment.

Randomization Plan: If applicable, a description of the randomization plan, including how randomization was achieved (e.g., random number generator or software).

Data Collection: Explanation of how data was collected, including data sources, measurement tools, and data recording methods.

Data Analysis:

Statistical Methods: Description of the statistical techniques and software used for data analysis, including any assumptions made.

Results: Presentation of the experimental results, including tables, figures, and statistical analyses. This section may include graphs, charts, and plots to illustrate trends and relationships.

Discussion of Findings: Interpretation of the results, addressing whether they support or reject the hypotheses. Discuss any unexpected results and their potential implications.

Conclusion:

Summary of the key findings and their implications for the research question.

Statement of Conclusions: A clear and concise statement about whether the experiment's objectives were achieved.

Practical Implications: Discuss the practical applications and significance of the findings in the context of mechanical engineering.

Recommendations:

Suggestions for future research or improvements to the experimental design.

Recommendations for implementing the findings in practical engineering applications.

References:

A list of all sources cited in the report, following a specific citation style (e.g., APA, IEEE, Chicago).

Appendices:

Any additional information that supplements the main report, such as raw data, detailed calculations, experimental protocols, or additional figures and tables.

Acknowledgments (optional):

Recognition of individuals, organizations, or institutions that contributed to the experiment or report.

It's essential to adhere to a clear and organized structure in your report to ensure that readers can follow your experiment's methodology and understand the significance of your findings. Additionally, make sure to use appropriate technical language and provide sufficient detail for peer researchers and engineers in the field.

3.5.9 SPLIT-PLOT AND BLOCKED EXPERIMENTS

Split-plot and blocked experiments are experimental design techniques used in various fields, including mechanical engineering, to efficiently study multiple factors and their interactions while minimizing experimental effort and resources. These techniques help researchers optimize processes, products, or systems by systematically varying factors and studying their effects. Let's understand what split-plot and blocked experiments are in the context of mechanical engineering:

Split-Plot Experiments:
- In a split-plot experiment, you have factors or variables that are divided into two or more levels of hierarchy, often referred to as "whole plots" and "subplots."
- Whole Plots: These represent the main experimental units, where changes to the factor levels are relatively more challenging, time-consuming, or expensive to implement.
- Subplots: These represent smaller-scale experimental units where changes to factor levels are easier or less costly.
- The key idea is to vary the whole-plot factors less frequently and the subplot factors more frequently to efficiently explore interactions and main effects.
- Split-plot experiments are used when certain factors are difficult to change frequently or when there is a natural hierarchy among factors.

Blocked Experiments:
- In a blocked experiment, you group experimental runs into blocks based on some characteristic that is not of primary interest but is known to influence the response.
- Blocking helps reduce the variability introduced by the blocking factor, making it easier to detect the effects of primary factors.
- Blocks are used to control for extraneous sources of variation or nuisance factors, ensuring that their effects do not confound the primary factors of interest.
- For example, in mechanical engineering, if you are testing the impact resistance of different materials, you may block the test runs based on the material supplier to account for potential variations between suppliers.

Applications in Mechanical Engineering:
- Split-plot and blocked experiments in mechanical engineering can be applied to various scenarios, such as:
- Testing the performance of different materials under various environmental conditions (whole plots: environmental conditions, subplots: materials).
- Evaluating the impact of machining parameters on surface finish (whole plots: machine settings, subplots: workpiece materials).
- Studying the effects of heat treatment on mechanical properties of alloys (whole plots: heat treatment process, subplots: alloy compositions).
- Optimizing manufacturing processes with factors like tool selection and cutting speed (whole plots: tool types, subplots: cutting conditions).

Benefits:
- Efficient resource utilization: By varying some factors less frequently (whole plots), you can save time, cost, and resources while still obtaining valuable insights into factor interactions.
- Better control: Blocking helps control for nuisance factors, improving the precision of estimated effects and making it easier to identify significant factors.
- Reduced experimental error: Split-plot and blocked designs can lead to more reliable and interpretable results compared to uncontrolled experiments.

These design techniques allow mechanical engineers to conduct experiments in a systematic and cost-effective manner, leading to better-informed decisions and optimized mechanical systems and processes. Properly designed and executed split-plot and blocked experiments can provide valuable insights for improving product performance, manufacturing processes, and system reliability.

STEPS

Developing a split-plot or blocked designed experiment in mechanical engineering involves several steps to plan, execute, and analyze the experiment effectively. These steps ensure that the experiment is designed to study the factors of interest while accounting for potential sources of variation or nuisance factors. Here are the key steps to develop a split-plot or blocked experiment in mechanical engineering:

1. **Define the Objectives**:
   - Clearly state the research objectives and what you aim to achieve through the experiment. Identify the specific factors and responses you want to study and optimize.
2. **Identify Factors and Levels**:
   - Determine the factors (independent variables) that may affect the mechanical system or process under investigation. Identify the levels or settings for each factor.
   - Categorize factors as whole-plot or subplot factors based on their influence and ease of manipulation.
3. **Select Response Variables**:
   - Choose response variables that represent the outcomes or performance metrics relevant to your objectives. These could include mechanical properties, product quality measures, or system performance indicators.
4. **Identify Potential Nuisance Factors**:
   - Identify any extraneous or nuisance factors that may introduce variability and affect the results. These could be environmental conditions, operator differences, or other sources of variability.
5. **Determine the Blocking Structure**:
   - Determine the criteria for grouping experimental runs into blocks. Blocks are used to control for the effects of nuisance factors.
   - Select blocking factors that are known to influence the response and group similar experimental runs together within blocks.
6. **Design the Experiment**:
   - Choose an appropriate experimental design methodology based on the factors, levels, and blocking structure. Common designs include split-plot designs, fractional factorial designs, or response surface designs.
   - Assign factor levels to whole-plot and subplot factors according to the design chosen. Randomize the order of experimental runs within blocks.
7. **Execute the Experiment**:
   - Conduct the experimental runs according to the designed plan, following the assigned factor settings.
   - Ensure that the blocking criteria are met, and blocks are executed in a randomized order.
   - Record data accurately for response variables and any other relevant observations.
8. **Data Analysis**:
   - Perform statistical analyses to examine the effects of factors on the response variables.
   - Assess the significance of main effects and interactions using appropriate statistical tests (e.g., ANOVA, regression analysis).
   - Evaluate the contribution of nuisance factors and their interactions with primary factors.

9. **Interpret Results**:
   - Interpret the results of the analysis to draw conclusions about the factors that significantly affect the response variables.
   - Identify any interactions or unexpected findings that may require further investigation or adjustments in the mechanical system or process.

10. **Optimization**:
   - If the goal is optimization, use the results to determine the optimal factor settings that maximize or minimize the response variables.
   - Perform sensitivity analyses to assess the robustness of the optimized settings to variations in factors.

11. **Report and Documentation**:
   - Prepare a detailed report that documents the experiment's design, execution, results, and conclusions.
   - Include any recommendations for process or system improvements based on the findings.

12. **Validation and Implementation**:
   - Validate the findings and recommendations through additional experiments or pilot studies if necessary.
   - Implement the optimized settings or process improvements in the mechanical system or manufacturing process.

13. **Continuous Monitoring and Improvement**:
   - Continuously monitor the mechanical system or process to ensure that the optimizations are maintained and assess the long-term impact on performance.

Throughout these steps, it's crucial to maintain rigor in experimental design, data collection, and analysis to ensure the reliability and validity of the results. Collaboration with statistical experts or engineers experienced in experimental design can be beneficial for complex split-plot or blocked experiments in mechanical engineering.

PITFALLS AND REMEDIES

Split-plot and blocked designed experiments in mechanical engineering can be powerful tools for studying complex systems and optimizing processes. However, like any experimental approach, they can also present challenges and pitfalls. Here are some common pitfalls and their corresponding remedies in split-plot and blocked experiments:

Pitfall 1: Inadequate Blocking
Issue: Ineffective blocking can lead to a failure to control for nuisance factors adequately, resulting in biased or confounded results.

Remedy: Carefully identify and select appropriate blocking factors that are known to influence the response. Ensure that blocking factors are homogeneous within blocks. Conduct a thorough pilot study to identify potential sources of variation and validate the choice of blocking factors.

Pitfall 2: Insufficient Randomization
Issue: Inadequate randomization within blocks can lead to biased or nonrepresentative results.

Remedy: Use randomization procedures to assign experimental runs to blocks and to randomize the order of runs within blocks. Implement a systematic randomization process to ensure that each factor combination is equally likely within blocks.

Pitfall 3: Incomplete Factorial Design

Issue: A factorial design may be incomplete, missing some factor combinations, potentially overlooking important interactions.

Remedy: Ensure that the experimental design includes all factor combinations, especially when studying interactions. Fractional factorial designs can be used if there are too many factor combinations to test. Consider adding center points or axial points to investigate curvature or nonlinearity in the response.

Pitfall 4: Ignoring Uncontrolled Factors

Issue: Failing to identify and account for uncontrolled factors (noise factors) can result in unexplained variability in the response.

Remedy: Conduct a noise factor analysis to identify and quantify the effects of uncontrolled factors on the response. If possible, include noise factors as covariates in the analysis to account for their influence.

Pitfall 5: Overlooking Interaction Effects

Issue: Ignoring interaction effects between factors can lead to suboptimal or misleading conclusions.

Remedy: Perform thorough statistical analyses to assess not only the main effects but also the interaction effects between factors. Visualize interactions using interaction plots or contour plots to gain a deeper understanding of the system's behavior.

Pitfall 6: Lack of Replication

Issue: Conducting too few replicate runs within each block can result in unreliable estimates and reduced statistical power.

Remedy: Ensure an adequate number of replicates within each block to obtain precise estimates and detect significant effects. Use power and sample size calculations to determine the appropriate number of replicates based on expected effect sizes and variability.

Pitfall 7: Inadequate Documentation

Issue: Poor documentation can lead to difficulties in reproducing the experiment or understanding the results later.

Remedy: Maintain detailed records of the experimental design, factor settings, data collection procedures, and any deviations from the plan. Create a comprehensive report that includes all relevant information, ensuring that the experiment can be replicated or extended in the future.

Pitfall 8: Over-Complexity

Issue: Overly complex experimental designs can be difficult to execute and analyze, leading to confusion and inefficiency.

Remedy: Aim for a balance between experimental complexity and the resources available. Simpler designs may be more practical and informative in some cases. Consult with experts in experimental design or statistics when designing complex experiments.

Pitfall 9: Ignoring Assumptions

Issue: Violating the assumptions of the statistical analysis can lead to incorrect conclusions.

Remedy: Validate the assumptions of the chosen statistical tests or models. For example, check for normality and homogeneity of variances. If assumptions are not met, consider appropriate transformations or nonparametric tests.

Pitfall 10: Lack of Follow-Up

Issue: Failing to act on the results and implement improvements can render the experiment meaningless.

Remedy: Use the findings to make informed decisions and implement optimizations or process improvements. Monitor the system or process to ensure that the benefits of the experiment are sustained.

By being aware of these pitfalls and taking appropriate measures to address them, engineers can maximize the value of split-plot and blocked designed experiments in mechanical engineering and avoid common pitfalls that may compromise the validity and utility of the results. Collaboration with experts in experimental design and statistics can also be valuable in addressing these challenges effectively.

EXAMPLE

In mechanical engineering, split-plot and blocked designs are used to study the effects of factors while accounting for certain restrictions or constraints in the experimental setup. These designs involve dividing the experiment into different subplots or blocks. Here's an example of a matrix for a split-plot and blocked designed experiment in mechanical engineering:

Objective: To investigate the effects of two factors, "Cutting Tool Material" and "Cutting Speed," on the wear rate of a cutting tool, while considering that different machine operators have different preferences for settings.

Factors:

1. **Cutting Tool Material (Factor A)**:
   - Level 1: High-Speed Steel (HSS)
   - Level 2: Carbide
2. **Cutting Speed (Factor B)**:
   - Level 1: Low Speed (200 RPM)
   - Level 2: High Speed (400 RPM)

Block Factor: 3. **Operator (Block C)**:

- Operator 1
- Operator 2
- Operator 3

Response Variable: Wear Rate (measured in cubic millimeters per minute, mm³/min)

Matrix for Split-Plot and Blocked Designed Experiment:
In this example, split-plot and blocked designs are used to assess the impact of cutting tool material and cutting speed on the wear rate of a cutting tool, while also considering the influence of different machine operators. The matrix is shown in Table 3.52.

In this matrix, each row represents a specific experimental run. The experiment is divided into blocks based on the operator, ensuring that each operator tests all combinations of cutting tool material and cutting speed. This design allows you to assess the main effects of cutting tool material and cutting speed while accounting for the potential influence of different operators.

By analyzing the data collected from these experimental runs, you can determine how the factors affect the wear rate of the cutting tool, considering both within-operator and between-operator variability, making the results more robust and applicable in real-world scenarios in mechanical engineering.

TABLE 3.52

Matrix for Split-Plot and Blocked Designed Experiment

Run	Cutting Tool Material (A)	Cutting Speed (B)	Operator (C)
1	HSS	Low Speed	Operator 1
2	Carbide	Low Speed	Operator 1
3	HSS	High Speed	Operator 1
4	Carbide	High Speed	Operator 1
5	HSS	Low Speed	Operator 2
6	Carbide	Low Speed	Operator 2
7	HSS	High Speed	Operator 2
8	Carbide	High Speed	Operator 2
9	HSS	Low Speed	Operator 3
10	Carbide	Low Speed	Operator 3
11	HSS	High Speed	Operator 3
12	Carbide	High Speed	Operator 3

REPORTING

Statistical tools and techniques play a crucial role in the analysis of split-plot and blocked designed experiments in mechanical engineering. These tools help researchers assess the effects of factors, identify interactions, and optimize processes or systems. Here are some common statistical tools used in split-plot and blocked experiments in mechanical engineering:

1. **Analysis of Variance (ANOVA):**
   - ANOVA is a fundamental tool for split-plot and blocked experiments. It assesses the significance of the main effects and interaction effects by partitioning the total variation in the response variable into different sources.
   - In split-plot experiments, ANOVA is used to analyze both whole-plot and subplot factors separately, providing insights into their effects on the response.
   - In blocked experiments, ANOVA helps assess the significance of factors and evaluate whether blocking has effectively controlled for nuisance factors.
2. **Regression Analysis:**
   - Regression models are used to establish relationships between independent variables (factors) and the response variable.
   - Multiple regression is commonly employed to model the effects of multiple factors on the response. Interaction terms can be included to capture interaction effects.
3. **Interaction Plots:**
   - Interaction plots visually display the interaction effects between two or more factors. These plots help interpret the nature of interactions, such as synergistic or antagonistic effects.
4. **Contour Plots:**
   - Contour plots are used in RSM experiments to visualize the response surface and identify optimal factor settings that maximize or minimize the response.
   - They are particularly useful for understanding complex interactions and finding optimal operating conditions.

5. **Main Effects Plots**:
   - Main effects plots illustrate the impact of individual factors on the response while holding other factors constant. They provide insights into which factors are most influential.
6. **Residual Analysis**:
   - Residual analysis checks the adequacy of the statistical model by examining the distribution of residuals (the differences between observed and predicted values).
   - It helps verify assumptions like normality and homoscedasticity and identifies outliers.
7. **Power and Sample Size Calculations**:
   - These calculations help determine the required sample size to achieve a desired level of statistical power, ensuring that the experiment can detect significant effects if they exist.
8. **ANOVA with Covariates**:
   - In blocked experiments, it may be necessary to include covariates (continuous nuisance factors) in the analysis to account for their influence on the response variable.
9. **Design of Experiments (DOE) Software**:
   - Specialized DOE software packages, such as Minitab, JMP, or Design-Expert, provide tools for designing experiments, performing statistical analyses, and generating graphical outputs.
10. **Statistical Hypothesis Testing**:
    - Hypothesis tests, such as t-tests and F-tests, are used to assess the significance of differences or effects observed in the experiment.
11. **Sensitivity Analysis**:
    - Sensitivity analyses help assess the robustness of optimized settings by varying factors within specified ranges and observing their impact on the response.
12. **Response Optimization Algorithms**:
    - Optimization algorithms, such as gradient descent or genetic algorithms, can be applied to find optimal factor settings that maximize or minimize the response.
13. **Control Charts and Process Capability Analysis**:
    - In follow-up stages, control charts and process capability analysis are used to monitor and control the process based on the optimized settings.

Selecting the appropriate statistical tools depends on the specific objectives of the experiment, the complexity of the system, and the types of data collected. Properly chosen and executed statistical analyses are essential for extracting meaningful insights from split-plot and blocked experiments in mechanical engineering and making informed decisions for process improvement and optimization.

A well-structured and informative report for a split-plot and blocked designed experiment in mechanical engineering should provide a comprehensive overview of the experiment, its objectives, methods, results, and conclusions. Here are the essential elements that should be included in such a report:

1. **Title and Cover Page**:
   - Title: A clear and concise title that reflects the experiment's purpose.
   - Author(s) and affiliation.
   - Date of the report.
2. **Abstract**:
   - A brief summary of the experiment's purpose, methods, key findings, and conclusions.
   - The abstract should be concise and provide a quick overview of the report's content.
3. **Table of Contents**:
   - List of sections and subsections with page numbers for easy navigation.

4. **List of Figures and Tables**:
   - A separate list of all figures and tables included in the report, along with their corresponding page numbers.
5. **Nomenclature and Abbreviations**:
   - Definitions of any specialized terminology, symbols, or abbreviations used throughout the report.
6. **Introduction:**
   - Background and context: Explain the problem or process being studied and why it is important.
   - Objectives: Clearly state the experiment's goals and what you aim to achieve.
   - Hypotheses or research questions: If applicable, provide the hypotheses or questions guiding the experiment.
7. **Literature Review**:
   - A brief review of relevant literature and prior research related to the experiment.
   - Discuss how the current study fits into the existing body of knowledge.
8. **Experimental Design**:
   - Description of the experimental design, including factors, levels, and their arrangement within whole plots and subplots.
   - Explanation of how blocking was implemented and the rationale behind it.
   - Details about the randomization process.
9. **Materials and Methods**:
   - Description of the equipment, materials, and instruments used in the experiment.
   - Experimental procedures: Step-by-step instructions on how the experiment was conducted, including any measurements, data collection methods, and data recording procedures.
   - Statistical analysis plan: Explain the statistical methods and tools used to analyze the data.
10. **Results**:
    - Presentation of the experimental data in a clear and organized manner.
    - Tables, figures, graphs, and charts: Use visual aids to present the results effectively.
    - Include summaries of key findings, including main effects and interaction effects.
11. **Discussion:**
    - Interpretation of the results: Explain the implications of the findings and how they relate to the experiment's objectives.
    - Address any unexpected results or trends.
    - Compare the results with prior literature and discuss their significance.
    - Consider limitations of the study and potential sources of error.
12. **Conclusion:**
    - Summarize the main findings and their relevance to the experiment's objectives.
    - Discuss the practical implications of the results.
    - Offer recommendations or insights for further research or process improvement.
13. **Acknowledgments:**
    - Recognize individuals, organizations, or funding sources that contributed to the experiment's success.
14. **References:**
    - Cite all relevant sources, research papers, and references used in the report.
15. **Appendices:**
    - Include any supplementary information that supports the main body of the report, such as raw data, statistical calculations, detailed experimental procedures, or additional figures and tables.

16. **Signatures and Approval**:
    - If applicable, include signatures and approvals from supervisors, team members, or stakeholders who have reviewed and endorsed the report.
17. **Glossary (optional)**:
    - Define any technical terms or acronyms that may be unfamiliar to the readers.
18. **Index (optional)**:
    - An index may be included for reports with extensive content to aid in locating specific information.

It's important to ensure that the report is well-organized, free from grammatical errors, and adheres to any specific formatting or citation style required by your institution or organization. Clarity and precision in presenting the experiment's details and results are key to effective communication in a split-plot and blocked designed experiment report in mechanical engineering.

4 Specific Industries

4.1 AGRICULTURE

Designed experiments are crucial in agriculture for optimizing crop production, improving agricultural practices, and enhancing overall farm efficiency. Here are some common types of designed experiments with examples of their applications in agriculture.

In agriculture designed experiments, various types of variables are considered to investigate crop growth, yield, soil quality, and the effects of different agricultural practices. These variables can be categorized into several common types, including:

1. **Independent Variables**:
   - **Crop Management Practices**: Variables related to agricultural practices and management decisions, such as planting density, irrigation frequency, fertilization rates, and pesticide application.
   - **Soil Characteristics**: Variables related to soil properties and conditions, including soil type, pH, organic matter content, nutrient levels, and moisture content.
   - **Crop Varieties**: Variables related to the choice of crop varieties or cultivars used in agricultural experiments, which can vary in terms of yield, disease resistance, and adaptability.
   - **Environmental Factors**: Variables related to weather and environmental conditions, including temperature, precipitation, sunlight, and wind speed.
   - **Experimental Treatments**: Variables representing specific treatments or interventions applied to study plots, such as the application of specific fertilizers, planting dates, or crop rotation schemes.
2. **Dependent Variables**:
   - **Crop Yield**: Variables related to the quantity of crops harvested per unit area, including crop weight, number of fruits, or grain production.
   - **Crop Quality**: Variables related to the quality of harvested crops, such as nutritional content, taste, texture, and shelf life.
   - **Soil Health Metrics**: Variables related to soil health and fertility, including soil organic matter, microbial activity, nutrient availability, and soil erosion rates.
   - **Pest and Disease Incidence**: Variables related to the prevalence and severity of pests, diseases, and weed infestations in agricultural fields.
   - **Environmental Impact**: Variables related to the environmental impact of agricultural practices, such as greenhouse gas emissions, water usage, and soil degradation.
3. **Categorical Variables**:
   - **Crop Rotations**: Categories representing different crop rotation schemes used in agricultural experiments to study the effects of crop sequencing on soil health and crop yields.
   - **Tillage Practices**: Categories representing different soil tillage practices, including no till, minimum tillage, and conventional tillage.
   - **Crop Varieties**: Categories representing different crop varieties or cultivars used in the experiment, particularly when comparing the performance of different varieties.

DOI: 10.1201/9781003528531-4

4. **Control Variables (Covariates)**:
   - These are variables that are held constant or controlled during experiments to eliminate their influence on the dependent variable, for example, maintaining consistent irrigation levels or controlling for weed pressure in different plots.
5. **Random Variables**:
   - In some cases, random variables may be introduced to account for variability or uncertainty in agricultural measurements or to control for natural variability in crop growth and environmental conditions.
6. **Interaction Variables**:
   - Interaction variables are used to investigate the combined effects of two or more independent variables on the dependent variable, helping to understand how different factors may interact to influence crop yields and soil health.
7. **Noise Variables (Error Variables)**:
   - These are uncontrolled or unmeasured variables that can introduce variability or errors into agricultural experiments. Statistical analysis and experimental controls are used to minimize the impact of noise variables.

Agriculture designed experiments are crucial for optimizing agricultural practices, improving crop yields, enhancing soil health, and mitigating environmental impacts. Careful consideration and control of variables are essential for obtaining meaningful results that can inform sustainable farming practices.

4.1.1 FACTORIAL EXPERIMENTS

Agricultural scientists may use factorial experiments to study the effects of multiple factors, such as irrigation frequency, fertilizer type, and planting density, on crop yield and quality.

Factorial experiments are commonly used in agriculture to study the effects of multiple factors on crop growth, yield, and other agricultural outcomes. Here is an example of a matrix for a factorial experiment designed experiment in agriculture:

Objective: To investigate the effects of two factors, fertilizer type (Factor A) and irrigation frequency (Factor B), on the yield of a specific crop.

Factors:

1. **Factor A**: fertilizer type
   - Fertilizer X (e.g., nitrogen-based)
   - Fertilizer Y (e.g., phosphorus-based)
2. **Factor B**: irrigation frequency
   - Low frequency (e.g., once a week)
   - High frequency (e.g., three times a week)

Response Variable: Crop yield (measured in bushels per acre)

Factorial Experiment Matrix: In a factorial experiment, all possible combinations of the factor levels are tested systematically (see Table 4.1).

In this matrix:

- Each row represents a specific combination of fertilizer type and irrigation frequency.
- Factor A (fertilizer type) has two levels: Fertilizer X and Fertilizer Y.
- Factor B (irrigation frequency) has two levels: low frequency and high frequency.
- Crop yield measurements are recorded for each combination.

TABLE 4.1
Factorial Experiment Matrix

Run	Fertilizer Type (A)	Irrigation Frequency (B)	Crop Yield (bushels/acre)
1	Fertilizer X	Low frequency	...
2	Fertilizer Y	Low frequency	...
3	Fertilizer X	High frequency	...
4	Fertilizer Y	High frequency	...

By conducting this factorial experiment and analyzing the results, you can determine how fertilizer type and irrigation frequency interact to affect crop yield. This information can help farmers make informed decisions about crop management practices to optimize yield and resource utilization.

4.1.2 RESPONSE SURFACE METHODOLOGY (RSM)

RSM can be applied in soil science to optimize the parameters for soil remediation processes. Researchers may vary factors like pH adjustment, organic matter addition, and microbial inoculation to enhance soil fertility.

RSM is commonly used in agriculture to optimize factors and predict the response of crops or plants. Here is an example of a matrix for a RSM designed experiment in agriculture:

Objective: To optimize the growth of a specific crop by varying factors such as fertilizer concentration (Factor A) and irrigation duration (Factor B).

Factors:

1. **Factor A**: fertilizer concentration (%)
   - Low (e.g., 5%)
   - Medium (e.g., 10%)
   - High (e.g., 15%)
2. **Factor B**: Irrigation duration (hours)
   - Short (e.g., 2 hours)
   - Medium (e.g., 4 hours)
   - Long (e.g., 6 hours)

Response Variable: Crop height (measured in centimeters (cm))

RSM Matrix: In an RSM experiment, you systematically vary the levels of factors within a defined range to create a response surface for analysis (Table 4.2).

In this matrix:

- Each row represents a specific combination of fertilizer concentration and irrigation duration.
- Factor A (fertilizer concentration) has three levels: low, medium, and high.
- Factor B (irrigation duration) has three levels: short, medium, and long.
- Crop height measurements are recorded for each combination.

RSM allows you to fit mathematical models to the data and identify optimal factor settings for maximizing crop height while considering the interactions between factors. It helps you understand the relationship between the factors and the response and enables you to make informed decisions for crop management and yield optimization.

TABLE 4.2
RSM Matrix

Run	Fertilizer Concentration (A)	Irrigation Duration (B)	Crop Height (cm)
1	Low	Short	...
2	Medium	Short	...
3	High	Short	...
4	Low	Medium	...
5	Medium	Medium	...
6	High	Medium	...
7	Low	Long	...
8	Medium	Long	...
9	High	Long	...

4.1.3 FRACTIONAL FACTORIAL EXPERIMENTS

In crop breeding, fractional factorial experiments can be used to assess the effects of various factors (e.g., temperature, humidity, light duration) on plant growth and development without testing all possible combinations.

Fractional factorial experiments are used to study the effects of factors on a response variable by conducting only a fraction of the full factorial experiment. Here is an example of a matrix for a fractional factorial experiment designed experiment in agriculture:

Objective: To investigate the impact of multiple factors on the yield of a specific crop, while reducing the number of experimental runs by conducting a fraction of the full factorial experiment.

Factors: Consider three factors (Factor A, Factor B, and Factor C), each with two levels (low and high):

1. **Factor A**: Fertilizer type (low and high)
2. **Factor B**: Irrigation frequency (low and high)
3. **Factor C**: Pest control (low and high)

Response Variable: Crop yield (measured in bushels per acre)

Fractional Factorial Experiment Matrix: In this example, we use a half-fractional design (2^{3-1}) to reduce the number of experimental runs (Table 4.3).
In this matrix:

- Each row represents a specific combination of factors at their respective levels.
- The combination of levels for each factor follows a fractional design pattern that allows for studying main effects and some interactions.
- Crop yield measurements are recorded for each combination.

Fractional factorial experiments help reduce the number of experimental runs while still providing insights into the effects of factors on the response variable. The design is chosen based on the specific research objectives and the need to balance experimental resources with the desire to investigate multiple factors and their interactions.

TABLE 4.3
Fractional Factorial Experiment Matrix

Run	Factor A	Factor B	Factor C	Crop Yield (Bushels/Acre)
1	Low	Low	Low	...
2	High	Low	Low	...
3	Low	High	Low	...
4	High	High	Low	...
5	Low	Low	High	...
6	High	Low	High	...
7	Low	High	High	...
8	High	High	High	...

4.1.4 CENTRAL COMPOSITE DESIGN (CCD)

CCD is employed in precision agriculture to optimize the parameters of a variable rate fertilizer application system. Farmers vary factors like fertilizer rate, Global Positioning System (GPS) guidance, and field slope to maximize nutrient use efficiency.

CCD is a RSM used to study the response of a system to multiple factors, including their interactions, while minimizing the number of experimental runs. Here is an example of a matrix for a CCD designed experiment in agriculture:

Objective: To optimize the growth of a specific crop by varying factors such as fertilizer concentration (Factor A) and irrigation frequency (Factor B).

Factors:

1. **Factor A**: Fertilizer concentration (%)
   - Low (−1)
   - Medium (0)
   - High (+1)
2. **Factor B**: Irrigation frequency (times per week)
   - Low (−1)
   - Medium (0)
   - High (+1)

Response Variable: Crop yield (measured in bushels per acre)

CCD Matrix: In CCD, the matrix includes a combination of factorial points, axial points, and center points to explore the response surface adequately (Table 4.4).
In this matrix:

- Each row represents a specific combination of fertilizer concentration and irrigation frequency.
- Factor A (fertilizer concentration) and Factor B (irrigation frequency) are varied at different levels, including low, high, and central (0) levels.
- Additional runs at axial points (−α and +α) and center points are included to explore the curvature of the response surface.

TABLE 4.4
CCD Matrix

Run	Fertilizer Concentration (A)	Irrigation Frequency (B)	Crop Yield (Bushels/Acre)
1	−1	−1	...
2	+1	−1	...
3	−1	+1	...
4	+1	+1	...
5	0	−1	...
6	0	+1	...
7	−1	0	...
8	+1	0	...
9	0	0	...
10	−α	0	...
11	+α	0	...
12	0	−α	...
13	0	+α	...

CCD allows you to fit a quadratic model to the data and analyze the interactions between factors. It helps identify the optimal factor settings for maximizing crop yield while considering the interactions between fertilizer concentration and irrigation frequency.

4.1.5 Taguchi Methods

Taguchi methods are used in agricultural engineering to optimize the parameters of a pesticide spraying system. Farmers may vary factors such as nozzle type, spray pressure, and wind speed to minimize pesticide drift and maximize coverage.

Taguchi methods are used in agriculture to optimize processes or systems with multiple factors, aiming to improve yield, quality, or other relevant outcomes. Here is an example of a matrix for Taguchi method designed experiment in agriculture:

Objective: To optimize the growth of a specific crop by varying factors such as planting density (Factor A), irrigation frequency (Factor B), and pest control method (Factor C).

Factors:

1. **Factor A**: Planting density (low, medium, and high)
2. **Factor B**: Irrigation frequency (low, medium, and high)
3. **Factor C**: Pest control method (Method 1, Method 2, and Method 3)

Response Variable: Crop yield (measured in bushels per acre)

Taguchi Method Matrix: In Taguchi methods, an orthogonal array is used to systematically vary factor levels to minimize experimental runs while capturing important information (Table 4.5).

TABLE 4.5
Taguchi Method Matrix

Run	Factor A	Factor B	Factor C	Crop Yield (Bushels/Acre)
1	Low	Low	Method 1	...
2	Medium	Medium	Method 2	...
3	High	High	Method 3	...
4	Low	Medium	Method 1	...
5	Medium	High	Method 2	...
6	High	Low	Method 3	...
7	Low	High	Method 1	...
8	Medium	Low	Method 2	...
9	High	Medium	Method 3	...
10	Low	Low	Method 2	...
11	Medium	Medium	Method 3	...
12	High	High	Method 1	...
...	...	...	...	...

In this matrix:

- Each row represents a specific combination of factors (planting density, irrigation frequency, and pest control method).
- The combination of factor levels is determined by the Taguchi orthogonal array.
- Crop yield measurements are recorded for each combination.

Taguchi methods allow you to identify the optimal combination of factor levels that maximizes the desired outcome (crop yield) while considering interactions between factors. The design is efficient in terms of the number of experimental runs required, making it a useful approach in agriculture optimization.

4.1.6 Mixture Experiments

In animal nutrition, mixture experiments help optimize animal feed formulations. Researchers vary proportions of feed ingredients like grains, protein sources, and additives to achieve desired nutrient profiles for livestock.

Mixture designed experiments are used in agriculture when multiple ingredients or components are mixed together to create a product or treatment. Here is an example of a matrix for a mixture designed experiment in agriculture.

Objective: To optimize a plant growth medium by varying the proportions of three ingredients: soil (Factor A), compost (Factor B), and sand (Factor C).

Factors:

1. **Factor A**: Proportion of soil (%)
   - Varying levels from 0% (absence of soil) to 100%
2. **Factor B**: Proportion of compost (%)
   - Varying levels from 0% (absence of compost) to 100%
3. **Factor C**: Proportion of sand (%)
   - Varying levels from 0% (absence of sand) to 100%

TABLE 4.6

Mixture Designed Experiment Matrix

Run	Soil (%)	Compost (%)	Sand (%)	Plant Growth (cm)
1	100	0	0	...
2	0	100	0	...
3	0	0	100	...
4	50	50	0	...
5	50	0	50	...
6	0	50	50	...
7	33.3	33.3	33.3	...
8	...	...	...	...

Response Variable: Plant growth (measured in height, cm)

Mixture Designed Experiment Matrix:

In a mixture design, you systematically vary the proportions of the components (factors) within a constrained space, often using a simplex lattice or other suitable design (Table 4.6).

In this matrix:

- Each row represents a specific combination of proportions for the three ingredients (soil, compost, and sand).
- The proportions of the ingredients are varied within the constraints of the design.
- Plant growth measurements (height) are recorded for each combination.

Mixture designed experiments help determine the optimal mix of ingredients to achieve desired outcomes, such as plant growth in this case. They also allow for understanding the effects of the proportions of different components on the response variable.

4.1.7 SEQUENTIAL EXPERIMENTATION

Farmers may use sequential experimentation to optimize crop planting dates based on local weather conditions and climate trends. They adjust planting schedules based on real-time weather forecasts and historical data to maximize crop yields.

Sequential experimentation involves conducting a series of experiments where each experiment is designed based on the results of the previous one. This allows for the gradual refinement of the process or system being studied. Here is a simplified example of a sequential experimentation matrix in agriculture:

Objective: To optimize the irrigation schedule for a specific crop to maximize yield.

Initial Experiment:

In the initial experiment, you might test two different irrigation schedules (Factor A) (Table 4.7).

Second Experiment:

Based on the initial experiment, you decide to further refine Schedule 1 by varying the irrigation frequency (Factor B) (Table 4.8).

Third Experiment:

Building on the results of the second experiment, you now focus on optimizing the amount of fertilizer (Factor C) used with the medium irrigation frequency (Table 4.9).

TABLE 4.7

Sequential Experiment Matrix (Initial Experiment)

Run	Irrigation Schedule (A)	Crop Yield (Bushels/Acre)
1	Schedule 1	...
2	Schedule 2	...

Let us assume that Schedule 1 produced a higher yield.

TABLE 4.8

Sequential Experiment Matrix (Second Experiment)

Run	Irrigation Frequency (B)	Crop Yield (Bushels/Acre)
1	Low	...
2	Medium	...
3	High	...

In this experiment, you find that a medium irrigation frequency (Factor B) leads to the highest yield.

TABLE 4.9

Sequential Experiment Matrix (Third Experiment)

Run	Fertilizer Amount (C)	Crop Yield (Bushels/Acre)
1	Low	...
2	Medium	...
3	High	...

Based on this experiment, you determine that a medium amount of fertilizer (Factor C) combined with medium irrigation frequency (Factor B) maximizes crop yield.

The sequential experimentation process continues until you reach the desired level of optimization or until no further improvements are observed. Each experiment is designed based on the insights gained from the previous one, allowing for a gradual refinement of the agricultural process.

The specific factors, levels, and results would vary depending on the crop, location, and other variables of interest in a real-world agriculture study.

4.1.8 Robust Parameter Design (RPD)

In organic farming, RPD can be applied to ensure the robustness of pest management practices against variations in pest populations and weather conditions, reducing the reliance on synthetic pesticides.

RPD is used in agriculture to optimize processes while considering variability in environmental conditions. Here is a simplified example of a matrix for a RPD experiment in agriculture.

Objective: To optimize the planting depth (Factor A) and irrigation frequency (Factor B) for a specific crop while considering variations in soil moisture levels (Factor C).

Factors:

1. **Factor A**: Planting depth (cm)
   - Low (e.g., 5 cm)
   - Medium (e.g., 10 cm)
   - High (e.g., 15 cm)
2. **Factor B**: Irrigation frequency (times per week)
   - Low (e.g., once a week)
   - Medium (e.g., three times a week)
   - High (e.g., daily)
3. **Factor C**: Soil moisture level (low, medium, and high)
   - Represents variations in soil moisture due to environmental conditions.

Response Variable: Crop yield (measured in bushels per acre)

RPD Matrix: In RPD, you systematically vary the factors while considering the impact of environmental variability (Table 4.10).

In this matrix:

- Each row represents a specific combination of planting depth, irrigation frequency, and soil moisture level.
- Factors A and B are systematically varied to explore the design space.
- Factor C represents variations in soil moisture that could occur naturally.
- Crop yield measurements are recorded for each combination.

RPD helps identify robust settings for factors that can produce consistent crop yields even in the presence of environmental variability. It allows for optimizing agricultural processes while accounting for real-world conditions.

4.1.9 RANDOMIZED EXPERIMENTS

In agricultural research, randomized experiments can be used to study the effects of different irrigation techniques on water use efficiency and crop health. Fields or plots are randomly assigned to different irrigation treatments to assess their impact.

TABLE 4.10
RPD Matrix

Run	Planting Depth (A)	Irrigation Frequency (B)	Soil Moisture Level (C)	Crop Yield (Bushels/Acre)
1	Low	Low	Low	...
2	Medium	Medium	Medium	...
3	High	High	High	...
4	Low	Medium	High	...
5	Medium	High	Low	...
6	High	Low	Medium	...
7	Low	High	Medium	...
8	Medium	Low	High	...
9	High	Medium	Low	...
10	...	...	...	...

TABLE 4.11
Randomized Experiment Matrix

Plot	Treatment A	Treatment B	Crop Yield (Bushels/Acre)
1	Fertilizer 1	Low	...
2	Fertilizer 2	High	...
3	Fertilizer 1	High	...
4	Fertilizer 2	Low	...
5	...	...	...

Randomized experiments are used in agriculture to investigate the impact of various treatments or factors on crop growth or other agricultural outcomes. In a randomized experiment, treatments are randomly assigned to experimental units to minimize bias and account for variability. Here is a simplified example of a matrix for a randomized experiment designed experiment in agriculture:

Objective: To test the effectiveness of two different types of fertilizer treatments (Treatment A and Treatment B) on crop yield.

Factors:

1. **Treatment A**: Type of fertilizer (Fertilizer 1 and Fertilizer 2)
2. **Treatment B**: Application rate (low and high)

Response Variable: Crop yield (measured in bushels per acre)

Randomized Experiment Matrix: In a randomized experiment, the treatments are randomly assigned to experimental plots to ensure that the results are not biased by factors other than the treatments themselves (Table 4.11).
    In this matrix:

- Each row represents an experimental plot.
- Treatment A and Treatment B are randomly assigned to each plot.
- Crop yield measurements are recorded for each plot.

Randomized experiments help ensure that the results are not biased by factors other than the treatments being studied. Randomization helps control for variability and allows for statistical analysis to determine whether there are significant differences in crop yield between the different fertilizer treatments and application rates.

4.1.10 Split-Plot and Blocked Experiments

When optimizing the parameters of a crop rotation system, split-plot or blocked experiments may be used to account for variations in soil types, pest pressures, and crop preferences.
    Split-plot and blocked experiments are used in agriculture when there are different levels of control or factors within the experimental design. Here are examples of matrices for split-plot and blocked experiment designed experiments in agriculture:

TABLE 4.12
Split-Plot Design Matrix

Whole Plot	Factor A (Irrigation Frequency)	Factor B (Fertilizer Type)	Crop Yield (Bushels/Acre)
1	Low	Organic	...
2	High	Organic	...
3	Low	Synthetic	...
4	High	Synthetic	...

TABLE 4.13
Split-Plot Design Matrix

Subplot	Factor C (Pest Control Method)	Crop Yield (Bushels/Acre)
1	Method 1	...
2	Method 2	...

Example 1: Split-Plot Design

Objective: To study the effect of two main factors, Factor A (irrigation frequency) and Factor B (fertilizer type), while considering a subfactor, Factor C (pest control method).

Factors:

1. **Factor A:** Irrigation frequency (low and high)
2. **Factor B:** Fertilizer type (organic and synthetic)
3. **Factor C:** Pest control method (Method 1 and Method 2)

Response Variable: Crop yield (bushels per acre)

Split-Plot Design Matrix: In a split-plot design, the main factors (Factor A and Factor B) are assigned to whole plots, while the subfactor (Factor C) is assigned to subplots within each whole plot (Table 4.12).

Within each whole plot, the subfactor Factor C (pest control method) is varied as follows (Table 4.13).

Example 2: Blocked Design

Objective: To study the effect of different soil types (Factor A) on crop yield, while considering the impact of weather variability as a blocking factor (Factor B).

Factors:

1. **Factor A:** Soil type (clay, loam, and sandy)
2. **Factor B:** Weather conditions (blocked factor: rainy and sunny)

Response Variable: Crop yield (bushels per acre)

Blocked Design Matrix: In a blocked design, the blocking factor (Factor B) is used to create separate blocks or groups of experimental units to account for variability (Table 4.14).

TABLE 4.14
Blocked Design Matrix

Block	Soil Type (Factor A)	Weather Conditions (Blocked Factor B)	Crop Yield (Bushels/Acre)
1	Clay	Rainy	...
1	Clay	Rainy	...
2	Clay	Sunny	...
2	Clay	Sunny	...
3	Loam	Rainy	...
3	Loam	Rainy	...
4	Loam	Sunny	...
4	Loam	Sunny	...
5	Sandy	Rainy	...
5	Sandy	Rainy	...
6	Sandy	Sunny	...
6	Sandy	Sunny	...

In this matrix:

- Each row represents an experimental unit.
- Soil type (Factor A) is the primary factor of interest.
- Weather conditions (Factor B) are used to create separate blocks, ensuring that variations in weather are accounted for when evaluating the effect of soil type on crop yield.

Blocked designs help control extraneous variability and ensure that the results are not confounded by the blocking factor, in this case, weather conditions.

4.2 BIOTECHNOLOGY

In biotechnology, designed experiments are essential for optimizing bioprocesses, developing new biopharmaceuticals, and advancing biotechnological innovations. Here are some common types of designed experiments with examples of their applications in biotechnology.

In biotechnology designed experiments, various types of variables are considered to investigate biological processes, optimize biotechnological procedures, and study the effects of different experimental conditions. These variables can be categorized into several common types, including:

1. **Independent Variables**:
   - **Biological Factors**: Variables related to biological systems, such as cell lines, microbial strains, genes, enzymes, or proteins used in biotechnological processes.
   - **Process Parameters**: Variables related to biotechnological process conditions, including temperature, pH, agitation speed, aeration rate, and fermentation time.
   - **Substrate Concentrations**: Variables related to the concentration of substrates, nutrients, or feedstock materials used in bioprocessing, such as glucose, nutrients, or carbon sources.
   - **Experimental Treatments**: Variables representing specific treatments or interventions applied to biological systems, such as the addition of inducers, inhibitors, or growth factors.
2. **Dependent Variables**:
   - **Bioprocess Yield**: Variables related to the quantity of bioproducts (e.g., biofuels, enzymes, and pharmaceuticals) produced during biotechnological processes.
   - **Biological Activity**: Variables related to the activity of enzymes, proteins, or cells, including enzymatic activity, cell growth, biomass accumulation, or product formation rates.

- **Quality Attributes**: Variables related to the quality of biotechnological products, such as purity, potency, stability, and specific characteristics (e.g., protein folding and glycosylation patterns).
- **Biomolecular Analyses**: Variables representing results from biomolecular analyses, including deoxyribonucleic acid (DNA) sequencing data, gene expression profiles, or protein quantification.

3. **Categorical Variables**:
   - **Strain or Cell Type**: Categories representing different strains or types of biological entities used in experiments, such as wild-type strains, mutants, or genetically modified organisms.
   - **Treatment Categories**: Categories representing different treatment conditions or experimental groups, such as control groups and treatment groups with specific interventions.
   - **Time Points**: Categories representing different time points during bioprocesses or experiments, particularly when studying dynamic processes.

4. **Control Variables (Covariates)**:
   - These are variables that are held constant or controlled during experiments to eliminate their influence on the dependent variable, for example, maintaining constant temperature or pH levels during bioprocessing.

5. **Random Variables**:
   - In some cases, random variables may be introduced to account for variability or uncertainty in biological measurements or to control for natural variability in biological systems.

6. **Interaction Variables**:
   - Interaction variables are used to investigate the combined effects of two or more independent variables on the dependent variable, helping to understand how different factors interact to influence biotechnological processes or biological responses.

7. **Noise Variables (Error Variables)**:
   - These are uncontrolled or unmeasured variables that can introduce variability or errors into biotechnology experiments. Statistical analysis and experimental controls are used to minimize the impact of noise variables.

Biotechnology designed experiments are essential for optimizing bioprocesses, enhancing product yields, and understanding biological systems at the molecular level. Careful consideration and control of variables are crucial for obtaining reliable and meaningful results in biotechnology research and development.

4.2.1 Factorial Experiments

Biotechnologists may use factorial experiments to optimize the parameters of a fermentation process for producing recombinant proteins. Factors such as pH, temperature, agitation rate, and nutrient concentrations can be varied to maximize protein yield.

Factorial experiments are commonly used in biotechnology to investigate the effects of multiple factors or variables on a biological process, such as cell growth, enzyme activity, or gene expression. Here is an example of a matrix for a factorial experiment designed experiment in biotechnology:

Objective: To study the impact of two factors, Factor A (temperature) and Factor B (pH), on the growth rate of a microorganism.

TABLE 4.15
Factorial Experiment Matrix

Run	Factor A (Temperature)	Factor B (pH)	Microorganism Growth Rate (OD/h)
1	Low (25°C)	Acidic (5.0)	...
2	High (37°C)	Acidic (5.0)	...
3	Low (25°C)	Alkaline (7.0)	...
4	High (37°C)	Alkaline (7.0)	...

Factors:

1. **Factor A**: Temperature (low and high)
   - Low temperature: 25°C
   - High temperature: 37°C
2. **Factor B**: pH (acidic and alkaline)
   - Acidic pH: 5.0
   - Alkaline pH: 7.0

Response Variable: Microorganism growth rate (measured in optical density units per hour (OD/h))

Factorial Experiment Matrix: In a 2 × 2 factorial design, each combination of Factor A and Factor B is systematically tested (Table 4.15).
   In this matrix:

- Each row represents an experimental run with a specific combination of temperature and pH.
- Factor A (temperature) has two levels (low and high).
- Factor B (pH) has two levels (acidic and alkaline).
- Microorganism growth rate measurements are recorded for each combination.

A factorial experiment allows you to assess the main effects of each factor (temperature and pH) and potential interactions between them. It helps identify how changes in these factors affect the growth rate of microorganisms and provides insights into optimizing biotechnological processes.

4.2.2 RSM

RSM is applied in bioprocess engineering to optimize the parameters of a bioreactor for microbial fermentation. Researchers vary factors like oxygen transfer rate, feed rate, and impeller speed to achieve desired biomass and product concentrations.

   RSM is frequently used in biotechnology to optimize multiple factors or variables while considering their interactions. Here is an example of a matrix for an RSM designed experiment in biotechnology:

Objective: To optimize the conditions for maximum enzyme activity (Factor A: temperature and Factor B: pH) while minimizing substrate cost (Factor C: concentration).

Factors:

1. **Factor A**: Temperature (°C)
   - Varying levels, e.g., 30, 40, and 50°C
2. **Factor B**: pH
   - Varying levels, e.g., 6.0, 7.0, and 8.0
3. **Factor C**: Substrate concentration (g/L)
   - Varying levels, e.g., 10, 20, and 30 g/L

Response Variables:

1. Enzyme activity (measured in units/mL)
2. Substrate cost (measured in currency units per experiment)

RSM Experiment Matrix: In an RSM experiment, you systematically vary the factors within a designed experimental space to fit a response surface model (Table 4.16).
In this matrix:

- Each row represents a specific combination of temperature, pH, and substrate concentration.
- The factors (temperature, pH, and substrate conc.) are varied systematically within the design space.
- Enzyme activity and substrate cost measurements are recorded for each combination.

RSM allows you to fit response surface models to the experimental data, analyze the effects of the factors, and find the optimal conditions that maximize enzyme activity while minimizing substrate cost, considering potential interactions between factors.

4.2.3 Fractional Factorial Experiments

In genetic engineering, fractional factorial experiments can be used to assess the effects of various factors (e.g., gene expression levels and culture conditions) on the production of biopharmaceuticals without testing all possible combinations.

TABLE 4.16
RSM Experiment Matrix

Run	Factor A (Temperature)	Factor B (pH)	Factor C (Substrate Conc.)	Enzyme Activity (Units/mL)	Substrate Cost (Currency Units)
1	30	6.0	10	...	...
2	40	6.0	20	...	...
3	50	6.0	30	...	...
4	30	7.0	20	...	...
5	40	7.0	10	...	...
6	50	7.0	30	...	...
7	30	8.0	30	...	...
8	40	8.0	30	...	...
9	50	8.0	20	...	...
10	...	...	...	...	...

TABLE 4.17
Fractional Factorial Experiment Matrix

Run	Factor A (Temperature)	Factor B (pH)	Factor C (Substrate Conc.)	Enzyme Activity (Units/mL)
1	Low (25°C)	Acidic (5.0)	Low (10 g/L)	...
2	High (37°C)	Acidic (5.0)	Low (10 g/L)	...
3	Low (25°C)	Alkaline (7.0)	Low (10 g/L)	...
4	High (37°C)	Alkaline (7.0)	Low (10 g/L)	...
5	Low (25°C)	Acidic (5.0)	High (30 g/L)	...
6	High (37°C)	Acidic (5.0)	High (30 g/L)	...
7	Low (25°C)	Alkaline (7.0)	High (30 g/L)	...
8	High (37°C)	Alkaline (7.0)	High (30 g/L)	...

Fractional factorial experiments are used in biotechnology to investigate the effects of multiple factors while reducing the number of experimental runs required. Here is an example of a matrix for a fractional factorial experiment designed experiment in biotechnology:

Objective: To study the impact of three factors, Factor A (temperature), Factor B (pH), and Factor C (substrate concentration), on enzyme activity.

Factors:

1. **Factor A**: Temperature (low and high)
2. **Factor B**: pH (acidic and alkaline)
3. **Factor C**: Substrate concentration (low and high)

Response Variable: Enzyme activity (measured in units/mL)

Fractional Factorial Experiment Matrix: In a fractional factorial design, you investigate a subset of the possible combinations of factors, allowing you to reduce the number of experimental runs (Table 4.17).

In this matrix:

- Each row represents an experimental run with a specific combination of temperature, pH, and substrate concentration.
- The fractional factorial design allows you to explore only a fraction of the total possible combinations, significantly reducing the number of runs while still capturing the main effects and certain interactions between factors.
- Enzyme activity measurements are recorded for each combination.

Fractional factorial experiments help efficiently explore the factor effects and interactions with fewer experimental runs, making them a valuable tool for optimizing biotechnological processes.

4.2.4 CCD

CCD is employed in vaccine development to optimize the parameters for a viral antigen purification process. Scientists vary factors like column size, flow rate, and resin type to maximize antigen purity and yield.

CCD is a popular experimental design in biotechnology for exploring the response surface and optimizing factors while considering quadratic effects and interactions. Here is an example of a matrix for a CCD designed experiment in biotechnology:

Objective: To optimize the conditions for maximum enzyme activity (Factor A: temperature and Factor B: pH) while minimizing substrate cost (Factor C: concentration).

Factors:

1. **Factor A**: Temperature (°C)
   - Varying levels, e.g., −1, 0, and +1 (corresponding to low, central, and high)
2. **Factor B**: pH
   - Varying levels, e.g., −1, 0, and +1 (corresponding to low, central, and high)
3. **Factor C**: Substrate concentration (g/L)
   - Varying levels, e.g., −1, 0, and +1 (corresponding to low, central, and high)

Response Variables:

1. Enzyme activity (measured in units/mL)
2. Substrate cost (measured in currency units per experiment)

CCD Experiment Matrix: In a CCD, you explore the factors over a central point and at various distances from the center to capture quadratic effects and interactions (Table 4.18).
In this matrix:

- Each row represents an experimental run with a specific combination of temperature, pH, and substrate concentration.
- The central point (Run 9) corresponds to the central levels of all factors.
- Runs 1–8 explore the factors at various distances from the center, allowing for the capture of quadratic effects and interactions.
- Enzyme activity and substrate cost measurements are recorded for each combination.

TABLE 4.18
CCD Experiment Matrix

Run	Factor A (Temperature)	Factor B (pH)	Factor C (Substrate Conc.)	Enzyme Activity (Units/mL)	Substrate Cost (Currency Units)
1	−1	−1	−1	...	...
2	1	−1	−1	...	...
3	−1	1	−1	...	...
4	1	1	−1	...	...
5	−1	−1	1	...	...
6	1	−1	1	...	...
7	−1	1	1	...	...
8	1	1	1	...	...
9	0	0	0	...	...
10	...	...	...	...	...

CCD allows you to fit response surface models to the experimental data, analyze the curvature of the response surface, and find the optimal conditions that maximize enzyme activity while minimizing substrate cost, considering quadratic effects and interactions between factors.

4.2.5 TAGUCHI METHODS

Taguchi methods are used in bioprocess optimization to minimize variability in bioreactor performance. Engineers may vary factors like nutrient concentrations and inoculum density to improve process robustness and reproducibility.

Taguchi methods are used to optimize processes and products by identifying the best combination of factors and levels while minimizing variability. Here is an example of a matrix for a Taguchi method designed experiment in biotechnology.

Objective: To optimize the conditions for maximum enzyme activity (Factor A: temperature and Factor B: pH) while minimizing substrate cost (Factor C: concentration).

Factors:

1. **Factor A**: Temperature (°C)
   - Varying levels, e.g., low and high
2. **Factor B**: pH
   - Varying levels, e.g., acidic and alkaline
3. **Factor C**: Substrate concentration (g/L)
   - Varying levels, e.g., low and high

Response Variables:

1. Enzyme activity (measured in units/mL)
2. Substrate cost (measured in currency units per experiment)

Taguchi Methods Experiment Matrix: In a Taguchi experiment, you systematically test various combinations of factors and levels, typically using orthogonal arrays to minimize the number of experimental runs (Table 4.19).

TABLE 4.19
Taguchi Method Experiment Matrix

Run	Factor A (Temperature)	Factor B (pH)	Factor C (Substrate Conc.)	Enzyme Activity (Units/mL)	Substrate Cost (Currency Units)
1	Low	Acidic	Low	...	...
2	High	Acidic	Low	...	...
3	Low	Alkaline	Low	...	...
4	High	Alkaline	Low	...	...
5	Low	Acidic	High	...	...
6	High	Acidic	High	...	...
7	Low	Alkaline	High	...	...
8	High	Alkaline	High	...	...

In this matrix:

- Each row represents a specific combination of temperature, pH, and substrate concentration.
- The factors and their levels are systematically tested using orthogonal arrays, which help reduce the number of runs while providing insights into the optimal settings.
- Enzyme activity and substrate cost measurements are recorded for each combination.

Taguchi methods allow you to identify the optimal factor settings (combination of temperature, pH, and substrate concentration) that maximize enzyme activity while minimizing variability and the impact of noise factors.

4.2.6 Mixture Experiments

In biopharmaceutical formulation, mixture experiments help optimize the composition of a protein formulation. Researchers vary proportions of buffer components, stabilizers, and preservatives to achieve desired stability and shelf life.

Mixture designed experiments are commonly used in biotechnology when you are dealing with formulations or mixtures of ingredients where the total composition sums to a fixed value (e.g., 100%). Here is an example of a matrix for a mixture designed experiment in biotechnology:

Objective: To optimize a protein formulation (Factor A: Protein X, Factor B: Protein Y, and Factor C: Protein Z) for maximum yield while minimizing cost.

Factors:

1. **Factor A**: Percentage of Protein X in the mixture
   - Varying levels, e.g., 10%, 20%, 30%, ..., 90%
2. **Factor B**: Percentage of Protein Y in the mixture
   - Varying levels, e.g., 10%, 20%, 30%, ... , 90%
3. **Factor C**: Percentage of Protein Z in the mixture
   - Varying levels, e.g., 10%, 20%, 30%, ..., 90%

Response Variables:

1. Yield of the protein formulation (measured in grams)
2. Production cost (measured in currency units per experiment)

Mixture Designed Experiment Matrix: In a mixture designed experiment, you systematically vary the proportions of the components within a fixed total composition (Table 4.20).
In this matrix:

- Each row represents a specific combination of the percentages of Protein X, Protein Y, and Protein Z in the formulation.
- The total composition is kept constant (e.g., 100%).
- Various proportions of the components are systematically tested to find the optimal mixture that maximizes yield while minimizing production cost.
- Yield and cost measurements are recorded for each combination.

TABLE 4.20
Mixture Designed Experiment Matrix

Run	Factor A (% Protein X)	Factor B (% Protein Y)	Factor C (% Protein Z)	Yield (Grams)	Cost (Currency Units)
1	10	10	80	...	...
2	20	20	60	...	...
3	30	40	30	...	...
4	40	50	10	...	...
5	...	...	...	...	...

Mixture designed experiments help you optimize complex formulations by exploring different proportions of components within a fixed total composition, making them valuable in biotechnology for developing cost-effective and high-yield products.

4.2.7 SEQUENTIAL EXPERIMENTATION

Biotechnologists may use sequential experimentation to optimize the parameters of a cell culture process for the production of monoclonal antibodies. They adjust factors like media composition and feeding strategies based on real-time cell growth and product titer data.

Sequential experimentation is often employed in biotechnology to optimize processes or formulations gradually. Here is an example of a matrix for a sequential experimentation designed experiment in biotechnology.

Objective: To optimize the growth medium for maximum cell biomass production (Factor A: Nutrient A and Factor B: Nutrient B).

Factors:

1. **Factor A**: Nutrient A concentration (g/L)
   - Varying levels, e.g., 1, 2, 3, and 4 g/L
2. **Factor B**: Nutrient B concentration (g/L)
   - Varying levels, e.g., 1, 2, 3, and 4 g/L

Response Variable: Cell biomass production (measured in grams per liter (g/L))

Sequential Experimentation Matrix: In a sequential experimentation approach, you gradually adjust the factor levels based on the results of previous experiments to converge toward the optimal conditions (Table 4.21).

In this matrix:

- Each row represents a specific combination of nutrient A and nutrient B concentrations.
- The factor levels are adjusted based on the results of previous experiments to gradually approach the optimal conditions for maximum cell biomass production.
- Biomass production measurements are recorded for each combination.

Sequential experimentation allows you to iteratively refine the factor settings, making informed decisions at each step based on the observed outcomes, with the goal of converging toward the best conditions for your biotechnological process.

TABLE 4.21

Sequential Experimentation Matrix

Run	Factor A (Nutrient A)	Factor B (Nutrient B)	Biomass Production (g/L)
1	1	1	...
2	2	1	...
3	1	2	...
4	3	1	...
5	2	2	...
6	1	3	...
7	4	1	...
8	3	2	...
9	2	3	...
10	...	...	...

4.2.8 RPD

In bioprocess scale-up, RPD can be applied to ensure the robustness of large-scale production processes against variations in equipment size, raw material quality, and environmental conditions.

RPD is used in biotechnology to optimize processes while making them less sensitive to variations or noise factors. Here is an example of a matrix for a RPD designed experiment in biotechnology:

Objective: To optimize the fermentation process for maximum yield of a bioproduct (Factor A: temperature and Factor B: pH) while ensuring robustness to variations in impeller speed (Factor C: impeller speed).

Factors:

1. **Factor A**: Temperature (°C)
   - Varying levels, e.g., 30°C, 35°C, and 40°C
2. **Factor B**: pH
   - Varying levels, e.g., 5.5, 6.0, and 6.5
3. **Factor C**: Impeller speed (rpm)
   - Varying levels, e.g., 500 rpm, 600 rpm, and 700 rpm

Response Variable: Yield of the bioproduct (measured in g/L)

RPD Matrix: In a RPD, you systematically investigate the main factors and noise factors while optimizing the process (Table 4.22).

In this matrix:

- Each row represents a specific combination of temperature, pH, and impeller speed.
- The main factors (temperature and pH) and a noise factor (impeller speed) are systematically varied to optimize the yield of the bioproduct.
- The goal is to find the optimal conditions that maximize yield while ensuring robustness to variations in impeller speed.

RPD allows you to consider both the main factors and noise factors, making the process less sensitive to variations and ensuring robust performance in biotechnological processes.

TABLE 4.22
RPD Matrix

Run	Factor A (Temperature)	Factor B (pH)	Factor C (Impeller Speed)	Yield (g/L)
1	30	5.5	500	...
2	30	5.5	600	...
3	30	5.5	700	...
4	35	6.0	500	...
5	35	6.0	600	...
6	35	6.0	700	...
7	40	6.5	500	...
8	40	6.5	600	...
9	40	6.5	700	...

TABLE 4.23
Randomized Experiment Matrix

Run	Factor A (Nutrient X)	Factor B (Nutrient Y)	Cell Growth (OD)
1	2	1	...
2	6	3	...
3	4	5	...
4	4	1	...
5	2	5	...
6	...	...	...

4.2.9 RANDOMIZED EXPERIMENTS

In pharmaceutical research, randomized experiments can be used to study the effects of different drug formulations on biological activity and therapeutic efficacy. Randomized trials assess the impact of formulation variations on drug performance.

Randomized experiments are used in biotechnology to evaluate the effects of various factors without bias and to account for the potential influence of uncontrolled variables. Here is an example of a matrix for a randomized experiment designed experiment in biotechnology.

Objective: To evaluate the effects of different nutrient formulations (Factor A: Nutrient X and Factor B: Nutrient Y) on cell growth.

Factors:

1. **Factor A**: Nutrient X concentration (g/L)
   - Varying levels, e.g., 2, 4, and 6 g/L
2. **Factor B**: Nutrient Y concentration (g/L)
   - Varying levels, e.g., 1, 3, and 5 g/L

Response Variable: Cell growth (measured in OD)

Randomized Experiment Matrix: In a randomized experiment, the factor combinations are selected randomly to ensure that any observed effects are not due to systematic biases (Table 4.23).

In this matrix:

- Each row represents a specific combination of nutrient X and nutrient Y concentrations.
- The factor combinations are selected randomly to ensure that the effects observed are not influenced by any systematic order or bias.
- Cell growth (measured as OD) is recorded for each combination.

Randomized experiments are valuable for unbiased assessment of the effects of factors, as they help eliminate the influence of uncontrolled variables or biases in the experimental setup, leading to more robust and reliable conclusions in biotechnology research.

4.2.10 Split-Plot and Blocked Experiments

When optimizing the parameters of a bioreactor scale-down model, split-plot or blocked experiments may be used to account for variations in scale-down equipment performance and representativeness.

Split-plot and blocked designs are used in biotechnology when certain factors cannot be easily changed or are treated as "hard-to-change" factors, leading to a two-level experimental design. Here are examples of matrices for split-plot and blocked designed experiments in biotechnology:

Example 1: Split-Plot Design

Objective: To optimize a fermentation process with two factors, Factor A (temperature) and Factor B (pH), where the temperature is a hard-to-change factor and pH can be easily changed.

Factors:

1. **Factor A**: Temperature (°C) – hard-to-change factor
   - Two levels: Low (30 °C) and high (35 °C)
2. **Factor B**: pH
   - Two levels: Low (5.5) and high (6.0)

Response Variable: Product yield (measured in g/L)

Split-Plot Design Matrix: In a split-plot design, you treat the hard-to-change factor as the whole plot and the easy-to-change factor as the subplot. The levels of the hard-to-change factor are not changed frequently (Table 4.24).

In this matrix:

- The whole-plot factor (temperature) is applied at the whole-plot level (e.g., the entire batch).
- The subplot factor (pH) is applied at the subplot level (e.g., within each batch).
- This design allows you to assess the impact of pH while considering the hard-to-change factor (temperature) as a higher-level constraint.

TABLE 4.24
Split-Plot Design Matrix

Run	Whole Plot (Temperature)	Subplot (pH)	Yield (g/L)
1	Low	Low	...
2	Low	High	...
3	High	Low	...
4	High	High	...

TABLE 4.25
Blocked Design Matrix

Block	Factor A (Nutrient X)	Factor B (Nutrient Y)	Cell Growth (OD)
1	2	1	...
1	4	1	...
2	2	3	...
2	4	3	...

Example 2: Blocked Design

Objective: To optimize a cell culture process with two factors, Factor A (Nutrient X) and Factor B (Nutrient Y), where variations in the equipment are suspected to influence the response.

Factors:

1. **Factor A**: Nutrient X concentration (g/L)
   - Two levels: Low (2 g/L) and high (4 g/L)
   - **Factor B**: Nutrient Y concentration (g/L)
   - Two levels: Low (1 g/L) and high (3 g/L)

Response Variable: Cell growth (measured in OD)

Blocked Design Matrix: In a blocked design, you divide the experiment into blocks to account for variations in equipment or other sources of variation (Table 4.25).
   In this matrix:
   - The experiment is divided into two blocks, each representing different experimental runs or batches.
   - Within each block, you test the combinations of Nutrient X and Nutrient Y concentrations.
   - This design helps account for potential variations in equipment or other sources of variation between the blocks.

Split-plot and blocked designs are useful in biotechnology when certain factors cannot be easily changed or when variations in equipment or batches need to be considered during experimentation.

4.3   ENERGY AND RENEWABLE RESOURCES

In the field of energy and renewable resources, designed experiments are essential for optimizing energy production processes, evaluating the efficiency of renewable technologies, and addressing sustainability challenges.

In experiments related to energy and renewable resources, various types of variables are considered to study energy generation, efficiency, and sustainable resource utilization. These variables can be categorized into several common types, including:

1. **Independent Variables**:
   - **Energy Source**: Variables related to the type of energy source being studied, such as solar, wind, hydroelectric, geothermal, biomass, or fossil fuels.
   - **Resource Availability**: Variables related to the availability of renewable resources, including sunlight, wind speed, water flow, or geothermal heat.

- **Technology Parameters**: Variables related to the design and specifications of energy conversion technologies, such as solar panel efficiency, turbine blade design, or geothermal well depth.
- **Operational Conditions**: Variables related to the operational parameters of energy systems, including temperature, pressure, flow rate, and control settings.
- **Policy and Regulatory Factors**: Variables related to government policies, incentives, subsidies, or regulations that may influence energy production, consumption, or investment decisions.

2. **Dependent Variables**:
   - **Energy Output**: Variables related to the amount of energy generated or extracted from renewable resources, often measured in kilowatt-hours (kWh) or other energy units.
   - **Efficiency Metrics**: Variables related to the efficiency of energy conversion processes, such as conversion efficiency, capacity factor, or energy return on investment (EROI).
   - **Environmental Impact**: Variables related to environmental outcomes, including greenhouse gas emissions, water usage, land use, and ecological impacts associated with energy production.
   - **Economic Variables**: Variables related to economic aspects of energy projects, such as cost per unit of energy, payback period, ROI, and levelized cost of electricity (LCOE).
   - **Energy Storage**: Variables related to the performance of energy storage technologies, including battery capacity, charging/discharging efficiency, and cycle life.

3. **Categorical Variables**:
   - **Technology Types**: Categories representing different types of energy conversion technologies or energy storage systems, such as photovoltaic panels, wind turbines, hydropower plants, or energy storage chemistries (e.g., lithium ion and flow batteries).
   - **Geographic Categories**: Categories representing different geographic regions or locations where energy projects are implemented, considering variations in resource availability and environmental conditions.
   - **Policy Categories**: Categories representing different energy policies, incentives, or regulatory frameworks that may vary by region or country.
   - **Market Segmentation**: Categories representing different market segments or consumer types, such as residential, commercial, industrial, or utility-scale energy consumers.

4. **Control Variables (Covariates)**:
   - These are variables that are held constant or controlled during experiments to eliminate their influence on the dependent variable, for example, maintaining consistent environmental conditions during solar panel testing.

5. **Random Variables**:
   - In some cases, random variables may be introduced to account for variability or uncertainty in measurements or to control for natural variations in resource availability.

6. **Interaction Variables**:
   - Interaction variables may be used to investigate how the combined effects of two or more independent variables impact energy generation, efficiency, or environmental outcomes.

7. **Noise Variables (Error Variables)**:
   - These are uncontrolled or unmeasured variables that can introduce variability or errors into experiments. Statistical analysis and experimental controls are used to minimize the impact of noise variables.

Energy and renewable resource experiments are crucial for optimizing energy technologies, assessing their sustainability, and informing energy policy decisions. Accurate and systematic consideration of variables is essential for obtaining meaningful results and improving the performance of renewable energy systems.

4.3.1 FACTORIAL EXPERIMENTS

Example: Energy engineers may use factorial experiments to optimize the parameters of a solar panel manufacturing process. Factors like material type, thickness, and temperature during production can be varied to maximize panel efficiency.

4.3.2 RSM

Example: RSM can be applied in wind turbine design to optimize blade shape and material properties. Researchers vary factors like blade curvature, length, and material composition to maximize energy capture and durability.

4.3.3 FRACTIONAL FACTORIAL EXPERIMENTS

Example: In biomass energy production, fractional factorial experiments can be used to assess the effects of various factors (e.g., feedstock type and pretreatment methods) on biofuel yield without testing all possible combinations.

4.3.4 CCD

Example: CCD is employed in geothermal energy exploration to optimize drilling parameters for geothermal wells. Engineers vary factors like drilling depth, mud composition, and injection rates to maximize heat extraction efficiency.

4.3.5 TAGUCHI METHODS

Example: Taguchi methods are used in energy-efficient building design to optimize heating, ventilation, and air conditioning (HVAC) systems. Engineers may vary factors like insulation materials, window types, and thermostat settings to minimize energy consumption while maintaining comfort.

4.3.6 MIXTURE EXPERIMENTS

Example: In biofuel production, mixture experiments help optimize the composition of biodiesel blends. Researchers vary proportions of different feedstock oils and additives to achieve desired fuel properties, such as cetane number and cold flow properties.

4.3.7 SEQUENTIAL EXPERIMENTATION

Example: Renewable energy project developers may use sequential experimentation to optimize the parameters of a solar farm's energy storage system. They adjust factors like battery size, charge–discharge cycles, and energy management strategies based on real-time energy generation and demand data.

4.3.8 RPD

Example: In hydropower generation, RPD can be applied to ensure the robustness of turbine designs against variations in water flow rates and sediment content, enhancing energy production reliability.

4.3.9 RANDOMIZED EXPERIMENTS

Example: In solar energy research, randomized experiments can be used to study the effects of different cleaning methods on solar panel efficiency. Panels are randomly assigned to different cleaning treatments to assess their impact on energy output.

4.3.10 SPLIT-PLOT AND BLOCKED EXPERIMENTS

Example: When optimizing the parameters of a biomass gasification process, split-plot or blocked experiments may be used to account for variations in feedstock quality, reactor design, and gasification conditions.

4.4 FOOD SCIENCE

In food science, designed experiments are essential for optimizing food production processes, improving product quality, and ensuring food safety. Here are some common types of designed experiments with examples of their applications in food science.

In food science designed experiments, various types of variables are considered to investigate food products, processes, and quality attributes. These variables can be categorized into several common types, including:

1. **Independent Variables**:
   - **Ingredients**: Variables related to the composition and type of ingredients used in food formulations, including proportions, quality, and source of ingredients.
   - **Processing Conditions**: Variables related to food processing parameters, such as temperature, pressure, cooking time, mixing speed, and fermentation conditions.
   - **Packaging and Storage Conditions**: Variables related to packaging materials, packaging techniques, and storage conditions (e.g., temperature, humidity, and light exposure) that can affect food shelf life and quality.
   - **Consumer Preferences**: Variables related to consumer preferences and sensory attributes, which can include factors like flavor, texture, appearance, and aroma.
   - **Nutritional Factors**: Variables related to the nutritional content of food products, including macronutrients (e.g., protein, carbohydrates, and fats), micronutrients (e.g., vitamins and minerals), and dietary fiber.
2. **Dependent Variables**:
   - **Food Quality Attributes**: Variables related to the quality and sensory attributes of food products, including taste, color, texture, aroma, shelf life, and overall acceptability.
   - **Nutritional Content**: Variables related to the nutritional composition of food products, such as protein content, vitamin content, fat content, and caloric value.
   - **Microbiological Safety**: Variables related to the presence of pathogens or spoilage microorganisms in food products, including microbial counts and foodborne pathogen detection.
   - **Texture Analysis**: Variables related to the mechanical properties of food products, such as hardness, cohesiveness, adhesiveness, and chewiness.
   - **Chemical Composition**: Variables related to the chemical composition of food products, including pH, acidity, moisture content, and flavor compounds.
3. **Categorical Variables**:
   - **Product Types**: Categories representing different types of food products or formulations, such as beverages, snacks, dairy products, or baked goods.
   - **Processing Methods**: Categories representing different food processing methods, such as frying, baking, pasteurization, and fermentation.
   - **Packaging Types**: Categories representing different types of packaging materials or packaging techniques used for food products.
   - **Consumer Demographics**: Categories representing different consumer groups, demographics, or market segments that may have varying preferences and needs.
4. **Control Variables (Covariates)**:
   - These are variables that are held constant or controlled during experiments to eliminate their influence on the dependent variable, for example, maintaining consistent cooking temperature or time during taste tests.

5. **Random Variables**:
   - In some cases, random variables may be introduced to account for variability or uncertainty in sensory evaluations or to control for natural variability in food properties.
6. **Interaction Variables**:
   - Interaction variables are used to investigate the combined effects of two or more independent variables on the dependent variable, helping to understand how different factors interact to influence food quality and consumer preferences.
7. **Noise Variables (Error Variables)**:
   - These are uncontrolled or unmeasured variables that can introduce variability or errors into food science experiments. Statistical analysis and experimental controls are used to minimize the impact of noise variables.

Food science designed experiments are crucial for optimizing food formulations, processes, and sensory attributes, as well as ensuring food safety and nutritional quality. Careful consideration and control of variables are essential for obtaining accurate and meaningful results in food research and development.

4.4.1 FACTORIAL EXPERIMENTS

Food scientists may use factorial experiments to optimize the parameters of a baking process for bread production. Factors like flour type, yeast concentration, dough mixing time, and baking temperature can be varied to achieve desired bread texture and taste.

Factorial design is a statistical technique often used in experimental research, including in the field of food science, to study the effects of multiple variables on a particular outcome or response. In factorial design, you manipulate two or more factors (independent variables) at different levels to understand how they interact and affect the response variable.

Let' us consider a simple example of a 2×2 factorial design in food science, where we are investigating the effects of two factors, A and B, on the taste of a new cookie recipe. Each factor has two levels, which means there are a total of four experimental conditions.

Factor A: Baking temperature

- Level 1: 350°F
- Level 2: 375°F

Factor B: Baking time

- Level 1: 10 minutes
- Level 2: 15 minutes

In this case, you would conduct experiments at all four possible combinations of these factors. You could set up your experiments like this (Table 4.26).

TABLE 4.26
Factorial Designed Experiment Matrix

Experiment	Factor A (Temperature)	Factor B (Time)	Taste Rating (Response)
1	Level 1	Level 1	
2	Level 1	Level 2	
3	Level 2	Level 1	
4	Level 2	Level 2	

In each experiment, you would bake a batch of cookies at the specified temperature and time and then have a panel of taste testers rate the taste. The taste ratings would be recorded in the "Taste Rating (Response)" column.

The goal of this factorial design is to determine the main effects of factors A and B and any potential interaction between them on the taste of the cookies. By systematically varying the levels of these factors and collecting data, you can analyze the results to make informed decisions about the best combination of baking temperature and time to achieve the desired taste for your cookies in food science experiments.

4.4.2 RSM

RSM is applied in food processing to optimize the parameters of a heat treatment process for canned vegetables. Researchers vary factors like temperature, pressure, and processing time to achieve desired microbial safety and product quality.

RSM is a statistical technique used in experimental research, including food science, to optimize processes and analyze the relationship between multiple factors and a response variable. In RSM, experiments are conducted at various combinations of factors, typically represented in a matrix, to create a response surface and find optimal conditions for the desired outcome.

Let us consider an example in food science where we want to optimize the recipe for a pasta sauce, specifically looking at the factors of tomato quantity (Factor A) and cooking time (Factor B). We will create a CCD, a common type of RSM experiment, with five levels for each factor, including the center point.

Factor A: Tomato quantity (grams)

- Level 1: 200 g
- Level 2: 300 g (center point)
- Level 3: 400 g
- Level 4: 500 g
- Level 5: 600 g

Factor B: Cooking time (minutes)

- Level 1: 10 minutes
- Level 2: 20 minutes (center point)
- Level 3: 30 minutes
- Level 4: 40 minutes
- Level 5: 50 minutes

You would then conduct experiments at these combinations of factor levels. To create a matrix, you can organize it as follows (Table 4.27).

TABLE 4.27

RSM Experiment Matrix

Experiment	Factor A (Tomato Quantity)	Factor B (Cooking Time)	Response (Taste)
1	Level 1 (200 g)	Level 1 (10 minutes)	
2	Level 5 (600 g)	Level 1 (10 minutes)	
3	Level 1 (200 g)	Level 5 (50 minutes)	
4	Level 5 (600 g)	Level 5 (50 minutes)	
5	Level 3 (400 g)	Level 2 (20 minutes)	
...	...	...	...

In each experiment, you would prepare the pasta sauce with the specified tomato quantity and cooking time and then have a panel of taste testers rate the taste. The taste ratings would be recorded in the "Response (Taste)" column.

The goal of this RSM designed experiment is to build a response surface or model based on these experimental data. This model can then be used to predict the optimal conditions (tomato quantity and cooking time) for the best-tasting pasta sauce. RSM helps food scientists optimize their processes and product formulations efficiently by systematically varying factors and analyzing the results.

4.4.3 Fractional Factorial Experiments

In dairy product development, fractional factorial experiments can be used to study the effects of various factors (e.g., milk composition and pasteurization conditions) on cheese quality without testing all possible combinations.

Fractional factorial design is a technique used in experimental research, including food science, to study the effects of factors while conducting only a fraction of the full set of possible experiments. This approach is useful when there are many factors to consider, and conducting a full factorial experiment would be impractical or time-consuming. Fractional factorial designs are based on a fraction (a subset) of the full factorial design matrix.

Let us consider an example in food science where you want to optimize a new salad dressing recipe with three factors: A, B, and C. Each factor has two levels (high and low), and you want to use a fractional factorial design to reduce the number of experiments.

Factor A: Amount of olive oil

- Level 1: Low (e.g., 20 mL)
- Level 2: High (e.g., 40 mL)

Factor B: Vinegar type

- Level 1: Red wine vinegar
- Level 2: Balsamic vinegar

Factor C: Seasoning

- Level 1: Low (e.g., 1 tsp)
- Level 2: High (e.g., 2 tsp)

A full factorial experiment would require $2^3 = 8$ experiments (two levels for each of the three factors). In a fractional factorial design, you can reduce the number of experiments by using a fraction, such as a half or quarter of the full factorial.

Let us create a matrix for a quarter-fractional factorial design (four experiments) where you systematically choose which experiments to run (Table 4.28).

TABLE 4.28

Fractional Factorial Designed Experiment Matrix

Experiment	Factor A (Olive Oil)	Factor B (Vinegar)	Factor C (Seasoning)	Taste Rating (Response)
1	Low	Red wine	Low	
2	High	Red wine	Low	
3	Low	Balsamic	High	
4	High	Balsamic	High	

In this matrix, you are conducting only a quarter of the full factorial experiments. You will prepare the salad dressings according to the specified levels of the factors, have a panel of taste testers rate them, and record the taste ratings in the "Taste Rating (Response)" column.

The analysis of these experiments would allow you to assess the main effects of the factors and any two-way interactions while reducing the number of experiments compared to a full factorial design. Fractional factorial designs are valuable for efficiently exploring factor effects and reducing experimentation time and resources in food science and other fields.

4.4.4 CCD

CCD is employed in beverage production to optimize the parameters of a juice pasteurization process. Engineers vary factors like temperature, holding time, and juice flow rate to maximize product shelf life and quality.

A CCD is a type of experimental design used in food science and various other fields to explore the response surface and optimize a process with multiple factors. It includes a central point, factorial points, and axial points to investigate factor effects and potential curvature in the response surface. Here is an example of a CCD matrix for a food science experiment involving three factors: A, B, and C.

Factor A: Temperature (°C)

- Low level: 50°C
- High level: 100°C

Factor B: Cooking time (minutes)

- Low level: 5 minutes
- High level: 30 minutes

Factor C: Salt (grams)

- Low level: 2 g
- High level: 10 g

In this example, we will assume a 2^3 full factorial design with a center point and a set of axial points for each factor.

1. Create the factorial points by varying each factor at its low and high levels (Table 4.29).
2. Add the center point experiments, typically at the midpoint of each factor's range (Table 4.30).
3. Finally, add axial point experiments that explore the edges of the design space. The axial points are typically set at ±1.68 times the distance from the center to the extreme levels (Table 4.31).

In this CCD matrix, you have designed a set of experiments to investigate the effects of temperature, cooking time, and salt on the taste of the food product. The taste ratings for each experiment would be recorded in the "Response (Taste)" column, and statistical analysis would be performed to optimize the process and identify significant factors and interactions.

4.4.5 TAGUCHI METHODS

Taguchi methods are used in food packaging to optimize packaging material properties. Scientists may vary factors like film thickness, barrier coatings, and sealing conditions to minimize oxygen permeation and extend the shelf life of packaged foods.

TABLE 4.29
CCD Designed Experiment Matrix

Experiment	Factor A (Temperature)	Factor B (Time)	Factor C (Salt)	Response (Taste)
1	Low	Low	Low	
2	High	Low	Low	
3	Low	High	Low	
4	High	High	Low	
5	Low	Low	High	
6	High	Low	High	
7	Low	High	High	
8	High	High	High	

TABLE 4.30
CCD Designed Experiment Matrix

Experiment	Factor A (Temperature)	Factor B (Time)	Factor C (Salt)	Response (Taste)
9	Midpoint	Midpoint	Midpoint	
10	Midpoint	Midpoint	Midpoint	
11	Midpoint	Midpoint	Midpoint	
12	Midpoint	Midpoint	Midpoint	

TABLE 4.31
CCD Designed Experiment Matrix

Experiment	Factor A (Temperature)	Factor B (Time)	Factor C (Salt)	Response (Taste)
13	Low	Midpoint	Midpoint	
14	High	Midpoint	Midpoint	
15	Midpoint	Low	Midpoint	
16	Midpoint	High	Midpoint	
17	Midpoint	Midpoint	Low	
18	Midpoint	Midpoint	High	

The Taguchi method is a robust optimization technique used in experimental design to improve product or process quality by identifying optimal settings for various factors while considering noise or variations. In the Taguchi method, experiments are organized into orthogonal arrays. Here is an example of a matrix for a Taguchi designed experiment in food science with three factors: A, B, and C.

Factor A: Ingredient type

- Level 1: Ingredient X
- Level 2: Ingredient Y

TABLE 4.32
Taguchi Method Designed Experiment Matrix

Experiment	Factor A (Ingredient)	Factor B (Mixing Time)	Factor C (Temperature)	Response (Quality)
1	Level 1	Level 1	Level 1	
2	Level 2	Level 1	Level 2	
3	Level 1	Level 2	Level 3	
4	Level 2	Level 2	Level 1	
5	Level 1	Level 3	Level 2	
6	Level 2	Level 3	Level 3	
7	Level 1	Level 1	Level 3	
8	Level 2	Level 2	Level 2	
9	Level 1	Level 3	Level 1	

Factor B: Mixing time (minutes)

- Level 1: 5 minutes
- Level 2: 10 minutes
- Level 3: 15 minutes

Factor C: Temperature (°C)

- Level 1: 150°C
- Level 2: 175°C
- Level 3: 200°C

Let us use an L9 orthogonal array, which is suitable for three factors with three levels each. An L9 array contains nine rows, indicating nine experimental runs (Table 4.32).

In this matrix, each row represents an experimental run with specific levels of factors A, B, and C. The "Response (Quality)" column is where you would record the quality assessment or performance measure of the food product for each experiment.

The Taguchi method focuses on identifying the optimal combination of factor levels that minimizes the impact of noise or variations in the system, ultimately leading to improved product quality or process performance. After conducting these experiments, statistical analysis, such as signal-to-noise ratio calculations, would be used to determine the best factor settings for achieving the desired quality in food science applications.

4.4.6 MIXTURE EXPERIMENTS

In confectionery production, mixture experiments help optimize the formulation of chocolate products. Researchers vary proportions of cocoa solids, cocoa butter, sweeteners, and emulsifiers to achieve desired sensory attributes and melt-in-mouth properties.

Mixture designed experimentation is used in food science when you are dealing with ingredients or components that make up a recipe or formulation. In such experiments, you typically have a mixture of ingredients, and you want to determine the optimal combination of these ingredients to achieve a desired outcome, such as taste, texture, or nutritional content. Here is an example of a matrix for a mixture designed experiment in food science:

TABLE 4.33
Mixture Designed Experiment Matrix

Experiment	Proportion of Oats (O)	Proportion of Nuts (N)	Proportion of Dried Fruits (F)	Taste Rating (Response)
1	0%	0%	100%	
2	0%	50%	50%	
3	50%	0%	50%	
4	33.33%	33.33%	33.33%	
5	75%	12.5%	12.5%	
6	12.5%	75%	12.5%	
7	12.5%	12.5%	75%	
8	25%	25%	50%	
9	25%	50%	25%	
10	50%	25%	25%	

Let us consider a food product like a granola bar, and you want to optimize the recipe by varying the proportions of three ingredients: oats (O), nuts (N), and dried fruits (F). You have a total of 100 g of ingredients to work with.

Factor A: Proportion of oats (O)

- Range: 0%–100% of 100 g
    Factor B: Proportion of nuts (N)
- Range: 0%–100% of 100 g
    Factor C: Proportion of dried fruits (F)
- Range: 0%–100% of 100 g

You can use a simplex-lattice design, which is suitable for mixture experiments. Here is an example of a 10-run simplex-lattice design matrix (Table 4.33).

In this matrix, each experiment represents a specific combination of ingredient proportions for your granola bar recipe. The "Taste Rating (Response)" column is where you would record the taste rating for each experiment.

The goal of the mixture designed experiment is to determine the optimal blend of ingredients that maximizes the desired taste or sensory characteristics of the granola bar. Statistical analysis, such as mixture design modeling, can help identify the ideal proportions of oats, nuts, and dried fruits to achieve the best product in food science applications.

4.4.7 SEQUENTIAL EXPERIMENTATION

Food scientists may use sequential experimentation to optimize the parameters of a food extrusion process for snack production. They adjust factors like moisture content and screw speed based on real-time product quality measurements to reduce waste and improve efficiency.

Sequential designed experimentation in food science typically involves a series of experiments where each new experiment is planned based on the results of previous ones. This iterative approach allows researchers to adapt and refine their experimental plan as they gather more information. Here is an example of how a sequential experimentation matrix might look for a food science study.

Suppose you are trying to optimize a bread recipe, and you want to experiment with three factors: A (flour type), B (yeast amount), and C (baking time). You start with an initial set of experiments and then adapt your plan based on the results.

TABLE 4.34

Sequential Designed Experiment Matrix

Experiment	Factor A (Flour)	Factor B (Yeast)	Factor C (Baking Time)	Taste Rating (Response)
1	Level 1	Level 1	Level 1	
2	Level 1	Level 1	Level 2	
3	Level 1	Level 2	Level 1	
4	Level 1	Level 2	Level 2	
5	Level 2	Level 1	Level 1	
6	Level 2	Level 1	Level 2	
7	Level 2	Level 2	Level 1	
8	Level 2	Level 2	Level 2	

Factor A: Flour type

- Level 1: All-purpose flour
- Level 2: Whole wheat flour

Factor B: Yeast amount (grams)

- Level 1: 5 g
- Level 2: 10 g

Factor C: Baking time (minutes)

- Level 1: 20 minutes
- Level 2: 30 minutes

Here is an example of a sequential experimentation matrix for the first set of experiments (Table 4.34).

After conducting these initial experiments and collecting taste ratings, you might decide to adjust your plan based on the results. For example, if you find that whole wheat flour consistently results in lower taste ratings, you might decide to exclude it from further experimentation. You might also refine the yeast amounts and baking times based on the initial feedback.

The key to sequential designed experimentation is the ability to adapt your experimental plan as you go, based on the information gathered in earlier experiments. This iterative process can be especially valuable when optimizing complex food products or processes in food science, as it allows for efficient exploration and adjustment of the factors involved.

4.4.8　RPD

In meat processing, RPD can be applied to ensure the robustness of a cooking process against variations in meat thickness and fat content, ensuring consistent product quality and safety.

RPD is a technique used in experimental design, including food science, to develop products or processes that are resistant to variations or sources of variability. In RPD, you aim to find optimal factor settings that minimize the sensitivity of the response variable to noise factors. Here is an example of how a matrix for RPD designed experimentation in food science might look.

Suppose you are working on a new chocolate chip cookie recipe and want to make it robust to variations in ingredients. You have identified three factors that can affect the cookie's taste: A (amount of sugar), B (baking temperature), and C (baking time).

TABLE 4.35
RPD Matrix

Experiment	Factor A (Sugar)	Factor B (Temperature)	Factor C (Time)	Taste Rating (Response)
1 (Control)	125 g	185°C	12.5 minutes	
2	125 g	175°C	10 minutes	
3	125 g	200°C	15 minutes	
4	150 g	185°C	10 minutes	
5	150 g	200°C	12.5 minutes	
6	100 g	185°C	15 minutes	
7	100 g	200°C	10 minutes	
8	100 g	175°C	12.5 minutes	
9	150 g	175°C	15 minutes	

Factor A: Amount of sugar (grams)

- Range: 100–150 g
    Factor B: Baking temperature (°C)
- Range: 175°C–200°C
    Factor C: Baking time (minutes)
- Range: 10–15 minutes

You will conduct a series of experiments to find factor settings that produce consistent taste results even when there are variations in the ingredients or process conditions. In RPD, you typically perform a series of experiments with a control or nominal setting and then vary the factors within a defined range.

Here is an example of a matrix for RPD designed experimentation (Table 4.35).

In this matrix, the first experiment (control) represents the nominal settings for sugar amount, baking temperature, and baking time. The other experiments vary these factors within their defined ranges. You would collect taste ratings for each batch of cookies and record them in the "Taste Rating (Response)" column.

The goal of RPD is to identify factor settings that make the cookie recipe less sensitive to variations in sugar, temperature, and time while maintaining a consistent taste. Statistical analysis, such as signal-to-noise ratio calculations or robust optimization methods, can help you determine the robust factor settings in food science applications.

4.4.9 Randomized Experiments

In sensory evaluation studies, randomized experiments can be used to assess the effects of different food formulations on consumer preferences. Panels of trained sensory evaluators are presented with randomized samples to evaluate product attributes.

In the context of experimental design in food science, a randomized complete block design (RCBD) is a commonly used approach. In an RCBD, experimental units are grouped into blocks, and treatments are randomly assigned within each block to reduce the impact of potential sources of variability. Here is an example of a matrix for a RCBD in food science.

Suppose you are conducting an experiment to determine the effects of different cooking temperatures (Factor A) and cooking times (Factor B) on the tenderness of steaks. You want to account for potential variability in meat quality, so you use a RCBD.

TABLE 4.36
Randomized Experiment Matrix (Block 1)

Experiment	Factor A (Temperature)	Factor B (Time)	Tenderness Rating (Response)
1	Level 1	Level 1	
2	Level 1	Level 2	
3	Level 1	Level 3	
4	Level 2	Level 1	
5	Level 2	Level 2	
6	Level 2	Level 3	

TABLE 4.37
Randomized Experiment Matrix (Block 2)

Experiment	Factor A (Temperature)	Factor B (Time)	Tenderness Rating (Response)
1	Level 1	Level 1	
2	Level 1	Level 2	
3	Level 1	Level 3	
4	Level 2	Level 1	
5	Level 2	Level 2	
6	Level 2	Level 3	

You have two blocks (Block 1 and Block 2), each representing a different batch of steaks with potentially varying qualities.
Factor A: Cooking temperature (°C)

- Level 1: 150°C
- Level 2: 160°C
- Level 3: 170°C

Factor B: Cooking time (minutes)

- Level 1: 10 minutes
- Level 2: 12 minutes
- Level 3: 14 minutes

Here is how the matrix for the RCBD might look (Tables 4.36 and 4.37).
Block 1
Block 2
In this matrix, each block represents a batch of steaks, and within each block, you randomly assign the combinations of cooking temperature and time (Factor A and Factor B). The "Tenderness Rating (Response)" column is where you record the tenderness ratings for each steak in the respective block.

Using a RCBD helps control for potential variability in meat quality between different batches of steaks, allowing you to better assess the effects of cooking temperature and time on tenderness in food science experiments.

4.4.10 SPLIT-PLOT AND BLOCKED EXPERIMENTS

When optimizing the parameters of a canning process for fruit preserves, split-plot or blocked experiments may be used to account for variations in fruit quality, jar size, and filling methods.

A split-plot design is often used in food science experiments when you have both main plots (whole plots) and subplots (smaller units within each whole plot). It is typically employed when different factors or treatments can be applied at different scales or levels. Here are examples of matrices for both split-plot and blocked experimentation in food science:

Example 1: Split-Plot Design

Suppose you are conducting an experiment to optimize a bread recipe. You are testing two factors: A (flour type) and B (yeast amount). However, due to limited oven space, you can only bake four loaves of bread at a time. Therefore, each baking batch represents a whole plot, and within each batch, you will test different yeast amounts as subplots.

Factor A: Flour type
  - Level 1: All-purpose flour
  - Level 2: Whole wheat flour

Factor B: Yeast amount (grams)
  - Level 1: 5 g
  - Level 2: 10 g

The matrix for a split-plot design is shown in Table 4.38.

In this split-plot design, each whole plot represents a baking batch with different flour types (Factor A), and within each batch, you vary the yeast amount (Factor B) for the subplots. Taste ratings are recorded for each loaf of bread.

TABLE 4.38
Split-Plot Experiment Matrix

Whole Plot	Subplot	Factor A (Flour)	Factor B (Yeast)	Taste Rating (Response)
1	1	Level 1	Level 1	
1	2	Level 1	Level 2	
1	3	Level 2	Level 1	
1	4	Level 2	Level 2	
2	1	Level 1	Level 1	
2	2	Level 1	Level 2	
2	3	Level 2	Level 1	
2	4	Level 2	Level 2	
3	1	Level 1	Level 1	
3	2	Level 1	Level 2	
3	3	Level 2	Level 1	
3	4	Level 2	Level 2	
4	1	Level 1	Level 1	
4	2	Level 1	Level 2	
4	3	Level 2	Level 1	
4	4	Level 2	Level 2	

TABLE 4.39
Blocked Experiment Matrix

Block	Experiment	Factor A (Preservative)	Tomato Quality Rating (Response)
Day 1	1	Level 1	
Day 1	2	Level 2	
Day 1	3	Level 1	
Day 1	4	Level 2	
Day 2	5	Level 1	
Day 2	6	Level 2	
Day 2	7	Level 1	
Day 2	8	Level 2	
Day 3	9	Level 1	
Day 3	10	Level 2	
Day 3	11	Level 1	
Day 3	12	Level 2	

Example 2: Blocked Experimentation

Blocked experimentation involves grouping experimental units into blocks to account for potential sources of variability. For instance, you might block based on time, location, or other relevant factors. Here is an example of a blocked design for testing the effectiveness of a food preservative on different batches of tomatoes.

Factor A: Food preservative (presence/absence)
- Level 1: Presence
- Level 2: Absence

You conduct the experiment over three different days, so you decide to block by day (Table 4.39).

In this blocked design, you group the experiments based on the day they were conducted (block). For each day, you test the effect of the preservative (Factor A) on tomato quality and record the quality ratings as the response. Blocking helps control for any day-to-day variations in environmental conditions or other factors that may affect tomato quality.

4.5   HEALTHCARE AND MEDICAL RESEARCH

Designed experiments are widely used in healthcare and medical research to evaluate treatments, medical interventions, and healthcare processes, as well as to study disease mechanisms and patient outcomes.

In healthcare and medical research designed experiments, various types of variables are considered to investigate health-related phenomena, interventions, and outcomes. These variables can be categorized into several common types, including:

1. **Independent Variables**:
   - **Treatment Variables**: Variables related to medical treatments, interventions, or therapeutic approaches, such as drug dosage, surgical procedure, therapy type, or medical device.
   - **Patient Characteristics**: Variables related to patient demographics and medical history, including age, gender, ethnicity, medical condition, comorbidities, and genetic factors.

- **Healthcare Practices**: Variables related to healthcare practices and protocols, including vaccination strategies, diagnostic criteria, and treatment guidelines.
- **Environmental Factors**: Variables related to environmental conditions that may impact health, such as air quality, temperature, pollution levels, or geographical location.
- **Behavioral Factors**: Variables related to patient behaviors and lifestyle choices, including diet, physical activity, smoking, alcohol consumption, and medication adherence.

2. **Dependent Variables:**
- **Health Outcomes**: Variables related to health outcomes, including measures of disease progression, symptom severity, survival rates, quality of life, or patient-reported outcomes.
- **Clinical Markers**: Variables related to clinical markers or biomarkers that indicate disease presence, progression, or treatment response, such as blood pressure, cholesterol levels, or biomarker concentrations.
- **Adverse Events**: Variables related to adverse events, side effects, or complications associated with medical treatments or interventions.
- **Patient Satisfaction**: Variables related to patient satisfaction with healthcare services, communication with healthcare providers, or overall healthcare experience.
- **Healthcare Utilization**: Variables related to healthcare utilization, such as hospital admissions, emergency room visits, healthcare costs, or healthcare resource consumption.

3. **Categorical Variables:**
- **Diagnostic Categories**: Categories representing different medical diagnoses or disease classifications.
- **Treatment Groups**: Categories representing different treatment arms or groups in clinical trials, including experimental and control groups.
- **Medical Specialties**: Categories representing different medical specialties or healthcare provider types (e.g., cardiology, oncology, and primary care).
- **Patient Risk Categories**: Categories representing different levels of patient risk or severity, such as low-risk, moderate-risk, and high-risk groups.

4. **Control Variables (Covariates):**
- These are variables that are held constant or controlled during experiments to eliminate their influence on the dependent variable, for example, controlling for age and gender in a clinical trial to ensure that treatment effects are not confounded by these factors.

5. **Random Variables:**
- Random variables may be introduced to account for variability or uncertainty in medical measurements or to control for natural variations in patient responses.

6. **Interaction Variables:**
- Interaction variables may be used to investigate how the combined effects of two or more independent variables influence health outcomes or treatment responses.

7. **Noise Variables (Error Variables):**
- These are uncontrolled or unmeasured variables that can introduce variability or errors into experiments. Statistical analysis and experimental controls are used to minimize the impact of noise variables.

Healthcare and medical research experiments are essential for advancing medical knowledge, improving patient care, and developing evidence-based healthcare practices. Rigorous consideration and control of variables are critical for obtaining accurate and meaningful results in medical research and healthcare interventions.

4.5.1 RANDOMIZED CONTROLLED TRIALS (RCTs)

Example: In clinical medicine, RCTs are used to assess the effectiveness of a new drug in treating a specific medical condition. Patients are randomly assigned to receive either the experimental drug or a placebo, and their health outcomes are compared.

4.5.2 CROSSOVER TRIALS

Example: Researchers in pharmacology may use crossover trials to study the effects of different drug dosages. Participants receive multiple treatments in random order, with washout periods in between, to minimize the influence of time on treatment effects.

4.5.3 CLUSTER RANDOMIZED TRIALS

Example: In public health, cluster randomized trials can be employed to evaluate the impact of a health intervention on entire communities. Communities or healthcare facilities are randomly assigned to receive the intervention or serve as controls.

4.5.4 FACTORIAL EXPERIMENTS

Example: In healthcare operations management, factorial experiments can be used to optimize hospital scheduling procedures. Factors like appointment duration, staff allocation, and appointment booking methods are varied to reduce patient wait times.

4.5.5 RSM

Example: RSM can be applied in pharmaceutical formulation to optimize the parameters of a tablet coating process. Researchers vary factors like coating thickness, drying temperature, and spray rate to achieve desired tablet characteristics.

4.5.6 QUASI-EXPERIMENTS

Example: In epidemiology, quasi-experiments can be used to assess the impact of a public health intervention, such as smoking bans in public places, on respiratory health outcomes by comparing data before and after the intervention.

4.5.7 SEQUENTIAL EXPERIMENTS

Example: In clinical research, sequential experimentation may be used in adaptive clinical trial designs. Researchers continuously analyze accumulating trial data to make real-time decisions on treatment arms, sample sizes, and study endpoints.

4.5.8 RPD

Example: In medical device manufacturing, RPD can be applied to ensure the robustness of product specifications against variations in manufacturing processes, reducing the likelihood of defects.

4.5.9 RANDOMIZED WITHDRAWAL TRIALS

Example: Randomized withdrawal trials are used in psychiatry to assess the long-term effectiveness of a medication. Patients initially receive the active drug, and then, a portion is switched to a placebo to evaluate relapse rates.

4.5.10 Simulation Experiments

Example: Healthcare administrators may use simulation experiments to optimize emergency department workflows. Various scenarios are simulated with different resource allocation strategies to improve patient throughput and reduce wait times.

4.6 INFORMATION TECHNOLOGY

In the field of information technology (IT), designed experiments are used to optimize software and hardware configurations, improve system performance, and enhance the user experience.

In IT research designed experiments, various types of variables are considered to investigate and optimize IT systems, software, networks, and services. These variables can be categorized into several common types, including:

1. **Independent Variables**:
   - **Software Parameters**: Variables related to software settings, configurations, and parameters, such as software version, algorithms, data structures, and coding practices.
   - **Hardware Specifications**: Variables related to hardware components and specifications, including central processing unit (CPU) speed, random access memory (RAM) capacity, storage type, and network bandwidth.
   - **Network Conditions**: Variables related to network characteristics, including latency, packet loss, bandwidth, and network topology.
   - **User Interface (UI) Design**: Variables related to UI elements, layouts, colors, fonts, and user interaction patterns.
   - **Security Settings**: Variables related to security configurations, such as encryption protocols, access control lists, and authentication methods.
   - **Environmental Conditions**: Variables related to physical or environmental factors that can impact IT systems, such as temperature, humidity, and power supply stability.
   - **Human Factors**: Variables related to user behavior, skills, and preferences, which can influence IT system usability and performance.
2. **Dependent Variables**:
   - **System Performance Metrics**: Variables related to the performance of IT systems, including response time, throughput, resource utilization, and system availability.
   - **Quality of Service (QoS)**: Variables related to network QoS parameters, such as latency, jitter, packet loss rate, and network reliability.
   - **Security Metrics**: Variables related to the security posture of IT systems, including vulnerability counts, intrusion detection alerts, and security incident rates.
   - **Usability Metrics**: Variables related to user experience and usability, including user satisfaction, task completion time, error rates, and ease of use.
   - **Software Reliability**: Variables related to software reliability measures, such as failure rates, error counts, and mean time between failures (MTBF).
   - **Cost and Resource Utilization**: Variables related to IT system costs, including hardware and software costs, energy consumption, and total cost of ownership (TCO).
3. **Categorical Variables**:
   - **Software Categories**: Categories representing different software types or application domains, such as operating systems, database management systems, web browsers, or mobile apps.
   - **Hardware Types**: Categories representing different hardware types, including server types, mobile device categories, and processor architectures.
   - **User Demographics**: Categories representing user characteristics, such as age groups, geographic locations, or user roles (e.g., administrator and end user).
   - **Security Threat Categories**: Categories representing different types of security threats, such as malware, phishing, denial-of-service attacks, or insider threats.

4. **Control Variables (Covariates)**:
   - These are variables that are held constant or controlled during experiments to eliminate their influence on the dependent variable, for example, controlling for network bandwidth when testing the impact of software configurations on network performance.
5. **Random Variables**:
   - Random variables may be introduced to account for variability or uncertainty in measurements or to control for natural variations in IT environments.
6. **Interaction Variables**:
   - Interaction variables may be used to investigate how the combined effects of two or more independent variables impact IT system performance, security, or usability.
7. **Noise Variables (Error Variables)**:
   - These are uncontrolled or unmeasured variables that can introduce variability or errors into experiments. Statistical analysis and experimental controls are used to minimize the impact of noise variables.

IT research experiments are essential for optimizing IT systems, enhancing security, improving user experience, and ensuring the reliability and efficiency of digital technologies. Careful consideration and control of variables are crucial for obtaining accurate and meaningful results in the field of IT.

4.6.1 FACTORIAL EXPERIMENTS

Example: IT professionals may use factorial experiments to optimize the performance of a database system. Factors such as database indexing, query optimization settings, and hardware resources can be varied to maximize query response times.

4.6.2 RSM

Example: RSM can be applied in software development to optimize the parameters of a machine learning algorithm. Researchers vary factors like learning rate, model complexity, and training dataset size to achieve the best predictive accuracy.

4.6.3 FRACTIONAL FACTORIAL EXPERIMENTS

Example: In network configuration, fractional factorial experiments can be used to study the effects of various factors (e.g., firewall rules and network topology) on network security and performance without testing all possible combinations.

4.6.4 CCD

Example: CCD is employed in cloud computing to optimize the parameters of a virtual machine provisioning process. Engineers vary factors like CPU allocation, memory size, and storage capacity to minimize resource wastage and meet service-level agreements.

4.6.5 TAGUCHI METHODS

Example: Taguchi methods are used in website design to optimize the user interface (UI) for an e-commerce platform. Designers may vary factors like button placement, color schemes, and navigation menus to improve user engagement and conversion rates.

4.6.6 MIXTURE EXPERIMENTS

Example: In software testing, mixture experiments help optimize the composition of test suites. Researchers vary proportions of test case types (e.g., unit tests, integration tests, and performance tests) to achieve effective software quality assurance.

4.6.7 Sequential Experimentation

Example: IT administrators may use sequential experimentation to optimize server load balancing algorithms. They adjust factors like server selection policies and load threshold settings based on real-time server performance data to ensure even resource distribution.

4.6.8 RPD

Example: In cybersecurity, RPD can be applied to ensure the robustness of a security system against variations in cyber threats and attack methods, reducing vulnerabilities and enhancing protection.

4.6.9 Randomized Experiments

Example: In web application development, randomized experiments can be used to evaluate the impact of different page layouts on user engagement. Randomized trials assign users to different versions of a website to assess their behavior and preferences.

4.6.10 Split-Plot and Blocked Experiments

Example: When optimizing the parameters of a data center cooling system, split-plot or blocked experiments may be used to account for variations in cooling equipment, server racks, and data center layouts.

4.7 LUBRICANTS

Design of experiment (DOE) methods are commonly used in the formulation of lubricants to optimize their properties and performance.

In lubricant designed experiments, various types of variables are considered to study the performance, properties, and characteristics of lubricants under different conditions and formulations. These variables can be categorized into several common types, including:

1. **Independent Variables**:
   - **Lubricant Formulation Variables**: Variables related to the composition and formulation of lubricants, such as base oil type, viscosity grade, additives (e.g., antiwear and antioxidants), and concentration of additives.
   - **Operating Conditions**: Variables related to the conditions under which lubricants are used, including temperature, pressure, speed, load, and environmental factors (e.g., humidity).
   - **Surface Materials**: Variables related to the materials and surface finishes of components in contact with lubricants, such as metals, alloys, ceramics, and coatings.
   - **Testing Parameters**: Variables related to laboratory or field testing parameters, such as test duration, test equipment specifications, and testing protocols.
2. **Dependent Variables**:
   - **Friction Coefficient**: Variables related to the frictional resistance between lubricated surfaces, which can impact energy efficiency and wear rates.
   - **Wear Rate**: Variables related to the amount of material loss or wear between lubricated surfaces, including wear scar size and wear volume.
   - **Viscosity**: Variables related to the viscosity of the lubricant, which affects its flow properties and film thickness.
   - **Oxidation Resistance**: Variables related to the ability of the lubricant to resist oxidation and maintain its performance over time.
   - **Temperature Rise**: Variables related to the temperature increase observed in lubricated components during operation.

3. **Categorical Variables**:
   - **Lubricant Types**: Categories representing different types of lubricants, such as mineral oils, synthetic oils, biobased lubricants, and greases.
   - **Additive Packages**: Categories representing different combinations and concentrations of additives used in lubricant formulations.
   - **Surface Treatments**: Categories representing different surface treatments or coatings applied to components to improve lubrication or wear resistance.
   - **Test Methods**: Categories representing different standardized test methods or test equipment used to evaluate lubricant performance.
4. **Control Variables (Covariates)**:
   - These are variables that are held constant or controlled during experiments to eliminate their influence on the dependent variable, for example, maintaining constant load or speed during friction and wear tests.
5. **Random Variables**:
   - In some cases, random variables may be introduced to account for variability or uncertainty in lubricant measurements or to control for natural variability in test conditions.
6. **Interaction Variables**:
   - Interaction variables are used to investigate the combined effects of two or more independent variables on the dependent variable, helping to understand how different factors interact to influence lubricant performance.
7. **Noise Variables (Error Variables)**:
   - These are uncontrolled or unmeasured variables that can introduce variability or errors into lubricant experiments. Statistical analysis and experimental controls are used to minimize the impact of noise variables.

Lubricant designed experiments are essential for improving lubrication technology, enhancing equipment reliability, and reducing friction and wear in various industries, including automotive, manufacturing, and aerospace. Careful consideration and control of variables are crucial for obtaining accurate and meaningful results in lubricant research and development.

4.7.1 FACTORIAL EXPERIMENTS

These are used to evaluate the main effects of multiple factors (ingredients or parameters) on lubricant properties. For example, you might investigate the impact of base oil type, additives, and temperature on lubricant viscosity and wear resistance.

In lubricant formulation, a factorial experiment can be used to assess the main effects and potential interactions of different factors (ingredients or parameters) on the properties of the lubricant. Here is an example of a matrix for a factorial designed experiment in lubricant formulation:

Objective: To investigate the influence of two factors, Factor A (base oil type) and Factor B (additive concentration), on the viscosity of a lubricant.

Factors:

1. **Factor A**: Base oil type
   - Two levels: Mineral oil (Level 1) and synthetic oil (Level 2)
2. **Factor B**: Additive concentration (% by weight)
   - Two levels: Low concentration (Level 1) and high concentration (Level 2)

TABLE 4.40
Factorial Experiment Matrix

Run	Factor A (Base Oil Type)	Factor B (Additive Concentration)	Viscosity (cSt)
1	Mineral oil (1)	Low concentration (1)	...
2	Mineral oil (1)	High concentration (2)	...
3	Synthetic oil (2)	Low concentration (1)	...
4	Synthetic oil (2)	High concentration (2)	...

Response Variable: Viscosity (measured in centistokes (cSt))

Factorial Experiment Matrix: A 2 × 2 factorial design involves four experimental runs, covering all possible combinations of the two factors at their respective levels (Table 4.40).
In this matrix:

- Each row represents a specific combination of base oil type and additive concentration.
- Viscosity measurements are recorded for each combination.

By analyzing the results of this factorial experiment, you can determine how each factor (base oil type and additive concentration) and their potential interactions affect the viscosity of the lubricant. This information is valuable for optimizing lubricant formulations to meet specific performance requirements.

4.7.2 RSM

RSM helps in optimizing lubricant formulations by exploring the interactions between different factors and responses. It allows you to find the optimal combination of factors for desired lubricant properties such as viscosity, friction, and wear.

RSM is a valuable technique for optimizing lubricant formulations by exploring the interactions between different factors and responses. Here is an example of a matrix for an RSM designed experiment in lubricant formulation:

Objective: To optimize a lubricant formulation for improved wear resistance, taking into account three factors: Factor A (base oil type), Factor B (additive concentration), and Factor C (temperature).

Factors:

1. **Factor A**: Base oil type
   - Varying levels: Mineral oil and synthetic oil
2. **Factor B**: Additive concentration (% by weight)
   - Varying levels: Low, medium, and high
3. **Factor C**: Temperature (°C)
   - Varying levels: 50°C, 75°C, and 100°C

Response Variable: Wear resistance (measured in wear scar diameter (WSD) in micrometers (μm))

TABLE 4.41
RSM Designed Experiment Matrix

Run	Factor A (Base Oil Type)	Factor B (Additive Concentration)	Factor C (Temperature)	Wear Resistance (WSD, μm)
1	Mineral oil	Low	50	...
2	Mineral oil	High	50	...
3	Synthetic oil	Low	50	...
4	Synthetic oil	High	50	...
5	Mineral oil	Medium	75	...
6	Synthetic oil	Medium	75	...
7	Mineral oil	Low	100	...
8	Synthetic oil	Low	100	...
9	Mineral oil	High	100	...
10	Synthetic oil	High	100	...
11	Mineral oil	Medium	50	...
12	Synthetic oil	Medium	50	...
13	Mineral oil	Medium	75	...
14	Synthetic oil	Medium	75	...
15	Mineral oil	Medium	100	...
16	Synthetic oil	Medium	100	...

RSM Designed Experiment Matrix: In this example, a CCD is used to create a matrix for RSM. CCD allows for the exploration of both linear and quadratic effects of the factors on the response (Table 4.41).

In this matrix:

- Each row represents a specific combination of base oil type, additive concentration, and temperature.
- Wear resistance (WSD) measurements are recorded for each combination.

By analyzing the results of this RSM designed experiment, you can model the relationships between the factors and the response, allowing you to optimize the lubricant formulation for the desired wear resistance while considering the interactions between these factors.

4.7.3 Mixture Designed Experiments

Lubricants often consist of a mixture of ingredients like base oils, additives, and modifiers. Mixture designs are suitable for optimizing the proportions of these components to achieve desired properties while minimizing cost or maximizing performance.

Mixture designed experiments are commonly used in lubricant formulation to optimize the proportions of different components, such as base oils, additives, and modifiers, while considering constraints and achieving specific performance targets. Here is an example of a matrix for a mixture designed experiment in lubricant formulation:

Objective: To optimize a lubricant formulation for enhanced oxidation stability while considering the proportions of three components: base oil A, base oil B, and an additive blend.

TABLE 4.42
Mixture Designed Experiment Matrix

Run	Component A (%)	Component B (%)	Component C (%)	Oxidation Stability (Hours)
1	20	80	5	...
2	40	60	5	...
3	60	40	5	...
4	80	20	5	...
5	20	80	10	...
6	40	60	10	...
7	60	40	10	...
8	80	20	10	...
9	20	80	15	...
10	40	60	15	...
11	60	40	15	...
12	80	20	15	...

Components:

1. **Component A**: Base oil A (proportion in %)
   - Varying proportions: 20%, 40%, 60%, and 80%
2. **Component B**: Base oil B (proportion in %)
   - Complementary proportions to component A to maintain 100%
3. **Component C**: Additive blend (proportion in %)
   - Varying proportions: 5%, 10%, and 15%

Response Variable: Oxidation stability (measured in hours)

Mixture Designed Experiment Matrix: A simplex-lattice design is a common choice for mixture experiments as it covers the entire space within the constraints of the proportions (Table 4.42).

In this matrix:

- Each row represents a specific combination of Component A, Component B, and Component C proportions.
- The proportions of Component A and Component B are complementary to each other (100% in total), and Component C is added in varying proportions.
- Oxidation stability measurements are recorded for each combination.

By analyzing the results of this mixture designed experiment, you can identify the optimal proportions of the components to achieve the desired oxidation stability in the lubricant formulation while adhering to the constraints of the mixture components' total percentage.

4.7.4 CCD

CCD is used to model complex relationships between factors and responses in lubricant formulation. It helps identify the optimal factor settings while assessing the curvature and interaction effects.

CCD is often used in lubricant formulation to optimize formulations while considering both linear and quadratic effects of factors. Here is an example of a matrix for a CCD designed experiment in lubricant formulation.

Objective: To optimize a lubricant formulation for improved wear resistance, taking into account three factors: Factor A (base oil type), Factor B (additive concentration), and Factor C (temperature).

Factors:

1. **Factor A**: Base oil type
   - Coded levels: −1 (mineral oil) and +1 (synthetic oil)
2. **Factor B**: Additive concentration (% by weight)
   - Coded levels: −1 (low), 0 (medium), and +1 (high)
3. **Factor C**: Temperature (°C)
   - Coded levels: −1 (low), 0 (medium), and +1 (high)

Response Variable: Wear resistance (measured in WSD in μm)

CCD Designed Experiment Matrix: In this example, a CCD with coded levels is used, allowing you to explore both linear and quadratic effects of the factors on the response (Table 4.43).
In this matrix:

- The coded levels for each factor allow for easy translation to actual levels within the experimentation range.
- The central points (9 to 14) are used for estimating curvature effects.
- Wear resistance (WSD) measurements are recorded for each combination of coded levels.

By analyzing the results of this CCD designed experiment, you can model the relationships between the factors and the response, allowing you to optimize the lubricant formulation for the desired wear resistance while considering linear and quadratic effects of the factors.

TABLE 4.43
CCD Designed Experiment Matrix

Run	Factor A (Coded Level)	Factor B (Coded Level)	Factor C (Coded Level)	Wear Resistance (WSD, μm)
1	−1	−1	−1	...
2	+1	−1	−1	...
3	−1	+1	−1	...
4	+1	+1	−1	...
5	−1	−1	+1	...
6	+1	−1	+1	...
7	−1	+1	+1	...
8	+1	+1	+1	...
9	−1.68	0	0	...
10	+1.68	0	0	...
11	0	−1.68	0	...
12	0	+1.68	0	...
13	0	0	−1.68	...
14	0	0	+1.68	...

4.7.5 TAGUCHI METHODS

Taguchi designs are valuable for robust lubricant formulation. They help in identifying the best combination of factors to produce lubricants that are less sensitive to variations in manufacturing or usage conditions.

Taguchi methods are often used in lubricant formulation to optimize product performance while considering factors and their interactions. Here is an example of a matrix for a Taguchi method designed experiment in lubricant formulation:

Objective: To optimize a lubricant formulation for enhanced engine protection, considering the factors: Factor A (base oil type), Factor B (additive concentration), and Factor C (temperature).

Factors:

1. **Factor A**: Base oil type
   - Levels: Mineral oil and synthetic oil
2. **Factor B**: Additive concentration (% by weight)
   - Levels: Low, medium, and high
3. **Factor C**: Temperature (°C)
   - Levels: 50°C, 75°C, and 100°C

Response Variable: Engine protection (measured as engine component wear rate in μm per hour)

Taguchi Method Designed Experiment Matrix: In the Taguchi method, you typically use an orthogonal array to efficiently explore factor combinations while minimizing the number of experimental runs. Here, we will use an L9 orthogonal array, which allows for nine experimental runs (Table 4.44).

In this matrix:

- Each row represents a specific combination of base oil type, additive concentration, and temperature.
- Engine protection measurements are recorded for each combination.

By analyzing the results of this Taguchi method designed experiment, you can identify the optimal combination of factors that provides the best engine protection while considering interactions between the factors and minimizing the number of experimental runs. Taguchi methods are known for their efficiency in experimentation and optimization.

TABLE 4.44
Taguchi Method Designed Experiment Matrix

Run	Factor A (Base Oil Type)	Factor B (Additive Concentration)	Factor C (Temperature)	Engine Protection (μm/Hour)
1	Mineral oil	Low	50	...
2	Mineral oil	Medium	75	...
3	Mineral oil	High	100	...
4	Synthetic oil	Low	75	...
5	Synthetic oil	Medium	100	...
6	Synthetic oil	High	50	...
7	Mineral oil	Medium	50	...
8	Synthetic oil	High	75	...
9	Mineral oil	Low	100	...

4.7.6 Sequential Experimentation

In this approach, you conduct a series of experiments, each based on the results of the previous one. It is useful for refining lubricant formulations iteratively, making adjustments as you gather more data.

Sequential experimentation is an approach where experiments are conducted in a sequence, and the results from one experiment influence the design of the next. Here is an example of a matrix for a sequential experimentation designed experiment in lubricant formulation:

Objective I: To optimize a lubricant formulation for maximum fuel efficiency while considering factors: Factor A (base oil type), Factor B (additive concentration), and Factor C (temperature).

Factors:

1. **Factor A**: Base oil type
   - Levels: Mineral oil and synthetic oil
2. **Factor B**: Additive concentration (% by weight)
   - Levels: Low, medium, and high
3. **Factor C**: Temperature (°C)
   - Levels: 50°C, 75°C, and 100°C

Response Variable: Fuel efficiency (measured in miles per gallon (MPG))

Sequential Experimentation Matrix:

In sequential experimentation, you would typically start with an initial set of experiments and then modify subsequent experiments based on the results obtained. Below is a simplified example.

Stage 1 – Initial Experiments:

For the first stage, a series of experiments is conducted to get a baseline understanding of the factors' effects (Table 4.45).

Stage 2 – Modified Experiments:

Based on the results of Stage 1, you can identify which factors or factor levels are more promising. You can modify your experiments to explore those areas further (Table 4.46).

TABLE 4.45

Sequential Experimentation Matrix (Stage 1 – Initial Experiments)

Run	Factor A (Base Oil Type)	Factor B (Additive Concentration)	Factor C (Temperature)	Fuel Efficiency (MPG)
1	Mineral oil	Low	50	...
2	Mineral oil	Medium	75	...
3	Synthetic oil	High	100	...

TABLE 4.46

Sequential Experimentation Matrix (Stage 2 – Modified Experiments)

Run	Factor A (Base Oil Type)	Factor B (Additive Concentration)	Factor C (Temperature)	Fuel Efficiency (MPG)
4	Synthetic oil	Medium	50	...
5	Synthetic oil	Low	75	...
6	Mineral oil	High	100	...

TABLE 4.47

Sequential Experimentation Matrix (Stage 3 – Further Refinement)

Run	Factor A (Base Oil Type)	Factor B (Additive Concentration)	Factor C (Temperature)	Fuel Efficiency (MPG)
7	Synthetic oil	Low	50	...
8	Synthetic oil	Medium	75	...
9	Mineral oil	Medium	100	...

TABLE 4.48

Sequential Experimentation Matrix (Initial Screening Experiments)

Run	Factor A (Base Oil Type)	Factor B (Additive Concentration)	Wear Resistance (WSD, μm)
1	Mineral oil	Low	...
2	Mineral oil	High	...
3	Synthetic oil	Low	...
4	Synthetic oil	High	...

Stage 3 – Further Refinement:
Continue refining your experiments based on the results obtained in Stage 2 (Table 4.47).

Objective II: To sequentially optimize a lubricant formulation for wear resistance, guided by initial screening experiments and subsequent confirmation experiments. Factors considered are Factor A (base oil type) and Factor B (additive concentration).

Factors:

1. **Factor A**: Base oil type
   - Levels: Mineral oil and synthetic oil
2. **Factor B**: Additive concentration (% by weight)
   - Initial Screening Levels: Low and high
   - Confirmation Levels: Optimal levels from initial screening

Response Variable: Wear resistance (measured in WSD in μm)

Sequential Experimentation Matrix:

1. **Initial Screening Experiments**:
   - Initial screening experiments with a limited set of combinations to identify promising factors and levels (Table 4.48) are conducted.
2. **Analysis and Selection**:
   - The results from the initial screening experiments are analyzed to identify which factors and levels show promise for improving wear resistance.
3. **Confirmation Experiments**:
   - Confirmation experiments based on the promising factors and levels identified in the initial screening (Table 4.49) are conducted.

TABLE 4.49

Sequential Experimentation Matrix (Confirmation Experiments)

Run	Factor A (Base Oil Type)	Factor B (Additive Concentration)	Wear Resistance (WSD, μm)
1	Mineral oil	Optimal	...
2	Synthetic oil	Optimal	...

In this sequential experimentation approach:

- Initial screening experiments are conducted to narrow down the factors and levels that are likely to yield better results.
- The results of the initial screening experiments are analyzed to select the most promising factor combinations.
- Confirmation experiments are then conducted with the selected factor combinations to validate and confirm the optimized lubricant formulation.

This iterative approach allows you to progressively improve the formulation while minimizing the number of experiments needed.

In sequential experimentation, you adapt your experimental design as you gather more information, with the goal of converging toward an optimal lubricant formulation that maximizes fuel efficiency. The matrix is continually modified based on the results from previous experiments.

4.7.7   RPD

RPD focuses on formulating lubricants that perform consistently under varying conditions. It helps in identifying factor settings that minimize the impact of external variations on lubricant performance.

RPD is used to develop products or processes that are insensitive to variations and can perform consistently under various conditions. In lubricant formulation, RPD can be applied to ensure that the formulation remains effective under a range of operating conditions and variations in raw materials. Here is an example of a matrix for an RPD designed experiment in lubricant formulation:

Objective: To develop a lubricant formulation that is robust to variations in base oil properties (Factor A) and operating temperature (Factor B) while optimizing wear resistance.

Factors:

1. **Factor A**: Base oil type
   - Levels: Mineral oil and synthetic oil
2. **Factor B**: Operating temperature (°C)
   - Levels: Low, medium, and high

Control Factors (Noise Factors):

1. **Noise Factor A**: Variation in base oil properties
   - Levels: Low variation and high variation
2. **Noise Factor B**: Variation in operating temperature
   - Levels: Low variation and high variation

TABLE 4.50

RPD Designed Experiment Matrix

Run	Factor A (Base Oil Type)	Factor B (Operating Temperature)	Noise Factor A (Base Oil Properties)	Noise Factor B (Operating Temperature)	Wear Resistance (WSD, μm)
1	Mineral oil	Low	Low variation	Low variation	...
2	Mineral oil	Medium	High variation	Low variation	...
3	Mineral oil	High	Low variation	High variation	...
4	Synthetic oil	Low	High variation	High variation	...
5	Synthetic oil	Medium	Low variation	Low variation	...
6	Synthetic oil	High	High variation	Low variation	...

Response Variable: Wear resistance (measured in WSD in μm)

RPD Designed Experiment Matrix:

In RPD, the goal is to evaluate the effects of control factors (base oil type and operating temperature) while considering variations in noise factors (variation in base oil properties and variation in operating temperature) (Table 4.50).

In this RPD matrix:

- Factor A represents the type of base oil.
- Factor B represents the operating temperature.
- Noise Factor A and Noise Factor B represent variations in base oil properties and operating temperature, respectively.
- Wear resistance measurements are recorded for each combination of control factors and noise factors.

The goal of RPD is to develop a lubricant formulation that performs consistently (i.e., robustly) across a range of conditions and variations in factors. Analysis of the data will help identify the factors and levels that lead to a robust formulation with minimal sensitivity to noise factors.

4.7.8 RANDOMIZED EXPERIMENTS

Randomized experiments are used to assess the effects of different factors in lubricant formulation without introducing bias. They help in obtaining unbiased estimates of factor effects on lubricant properties.

In a randomized design experiment, the factors and their levels are randomly assigned to experimental runs to minimize the effects of uncontrolled variability. Here is an example of a matrix for a randomized designed experiment in lubricant formulation:

Objective: To evaluate the effects of various factors on the wear resistance of a lubricant formulation while minimizing bias and controlling for uncontrolled variability.

Factors:

1. **Factor A**: Base oil type
   - Levels: Mineral oil and synthetic oil
2. **Factor B**: Additive concentration (% by weight)
   - Levels: Low, medium, and high
3. **Factor C**: Operating temperature (°C)
   - Levels: 50°C, 75°C, and 100°C

TABLE 4.51
Randomized Designed Experiment Matrix

Run	Factor A (Base Oil Type)	Factor B (Additive Concentration)	Factor C (Operating Temperature)	Wear Resistance (WSD, μm)
1	Mineral oil	Low	50	...
2	Synthetic oil	High	100	...
3	Mineral oil	Medium	75	...
4	Synthetic oil	Low	75	...
5	Synthetic oil	Medium	50	...
6	Mineral oil	High	100	...
7	Synthetic oil	Medium	100	...
8	Mineral oil	High	50	...
9	Mineral oil	Low	100	...
10	Synthetic oil	Low	50	...
11	Mineral oil	Medium	100	...
12	Synthetic oil	High	75	...

Response Variable: Wear resistance (measured in WSD in μm)

Randomized Designed Experiment Matrix:

In a randomized design, the assignment of factor levels to experimental runs is done randomly to minimize systematic bias.

Here is a simplified example of a randomized design matrix with 12 experimental runs (Table 4.51).

In this randomized design:

- The levels of factors A, B, and C are randomly assigned to each experimental run.
- The randomization helps minimize the effects of any uncontrolled variability or bias in the experiment.

Randomized designs are particularly useful when you want to ensure that the effects you observe are not influenced by the order or arrangement of the experimental runs, providing more robust and unbiased results.

4.7.9 Split-Plot and Blocked Designs

When certain factors in lubricant formulation are hard to change or when there are sources of variability that need to be accounted for (e.g., equipment variations), split-plot and blocked designs can be useful.

In a split-plot and blocked designed experiment, the factors are divided into main plots (large-scale factors) and subplots (small-scale factors) to investigate their effects while controlling for specific sources of variation. Here are examples of matrices for split-plot and blocked designed experiments in lubricant formulation:

Example 1: Split-Plot Design

Objective: To study the effects of two main factors (Factor A and Factor B) and their interaction on the performance of a lubricant formulation. Factor C is a smaller-scale factor that is difficult or expensive to change frequently, so it is allocated to subplots.

TABLE 4.52
Split-Plot Designed Experiment Matrix

Run	Whole Plot (Factor A)	Whole Plot (Factor B)	Subplot (Factor C)	Wear Resistance (WSD, μm)
1	Mineral oil	Low	50	...
2	Synthetic oil	Low	75	...
3	Mineral oil	High	100	...
4	Synthetic oil	High	75	...
5	Mineral oil	Low	100	...
6	Synthetic oil	High	50	...
7	Mineral oil	High	75	...
8	Synthetic oil	Low	50	...
9	Mineral oil	Low	75	...
10	Synthetic oil	Low	100	...
11	Mineral oil	High	50	...
12	Synthetic oil	High	100	...

Factors:

1. **Factor A**: Base oil type
   - Levels: Mineral oil and synthetic oil
2. **Factor B**: Additive concentration (% by weight)
   - Levels: Low and high
3. **Factor C**: Operating temperature (°C) (small-scale factor, difficult to change frequently)
   - Levels: 50°C, 75°C, and 100°C

Response Variable: Wear resistance (measured in WSD in μm)

Split-Plot Designed Experiment Matrix:

In a split-plot design, the main factors (Factor A and Factor B) are assigned at the whole-plot level, while the subplot factor (Factor C) is assigned at the subplot level.

A simplified example of a split-plot design matrix with 12 experimental runs is shown in Table 4.52. In this matrix:

- Whole-plot factors (Factor A and Factor B) are assigned at the whole-plot level.
- Subplot factor (Factor C) is assigned at the subplot level.
- Wear resistance measurements are recorded for each combination of factors.

Example 2: Blocked Design

Objective: To investigate the effects of base oil type (Factor A) and additive concentration (Factor B) on wear resistance while controlling for potential variations between different batches of base oils.

Factors:

1. **Factor A**: Base oil type
   - Levels: Mineral oil and synthetic oil
2. **Factor B**: Additive concentration (% by weight)
   - Levels: Low and high

Response Variable: Wear resistance (measured in WSD in μm)

Blocked Designed Experiment Matrix:

In a blocked design, the experimental runs are divided into blocks, and each block represents a different batch of base oil to account for batch-to-batch variation.

Here is a simplified example of a blocked design matrix with 12 experimental runs (three blocks of four runs each) (Tables 4.53–4.55).

Block 1 (Batch 1 Base Oil)
Block 2 (Batch 2 Base Oil)
Block 3 (Batch 3 Base Oil)

In this blocked design:

- Each block represents a different batch of base oil to account for batch-to-batch variations.
- Wear resistance measurements are recorded for each combination of factors within each block.

The blocked design helps control for potential variations between different batches of base oil, ensuring that the effects of Factor A and Factor B on wear resistance are evaluated while minimizing batch-related variations.

TABLE 4.53

Blocked Designed Experiment Matrix (Batch 1 Base Oil)

Run	Factor A (Base Oil Type)	Factor B (Additive Concentration)	Wear Resistance (WSD, μm)
1	Mineral oil	Low	...
2	Synthetic oil	Low	...
3	Mineral oil	High	...
4	Synthetic oil	High	...

TABLE 4.54

Blocked Designed Experiment Matrix (Batch 2 Base Oil)

Run	Factor A (Base Oil Type)	Factor B (Additive Concentration)	Wear Resistance (WSD, μm)
5	Mineral oil	Low	...
6	Synthetic oil	Low	...
7	Mineral oil	High	...
8	Synthetic oil	High	...

TABLE 4.55

Blocked Designed Experiment Matrix (Batch 3 Base Oil)

Run	Factor A (Base Oil Type)	Factor B (Additive Concentration)	Wear Resistance (WSD, μm)
9	Mineral oil	Low	...
10	Synthetic oil	Low	...
11	Mineral oil	High	...
12	Synthetic oil	High	...

4.7.10 FRACTIONAL FACTORIAL EXPERIMENTS

These designs are employed to evaluate a subset of factors when the full factorial design is too resource-intensive. They help in identifying the most influential factors in lubricant formulation.

Fractional factorial design experiments are used when you want to study the effects of multiple factors at different levels while reducing the number of experimental runs. Here are examples of matrices for fractional factorial designed experiments in lubricant formulation.

Example 1: 2^3 Fractional Factorial Design

Objective: To investigate the effects of three factors on the performance of a lubricant formulation, while reducing the number of experimental runs by using a 2^3 fractional factorial design.

Factors:

1. **Factor A**: Base oil type
   - Levels: Mineral oil (M) and synthetic oil (S)
2. **Factor B**: Additive concentration (% by weight)
   - Levels: Low (L) and high (H)
3. **Factor C**: Operating temperature (°C)
   - Levels: Low (L) and high (H)

Response Variable: Wear resistance (measured in WSD in μm)

2^3 Fractional Factorial Designed Experiment Matrix:

In this 2^3 fractional factorial design, only a fraction of the possible experimental runs is conducted to study the main effects and selected interactions.

A simplified example of a 2^3 fractional factorial design matrix with four experimental runs is shown in Table 4.56.

In this 2^3 design:

- The main effects of Factor A, Factor B, and Factor C are studied.
- Selected interactions (e.g., AB, AC, and BC) can be estimated.
- The design requires only four runs instead of the full 2^3 = 8 runs, reducing the experimental effort.

TABLE 4.56

2^3 Fractional Factorial Designed Experiment Matrix

Run	Factor A (Base Oil Type)	Factor B (Additive Concentration)	Factor C (Operating Temperature)	Wear Resistance (WSD, μm)
1	Mineral oil (M)	Low (L)	Low (L)	...
2	Synthetic oil (S)	Low (L)	High (H)	...
3	Mineral oil (M)	High (H)	Low (L)	...
4	Synthetic oil (S)	High (H)	High (H)	...

Example 2: 2^4-1 Fractional Factorial Design

Objective: To investigate the effects of four factors on the performance of a lubricant formulation, while reducing the number of experimental runs using a 2^4-1 fractional factorial design.

Factors:

1. **Factor A**: Base oil type
   - Levels: Mineral oil (M) and synthetic oil (S)
2. **Factor B**: Additive concentration (% by weight)
   - Levels: Low (L) and high (H)
3. **Factor C**: Operating temperature (°C)
   - Levels: Low (L) and high (H)
4. **Factor D**: Additive type
   - Levels: Type 1 (T1) and Type 2 (T2)

Response Variable: Wear resistance (measured in WSD in μm)

2^4-1 Fractional Factorial Designed Experiment Matrix:

In this 2^4-1 fractional factorial design, only a fraction of the possible experimental runs is conducted while studying main effects and selected interactions. The "-1" indicates that one run is omitted.

A simplified example of a 2^4-1 fractional factorial design matrix with 15 experimental runs is shown in Table 4.57.

In this 2^4-1 design:

- Main effects of all four factors (A, B, C, and D) are studied.
- Selected interactions can be estimated.
- The design requires only 15 runs instead of the full 2^4 = 16.

The choice of the designed experimental method depends on the specific objectives, resources, and complexity of the lubricant formulation process. These methods help lubricant manufacturers optimize formulations for performance, cost-effectiveness, and robustness in various applications.

TABLE 4.57

2^4-1 Fractional Factorial Designed Experiment Matrix

Run	Factor A (Base Oil Type)	Factor B (Additive Concentration)	Factor C (Operating Temperature)	Factor D (Additive Type)	Wear Resistance (WSD, μm)
1	Mineral oil (M)	Low (L)	Low (L)	Type 1 (T1)	...
2	Synthetic oil (S)	Low (L)	High (H)	Type 1 (T1)	...
3	Mineral oil (M)	High (H)	Low (L)	Type 1 (T1)	...
4	Synthetic oil (S)	High (H)	High (H)	Type 1 (T1)	...
5	Mineral oil (M)	Low (L)	Low (L)	Type 2 (T2)	...
6	Synthetic oil (S)	Low (L)	High (H)	Type 2 (T2)	...
7	Mineral oil (M)	High (H)	Low (L)	Type 2 (T2)	...
8	Synthetic oil (S)	High (H)	High (H)	Type 2 (T2)	...
9	Mineral oil (M)	Low (L)	Low (L)	Type 1 (T1)	...
10	Synthetic oil (S)	Low (L)	High (H)	Type 1 (T1)	...
11	Mineral oil (M)	High (H)	Low (L)	Type 1 (T1)	...
12	Synthetic oil (S)	High (H)	High (H)	Type 1 (T1)	...
13	Mineral oil (M)	Low (L)	Low (L)	Type 2 (T2)	...
14	Synthetic oil (S)	Low (L)	High (H)	Type 2 (T2)	...
15	Mineral oil (M)	High (H)	Low (L)	Type 2 (T2)	...

4.8   MANUFACTURING

In manufacturing, designed experiments are critical for optimizing processes, improving product quality, and reducing defects. Here are some common types of designed experiments with examples of their applications in manufacturing.

In manufacturing designed experiments, various types of variables are considered to improve and optimize manufacturing processes, product quality, and efficiency. These variables can be categorized into several common types, including:

1. **Independent Variables**:
   - **Process Parameters**: Variables related to manufacturing process conditions and parameters, such as temperature, pressure, speed, feed rate, and machine settings.
   - **Material Properties**: Variables related to the properties of raw materials or components used in manufacturing, including composition, hardness, viscosity, and size.
   - **Design Factors**: Variables related to the design of products or manufacturing systems, such as product dimensions, tolerances, and geometries.
   - **Environmental Conditions**: Variables related to environmental factors that may impact manufacturing, including humidity, lighting, and air quality.
   - **Equipment and Tooling**: Variables related to the type, condition, and specifications of manufacturing equipment, machinery, and tooling.
2. **Dependent Variables**:
   - **Product Quality Metrics**: Variables related to the quality and performance of manufactured products, including dimensions, tolerances, surface finish, strength, and durability.
   - **Production Yield**: Variables related to the quantity of defect-free products produced relative to the total number of products manufactured.
   - **Production Efficiency**: Variables related to the efficiency and productivity of manufacturing processes, such as cycle time, throughput, and downtime.
   - **Cost and Resource Utilization**: Variables related to production costs, resource consumption (e.g., energy and materials), and waste generation.
   - **Environmental Impact**: Variables related to the environmental impact of manufacturing processes, including emissions, waste disposal, and sustainability metrics.
3. **Categorical Variables**:
   - **Product Types**: Categories representing different types or models of products being manufactured within the same facility or process.
   - **Production Shifts**: Categories representing different production shifts or time periods when manufacturing occurs.
   - **Machine Operators**: Categories representing different machine operators or personnel responsible for manufacturing tasks.
   - **Batch Numbers**: Categories representing different production batches or lots, particularly in industries like pharmaceuticals and chemicals.
4. **Control Variables (Covariates)**:
   - These are variables that are held constant or controlled during experiments to eliminate their influence on the dependent variable, for example, maintaining consistent temperature or humidity levels during manufacturing trials.
5. **Random Variables**:
   - In some cases, random variables may be introduced to account for variability or uncertainty in manufacturing measurements or to control for natural variability in production processes.
6. **Interaction Variables**:
   - Interaction variables are used to investigate the combined effects of two or more independent variables on the dependent variable, helping to understand how different factors may interact to influence manufacturing outcomes.

7. **Noise Variables (Error Variables)**:
   - These are uncontrolled or unmeasured variables that can introduce variability or errors into the manufacturing process or experimental results. Statistical analysis and experimental controls are used to minimize the impact of noise variables in manufacturing experiments.

Manufacturing designed experiments are essential for process optimization, quality control, cost reduction, and continuous improvement in various industries, including automotive, electronics, pharmaceuticals, and consumer goods. Careful consideration and control of variables are crucial for achieving desired manufacturing outcomes.

4.8.1 FACTORIAL EXPERIMENTS

In automotive manufacturing, factorial experiments can be used to optimize the parameters of a welding process. Engineers vary factors like welding current, electrode material, and welding speed to achieve strong and reliable welds.

Factorial experiments are commonly used in manufacturing to study the main effects and interactions of multiple factors on a particular response variable. Here is an example of a matrix for a factorial experiment in manufacturing.

Objective: To optimize the manufacturing process of a metal alloy by studying the effects of two factors, Factor A (temperature) and Factor B (pressure), on the alloy's tensile strength.

Factors:

1. **Factor A**: Temperature (in degrees Celsius)
   - Low (e.g., 900°C)
   - Medium (e.g., 1000°C)
   - High (e.g., 1100°C)
2. **Factor B**: Pressure (in megapascals (MPa))
   - Low (e.g., 20 MPa)
   - Medium (e.g., 30 MPa)
   - High (e.g., 40 MPa)

Response Variable: Tensile strength (measured in MPa)

Factorial Experiment Matrix: In a 3 × 3 full factorial design, all possible combinations of the two factors and their levels are tested. The matrix is shown in Table 4.58.

TABLE 4.58
Factorial Experiment Matrix

Run	Temperature (A)	Pressure (B)	Tensile Strength (MPa)
1	Low	Low	...
2	Low	Medium	...
3	Low	High	...
4	Medium	Low	...
5	Medium	Medium	...
6	Medium	High	...
7	High	Low	...
8	High	Medium	...
9	High	High	...

In this matrix, each row represents a specific combination of Factor A (temperature) and Factor B (pressure), and the resulting tensile strength of the metal alloy is recorded. By systematically varying the levels of these factors and analyzing the results, you can identify the optimal conditions for maximizing tensile strength in the manufacturing process.

Factorial experiments are valuable for understanding the relationships between factors and responses, helping manufacturers make data-driven decisions to improve product quality and process efficiency.

4.8.2 RSM

RSM is commonly applied in semiconductor manufacturing to optimize the parameters for photolithography processes. Engineers vary factors like exposure time, mask alignment, and resist thickness to achieve desired feature sizes on semiconductor wafers.

RSM is also used in manufacturing to optimize processes with multiple factors. Here is an example of a matrix for an RSM designed experiment in manufacturing.

Objective: To optimize the production of a polymer material by studying the effects of three factors, Factor A (temperature), Factor B (pressure), and Factor C (mixing time), on the polymer's tensile strength.

Factors:

1. **Factor A**: Temperature (in degrees Celsius)
   - Low (e.g., 150°C)
   - Medium (e.g., 175°C)
   - High (e.g., 200°C)
2. **Factor B**: Pressure (in pounds per square inch (psi))
   - Low (e.g., 200 psi)
   - Medium (e.g., 250 psi)
   - High (e.g., 300 psi)
3. **Factor C**: Mixing time (in minutes)
   - Low (e.g., 10 minutes)
   - Medium (e.g., 15 minutes)
   - High (e.g., 20 minutes)

Response Variable: Tensile strength (measured in MPa)

RSM Experiment Matrix: In an RSM designed experiment, you often use a CCD or a Box–Behnken design to explore the response surface. A simplified example with a CCD is shown in Table 4.59.

In this matrix:

- Columns represent the factors (temperature, pressure, and mixing time) and the response variable (tensile strength).
- The levels of each factor are denoted as −1 (low), 0 (central point), and 1 (high), allowing for a quadratic fit to the response surface.
- The central point run (Run 9) provides replication at the center of the design to estimate experimental error.

By conducting experiments based on this design and analyzing the results, you can fit a mathematical model (e.g., a quadratic equation) to the response surface. This model helps you identify

TABLE 4.59
RSM Experiment Matrix

Run	Temperature (A)	Pressure (B)	Mixing Time (C)	Tensile Strength (MPa)
1	−1	−1	−1	...
2	−1	−1	1	...
3	−1	1	−1	...
4	−1	1	1	...
5	1	−1	−1	...
6	1	−1	1	...
7	1	1	−1	...
8	1	1	1	...
9	0	0	0	...

the optimal conditions for maximizing tensile strength and gain insights into the interactions between the factors.

RSM is a powerful tool for process optimization in manufacturing, as it allows you to navigate complex multifactor systems efficiently and achieve desired product quality and performance.

4.8.3 FRACTIONAL FACTORIAL EXPERIMENTS

In electronics manufacturing, fractional factorial experiments are often used to study the effects of various factors (e.g., component placement accuracy and solder paste quality) on circuit board assembly quality without testing all possible combinations.

Fractional factorial experiments are often used in manufacturing to explore the effects of multiple factors while conducting fewer experimental runs. Here is an example of a matrix for a fractional factorial experiment in manufacturing.

Objective: To optimize the production process of a semiconductor by studying the effects of four factors (factors A, B, C, and D) on the defect rate of manufactured chips.

Factors:

1. **Factor A**: Temperature (low/high)
2. **Factor B**: Pressure (low/high)
3. **Factor C**: Gas flow rate (low/high)
4. **Factor D**: Exposure time (short/long)

Response Variable: Defect rate (%)

Fractional Factorial Experiment Matrix: In a fractional factorial design, you selectively choose a fraction of the runs from a full factorial design to reduce the number of experimental trials. An example using a quarter fraction of a 2^4 full factorial design is shown in Table 4.60.

In this matrix, each row represents a specific combination of factors A, B, C, and D, and the corresponding defect rate is recorded. The design allows you to investigate the main effects and two-way interactions of the factors while conducting only a fraction of the full set of experiments. Fractional factorial designs are useful for efficiently identifying key factors affecting the process while minimizing resource and time requirements.

TABLE 4.60
Fractional Factorial Experiment Matrix

Run	Factor A	Factor B	Factor C	Factor D	Defect Rate (%)
1	Low	Low	Low	Low	...
2	High	Low	Low	Low	...
3	Low	High	Low	Low	...
4	High	High	Low	Low	...
5	Low	Low	High	Low	...
6	High	Low	High	Low	...
7	Low	High	High	Low	...
8	High	High	High	Low	...
9	Low	Low	Low	High	...
10	High	Low	Low	High	...
11	Low	High	Low	High	...
12	High	High	Low	High	...
...	...	...	...	...	...

4.8.4   CCD

CCD is employed in aerospace manufacturing to optimize the parameters of a machining process for aircraft components. Engineers vary factors like cutting speed, tool geometry, and coolant flow to improve machining efficiency and component quality.

CCD is used in manufacturing to explore the response surface and optimize processes with multiple factors. Here is an example of a matrix for a CCD designed experiment in manufacturing.

Objective: To optimize the production of a chemical compound by studying the effects of three factors (factors A, B, and C) on the compound's yield.

Factors:

1. **Factor A**: Temperature (in degrees Celsius)
   - Low (−1.68)
   - Medium (0.00)
   - High (1.68)
2. **Factor B**: Catalyst concentration (in moles/liter)
   - Low (−1.68)
   - Medium (0.00)
   - High (1.68)
3. **Factor C**: Reaction time (in hours)
   - Low (−1.68)
   - Medium (0.00)
   - High (1.68)

Response Variable: Yield of compound (measured in grams)

CCD Experiment Matrix: A CCD typically includes a factorial design, axial points, and center points to capture curvature in the response surface. An example with a 2^3 CCD is shown in Table 4.61.

TABLE 4.61
CCD Experiment Matrix

Run	Temperature (A)	Catalyst Concentration (B)	Reaction Time (C)	Yield of Compound (g)
1	−1.68	−1.68	−1.68	...
2	1.68	−1.68	−1.68	...
3	−1.68	1.68	−1.68	...
4	1.68	1.68	−1.68	...
5	−1.68	−1.68	1.68	...
6	1.68	−1.68	1.68	...
7	−1.68	1.68	1.68	...
8	1.68	1.68	1.68	...
9	0.00	0.00	0.00	...
10	−1.68	0.00	0.00	...
11	1.68	0.00	0.00	...
12	0.00	−1.68	0.00	...
13	0.00	1.68	0.00	...
14	0.00	0.00	−1.68	...
15	0.00	0.00	1.68	...
16	0.00	0.00	0.00	...

In this matrix:

- Columns represent the factors (temperature, catalyst concentration, and reaction time) and the response variable (yield of compound).
- The levels of each factor are represented as −1.68 (low), 1.68 (high), and 0.00 (center point), allowing for a quadratic fit to the response surface.
- Runs 1 to 8 are the factorial points, Runs 9 to 16 are the axial points, and Run 9 is the center point.
- The axial points and center point help capture the curvature in the response surface, enabling the identification of optimal conditions for maximizing the yield of the compound.

CCD designs are useful for efficiently exploring the impact of multiple factors on a response and identifying optimal process settings in manufacturing.

4.8.5 Taguchi Methods

Taguchi methods are used in lean manufacturing to optimize production processes. In a metal stamping operation, engineers may apply Taguchi techniques to identify settings that minimize defects and reduce material waste.

Taguchi methods are widely used in manufacturing for robust optimization of processes while considering factors like noise. Here is an example of a matrix for a Taguchi method designed experiment in manufacturing:

Objective: To optimize the quality of a metal casting process by studying the effects of four factors (factors A, B, C, and D) on the casting's surface finish.

TABLE 4.62
Taguchi Method Matrix

Run	Factor A	Factor B	Factor C	Factor D	Surface Finish (μm)
1	Low	Low	Short	A	...
2	High	Low	Short	A	...
3	Low	High	Short	A	...
4	High	High	Short	A	...
5	Low	Low	Long	A	...
6	High	Low	Long	A	...
7	Low	High	Long	A	...
8	High	High	Long	A	...
9	Low	Low	Short	B	...
10	High	Low	Short	B	...
11	Low	High	Short	B	...
12	High	High	Short	B	...
13	Low	Low	Long	B	...
14	High	Low	Long	B	...
15	Low	High	Long	B	...
16	High	High	Long	B	...

Factors:

1. **Factor A**: Pouring temperature (low/high)
2. **Factor B**: Mold temperature (low/high)
3. **Factor C**: Cooling time (short/long)
4. **Factor D**: Mold release agent (A, B, and C)

Response Variable: Surface finish (measured in μm)

Taguchi Method Matrix: Taguchi experiments use orthogonal arrays to systematically test combinations while minimizing the number of runs. An example using a Taguchi L16 orthogonal array for four factors is shown in Table 4.62.

In this matrix:

- Each row represents a specific combination of factors A, B, C, and D.
- The levels of each factor are represented as low/high for factors A, B, and C and A, B, and C for Factor D.
- The L16 orthogonal array allows you to test 16 different combinations with only 16 experimental runs.
- Surface finish measurements are recorded for each run.

Taguchi methods focus on robustness and the identification of factor settings that minimize the impact of noise. By analyzing the results from this matrix and performing signal-to-noise ratio analysis, you can determine the optimal factor settings for achieving the desired surface finish while minimizing sensitivity to variations in factors and noise.

4.8.6 MIXTURE EXPERIMENTS

In polymer manufacturing, mixture experiments help optimize the composition of polymer blends. Engineers may vary proportions of polymer resins and additives to achieve desired material properties for processes.

Mixture experiments are commonly used in manufacturing when a product or process involves a combination of ingredients or components. Here is an example of a matrix for a mixture designed experiment in manufacturing:

Objective: To optimize the formulation of a paint by studying the effects of three components (Component A, Component B, and Component C) on the paint's color intensity.

Components:

1. **Component A**: Pigment A (measured in grams)
2. **Component B**: Pigment B (measured in grams)
3. **Component C**: Solvent (measured in milliliters)

Response Variable: Color intensity (measured using a colorimeter)

Mixture Designed Experiment Matrix: In a mixture design, the total composition is constrained to sum to a constant (e.g., 100%). An example using a simplex centroid design for three components is shown in Table 4.63.

In this matrix:

- Each row represents a specific combination of Components A, B, and C.
- The total composition (sum of components) is constrained to be constant (e.g., 100 g or 100 mL).
- The simplex centroid design allows you to explore combinations within the mixture space efficiently.
- Color intensity measurements are recorded for each run.

Mixture designs are useful for optimizing formulations and processes involving various ingredients or components, where the total composition must remain constant. By analyzing the results from this matrix, you can determine the optimal component proportions for achieving the desired color intensity in the paint formulation.

TABLE 4.63
Mixture Designed Experiment Matrix

Run	Component A (g)	Component B (g)	Component C (mL)	Color Intensity
1	20	40	40	...
2	40	20	40	...
3	20	20	60	...
4	30	30	40	...
5	10	60	30	...
6	30	10	60	...
7	10	10	80	...
8	25	35	40	...

4.8.7 SEQUENTIAL EXPERIMENTATION

As an example, manufacturers may use sequential experimentation to optimize the parameters of a 3D printing process. After each print, they adjust settings like layer thickness and print speed based on the previous results to reduce defects and improve efficiency.

Sequential experimentation is a dynamic approach to optimizing processes in manufacturing, where the design evolves based on the results of previous runs. Unlike traditional fixed designs, it allows for flexibility in adjusting factors based on ongoing data analysis. Here is an example of how a matrix for a sequential experimentation designed experiment might evolve.

Objective: To optimize the cutting speed of a computer numerical control (CNC) machining process while minimizing tool wear and maximizing product quality.

Factors:

1. **Factor A**: Cutting speed (in meters per minute (m/min))
   - Low range (e.g., 100 m/min)
   - Medium range (e.g., 200 m/min)
   - High range (e.g., 300 m/min)
2. **Factor B**: Tool material (Tool A and Tool B)
   - Tool A: Carbide
   - Tool B: High-speed steel (HSS)

Response Variables:

- Tool wear (measured in µm)
- Surface finish (measured in Ra)

Sequential Experimentation Matrix (Initial Design): In a sequential experimentation design, you typically start with an initial set of runs and then adjust the design based on interim results. An example of the initial design is shown in Table 4.64.

Sequential Experimentation Matrix (Adjustment Based on Initial Results): After analyzing the initial results, you may decide to explore specific factor combinations in more detail or make adjustments to the design. For instance, if you find that a particular cutting speed shows promise, you may conduct additional runs at different levels within that range. The matrix evolves based on ongoing learning and optimization goals (Table 4.65).

TABLE 4.64

Sequential Experimentation Matrix (Initial Design)

Run	Cutting Speed (A)	Tool Material (B)	Tool Wear (µm)	Surface Finish (Ra)
1	Low	Tool A	...	...
2	Medium	Tool A	...	...
3	High	Tool A	...	...
4	Low	Tool B	...	...
5	Medium	Tool B	...	...
6	High	Tool B	...	...

TABLE 4.65

Sequential Experimentation Matrix (Adjustment Based on Initial Results)

Run	Cutting Speed (A)	Tool Material (B)	Tool Wear (μm)	Surface Finish (Ra)
1	Low	Tool A	...	...
2	Medium	Tool A	...	...
3	High	Tool A	...	...
4	Low	Tool B	...	...
5	Medium	Tool B	...	...
6	High	Tool B	...	...
7	Medium	Tool A	...	...
8	Medium	Tool A	...	...

In sequential experimentation, you continue to modify the design and conduct additional runs to converge toward the optimal settings while considering the trade-offs between factors and objectives. This iterative process allows for efficient optimization and adaptation as more information becomes available.

4.8.8 RPD

In automotive manufacturing, RPD can be applied to ensure the robustness of an assembly process against variations in component dimensions and tolerances, reducing the likelihood of assembly defects.

RPD aims to optimize processes while making them less sensitive to variation or noise factors. Here is an example of a matrix for a RPD experiment in manufacturing:

Objective: To optimize the injection molding process for producing plastic components with minimal warpage while considering variations in material properties and environmental conditions.

Factors:

1. **Factor A**: Melt temperature (in degrees Celsius)
   - Low (e.g., 200°C)
   - Medium (e.g., 220°C)
   - High (e.g., 240°C)
2. **Factor B**: Mold temperature (in degrees Celsius)
   - Low (e.g., 40°C)
   - Medium (e.g., 60°C)
   - High (e.g., 80°C)
3. **Factor C**: Cooling time (in seconds)
   - Short (e.g., 10 seconds)
   - Medium (e.g., 15 seconds)
   - Long (e.g., 20 seconds)

Noise Factors:

1. **Noise Factor D**: Ambient humidity (low/high)
2. **Noise Factor E**: Material variability (low/high)

Response Variable: Warpage (measured in millimeters (mm))

RPD Matrix: In RPD, you typically use an orthogonal array to conduct experiments that consider both factors and noise factors efficiently. An example using an L27 orthogonal array is shown in Table 4.66.

TABLE 4.66
RPD Matrix

Run	Factor A	Factor B	Factor C	Noise Factor D	Noise Factor E	Warpage (mm)
1	Low	Low	Short	Low	Low	...
2	High	Low	Short	Low	Low	...
3	Low	High	Short	Low	Low	...
4	High	High	Short	Low	Low	...
5	Low	Low	Medium	Low	Low	...
6	High	Low	Medium	Low	Low	...
7	Low	High	Medium	Low	Low	...
8	High	High	Medium	Low	Low	...
9	Low	Low	Long	Low	Low	...
10	High	Low	Long	Low	Low	...
11	Low	High	Long	Low	Low	...
12	High	High	Long	Low	Low	...
...	...	...	...	...	...	...

In this matrix:

- Each row represents a specific combination of factors and noise factors.
- Factor levels for factors A, B, and C are varied systematically based on the orthogonal array.
- Noise factors, such as ambient humidity and material variability, are included to account for variation.
- Warpage measurements are recorded for each run.

RPD designs aim to find factor settings that result in minimal sensitivity to noise factors while achieving the desired process outcome, in this case minimal warpage in the plastic components. Statistical analysis methods are then applied to identify the optimal settings that meet robustness and quality criteria.

4.8.9 RANDOMIZED EXPERIMENTS

In quality control, randomized experiments can be used to study the effects of changes in manufacturing processes on product quality. Randomized trials may assess the impact of variations in raw material suppliers on final product specifications.

Randomized experiments in manufacturing are designed to study the effects of factors on a process while minimizing bias and external influences. Here is an example of a matrix for a randomized experiment designed experiment in manufacturing.

Objective: To investigate the effect of cutting tool material (Factor A) and cutting speed (Factor B) on the surface finish of machined metal parts.

Factors:

1. **Factor A**: Cutting tool material
   - Tool A: Carbide
   - Tool B: HSS

TABLE 4.67
Randomized Experiment Matrix

Run	Factor A	Factor B	Surface Finish (Ra)
1	Tool A	High	...
2	Tool B	Low	...
3	Tool A	Medium	...
4	Tool B	High	...
5	Tool A	Low	...
6	Tool B	Medium	...
7	Tool A	High	...
8	Tool B	Low	...
9	Tool A	Medium	...
10	Tool B	High	...
11	Tool A	Low	...
12	Tool B	Medium	...
...	...	...	...

2. **Factor B**: Cutting speed (in m/min)
   - Low (e.g., 100 m/min)
   - Medium (e.g., 200 m/min)
   - High (e.g., 300 m/min)

Response Variable: Surface finish (measured in Ra)

Randomized Experiment Matrix: In a randomized experiment, the order in which factors are tested is randomized to reduce the impact of external variables. An example matrix for such an experiment is shown in Table 4.67.
   In this matrix:

- Each row represents a specific combination of factors (tool material and cutting speed).
- The order of the runs is randomized to minimize the influence of uncontrolled variables.
- Surface finish measurements (Ra) are recorded for each run.

Randomized experiments are useful for estimating the effects of factors while reducing the potential bias from external factors. Statistical analysis of the data allows you to identify which factors significantly influence the surface finish of the machined parts and determine optimal conditions for the manufacturing process.

4.8.10 SPLIT-PLOT AND BLOCKED EXPERIMENTS

A common example in manufacturing is optimizing the parameters of a heat treatment process in metallurgy, and split-plot or blocked experiments may be used to account for variations in furnace zones, cooling rates, and part positions.
   Split-plot and blocked designs are useful in manufacturing when factors have different levels of difficulty to change, such as in settings where changing some factors is more time-consuming or expensive than others. Here are two examples of matrices for split-plot and blocked designed experiments in manufacturing:

TABLE 4.68
Split-Plot Design Matrix (Whole Plots)

Run	Factor A	Factor B	Quenching Medium	Hardness (HRc)
1	Low	Low	Water	...
2	High	Low	Water	...
3	Low	High	Water	...
4	High	High	Water	...

TABLE 4.69
Split-Plot Design Matrix (Subplots)

Subplot	Factor C	Hardness (HRc)
1	Water	...
2	Oil	...
3	Water	...
4	Oil	...

Example 1: Split-Plot Design

Objective: To optimize a heat treatment process for a new alloy by varying factors A, B, and C.

Factors:

1. **Factor A:** Temperature (low/high)
2. **Factor B:** Time (low/high)
3. **Factor C:** Quenching medium (water and oil)

Response Variable: Hardness (measured in Rockwell hardness units (HRc))

Split-Plot Design Matrix: In a split-plot design, you have one set of factors (whole plots) that can be changed relatively easily and another set of factors (subplots) that are more challenging to change.
Whole plots (factors A and B) (Table 4.68)
Subplots (Factor C) (Table 4.69)
In this design:

- Factors A and B are the whole plots, which are easier to change.
- Factor C (quenching medium) is the subplot, which is more challenging to change and is nested within whole plots.
- The hardness measurements are recorded for each combination.

Example 2: Blocked Design

Objective: To optimize a chemical mixing process by varying factors X, Y, and Z.

Factors:

1. **Factor X:** Mixing time (low/high)
2. **Factor Y:** Temperature (low/high)
3. **Factor Z:** Concentration of reactant (low/high)

TABLE 4.70

Blocked Design Matrix (Low Temperature)

Run	Factor X	Factor Y	Factor Z	Reaction Yield (%)
1	Low	Low	Low	...
2	Low	Low	High	...
3	Low	High	Low	...
4	Low	High	High	...

TABLE 4.71

Blocked Design Matrix (High Temperature)

Run	Factor X	Factor Y	Factor Z	Reaction Yield (%)
5	Low	Low	Low	...
6	Low	Low	High	...
7	Low	High	Low	...
8	Low	High	High	...

Response Variable: Reaction yield (%)

Blocked Design Matrix:
In a blocked design, the experiments are grouped into blocks, and each block tests a specific combination of factor levels (Tables 4.70 and 4.71).
Block 1 (low temperature)
Block 2 (high temperature)
And so on for additional blocks.
In this design:

- Each block represents a specific temperature condition.
- Within each block, the factors X, Y, and Z are varied.
- Reaction yields are recorded for each combination within each block.

Blocked designs are useful when factors have different levels of difficulty to change, and they help account for potential interactions between factors within each block.

4.9   MATERIALS SCIENCE

In materials science, designed experiments are crucial for developing new materials, optimizing fabrication processes, and characterizing material properties.

In materials science designed experiments, various types of variables are considered to investigate the properties, performance, and behavior of materials. These variables can be categorized into several common types, including:

1. **Independent Variables:**
    - **Material Composition:** Variables related to the chemical composition and formulation of materials, including the type and proportion of elements or compounds present.
    - **Processing Parameters:** Variables related to the processing conditions used to manufacture or modify materials, such as temperature, pressure, heat treatment, alloying techniques, and synthesis methods.

- **Material Structure**: Variables related to the microstructure and crystallography of materials, including grain size, phase composition, crystal defects, and orientation.
- **Environmental Factors**: Variables related to the environmental conditions in which materials are exposed or tested, such as temperature, humidity, radiation, and corrosive agents.
- **Testing Conditions**: Variables related to the conditions under which materials are tested or evaluated, including load, strain rate, frequency, and testing apparatus.

2. **Dependent Variables**:
   - **Material Properties**: Variables related to the physical, mechanical, thermal, electrical, and optical properties of materials, including hardness, tensile strength, thermal conductivity, electrical resistivity, and refractive index.
   - **Material Performance**: Variables related to the performance of materials in specific applications or environments, such as wear resistance, fatigue life, corrosion resistance, and thermal stability.
   - **Material Behavior**: Variables related to the behavior of materials under different conditions, including deformation, creep, fracture toughness, and stress–strain relationships.
   - **Material Characterization**: Variables related to the results of characterization techniques, such as X-ray diffraction patterns, electron microscopy images, and spectroscopic data.

3. **Categorical Variables**:
   - **Material Types**: Categories representing different types of materials, such as metals, polymers, ceramics, composites, and nanomaterials.
   - **Processing Methods**: Categories representing different manufacturing or processing methods used to create or modify materials, such as casting, extrusion, sintering, and doping.
   - **Environmental Categories**: Categories representing different environmental conditions or exposure scenarios that materials may encounter, such as dry, wet, corrosive, or high-temperature environments.
   - **Testing Methods**: Categories representing different testing techniques or methods employed to evaluate material properties, including tension tests, hardness tests, and microscopy.

4. **Control Variables (Covariates)**:
   - These are variables that are held constant or controlled during experiments to eliminate their influence on the dependent variable, for example, maintaining consistent processing temperatures during material synthesis.

5. **Random Variables**:
   - In some cases, random variables may be introduced to account for variability or uncertainty in material measurements or to control for natural variability in material properties.

6. **Interaction Variables**:
   - Interaction variables are used to investigate the combined effects of two or more independent variables on the dependent variable, helping to understand how different factors interact to influence material behavior.

7. **Noise Variables (Error Variables)**:
   - These are uncontrolled or unmeasured variables that can introduce variability or errors into materials science experiments. Statistical analysis and experimental controls are used to minimize the impact of noise variables.

Materials science designed experiments are essential for optimizing material properties, developing new materials, and understanding material behavior under different conditions. Careful consideration and control of variables are crucial for obtaining accurate and meaningful results in materials research and development.

4.9.1 FACTORIAL EXPERIMENTS

Example: Materials scientists may use factorial experiments to optimize the parameters of a heat treatment process for a specific alloy. Factors like temperature, time, and cooling rate can be varied to achieve desired material properties, such as hardness and strength.

4.9.2 RSM

Example: RSM can be applied in metallurgy to optimize the parameters for a casting process. Researchers vary factors like melt temperature, mold temperature, and pouring rate to achieve desired casting quality and minimize defects.

4.9.3 FRACTIONAL FACTORIAL EXPERIMENTS

Example: In composite material development, fractional factorial experiments can be used to study the effects of various factors (e.g., fiber type and resin composition) on material strength and stiffness without testing all possible combinations.

4.9.4 CCD

Example: CCD is employed in ceramics processing to optimize the parameters of a sintering process. Engineers vary factors like sintering temperature, dwell time, and atmosphere composition to achieve the desired density and microstructure.

4.9.5 TAGUCHI METHODS

Example: Taguchi methods are used in semiconductor fabrication to optimize the parameters of photolithography processes. Researchers may vary factors like exposure dose, focus, and photoresist type to minimize defects and improve feature resolution.

4.9.6 MIXTURE EXPERIMENTS

Example: In polymer material development, mixture experiments help optimize the composition of polymer blends. Researchers vary proportions of polymer types, additives, and fillers to achieve desired material properties, such as thermal conductivity and electrical insulation.

4.9.7 SEQUENTIAL EXPERIMENTATION

Example: Materials scientists may use sequential experimentation to optimize the parameters of a deposition process for thin films. They adjust factors like deposition rate, substrate temperature, and gas flow based on real-time film thickness and quality measurements.

4.9.8 RPD

Example: In material testing, RPD can be applied to ensure the robustness of test methods against variations in sample preparation, loading conditions, and environmental factors, enhancing the reliability of material property measurements.

4.9.9 RANDOMIZED EXPERIMENTS

Example: In metallurgical research, randomized experiments can be used to study the effects of different alloy compositions on mechanical properties. Samples are randomly assigned to different alloy groups to assess material performance.

4.9.10 SPLIT-PLOT AND BLOCKED EXPERIMENTS

Example: When optimizing the parameters of a powder metallurgy process, split-plot or blocked experiments may be used to account for variations in powder characteristics, compaction methods, and sintering conditions.

4.10 PHARMACEUTICALS

In the pharmaceutical industry, designed experiments are essential for optimizing drug formulations, improving manufacturing processes, and ensuring product quality and safety. Here are some common types of designed experiments with examples of their applications in pharmaceuticals.

In pharmaceutically designed experiments, various types of variables are considered to investigate drug formulations, pharmaceutical processes, and their effects on drug efficacy, safety, and quality. These variables can be categorized into several common types, including:

1. **Independent Variables:**
   - **Drug Formulation Variables**: Variables related to the composition and formulation of pharmaceutical products, including drug dosage, active ingredient concentration, excipient types, and drug release profiles.
   - **Process Variables**: Variables related to manufacturing processes and conditions, such as mixing speed, temperature, pressure, drying time, and sterilization methods.
   - **Drug Delivery Variables**: Variables related to drug delivery systems and administration methods, including dosage forms (e.g., tablets, capsules, and injections), release mechanisms, and dosing regimens.
   - **Experimental Conditions**: Variables related to experimental conditions, such as test environments, test equipment, and testing protocols.
2. **Dependent Variables:**
   - **Pharmacological Responses**: Variables related to the pharmacological responses of the drug, including efficacy, potency, onset of action, duration of action, and therapeutic outcomes.
   - **Pharmaceutical Properties**: Variables related to pharmaceutical properties, such as drug dissolution rate, solubility, stability, and bioavailability.
   - **Pharmaceutical Quality**: Variables related to pharmaceutical quality attributes, including purity, identity, content uniformity, and physical characteristics (e.g., particle size and crystallinity).
   - **Toxicological Responses**: Variables related to toxicological responses, such as cytotoxicity, genotoxicity, and adverse effects.
   - **Pharmacokinetic Parameters**: Variables related to drug absorption, distribution, metabolism, and excretion (ADME), including drug plasma concentrations over time.
3. **Categorical Variables:**
   - **Drug Classes**: Categories representing different classes of drugs (e.g., antibiotics, analgesics, and antipsychotics) or drug categories (e.g., small molecules and biologics).
   - **Dosage Forms**: Categories representing different pharmaceutical dosage forms, such as tablets, capsules, oral solutions, and transdermal patches.
   - **Patient Populations**: Categories representing different patient groups, such as age groups, disease states, or genetic subpopulations.
4. **Control Variables (Covariates):**
   - These are variables that are held constant or controlled during experiments to eliminate their influence on the dependent variable, for example, controlling for temperature and humidity during drug stability testing.

5. **Random Variables**:
   - In some cases, random variables may be introduced to account for variability or uncertainty in pharmaceutical measurements or to control for natural variability in biological systems (e.g., interindividual variability in clinical trials).
6. **Interaction Variables**:
   - Interaction variables are used to investigate the combined effects of two or more independent variables on the dependent variable, helping to understand how different factors interact to influence drug properties or responses.
7. **Noise Variables (Error Variables)**:
   - These are uncontrolled or unmeasured variables that can introduce variability or errors into the experimental results. Statistical analysis and experimental controls are used to minimize the impact of noise variables in pharmaceutical experiments.

Pharmaceutical designed experiments are essential for drug development, formulation optimization, quality control, and regulatory compliance. Careful control and manipulation of variables are critical for ensuring the safety, efficacy, and quality of pharmaceutical products.

4.10.1 FACTORIAL EXPERIMENTS

Pharmaceutical scientists may use factorial experiments to optimize the composition of a tablet formulation. They vary factors such as drug concentration, excipient type, and compression force to determine the best combination for tablet quality and dissolution rate.

Factorial experiments are commonly used in the pharmaceutical industry to study the effects of multiple factors on drug formulations, processes, and outcomes. Here is an example of a matrix for a factorial experiment in the pharmaceutical industry.

Objective: To investigate the effects of two factors, Factor A (active ingredient concentration) and Factor B (mixing time), on the dissolution rate of a pharmaceutical tablet.

Factors:

1. **Factor A**: Active ingredient concentration (%)
   - Low (e.g., 5%)
   - Medium (e.g., 10%)
   - High (e.g., 15%)
2. **Factor B**: Mixing time (minutes)
   - Short (e.g., 5 minutes)
   - Medium (e.g., 10 minutes)
   - Long (e.g., 15 minutes)

Response Variable: Dissolution rate (measured in mg/min)

Factorial Experiment Matrix: In a 2 × 3 factorial experiment, each combination of Factor A and Factor B is tested. The matrix is shown in Table 4.72.

In this matrix, each row represents a specific combination of active ingredient concentration and mixing time, and the corresponding dissolution rate is recorded for each combination. This factorial experiment allows researchers in the pharmaceutical industry to assess the main effects of each factor (Factor A and Factor B) and their interactions on the dissolution rate of the tablet.

The goal is to optimize the drug formulation process to achieve the desired dissolution rate, and factorial experiments are valuable for understanding how different factors influence pharmaceutical outcomes.

TABLE 4.72
Factorial Experiment Matrix

Run	Active Ingredient Concentration (%)	Mixing Time (minutes)	Dissolution Rate (mg/min)
1	Low	Short	...
2	Low	Medium	...
3	Low	Long	...
4	Medium	Short	...
5	Medium	Medium	...
6	Medium	Long	...
7	High	Short	...
8	High	Medium	...
9	High	Long	...

4.10.2 RSM

RSM is applied in drug formulation to optimize the parameters for a controlled-release drug delivery system. Researchers vary factors like polymer composition, drug loading, and coating thickness to achieve desired release kinetics.

RSM is commonly used in the pharmaceutical industry to optimize drug formulations and processes by studying the response of interest over a range of factor levels. Here is an example of a matrix for an RSM designed experiment in the pharmaceutical industry.

Objective: To optimize the formulation of a tablet by studying the effects of two factors, Factor A (binder concentration) and Factor B (compression force), on tablet hardness.

Factors:

1. **Factor A**: Binder concentration (% w/w)
   - Low (e.g., 2%)
   - Medium (e.g., 4%)
   - High (e.g., 6%)

2. **Factor B**: Compression force (kN)
   - Low (e.g., 5 kN)
   - Medium (e.g., 10 kN)
   - High (e.g., 15 kN)

Response Variable: Tablet hardness (measured in N/mm^2)

RSM Experiment Matrix: In an RSM experiment, a matrix of factor combinations is created to model the response surface. The matrix is shown in Table 4.73.

In this matrix, each row represents a specific combination of binder concentration and compression force, and the corresponding tablet hardness is recorded. The goal of the RSM experiment is to model the response surface and find the optimal combination of factors that yield the desired tablet hardness.

RSM techniques, such as regression analysis and contour plots, are used to analyze the data and optimize drug formulation processes in the pharmaceutical industry.

TABLE 4.73
RSM Experiment Matrix

Run	Binder Concentration (%)	Compression Force (kN)	Tablet Hardness (N/mm²)
1	Low	Low	...
2	Low	Medium	...
3	Low	High	...
4	Medium	Low	...
5	Medium	Medium	...
6	Medium	High	...
7	High	Low	...
8	High	Medium	...
9	High	High	...
10	Low	Low	...
11	Low	Medium	...
12	Low	High	...
13	Medium	Low	...
14	Medium	Medium	...
15	Medium	High	...
16	High	Low	...
17	High	Medium	...
18	High	High	...

4.10.3 Fractional Factorial Experiments

In vaccine development, fractional factorial experiments can be used to assess the effects of various factors (e.g., antigen concentration and adjuvant type) on vaccine efficacy without testing all possible combinations.

Fractional factorial experiments are used in the pharmaceutical industry to efficiently screen a large number of factors and their interactions while conducting fewer experiments. Here is an example of a matrix for a fractional factorial experiment in the pharmaceutical industry:

Objective: To screen the effects of five factors (Factor A, Factor B, Factor C, Factor D, and Factor E) on the dissolution rate of a pharmaceutical tablet.

Factors:

1. **Factor A**: Active ingredient type (A1/A2)
2. **Factor B**: Binder type (B1/B2)
3. **Factor C**: Lubricant type (C1/C2)
4. **Factor D**: Compression force (low/high)
5. **Factor E**: Tablet shape (E1/E2)

Response Variable: Dissolution rate (measured in mg/min)

Fractional Factorial Experiment Matrix: In a fractional factorial experiment, a fraction of the full factorial design is performed, allowing you to study the main effects and some interactions while

TABLE 4.74
Fractional Factorial Experiment Matrix

Run	Factor A	Factor B	Factor C	Factor D	Factor E	Dissolution Rate (mg/min)
1	A1	B1	C1	Low	E1	...
2	A2	B1	C1	Low	E2	...
3	A1	B2	C1	Low	E2	...
4	A2	B2	C1	Low	E1	...
5	A1	B1	C2	Low	E2	...
6	A2	B1	C2	Low	E1	...
7	A1	B2	C2	Low	E1	...
8	A2	B2	C2	Low	E2	...

reducing the number of experiments. In this example, we will use a half-fraction design, which tests half of the possible combinations. The matrix is shown in Table 4.74.

In this matrix, each row represents a specific combination of Factor A, Factor B, Factor C, Factor D, and Factor E, and the corresponding dissolution rate is recorded. By conducting a fractional factorial experiment, you can efficiently explore the effects of these factors on the pharmaceutical tablet's dissolution rate while significantly reducing the number of experimental runs compared to a full factorial design.

Fractional factorial experiments are valuable for initial screening of factors to identify which ones have a significant impact on the response variable, helping researchers focus on further investigations or optimization of critical factors in drug formulation processes.

4.10.4   CCD

CCD is employed in pharmaceutical manufacturing to optimize the parameters of a tablet coating process. Engineers vary factors like coating solution viscosity, drying temperature, and curing time to achieve uniform and efficient tablet coating.

CCD is a widely used experimental design in the pharmaceutical industry to study the effects of multiple factors and their interactions while allowing for the estimation of curvature in response surfaces. Here is an example of a matrix for a CCD designed experiment in the pharmaceutical industry.

Objective: To optimize the drug formulation for a pharmaceutical tablet by studying the effects of two factors, Factor A (binder concentration) and Factor B (compression force), on tablet hardness.

Factors:

1. **Factor A**: Binder concentration (%)
   - Low (−1)
   - Medium (0)
   - High (+1)
2. **Factor B**: Compression force (kN)
   - Low (−1)
   - Medium (0)
   - High (+1)

Response Variable: Tablet hardness (measured in N/mm²)

TABLE 4.75
CCD Experiment Matrix

Run	Binder Concentration (A)	Compression Force (B)	Tablet Hardness (N/mm²)
−3	Low	Medium	...
−2	High	Medium	...
−1	Medium	Low	...
0	Medium	Medium	...
1	Medium	High	...
2	Low	Medium	...
3	Low	Low	...
4	Medium	Medium	...
5	Low	High	...

CCD Experiment Matrix: In a CCD, a set of experimental runs is designed to estimate the main effects, interactions, and curvature in the response surface. The matrix is shown in Table 4.75.

In this matrix, each row represents a specific combination of binder concentration (Factor A) and compression force (Factor B), and the corresponding tablet hardness is recorded. The values for Factor A and Factor B are coded as −1, 0, and +1, representing low, medium, and high levels, respectively. The negative and positive levels are used to explore the effects of factors beyond the mid-level settings (0).

The CCD allows for a thorough exploration of the response surface and helps pharmaceutical researchers identify the optimal combination of factors (binder concentration and compression force) to achieve the desired tablet hardness.

Statistical techniques, such as regression analysis and contour plots, are used to analyze the data and model the response surface to guide the formulation optimization process in the pharmaceutical industry.

4.10.5 TAGUCHI METHODS

Taguchi methods are used in analytical chemistry to optimize analytical methods for drug quality control. Scientists may vary factors like sample preparation conditions and instrument parameters to minimize measurement errors and ensure product quality.

Taguchi methods, developed by Japanese engineer and statistician Genichi Taguchi, are often used in the pharmaceutical industry to optimize processes and formulations while minimizing variability and ensuring robustness. Here is an example of a matrix for Taguchi method designed experiment in the pharmaceutical industry.

Objective: To optimize the tablet coating process for a pharmaceutical product while minimizing the effects of variations in temperature and humidity.

Factors:

1. **Factor A**: Coating time (levels: short, medium, and long)
2. **Factor B**: Inlet air temperature (levels: low, medium, and high)
3. **Factor C**: Relative humidity (levels: low, medium, and high)

Response Variable: Coating thickness (measured in μm)

TABLE 4.76
Taguchi Method Experiment Matrix

Run	Coating Time (A)	Inlet Air Temperature (B)	Relative Humidity (C)	Coating Thickness (μm)
1	Short	Low	Low	…
2	Short	Medium	Medium	…
3	Short	High	High	…
4	Medium	Low	Medium	…
5	Medium	Medium	High	…
6	Medium	High	Low	…
7	Long	Low	High	…
8	Long	Medium	Low	…
9	Long	High	Medium	…

Taguchi Method Experiment Matrix: Taguchi methods typically use an orthogonal array to systematically test different combinations of factors at different levels. The matrix is shown in Table 4.76.

In this matrix, each row represents a specific combination of coating time (Factor A), inlet air temperature (Factor B), and relative humidity (Factor C), along with the resulting coating thickness. The Taguchi methods aim to identify the optimal factor settings that minimize the effects of noise (variations in temperature and humidity) and result in a robust coating process with consistent quality.

Taguchi methods also involve signal-to-noise (S/N) ratios and other statistical techniques to assess the robustness of the designed experiment and select the best factor settings that minimize the impact of variability on the pharmaceutical product's quality.

4.10.6 Mixture Experiments

In drug formulation, a common mixture experiment is used to help optimize the composition of topical creams. Researchers vary proportions of active ingredients, excipients, and penetration enhancers to achieve desired drug release and skin permeation profiles.

Mixture designed experiments are commonly used in the pharmaceutical industry to optimize the composition of drug formulations where multiple ingredients are involved. Here is an example of a matrix for a mixture designed experiment in the pharmaceutical industry:

Objective: To optimize the composition of a tablet formulation by studying the effects of three ingredients: Factor A (active ingredient), Factor B (binder), and Factor C (filler).

Ingredients:

1. **Factor A**: Active ingredient (% by weight)
   - Varies from 0% to 100%
2. **Factor B**: Binder (% by weight)
   - Varies from 0% to 30%
3. **Factor C**: Filler (% by weight)
   - Varies from 0% to 70%

Response Variable: Tablet hardness (measured in N/mm²)

TABLE 4.77
Mixture Designed Experiment Matrix

Run	Active Ingredient (%)	Binder (%)	Filler (%)	Tablet Hardness (N/mm²)
1	10	5	85	...
2	30	10	60	...
3	20	15	65	...
4	50	20	30	...
5	5	25	70	...
6	40	30	30	...
7	70	10	20	...
8	15	5	80	...
9	25	10	65	...

Mixture Designed Experiment Matrix: In a mixture designed experiment, the total weight of the mixture remains constant, and the proportions of each ingredient are varied systematically. The matrix is shown in Table 4.77.

In this matrix, each row represents a specific combination of active ingredient, binder, and filler proportions, and the corresponding tablet hardness is recorded. The proportions of the three ingredients vary to explore the effects on tablet hardness while keeping the total weight constant.

Mixture designed experiments allow pharmaceutical researchers to optimize drug formulations by determining the ideal combination of ingredients to achieve the desired tablet hardness or other product characteristics. Statistical techniques are used to analyze the data and model the relationships between ingredient proportions and the response variable.

4.10.7 SEQUENTIAL EXPERIMENTATION

Pharmaceutical researchers may use sequential experimentation to optimize the parameters of a pharmaceutical crystallization process. They adjust factors like solvent composition and cooling rate based on real-time crystallization data to achieve desired crystal properties.

Sequential experimentation is often used in the pharmaceutical industry to optimize drug formulations, processes, or clinical trials by iteratively conducting experiments based on the results of previous ones. Here is an example of a matrix for a sequential experimentation designed experiment in the pharmaceutical industry:

Objective: To optimize the formulation of a pharmaceutical tablet for immediate release by studying the effects of two factors, Factor A (binder concentration) and Factor B (compression force), on tablet hardness.

Factors:

1. **Factor A**: Binder concentration (%)
   - Low (e.g., 2%)
   - Medium (e.g., 4%)
   - High (e.g., 6%)
2. **Factor B**: Compression force (kN)
   - Low (e.g., 5 kN)
   - Medium (e.g., 10 kN)
   - High (e.g., 15 kN)

Response Variable: Tablet hardness (measured in N/mm²)

TABLE 4.78
Sequential Experimentation Matrix

Run	Binder Concentration (%)	Compression Force (kN)	Tablet Hardness (N/mm²)
1	Low	Low	...
2	Low	Medium	...
3	Low	High	...
4	Medium	Low	...
5	Medium	Medium	...
6	Medium	High	...
7	High	Low	...
8	High	Medium	...
9	High	High	...

Sequential Experimentation Matrix: In sequential experimentation, you start with an initial set of experiments and then adjust the factor levels based on the results obtained. The matrix might look like this for an initial set of experiments is shown in Table 4.78.

After analyzing the results of these initial experiments, you may choose to conduct additional experiments in areas where further optimization is needed. For example, if the initial results suggest that tablet hardness can be improved by increasing binder concentration, you might conduct additional experiments with higher binder concentrations.

The key feature of sequential experimentation is that it allows you to adapt and refine your experiments based on the knowledge gained from earlier experiments, ultimately leading to an optimized formulation while conserving resources and time.

In practice, you may perform several iterations of experiments, each building upon the results of the previous ones until the desired formulation or process is achieved.

4.10.8 RPD

In clinical trial design, RPD may be applied to ensure the robustness of study protocols against variations in patient characteristics and data collection procedures, enhancing the reliability of clinical trial results.

RPD is used in the pharmaceutical industry to optimize drug formulations, manufacturing processes, and clinical trials by making them less sensitive to variability and environmental factors. Here is an example of a matrix for an RPD designed experiment in the pharmaceutical industry.

Objective: To optimize the formulation of a pharmaceutical tablet by studying the effects of two factors, Factor A (binder concentration) and Factor B (compression force), on tablet hardness while considering potential sources of variation.

Factors:

1. **Factor A**: Binder concentration (%)
    * Low (e.g., 2%)
    * Medium (e.g., 4%)
    * High (e.g., 6%)
2. **Factor B**: Compression force (kN)
    * Low (e.g., 5 kN)
    * Medium (e.g., 10 kN)
    * High (e.g., 15 kN)

TABLE 4.79
RPD Matrix

Run	Binder Concentration (%)	Compression Force (kN)	Ambient Temperature (°C)	Relative Humidity (%)	Tablet Hardness (N/mm²)
1	Low	Low	20	40	...
2	Low	Medium	25	45	...
3	Low	High	22	50	...
4	Medium	Low	23	42	...
5	Medium	Medium	21	48	...
6	Medium	High	24	46	...
7	High	Low	19	43	...
8	High	Medium	26	41	...
9	High	High	27	49	...

Noise Factors:

1. **Noise Factor C**: Ambient temperature (°C)
   - Varies within its normal operating range.
2. **Noise Factor D**: Relative humidity (%)
   - Varies within its normal operating range.

Response Variable: Tablet hardness (measured in N/mm²)

RPD Matrix: In an RPD designed experiment, you consider the impact of noise factors or sources of variation that could affect the response. The matrix is shown in Table 4.79.

In this matrix, each row represents a specific combination of binder concentration, compression force, ambient temperature, relative humidity, and the resulting tablet hardness. Noise factors such as ambient temperature and relative humidity are included to assess the robustness of the tablet hardness under varying environmental conditions.

The goal of RPD is to identify factor settings that result in a pharmaceutical tablet with consistent hardness across different operating conditions, making the formulation less sensitive to sources of variation and ensuring product quality and reliability. Statistical analysis and optimization techniques are used to achieve this robustness.

4.10.9 RANDOMIZED EXPERIMENTS

In clinical research, randomized experiments are used to study the effects of different drug treatments on patient outcomes. Patients are randomly assigned to different treatment groups to assess drug efficacy and safety.

Randomized experiments are commonly used in the pharmaceutical industry to assess the efficacy and safety of new drug formulations or treatments. In these experiments, patients or subjects are randomly assigned to different treatment groups. Here is an example of a matrix for randomized experiment designed experiment in the pharmaceutical industry:

Objective: To evaluate the effectiveness of a new drug treatment for a specific medical condition compared to a control group.

TABLE 4.80
Randomized Experiment Matrix

Patient ID	Treatment Group	Health Measure 1	Health Measure 2	...	Health Measure N
1	Group A	...	...	...	...
2	Group B	...	...	...	...
3	Group A	...	...	...	...
4	Group A	...	...	...	...
5	Group B	...	...	...	...
6	Group A	...	...	...	...
7	Group B	...	...	...	...
...	...	...	...	...	...
N	Group B	...	...	...	...

Groups:

1. **Treatment Group A**: Patients receive the new drug (drug A).
2. **Treatment Group B**: Patients receive a placebo (control).

Response Variables: Various health-related measures, such as symptom relief, adverse effects, or overall patient outcomes, depending on the study's objectives.

Randomized Experiment Matrix: In a randomized experiment, patients or subjects are randomly assigned to different treatment groups to ensure unbiased and statistically valid results. The matrix is shown in Table 4.80.

In this matrix, each row represents a patient or subject in the study, with a unique identifier (Patient ID). Patients are randomly assigned to either Treatment Group A (receiving the new drug) or Treatment Group B (receiving a placebo). Various health measures, such as symptom relief, adverse effects, or other relevant outcomes, are assessed for each patient.

Randomization helps ensure that any observed differences in health outcomes between the treatment groups are not due to bias or confounding factors. Statistical analysis, such as hypothesis testing or regression modeling, is performed to determine the effectiveness and safety of the new drug compared to the control group.

Randomized experiments are a fundamental approach in clinical trials and drug development to provide scientifically sound evidence for regulatory approval and clinical decision-making in the pharmaceutical industry.

4.10.10   Split-Plot and Blocked Experiments

When optimizing the parameters of a pharmaceutical packaging process, split-plot or blocked experiments may be used to account for variations in packaging materials, equipment performance, and environmental conditions.

Split-plot and blocked designed experiments are used in the pharmaceutical industry to optimize processes, formulations, or clinical trials when certain factors are difficult or expensive to change. These designs help account for the variability introduced by these factors. Here are some examples of matrices for split-plot and blocked designed experiments:

TABLE 4.81

Split-Plot Designed Experiment Matrix

Run	Whole Plot (Factor B)	Subplot (Factor A)	Tablet Hardness (N/mm²)
1	10	Low	...
2	10	Medium	...
3	10	High	...
4	15	Low	...
5	15	Medium	...
6	15	High	...
7	20	Low	...
8	20	Medium	...
9	20	High	...

Example 1: Split-Plot Designed Experiment

Objective: To optimize the manufacturing process of a pharmaceutical tablet, considering the effects of two factors: Factor A (binder concentration) and Factor B (compression force). However, changing compression force is time-consuming and costly, so it is treated as a "hard-to-change" factor.

Factors:

1. **Factor A:** Binder concentration (%)
   - Low (e.g., 2%)
   - Medium (e.g., 4%)
   - High (e.g., 6%)
2. **Factor B:** Compression force (kN)
   - Hard to change (three levels, e.g., 10 kN, 15 kN, and 20 kN)

Response Variable: Tablet hardness (measured in N/mm²)

Split-Plot Designed Experiment Matrix: In a split-plot design, the "hard-to-change" factor (Factor B) is applied to whole plots, and the other factor (Factor A) is applied to subplots within each whole plot. The matrix is shown in Table 4.81.

In this matrix, each row represents a specific combination of Factor A (binder concentration) applied to subplots and Factor B (compression force) applied to whole plots. The split-plot design allows for efficient experimentation when one factor is "hard to change" by reducing the number of whole-plot changes, which are time-consuming or expensive.

Example 2: Blocked Designed Experiment

Objective: To evaluate the effectiveness of a new drug treatment for a specific medical condition in a clinical trial, while considering the influence of potential confounding factors, such as age and gender.

Groups:

1. **Treatment Group A:** New drug treatment.
2. **Treatment Group B:** Placebo (control).

TABLE 4.82
Blocked Designed Experiment Matrix

Patient ID	Treatment Group	Age Group	Gender	Health Improvement
1	Group A	30–40	Male	...
2	Group B	40–50	Female	...
3	Group A	30–40	Female	...
4	Group A	40–50	Male	...
5	Group B	30–40	Female	...
6	Group A	50–60	Male	...
7	Group B	40–50	Male	...
...	...	...	...	...
N	Group A	30–40	Female	...

Blocks: Patients are divided into blocks based on age and gender to account for potential confounding effects.

Response Variable: Health improvement or symptom relief (measured on a standardized scale).

Blocked Designed Experiment Matrix: In a blocked design, patients are divided into blocks based on specific characteristics (e.g., age and gender) to account for potential confounding effects. The matrix is shown in Table 4.82.

In this matrix, each row represents a patient in the clinical trial, with information on their assigned treatment group, age group, gender, and health improvement outcome. Patients are divided into blocks based on age and gender to account for potential confounding factors in the evaluation of the treatment's effectiveness.

Blocked designs help ensure that the treatment groups are comparable within each block, reducing the potential for bias and confounding effects when assessing treatment outcomes. This improves the reliability and validity of the clinical trial results in the pharmaceutical industry.

4.11　SOCIAL SCIENCES

Designed experiments are less common in social sciences compared to fields like engineering or biology due to the inherent complexity of human behavior and the ethical considerations involved. However, social scientists still use certain types of designed experiments for specific research purposes.

In social science designed experiments, various types of variables are considered to investigate human behavior, attitudes, and social phenomena. These variables can be categorized into several common types, including:

1. **Independent Variables:**
   - **Treatment Variables:** Variables related to interventions or treatments applied to participants, groups, or situations, such as educational programs, therapy sessions, or policy changes.
   - **Demographic Variables:** Variables related to participant characteristics, such as age, gender, ethnicity, socioeconomic status, education level, and marital status.
   - **Environmental Factors:** Variables related to the physical or social environment in which the study takes place, including location, climate, urban/rural setting, and community characteristics.
   - **Experimental Conditions:** Variables related to the conditions under which the experiment is conducted, such as the timing of data collection, instructions given to participants, or manipulations of stimuli.

- **Independent Variables in Surveys**: In survey-based research, independent variables may include questions related to opinions, behaviors, or experiences, often measured on Likert scales or as categorical responses.

2. **Dependent Variables**:
   - **Behavioral Variables**: Variables related to observable actions or behaviors, such as voting behavior, consumer choices, social interactions, or performance on tasks.
   - **Attitudinal Variables**: Variables related to attitudes, beliefs, perceptions, or opinions, often measured through surveys or questionnaires.
   - **Psychological Variables**: Variables related to cognitive or emotional processes, including variables such as self-esteem, motivation, stress levels, or mental health outcomes.
   - **Social Outcomes**: Variables related to social outcomes or consequences, such as crime rates, educational attainment, employment status, or well-being indicators.
   - **Dependent Variables in Surveys**: In survey-based research, dependent variables may include responses to questions about preferences, satisfaction, or perceptions.

3. **Categorical Variables**:
   - **Group Membership**: Categories representing different groups or conditions, such as experimental group vs. control group, political affiliation, or treatment type.
   - **Demographic Categories**: Categories representing different demographic groups or categories, such as age groups, gender categories, or income brackets.
   - **Geographic Categories**: Categories representing different geographic regions or areas, which can be important in studies of regional disparities or cultural differences.
   - **Policy Categories**: Categories representing different policy or program conditions, which can be used to study the impact of policies or interventions.

4. **Control Variables (Covariates)**:
   - These are variables that are held constant or controlled during experiments to eliminate their influence on the dependent variable, for example, controlling for age or gender in a study examining the impact of an educational program.

5. **Random Variables**:
   - In some cases, random variables may be introduced to account for variability or uncertainty in social science measurements or to control for natural variability in human behavior.

6. **Interaction Variables**:
   - Interaction variables are used to investigate the combined effects of two or more independent variables on the dependent variable, helping to understand how different factors interact to influence social outcomes.

7. **Noise Variables (Error Variables)**:
   - These are uncontrolled or unmeasured variables that can introduce variability or errors into social science experiments. Statistical analysis and experimental controls are used to minimize the impact of noise variables.

Social science designed experiments are essential for understanding human behavior, evaluating the effectiveness of interventions, and exploring complex social phenomena. Careful consideration and control of variables are crucial for obtaining accurate and meaningful results in social science research.

4.11.1 RCTs

Example: In education research, an RCT can be used to evaluate the effectiveness of a new teaching method. Students are randomly assigned to either the experimental group (receiving the new method) or the control group (receiving the traditional method), and their academic performance is compared.

4.11.2   FIELD EXPERIMENTS

Example: In economics, a field experiment can be conducted to study the impact of a policy intervention on consumer behavior. Researchers may randomly offer discounts on certain products in a retail store to assess the effects on sales and customer choices.

4.11.3   LAB EXPERIMENTS

Example: In psychology, a lab experiment can be used to investigate the effects of social influence on decision-making. Participants are randomly assigned to different conditions where they receive varying levels of peer pressure, and their decisions are observed and analyzed.

4.11.4   QUASI-EXPERIMENTS

Example: In public health, a quasi-experiment can be used to assess the impact of a smoking cessation program. Researchers compare health outcomes (e.g., lung function) in individuals who voluntarily participate in the program to those who do not, controlling for relevant variables.

4.11.5   SURVEY EXPERIMENTS

Example: In political science, a survey experiment can be conducted to assess the impact of campaign messages on voter preferences. Respondents are randomly exposed to different campaign messages, and their subsequent candidate preferences are recorded.

4.11.6   NATURAL EXPERIMENTS

Example: In sociology, a natural experiment can be used to study the effects of a sudden policy change. Researchers may examine the impact of a new law legalizing same-sex marriage by comparing social and demographic outcomes before and after its implementation.

4.11.7   CROSSOVER EXPERIMENTS

Example: In social work, a crossover experiment can be conducted to assess the effectiveness of two different counseling approaches. Participants receive both types of counseling in random order, and their well-being is measured before and after each treatment.

4.11.8   LONGITUDINAL EXPERIMENTS

Example: In developmental psychology, a longitudinal experiment can be used to study the long-term effects of early interventions on child development. Children from different backgrounds are randomly assigned to receive an intervention, and their progress is tracked over several years.

4.11.9   POLICY EXPERIMENTS

Example: In public policy research, a policy experiment can be conducted to evaluate the effects of a new welfare reform. The reform is implemented in selected regions, while others serve as control groups, and the impact on employment and poverty rates is analyzed.

4.11.10   ONLINE EXPERIMENTS

Example: In communication studies, an online experiment can be used to investigate the effects of social media content on public opinion. Participants are exposed to different types of content in a controlled online environment, and their reactions are measured.

4.12 TRANSPORTATION AND AEROSPACE

Designed experiments are valuable tools in the fields of transportation and aerospace for optimizing processes, enhancing safety, and improving the performance of vehicles, systems, and components.

In transportation and aerospace research designed experiments, various types of variables are considered to study and optimize systems, vehicles, and operations related to transportation and aerospace engineering. These variables can be categorized into several common types, including:

1. **Independent Variables**:
   - **Vehicle Characteristics**: Variables related to the design, specifications, and features of vehicles or aircraft, such as size, weight, shape, engine type, propulsion system, and wing design.
   - **Operational Parameters**: Variables related to the operation of transportation systems, including speed, altitude, route, fuel type, cargo load, and payload capacity.
   - **Environmental Conditions**: Variables related to external conditions that affect transportation and aerospace systems, such as weather, temperature, humidity, wind speed, and air pressure.
   - **Maintenance and Repairs**: Variables related to maintenance schedules, repair procedures, and equipment reliability, which can impact system performance and safety.
   - **Control Systems**: Variables related to the control and automation systems used in transportation and aerospace, including avionics, navigation systems, and vehicle guidance.
2. **Dependent Variables**:
   - **Performance Metrics**: Variables related to the performance of transportation and aerospace systems, including speed, fuel efficiency, range, payload capacity, and flight duration.
   - **Safety and Reliability Metrics**: Variables related to safety and reliability measures, including accident rates, failure modes, downtime, and maintenance intervals.
   - **Environmental Impact**: Variables related to environmental consequences, such as emissions of pollutants, noise levels, and ecological impacts of transportation and aerospace activities.
   - **Economic Metrics**: Variables related to economic aspects, including cost per unit distance, profitability, ROI, and total operating costs.
   - **Customer Satisfaction**: Variables related to customer satisfaction with transportation services, including passenger comfort, on-time performance, and service quality.
3. **Categorical Variables**:
   - **Vehicle Types**: Categories representing different types of vehicles or aircraft, such as cars, trucks, buses, airplanes, helicopters, or spacecraft.
   - **Route Categories**: Categories representing different transportation routes or flight paths, considering variations in distance, terrain, and traffic conditions.
   - **Weather Categories**: Categories representing different weather conditions, such as clear skies, rain, snow, fog, or thunderstorms.
   - **Aircraft Models**: Categories representing different aircraft models and configurations, which can impact performance and efficiency.
4. **Control Variables (Covariates)**:
   - These are variables that are held constant or controlled during experiments to eliminate their influence on the dependent variable, for example, controlling for altitude when testing the fuel efficiency of an aircraft.
5. **Random Variables**:
   - Random variables may be introduced to account for variability or uncertainty in measurements or to control for natural variations in transportation and aerospace conditions.

6. **Interaction Variables**:
   - Interaction variables may be used to investigate how the combined effects of two or more independent variables influence transportation and aerospace performance or safety.
7. **Noise Variables (Error Variables)**:
   - These are uncontrolled or unmeasured variables that can introduce variability or errors into experiments. Statistical analysis and experimental controls are used to minimize the impact of noise variables.

Transportation and aerospace research experiments are critical for optimizing systems, improving safety, reducing environmental impact, and enhancing the efficiency of transportation modes and aerospace technologies. Careful consideration and control of variables are essential for obtaining accurate and meaningful results in these fields.

4.12.1 FACTORIAL EXPERIMENTS

Example: Aerospace engineers may use factorial experiments to optimize the design of an aircraft wing. Factors such as wing shape, winglet size, and airfoil profiles can be varied to maximize lift and minimize drag.

4.12.2 RSM

Example: RSM can be applied in vehicle manufacturing to optimize the parameters of a welding process for automotive components. Researchers vary factors like welding current, electrode material, and joint geometry to achieve desired weld strength.

4.12.3 FRACTIONAL FACTORIAL EXPERIMENTS

Example: In spacecraft propulsion, fractional factorial experiments can be used to study the effects of various factors (e.g., propellant type and nozzle design) on rocket engine performance without testing all possible combinations.

4.12.4 CCD

Example: CCD is employed in aviation to optimize the parameters of an aircraft engine's combustion process. Engineers vary factors like fuel injection rate, air–fuel ratio, and ignition timing to improve fuel efficiency and reduce emissions.

4.12.5 TAGUCHI METHODS

Example: Taguchi methods are used in automotive design to optimize the parameters of a vehicle suspension system. Engineers may vary factors like shock absorber settings, tire pressure, and suspension geometry to enhance ride comfort and handling.

4.12.6 MIXTURE EXPERIMENTS

Example: In aviation fuel formulation, mixture experiments help optimize the composition of aviation fuels. Researchers vary proportions of different hydrocarbon fractions and additives to meet performance and safety standards.

4.12.7 SEQUENTIAL EXPERIMENTATION

Example: Aerospace engineers may use sequential experimentation to optimize the parameters of a rocket launch trajectory. They adjust factors like thrust profiles and staging timings based on real-time telemetry data to achieve mission objectives.

4.12.8 RPD

Example: In automotive manufacturing, RPD can be applied to ensure the robustness of assembly processes against variations in component dimensions and tolerances, reducing defects and improving reliability.

4.12.9 RANDOMIZED EXPERIMENTS

Example: In aviation safety research, randomized experiments can be used to assess the effectiveness of different pilot training programs in improving cockpit crew performance. Randomized trials assign pilots to different training methods to evaluate their impact on safety outcomes.

4.12.10 SPLIT-PLOT AND BLOCKED EXPERIMENTS

Example: When optimizing the parameters of a railroad track maintenance process, split-plot or blocked experiments may be used to account for variations in track sections, equipment performance, and environmental conditions.

5 Statistic Tools Used for Analysis of Precision, Accuracy, Repeatability, and Reproducibility

There are many statistical tools you can use to improve the precision, accuracy, repeatability, and reproducibility of data in your laboratory. Here are some effective tools for each metric:

Precision:
- **Standard Deviation (SD)**: Measures the spread of data points around the mean; a lower SD indicates higher precision.
- **Variance (Var) also Known as Analysis of Variance (ANOVA)**: Similar to SD but squared, useful for statistical calculations. ANOVA: Compares data from different groups to identify statistically significant differences, useful for evaluating factors affecting variability.
- **Relative SD (RSD)**: Expresses SD as a percentage of the mean and helpful for comparing precision across different measurements.
- **Coefficient of Variation (CV)**: Similar to RSD but normalized by the mean and allows for better comparison across measurements with different units.
- **Control Charts**: Monitor data over time to identify trends and potential inconsistencies, helping maintain precision.

Accuracy:
- **Bias**: Difference between the mean of your measurements and the true value. Minimized through calibration and using reference standards.
- **Calibration Curves**: Relate instrument response to known concentrations of analytes, allowing for accurate quantification.
- **Standard Reference Materials (SRMs)**: Certified materials with known compositions, used to verify accuracy and compare results across laboratories.
- **Recovery Experiments**: Spike known amounts of analytes into samples and measure their recovery, assessing accuracy of the analytical method.

Repeatability:
- **Intra-Assay Variation**: Measures variation within a single analytical run, assessed by replicate measurements of the same sample.
- **Control Charts for Replicates**: Monitor replicate data for trends and outliers, ensuring consistent performance within an assay.
- **Method Validation**: Evaluate the repeatability of your analytical method through formal testing and statistical analysis.

Reproducibility:
- **Inter-assay Variation**: Measures variation between different analytical runs or between different analysts.

DOI: 10.1201/9781003528531-5

- **Control Charts for Inter-assay Data**: Monitor data across different runs or analysts to identify inconsistencies and ensure method reproducibility.
- **Ring Tests**: Collaborative experiments where multiple laboratories analyze the same samples, assessing the reproducibility of the method across different settings.

Additional tools:

- **Regression Analysis**: Relates analytical response to potential influencing factors, providing insights into sources of variation and improving prediction accuracy.
- **Multivariate Statistics**: Analyze data with multiple variables to understand complex relationships and identify hidden patterns impacting variability.
- **Grubbs' Test**: Used to identify and remove outliers from a dataset. Outliers can significantly affect precision and accuracy.
- **Reproducibility and Repeatability (R&R) Studies**: R&R studies, also known as gage R&R (GR&R) studies, assess the variability contributed by different sources, including operators and equipment. They help distinguish between repeatability (within-run variability) and reproducibility (between-run variability).
- **Youden Plot**: Youden plots are used for assessing measurement system performance, especially for comparing different instruments or methods. They graphically represent precision, bias, and linearity.

By regularly employing these statistical tools, you can gain valuable insights into your analytical processes, identify and address sources of errors, and ultimately improve the overall quality and reliability of your data in the laboratory.

5.1 HOW TO DO THE STATISTICS FOR PRECISION

5.1.1 SD

Calculating the SD for your data in an analytical laboratory involves several steps to assess the variability or spread of your measurements. Here are the steps to perform a SD analysis:

1. **Prepare your data:**
   - Collect your data points from your analytical measurements. Make sure they are all from the same sample or experimental condition.
   - Ensure your data points are accurate and have the correct units.
2. **Calculate the mean:**
   - Add up all your data points and divide by the total number of points. This gives you the average value of your dataset.

$$\text{Mean}\,(\mu) = \Sigma\,(\text{Sum of Data Points})\,/\,\text{Number of Data Points}$$

3. **Calculate and square the deviations from the mean:**
   - For each data point, subtract the mean you just calculated. Square the result. Repeat this for all your data points.
   - For each data point, calculate the deviation from the mean. Deviation is the difference between each data point and the mean:

$$\text{Deviation} = \text{Data Point} - \text{Mean}$$

   - Square each of the deviations. Squaring ensures that both positive and negative deviations contribute to the variability:

$$\text{Squared Deviation} = (\text{Deviation})^2$$

4. **Calculate the variance:**
   - Add up all the squared deviations you calculated in step 3. Divide the sum by the total number of data points, minus one (n − 1). This accounts for the estimation of the population mean using the sample mean.
   - Calculate the variance by finding the average of the squared deviations. This represents the average squared distance from the mean:

$$\text{Variance}(\sigma \wedge 2) = \Sigma(\text{Squared Deviations}) / (\text{Number of Data Points} - 1)$$

   Note: The denominator is typically (n − 1) for sample data, where "n" is the number of data points. For population data, it would be "n" without the subtraction.
5. **Calculate the SD:**
   - Take the square root of the variance you calculated in step 4. This gives you the final SD of your dataset.
   - Calculate the variance by finding the average of the squared deviations. This represents the average squared distance from the mean:

$$\text{Variance}(\sigma \wedge 2) = \Sigma(\text{Squared Deviations}) / (\text{Number of Data Points} - 1)$$

   Note: The denominator is typically (n − 1) for sample data, where "n" is the number of data points. For population data, it would be "n" without the subtraction.
6. **Interpret the Results**

   The SD represents the typical or average deviation of individual data points from the mean. A larger SD indicates greater variability in the data, while a smaller SD indicates less variability. It is often used to assess the precision and consistency of measurements in analytical chemistry.
7. **Communicate the Findings**

   In your analytical laboratory report or documentation, include the calculated SD along with the mean and any other relevant statistical parameters. Interpret the results in the context of your analytical experiments, highlighting the precision and variability of your measurements.
8. **Quality Control**

   Use the SD as a quality control measure to monitor the consistency and reliability of your analytical processes. If the SD exceeds acceptable limits, investigate potential sources of variability and take corrective actions as needed.

Additional considerations:

- For smaller datasets, consider using Bessel's correction instead of dividing by n − 1. This correction adjusts for the underestimation of population variance by the sample variance based on the sample size.
- Round the final SD to an appropriate number of decimal places based on the precision of your data.
- Use statistical software or calculators for faster and more efficient calculations, especially for large datasets.
- Interpret the SD in the context of your analytical method and acceptable tolerance for variation. A low SD indicates high precision, meaning your data points are tightly clustered around the mean. Conversely, a high SD indicates low precision, with data points spread out more widely.

5.1.2 Variance (Var)/ANOVA

Variance analysis, also known as ANOVA, is a statistical technique used to assess the sources of variation in a dataset and determine whether there are statistically significant differences between groups or conditions. In an analytical laboratory, it can be valuable for comparing different experimental setups, instruments, or methods. Here are the steps to perform a variance analysis in an analytical laboratory:

Step 1: Define Your Hypotheses
- Clearly define the null hypothesis (H0) and alternative hypothesis (Ha) based on the specific research question or comparison you want to make. The null hypothesis typically states that there are no significant differences between groups, while the alternative hypothesis suggests that differences exist.

Step 2: Organize Your Data
- Organize your data into groups or categories based on the factor(s) you want to analyze. Each group represents a distinct condition, treatment, or category.

Step 3: Calculate Descriptive Statistics
- Calculate summary statistics for each group, including the mean, SD, and sample size (n). These statistics provide initial insights into the data.

Step 4: Perform the ANOVA
- Use the appropriate ANOVA test based on the design of your experiment:
  - **One-Way ANOVA:** Use when you have one independent variable (factor) with multiple levels or groups.
  - **Two-Way ANOVA:** Use when you have two independent variables (factors) and want to assess their main effects and interaction.
  - **ANOVA with Replication:** Use when you have multiple measurements (replicates) within each group.
- Calculate the F-statistic, which measures the ratio of between-group variability to within-group variability. The F-statistic follows an F-distribution.
- Determine the degrees of freedom for the numerator (df1) and denominator (df2) for the F-distribution.
- Use a significance level (alpha, typically set at 0.05) to determine the critical F-value from the F-distribution table.

Step 5: Calculate the P-Value
- Calculate the p-value associated with the F-statistic. The p-value represents the probability of obtaining the observed results if the null hypothesis is true.
- Compare the p-value to the chosen significance level (alpha) to make a statistical decision:

    If $p \leq$ alpha, reject the null hypothesis (indicating significant differences).

    If $p >$ alpha, fail to reject the null hypothesis (indicating no significant differences).

Step 6: Post hoc Tests (If Applicable)
- If you reject the null hypothesis in a one-way or two-way ANOVA, perform post hoc tests to identify which specific groups differ significantly from each other. Common post hoc tests include Tukey's honestly significant difference (HSD), Bonferroni, or Dunnett's test.

Step 7: Interpret the Results
- Interpret the results in the context of your research question. If you reject the null hypothesis, explain the practical implications of the differences found.

Step 8: Report the Findings
- Include the results of the ANOVA, including F-statistic, degrees of freedom, p-value, and any post hoc test results, in your analytical laboratory report or documentation.

Step 9: Quality Control and Continuous Improvement
- Use variance analysis as a tool for quality control in your laboratory. Monitor and analyze variation over time to identify areas for improvement and ensure consistent analytical results.

Variance analysis is a powerful tool for comparing groups or conditions in an analytical laboratory setting. It helps identify factors that may contribute to variability in your measurements and supports data-driven decision-making for process optimization and quality assurance.

5.1.3   RSD

The RSD analysis is a statistical measure used to assess the precision or variability of data relative to its mean. It is commonly used in analytical laboratories to evaluate the consistency of measurements. Below are the steps to perform an RSD analysis for improved precision in an analytical laboratory:

Step 1: Data Collection: Collect the data or measurements you want to analyze. Ensure that you have a complete dataset with replicates or multiple measurements for each sample or condition.

Step 2: Calculate the Mean (Average): Calculate the mean (average) of your dataset. Sum up all the data points and divide by the total number of data points:

$$\text{Mean}(\mu) = \Sigma(\text{Sum of Data Points}) / \text{Number of Data Points}$$

Step 3: Calculate the SD: Calculate the SD of your dataset. This quantifies the spread or variability of the data:

$$\text{Standard Deviation}(\sigma) = \sqrt{(\Sigma(\text{Data Point} - \text{Mean})^2 / (\text{Number of Data Points} - 1))}$$

Step 4: Calculate the RSD: Calculate the RSD by dividing the SD by the mean and multiplying by 100 to express it as a percentage:

$$\text{RSD}(\%) = (\text{Standard Deviation} / \text{Mean}) * 100$$

Step 5: Interpret the RSD
- A lower RSD indicates higher precision and lower relative variability.
- A higher RSD suggests lower precision and greater relative variability.
- The specific acceptable range for RSD may depend on your laboratory's quality control standards, industry guidelines, or analytical method requirements.

Step 6: Quality Control and Interpretation
- Evaluate the RSD in the context of your analytical objectives and laboratory quality control standards.
- Compare the calculated RSD to predetermined acceptance criteria or acceptable RSD ranges.
- Determine whether the precision meets the required standards for your specific analytical application.

Step 7: Repeat as Necessary
- If precision is not within acceptable limits, investigate potential sources of variability in your analytical process or measurement system.

- Consider making adjustments, such as improving instrument calibration, optimizing sample preparation, or enhancing operator training, to reduce variability and improve precision.

Step 8: Record and Report

- Include the RSD values along with the mean and other relevant statistical parameters in your laboratory report or documentation.
- Keep records of RSD values for quality control and continuous improvement purposes.

The RSD analysis is a valuable tool for assessing the precision of analytical measurements. By calculating and monitoring RSD values, you can identify areas for improvement in your laboratory processes and ensure that your measurements are consistent and reliable.

5.1.4 CV

The CV analysis is a statistical measure used to assess the relative variability or precision of data, particularly when comparing datasets with different units or scales. It is commonly used in analytical laboratories to evaluate the consistency of measurements. Here are the steps to perform a CV analysis for improved precision in an analytical laboratory:

Step 1: Data Collection: Collect the data or measurements you want to analyze. Ensure that you have a complete dataset, which may include replicates or multiple measurements for each sample or condition.

Step 2: Calculate the Mean (Average): Calculate the mean (average) of your dataset. Sum up all the data points and divide by the total number of data points:

$$\text{Mean}\,(\mu) = \Sigma\,(\text{Sum of Data Points}) / \text{Number of Data Points}$$

Step 3: Calculate the SD: Calculate the SD of your dataset. This quantifies the spread or variability of the data:

$$\text{Standard Deviation}(\sigma) = \sqrt{\,(\Sigma(\text{Data Point} - \text{Mean})\,^\wedge\,2\,/\,(\text{Number of Data Points} - 1))}$$

Step 4: Calculate the CV: Calculate the CV by dividing the SD by the mean and multiplying by 100 to express it as a percentage:

$$\text{CV}\,(\%) = (\text{Standard Deviation}\,/\,\text{Mean}) * 100$$

Step 5: Interpret the CV

- The CV expresses the relative variability of the data as a percentage of the mean.
- A lower CV indicates higher precision and lower relative variability.
- A higher CV suggests lower precision and higher relative variability.
- The specific acceptable range for CV may depend on your laboratory's quality control standards, industry guidelines, or analytical method requirements.

Step 6: Quality Control and Interpretation

- Evaluate the CV in the context of your analytical objectives and laboratory quality control standards.
- Compare the calculated CV to predetermined acceptance criteria or acceptable CV ranges.
- Determine whether the precision meets the required standards for your specific analytical application.

Step 7: Repeat as Necessary
- If precision is not within acceptable limits, investigate potential sources of variability in your analytical process or measurement system.
- Consider making adjustments, such as improving instrument calibration, optimizing sample preparation, or enhancing operator training, to reduce variability and improve precision.

Step 8: Record and Report
- Include the CV values along with the mean and other relevant statistical parameters in your laboratory report or documentation.
- Keep records of CV values for quality control and continuous improvement purposes.

The CV analysis is a valuable tool for assessing the relative precision of analytical measurements, especially when dealing with data from different scales or units. By calculating and monitoring CV values, you can identify areas for improvement in your laboratory processes and ensure that your measurements are consistent and reliable.

5.1.5 CONTROL CHARTS

Establishing control charts in an analytical laboratory is a crucial step in quality control and process monitoring. Control charts help assess the stability and precision of measurements over time, allowing you to detect deviations or trends that may indicate issues with the analytical process. Here are the steps to establish control charts for improved precision in an analytical laboratory:

Step 1: Define the Objective and Select the Control Chart Type
- Determine the objective of your control chart. Are you monitoring the mean (X-bar chart), the range (R chart), or both (X-bar and R chart)?
- Choose the appropriate control chart type based on your objectives and the nature of the data.

Step 2: Collect Data
- Collect a set of data points over time. Ensure that the data are representative of the process or measurement you want to monitor.

Step 3: Calculate the Control Chart Constants
- Depending on the control chart type, calculate the control chart constants, such as A2, D3, D4, B3, and B4. These constants are used to establish control limits and assess data points' variation.

Step 4: Calculate the Control Limits
- Calculate the control limits for the control chart. Control limits consist of the center-line (CL), upper control limit (UCL), and lower control limit (LCL). These limits are essential for identifying variations in the process.
- For X-bar and R charts, calculate the CL, UCL, and LCL for both the X-bar and R charts separately.

Step 5: Plot the Data
- Plot the data points on the control chart. Place the data points along the X-bar and R charts according to their respective time order.

Step 6: Calculate the Statistics
- Calculate the X-bar (average) and R (range) for each subgroup of data points, if applicable. Subgroups can be determined based on time intervals, batches, or other relevant factors.
- Compute the X-bar and R statistics to monitor the central tendency and dispersion of the process, respectively.

Step 7: Plot the Control Limits
- Plot the calculated control limits on the control chart. This visually shows the acceptable range of variation for the process.
- The control limits should be placed above and below the X-bar and R charts, parallel to the CL.

Step 8: Interpret the Control Chart
- Regularly update the control chart with new data points. Examine the chart for trends, patterns, or points that fall outside the control limits.
- Interpret the control chart by assessing whether variations are within normal process variability or if they indicate special causes of variation that require investigation.

Step 9: Take Corrective Action
- If the control chart indicates the presence of special causes of variation, investigate and address the root causes.
- Implement corrective actions to improve precision and maintain process stability.

Step 10: Continuous Monitoring and Improvement
- Continuously monitor the process using control charts. Regularly update the charts with new data.
- Use control charts as a tool for ongoing process improvement, identifying opportunities to reduce variability and enhance precision.

By following these steps, you can establish control charts in your analytical laboratory to monitor precision, detect variations, and ensure the consistency and reliability of your measurements over time.

5.2　HOW TO DO THE STATISTICS FOR ACCURACY

5.2.1　BIAS ANALYSIS

Establishing a bias analysis in an analytical laboratory is essential for assessing and improving the accuracy of measurements. Bias analysis helps identify and quantify systematic errors in analytical methods or instruments. Here are the steps to establish a bias analysis for improved accuracy:

Step 1: Define the Objective
- Clearly define the objective of the bias analysis. Determine what aspect of accuracy you want to assess or improve, such as method accuracy, instrument bias, or calibration accuracy.

Step 2: Select Reference Materials or Methods
- Identify appropriate reference materials, certified reference materials (CRMs), or reference methods that are well-characterized and traceable to national or international standards.
- Ensure that the reference materials or methods are relevant to your analytical measurements.

Step 3: Prepare Samples
- Prepare a set of test samples or specimens that cover the analytical range of interest.
- Ensure that the samples are representative and homogeneous.

Step 4: Conduct Measurements
- Perform measurements on the test samples using the analytical method or instrument you want to assess.
- Record the measurement results.

Step 5: Obtain Reference Measurements
- Obtain reference measurements for the same set of test samples using the reference materials or methods selected in Step 2.
- Ensure that the reference measurements are conducted under the same conditions as the test measurements.

Step 6: Calculate Bias
- Calculate the bias (systematic error) for each measurement as the difference between the test measurement result and the reference measurement result.
- Express bias as an absolute value or a percentage of the reference value.

Step 7: Statistical Analysis
- Perform statistical analysis to determine the mean bias and its confidence interval.
- Evaluate the significance of bias by comparing it to predetermined acceptable limits or specifications.

Step 8: Interpret the Results
- Interpret the bias analysis results in the context of your laboratory's quality control standards, industry guidelines, or regulatory requirements.
- Determine whether the observed bias is within acceptable limits or if corrective actions are needed.

Step 9: Identify and Address Sources of Bias
- If bias is detected and considered significant, investigate potential sources of bias in the analytical method or instrument.
- Implement corrective actions to reduce or eliminate bias, such as adjusting calibration, improving sample handling, or recalibrating instruments.

Step 10: Continuous Monitoring and Improvement
- Continuously monitor and reassess bias over time to ensure ongoing accuracy and reliability of measurements.
- Use bias analysis as a tool for continuous improvement in your laboratory's measurement processes.

Step 11: Documentation and Reporting
- Document the bias analysis process, including the test samples, reference measurements, calculations, and results.
- Include bias analysis results and any corrective actions taken in your laboratory reports and documentation.

Establishing a bias analysis process is essential for maintaining the accuracy and traceability of measurements in an analytical laboratory. It helps ensure that systematic errors are identified, quantified, and addressed, leading to improved accuracy and reliability of analytical results.

5.2.2 Calibration Curves

Establishing a calibration curve is a critical step in analytical chemistry to determine the relationship between the concentration of an analyte and the instrument's response. This curve is used to accurately quantify unknown concentrations based on measured responses. Here are the steps to establish a calibration curve analysis for improved accuracy in an analytical laboratory:

Step 1: Define the Objective
- Clearly define the objective of the calibration curve. Determine which analyte you want to quantify and the specific analytical method or instrument you will use.

Step 2: Select Analyte Standards
- Choose a set of analyte standards with known concentrations that span the expected concentration range of your samples.
- Ensure that the standards are well-prepared and accurately characterized.

Step 3: Prepare Standard Solutions
- Prepare a series of standard solutions by diluting the analyte standards to create a range of known concentrations.
- Use appropriate solvents or diluents to ensure compatibility with your analytical method.

Step 4: Perform Measurements
- Measure the instrument response (e.g., peak area, absorbance, and intensity) for each standard solution using the chosen analytical instrument.
- Record the response values for each concentration.

Step 5: Plot the Calibration Curve
- Create a plot of the instrument response (y-axis) versus the known analyte concentrations (x-axis).
- Typically, a linear calibration curve is sought, but other functional relationships (e.g., quadratic) may be used if appropriate.
- Use regression analysis to fit a curve to the data points, and calculate the equation of the curve.

Step 6: Evaluate Linearity and Regression Statistics
- Assess the linearity of the calibration curve by calculating correlation coefficients (R-squared) and residuals.
- A high R-squared value (close to 1) indicates good linearity.
- Analyze residuals to ensure that the model fits the data well.

Step 7: Determine Limits of Detection and Quantification
- Calculate the limit of detection (LOD) and limit of quantification (LOQ) for the method based on the calibration curve and instrument noise.
- LOD is the lowest concentration that can be reliably detected, while LOQ is the lowest concentration that can be quantified with acceptable precision.

Step 8: Validate the Calibration Curve
- Validate the calibration curve by assessing its accuracy, precision, and robustness.
- Perform validation experiments using known samples to confirm that the curve accurately quantifies the analyte.

Step 9: Document the Calibration Curve
- Record all details of the calibration curve, including the equation, regression statistics, and validation results.
- Include this information in your laboratory's standard operating procedures (SOPs) and documentation.

Step 10: Use the Calibration Curve for Sample Analysis
- Once the calibration curve is established and validated, use it to analyze unknown samples.
- Measure the instrument response for the unknown samples and use the calibration curve equation to calculate their concentrations.

Step 11: Monitor and Update
- Continuously monitor the performance of the calibration curve over time.
- Periodically re-establish the calibration curve using fresh standards to ensure accuracy and reliability.

Establishing a calibration curve is essential for quantitative analytical chemistry. It provides a reliable method for accurately determining the concentration of analytes in unknown samples, contributing to improved accuracy and precision in your laboratory's analytical measurements.

5.2.3　SRMs

Establishing the use of SRMs is a crucial step in analytical laboratory work to ensure the accuracy and reliability of measurements. SRMs are certified materials with well-characterized properties, and they serve as benchmarks for verifying the accuracy of analytical methods and instruments. Here are the steps to establish an SRM analysis for improved accuracy in an analytical laboratory:

Step 1: Identify the Analyte or Property of Interest
- Determine the specific analyte or property you want to measure or analyze using the SRM.
- SRMs are available for a wide range of substances and properties, including chemical composition and physical properties.

Step 2: Select an Appropriate SRM
- Choose an SRM that is relevant to your analytical needs and is certified for the analyte or property of interest.
- Ensure that the SRM's certification matches your laboratory's requirements and measurement range.

Step 3: Acquire the SRM
- Obtain the selected SRM from a recognized supplier or authority, such as the National Institute of Standards and Technology (NIST) in the United States or equivalent organizations in other countries.
- Ensure that the SRM is properly stored and handled according to recommended procedures.

Step 4: Prepare SRM Samples
- If the SRM is not provided in a suitable form for your analysis, prepare SRM samples according to the instructions provided with the SRM certificate.
- Take care to follow recommended procedures for sample preparation to avoid introducing errors.

Step 5: Perform Measurements
- Perform measurements on the SRM samples using your analytical method or instrument.
- Record the measurement results, including instrument parameters, conditions, and any necessary calibration data.

Step 6: Compare Results to SRM Certificate
- Compare your measurement results to the certified values and uncertainties provided in the SRM certificate.
- Assess the accuracy and precision of your measurements by evaluating how closely your results match the SRM values.

Step 7: Calculate Bias and Uncertainty
- Calculate the bias (systematic error) by determining the difference between your measured values and the SRM-certified values.
- Calculate the uncertainty associated with your measurements based on your laboratory's practices and any additional factors specific to your analysis.

Step 8: Validate and Verify Your Analytical Method
- Use the SRM analysis to validate and verify the accuracy and precision of your analytical method or instrument.
- Ensure that your method meets acceptable criteria for accuracy, precision, and bias.

Step 9: Document and Report
- Document all details of the SRM analysis, including measurement results, bias calculations, uncertainties, and any corrective actions taken.
- Include this information in your laboratory's SOPs and documentation.

Step 10: Continuous Monitoring and Improvement
- Continuously monitor the performance of your analytical method or instrument using SRMs.
- Periodically re-analyze SRMs to ensure that your measurements remain accurate and reliable.

The use of SRMs is essential for establishing and maintaining the accuracy of analytical measurements in a laboratory. By following these steps and regularly incorporating SRM analysis into your quality control procedures, you can ensure that your laboratory produces accurate and reliable data.

5.2.4 RECOVERY EXPERIMENTS

A recovery experiment, also known as a spike recovery experiment, is a valuable procedure in analytical chemistry used to assess the accuracy and reliability of an analytical method. It involves the addition of a known quantity of the target analyte to a sample, followed by the analysis of the spiked sample to determine the extent to which the added analyte is recovered. Here are the steps to establish a recovery experiment analysis for improved accuracy in an analytical laboratory:

Step 1: Define the Objective
- Clearly define the objective of the recovery experiment. Determine which analyte you want to assess, and specify the target recovery percentage you aim to achieve.

Step 2: Prepare the Spiking Solution
- Prepare a spiking solution containing a known quantity of the target analyte. Ensure that the spiking solution is accurately prepared and stable.
- Document the concentration and composition of the spiking solution.

Step 3: Prepare Sample Matrices
- Obtain a set of sample matrices that closely resemble the type of samples you typically analyze in your laboratory.
- Ensure that the sample matrices are representative and free from interference with the target analyte.

Step 4: Spike the Samples
- Add a known volume or mass of the spiking solution (from Step 2) to each sample matrix. This is typically done in duplicate or triplicate for each sample.
- The amount added should be within the expected range of analyte concentrations in your samples.

Step 5: Perform Measurements
- Analyze the spiked samples using your analytical method or instrument. Perform the analysis in the same manner as you would for routine sample analysis.
- Record the measurement results, including instrument parameters and conditions.

Step 6: Calculate Recovery
- Calculate the recovery of the target analyte for each spiked sample. This is done by comparing the measured concentration of the spiked analyte to the expected concentration based on the amount added.
- Calculate recovery as a percentage using the formula:

$$\text{Recovery}(\%) = (\text{Measured Concentration / Expected Concentration}) * 100.$$

Step 7: Statistical Analysis
- Perform statistical analysis to assess the accuracy and precision of the recovery experiment.
- Calculate the mean recovery and the SD of recovery values for the spiked samples.
- Evaluate the significance of deviations from the target recovery percentage.

Step 8: Interpret the Results
- Interpret the recovery experiment results in the context of your laboratory's quality control standards and industry guidelines.
- Assess whether the method is accurate and provide consistent recovery within an acceptable range.

Step 9: Identify and Address Issues
- If recovery results are outside acceptable limits, investigate potential sources of error or bias in your analytical method.
- Implement corrective actions to improve method accuracy and reliability.

Step 10: Document and Report
- Document all details of the recovery experiment, including spiking levels, measurement results, statistical analysis, and any corrective actions taken.
- Include this information in your laboratory's SOPs and documentation.

Step 11: Continuous Monitoring and Improvement
- Continuously monitor the accuracy and precision of your analytical method by periodically performing recovery experiments.
- Use recovery experiments as a tool for ongoing method improvement and validation.

Recovery experiments are essential for ensuring the accuracy and reliability of analytical methods in your laboratory. By following these steps and regularly incorporating recovery experiments into your quality control procedures, you can identify and address issues that may affect the accuracy of your analytical measurements.

5.3 HOW TO DO THE STATISTICS FOR REPEATABILITY AND REPRODUCIBILITY

5.3.1 INTRA-ASSAY VARIATION

Intra-assay variation analysis, also known as within-assay or within-run variation analysis, is a crucial procedure in analytical laboratories to assess the repeatability or precision of an analytical method within a single run or batch of samples. Here are the steps to establish an intra-assay variation analysis for improved repeatability in an analytical laboratory:

Step 1: Define the Objective
- Clearly define the objective of the intra-assay variation analysis. Determine which analyte or parameter you want to assess for repeatability and precision within a single run or batch.

Step 2: Prepare Sample Replicates
- Select a set of representative samples or control materials that contain the analyte of interest at known concentrations.
- Prepare multiple replicates (usually at least three) of each sample to be analyzed within the same run or batch.

Step 3: Perform Measurements
- Analyze the prepared sample replicates using your analytical method or instrument within the same run or batch.
- Ensure that the measurements are performed under consistent conditions, including instrument settings and calibration.

Step 4: Record Measurement Results
- Record the measurement results for each replicate, including raw data, instrument parameters, and any relevant conditions.

Step 5: Calculate Variation Metrics
- Calculate variation metrics, such as the mean, SD, and CV, for the measurement results of each sample replicate.
- The mean represents the central tendency, the SD quantifies the spread, and the CV expresses the relative variation as a percentage of the mean.

Step 6: Statistical Analysis
- Perform statistical analysis to assess the repeatability of the intra-assay measurements.
- Calculate the mean, SD, and CV for each set of sample replicates.
- Evaluate the significance of variation by comparing CV values to predetermined acceptance criteria or quality control standards.

Step 7: Interpret the Results
- Interpret the intra-assay variation analysis results in the context of your laboratory's quality control standards and industry guidelines.
- Assess whether the method demonstrates acceptable repeatability within the same run or batch.

Step 8: Identify and Address Issues
- If variation results are outside acceptable limits, investigate potential sources of error or bias in your analytical method or instrument within the same run or batch.
- Implement corrective actions to improve repeatability.

Step 9: Document and Report
- Document all details of the intra-assay variation analysis, including measurement results, variation metrics, statistical analysis, and any corrective actions taken.
- Include this information in your laboratory's SOPs and documentation.

Step 10: Continuous Monitoring and Improvement
- Continuously monitor the repeatability of your analytical method by periodically performing intra-assay variation analyses within the same run or batch.
- Use intra-assay variation analysis as a tool for ongoing method validation and improvement.

Intra-assay variation analysis is essential for assessing the precision and repeatability of measurements within a single run or batch in an analytical laboratory. By following these steps and regularly conducting intra-assay variation analyses, you can identify and address issues that may affect the repeatability and reliability of your measurements.

5.3.2   Control Charts for Replicates

Control charts for replicate analysis are a valuable tool in analytical laboratories for monitoring the repeatability and precision of measurements taken on replicate samples over time. These charts help identify trends or deviations in replicate data. Here are the steps to establish control charts for replicates analysis for improved repeatability in an analytical laboratory:

Step 1: Define the Objective
- Clearly define the objective of using control charts for replicates analysis. Determine which parameter or analyte you want to monitor for repeatability and precision.

Step 2: Collect Replicate Data
- Collect a series of replicate measurements for the same sample over multiple runs or time intervals.
- Ensure that the replicate data are collected under consistent conditions, including instrument settings and calibration.

Step 3: Calculate Central Tendency and Variability
- Calculate summary statistics for each set of replicate measurements, including the mean (average) and SD or other measures of variability.
- These statistics represent the central tendency and variability of the data for each set of replicates.

Step 4: Choose the Type of Control Chart
- Select the appropriate type of control chart based on the nature of your data and the objective of your analysis. Common control charts for replicates analysis include the individual moving range (I-MR) chart and the X-bar and R chart.
- The I-MR chart is used when you have individual measurements, while the X-bar and R chart are used when you have subgroup measurements with multiple replicates.

Step 5: Calculate Control Limits
- Calculate control limits for the selected control chart. Control limits consist of the CL, UCL, and LCL.
- For an I-MR chart, the control limits are typically calculated based on the SD of the replicate measurements.
- For an X-bar and R chart, the control limits are calculated separately for the X-bar (mean) and R (range) charts.

Step 6: Plot the Control Chart
- Create the control chart by plotting the replicate data points and control limits on the chart.
- Use appropriate software or tools to generate the control chart, which should display the CL, UCL, LCL, and data points over time.

Step 7: Interpret the Control Chart
- Regularly update the control chart with new replicate data points. Examine the chart for trends, patterns, or points that fall outside the control limits.
- Interpret the control chart to assess whether the variations in replicate measurements are within normal process variability or if they indicate special causes of variation.

Step 8: Take Corrective Action
- If the control chart indicates the presence of special causes of variation, investigate and address the root causes.
- Implement corrective actions to improve measurement repeatability and precision.

Step 9: Document and Report
- Document all details of the control chart analysis, including the control chart plots, control limits, and any corrective actions taken.
- Include this information in your laboratory's SOPs and documentation.

Step 10: Continuous Monitoring and Improvement
- Continuously monitor the repeatability and precision of replicate measurements using control charts.
- Periodically update the control chart and use it as a tool for ongoing process improvement and validation.

Control charts for replicates analysis are essential for monitoring the repeatability of measurements in an analytical laboratory. By following these steps and regularly using control charts, you can identify and address issues that may affect the repeatability and reliability of your measurements.

5.3.3 Method Validation

Method validation is a critical process in analytical chemistry that ensures the reliability, repeatability, and accuracy of an analytical method. It involves a series of experiments and assessments to demonstrate that a method is fit for its intended purpose. Here are the steps to establish method validation analysis for improved repeatability in an analytical laboratory:

Step 1: Define the Objective
- Clearly define the objective of method validation. Determine the specific analyte or parameter you want to validate, the intended use of the method, and the regulatory or quality standards that must be met.

Step 2: Review Method Documentation
- Review and gather all available documentation related to the analytical method, including the method's SOP, instrument manuals, and any relevant literature.

Step 3: Select Validation Parameters
- Identify the key parameters to be validated. These parameters may include accuracy, precision, linearity, specificity, LOD, LOQ, and robustness.
- Select the parameters based on the requirements of the analytical method and the regulatory guidelines that apply to your laboratory.

Step 4: Prepare Validation Samples
- Prepare a set of validation samples that represent a range of analyte concentrations or conditions.
- Ensure that these samples are well-characterized and accurately prepared and that they cover the intended working range of the method.

Step 5: Perform Validation Experiments
- Conduct a series of validation experiments, each designed to assess a specific parameter. These experiments may include:
  - Accuracy experiments: Compare method results to a reference method or known reference materials.
  - Precision experiments: Evaluate repeatability (within-run) and intermediate precision (between-run).
  - Linearity experiments: Assess the linearity of the method's response over a range of concentrations.
  - Specificity experiments: Verify the method's ability to distinguish the analyte from potential interferences.
  - LOD and LOQ determination: Calculate the method's LOD and quantification.
  - Robustness experiments: Assess the method's robustness by deliberately varying method parameters.

Step 6: Analyze and Record Data
- Analyze the validation samples using the analytical method, recording all measurement results and relevant data.
- Maintain detailed records of experimental conditions, sample preparation, and instrument settings.

Step 7: Perform Data Analysis
- Analyze the data collected during validation experiments using appropriate statistical methods.
- Calculate validation parameters such as recovery, precision (SD and RSD), regression coefficients, and other relevant metrics.

Step 8: Compare Results to Acceptance Criteria
- Compare the validation results to predetermined acceptance criteria or criteria specified in regulatory guidelines.
- Determine whether the method meets the required criteria for accuracy, precision, linearity, specificity, LOD, LOQ, and robustness.

Step 9: Document and Report
- Document all details of the method validation process, including experimental procedures, data analysis, and results.
- Prepare a comprehensive method validation report that summarizes the findings, compliance with acceptance criteria, and any recommendations or corrective actions needed.

Step 10: Peer Review and Approval
- Subject the method validation report to peer review within the laboratory to ensure accuracy and completeness.
- Obtain approval from relevant personnel or authorities, including laboratory management and quality assurance.

Step 11: Continuous Monitoring and Maintenance
- Continuously monitor the performance of the validated method by implementing routine quality control procedures.
- Periodically revalidate the method when significant changes occur, such as instrument replacement or modification.

Method validation is an ongoing process that ensures the repeatability and reliability of analytical methods in an analytical laboratory. By following these steps and adhering to regulatory guidelines, you can establish and maintain validated methods that provide accurate and precise results for your laboratory's analytical needs.

5.3.4 RING TESTS

Ring tests, also known as proficiency testing or interlaboratory comparison, are an important tool in analytical laboratories for assessing the repeatability and accuracy of analytical methods across different laboratories. These tests involve multiple laboratories analyzing the same samples to evaluate their performance. Here are the steps to establish ring tests data analysis for improved repeatability in an analytical laboratory:

Step 1: Define the Objective
- Clearly define the objective of conducting ring tests. Determine which parameters or analytes you want to assess for repeatability and accuracy across multiple laboratories.

Step 2: Select Ring Test Participants
- Identify and invite laboratories to participate in the ring test. Ensure that a diverse group of laboratories with different capabilities and expertise is included.
- Specify the number of participating laboratories based on the requirements of the ring test.

Step 3: Prepare Test Samples
- Prepare a set of test samples that contain the analytes of interest at known concentrations. These samples should be representative of the types of samples routinely analyzed in your laboratory.
- Ensure that the samples are well-characterized and stable.

Step 4: Distribute Test Samples
- Distribute the test samples to all participating laboratories, along with detailed instructions on sample handling, storage, and analysis.
- Specify the deadline for completing the analysis and returning the results.

Step 5: Conduct the Analysis
- Each participating laboratory analyzes the test samples using their own analytical methods and equipment following their standard procedures.
- Ensure that laboratories maintain consistency in sample handling, analysis conditions, and calibration.

Step 6: Collect the Data
- Collect the analysis results from all participating laboratories, including raw data, measurement uncertainties, and any other relevant information.
- Compile the data in a standardized format for analysis.

Step 7: Calculate Statistics
- Perform statistical analysis on the collected data to assess the repeatability and accuracy of the participating laboratories' results.
- Calculate summary statistics, such as mean, SD, RSD, and z-scores.

Step 8: Identify Outliers
- Identify any outlier results that significantly deviate from the expected values or from the results of other laboratories.
- Investigate the potential causes of outliers, such as method deviations or instrument issues.

Step 9: Generate a Ring Test Report
- Prepare a comprehensive ring test report that summarizes the findings, including statistical analyses, outlier identification, and any corrective actions taken.
- Include a comparison of the results from different laboratories and an assessment of the overall repeatability and accuracy.

Step 10: Share and Review Results
- Share the ring test report with all participating laboratories and stakeholders.
- Encourage laboratories to review their results and implement corrective actions if needed.

Step 11: Continuous Improvement
- Use the findings from the ring test to identify areas for improvement in your laboratory's analytical methods, procedures, and quality control processes.
- Participate in future ring tests to continuously assess and improve your laboratory's performance.

Ring tests provide valuable insights into the repeatability and accuracy of analytical methods across different laboratories. By following these steps and actively participating in ring tests, your laboratory can improve its analytical performance, identify areas for improvement, and ensure the reliability of your analytical results.

5.3.5 REGRESSION ANALYSIS

Regression analysis is a statistical technique used to model relationships between variables, make predictions, and assess the precision, accuracy, reproducibility, and repeatability of analytical measurements in an analytical laboratory. Here are the steps to establish regression analysis for improved precision, accuracy, reproducibility, and repeatability in an analytical laboratory:

Step 1: Define the Objective
- Clearly define the objective of the regression analysis. Determine the specific parameters or variables you want to study, model, or predict, and specify the research question or hypothesis.

Step 2: Collect Data
- Collect the data relevant to your research objective. Ensure that the data are accurate, complete, and representative of the population or system you are studying.
- Maintain detailed records of the data collection process, including sample preparation and measurement conditions.

Step 3: Choose the Regression Model
- Select the appropriate regression model that fits your data and research question. Common types of regression models include linear regression, multiple regression, polynomial regression, and logistic regression.
- Consider the nature of your data (continuous or categorical), the relationship you expect to find, and the assumptions of the chosen regression model.

Step 4: Prepare Data
- Prepare the data for analysis by cleaning and transforming it if necessary. Handle missing data, outliers, and any other data issues that may affect the accuracy of the regression analysis.

- Normalize or standardize variables as needed to ensure that they are on the same scale.

Step 5: Perform Regression Analysis
- Use appropriate software or statistical tools to perform the regression analysis. Fit the chosen regression model to your data.
- Calculate regression coefficients, including intercepts and slopes, and assess their statistical significance.

Step 6: Evaluate Model Fit
- Evaluate the goodness of fit of the regression model by examining measures such as the coefficient of determination (R-squared), adjusted R-squared, and residual analysis.
- Assess whether the model adequately explains the variability in the data.

Step 7: Interpret Results
- Interpret the results of the regression analysis. Understand the meaning and implications of the regression coefficients and their statistical significance.
- Assess how well the regression model predicts the dependent variable or explains the relationship between variables.

Step 8: Validate and Cross-Validate
- Validate the regression model using appropriate validation techniques, such as hold-out validation, k-fold cross-validation, or bootstrapping.
- Assess the model's performance on new or unseen data to ensure its generalizability.

Step 9: Report and Document
- Prepare a comprehensive report that documents the regression analysis, including the research objective, data collection process, regression model, results, and interpretation.
- Include graphical representations of the regression analysis, such as scatterplots and regression plots.

Step 10: Continuous Monitoring and Improvement
- Continuously monitor the precision, accuracy, reproducibility, and repeatability of analytical measurements by regularly updating and revalidating the regression model.
- Use the regression model as a tool for ongoing quality control and process improvement.

Regression analysis is a powerful tool for assessing and improving the precision, accuracy, reproducibility, and repeatability of analytical measurements in an analytical laboratory. By following these steps and using regression analysis effectively, you can gain valuable insights into your data, make informed decisions, and enhance the reliability of your laboratory's analytical results.

5.3.6 Multivariate Statistics

Multivariate statistical analysis is a powerful approach used in analytical laboratories to simultaneously analyze multiple variables and their relationships to improve precision, accuracy, reproducibility, and repeatability. Here are the steps to establish multivariate statistical analysis for these purposes in an analytical laboratory.

Step 1: Define the Objective
- Clearly define the objective of the multivariate statistical analysis. Determine which variables or parameters you want to study, how they relate to each other, and what specific goals you aim to achieve.

Step 2: Data Collection
- Collect the data relevant to your research objective. Ensure that the data are accurate, complete, and representative of the system or process you are studying.
- Record detailed information about the data collection process, including sample preparation, measurement conditions, and any relevant contextual factors.

Step 3: Data Preparation
- Prepare the data for multivariate analysis. This may involve cleaning the data, handling missing values, and transforming variables to meet the assumptions of the chosen multivariate technique.
- Normalize or standardize variables as needed to ensure they are on the same scale.

Step 4: Choose the Multivariate Technique
- Select the appropriate multivariate technique that best suits your research objective and the nature of your data. Common multivariate techniques include principal component analysis (PCA), factor analysis, discriminant analysis, cluster analysis, and partial least squares regression (PLS).
- Consider the specific goals of your analysis and whether you are interested in dimension reduction, pattern recognition, classification, or regression.

Step 5: Perform Multivariate Analysis
- Use specialized software or statistical tools to perform the chosen multivariate analysis. Apply the technique to your prepared data to extract relevant information and patterns.
- Generate multivariate models, loadings plots, scores plots, or other relevant outputs based on the chosen technique.

Step 6: Interpret Results
- Interpret the results of the multivariate analysis to gain insights into the relationships between variables and patterns within your data.
- Identify any underlying structures, clusters, trends, or outliers that may affect precision, accuracy, reproducibility, or repeatability.

Step 7: Validation and Cross-Validation
- Validate the multivariate models and results using appropriate validation techniques, such as cross-validation or bootstrapping.
- Ensure that the models generalize well to new data and maintain their predictive power.

Step 8: Report and Document
- Prepare a comprehensive report that documents the multivariate statistical analysis, including the research objective, data preparation, chosen technique, results, and interpretation.
- Include visual representations of the multivariate analysis, such as biplots, scatterplots, or clustering dendrograms.

Step 9: Continuous Monitoring and Improvement
- Continuously monitor the precision, accuracy, reproducibility, and repeatability of analytical measurements by regularly updating and validating the multivariate models.
- Use multivariate statistical analysis as a tool for ongoing quality control and process improvement.

Multivariate statistical analysis allows you to uncover hidden patterns and relationships in complex data, leading to improved precision, accuracy, reproducibility, and repeatability in analytical measurements. By following these steps and effectively using multivariate techniques, you can enhance the reliability and insightfulness of your laboratory's analytical results.

5.3.7 Grubbs' Test

Grubbs' test, also known as the Grubbs' outlier test, is a statistical test used to detect and remove outliers in a dataset. Outliers can impact the precision, accuracy, reproducibility, and repeatability of analytical measurements. Here are the steps to perform Grubbs' test on your data in an analytical laboratory:

Step 1: Define the Objective
- Clearly define the objective of using Grubbs' test. Determine which variables or measurements you want to assess for the presence of outliers and why it is important for your analytical process.

Step 2: Data Collection
- Collect the dataset that you want to analyze for outliers. Ensure that the data are complete and accurate.
- Organize the data in a clear and structured format, including any relevant contextual information.

Step 3: Calculate Descriptive Statistics
- Calculate basic descriptive statistics for your dataset, including the mean and SD. These statistics will help in identifying outliers.
- Determine the critical value (α) based on the desired significance level (e.g., 0.05) for the Grubbs' test.

Step 4: Perform Grubbs' Test
- Set up the null and alternative hypotheses for the Grubbs' test:
  - Null Hypothesis (H0): There are no outliers in the dataset.
  - Alternative Hypothesis (Ha): There is at least one outlier in the dataset.
    Calculate the test statistic (G) for each data point using the formula for Grubbs' test:

$$G = \left(| X - \bar{X} | \right) / S$$

  where
    X is the data point being tested.
    $\bar{X}$ is the sample mean.
    S is the sample SD.
- Calculate the critical value (G_critical) using the formula for Grubbs' critical value based on the significance level (α) and the number of data points (n) in your dataset.
- Compare the calculated test statistic (G) for each data point to the critical value (G_critical). If the calculated G is greater than G_critical, the data point is considered an outlier.

Step 5: Identify Outliers
- Identify the data points that are identified as outliers based on the Grubbs' test results.

Step 6: Document and Report
- Document the results of the Grubbs' test, including which data points are identified as outliers and their corresponding test statistics.
- Prepare a report summarizing the findings and the impact of outliers on precision, accuracy, reproducibility, and repeatability in your analytical measurements.

Step 7: Decide on Handling Outliers
- Decide on how to handle the identified outliers. Depending on the context and nature of the data, you may choose to:
  - Remove the outliers from the dataset.
  - Investigate and correct potential errors that led to outliers.
  - Report the outliers separately if they are valid data points with a specific meaning.

Step 8: Continuous Monitoring
- Implement Grubbs' test as a part of your quality control process and regularly monitor for outliers in your analytical data.
- Use the insights gained from Grubbs' test to improve data quality and enhance precision, accuracy, reproducibility, and repeatability in your analytical laboratory.

Grubbs' test is a valuable tool for identifying and managing outliers in your data, which can significantly impact the quality and reliability of analytical measurements.

5.3.8 GR&R

GR&R is a statistical method used to assess the precision, accuracy, reproducibility, and repeatability of measurement systems, such as analytical instruments, in an analytical laboratory. Here are the steps to perform a GR&R study for improved precision, accuracy, reproducibility, and repeatability:

Step 1: Define the Objective
- Clearly define the objective of the GR&R study. Determine which measurement system or instrument you want to assess and why it is important for your analytical process.

Step 2: Select the Measurement Device
- Choose the measurement device or instrument that you want to evaluate. Ensure that it is representative of the instruments used in your laboratory for the specific measurement.

Step 3: Define the Characteristics
- Define the specific characteristics or parameters that you will measure using the selected instrument. These characteristics should be relevant to your analytical process and research objectives.

Step 4: Collect Samples or Artifacts
- Prepare a set of samples or artifacts that will be used for the measurements. These should cover a range of values and conditions for the selected characteristics.
- Ensure that the samples are stable, well-characterized, and representative of the types of samples typically analyzed in your laboratory.

Step 5: Select Operators
- Choose a group of operators or individuals who will perform the measurements using the selected instrument. Ensure that the operators are trained and competent in using the instrument.

Step 6: Design the Experiment
- Create a measurement plan that outlines how the GR&R study will be conducted. Specify the number of trials, the order of measurements, and any other relevant details.
- Randomize the order of measurements to reduce potential bias.

Step 7: Perform Measurements
- Conduct the measurements according to the measurement plan. Each operator should independently measure the selected characteristics on the samples or artifacts.
- Ensure that the measurements are performed under consistent and controlled conditions.

Step 8: Record Data
- Record all measurement results, including values obtained by each operator for each sample or artifact.
- Organize the data in a structured format, such as a data table.

Step 9: Calculate Variance Components
- Use statistical software or tools to calculate the variance components for the GR&R study. This involves partitioning the total variability into components due to repeatability (within-operator variation) and reproducibility (between-operator variation).

Step 10: Assess the Results
- Evaluate the results of the GR&R study. Calculate the percent contribution of repeatability and reproducibility to the total variability.
- Determine the precision, accuracy, and capability of the measurement system based on the calculated variance components.

Step 11: Interpret the Results
- Interpret the results in the context of your laboratory's analytical process and quality requirements. Assess whether the measurement system meets the desired precision, accuracy, reproducibility, and repeatability standards.
- Identify areas for improvement and take corrective actions if needed.

Step 12: Document and Report
- Prepare a comprehensive report that documents the GR&R study, including the objective, measurement plan, results, and interpretation.
- Include recommendations for improving the measurement system if deficiencies are identified.

Step 13: Continuous Monitoring and Improvement
- Implement GR&R as a part of your ongoing quality control process. Regularly assess and monitor the measurement system's performance.
- Use the insights gained from GR&R studies to enhance precision, accuracy, reproducibility, and repeatability in your analytical laboratory.

GR&R studies are essential for ensuring the reliability of measurement systems in analytical laboratories. By following these steps and conducting regular GR&R assessments, you can identify and address issues that may affect the quality of your measurements and improve the overall analytical process.

5.3.9 YOUDEN PLOT

A Youden plot is a graphical tool used in analytical chemistry to assess the precision, accuracy, reproducibility, and repeatability of a measurement method or instrument. It helps in identifying potential sources of variation and determining the optimal conditions for measurements. Here are the steps to create a Youden plot for improved precision, accuracy, reproducibility, and repeatability in an analytical laboratory:

Step 1: Define the Objective
- Clearly define the objective of creating a Youden plot. Determine which measurement method, instrument, or analytical condition you want to assess and improve.

Step 2: Select the Analyte or Parameter
- Choose the specific analyte, parameter, or characteristic that you will measure or analyze using the selected measurement method or instrument.

Step 3: Prepare Samples or Standards
- Prepare a set of samples or standards that cover a range of concentrations or values for the selected analyte or parameter.
- Ensure that the samples or standards are well-characterized and representative of the types of samples routinely analyzed in your laboratory.

Step 4: Collect Data
- Perform measurements or analyses on the prepared samples or standards using the measurement method or instrument you want to assess.
- Record all measurement results, including values obtained for each sample or standard.

Step 5: Calculate Variability
- Calculate the variability in the measurements for each sample or standard. This can be done by calculating the SD or RSD for each concentration or value.
- Assess the precision and repeatability of the measurements.

Step 6: Create a Youden Plot
- Create a Youden plot, also known as a Youden pair plot or concentration–response plot. This plot typically has concentration or value on the x-axis and measurement results (e.g., instrument response) on the y-axis.
- Plot the data points for each sample or standard on the Youden plot.

Step 7: Interpret the Youden Plot
- Interpret the Youden plot to assess the accuracy, precision, and linearity of the measurement method or instrument.
- Look for trends, patterns, or deviations from linearity that may indicate measurement bias or nonlinearity.

Step 8: Identify Optimal Conditions
- Use the Youden plot to identify the optimal conditions for measurements. Determine the concentration or value range where the measurement method or instrument performs best in terms of accuracy and precision.
- Consider adjustments or calibration if necessary.

Step 9: Document and Report
- Prepare a report that documents the Youden plot analysis, including the objective, data, plot, interpretation, and recommendations for improvement.
- Include any corrective actions or adjustments to enhance precision, accuracy, reproducibility, and repeatability.

Step 10: Continuous Monitoring and Improvement
- Implement regular Youden plot assessments as part of your laboratory's quality control process. Continuously monitor the performance of the measurement method or instrument.
- Use the insights gained from Youden plot analyses to improve measurement conditions and maintain quality standards.

Youden plots are valuable tools for evaluating and optimizing measurement methods and instruments in analytical laboratories. By following these steps and regularly utilizing Youden plots, you can enhance the precision, accuracy, reproducibility, and repeatability of your laboratory's measurements and analyses.

5.4 EXAMPLES OF STATISTICS IN USE

The following are examples of various statistical tools used to develop increased precision, accuracy, repeatability, and reproducibility in analytical tests used in laboratories. Each method will have examples, and each can be used for improved precision, accuracy, repeatability, and reproducibility of the data.

5.4.1 SD ANALYSIS

Example: An analytical lab measures the concentration of a chemical in water samples using a chromatography technique.

Goal: Improve the precision, accuracy, repeatability, and reproducibility of the measurements.

Steps:

1. Identify potential sources of variation: Brainstorm factors that could affect the measurements, such as
   - Instrumentation: Calibration, stability, and temperature fluctuations.
   - Reagents: Purity, storage conditions, and batch-to-batch variability.
   - Analysis technique: Injection volume, flow rate, and data acquisition settings.
   - Analyst technique: Pipetting accuracy and handling procedures.
   - Environmental factors: Temperature, humidity, and vibration.
2. Perform SD analysis for different levels of variation:
   - Repeatability: Measure the same sample multiple times under identical conditions (same analyst, instrument, reagents, etc.). Calculate the SD of these results. This reflects the short-term precision of the method.
   - Intermediate precision: Measure the same sample multiple times over a longer period (different days, analysts, reagent batches, etc.). Calculate the SD of these results. This reflects the within-laboratory reproducibility.

- Reproducibility: Measure the same sample in different laboratories using the same method. Calculate the SD of these results. This reflects the between-laboratory reproducibility.

3. Analyze the results: Compare the SDs from each level of variation. A large SD indicates high variability and potential issues with precision or accuracy.
4. Identify improvement targets: Based on the SD analysis, focus on the factors contributing to the largest variations. You may implement changes like:
   - Calibrating instruments more frequently.
   - Using high-quality, consistent reagents.
   - Standardizing analyst procedures.
   - Controlling environmental factors.
   - Validating and optimizing the analysis technique.
5. Repeat the analysis: After implementing changes, repeat the SD analysis for each level of variation. Monitor the reduction in SD to assess the effectiveness of the improvements.

Benefits:

- Improved data quality: Lower SDs lead to more reliable and precise data, reducing uncertainty and improving confidence in results.
- Enhanced accuracy: Identifying and reducing bias sources can lead to more accurate measurements that are closer to the true values.
- Streamlined workflows: Standardized procedures and reduced variability can improve efficiency and productivity in the lab.
- Meeting regulatory requirements: Many analytical labs need to demonstrate adherence to quality standards for data accuracy and precision.

SD analysis is a common statistical technique used to assess precision, accuracy, repeatability, and reproducibility in analytical laboratories. Here are some examples of how SD analysis can be applied in different examples:

Precision Assessment
   Example: Evaluating the precision of a balance used for weighing substances.
   Procedure: Weigh the same substance multiple times (e.g., 10 times) using the same balance under the same conditions.
   Analysis: Calculate the SD of the measured weights. A smaller SD indicates higher precision.

Accuracy Assessment
   Example: Assessing the accuracy of a spectrophotometer for measuring the absorbance of a known concentration of a chemical compound.
   Procedure: Measure the absorbance of a standard solution with a known concentration multiple times (e.g., 5 times).
   Analysis: Calculate the mean absorbance and the SD of the measurements. Compare the mean absorbance to the known concentration to assess accuracy.

Repeatability Assessment
   Example: Evaluating the repeatability of a pH meter within a single laboratory.
   Procedure: Measure the pH of the same buffer solution multiple times (e.g., five times) using the same pH meter.
   Analysis: Calculate the SD of the pH measurements. A smaller SD indicates higher repeatability.

Reproducibility Assessment
   Example: Assessing the reproducibility of a chromatography method among different laboratories.

Procedure: Multiple laboratories analyze the same set of samples using the same method.

Analysis: Calculate the SD of the results obtained by different laboratories for each sample. A smaller SD indicates better reproducibility.

Linearity Assessment

Example: Evaluating the linearity of an analytical instrument's response to varying concentrations of a substance.

Procedure: Analyze a series of standard solutions with known concentrations at different levels.

Analysis: Plot the instrument's response (e.g., peak area) against the known concentrations and calculate the SD of the residuals (the differences between observed and expected responses). A small SD of residuals indicates good linearity.

Matrix Effect Assessment

Example: Assessing the impact of different sample matrices on the accuracy of an analytical method.

Procedure: Analyze a set of samples with the same analyte concentration but different matrices (e.g., different types of soil samples).

Analysis: Calculate the SD of the measured analyte concentrations for each matrix type. Differences in SDs indicate matrix effects.

In each of these examples, SD analysis helps assess the variation and quality of analytical measurements. Smaller SDs generally indicate higher precision, repeatability, and reproducibility. By calculating and analyzing SDs, laboratories can identify areas for improvement and ensure the reliability of their analytical results.

5.4.2   ANOVA

SD analysis is a valuable tool for assessing precision and repeatability, but ANOVA takes things further by statistically identifying and quantifying the impact of multiple factors on analytical outcomes. This allows for a more comprehensive understanding of variability and targeted improvement strategies.

Example: An analytical lab measures the vitamin C content in fruit juice samples using high-performance liquid chromatography (HPLC). They want to investigate the influence of three factors on measurement variability:

- Fruit type: Orange, grapefruit, and apple
- Analyst: Analyst A and Analyst B
- Instrument: Instrument 1 and Instrument 2

Steps:

1. Design the experiment: Choose an appropriate ANOVA design based on the factors and number of levels. In this case, a three-way ANOVA with fixed effects is suitable.
2. Perform the analysis: Run the experiment, collecting replicate measurements for each combination of factors.
3. Analyze the data: Use statistical software to perform the ANOVA. The analysis will generate an ANOVA table with F-statistics and p-values for each factor and their interactions.

Interpreting the Results:

- Significant F-statistic: Indicates that the factor has a statistically significant effect on the variability of the measurements. A higher F-value suggests a stronger effect.

TABLE 5.1
Example ANOVA Table

Source of Variation	Degrees of Freedom	Mean Square	F-Statistic	p-Value
Fruit type	2	100	5.00	0.01
Analyst	1	40	2.00	0.10
Instrument	1	20	1.00	0.30
Fruit type x analyst	2	15	0.75	0.45
Error	15	5	-	-

- Nonsignificant F-statistic: Suggests that the factor does not have a significant impact on variability.
- Interactions: ANOVA can also reveal how factors interact with each other. Significant interaction terms indicate that the effect of one factor depends on the level of another factor (Table 5.1).

Insights and Improvement Strategies:

- In this example, fruit type has a significant influence on measurement variability. Further analysis could explore specific differences between fruits.
- While analyst and instrument effects are not significant individually, the interaction term might warrant investigation. Perhaps specific analyst–instrument combinations lead to higher variability.
- Based on the findings, the lab could focus on standardizing sample preparation for different fruit types and potentially investigating factors contributing to the analyst–instrument interaction.

Benefits of ANOVA in Analytical Labs:

- Identifies key sources of variability: Pinpoints factors with the greatest impact on precision and accuracy.
- Quantifies the effect of different factors: Allows for objective comparison of factor contributions.
- Guides targeted improvement strategies: Enables efficient efforts to optimize analytical performance.
- Supports data-driven decision-making: Provides robust statistical evidence for process improvement.

ANOVA is a powerful statistical technique used to assess variation and compare means in different groups or conditions. Here are some examples of how ANOVA can be applied in analytical laboratories to evaluate precision, accuracy, repeatability, and reproducibility:

Precision Assessment – One-Way ANOVA
Example: Evaluating the precision of three different laboratory balances used for weighing substances.
Procedure: Weigh the same substance multiple times (e.g., 10 times) using each of the three balances.
Analysis: Perform a one-way ANOVA to compare the means of the measurements obtained from the three balances. If the p-value is significant, it suggests differences in precision among the balances.

Accuracy Assessment – Two-Way ANOVA

Example: Assessing the accuracy of a spectrophotometer for measuring the absorbance of a known concentration of a chemical compound under different temperature conditions.

Procedure: Measure the absorbance of a standard solution with a known concentration at two different temperature settings (e.g., room temperature and elevated temperature).

Analysis: Conduct a two-way ANOVA with temperature and concentration as factors to assess whether temperature significantly affects accuracy.

Repeatability Assessment – One-Way ANOVA

Example: Evaluating the repeatability of pH measurements using three different pH meters within a single laboratory.

Procedure: Measure the pH of the same buffer solution multiple times (e.g., five times) using each of the three pH meters.

Analysis: Perform a one-way ANOVA to compare the means of the pH measurements obtained from the three meters. If the p-value is significant, it indicates differences in repeatability among the meters.

Reproducibility Assessment – Two-Way ANOVA

Example: Assessing the reproducibility of a chromatography method among different laboratories and under varying column conditions.

Procedure: Multiple laboratories analyze the same set of samples using the same method with two different column types (e.g., Column A and Column B).

Analysis: Conduct a two-way ANOVA with laboratories and column types as factors to assess whether differences in laboratories and column types significantly affect reproducibility.

Linearity Assessment – Simple Linear Regression ANOVA

Example: Evaluating the linearity of an analytical instrument's response to varying concentrations of a substance.

Procedure: Analyze a series of standard solutions with known concentrations at different levels and measure the instrument's response (e.g., peak area).

Analysis: Perform a simple linear regression analysis and examine the ANOVA table to assess whether the relationship between concentration and instrument response is significant.

Matrix Effect Assessment – Two-Way ANOVA

Example: Assessing the impact of different sample matrices on the accuracy of an analytical method.

Procedure: Analyze a set of samples with the same analyte concentration but different matrices (e.g., different types of food samples) using two different instruments (Instrument A and Instrument B).

Analysis: Conduct a two-way ANOVA with sample matrices and instruments as factors to assess whether both factors significantly affect accuracy.

In each of these examples, ANOVA is used to examine the variation in measurement results and determine whether factors such as instruments, conditions, or sample types significantly impact precision, accuracy, repeatability, or reproducibility in analytical laboratories. ANOVA helps identify sources of variability and provides valuable insights for quality control and improvement efforts.

5.4.3 RSD

RSD, also known as CV, is a powerful tool for understanding and comparing variability in analytical measurements. It expresses the SD as a percentage of the mean, making it independent of the units of measurement and facilitating comparisons across different analytes and methods.

Benefits of using RSD

- Simple to calculate and interpret.
- Unitless, allowing comparison across different analytes and methods.
- Provides insights into different types of variability (repeatability, reproducibility, accuracy).
- Helps monitor and improve analytical performance.

Limitations of RSD

- Not sensitive to systematic errors that bias all measurements in the same direction.
- Assumes normality of data, which may not always be the case in analytical measurements.

RSD is a valuable tool for analytical labs to assess and improve precision, accuracy, repeatability, and reproducibility. By analyzing RSD at different levels (repeat, intermediate, and reproducibility), labs can identify sources of variability and implement targeted strategies for optimization.

Here is how RSD can be used to analyze precision, accuracy, repeatability, and reproducibility:

Precision

Precision Repeatability RSD (RSDr): Calculated using replicate measurements from the same sample, same analyst, same instrument, and within a short timeframe. A low RSDr indicates high repeatability, meaning consistent results within a single analysis run.

Example: Evaluating the precision of a balance used for weighing substances.

Procedure: Weigh the same substance multiple times (e.g., 10 times) using the same balance under the same conditions.

Analysis: Calculate the RSD for the repeated measurements. A smaller RSD indicates higher precision.

Intermediate Precision RSD (RSDi): Calculated using replicate measurements from the same sample within a single laboratory but over a longer timeframe, potentially involving different analysts, instruments, or reagent batches. A low RSDi compared to RSDr indicates good within-laboratory reproducibility.

Accuracy

RSD Compared to a Known Reference Value: If the true value of the analyte is known (e.g., from a CRM), the RSD can be calculated by comparing the measured values to the reference value. A low RSD suggests good accuracy, meaning measurements are close to the true value.

Example: Assessing the accuracy of a spectrophotometer for measuring the absorbance of a known concentration of a chemical compound.

Procedure: Measure the absorbance of a standard solution with a known concentration multiple times (e.g., five times).

Analysis: Calculate the RSD for the measurements. Assess how closely the RSD aligns with acceptable accuracy criteria.

Reproducibility

Calculated using replicate measurements from the same sample analyzed in different laboratories. A low RSDr indicates good between-laboratory reproducibility, meaning consistent results across different laboratories using the same method.

Example: Assume an analytical lab measures the lead concentration in water samples using atomic absorption spectroscopy. They perform:

- Five replicate measurements on a single sample (RSDr=8%). This indicates good repeatability.

- Ten measurements on the same sample over a week (RSDi = 12%). This suggests lower within-laboratory reproducibility compared to repeatability, highlighting potential variability from other factors within the lab.
- Measurements on the same sample in three different labs (RSDr = 15%). This indicates acceptable between-laboratory reproducibility, meaning the labs obtain relatively consistent results for the analysis.

Another Example: Assessing the reproducibility of a chromatography method among different laboratories.

Procedure: Multiple laboratories analyze the same set of samples using the same method.

Analysis: Calculate the RSD for the results obtained by different laboratories for each sample. A smaller RSD indicates better reproducibility.

The acceptable RSD values for analytical methods can vary depending on various factors like regulatory requirements, analyte complexity, and instrumentation. Always consult relevant guidelines and best practices for your specific field when interpreting RSD values.

It is most useful when applied to express the SD as a percentage of the mean. It is commonly used to assess precision, accuracy, repeatability, and reproducibility in analytical laboratories.

Repeatability

Example: Evaluating the repeatability of pH measurements using the same pH meter within a single laboratory.

Procedure: Measure the pH of the same buffer solution multiple times (e.g., five times) using the same pH meter.

Analysis: Calculate the RSD for the pH measurements. A smaller RSD indicates higher repeatability.

Linearity Assessment – RSD for Linearity

Example: Evaluating the linearity of an analytical instrument's response to varying concentrations of a substance.

Procedure: Analyze a series of standard solutions with known concentrations at different levels.

Analysis: Calculate the RSD of residuals (differences between observed and expected responses) in a linearity plot. Assess how closely the RSD aligns with acceptable linearity criteria.

Matrix Effect Assessment – RSD for Matrix Effect

Example: Assessing the impact of different sample matrices on the accuracy of an analytical method.

Procedure: Analyze a set of samples with the same analyte concentration but different matrices (e.g., different types of food samples) using the same instrument.

Analysis: Calculate the RSD for the measured analyte concentrations for each matrix type. Differences in RSD values indicate matrix effects.

In these examples, the RSD is used to quantify the variation in measurement results as a percentage of the mean. A smaller RSD indicates better precision, repeatability, and reproducibility, while aligning with acceptable accuracy and linearity criteria. RSD is a valuable tool for assessing the quality and reliability of analytical measurements in laboratories.

5.4.4 CONTROL CHARTS

Control charts are graphical tools used to monitor and visualize trends in data, helping to identify out-of-control situations that indicate potential issues with precision, accuracy, repeatability, and reproducibility in analytical labs. Different chart types are used for different purposes:

Precision

X-Bar Chart: Tracks the means of replicate measurements over time. UCL and LCL are set based on the expected variability. Points falling outside the limits indicate potential problems with precision (systematic bias or increased randomness).

Range Chart: Plots the range (difference between highest and lowest values) of replicate measurements. Out-of-control points suggest changes in variability within the analysis run.

Accuracy

Shewhart Chart with Reference Values: Plot individual measurements against a known reference value. UCL and LCL are set based on the desired accuracy. Points outside the limits indicate potential bias in the measurements.

Levey–Jennings Chart: Plots standardized residuals (measurement minus mean divided by SD) over time. Useful for monitoring systematic trends in accuracy over time.

Repeatability and Reproducibility

R Charts: Plot the range of replicate measurements across different analytical runs or batches. Useful for monitoring repeatability within a lab and reproducibility between labs.

Cumulative Sum (CUSUM) Charts: These accumulate positive or negative deviations from the target value, highlighting small shifts in accuracy that might not be evident on an X-bar chart.

Example: Assume an analytical lab measures the pH of a buffer solution using a pH meter. They set control limits for the X-bar and range charts based on historical data. Here is how these charts could help analyze different aspects of performance:

- Increased variability: If the range chart shows out-of-control points, it suggests an increase in variability within the analysis run. This could be due to calibration issues, instrument drift, or changes in the sample matrix.
- Systematic bias: If the X-bar chart shows points consistently above or below the UCL or LCL, it indicates a potential systematic bias in the measurements. This could be due to incorrect reference materials, faulty reagents, or instrument calibration errors.
- Improved performance: Over time, if control charts consistently show points within the control limits, it indicates stable and predictable performance, suggesting good precision and accuracy within the lab.

Benefits of Using Control Charts

- Early detection of potential problems: Out-of-control points prompt timely investigations and corrective actions.
- Visualization of trends: Allows easy identification of shifts in variability or bias over time.
- Data-driven decision-making: Provides quantitative evidence for process improvement strategies.
- Continuous monitoring: Ensures consistent laboratory performance and data quality.

Limitations of Control Charts

- Require historical data to establish control limits.
- Need appropriate chart selection and interpretation skills.
- May not detect all types of errors or subtle changes.

Here are other examples of control charts for different quality attributes in analytical labs:

Precision Control Chart (X-Bar and R Chart)
   Example: Monitoring the precision of pH measurements using the same pH meter.
   Data Collection: Measure the pH of a buffer solution at regular intervals (e.g., hourly) using the same pH meter.

Control Chart Type: X-bar and R chart
   Interpretation: Plot the average pH value (X-bar) and the range of pH measurements (R) on the control chart. Set control limits based on historical data or acceptable precision criteria. Monitor for trends, shifts, or points outside control limits, which may indicate a loss of precision.

Accuracy Control Chart (Bias Chart)
   Example: Monitoring the accuracy of a spectrophotometer for measuring the absorbance of a known concentration of a chemical compound.
   Data Collection: Measure the absorbance of a standard solution with a known concentration at regular intervals.

Control Chart Type: Bias chart
   Interpretation: Calculate the bias (difference between measured and expected values) for each measurement and plot it on the control chart. Set control limits based on acceptable accuracy criteria. Monitor for bias trends or points outside control limits, which may indicate accuracy issues.

Repeatability Control Chart (Individuals Chart)
   Example: Monitoring the repeatability of a balance used for weighing substances.
   Data Collection: Weigh the same substance multiple times (e.g., daily) using the same balance.
   Control Chart Type: Individuals chart
   Interpretation: Plot the individual measurement values on the control chart. Set control limits based on acceptable repeatability criteria. Monitor for outliers or excessive variability in measurements, which may indicate repeatability problems.

Reproducibility Control Chart (P Chart or NP Chart)
   Example: Assessing the reproducibility of a chromatography method among different laboratories.
   Data Collection: Multiple laboratories analyze the same set of samples using the same method.
   Control Chart Type: P chart (proportion of acceptable results) or NP chart (number of acceptable results).
   Interpretation: Calculate the proportion or count of acceptable results for each laboratory and plot it on the control chart. Set control limits based on acceptable reproducibility criteria. Monitor for variations among laboratories that may indicate reproducibility issues.

   These are just a few examples of control charts used to monitor and control precision, accuracy, repeatability, and reproducibility in analytical laboratories. The choice of control chart type and parameters may vary based on the specific quality attributes being monitored and the laboratory's quality control objectives. Control charts help identify deviations from acceptable performance and facilitate timely corrective actions to maintain measurement quality.

5.4.5 Bias Analysis

Bias refers to the systematic difference between a measured value and the true value, regardless of the variability (random error) in the measurements. While precision and reproducibility focus on consistency of results, bias tells us how close those results are to the actual value. Analyzing bias is crucial for ensuring accurate and reliable data in analytical labs.
   Here is how bias analysis relates to different aspects of analytical performance:

Accuracy: Bias directly affects accuracy. A significant bias will lead to inaccurate results, even if the measurements are precise and reproducible. Bias analysis helps identify and quantify the degree of inaccuracy due to systematic errors.

Precision and Repeatability: High precision and repeatability do not guarantee accuracy if there is a bias present. For example, imagine darts consistently hitting the same spot off-center on a bullseye. Their grouping is precise and repeatable, but they are all biased away from the bullseye's center. While analyzing bias, it is important to consider precision and repeatability. High variability can mask underlying bias or make it difficult to quantify its extent. Analyzing both aspects together provides a more comprehensive picture of analytical performance.

Reproducibility: Bias can be present within a single lab (systematic error specific to the lab) or affect multiple labs using the same method (systematic error in the method itself). Reproducibility studies can help identify whether bias is lab-specific or inherent to the method.

Methods for Bias Analysis

Comparison to Reference Materials: Analyzing CRMs with known true values helps quantify bias directly. This is the most reliable method but may not be available for all analytes or at desired concentrations.

Comparison to Alternative Methods: Comparing results with established alternative methods can reveal potential bias in either method. However, both methods might share the same bias if they originate from common factors like sample preparation or calibration standards.

Spike Recovery Experiments: Adding known amounts of the analyte to samples and measuring the recovery help assess bias in sample matrices and extraction procedures.

Blank Analysis: Analyzing blanks (samples without the analyte) helps detect contamination or background contributions that can lead to bias.

Benefits of Bias Analysis

- Provides quantitative information about the magnitude and direction of bias.
- Helps identify sources of systematic errors affecting accuracy.
- Supports validation and optimization of analytical methods.
- Improves data quality and reliability for decision-making.

Limitations of Bias Analysis

- Identifying the specific source of bias can be challenging and may require further investigation.
- Reference materials or alternative methods might not be readily available or perfect themselves.
- Statistical tools are sometimes needed to quantify and interpret bias.

Example: An analytical lab measures the vitamin C content in orange juice using HPLC. Bias analysis using a CRM reveals a consistent 5% underestimation of vitamin C compared to the true value. This indicates a systematic bias in the measurement, even though the results within runs and between batches might be precise and reproducible. Further investigation might focus on factors like calibration standards, extraction procedures, or instrument settings to identify and address the source of bias.

Here are other examples of bias analysis for different quality attributes:

Accuracy Bias Analysis

Example: Assessing the accuracy of a method for determining the concentration of a specific chemical compound.

Procedure: Analyze a set of samples with known concentrations of the compound using the method under evaluation.

Analysis: Calculate the bias for each sample by comparing the measured concentrations to the known concentrations. Bias is typically expressed as a percentage.

Interpretation: Determine whether the bias is within an acceptable range (e.g., ±5% of the known concentration). If bias exceeds acceptable limits, corrective actions may be necessary.

Precision Bias Analysis

Example: Evaluating the precision of a balance used for weighing substances.

Procedure: Weigh the same substance multiple times using the same balance.

Analysis: Calculate the SD or RSD of the repeated measurements to assess precision.

Interpretation: Compare the calculated precision (SD or RSD) to acceptable criteria. Higher precision is indicated by a lower SD or RSD.

Repeatability and Reproducibility Bias Analysis (ANOVA)

Example: Assessing the repeatability and reproducibility of a method among different operators and laboratories.

Procedure: Multiple operators and laboratories analyze the same set of samples using the method.

Analysis: Conduct an ANOVA to partition the total variability into components due to repeatability (within-operator) and reproducibility (between-operator or between-laboratory).

Interpretation: Evaluate the proportions of variability attributed to repeatability and reproducibility. A lower proportion of variability due to reproducibility indicates better method performance.

Matrix Effect Bias Analysis for Accuracy

Example: Assessing the impact of different sample matrices on the accuracy of an analytical method.

Procedure: Analyze a set of samples with known analyte concentrations but different matrices (e.g., soil, water, and food) using the same method.

Analysis: Calculate the bias for each matrix type by comparing the measured analyte concentrations to the known concentrations.

Interpretation: Determine whether bias differences among matrix types are within acceptable limits. Corrective actions may be needed if significant bias is observed.

Linearity Bias Analysis

Example: Evaluating the linearity of an analytical instrument's response to varying concentrations of a substance.

Procedure: Analyze a series of standard solutions with known concentrations at different levels.

Analysis: Calculate the bias for each concentration level by comparing the observed instrument responses to expected responses based on linearity.

Interpretation: Assess whether the bias at different concentration levels is within acceptable linearity criteria. Corrective actions may be required if bias exceeds limits.

Bias analysis helps identify and quantify systematic errors in measurement methods, instruments, or procedures. It plays a crucial role in ensuring the reliability and accuracy of analytical results in laboratories. Corrective actions can be taken based on the results of bias analysis to improve measurement quality.

5.4.6 Calibration Curves

Calibration curves are graphical representations of the relationship between a known analyte concentration (standard solutions) and the corresponding instrument response (signal intensity, peak area, etc.). They are essential for quantifying unknown analyte concentrations in actual samples.

Calibration curves are integral to analytical labs, playing a crucial role in precision, accuracy, and repeatability within a single lab. However, their direct influence on reproducibility is limited due to variations in laboratory practices and sample matrices. By establishing accurate, precise, and linear calibration curves and considering potential matrix effects, labs can strive for optimal data quality and reliable analysis results.

Calibration curves are essential tools in analytical laboratories for assessing and improving precision, accuracy, repeatability, and reproducibility. Here are examples of calibration curves in different examples:

Accuracy Calibration Curve

Accuracy refers to how close measured values are to the true value. Calibration curves influence accuracy through two aspects:

Curve Accuracy: If the curve itself is inaccurate due to issues with standard preparation, instrument calibration, or curve fitting, it will translate to systematic errors in all concentration estimations, impacting accuracy.

LOD and Quantitation (LOD and LOQ): Calibration curves define the lowest concentration detectable and quantifiable with acceptable accuracy. Measurements below the LOQ have higher uncertainty and become less reliable, affecting accuracy for low-concentration samples.

Example: Assessing the accuracy of an analytical method for determining the concentration of a chemical compound.

Procedure: Prepare a series of standard solutions with known concentrations of the compound. Analyze each standard solution using the method under evaluation.

Analysis: Plot the measured concentrations (y-axis) against the known concentrations (x-axis) to create a calibration curve.

Interpretation: Assess how closely the calibration curve aligns with a 45-degree diagonal line. Deviations from the diagonal line indicate accuracy issues.

Precision Calibration Curve:

Attribute: Precision

Example: Evaluating the precision of an analytical instrument (e.g., a spectrophotometer) for measuring absorbance.

Procedure: Prepare a single standard solution with a known concentration and analyze it multiple times (e.g., 10 times) using the instrument.

Analysis: Plot the measured absorbance values (y-axis) against the replicate measurements (x-axis) to create a calibration curve.

Interpretation: Assess the scatter of data points around the calibration curve. Smaller scatter indicates higher precision.

Repeatability Calibration Curve

A well-constructed calibration curve with high linearity contributes to improved precision and repeatability. This means repeated measurements of the same concentration will give closer results with less dispersion. Nonlinear curves can lead to errors in estimating unknown concentrations, impacting both precision and repeatability.

Example: Monitoring the repeatability of a pH meter within a single laboratory.

Procedure: Measure the pH of a buffer solution multiple times (e.g., five times) using the same pH meter.

Analysis: Plot the measured pH values (y-axis) against the measurement order (e.g., 1st, 2nd, and 3rd measurement) to create a calibration curve.

Interpretation: Assess the spread of data points along the calibration curve. Smaller spread indicates higher repeatability.

Reproducibility Calibration Curve

Reproducibility refers to the consistency of results between different laboratories using the same method. While a well-defined calibration curve within a lab can improve precision and accuracy, it does not directly guarantee reproducibility for several reasons:

- Differences in instrumentation, reagents, and procedures across labs can introduce additional variability even with identical calibration curves.
- Matrix effects from different sample types can interact differently with the analytical method, impacting the calibration curve and leading to discrepancies in results between labs.

Example: Assessing the reproducibility of a chromatography method among different laboratories.

Procedure: Multiple laboratories analyze the same set of samples with known concentrations using the method.

Analysis: Plot the measured concentrations (y-axis) from each laboratory against the known concentrations (x-axis) to create a calibration curve.

Interpretation: Examine the agreement among the calibration curves from different laboratories. Consistency in calibration curves indicates good reproducibility.

Linearity Calibration Curve

Example: Evaluating the linearity of an analytical instrument's response to varying concentrations of a substance.

Procedure: Prepare a series of standard solutions with known concentrations at different levels. Analyze each standard solution using the instrument.

Analysis: Plot the measured instrument responses (e.g., peak areas) (y-axis) against the known concentrations (x-axis) to create a calibration curve.

Interpretation: Assess whether the calibration curve is a straight line within acceptable linearity criteria. Deviations from linearity indicate issues with linearity.

Other Examples

HPLC analysis of Vitamin C: A well-fitting linear calibration curve with high precision and repeatability allows accurate quantification of vitamin C in juice samples within the established LOD and LOQ. However, this does not guarantee consistent results for measuring vitamin C in fruit tissues due to potential matrix effects.

Atomic Absorption Spectroscopy for Trace Metals: A calibration curve showing nonlinearity at low concentrations can lead to inaccurate estimations of trace metals in environmental samples, affecting both accuracy and potentially precision.

Calibration curves provide visual representations of measurement performance and serve as valuable tools for diagnosing and improving the quality attributes of analytical methods and instruments in laboratories. They help ensure that measurements are accurate, precise, repeatable, and reproducible.

5.4.7 STANDARD REFERENCE ANALYSIS

By analyzing CRMs or other well-characterized reference standards, labs can gain valuable insights into their analytical performance and identify areas for improvement.

Here are some examples of how standard reference analysis can be used for different aspects of lab performance:

1. **Precision**:
   - Repeatability analysis: Analyze several replicates of a CRM within a single run or batch. Calculate the SD or RSD to assess the repeatability of the measurements, i.e., how consistent the results are within a single analysis run.

2. **Accuracy**:
   - Recovery analysis: Spike known amounts of the analyte into blank matrices and analyze them alongside CRMs. Compare the measured concentrations to the expected values (spiked amount+CRM value) to calculate the recovery. A high percentage recovery indicates good accuracy, meaning the analytical method correctly quantifies the analyte.
3. **Reproducibility**:
   - Interlaboratory study: Analyze the same CRM in multiple laboratories using the same analytical method. Compare the results across labs to assess the reproducibility, i.e., how consistent the results are between different laboratories. This can be done through statistical analysis like ANOVA to quantify the variability between labs.
4. **LOD and Limit of Quantitation (LOQ)**:
   - Analyze CRMs containing progressively lower concentrations of the analyte. The LOD is the lowest concentration that can be reliably detected, while the LOQ is the lowest concentration that can be accurately quantified. Standard reference analysis helps validate the established LOD and LOQ for the analytical method.

Benefits of Standard Reference Analysis:

- Provides objective assessment of precision, accuracy, repeatability, and reproducibility.
- Identifies potential biases and systematic errors in the analytical method.
- Contributes to method validation and optimization.
- Enhances data quality and reliability for confident decision-making.

Limitations of Standard Reference Analysis:

- CRMs may not be available for all analytes or at all desired concentrations.
- Analyzing CRMs adds cost and time to the analytical workflow.
- Proper interpretation of results requires statistical skills and understanding of potential matrix effects.

Examples

- Inductively coupled plasma mass spectrometry (ICP-MS) analysis of trace metals in water samples: Analyzing a CRM containing trace metals can assess the accuracy and LOD/LOQ for individual elements. This helps ensure reliable detection and quantification of trace metals in environmental samples.
- HPLC analysis of Vitamin C in fruits: Analyzing a CRM containing vitamin C can validate the accuracy of the calibration curve and quantify vitamin C content in different fruit types with confidence.

SRMs are CRMs provided by organizations like NIST to assess and improve precision, accuracy, repeatability, and reproducibility in analytical laboratories. Here are examples of how SRMs can be used for these purposes:

1. **Accuracy Assessment with SRMs:**
   - **Attribute:** Accuracy
   - **Example:** Evaluating the accuracy of an analytical method for determining the concentration of a specific chemical compound.
   - **Procedure:** Analyze an SRM with a known concentration of the compound using the method under evaluation.

- **Analysis:** Compare the measured concentration of the SRM to its certified or reference value. Calculate the bias (difference) as a measure of accuracy.
- **Interpretation:** Determine whether the bias is within acceptable accuracy criteria (e.g., ±5% of the reference value). Corrective actions may be required if bias exceeds acceptable limits.

2. **Precision Assessment with SRMs:**
   - **Attribute:** Precision
   - **Example:** Assessing the precision of an analytical instrument (e.g., a spectrophotometer) for measuring absorbance.
   - **Procedure:** Analyze an SRM multiple times (e.g., 5 times) using the instrument.
   - **Analysis:** Calculate the SD or RSD of the repeated measurements to assess precision.
   - **Interpretation:** Compare the calculated precision (SD or RSD) to acceptable precision criteria. A smaller SD or RSD indicates higher precision.

3. **Repeatability Assessment with SRMs:**
   - **Attribute:** Repeatability
   - **Example:** Monitoring the repeatability of a balance used for weighing substances.
   - **Procedure:** Weigh an SRM multiple times (e.g., 10 times) using the same balance.
   - **Analysis:** Calculate the SD or RSD of the repeated measurements to assess repeatability.
   - **Interpretation:** Assess the spread of data points around the reference value. Smaller SD or RSD indicates higher repeatability.

4. **Reproducibility Assessment with SRMs (Interlaboratory Comparison):**
   - **Attribute:** Reproducibility
   - **Example:** Assessing the reproducibility of a chromatography method among different laboratories.
   - **Procedure:** Multiple laboratories analyze the same SRM samples with known properties using the same method.
   - **Analysis:** Collect data from different laboratories and calculate the mean, SD, and RSD for the results.
   - **Interpretation:** Examine the agreement among the laboratories' results. Smaller RSD values indicate better reproducibility among laboratories.

SRMs play a critical role in ensuring the reliability and quality of analytical measurements. They serve as benchmarks for accuracy and precision, allowing laboratories to validate their methods, instruments, and procedures. The use of SRMs helps identify and rectify issues related to accuracy, precision, repeatability, and reproducibility, ultimately improving measurement quality.

5.4.8 RECOVERY EXPERIMENTS

By analyzing how much of a known amount of analyte can be reliably recovered from a sample matrix, labs can gain valuable insights into various aspects of their analytical performance. Here are some examples of how recovery experiments can be used to analyze:

1. **Precision:**
   - Intra-assay precision: Analyze multiple replicates of the same spiked sample within a single analytical run. Calculate the SD or RSD of the recovery values to assess the repeatability of the extraction and quantification process within a single run.
   - Inter-assay precision: Analyze replicates of the same spiked sample across different analytical runs or batches. Calculate the SD or RSD of the recovery values to assess the intermediate precision of the entire analytical process, including potential variability from different analysts, instruments, or reagent batches.

2. **Accuracy:**
   - Average recovery: Calculate the average percentage recovery across all replicates in the experiment. Compare this value to the expected recovery (100%) to assess the overall bias in the extraction and quantification procedures. A significant deviation from 100% might indicate systematic errors in the method.
   - Matrix effects: Analyze recovery experiments using different sample matrices (e.g., different fruit types for vitamin C analysis). Significant differences in recovery between matrices indicate potential matrix effects that can impact the accuracy of the analysis for real samples.
3. **Repeatability and Reproducibility:**
   - Recovery experiment across labs: Analyze the same spiked sample in multiple laboratories using the same analytical method. Compare the average recovery values and RSDs across labs to assess the reproducibility of the entire process. This can help identify potential variability due to differences in instrumentation, reagents, or laboratory practices.
   - Spike levels: Perform recovery experiments with different spiking concentrations of the analyte. Consistent recovery across a range of concentrations suggests good repeatability and accuracy across the working range of the method.

Benefits of Recovery Experiments:

- Provide valuable insights into the efficiency and accuracy of extraction and quantification processes.
- Help identify potential sources of bias, variability, and matrix effects.
- Contribute to method validation and optimization.
- Enhance data quality and reliability for accurate analysis of real samples.

Limitations of Recovery Experiments:

- Require additional time and resources to perform.
- May not be applicable to all analytes or sample matrices.
- Interpretation of results can be complex and requires statistical expertise.

Examples

- Analysis of pesticide residues in vegetables: Recovery experiments with spiked vegetable samples help assess the efficiency of pesticide extraction and potential matrix effects for different types of vegetables. This information is crucial for accurate quantification of pesticide residues in real samples.
- Determination of vitamin C content in fruit juices: Recovery experiments with spiked juices can reveal biases in the extraction and quantification procedures and highlight potential matrix effects from different fruit types. This helps ensure accurate assessment of vitamin C content in various juices.

Recovery experiments are conducted in analytical laboratories to assess precision, accuracy, repeatability, and reproducibility by determining how well an analytical method can recover known amounts of an analyte from a sample matrix. Here are examples of recovery experiments and their analysis for various quality attributes:

1. **Accuracy Assessment with Recovery Experiments**:
   - **Attribute:** Accuracy
   - **Example:** Evaluating the accuracy of a method for quantifying a specific chemical compound in a complex sample matrix (e.g., soil).

- **Procedure:** Spiking known amounts of the analyte into the sample matrix at different concentrations (e.g., low, medium, and high). Analyze the spiked samples using the method.
- **Analysis:** Calculate the recovery percentage for each spike level by comparing the measured concentration to the expected concentration.
- **Interpretation:** Determine whether the recovery percentages are within an acceptable range (e.g., 90%–110%) for each spike level. Deviations may indicate accuracy issues.

2. **Precision Assessment with Recovery Experiments:**
   - **Attribute:** Precision
   - **Example:** Evaluating the precision of an analytical instrument (e.g., an atomic absorption spectrometer) for measuring trace metals in water samples.
   - **Procedure:** Prepare multiple replicate samples spiked with a known concentration of the target metal. Analyze each spiked sample multiple times.
   - **Analysis:** Calculate the recovery percentage and SD for each replicate set.
   - **Interpretation:** Evaluate the precision by assessing the SD of recovery percentages. Lower SD indicates higher precision.

3. **Repeatability Assessment with Recovery Experiments:**
   - **Attribute:** Repeatability
   - **Example:** Monitoring the repeatability of a liquid chromatography method for pharmaceutical analysis.
   - **Procedure:** Prepare multiple identical samples spiked with the same analyte concentration. Analyze each sample using the method.
   - **Analysis:** Calculate the recovery percentage for each sample.
   - **Interpretation:** Assess the repeatability by comparing the recovery percentages of replicate samples. Smaller variability indicates higher repeatability.

4. **Reproducibility Assessment with Recovery Experiments (Interlaboratory Comparison):**
   - **Attribute:** Reproducibility
   - **Example:** Assessing the reproducibility of an analytical method for pesticide residues in food among different laboratories.
   - **Procedure:** Multiple laboratories independently spike samples with known pesticide concentrations and analyze them using the same method.
   - **Analysis:** Collect data from different laboratories, calculate recovery percentages, and assess the reproducibility through interlaboratory comparisons.
   - **Interpretation:** Evaluate the agreement among the laboratories' recovery percentages. Smaller variability indicates better reproducibility among laboratories.

Recovery experiments provide insights into how well an analytical method can recover analytes from real sample matrices. Analysis of the recovery percentages and their variability helps assess and improve the accuracy, precision, repeatability, and reproducibility of the method, ensuring reliable and trustworthy analytical results.

5.4.9 Intra-Assay Variation Analysis

Intra-assay variation analysis in analytical labs focuses on evaluating the consistency and reliability of measurements within a single analytical run. It provides valuable insights into the precision, accuracy, repeatability, and reproducibility of the analytical method under controlled conditions. Here are some examples of how intra-assay variation analysis can be applied:

1. **Precision:**
   - Repeated measures of a reference standard: Analyzing multiple replicates of a CRM within a single run provides data on repeatability. Calculate the SD or RSD of the results to assess the spread of values and the overall precision of the measurements. Lower SD/RSD indicates better precision.

- Linearity study: Analyzing several dilutions of a standard solution across a concentration range within a single run allows investigation of the linearity of the calibration curve. Nonlinearity can indicate poor precision at certain concentration ranges.

2. **Accuracy:**
   - Recovery experiment: Spiking known amounts of the analyte into blank samples and analyzing them alongside a CRM within the same run help assess accuracy. Compare the measured concentrations of the spiked samples to the expected values (spiked amount+CRM value) to calculate the recovery. Consistent recovery close to 100% indicates good accuracy.
   - Blank analysis: Analyzing blank samples within the run helps detect background contamination that can bias the measurements and impact accuracy.

3. **Repeatability:**
   - Multiple injections of the same sample: Analyzing the same sample solution multiple times within a single run provides data on repeatability, i.e., how consistent the results are for the same sample under the same conditions. Calculate the SD/RSD of the values to assess the variability within the run.
   - Comparison of replicates across different analysts: If multiple analysts perform the same analysis within a single run using the same instrument and reagents, comparing the SD/RSD of their replicate measurements helps assess method repeatability and identify potential analyst-related variability.

4. **Reproducibility (Limited Scope):**
   While intra-assay analysis focuses on a single run, it can also offer preliminary insights into reproducibility, which refers to the consistency of results between different runs or batches. For example:
   - Comparing SD/RSD values from multiple intra-assay analyses performed on different days or instruments can hint at potential inter-assay variability.
   - Analyzing replicates of the same spiked sample across multiple runs within a short timeframe can provide early indications of repeatability and intermediate precision.

However, full evaluation of reproducibility typically requires analyzing the same samples across different labs or extended periods, which extends beyond the scope of a single intra-assay analysis.

Benefits of Intra-Assay Variation Analysis:

- Provides quick and efficient assessment of measurement consistency within a single run.
- Helps identify potential sources of variability and bias.
- Contributes to method validation and optimization.
- Enhances data quality and confidence in analytical results.

Limitations of Intra-Assay Variation Analysis:

- Cannot fully evaluate reproducibility due to limited scope within a single run.
- Requires careful control of experimental conditions to minimize external variation.
- Statistical analysis skills may be needed for interpretation of results.

Examples

- HPLC analysis of pharmaceuticals: Analyzing replicate injections of a standard solution within a single run helps confirm precision and linearity of the calibration curve for accurate quantification of pharmaceutical compounds in real samples.

- Enzyme-linked immunosorbent assay (ELISA) for protein analysis: Repeated measurements of a control sample in an ELISA run assess the repeatability and background levels of the assay, ensuring reliable quantification of proteins in unknown samples.

Intra-assay variation analysis assesses the precision, accuracy, repeatability, and reproducibility of an analytical method within a single laboratory. Here are examples of intra-assay variation analysis for various quality attributes:

1. **Precision Assessment with Intra-Assay Variation**:
   - **Attribute:** Precision
   - **Example:** Evaluating the precision of a liquid chromatography-mass spectrometry (LC-MS) method for quantifying drug metabolites in blood samples.
   - **Procedure:** Analyze a set of replicate blood samples (e.g., 10 replicates) spiked with known concentrations of drug metabolites using the same LC-MS system.
   - **Analysis:** Calculate the CV or RSD for the measurements to assess precision.
   - **Interpretation:** A lower CV or RSD indicates higher precision.
2. **Accuracy Assessment with Intra-Assay Variation**:
   - **Attribute:** Accuracy
   - **Example:** Evaluating the accuracy of an ELISA for detecting a specific biomarker in serum samples.
   - **Procedure:** Analyze a set of serum samples spiked with known concentrations of the biomarker using the ELISA method.
   - **Analysis:** Calculate the accuracy by comparing the measured biomarker concentrations to the expected concentrations.
   - **Interpretation:** Determine whether the measured concentrations closely match the expected concentrations to assess accuracy.
3. **Repeatability Assessment with Intra-Assay Variation**:
   - **Attribute:** Repeatability
   - **Example:** Monitoring the repeatability of a pH meter within a single laboratory for measuring pH in water samples.
   - **Procedure:** Measure the pH of identical water samples (e.g., five replicates) using the same pH meter.
   - **Analysis:** Calculate the SD or RSD for the repeated pH measurements to assess repeatability.
   - **Interpretation:** A smaller SD or RSD indicates higher repeatability.
4. **Reproducibility Assessment with Intra-Assay Variation**:
   - **Attribute:** Reproducibility
   - **Example:** Assessing the reproducibility of an ICP-MS method for analyzing trace elements in soil samples.
   - **Procedure:** Analyze a set of soil samples independently (e.g., five replicates) using the same ICP-MS instrument within the same laboratory.
   - **Analysis:** Calculate the mean and SD of measurements for each sample to assess reproducibility.
   - **Interpretation:** Examine the agreement among the measurements of replicate samples to assess reproducibility.

Intra-assay variation analysis involves conducting measurements within a single laboratory using the same method and equipment. It helps identify and quantify precision, accuracy, repeatability, and reproducibility issues that may arise during routine laboratory operations. Corrective actions can be taken based on the results to improve the quality of analytical measurements.

5.4.10 RING TESTS

Ring tests, also known as proficiency testing or interlaboratory studies, are valuable tools for analytical labs to assess their precision, accuracy, reproducibility, and overall performance compared to other labs using the same analytical method. By analyzing the same reference material or spiked samples under controlled conditions, laboratories gain valuable insights into their strengths and weaknesses. Here are some examples of how ring test analysis can be used to evaluate different aspects of laboratory performance:

1. **Precision:**
   - Intra-laboratory precision: Analyze several replicates of the ring test sample within a single run to estimate repeatability. Calculate the SD or RSD of the results to assess the consistency of measurements within your lab.
   - Interlaboratory precision: Compare the SD/RSD values of your lab with the overall averages across all participating labs. Smaller deviations from the average indicate better intermediate precision, signifying consistency within your analytical process compared to others.
2. **Accuracy:**
   - Target value comparison: Compare your measured concentration of the analyte in the ring test sample to the provided target value (true value or expected concentration). The difference between your result and the target value reflects your bias.
   - Z-score calculation: Many ring tests provide Z-scores, which statistically assess the deviation of your result from the target value and overall lab population. A Z-score within a predefined acceptable range indicates acceptable accuracy of your measurements.
3. **Reproducibility:**
   - Ring test results across rounds: If you participate in multiple rounds of the same ring test over time, comparing your results and Z-scores can reveal your reproducibility, i.e., the consistency of your performance over time and different batches.
   - Comparison with historical data: Analyze trends in your ring test performance over time. Consistent improvement in precision, accuracy, and Z-scores demonstrates progress in method optimization and overall reproducibility.
4. **Identification of Bias and Variability:**
   - Ring test results can help identify potential sources of bias in your analytical method. Significant and consistent deviations from the target value might indicate issues with calibration, extraction procedures, or instrument calibration.
   - High SD/RSD values compared to other labs might point to variability within your analytical process. Investigating factors like reagent quality, analyst technique, or instrument maintenance can help address these issues and improve precision.

Benefits of Ring Test Analysis:

- Provides independent assessment of precision, accuracy, and reproducibility.
- Identifies potential biases and sources of variability.
- Contributes to method validation and optimization.
- Improves data quality and confidence in analytical results.
- Benchmarks lab performance against others in the field.

Limitations of Ring Test Analysis:

- May require additional time and resources to participate.
- Results may not be specific to every analyte or sample matrix.
- Interpreting Z-scores and statistical analysis might require some expertise.

Examples

Water quality analysis: Participating in ring tests for analyzing specific pollutants in water samples allows labs to assess their accuracy and reproducibility compared to other labs across different regions. This helps ensure reliable data for environmental monitoring and regulatory compliance.

- Clinical laboratory testing: Ring tests for analyzing blood glucose or cholesterol levels provide crucial insights into the accuracy and reproducibility of diagnostic tests, ensuring reliable results for patient care.
1. **Precision Ring Test**:
    - **Attribute:** Precision
    - **Example:** Assessing the precision of an analytical method for measuring the acidity (pH) of water samples.
    - **Procedure:** Multiple laboratories receive identical water samples and analyze them using the same pH meter.
    - **Analysis:** Calculate the mean, SD, and RSD of pH measurements from participating laboratories.
    - **Interpretation:** Evaluate the RSD values to assess the precision of the method across laboratories. Smaller RSD indicates better precision.
2. **Accuracy Ring Test**:
    - **Attribute:** Accuracy
    - **Example:** Evaluating the accuracy of a spectrophotometric method for quantifying a specific analyte (e.g., glucose) in blood samples.
    - **Procedure:** Laboratories receive blood samples with known analyte concentrations and analyze them using the method.
    - **Analysis:** Compare the measured analyte concentrations from each laboratory to the known concentrations and calculate the bias (difference) as a measure of accuracy.
    - **Interpretation:** Determine whether the bias values are within acceptable accuracy criteria. Deviations may indicate accuracy issues.
3. **Repeatability Ring Test**:
    - **Attribute:** Repeatability
    - **Example:** Monitoring the repeatability of a gas chromatography-mass spectrometry (GC-MS) method for analyzing volatile organic compounds (VOCs) in air samples.
    - **Procedure:** Participating laboratories receive identical air samples and independently analyze them using the same GC-MS system.
    - **Analysis:** Calculate the mean and SD of VOC concentrations from different laboratories to assess repeatability.
    - **Interpretation:** Evaluate the spread of data points to assess the repeatability of the method across laboratories. Smaller variability indicates higher repeatability.
4. **Reproducibility Ring Test**:
    - **Attribute:** Reproducibility
    - **Example:** Assessing the reproducibility of an X-ray diffraction (XRD) method for identifying and quantifying mineral phases in geological samples.
    - **Procedure:** Laboratories receive identical geological samples and analyze them using their respective XRD instruments.
    - **Analysis:** Collect data from participating laboratories, calculate the mean, and assess the agreement among the results to assess reproducibility.
    - **Interpretation:** Examine the consistency of results among laboratories to assess the reproducibility of the method.

Ring tests are valuable for identifying variations in measurement quality across different laboratories and are often used to establish confidence in analytical methods. They help laboratories detect and address precision, accuracy, repeatability, and reproducibility issues and promote harmonization of measurement practices among participating laboratories.

5.4.11 GRUBBS' TEST

While Grubbs' test does not directly measure all aspects of precision, accuracy, repeatability, and reproducibility in analytical labs, it plays a crucial role in data analysis by identifying outliers within a dataset. This, in turn, indirectly contributes to understanding and improving these key performance indicators. Here is how Grubbs' test comes into play:

Understanding Outliers:

- Outliers are data points that significantly deviate from the expected pattern or distribution of the rest of the data. They can arise due to various reasons like analytical errors, sample contamination, or unusual sample matrix effects.
- Identifying and removing outliers is crucial for accurate data analysis and interpretation. Outliers can distort statistical calculations, masking or skewing real trends and impacting the assessment of precision, accuracy, repeatability, and reproducibility.

Grubbs' Test in Action:

- Grubbs' test is a statistical test specifically designed to detect and remove outliers within a normally distributed dataset. It compares each data point to the mean and SD of the dataset and calculates a test statistic.
- If the test statistic for a particular data point falls beyond a critical value (calculated based on sample size and desired confidence level), that point is considered an outlier and can be removed from the analysis.

Impact on Performance Indicators:

- Precision: By removing outliers that inflate the SD, Grubbs' test can provide a more accurate representation of the spread of data, potentially leading to a slightly improved estimation of precision.
- Accuracy: While Grubbs' test does not directly measure accuracy, removing outliers that significantly deviate from the expected value can help provide a clearer picture of the overall trend and potentially reduce bias in the data.
- Repeatability and Reproducibility: Grubbs' test focuses on single datasets, but it can indirectly support investigations into repeatability and reproducibility. Consistent presence of outliers across multiple datasets might indicate systematic issues affecting these aspects, prompting further investigation into analytical procedures or instrument performance.

Benefits of Grubbs' Test:

- Simple and easy-to-use statistical test.
- Effective in identifying and removing outliers.
- Helps improve data quality and accuracy.
- Supports assessment of other performance indicators indirectly.

Limitations of Grubbs' Test:

- Assumes a normally distributed dataset.
- Not suitable for identifying multiple outliers in small datasets.
- Overuse can lead to discarding valid data points.

Examples

- HPLC analysis of pharmaceutical residues: Analyzing multiple replicates of a sample might reveal an outlier due to potential sample contamination. Removing this outlier using Grubbs' test can lead to a more accurate estimation of the average residue concentration and improve precision.
- Spectroscopic analysis of environmental samples: Outliers in metal concentration measurements detected by Grubbs' test could point to localized contamination or analytical errors. Removing them ensures a more accurate portrayal of the average metal concentration, reducing bias and potentially improving accuracy.

Grubbs' test, while not directly measuring precision, accuracy, repeatability, and reproducibility, plays a valuable role in analytical labs by identifying and removing outliers. This, in turn, helps ensure data quality, reduce bias, and potentially improve estimations of these key performance indicators. However, it is crucial to use Grubbs' test cautiously, considering its limitations and ensuring its appropriate application for optimal data analysis and interpretation in your specific research context.

1. **Outlier Detection in Precision Analysis:**
   - **Attribute:** Precision
   - **Example:** Evaluating the precision of a balance used for weighing substances in a laboratory.
   - **Procedure:** Weigh a set of replicates of the same substance using the same balance. Calculate the mean and SD of the measurements.
   - **Analysis:** Apply Grubbs' test to identify any outliers (extreme measurements) that may indicate errors or inconsistencies in the precision data.
   - **Interpretation:** If outliers are detected, investigate the potential causes and consider repeating the measurements to improve precision.
2. **Outlier Detection in Accuracy Analysis:**
   - **Attribute:** Accuracy
   - **Example:** Assessing the accuracy of an analytical method for determining the concentration of a specific chemical compound.
   - **Procedure:** Analyze a series of standard solutions with known concentrations using the method.
   - **Analysis:** Apply Grubbs' test to identify outliers in the dataset. Outliers may indicate inaccurate measurements or calibration issues.
   - **Interpretation:** Investigate the outliers to determine the reasons for inaccuracy and take corrective actions to improve accuracy.
3. **Outlier Detection in Repeatability Assessment:**
   - **Attribute:** Repeatability
   - **Example:** Monitoring the repeatability of a pH meter for measuring pH in water samples.
   - **Procedure:** Measure the pH of replicate water samples using the same pH meter.
   - **Analysis:** Use Grubbs' test to identify outliers in the dataset that may be indicative of inconsistencies in repeatability.
   - **Interpretation:** Investigate the outliers to understand the sources of variation and consider recalibration or maintenance of the pH meter.

4. **Outlier Detection in Reproducibility Assessment**:
   - **Attribute:** Reproducibility
   - **Example:** Assessing the reproducibility of a method for analyzing metal content in soil samples among different laboratories.
   - **Procedure:** Multiple laboratories independently analyze the same set of soil samples using the same method.
   - **Analysis:** Apply Grubbs' test to identify outliers in the results, which may suggest discrepancies among laboratories.
   - **Interpretation:** Investigate the outliers to determine the causes of differences and work toward improving method harmonization and reproducibility.

While Grubbs' test is primarily used for identifying outliers, it indirectly contributes to improving precision, accuracy, repeatability, and reproducibility in analytical laboratories by highlighting data points that may require further investigation and corrective actions. Detecting and addressing outliers can lead to enhanced data quality and measurement reliability.

5.4.12 GR&R

GR&R studies are powerful tools for analytical laboratories to assess the precision, accuracy, repeatability, and reproducibility of their measurement systems. By analyzing how consistently and accurately various gauges or instruments measure a specific characteristic, labs can identify areas for improvement and ensure reliable data quality. Here are some examples of gauge R&R studies in analytical labs:

1. **Measuring pH of Solutions:**
   - **Precision:** Analyze multiple replicates of a standard solution with known pH using a pH meter. Calculate the SD or RSD of the results to assess the repeatability of the instrument within a single run.
   - **Accuracy:** Compare the average measured pH to the known value of the standard solution. The difference reflects the bias of the instrument.
   - **Repeatability and Reproducibility:** Analyze the same standard solution across multiple days or by different analysts to assess the inter-assay precision and potential operator/day variability.
2. **Weighing Samples for Analysis:**
   - **Repeatability:** Weigh the same sample multiple times on the same balance within a single run. Calculate the SD/RSD to assess the repeatability of the balance.
   - **Accuracy:** Compare the average weight of the sample to its known true weight (e.g., using a CRM). The difference reflects the bias of the balance.
   - **Reproducibility:** Weigh the same sample on different balances or across different time periods to assess the inter-instrument and long-term reproducibility of the weighing process.
3. **Calibrating Pipettes:**
   - **Repeatability:** Dispense volumes of water from the same pipette multiple times within a single run and measure the delivered volume each time. Calculate the SD/RSD to assess the repeatability of the pipette.
   - **Accuracy:** Compare the average dispensed volume to the target volume (usually marked on the pipette). The difference reflects the accuracy of the pipette calibration.
   - **Reproducibility:** Dispense volumes from the same pipette across different days or analysts to assess the inter-operator and long-term reproducibility of the pipetting technique and calibration.

Benefits of GR&R Studies:

- Identify sources of variability and bias in measurement systems.
- Improve the precision and accuracy of analytical results.
- Strengthen data quality and confidence in measurements.
- Optimize instrument calibration and maintenance procedures.
- Demonstrate laboratory competency and compliance with regulations.

Limitations of GR&R Studies:

- Require time and resources to conduct effectively.
- Statistical expertise might be needed for data analysis and interpretation.
- Can be limited to specific instruments or procedures within a lab.

Example:
Evaluating the precision, repeatability, and reproducibility of a pH meter used for measuring pH in various laboratory applications.

Procedure:

1. Select a representative set of pH samples that cover the expected pH range and characteristics of typical laboratory measurements.
2. Prepare multiple samples of each pH value to ensure sufficient data points for analysis.
3. Identify multiple operators who will perform the pH measurements. Ensure they represent the range of skill levels typically encountered in the laboratory.
4. Randomly assign the samples to the operators, making sure that each sample is measured by multiple operators.
5. Each operator measures the pH of the assigned samples using the pH meter, repeating the measurements multiple times.
6. Record the measurement results in a structured data format.

GR&R Analysis:

1. Calculate the repeatability (within-operator variation) and reproducibility (between-operator variation) components of variance using statistical analysis.
2. Express the results as percentages of total variation, which indicate the contribution of repeatability and reproducibility to the overall measurement variation.
3. Evaluate the results to determine the contribution of measurement system variation to the total variation in pH measurements.
4. Calculate the overall measurement error and assess whether it meets acceptable criteria for precision, repeatability, and reproducibility.

Interpretation:

- Precision: GR&R analysis quantifies the precision of the pH measurements by separating the variation due to repeatability (within-operator) and reproducibility (between-operator). A smaller percentage of reproducibility variation indicates higher precision.
- Repeatability: The repeatability component measures the variation in measurements made by the same operator on the same samples. Lower repeatability percentages suggest better repeatability.

- Reproducibility: The reproducibility component measures the variation in measurements made by different operators on the same samples. Lower reproducibility percentages indicate better reproducibility.
- Overall Assessment: GR&R analysis provides an overall assessment of the measurement system's capability to deliver consistent and reproducible pH measurements. It allows you to identify sources of variation and take corrective actions if necessary.

By conducting gauge R&R tests for measurement systems like pH meters, laboratories can ensure that the instruments deliver reliable and consistent results, contributing to improved precision, repeatability, and reproducibility of measurements in analytical processes.

GR&R studies are valuable tools for analytical labs to ensure the reliability and accuracy of their data. By analyzing precision, accuracy, repeatability, and reproducibility, labs can identify weaknesses in their measurement systems and implement improvements to optimize performance and achieve consistent, high-quality analytical results. Remember, consulting relevant guidelines and best practices in your field is crucial for designing and interpreting GR&R studies effectively in your specific analytical context.

5.4.13 YOUDEN PLOTS

Youden plots are powerful tools for visualizing and analyzing data from interlaboratory studies or ring tests in analytical labs. They offer a simple yet effective way to assess precision, accuracy, repeatability, and reproducibility by visually representing the relationships between results from different labs. Here are some examples of how Youden plots can be used:

1. **Assessing Overall Performance:**
   - Plot the measured values from each lab for two replicates of the same sample on the x- and y-axes. The ideal example would be a tight cluster of points around the origin (target value), indicating good precision and accuracy across all labs.
   - Deviations from the ideal example can reveal:
     - Systematic bias: If points are consistently above or below the origin, it suggests a bias for overestimation or underestimation in some labs.
     - Poor precision: A wide scatter of points, regardless of position, indicates poor repeatability within individual labs.
     - Interlaboratory variability: Different clusters of points suggest inconsistencies between labs, impacting both precision and reproducibility.
2. **Identifying Outliers:**
   - Youden plots readily highlight potential outliers, data points significantly deviating from the main cluster. This can help investigate and address potential analytical errors, sample issues, or instrument malfunctions in specific labs.
3. **Comparing Reproducibility across Time:**
   - Analyzing Youden plots from multiple rounds of the same ring test over time can show trends in performance. Ideally, the clusters of points should become tighter and closer to the origin, indicating improved reproducibility and overall laboratory performance.

Examples of Youden Plot Applications:

- Water quality analysis: Comparing nitrate or phosphate levels measured by different labs in multiple water samples using a Youden plot reveals potential regional variations, systematic biases in specific labs, or variability in analytical methods.

- Clinical laboratory testing: Youden plots for analyzing cholesterol or blood sugar levels across different labs help identify consistent biases or outliers that could impact patient care and highlight areas for interlaboratory standardization.

Here are some examples of Youden plots for different examples:

Accuracy Assessment:
   Example: Evaluating the accuracy of an analytical method for the determination of a specific chemical compound.
   Samples: A series of reference standards with known concentrations of the compound.
   Youden Plot: Plot the known concentrations on the x-axis and the measured concentrations on the y-axis. Assess how closely the measured values align with the true concentrations. Deviations from a 45-degree diagonal line indicate accuracy issues.

Precision Assessment:
   Example: Assessing the precision of a pipetting technique used for preparing dilutions.
   Samples: Prepare multiple dilutions of a standard solution and pipette each dilution into separate containers.
   Youden Plot: Plot the expected volumes (based on dilution calculations) on the x-axis and the measured volumes (from pipetting) on the y-axis. Scatter in the data points indicates imprecision.

Repeatability Assessment:
   Example: Evaluating the repeatability of a pH meter within a single laboratory.
   Samples: Use a buffer solution at a fixed pH value.
   Youden Plot: Measure the pH of the buffer solution multiple times with the same pH meter. Plot the expected pH value on the x-axis and the measured pH values on the y-axis. Variability in the measurements indicates repeatability issues.

Reproducibility Assessment:
   Example: Assessing the reproducibility of a chromatography method among multiple laboratories.
   Samples: Prepare a set of standard solutions and distribute them to different laboratories for analysis.
   Youden Plot: Each laboratory measures the concentrations of compounds in the standard solutions using the same method. Plot the expected concentrations on the x-axis and the measured concentrations from each laboratory on the y-axis. Variability among laboratories suggests reproducibility challenges.

Linearity Assessment:
   Example: Evaluating the linearity of an analytical instrument's response.
   Samples: Analyze a series of standard solutions with varying concentrations.
   Youden Plot: Plot the known concentrations on the x-axis and the instrument responses on the y-axis. Deviations from a straight line indicate nonlinearity in the instrument's response.

Matrix Effect Assessment:
   Example: Assessing the impact of different sample matrices on the accuracy of an analytical method.
   Samples: Analyze a set of samples with the same analyte concentration but different matrices (e.g., different types of food samples).

Youden Plot: Plot the expected analyte concentrations on the x-axis and the measured concentrations for each matrix type on the y-axis. Differences in measured values for the same concentration indicate matrix effects.

These examples demonstrate how Youden plots can be applied to assess different aspects of measurement quality in analytical laboratories, including accuracy, precision, repeatability, reproducibility, linearity, and the impact of sample matrices. By visually examining the data in a Youden plot, laboratory analysts can identify potential issues and make improvements to ensure the reliability of their analytical results.

Benefits of Youden Plots:

- Provides a simple and intuitive visual representation of interlaboratory data.
- Aids in identifying bias, outliers, and overall performance trends.
- Facilitates comparisons and communication between labs.
- Supports optimization of analytical methods and overall laboratory quality.

Limitations of Youden Plots:

- Requires data from at least two replicates per lab.
- Assumptions about data normality can impact interpretation.
- Interpretation might require some statistical knowledge.

Youden plots offer a valuable tool for analytical labs participating in interlaboratory studies or ring tests. By visually analyzing data related to precision, accuracy, repeatability, and reproducibility, labs can gain valuable insights into their performance, identify areas for improvement, and contribute to improving data quality and reliability within the broader analytical community. Remember, consulting relevant guidelines and best practices in your field is crucial for interpreting and utilizing Youden plots effectively in your specific analytical context.

References and Reviews

Bailey, R.A. *The Design of Experiments: Statistical Principles for Practical Applications.* Cambridge University Press, 2008 – This book provides a comprehensive introduction to experimental design principles and statistical methods for practical applications in science and engineering.

Belle, G. van, L.D. Fisher, and P.J. Heagerty. *Design and Analysis of Experiments in the Health Sciences.* Wiley, 2004 – Focused on applications in health sciences, this book covers experimental design and analysis techniques relevant to medical and biological research.

Box, G.E.P., J. Stuart Hunter, and W.G. Hunter. *Statistics for Experimenters: Design, Innovation, and Discovery.* Wiley, 2005 – This classic text presents experimental design principles and techniques for innovation and discovery in science and engineering.

Cunningham, D.W. *Experimental Design: From User Studies to Psychophysics.* Academic Press, 2016 – This book covers experimental design principles and techniques, with a focus on applications in user studies and psychophysics.

de Bruijn, A., and G.M. Willems. *Experimental Design for Formulation.* CRC Press, 2000 – Focused on the formulation process, this book presents experimental design techniques for optimizing formulations in science and engineering.

Kleinbaum, D.G., L.L. Kupper, and K.E. Muller. *Applied Regression Analysis and Other Multivariable Methods.* Cengage Learning, 2013 – This book covers regression analysis and multivariable methods with applications in engineering, science, and public health.

Kume, H. *Statistical Methods for Quality Improvement.* CRC Press, 2012 – This book covers statistical methods for quality improvement in engineering and manufacturing processes, with a focus on practical applications.

Lawson, J. *Design and Analysis of Experiments with R.* CRC Press, 2018 – This book combines the principles of experimental design with practical examples using the R programming language, making it suitable for both students and researchers.

Mason, R.L., R.F. Gunst, and J.L. Hess. *Statistical Design and Analysis of Experiments, with Applications to Engineering and Science.* Wiley, 2003 – This book covers experimental design and analysis techniques with applications in engineering and science, emphasizing practical problem-solving.

Meeker, W.Q., and L.A. Escobar. *Statistical Methods for Reliability Data.* Wiley, 1998 – Focused on reliability data analysis, this book presents statistical methods for assessing and improving the reliability of engineering systems.

Montgomery, D.C. *Design and Analysis of Experiments.* Wiley, 2017 – This classic text covers the principles of experimental design and analysis, emphasizing practical applications in science and engineering.

Montgomery, D.C., and G.C. Runger. *Applied Statistics and Probability for Engineers.* Wiley, 2018 – This textbook covers probability and statistics concepts relevant to engineering applications, including experimental design and analysis.

Montgomery, D.C., and G.C. Runger. *Engineering Statistics.* Wiley, 2010 – This comprehensive textbook covers statistical methods relevant to engineering applications, including experimental design, regression analysis, and quality control.

Myers, R.H. *Statistical Methods for Engineers and Scientists: An Introduction with Applications Using R.* CRC Press, 2016 – This book provides an introduction to statistical methods tailored for engineers and scientists, with examples and applications using the R software.

NIST/SEMATECH, *Engineering Statistics Handbook.* National Institute of Standards and Technology, Ongoing – This handbook provides a comprehensive resource on statistical methods and techniques relevant to engineering applications.

Oehlert, G.W. *Practical Design and Analysis of Experiments.* SAS Institute, 2000 – This book provides a practical introduction to experimental design and analysis techniques, with examples using SAS software.

Pahl, G., and W. Beitz. *Engineering Design: A Systematic Approach.* Springer, 2006 – This book covers a systematic approach to engineering design, including experimental design principles and methods.

Rao, K. Ramachandra, and S. Gangopadhyay. *Statistical Techniques for Transportation Engineering.* CRC Press, 2008 – This book covers statistical techniques applicable to transportation engineering, including experimental design and analysis.

Sall, J., A. Lehman, M. Stephens, and L. Creighton. *Introduction to Design of Experiments with JMP Examples.* SAS Institute, 2012 – Focused on practical examples using JMP software, this book introduces experimental design principles and techniques.

Subramanian, E. *Quality Improvement through Statistical Methods.* Springer, 2012 – Focused on quality improvement, this book presents statistical methods and techniques applicable to engineering and manufacturing processes.

Tobias, R.H., and K.S. Krishnamoorthi. *Statistical Methods in Engineering and Quality Assurance.* McGraw-Hill Education, 1997 – This book covers statistical methods applicable to engineering and quality assurance, with an emphasis on practical problem-solving techniques.

Vining, G. Geoffrey, and S.M. Kowalski. *Statistical Methods for Engineers.* Cengage Learning, 2011 – This book provides a comprehensive overview of statistical methods commonly used in engineering applications, with a focus on experimental design and analysis.

Wu, C.F. Jeff and M.S. Hamada. *Experiments: Planning, Analysis, and Optimization.* Wiley, 2011 – This book provides a comprehensive treatment of experimental design and analysis techniques, with a focus on optimization in engineering and scientific experiments.

Index

For Product Safety Concerns and Information please contact our EU
representative GPSR@taylorandfrancis.com
Taylor & Francis Verlag GmbH, Kaufingerstraße 24, 80331 München, Germany

www.ingramcontent.com/pod-product-compliance
Ingram Content Group UK Ltd.
Pitfield, Milton Keynes, MK11 3LW, UK
UKHW051129130726
473146UK00025B/927